DISTANCE AND MIDPOINT FORMULAS

The **distance** between $P(x_1, y_1)$ and $Q(x_2, y_2)$ is

$$PQ = \sqrt{(x_2 - x_1)^2 + (y_2 - y_1)^2}$$

and the coordinates of the **midpoint** of line segment $\overline{PQ}$ are

$$\left(\frac{x_1 + x_2}{2}, \frac{y_1 + y_2}{2}\right)$$

EQUATION OF A CIRCLE

The equation of a circle with center (h, k) and radius r is

$$(x - h)^2 + (y - k)^2 = r^2$$

LINES

1. The **slope** of a line through $P(x_1, y_1)$ and $Q(x_2, y_2)$ is

$$m = \frac{y_2 - y_1}{x_2 - x_1}$$

2. The **slope-intercept form** of a line with slope m and y-intercept b is

$$y = mx + b$$

3. The **point-slope form** of a line through $P(x_1, y_1)$ with slope m is

$$y - y_1 = m(x - x_1)$$

VERTEX FORMULA

The graph of $f(x) = ax^2 + bx + c$ is a *parabola*.
The coordinates of its **vertex** are

$$\left(-\frac{b}{2a}, f\left(-\frac{b}{2a}\right)\right)$$

LOGARITHMS

1. $y = \log_b x$ is equivalent to $b^y = x$

2. $\log_b 1 = 0$

3. $\log_b b = 1$

4. $\log_b b^x = x$

5. $b^{\log_b x} = x$

6. $\log_b xy = \log_b x + \log_b y$

7. $\log_b \dfrac{x}{y} = \log_b x - \log_b y$

8. $\log_b x^n = n \log_b x$

9. $\log_b x = \dfrac{\log_a x}{\log_a b}$

GRAPHS OF TEN BASIC FUNCTIONS

1. **Constant Function**

$f(x) = k$, $(0, k)$

2. **Identity Function**

$f(x) = x$

3. **Absolute Value Function**

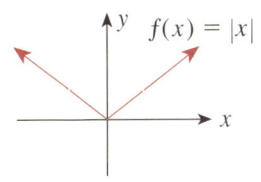

$f(x) = |x|$

4. **Squaring Function**

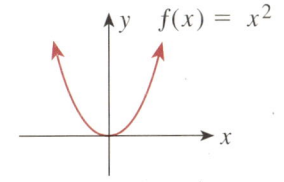

$f(x) = x^2$

5. **Cubing Function**

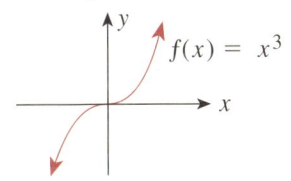

$f(x) = x^3$

6. **Reciprocal Function**

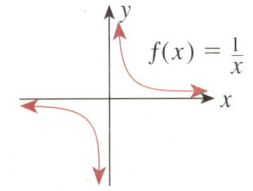

$f(x) = \frac{1}{x}$

7. **Square Root Function**

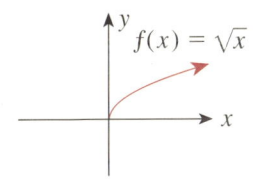

$f(x) = \sqrt{x}$

8. **Cube Root Function**

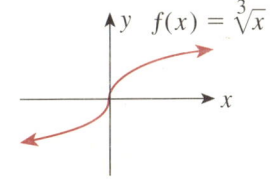

$f(x) = \sqrt[3]{x}$

9. **Exponential Function**

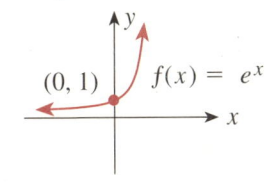

$f(x) = e^x$, $(0, 1)$

10. **Logarithmic Function**

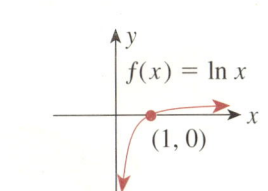

$f(x) = \ln x$, $(1, 0)$

BINOMIAL THEOREM

$$(A + B)^n = \sum_{i=0}^{n} \binom{n}{i} A^{n-i} B^i$$

College Algebra
with Graphing Technology

College Algebra
with Graphing Technology

David E. Stevens

Wentworth Institute of Technology

West Publishing Company

Minneapolis/St. Paul ■ New York ■ Los Angeles ■ San Francisco

Production Credits

Text Design: Katherine Townes/TECHarts
Copyediting: Katherine Townes/TECHarts
Composition: G & S Typesetters, Inc.
Illustration: Scientific Illustrators
Cover Image: Graeme Outerbridge
Cover Design: Diane Beasley Design

Production, Prepress, Printing and Binding by West Publishing Company.

PRINTED ON 10% POST CONSUMER RECYCLED PAPER   PRINTED WITH SOY INK™

British Library Cataloguing-in-Publication Data. A catalogue record for this book is available from the British Library.

Copyright © 1996 by West Publishing Company
610 Opperman Drive
P. O. Box 64526
St. Paul, MN 55164-0526

Printed in the United States of America

03 02 01 00 99 98 97 96   8 7 6 5 4 3 2 1 0

Library of Congress Cataloging-in-Publication Data

Stevens, David E.
College algebra with graphing technology / David E. Stevens.
p. cm.
Includes index.
ISBN 0-314-06241-6 (hardcover)
1. Algebra—Data processing. 2. Graphic calculators. I. Title.
QA154.2.S754 1996
512.9—dc20   95-23113
                                                CIP

CONTENTS

CONTENTS ◆ Graphing Technology Usage

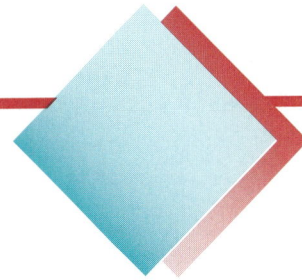

CONTENTS ◆ Applied Problems and Models

PREFACE

◆◆ **Intent**

A course in college algebra is a prerequisite for more advanced courses in mathematics as well as a prerequisite for other college courses in astronomy, biology, physics, chemistry, computer science, business, and engineering. The goal of this text is to effectively prepare students for these courses and to illustrate through real-life applied problems that knowledge of college algebra is fundamental to these disciplines. Unlike a traditional text, however, this text integrates *graphing technology* (graphing calculator or graphing software) into the college algebra course. In each chapter after Chapter 1, graphing technology is used as a true teaching aid, without sacrificing mathematical thought or rigor. Whenever a graphical interpretation of the mathematics is appropriate, graphing technology is included. See *Table of Contents: Graphing Technology Usage* on page ix.

◆◆ **Approach**

For their success in a college algebra course, it is essential that students be active rather than passive readers of the text. Therefore, I have written this text using an *interactive approach* with graphing technology. Each key mathematical concept is supported by a step-by-step text example, with marginal annotations and explanatory notes, and each text example is followed by a practice problem for the student to work. The practice problem may ask the reader to check the preceding text example by using graphing technology, to work the preceding text example using an alternative algebraic approach, to extend the preceding text example by asking for additional information, or to try an entirely different problem that has similar mathematical steps. In effect, the practice problems require the student to become involved with the mathematics and thus constitute a built-in workbook for the student. *A complete, detailed solution of each practice problem is given in the back of the text.*

◆◆ **Features**

Written in a warm and user-friendly style, this text addresses the concerns of writing across the curriculum, group learning, critical thinking, and the use of modern technology in the math classroom. The following features distinguish this text from other texts in the market.

◆ Applied Problems

Student interest in mathematics is undoubtedly greatest when applied problems are integrated directly within the curriculum. For this reason, applied problems from the fields of science, engineering, and business are introduced at every reasonable opportunity.

Chapter opening applications   To arouse student interest, each chapter opens with an applied problem and a related photograph. The solution to the problem is presented within the chapter, after the necessary mathematics has been developed.

Application subsections   Separate material has been included at the end of most sections to apply the mathematics to real-life situations. The logo ◆ identifies these application subsections. See *Table of Contents: Applied Problems and Models* on page xii.

◆ Exercise Sets

The heart of any math textbook is its end-of-section exercise sets. It is here that students are given an opportunity to practice the mathematical concepts that have been developed. The exercise sets in this text are presented in two parts: *Basic Skills* and *Critical Thinking*. Applied problems appear in both parts.

Basic Skills   These exercises are routine in nature, and most of them mimic the text examples that are worked out in each section. The reader is encouraged to use graphing technology to confirm results and to solve problems to which standard algebraic methods do not apply.

Critical Thinking   These exercises require the student to think critically and to transcend routine application of the basic skills to the next level of difficulty. Exercises in this group may require the student to draw upon skills developed in earlier chapters or to use graphing technology to explore new ideas.

Some of the exercise sets also contain problems that are *calculus related*. Designed for students who are taking a college algebra course as a prerequisite to calculus, these exercises illustrate the algebraic support that is needed in calculus. The logo $\boxed{\frac{\Delta y}{\Delta x}}$ identifies these problems.

◆ Chapter Reviews

To help students prepare for chapter exams, each chapter in this text concludes with an extensive chapter review, which is divided into two parts: *Questions for Writing and Group Discussion* and *Review Exercises*.

Questions for Writing and Group Discussion   In keeping with the interactive approach, these questions allow students to state in their own words what they have learned in the chapter. Since many of these questions have open-ended answers, they are well suited for class or group discussions and are extremely valuable to those who believe in cooperative or collaborative learning.

Review Exercises   These exercises reinforce the ideas discussed in the chapter and allow the instructor to indicate to the student the types of problems that may appear on a chapter test.

◆ Cumulative Reviews

To help students pull together ideas from several chapters, cumulative review exercises are placed strategically after Chapters 3, 6, and 8. The problems in these exercises are ungraded by difficulty and are presented in a random order. Some problems are basic and similar to those already studied; others are more challenging and require creative thinking.

◆ Pedagogy

Every effort has been made to make this a text from which students can learn and succeed. The following pedagogical features attest to this fact.

Caution notes   flagged by the symbol **CAUTION**, help eliminate misconceptions and bad mathematical habits by pointing out the errors most commonly made by students.

Introductory comments   at the start of each section introduce vocabulary and inform the reader of the purpose of the section.

Boxed definitions, formulas, laws, and properties   state key mathematical ideas and provide the reader with quick, easy access to this information.

Step-by-step procedural boxes   indicate the sequence of steps that a student can follow for tasks such as simplifying algebraic expressions, solving certain types of equations and inequalities, or finding the inverse of a function or a matrix.

◆◆ Development

The prerequisite for a course in college algebra is two years of high school algebra or a semester of college intermediate algebra. Chapter 1, *Prerequisites of College Algebra*, provides a quick review of the topics in intermediate algebra, such as exponents, radicals, factoring, algebraic fractions, equations, inequalities, and formulas. Complex numbers are introduced in this chapter in preparation for their use with quadratic functions in Chapter 3 and polynomial functions in Chapter 4.

Chapter 2, *Functions and Graphs*, introduces the coordinate plane and the language of functions and graphs. Graphing technology is introduced in Section 2.2 and is used extensively throughout the remainder of the text to verify results, solve equations, and explore new ideas. Section 2.5 lists eight basic functions and their graphs (constant, identity, absolute value, squaring, cubing, reciprocal, square root, and cube root), and then applies the vertical and horizontal shift rules, the *x*- and *y*-axes reflection rules, and the vertical stretch and compress rules to sketch the graphs of several related functions. These eight basic functions and their graphs are then used to discuss composition of functions, inverse functions, applied functions, and variation.

Chapter 3, *Linear and Quadratic Functions*, uses the ideas of shifting, reflecting, stretching, and compressing to develop the linear and quadratic functions from the identity and squaring functions. The vertex formula for a parabola is developed and used to solve some max–min applied problems.

Section 3.4 discusses the algebraic and graphical methods of solving quadratic equations and inequalities, and Section 3.5 gives an introduction to solving a system of two equations in two unknowns by looking at the intersection points of their graphs.

Chapter 4, *Polynomial and Rational Functions*, focuses on the relationship between factors, roots, zeros, and *x*-intercepts. Graphing technology and the rational zero theorem are used to find the real zeros of a polynomial function. This chapter also discusses algebraic and graphical methods for solving polynomial equations and inequalities and rational equations and inequalities. With the aid of graphing technology, it is possible to solve max–min applied problems that deal with polynomial and rational functions.

Chapter 5, *Exponential and Logarithmic Functions*, discusses the properties of real exponents, defines the exponential function, and develops the logarithmic function as the inverse of the exponential function. By letting the number of compounding periods in the compound interest formula increase without bound, the reader is shown how the number e develops in a real-life situation. The properties of logarithms are used in graphing functions that contain logarithmic expressions (Section 5.3) and also in solving exponential and logarithmic equations (Section 5.4). Graphing technology is used to support the work and to help solve equations that are not solvable by ordinary algebraic methods.

Chapter 6, *Conic Sections*, discusses the conics and their related quadratic equations in two unknowns. This chapter states the geometric properties of the circle, parabola, ellipse, and hyperbola, as well as the distinguishing characteristics of their equations. The reflection properties of the conics are applied to various technical applications. Throughout the chapter, graphing technology is used to support algebraic manipulations.

Chapters 7 and 8 include other topics of interest in a college algebra course. Chapter 7, *Linear Systems and Matrices*, extends the discussion of 2×2 systems (from Section 3.5) to a procedure for solving $n \times n$ linear systems by the elimination method. Matrices are first introduced as an aid for solving a system of linear equations and are then applied to problems that require using the matrix operations of addition and multiplication. Determinants and inverses of matrices are evaluated algebraically and by using the matrix features of a graphing calculator. Section 7.4 introduces linear systems of inequalities and applies the results to linear programming problems in two unknowns.

Chapter 8, *Sequences and Series*, offers an introduction to these topics for the precalculus student. Section 8.2 introduces the idea of proof by mathematical induction. In this section, the reader is encouraged to use pattern recognition to guess a formula for the sum of a series and then to prove this guess by mathematical induction. Similarly, pattern recognition and mathematical induction are used to develop a formula for the general element of an arithmetic sequence, for the general element of a geometric sequence, and for the expansion of $(A + B)^n$. The sequence and series features of a graphing calculator are used to support the work.

◆◆ Supplements

The following supplements are available for users of this text.

Instructor's Solution Manual, by Eleanor Canter of Wentworth Institute of Technology, includes detailed solutions to all the even-numbered exercises.

Student's Solution Manual, by Eleanor Canter of Wentworth Institute of Technology, provides detailed solutions for the odd-numbered exercises from the text.

Instructor's Manual with Test Bank, by Cheryl Roberts of Northern Virginia Community College, includes sample syllabi, suggested course schedules, chapter outlines with references to videos, homework assignments, chapter tests, and a test bank of multiple-choice questions and open-ended problems.

Graphing Technology Laboratory Manual, by David Lawrence of Southwestern Oklahoma State University, includes keystroke instructions for various types of graphing calculators—Texas Instruments, Casio, Sharp, and Hewlett-Packard, as well as Derive software.

DERIVE© Laboratory Manual, by Lloyd R. Jaisingh of Morehead State University contains sixteen laboratory experiments using the computer algebra system DERIVE©.

WESTEST, a computer-generated testing program, includes algorithmically generated questions and is available in both Macintosh and PC versions.

West Math Tutor Software by Mathens, an algorithmically based tutorial, is available for Macintosh and PC platforms. The instructor's version contains the complete bank of algorithms so that an instructor can create assignments, save them to disk for student use, and grade automatically when assignments are turned in. The student version is customized for each chapter of the appropriate text. The package is available to instructors, and a disk is available to students. It is also available as a site license for a math laboratory.

Videos, which are produced specifically for this text.

Fifty Transparency Masters illustrate the important figures, rules, and procedures in the text.

Please ask your West representative about qualifications for these supplements.

◆◆ Acknowledgments

Most of the material in this text has been class-tested with several hundred students at Wentworth Institute of Technology. I thank these students for their helpful comments and critiques. Special thanks go to my friend and colleague Eleanor Canter for her work in checking the answers and writing complete solutions to the more than 4500 exercises in this text. I also express my sincere thanks to the following reviewers—their ideas were extremely helpful in shaping this text into its present form.

Daniel D. Anderson,
University of Iowa

Charles D. Bedal,
Chandler Gilbert Community College

Bobbi Parrino Cook,
Indian River Community College

Gladys C. Cummings,
St. Petersburg Junior College

Sue Ehlers,
Anoka–Ramsey Community College

Susan S. Garstka,
Morraine Valley Community College

Norma F. James,
New Mexico State University

Larry Kurtz,
University of Montevallo

William Livingston,
Missouri Southern State College

Steve Martin,
Richard Bland College

Lawrence P. Merbach,
North Dakota State College of Science

Peggy I. Miller,
University of Nebraska–Kearney

Juan A. Gatica,
University of Iowa

William L. Grimes,
Central Missouri State University

Brian Hayes,
Triton College

Elizabeth Hodes,
Santa Barbara City College

Charlotte Newsom,
Tidewater Community College

Smruti Patel,
Hutchinson Community College

Jan Vandever,
South Dakota State University

Jamie Whitehead Ashby,
Texarkana College

The production of a textbook is a team effort between the editorial staff and the author. My editor, Nancy Hill-Whilton, offered the support and guidance that I needed to complete this project. Denise Bayko organized our reviewers' comments into a format that revealed where extra work was needed. Kathi Townes copyedited the manuscript and encouraged me to provide additional information that would benefit the reader, and Sandy Gangelhoff kept the project moving through the various stages of production. I thank each of you for the encouragement and enthusiasm that you provided in the preparation of this book.

D. E. Stevens
Boston, Massachusetts
1995

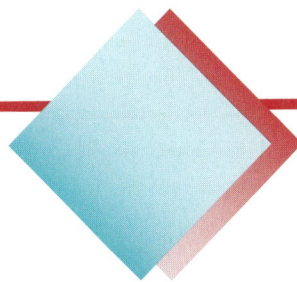

TO THE STUDENT

In order to use this text effectively, you should purchase a graphing calculator. As an alternative to a graphing calculator, you can use a computer equipped with graphing software. Your instructor has information on the type of calculator that you should purchase and the type of software that is available from West.

◆◆ Get Involved

The key to success in a college algebra course is to *get involved* with the mathematics. Do the homework exercises the day that your instructor assigns them. Before beginning the homework exercises, read the appropriate section in the text. Read carefully each text example in the section and try the practice problem that follows the text example. (A complete, detailed solution of each practice problem is given in the back of the text.) Form a study group with other members of your class and discuss the homework exercises with the group. You'll be surprised how much you can learn from each other.

◆◆ Ask Questions

You are in college to learn. You have an inquiring mind, and inquiring minds want to know. So, be sure to *ask questions*. Make note of the homework exercises that you did not understand, and come to class prepared to ask specific questions. Remember, there is no such thing as a "dumb" question. In fact, many other members of your class probably want to ask the same question that is causing you difficulty. Your instructor will welcome your participation in the discussion.

◆◆ Have Confidence

In order to do well in a mathematics test, you must *have confidence* in your ability to do the work. Building confidence requires study, intense work, and perseverance. If you have put in an honest effort, then you will never feel intimidated, defensive, or flustered during a test situation. You can enter the classroom knowing that you have given your very best.

◆◆ Never Give Up

Life is exciting when you're in college. You are full of optimism and enthusiasm for the opportunities that lie ahead. In the future, you will look back at your college experience as one of the best times of your life. So laugh, enjoy yourself, and *never give up!*

> Achievement consists of never giving up . . . If there is no dark and dogged will, there will be no shining accomplishment; if there is no dull and determined effort, there will be no brilliant achievement.

HSUN TZU
Chinese philosopher

College Algebra
with Graphing Technology

The distance from Earth to Alpha Centauri (the star closest to our Sun) is approximately 25,500,000,000,000 miles. How many years does it take light from this star to reach Earth? Assume that light travels at 186,000 miles per second.

For the solution, see Example 2 in Section 1.2.

Prerequisites of College Algebra

1.1 Real Numbers

◆ **Introductory Comments**

A chef requires 3 eggs and $\frac{2}{3}$ cup of sugar for a cake recipe. A meteorologist reports that the temperature is $-4°C$ and the barometric pressure is 29.35 inches. A student determines that the side of a right triangle is $\sqrt{5}$ units and the area of a circle is π square units. Each of the numbers

$$3 \qquad \frac{2}{3} \qquad -4 \qquad 29.35 \qquad \sqrt{5} \qquad \text{and} \qquad \pi$$

is an element of the **set of real numbers**.* These are the type of numbers that we work with every day. The set of real numbers has five important subsets:

1. Natural numbers $\{1, 2, 3, 4, \ldots\}$
2. Whole numbers $\{0, 1, 2, 3, \ldots\}$
3. Integers $\{\ldots, -3, -2, -1, 0, 1, 2, 3, \ldots\}$
4. Rational numbers

{All real numbers of the form $\frac{p}{q}$, where p and q are integers and $q \neq 0$.}
or
{All decimal numbers that either terminate or repeat the same block of digits.}

Examples:
$$\frac{3}{5}, \quad \frac{-8}{3}, \quad \frac{16}{1}, \quad \frac{0}{4}, \quad 0.75, \quad -5.343434\ldots$$

5. Irrational numbers

{All real numbers that are not rational.}
or
{All decimal numbers that neither terminate nor repeat the same block of digits.}

Examples:
$$\sqrt{2}, \quad -\sqrt{3}, \quad \pi, \quad \sqrt[3]{6}, \quad 3.050050005\ldots$$

In algebra, real numbers such as 3, $\frac{2}{3}$, -4, 29.35, $\sqrt{5}$, and π are called **constants**, since each is a fixed number. Letters from the beginning of the alphabet, like $a, b, c,$ and k, are often used as symbols for yet unspecified constants. A **variable** is a symbol that represents *any* member from a given set of numbers. Letters from the end of the alphabet, like $x, y, z,$ and t, are often used as variables. By assuming that a variable represents any member from the set of real numbers, we can manipulate and simplify any expression that contains constants and variables by using the *fundamental properties of real numbers*.

---

*The concept of a set is often used in mathematics. A *set* is a collection of objects and these objects are called the *elements* of the set. A *subset* of a given set is formed by selecting particular elements of the set. Braces { } are used to enclose the elements of sets and subsets.

◆ **Fundamental Properties of Real Numbers**

For any real numbers *a*, *b*, and *c*,

Property	Example
Commutative Property of Addition	
$a + b = b + a$	$2x + 4 = 4 + 2x$
Commutative Property of Multiplication	
$ab = ba$	$(x + y)2 = 2(x + y)$
Associative Property of Addition	
$(a + b) + c = a + (b + c)$	$(x + 2) + 3 = x + (2 + 3)$ $= x + 5$
Associative Property of Multiplication	
$(ab)c = a(bc)$	$3(2x) = (3 \cdot 2)x = 6x$
Distributive Property (left)	
$a(b + c) = ab + ac$	$7(x + y) = 7x + 7y$
Distributive Property (right)	
$(a + b)c = ac + bc$	$5x + 3x = (5 + 3)x = 8x$
Additive Identity Property	
$a + 0 = 0 + a = a$	$(4x + 3y) + 0 = 4x + 3y$
Multiplicative Identity Property	
$a(1) = (1)a = a$	$1(-3y) = -3y$
Additive Inverse Property	
$a + (-a) = (-a) + a = 0$	$5x + (-5x) = 0$
Multiplicative Inverse Property	
$a \cdot \dfrac{1}{a} = \dfrac{1}{a} \cdot a = 1, \quad a \neq 0$	$(x + 4) \cdot \dfrac{1}{x + 4} = 1, \quad x \neq -4$

The difference of two real numbers can be defined in terms of addition. To subtract b from a, we add a with the additive inverse, or opposite, of b.

◆ Subtraction

For any real numbers a and b,

$$a - b = a + (-b).$$

The quotient of two real numbers, denoted $a \div b$ or $\dfrac{a}{b}$, can be defined in terms of multiplication. To divide a by b, we multiply a by the multiplicative inverse (or *reciprocal*) of b. Since 0 does not have a multiplicative inverse, division by zero is *undefined*.

◆ Division

For any real numbers a and b, except $b = 0$,

$$a \div b = \frac{a}{b} = a \cdot \frac{1}{b} = \frac{1}{b} \cdot a.$$

Note that if a is a negative number, then its additive inverse, $-a$, is a positive number. For example, if $a = -4$, then $-a = -(-4) = 4$. Here are some other *properties of negation* that we use frequently throughout this text.

◆ Properties of Negation

For any real numbers, a, b, and c,

Property	Example
$(-1)a = -a$	$-1(3xy) = -3xy$
$-(-a) = a$	$-(-2x) = 2x$
$(-a)b = a(-b) = -(ab)$	$(-4x)2 = 4x(-2) = -(4x \cdot 2) = -8x$
$(-a)(-b) = ab$	$(-3x)(-y) = 3xy$
$\dfrac{-a}{b} = \dfrac{a}{-b} = -\dfrac{a}{b}, \ b \neq 0$	$\dfrac{-2x}{3} = \dfrac{2x}{-3} = -\dfrac{2x}{3}$
$\dfrac{-a}{-b} = \dfrac{a}{b}, \ b \neq 0$	$\dfrac{-(x+3)}{-y} = \dfrac{x+3}{y}$
$-(a+b) = -a - b$	$-(-2 + 7x) = 2 - 7x$
$-(a-b) = b - a$	$-(2x - 3) = 3 - 2x$

◆ The Real Number Line and Inequality Symbols

A geometric interpretation of the real numbers can be shown on the *real number line,* as illustrated in Figure 1.1. For each point on this line there corresponds exactly one real number, and for each real number there corresponds exactly one point on this line. This type of relationship is called a **one-to-one correspondence**.

FIGURE 1.1

On the real number line there exists a *one-to-one correspondence* between the set of real numbers and the set of points on the line.

The real number associated with a point on the real number line is called the **coordinate** of the point. In Figure 1.1, $-\frac{13}{4}$ is the coordinate of point A, $-\sqrt{2}$ is the coordinate of point B, $\frac{5}{3}$ is the coordinate of point C, and π is the coordinate of point D.

The real number line gives us a convenient way to compare two distinct real numbers a and b. If a is to the *right* of b on the real number line, then *a* **is greater than** *b*, and we write

$$a > b$$

If a is to the *left* of b on the real number line, then *a* **is less than** *b*, and we write

$$a < b$$

Comparing the coordinates of points C and D in Figure 1.1, we may write either

$$\pi > \frac{5}{3} \quad \text{or} \quad \frac{5}{3} < \pi$$

In general, for real numbers a and b,

$$a < b \quad \text{if and only if} \quad b > a$$

Note: The phrase *if and only if* occurs frequently in mathematics. In the preceding statement, it implies two statements:

1. If $a < b$, then $b > a$ and   **2.** If $b > a$, then $a < b$.

The symbols $>$ and $<$ are called **inequality symbols**, and statements such as $a > b$ and $a < b$ are called **inequalities**. Two other inequality symbols are used frequently:

$\leq$ read "less than or equal to" and $\geq$ read "greater than or equal to."

Inequalities can be used to indicate whether a number is *positive, negative, nonnegative,* or *nonpositive.* Some inequalities and their meanings are shown in Table 1.1.

TABLE 1.1
Some inequalities and their meanings

Inequality	Meaning
$a > 0$ or $0 < a$	a is positive.
$a < 0$ or $0 > a$	a is negative.
$a \geq 0$ or $0 \leq a$	a is nonnegative.
$a \leq 0$ or $0 \geq a$	a is nonpositive.

FIGURE 1.2

Three distinct points on the real number line with $a < b < c$.

Figure 1.2 shows three distinct real numbers a, b, and c on a real number line. To indicate that b is between a and c on this line, we can write either

$$a < b < c$$ or $$c > b > a$$

Each of these expressions is called a **double inequality**. When you use a statement such as $a < b < c$ or $c > b > a$, be sure that all the inequality symbols point in the same direction. For example, the expression $a < c > b$ is meaningless.

EXAMPLE 1   Write each statement as an inequality.

(a) a is at most 6.   **(b)** b is at least -2.

(c) c is nonnegative and less than 10.

◆ **Solution**

(a) a is at most 6   is written as

$$a \leq 6$$

(b) b is at least -2 is written as

$$b \geq -2$$

(c) c is nonnegative *and* less than 10   is written as

$$0 \leq c \quad and \quad c < 10$$

or, more compactly, $0 \leq c < 10$.   ◆

PROBLEM 1   Write each statement as an inequality.

(a) a is not more than 8.   **(b)** b is negative and at least -4.   ◆

◆ **Absolute Value**

The distance between zero and a number a on the real number line, without regard to direction, is called the *absolute value* of a and is denoted $|a|$. Because distance is independent of direction and is always nonnegative, the absolute value of any real number is also nonnegative, that is, $|a| \geq 0$. A more formal definition of **absolute value** follows.

◆ **Absolute Value**

For any real number a,

$$|a| = \begin{cases} a & \text{if } a \geq 0 \\ -a & \text{if } a < 0. \end{cases}$$

To find $|4|$ from this definition, we use $|a| = a$, since $4 \geq 0$. Thus,

$$|4| = 4.$$

To find $|-4|$ from this definition, we use $|a| = -a$, since $-4 < 0$. Thus,

$$|-4| = -(-4) = 4.$$

In general, we have $|a| = |-a|$ for any real number a. We now give two *properties of absolute value*, which involve products and quotients.

◆ **Properties of Absolute Value**

For any real numbers a and b,

Property	*Example*
Product Property $|ab| = |a||b|$	$|2x| = |2||x| = 2|x|$
Quotient Property $\left|\dfrac{a}{b}\right| = \dfrac{|a|}{|b|}, \; b \neq 0$	$\left|\dfrac{x+2}{-4}\right| = \dfrac{|x+2|}{|-4|} = \dfrac{|x+2|}{4}$

If points A and B on the real number line have coordinates a and b, respectively, then the **distance** between points A and B, denoted AB, is defined as follows.

◆ **Distance Between Two Points on the Real Number Line**

For any points A and B with coordinates a and b, respectively,

$$AB = |a - b|.$$

FIGURE 1.3

The distance between -8 and 5 is 13 units.

For example, if the coordinate of point A is -8 and the coordinate of point B is 5, then

$$AB = |-8 - 5| = |-13| = 13 \text{ units},$$

as shown in Figure 1.3. Note that $BA = |5 - (-8)| = |13| = 13$ units. In general, for any points A and B, the distance is the same in both directions, that is, $AB = BA$.

EXAMPLE 2   Write each statement as an inequality with absolute value notation.

(a) The distance between a and 2 is at most 3.

(b) b is more than 3 units from -5.

◆ **Solution**

(a) The distance between a and 2 is written as $|a - 2|$, or as $|2 - a|$. Hence,

the distance between a and 2 is at most 3   is written as

$$|a - 2| \leq 3.$$

(b) If b is more than 3 units from -5, then

the distance from b to –5 is greater than 3,   and we write

$|b - (-5)|$ $>$ 3 or $|b + 5| > 3.$ ◆

PROBLEM 2   Express "the distance between c and -1 is at least 5" as an inequality with absolute value notation.   ◆

◆ **Interval Notation**

Any unbroken portion of the real number line is called an **interval**. We may describe the set of all real numbers in an interval by using either *set-builder notation* or *interval notation*. For example, consider the interval shown on the real number line in Figure 1.4. The *closed* dot indicates that -3 *is* contained in the interval, and the *open* circle indicates that 5 is *not* contained in the interval. Hence, this figure represents the set of all real numbers greater than or equal to -3 and less than 5. We may describe the set of real numbers x in this interval by writing

$$\text{Set-builder notation:} \quad \{x \mid -3 \leq x < 5\}$$

or

$$\text{Interval notation:} \quad [-3, 5)$$

bracket   parenthesis

In interval notation, a bracket indicates than an endpoint is included in the interval and a parenthesis indicates that the endpoint is excluded from the interval.

An interval that includes both its endpoints is called a **closed interval** and an interval that excludes both its endpoints is called an **open interval**. An

FIGURE 1.4

All real numbers x such that x is greater than or equal to -3 and less than 5.

interval that contains one of its endpoints but not the other, such as $[-3, 5)$, is called a **half-open interval**.

CAUTION

When using interval notation, be sure the lesser of the two numbers is written first. To express the interval shown in Figure 1.4 as

$$(5, -3] \quad \text{is WRONG!}$$

Remember, always record the numbers in the order they appear on the number line from left to right.

In interval notation, the symbol ∞ (*infinity*) indicates that an interval has no right-hand boundary and the symbol $-\infty$ (*negative infinity*) indicates that an interval has no left-hand boundary. We refer to an interval with no right-hand boundary, or no left-hand boundary, or neither, as an **unbounded interval**. The symbols ∞ and $-\infty$ are *not* real numbers and therefore cannot be included in an interval. Thus, in an unbounded interval, $-\infty$ is always preceded by a parenthesis and ∞ is always followed by a parenthesis.

In summary, we list nine types of intervals. In each case, a and b are constants and x is a variable. A closed dot indicates that the point is included as part of the interval, and an open circle indicates that the point is not part of the interval.

◆ **Types of Intervals**

Type of Interval		Set-Builder Notation	Interval Notation
Unbounded intervals			
	x	$\{x \mid x > a\}$	(a, ∞)
	x	$\{x \mid x \geq a\}$	$[a, \infty)$
	x	$\{x \mid x < a\}$	$(-\infty, a)$
	x	$\{x \mid x \leq a\}$	$(-\infty, a]$
	x	$\{x \mid x \text{ is real}\}$	$(-\infty, \infty)$
Open interval			
	x	$\{x \mid a < x < b\}$	(a, b)
Closed interval			
	x	$\{x \mid a \leq x \leq b\}$	$[a, b]$
Half-open interval			
	x	$\{x \mid a < x \leq b\}$	$(a, b]$
	x	$\{x \mid a \leq x < b\}$	$[a, b)$

Note: Throughout the remainder of this text, we shall use interval notation, instead of set-builder notation, to describe the set of all real numbers in an interval.

EXAMPLE 3   Use interval notation to describe the set of all real numbers designated on each real number line.

(a)

(b)

◆ Solution

(a) This real number line shows all real numbers x such that x is less than or equal to -2. We may describe this set of real numbers by writing

$$(-\infty, -2].$$

(b) This real number line shows all real numbers x that are greater than -2 and less than 2. We may describe this set of real numbers by writing

$$(-2, 2). \qquad\qquad ◆$$

PROBLEM 3   Indicate each interval on a real number line.

(a) $(3, \infty)$ **(b)** $[4, 9)$

◆

The **union** of two sets A and B is denoted $A \cup B$, which we read as "A union B." The union operation denotes the set of all elements that are members of set A, or members of set B, or members of both sets A and B. In the next example, we use the union operation in conjuction with interval notation to describe the set of all real numbers designated by two or more intervals on a real number line.

EXAMPLE 4   Use interval notation to describe the set of all real numbers designated by each real number line.

(a)

(b)

◆ Solution

(a) This real number line shows two intervals that represent all real numbers x such that x is less than or equal to -3 *or* greater than or equal to 3. We may describe the set of real numbers in these two intervals by writing

$$(-\infty, -3] \cup [3, \infty).$$

(b) This real number line shows three intervals that represent all real numbers x except -1 and 7. We may describe the set of real numbers in these three intervals by writing

$$(-\infty, -1) \cup (-1, 7) \cup (7, \infty).$$

◆

PROBLEM 4 Indicate the given intervals on a real number line.

(a) $[-2, -1) \cup (-1, 2]$ **(b)** $(-\infty, -3] \cup [3, 5) \cup (5, \infty)$ ◆

Exercises 1.1

◆ **Basic Skills**

In Exercises 1–10, write each statement as an inequality.

1. x is negative.
2. y is positive.
3. a is at most 7.
4. b is at least -9.
5. p is greater than 2 and less than or equal to 10.
6. q is less than or equal to 4 and greater than -1.
7. c is positive and less than 8.
8. d is negative and more than -5.
9. t is nonpositive and at least -2.
10. k is nonnegative and at most 4.

In Exercises 11–20, find AB, the distance from A to B on the real number line. Write each answer without absolute value notation.

11. The coordinate of point A is -9 and the coordinate of point B is 6.
12. The coordinate of point A is 13 and the coordinate of point B is -2.
13. The coordinate of point A is -1.9 and the coordinate of point B is -3.8.
14. The coordinate of point A is 0.16 and the coordinate of point B is 0.09.
15. The coordinate of point A is $\frac{7}{8}$ and the coordinate of point B is $-\frac{1}{3}$.
16. The coordinate of point A is $-\frac{3}{4}$ and the coordinate of point B is $-\frac{9}{10}$.
17. The coordinate of point A is $-3\frac{1}{2}$ and the coordinate of point B is $5\frac{2}{3}$.
18. The coordinate of point A is 19 and the coordinate of point B is $6\frac{5}{8}$.
19. The coordinate of point A is $\sqrt{2}$ and the coordinate of point B is π.

20. The coordinate of point A is $-\sqrt{2}$ and coordinate of point B is $\sqrt{3}$.

In Exercises 21–30, write each statement using an inequality with absolute value notation.

21. The distance between a and 7 is at least 3.
22. The distance between b and 2 is at most 7.
23. The distance between x and -4 is less than 2.
24. The distance between y and -6 is greater than 1.
25. a is more than 4 units from -1.
26. c is less than 6 units from 0.
27. y is closer to 1 than to 0.
28. y is closer to -3 than to -2.
29. b is farther from -2 than from 5.
30. a is farther from 5 than from -1.

In Exercises 31–50, use interval notation to describe the set of all real numbers designated on each real number line.

31.
32.
33.
34.
35.
36.

37.

38.

39.

40.

41.

42.

43.

44.

45.

46.

47.

48.

49.

50.

In Exercises 51–60, indicate the given intervals on a real number line.

51. $(-3, 0]$

52. $[-6, -3]$

53. $(-\infty, \infty)$

54. $[2, \infty)$

55. $[0, 1) \cup (3, \infty)$

56. $(-\infty, 1) \cup [2, 5]$

57. $(-\infty, 7) \cup (7, \infty)$

58. $[-1, 0) \cup (0, 2]$

59. $(-\infty, 0] \cup [1, 2) \cup (2, \infty)$

60. $(0, 1) \cup (1, 3) \cup (3, \infty)$

◆ **Critical Thinking**

In Exercises 61–66, rewrite each expression so that it does not contain absolute value bars.

61. $|\pi - x|$ if $x \geq \pi$

62. $|\pi - x|$ if $x < \pi$

63. $|x - 3| + |x - 4|$ if $3 < x < 4$

64. $|x - 3| - |x - 4|$ if $x < 3$

65. $|x| < 5$

66. $|y| > 3$

67. What meaning (if any) can be assigned to the interval notation $[a, a]$? to (a, a)?

68. Explain why each of the given interval notations is meaningless.

(a) $(0, -6)$

(b) $[-2, -\infty)$

(c) $[-\infty, 5)$

(d) $[1, \infty]$

69. Use interval notation to describe the set of all real numbers that are less than their reciprocals.

70. Use interval notation to describe the set of all real numbers that are less than or equal to their squares.

71. Use a calculator to help list the following real numbers in order from smallest to largest:

$$3.145 \qquad \pi \qquad \frac{22}{7} \qquad 3.2 \qquad \sqrt{10} \qquad \frac{157}{50}$$

72. Use a calculator to help list the following real numbers in order from largest to smallest:

$$\frac{7}{5} \qquad \sqrt{2} \qquad 1.414 \qquad \frac{71}{50} \qquad \frac{8\pi}{17} \qquad 1.5$$

1.2 **Integer Exponents and Scientific Notation**

◆ **Introductory Comments**

When we add or subtract two or more numbers, the numbers are called *terms*. To describe the repeated addition of a real number a, we use multiplication:

$$a + a + a + \cdots + a = na.$$

n terms

When we multiply two or more numbers, the numbers are called *factors*. To describe the repeated multiplication of a real number *a*, we use a positive integer exponent:

◆◆ **Definition of a^n**

For any real number *a* and any positive integer *n*,

$$a^n = (a)(a)(a) \cdots (a)$$

n factors

The expression a^n is read "*a* to the *n*th power" and is referred to as the **exponential form** of the repeated multiplication. In the expression a^n, the number *a* is called the **base** and *n* is the **exponent**, or **power**.

Note the difference in the meanings of these exponential forms:

$$-2^4 = -(2)(2)(2)(2) = -16$$
whereas $$(-2)^4 = (-2)(-2)(-2)(-2) = 16$$

Also,

$$2y^4 = 2(y)(y)(y)(y)$$
whereas $$(2y)^4 = (2y)(2y)(2y)(2y) = 16y^4$$

Our definition of a^n gives us the following **properties of positive integer exponents**.

◆◆ **Properties of Positive Integer Exponents**

For any positive integers *m* and *n* and real numbers *a* and *b* that yield nonzero bases and denominators,

Property	*Example*
Product Property $a^m a^n = a^{m+n}$	$(-2x^2)(3x^5) = (-2 \cdot 3)x^{2+5}$ $= -6x^7$
Power Property $(a^m)^n = a^{mn}$	$[(x + y)^2]^3 = (x + y)^{2 \cdot 3}$ $= (x + y)^6$
Quotient Property $\dfrac{a^m}{a^n} = \begin{cases} a^{m-n} & \text{if } m > n \\ \dfrac{1}{a^{n-m}} & \text{if } m < n \\ 1 & \text{if } m = n \end{cases}$	$\dfrac{3x^5y^3z^2}{6x^2y^4z^2} = \dfrac{3}{6} \cdot x^{5-2} \cdot \dfrac{1}{y^{4-3}} \cdot 1$ $= \dfrac{x^3}{2y}$
Power of a Product Property $(ab)^n = a^n b^n$	$(-3xy)^2 = (-3)^2 x^2 y^2 = 9x^2 y^2$
Power of a Quotient Property $\left(\dfrac{a}{b}\right)^n = \dfrac{a^n}{b^n}$	$\left(\dfrac{x^2 + 2}{y}\right)^3 = \dfrac{(x^2 + 2)^3}{y^3}$

CAUTION

It is not possible to manipulate a power of a sum or difference in the same manner as a power of a product or quotient; that is,

$$(a + b)^n \neq a^n + b^n \quad \text{and} \quad (a - b)^n \neq a^n - b^n.$$

Note that

$$(3 \cdot 2)^2 = 3^2 \cdot 2^2 \quad \text{whereas} \quad (3 + 2)^2 \neq 3^2 + 2^2$$
$$36 = 36 \qquad\qquad\qquad 25 \neq 13$$

◆ Integer Exponents

Our definition of a^n applies only to *positive* integer exponents n. How should we define a^0 or a^{-n}? Certainly, we would like the properties of positive integer exponents to hold for zero and negative integer exponents as well. If the product property $a^m a^n = a^{m+n}$ is to hold for the zero exponent, then

$$a^0 \cdot a^n = a^{0+n} = a^n.$$

The only way $a^0 \cdot a^n$ can equal a^n is for a^0 to be the *identity element for multiplication,* that is, $a^0 = 1$. If the product property is to hold for negative integer exponents, then

$$a^{-n} \cdot a^n = a^{-n+n} = a^0 = 1.$$

The only way $a^{-n} \cdot a^n$ can equal 1 is for a^{-n} to be the *multiplicative inverse,* or *reciprocal,* of a^n, that is, $a^{-n} = \dfrac{1}{a^n}$.

◆ **Definitions of a^0 and a^{-n}**

For any real number a, except 0, and any integer n,

$$a^0 = 1 \quad \text{and} \quad a^{-n} = \frac{1}{a^n}.$$

Here are some illustrations of evaluating zero and negative integer exponents:

$$5^0 = 1 \qquad 3^{-2} = \frac{1}{3^2} = \frac{1}{9} \qquad \left(-\frac{1}{2}\right)^{-4} = \frac{1}{(-\frac{1}{2})^4} = \frac{1}{\frac{1}{16}} = 16$$

We can use the power key, $\boxed{\wedge}$ or $\boxed{y^x}$, on a calculator to raise a number to an integer power. Figure 1.5 shows a typical display on a calculator for evaluating 5^0, 3^{-2}, and $(-\frac{1}{2})^{-4}$ and confirms our work.

Now, consider the quotient a^{-m}/b^{-n}, where m and n are positive integers. To write a^{-m}/b^{-n} *without* negative exponents, we can proceed as follows:

$$\frac{a^{-m}}{b^{-n}} = \frac{1/a^m}{1/b^n} = \frac{1}{a^m} \cdot \frac{b^n}{1} = \frac{b^n}{a^m}.$$

```
5^0
                1
3^-2▸Frac
              1/9
(-1/2)^-4
               16
```

FIGURE 1.5

Typical display of evaluating 5^0, 3^{-2}, and $(-\frac{1}{2})^{-4}$

Notice that the factor a^{-m} in the numerator becomes a^m in the denominator, and the factor b^{-n} in the denominator becomes b^n in the numerator. In general, a factor may be moved from the numerator to the denominator— or from the denominator to the numerator—by changing the sign of the exponent.

CAUTION Only a *factor* of the numerator or denominator may be moved by changing the sign of its exponent. A *term* of the numerator or denominator cannot be moved in this fashion. For example, to write

up

$$\frac{3 \cdot \boxed{2^{-1}}}{\boxed{5^{-1}}}$$

down

as $\quad \dfrac{3 \cdot 5}{2} = \dfrac{17}{2}$ is CORRECT!

However, to write

up

$$\frac{3 + \boxed{2^{-1}}}{\boxed{5^{-1}}}$$

down

as $\quad \dfrac{3 + 5}{2} = 4$ is WRONG!

To evaluate $(3 + 2^{-1})/5^{-1}$ correctly, we proceed as follows:

$$\frac{3 + 2^{-1}}{5^{-1}} = \frac{3 + \frac{1}{2}}{\frac{1}{5}} = \frac{\frac{7}{2}}{\frac{1}{5}} = \frac{35}{2}$$

The display in Figure 1.6 confirms our work.

It can be shown that each of the five properties of positive integer exponents are valid for *all* integer exponents. In fact, since we now know the meaning of zero and negative integer exponents, the quotient property for exponents can be written simply as

$$\frac{a^m}{a^n} = a^{m-n}, \quad a \neq 0.$$

The important definitions and properties of integer exponents are summarized next. We refer to them as the **laws of exponents**.

```
(3+2^-1)/5^-1
                17.5
Ans▶Frac
                35/2
```

FIGURE 1.6

Typical display of evaluating $(3 + 2^{-1})/5^{-1}$.

◆ **Laws of Exponents**

For any integers m and n and real numbers a and b that yield nonzero bases and denominators,

1. $a^0 = 1$

2. $a^{-n} = \dfrac{1}{a^n}$

3. $a^m a^n = a^{m+n}$

4. $(a^m)^n = a^{mn}$

5. $\dfrac{a^m}{a^n} = a^{m-n}$

6. $(ab)^n = a^n b^n$

7. $\left(\dfrac{a}{b}\right)^n = \dfrac{a^n}{b^n}$

8. $\dfrac{a^{-m}}{b^{-n}} = \dfrac{b^n}{a^m}$

EXAMPLE 1   Simplify each expression. Express the answer with positive exponents.

(a) $(x + y)^{-4}(x + y)^3$ **(b)** $\left(\dfrac{3}{y}\right)^{-2}$

(c) $(-3x^{-3}y^4)^{-2}$ **(d)** $\dfrac{4^{-2}x^{-3}y^4}{4x^{-4}y^{-3}}$

◆ **Solution**

(a) $(x + y)^{-4}(x + y)^3 = (x + y)^{-1}$ **Add exponents**

$$= \frac{1}{x + y}$$ **Change to a positive exponent**

(b) $\left(\dfrac{3}{y}\right)^{-2} = \dfrac{3^{-2}}{y^{-2}}$ **Apply $(a/b)^n = a^n/b^n$ (law 7)**

$$= \frac{y^2}{3^2} \quad \text{or} \quad \frac{y^2}{9}$$ **Change to positive exponents**

(c) $(-3x^{-3}y^4)^{-2} = (-3)^{-2}(x^{-3})^{-2}(y^4)^{-2}$ **Apply $(ab)^n = a^n b^n$ (law 6)**

$$= (-3)^{-2}x^6 y^{-8}$$ **Multiply exponents**

$$= \frac{x^6}{(-3)^2 y^8} \quad \text{or} \quad \frac{x^6}{9y^8}$$ **Change to positive exponents**

(d) $\dfrac{4^{-2}x^{-3}y^4}{4x^{-4}y^{-3}} = 4^{-2-1}x^{-3-(-4)}y^{4-(-3)}$ **Subtract exponents**

$$= 4^{-3}x^1 y^7$$ **Simplify**

$$= \frac{xy^7}{4^3} \quad \text{or} \quad \frac{xy^7}{64}$$ **Change to a positive exponent** ◆

Several alternate procedures may be used to simplfy the expressions in Example 1. For example, we may first change to positive exponents, and then apply the properties of positive integer exponents.

PROBLEM 1   Simplify the expression in Example 1(d) by first applying the property $a^{-m}/b^{-n} = b^n/a^m$ to eliminate the negative exponents. Then simplify by using the properties of positive integer exponents. You should obtain the same result as in Example 1. ◆

◆ **Application: Scientific Notation and Conversion Factors**

The laws of exponents may be used in conjunction with *scientific notation* to perform arithmetic calculations containing large and small numbers.

◆ **Scientific Notation**

A number is written in **scientific notation** if it is of the form

$$k \times 10^n \qquad \text{where } 1 \le |k| < 10 \text{ and } n \text{ is an integer.}$$

To change an ordinary decimal number to a number in scientific notation, or vice versa, remember the following three facts:

1. A decimal number greater than 10 has an exponent on base 10 that is a positive integer.
2. A decimal number between 0 and 1 has an exponent on base 10 that is a negative integer.
3. A decimal number between 1 and 10 has an exponent on base 10 that is 0.

Here are some examples of changing an ordinary decimal number to a number in scientific notation, and vice versa:

$$0.00000065 = 6.5 \times 10^{-7} \qquad 8{,}240{,}000{,}000 = 8.24 \times 10^9$$

7 places   9 places

$$3.2 \times 10^3 = 3200. \qquad 9.84 \times 10^{-4} = 0.000984$$

3 places   4 places

$$8.75 = 8.75 \times 10^0 \qquad 1.56 \times 10^0 = 1.56$$

To perform arithmetic calculations containing large and small numbers, we may first write the numbers in scientific notation and then apply the laws of exponents. For example, to evaluate

$$\frac{(0.0000650)(0.00000008)}{1{,}300{,}000}$$

we proceed as follows:

$$\frac{(0.0000650)(0.00000008)}{1{,}300{,}000} = \frac{(6.50 \times 10^{-5})(8 \times 10^{-8})}{1.3 \times 10^6}$$

$$= \frac{(6.50)(8)}{(1.3)} \times \frac{(10^{-5})(10^{-8})}{10^6} \qquad \text{\color{red}\textbf{Rearrange the factors}}$$

$$= 40 \times 10^{-19} \qquad \text{\color{red}\textbf{Evaluate and simplify exponents}}$$

$$= (4 \times 10^1) \times 10^{-19} = 4 \times 10^{-18} \qquad \text{\color{red}\textbf{Write in scientific notation}}$$

$$= 0.000000000000000004 \qquad \text{\color{red}\textbf{Write as a decimal number}}$$

We can use the scientific notation key, $\boxed{\text{EE}}$ or $\boxed{\text{EXP}}$, on a calculator to perform arithmetic calculations containing large and small numbers. Figure 1.7 shows a typical display on a calculator for evaluating the previous expression. This display confirms our work.

Scientist and engineers often use scientific notation along with *conversion factors* in their work. A **conversion factor** is formed by two equal values, one of the values placed in the numerator and the other value in the denominator of a fraction. Hence, a conversion factor is equal to 1. A conversion factor is considered an exact number and therefore does not affect the number of sig-

FIGURE 1.7

Typical display of evaluating

$$\frac{(0.0000650)(0.00000008)}{1{,}300{,}000}$$

nificant digits that we maintain in a calculation with measured data. (See Appendix for a discussion of significant digits).

EXAMPLE 2 *Chapter Opening Problem*

The distance from Earth to Alpha Centauri (the star closest to our Sun) is approximately 25,500,000,000,000 miles. How many years does it take light from this star to reach Earth? Assume that light travels at 186,000 miles per second.

◆ **Solution**   First, we express these large numbers in scientific notation and find the time in second(s) that it takes light from this star to reach Earth:

$$\frac{25,500,000,000,000 \text{ mi}}{186,000 \text{ mi/s}} = \frac{2.55 \times 10^{13}}{1.86 \times 10^5} \text{ s}$$

Now, we apply three conversion factors and change from seconds to years (yr):

$$\frac{2.55 \times 10^{13}}{1.86 \times 10^5} \text{ s} \cdot \frac{1 \text{ hr}}{3600 \text{ s}} \cdot \frac{1 \text{ day}}{24 \text{ hr}} \cdot \frac{1 \text{ yr}}{365 \text{ days}}$$

We can evaluate this expression by using the scientific notation key on a calculator. The calculator display in Figure 1.8 indicates that it takes approximately 4.35 years for light to reach Earth from Alpha Centauri. We round this answer to three significant digits since the measured data, 2.55×10^{13} mi and 1.86×10^5 mi, are given to three significant digits.   ◆

PROBLEM 2   The distance from Earth to our Sun is approximately 93,000,000 mi. How many minutes does it take light from the Sun to reach Earth? Assume that light travels at 186,000 mi/s.   ◆

In engineering, prefixes indicating powers of ten, such as *mega-, kilo-, centi-, milli-,* and *micro-,* are attached to the basic units of measurement—meters, grams, volts, watts, and so on, in order to work with very large or small numbers. Listed in Table 1.2 are the symbols and meanings of these commonly used prefixes.

FIGURE 1.8

Typical display of the evaluation in Example 2.

```
2.55E13/(1.86E5*
3600*24*365)
        4.347310191
```

TABLE 1.2

Symbols and meanings of the most commonly used prefixes

Prefixes	Symbol	Meaning
mega-	M	10^6
kilo-	k	10^3
centi-	c	10^{-2}
milli-	m	10^{-3}
micro-	μ	10^{-6}

Numerical quantities are usually substituted in their basic unit size into a formula. As illustrated in the next example, if a prefix is attached to a basic unit, then the prefix must be converted to a power of ten before substituting the unit into the formula. If necessary, the answer in basic units may then be converted to a specific prefix.

EXAMPLE 3 The formula $P = V^2/R$ gives the power P in watts (W) dissipated by a resistance of R ohms (Ω) when the voltage drop across the resistance is V volts (V). Find the power P in *milliwatts* (mW) if $V = 1.2$ kV and $R = 20$ MΩ.

◆ **Solution**   Converting to powers of ten, we have

$$V = 1.2 \text{ kV} = 1.2 \times 10^3 \text{ V} \quad \text{and} \quad R = 20 \text{ M}\Omega = 20 \times 10^6 \ \Omega.$$

Thus,

$$P = \frac{V^2}{R} = \frac{(1.2 \times 10^3)^2}{20 \times 10^6} = \frac{(1.2)^2 \times (10^3)^2}{20 \times 10^6} \text{ W}$$

$$= \frac{1.44 \times 10^6}{20 \times 10^6} \text{ W}$$

$$= 0.072 \times 10^0 \text{ W}$$

$$= 0.072 \text{ W}$$

$$= 72 \times 10^{-3} \text{ W} = 72 \text{ mW}$$

The calculator display in Figure 1.9 supports our work.   ◆

FIGURE 1.9

Typical display of the evaluation in Example 3.

PROBLEM 3   Use the power formula in Example 3 to calculate P in megawatts (MW) if $R = 30 \ \Omega$ and $V = 600,000$ V.   ◆

Exercises 1.2

◆ **Basic Skills**

In Exercises 1–20, use the definitions of a^n, a^{-n}, and a^0 to evaluate each expression. Use a calculator to verify each answer.

1. $(-6)^2$
2. 5^3
3. -4^2
4. -2^5
5. 4^{-3}
6. 3^{-4}
7. -8^{-2}
8. -2^{-3}
9. $\left(\dfrac{2}{3}\right)^{-4}$
10. $\left(\dfrac{2}{5}\right)^{-1}$
11. $(-9)^{-2}$
12. $\left(\dfrac{-3}{4}\right)^{-3}$
13. $\dfrac{(-\frac{1}{3})^2}{2^{-1}}$
14. $\dfrac{4^{-1}}{-(\frac{2}{3})^4}$

15. $\dfrac{2^{-1} - 3^0}{2^0 + 3^0}$
16. $\dfrac{3^0}{2^0 + 3^{-2}}$
17. $3^{-1} + 4^{-1}$
18. $\dfrac{8^{-2} - 4^{-1}}{2^{-1}}$
19. $\dfrac{3 \cdot 2^{-3}}{5 \cdot 4^{-2}}$
20. $\dfrac{3 + 2^{-3}}{5 + 4^{-2}}$

In Exercises 21–40, use the laws of exponents to simplify each expression. Express the answer with positive exponents only.

21. $-8x^{-1}y^{-3}$
22. $2^{-1}xy^{-4}$
23. $-5(1 - 2x)^{-6}(-2)$
24. $-3(x^2 + 1)^{-4}(2x)$
25. $(-2x^2y^3)(5xy^5)^2$
26. $(2x)^4(-x^2y)(-3xy^{-1})$
27. $(-2y^{-1})^4$
28. $(-3x^2y^{-2})^3$

29. $\dfrac{8x^4yz^3}{4xy^{-3}z^3}$

30. $\dfrac{9x^{-1}y^2z}{12x^3y^{-2}}$

31. $(x + 3)^2(x + 3)^{-3}$

32. $(x + 1)(x + 1)^4$

33. $[(x - 2)^2]^4$

34. $[(t - 6)^{-1}]^{-2}$

35. $\dfrac{(2y + 3)^{-5}(2y + 3)}{(2y + 3)^{-6}}$

36. $\dfrac{(2a - 5)(2a + 5)^{-3}}{(2a + 5)^3(2a - 5)^2}$

37. $\dfrac{(3m^2n)^{-2}}{2m^{-3}}$

38. $\dfrac{(3x^{-1}y^2)^{-4}}{(-x^4y^3)^{-1}}$

39. $\dfrac{(2p)^{-3}(q - r)^3}{(q - r)^{-4}(4p^2)^{-3}}$

40. $\dfrac{[a^{-2}(b - c)^2]^{-1}}{a(b - c)^{-3}}$

In Exercises 41–46, express each number as an ordinary decimal number.

41. 6.9×10^{-9}

42. 2.33×10^6

43. -1750×10^8

44. 2200×10^{-5}

45. 0.00392×10^{-1}

46. -0.0698×10^0

In Exercises 47–52, express each number in scientific notation.

47. $54,300$

48. 0.00000294

49. 0.13×10^{-6}

50. 1730×10^5

51. 240×10^3

52. 0.005×10^{-8}

In Exercises 53–60, perform the indicated operations. Write each answer in scientific notation.

53. $(8,000,000,000)(0.00000025)$

54. $(170,000,000)(-20,000,000,000,000)$

55. $\dfrac{0.0000496}{16,000,000,000,000}$

56. $\dfrac{486,000,000}{0.000000006}$

57. $(0.0000000002)^{-5}$

58. $(25,000,000,000)^{-2}$

59. $(300,000,000)(0.00002)^3$

60. $\dfrac{(8,000,000)^2(2000)}{(0.0004)^4}$

In Exercises 61–64, use the power key, $\boxed{\wedge}$ or $\boxed{y^x}$, on a calculator to evaluate each expression. Record each answer as an ordinary decimal number rounded to four significant digits (see Appendix for a discussion of significant digits).

61. $(1.0025)^{45}$

62. $(-0.936)^{12}$

63. $(0.0287)^{-3}$

64. $(1.806)^{-10}$

In Exercises 65–68, use the scientific notation key, $\boxed{EE}$ or $\boxed{EXP}$, on a calculator to evaluate each expression. Record the answer in the scientific notation form $k \times 10^n$, rounding k to three significant digits (see Appendix for a discussion of significant digits).

65. $(2.731 \times 10^{-11})(3.924 \times 10^{-13})$

66. $\dfrac{(8.75 \times 10^{12})(167 \times 10^{14})}{0.000000000796}$

67. $\dfrac{(0.0000212)(0.000000567)^2}{3.42 \times 10^{12}}$

68. $(8.92 \times 10^{15})^2(0.0000000274)^3$

In Exercises 69–74, use scientific notation to help solve each problem.

69. The distance from Earth to the planet Pluto is approximately 3,700,000,000 mi. If a space ship leaves from Earth and averages 35,000 mi/hr, how many years will it take the ship to reach Pluto?

70. *Halley's comet* travels approximately 43,300,000,000 km in one orbit around the Sun and averages 65,100 km/hr. How many years does it take Halley's comet to orbit the Sun once?

71. The formula $I = V/R$ gives the current I in amps (A) through a resistance of R ohms (Ω) when the voltage drop across the resistance is V volts (V). Calculate the current in microamps (μA) if the resistance is 330 MΩ and the voltage drop is 2.2 kV.

72. When two resistances R_1 and R_2 are connected in parallel, their total resistance R_t is given by

$$R_t = \dfrac{R_1 R_2}{R_1 + R_2}.$$

If $R_1 = 4.2$ MΩ and $R_2 = 680$ kΩ, find the total resistance R_t in kΩ.

73. The land area of the earth is approximately 1.484×10^8 square kilometers (sq km). In 1980, the world population was approximately 4,483,000,000. At that time, what was the amount of land area in square meters (sq m) per person?

74. If our national debt is approximately one trillion dollars and the population of the United States is approximately 250,000,000, what is the amount of debt per person?

━━━ ◆ **Critical Thinking** ━━━━━━━━━━━━━━━━━━━━━━━━━━

75. Given that x, y, and z are integers and $a \neq 0$, simplify $a^{x-y} \cdot a^{y-z} \cdot a^{z-x}$.

76. Given that n is an integer and $b \neq 0$, rewrite each expression as a power of b:

    (a) $(b^{2-n} \cdot b^{n-4})^{-3}$ (b) $\dfrac{(b^2)^n}{b^n \cdot b^{n-2}}$

77. We have defined a^0 to be equal to 1, provided $a \neq 0$. If we erroneously define 0^0 to be equal to 1, then its reciprocal $1/0^0$ must also be equal to 1 and, by the laws of exponents, we could write

$$1 = 0^0 = \frac{1}{0^0} = \frac{a^0}{0^0} = \left(\frac{a}{0}\right)^0.$$

Do you see the fallacy in this argument? Explain.

78. What is the value of $(-1)^n$ when n is an even integer? an odd integer? 0?

79. Suppose you have a home mortgage of $80,000 at an interest rate of 9% per year over 30 years. The monthly payment may be calculated by evaluating

$$\frac{(7.5 \times 10^{-3})(8 \times 10^4)(1.0075)^{360}}{(1.0075)^{360} - 1}$$

(a) Determine, to the nearest cent, the amount of the monthly payment.
(b) Determine the total amount you will pay to the lender.

80. You may exceed the computational range of your calculator when working with the power key or scientific notation key. Most calculators will not display positive values less than 1×10^{-99} or greater than $9.999999999 \times 10^{99}$. Determine the smallest and largest integer n for which you are able to obtain a value for 25^n on the display of your calculator.

1.3 Rational Exponents and Radical Expressions

◆ **Introductory Comments**

In Section 1.2 we defined the expression a^n for integer exponents. We now extend our definition to include exponents that are rational numbers; that is, we now assign meaning to expressions such as these:

$$49^{1/2} \qquad 32^{2/5} \qquad 16^{-3/4}$$

In order to make the transition from integer exponents to rational exponents, we must first discuss the meaning of the n*th root* of a real number.

◀ **Definition of *n*th root**

> If a and b are any real numbers and $n \geq 2$ is an integer such that $b^n = a$, then b is an **nth root** of a.

We now describe the nature of the nth root(s) of a for n even or odd and a positive or negative.

◀ **Nature of the *n*th root(s) of *a***

n	a	nth root(s) of a	*Example*
even	positive	two real roots, one positve and one negative	The two real fourth roots of 16 are ± 2, since $(\pm 2)^4 = 16$.
even	negative	no real root	No real square (2nd) root of -9 exists, because no real number squared is -9.
odd	positive	one real root, a positive root	The one real cube (3rd) root of 8 is 2, since $2^3 = 8$.
odd	negative	one real root, a negative root	The one real cube root of -8 is -2, since $(-2)^3 = -8$.

Note: The *n*th root of zero is 0, since $0^n = 0$ for any integer $n \geq 2$.

To avoid ambiguity when working with even roots of positive numbers, we define the *principal n*th root of a number.

◆ **Principal *n*th Root of a Number**

> If a is any real number that has at least one real *n*th root, then the *n*th root that has the same sign as a is called the **principal *n*th root of a** and is denoted by $\sqrt[n]{a}$.

In the expression $\sqrt[n]{a}$, the symbol $\sqrt{}$ is called the **radical**, n is the **index**, and a is the **radicand.** We usually write the principal square root as $\sqrt{a}$ instead of $\sqrt[2]{a}$, but for $n \geq 3$ we must show the index. Here are some illustrations of evaluating *n*th roots:

$$\sqrt[4]{16} = 2 \qquad -\sqrt[4]{16} = -2 \qquad \sqrt{-9} \text{ is not a real number}$$
$$\sqrt[3]{8} = 2 \qquad \sqrt[3]{-8} = -2 \qquad \sqrt[5]{0} = 0$$

◆ **Rational Exponents**

We now define a rational exponent, $a^{1/n}$, where n is a positive integer. Certainly, we would like a definition of $a^{1/n}$ to be consistent with the laws of exponents listed in Section 1.2. If the power property for exponents is to hold true, then we must have

$$(a^{1/n})^n = a^{(1/n)(n)} = a^1 = a.$$

The only way $(a^{1/n})^n$ can equal a is for $a^{1/n}$ to be an *n*th root of a. We define the expression $a^{1/n}$ as the *exponential form of* $\sqrt[n]{a}$.

◆ **Definition of $a^{1/n}$**

> For any integer $n \geq 2$, where $a \geq 0$ when n is even,
>
> $$a^{1/n} = \sqrt[n]{a}.$$

Notice the difference between these examples of the definition of $a^{1/n}$:

$$49^{1/2} = \sqrt{49} = 7 \qquad \text{and} \qquad -49^{1/2} = -\sqrt{49} = -7$$
$$27^{1/3} = \sqrt[3]{27} = 3 \qquad \text{and} \qquad (-27)^{1/3} = \sqrt[3]{-27} = -3$$

How might we define $a^{m/n}$, where m and n are positive integers with m/n reduced to lowest terms and $a \geq 0$ when n is even? Again, if the power property for exponents is to hold true, we must define $a^{m/n}$ as follows:

◆ **Definition of $a^{m/n}$**

If $a^{1/n}$ is any real number and m and n are positive integers such that m/n is reduced to lowest terms, then

$$a^{m/n} = (a^{1/n})^m = \left(\sqrt[n]{a}\right)^m$$

or

$$a^{m/n} = (a^m)^{1/n} = \sqrt[n]{a^m}.$$

Thus, to evaluate an expression that contains a rational exponent, such as $9^{3/2}$, we can use either of two methods of reasoning:

1. $9^{3/2} = (9^{1/2})^3 = \left(\sqrt{9}\right)^3 = (3)^3 = 27$ or

2. $9^{3/2} = (9^3)^{1/2} = \sqrt{9^3} = \sqrt{729} = 27.$

As you can see, to evaluate $a^{m/n}$ the first method is easier to apply: *first take the root, then raise to the power.* Here are two more examples of evaluating expressions with rational exponents:

$$32^{2/5} = (32^{1/5})^2 = \left(\sqrt[5]{32}\right)^2 = (2)^2 = 4$$

$$16^{-3/4} = \frac{1}{16^{3/4}} = \frac{1}{(16^{1/4})^3} = \frac{1}{\left(\sqrt[4]{16}\right)^3} = \frac{1}{2^3} = \frac{1}{8}$$

We can use the power key, $\boxed{\wedge}$ or $\boxed{y^x}$, on a calculator to raise a number to a rational power. Figure 1.10 shows a typical display on a calculator for evaluating $32^{2/5}$ and $16^{-3/4}$ and confirms our work.

```
32^(2/5)
                 4
16^(-3/4)▶Frac
               1/8
```

FIGURE 1.10

Typical display of evaluating $32^{2/5}$ and $16^{-3/4}$.

◆ **Exponential and Radical Expressions**

To change a radical expression to an exponential expression, or vice versa, we use the definition of a rational exponent, $a^{m/n} = \sqrt[n]{a^m}$. The procedure is shown in the next example.

EXAMPLE 1

(a) Change to an exponential expression with a negative exponent:

$$\frac{1}{\sqrt{x^2 + 9}}$$

(b) Change to a radical expression: $(3x)^{2/3}$

◆ **Solution**

(a) The index of the radical becomes the denominator of the exponent. Hence,

$$\frac{1}{\sqrt{x^2 + 9}} = \frac{1}{(x^2 + 9)^{1/2}} = (x^2 + 9)^{-1/2}$$

(b) The denominator of the exponent becomes the index of the radical, and the numerator of the exponent is then the power of the radicand. Hence,

$$(3x)^{2/3} = \sqrt[3]{(3x)^2} = \sqrt[3]{9x^2} \qquad ◆$$

PROBLEM 1

(a) Change to an exponential expression: $\sqrt[5]{(x^2 + 2)^2}$

(b) Change to a radical expression: $(2x^2)^{3/7}$ ◆

To simplify an expression containing rational exponents, we use the laws of exponents (Section 1.2). First, we look at problems in which each *variable base represents a positive real number.*

EXAMPLE 2   Simplify each exponential expression. Assume all variables represent *positive* real numbers.

(a) $(9x^2)^{1/2}$ **(b)** $(5x^{1/4})(-4x^{5/6})$ **(c)** $\left(\dfrac{x^{-3}}{-8y}\right)^{-2/3}$

◆◆ **Solution**

(a) $(9x^2)^{1/2} = 9^{1/2}(x^2)^{1/2}$ **Apply** $(ab)^n = a^n b^n$

$\qquad\qquad = 9^{1/2}x^{(2)(1/2)}$ **Apply** $(a^m)^n = a^{mn}$

$\qquad\qquad = 3x$ **Simplify**

(b) $(5x^{1/4})(-4x^{5/6}) = (5)(-4)x^{1/4}x^{5/6}$ **Regroup**

$\qquad\qquad\qquad = -20x^{1/4+5/6}$ **Apply** $a^m a^n = a^{m+n}$

$\qquad\qquad\qquad = -20x^{3/12+10/12}$ **Add the exponents and simplify**
$\qquad\qquad\qquad = -20x^{13/12}$

(c) $\left(\dfrac{x^{-3}}{-8y}\right)^{-2/3} = \dfrac{(x^{-3})^{-2/3}}{(-8)^{-2/3}y^{-2/3}}$ **Apply** $(a/b)^n = a^n/b^n$ **and** $(ab)^n = a^n b^n$

$\qquad\qquad\qquad = \dfrac{x^2}{(-8)^{-2/3}y^{-2/3}}$ **Apply** $(a^m)^n = a^{mn}$

$\qquad\qquad\qquad = (-8)^{2/3}x^2 y^{2/3}$ **Change to positive exponents**

$\qquad\qquad\qquad = 4x^2 y^{2/3}$ **Simplify** ◆

When we simplify the expression $(9x^2)^{1/2}$ in Example 2(a), if we do *not* restrict x to *positive* real numbers, then we *cannot* state that $(9x^2)^{1/2}$ and $3x$ are equal. For example, suppose $x = -2$. Then

$$(9x^2)^{1/2} = [9(-2)^2]^{1/2} \qquad \text{whereas} \qquad 3x = 3(-2)$$
$$= [9(4)]^{1/2} \qquad\qquad\qquad\qquad = -6$$
$$= 36^{1/2} = 6$$

If we allow x to be *any* real number, then we must write

$$(9x^2)^{1/2} = 3|x|.$$

In general, we have the following definition.

◄ **Definition of** $(a^n)^{1/n}$

If n is any positive integer and a is *any* real number, then

$$(a^n)^{1/n} = \sqrt[n]{a^n} = \begin{cases} a & \text{if } n \text{ is odd} \\ |a| & \text{if } n \text{ is even.} \end{cases}$$

PROBLEM 2   Simplify $[(-2)^4 x^4]^{1/4}$, assuming each of the following descriptions of x.

(a) x is a *positive* real number.   **(b)** x is *any* real number.   ◆

◆   **Simplifying Radical Expressions**

We can use the laws of exponents (Section 1.2) in conjunction with the definition of a rational exponent to derive the following *properties of radicals*.

◄ **Properties of Radicals**

For any real numbers a and b and positive integers $m \geq 2$ and $n \geq 2$ such that $\sqrt[n]{a}$ and $\sqrt[n]{b}$ are real numbers,

Property	**Example**
1. $\left(\sqrt[n]{a}\right)^n = a$	$\left(\sqrt[3]{10}\right)^3 = 10$
2. $\sqrt[m]{\sqrt[n]{a}} = \sqrt[mn]{a}$	$\sqrt[3]{\sqrt[2]{7}} = \sqrt[3 \cdot 2]{7} = \sqrt[6]{7}$
3. $\sqrt[n]{ab} = \sqrt[n]{a} \cdot \sqrt[n]{b}$	$\sqrt{25 \cdot 2} = \sqrt{25} \cdot \sqrt{2} = 5\sqrt{2}$
4. $\sqrt[n]{\dfrac{a}{b}} = \dfrac{\sqrt[n]{a}}{\sqrt[n]{b}}$	$\sqrt[3]{\dfrac{-27}{64}} = \dfrac{\sqrt[3]{-27}}{\sqrt[3]{64}} = \dfrac{\sqrt[3]{(-3)^3}}{\sqrt[3]{4^3}} = \dfrac{-3}{4}$

CAUTION   It is not possible to take the nth root of a sum or difference in the same manner as taking the nth root of a product or quotient (as in property 3 or 4), that is,

$$\sqrt[n]{a + b} \neq \sqrt[n]{a} + \sqrt[n]{b} \quad \text{and} \quad \sqrt[n]{a - b} \neq \sqrt[n]{a} - \sqrt[n]{b}.$$

Notice that

$$\sqrt{9 \cdot 16} = \sqrt{9} \cdot \sqrt{16}, \quad \text{whereas} \quad \sqrt{9 + 16} \neq \sqrt{9} + \sqrt{16}$$
$$12 = 12 \qquad\qquad\qquad\qquad 5 \neq 7$$

We can use the properties of radicals to write a radical expression in *simplified form*.

◆ Simplified Form
of a Radical

A radical expression is in **simplified form** if both of the following conditions are satisfied.

Example

1. The radicand contains no factor to a power m greater than or equal to the index n of the radical; that is, $\sqrt[n]{a^m}$ has $m < n$.

$$\sqrt{75} = \sqrt{5^2 \cdot 3} = \sqrt{5^2} \cdot \sqrt{3} = 5\sqrt{3}$$

Simplified form

2. The power of the radicand m and the index n of the radical have no common factor other than 1; that is, for $\sqrt[n]{a^m}$, the exponent m/n is reduced to lowest terms.

$$\sqrt{5^2} = 5^{2/4} = 5^{1/2} = \sqrt{5}$$

Simplified form

EXAMPLE 3   Write each radical expression in simplified form. Assume all variables represent *positive* real numbers.

(a) $\sqrt{18x^3}$

(b) $\sqrt{\sqrt[3]{16x^2}}$

(c) $\sqrt{\dfrac{8x^5}{y^6}}$

(d) $\sqrt[3]{3x^2y} \cdot \sqrt[3]{9xy^2}$

◆ Solution

(a) $\sqrt{18x^3} = \sqrt{(3)^2 2 \cdot (x^2)x}$ Express each factor in terms of perfect squares

$\qquad\qquad = \sqrt{(3x)^2 \cdot 2x}$ Group the perfect square factors

$\qquad\qquad = \sqrt{(3x)^2} \cdot \sqrt{2x}$ Apply $\sqrt[n]{ab} = \sqrt[n]{a}\,\sqrt[n]{b}$

$\qquad\qquad = 3x\sqrt{2x}$ Simplify using $\sqrt[n]{a^n} = a$

(b) $\sqrt{\sqrt[3]{16x^2}} = \sqrt[6]{16x^2}$ Apply $\sqrt[m]{\sqrt[n]{a}} = \sqrt[mn]{a}$

$\qquad\qquad = \sqrt[6]{(4x)^2}$ Write the radicand as a perfect square

$\qquad\qquad = (4x)^{2/6}$ Change to exponential form

$\qquad\qquad = (4x)^{1/3} = \sqrt[3]{4x}$ Reduce the order of the radical

(c) $\sqrt{\dfrac{8x^5}{y^6}} = \dfrac{\sqrt{8x^5}}{\sqrt{y^6}}$ **Apply $\sqrt[n]{a/b} = \sqrt[n]{a}/\sqrt[n]{b}$**

$$= \dfrac{\sqrt{(2)(2)^2 \cdot (x)(x^2)^2}}{\sqrt{(y^3)^2}}$$ **Rewrite numerator and denominator in terms of perfect squares**

$$= \dfrac{\sqrt{(2x^2)^2} \cdot \sqrt{2x}}{\sqrt{(y^3)^2}}$$ **Regroup and apply $\sqrt[n]{ab} = \sqrt[n]{a}\,\sqrt[n]{b}$**

$$= \dfrac{2x^2\sqrt{2x}}{y^3}$$ **Simplify using $\sqrt[n]{a^n} = a$**

(d) $\sqrt[3]{3x^2y} \cdot \sqrt[3]{9xy^2} = \sqrt[3]{(3x^2y)(9xy^2)}$ **Apply $\sqrt[n]{a}\,\sqrt[n]{b} = \sqrt[n]{ab}$**

$$= \sqrt[3]{27x^3y^3}$$ **Multiply**

$$= \sqrt[3]{(3xy)^3}$$ **Rewrite as a perfect cube**

$$= 3xy$$ **Simplify using $\sqrt[n]{a^n} = a$**   ◆

PROBLEM 3   Write each radical expression in simplified form. Assume all variables represent *positive* real numbers.

(a) $\sqrt[3]{32(x+y)^3}$ **(b)** $\sqrt[4]{\sqrt[3]{x^2y^6}}$ ◆

◆ Rationalizing Denominators and Numerators

In algebra we must in some instances remove the radical symbol from the denominator or numerator of a fraction in order to perform further simplifications. This process is called **rationalizing the denominator or numerator**. To rationalize a denominator (or a numerator) that contains a radical with an index of n, we multiply numerator and denominator of the expression by a **rationalizing factor**—a factor that produces a perfect nth power in the radicand.

EXAMPLE 4

(a) Rationalize the denominator: $\dfrac{2\sqrt{3}}{\sqrt{2}}$

(b) Rationalize the numerator: $\dfrac{\sqrt[3]{5x^2}}{\sqrt[3]{2y}}$

◆ Solution

(a) $\dfrac{2\sqrt{3}}{\sqrt{2}} = \dfrac{2\sqrt{3}}{\sqrt{2}} \cdot \boxed{\dfrac{\sqrt{2}}{\sqrt{2}}} = \dfrac{2\sqrt{3 \cdot 2}}{\sqrt{2^2}} = \dfrac{2\sqrt{6}}{2} = \sqrt{6}$

This rationalizing factor produces a perfect square in the denominator.

(b) $\dfrac{5x^2}{2y} = \dfrac{5x^2}{2y} \cdot \boxed{\dfrac{5^2 x}{5^2 x}} = \dfrac{(5x)^3}{50xy} = \dfrac{5x}{50xy}$

This rationalizing factor produces a perfect cube in the numerator.

PROBLEM 4   Rationalize the denominator of the radical expression in Example 4(b), and simplify.

Exercises 1.3

◆ Basic Skills

In Exercises 1–20, evaluate each exponential expression. Use a calculator to verify each answer.

1. $4^{1/2}$ **2.** $8^{1/3}$

3. $-64^{1/6}$ **4.** $-81^{1/2}$

5. $49^{3/2}$ **6.** $4^{5/2}$

7. $(-125)^{2/3}$ **8.** $(-64)^{2/3}$

9. $100^{-3/2}$ **10.** $27^{-2/3}$

11. $-81^{-3/4}$ **12.** $-36^{-3/2}$

13. $\left(-\dfrac{1}{8}\right)^{2/3}$ **14.** $\left(\dfrac{1}{16}\right)^{1/4}$

15. $(2.25)^{-1/2}$ **16.** $(-125)^{1/3} \cdot 25^{-3/2}$

17. $\dfrac{4^{-3/2} \cdot 32^{3/5}}{-27^{0/3}}$ **18.** $\dfrac{6^{-2}(0.04)^{1/2}}{(-3)^{-1}}$

19. $(2^{-2} - 8^{-2/3} + 4^{-3/2})^{-1}$

20. $4^{-3/2} - 4^{-1/2} + 4^0 - 4^{1/2} + 4^{3/2}$

In Exercises 21–28, use a calculator to evaluate each expression. Record each answer as a decimal number rounded to four significant digits. (See Appendix for a review of significant digits.)

21. $(27.5)^{1/4}$ **22.** $(-0.654)^{1/3}$

23. $(287)^{-6/5}$ **24.** $(2.345 \times 10^{-3})^{3/2}$

25. $\sqrt{275}$ **26.** $\sqrt[3]{-8.32}$

27. $\sqrt[4]{0.1844}$ **28.** $\sqrt{12.6 \times 10^8}$

In Exercises 29–36, rewrite each radical expression as an exponential expression. For Exercises 33 and 34, express the answer with a negative exponent.

29. $\sqrt{2a}$ **30.** $\sqrt[4]{(x^2 + 1)^3}$

31. $\sqrt[7]{(y^2 - 3)^4}$ **32.** $\sqrt[10]{(2y)^5}$

33. $\dfrac{x}{\sqrt{x^2 + y^2}}$ **34.** $\dfrac{3}{\sqrt[3]{x^2}}$

35. $x\sqrt{x}$ **36.** $\sqrt[3]{y} \cdot \sqrt{y}$

In Exercises 37–44, rewrite each exponential expression as a radical expression.

37. $x^{1/2}$ **38.** $y^{3/4}$

39. $(3m^2)^{3/4}$ **40.** $(-2x)^{2/3}$

41. $(x + y)^{4/5}$ **42.** $[2(x - 3)^2]^{2/5}$

43. $2x^{-2/3}$ **44.** $(3p + q)^{-1/6}$

In Exercises 45–58, simplify each exponential expression. Assume the variables represent positive real numbers. Express the answer without zero or negative exponents.

45. $(4x^2)^{1/2} + (32x^5)^{1/5}$ **46.** $[16(x + 1)^4]^{-1/4}$

47. $\left(\dfrac{x^3}{8}\right)^{-2/3}$ **48.** $\left(\dfrac{x^5 y^{10}}{32}\right)^{2/5}$

49. $(x + y)^{4/3}(x + y)^{3/2}$

50. $(x^2 + 1)^{-1/2}(x^2 + 1)^{3/2}$

51. $(-64x^3 y^{-3})^{-2/3}$ **52.** $(25x^4 y^{-6})^{3/2}$

53. $\dfrac{x^2(x^2 + 4)^{1/2}}{x^{2/3}(x^2 + 4)^{3/4}}$ **54.** $\dfrac{[36(m + 8)^{-2}]^{-1/2}}{(m + 8)^{2/3}}$

55. $\dfrac{(2^{-4/3} x^{2/3})^{3/4}}{x^{1/2} y^{-2}}$ **56.** $\dfrac{(3^{2/3} x^{-2} y^{1/2})^{-3/2}}{25x^{1/3} y^2}$

57. $\frac{1}{2}(x^2 - 1)^{-1/2}(2x)$

58. $\frac{1}{3}(3x - 6x^3)^{-2/3}(3 - 18x^2)$

59. Rework Exercise 45 assuming x is *any* real number.

60. Rework Exercise 46 assuming x is *any* real number.

In Exercises 61–76, write each radical expression in simplified form. Assume the variables represent positive real numbers.

61. $\sqrt{24}$

62. $\sqrt[3]{72}$

63. $\sqrt{(x + 1)^2}$

64. $\sqrt[4]{16x^4} + \sqrt[3]{8x^3}$

65. $\sqrt{x^9}$

66. $\sqrt[3]{x^8}$

67. $\sqrt[3]{16x^4y^3}$

68. $\sqrt{125x^7y^2z^3}$

69. $\sqrt{49(a + b)^2(a^2 + b^2)}$

70. $\sqrt[3]{24(x^3 - 8)(x - 8)^3}$

71. $\sqrt[3]{\dfrac{54x^3y^4}{125z^3}}$

72. $\sqrt{\dfrac{b^2 - 4ac}{4a^2}}$

73. $\sqrt{\sqrt[3]{25x^2y^4}}$

74. $\sqrt[4]{\sqrt[3]{36x^2}}$

75. $\dfrac{\sqrt{40x^4yz}}{2\sqrt{5xy^3z}}$

76. $\sqrt[3]{6x^2y^2} \cdot \sqrt[3]{9xy^2}$

77. Rework Exercise 63 assuming x can be *any* real number.

78. Rework Exercise 64 assuming x can be *any* real number.

In Exercises 79–82, rationalize the denominator. Assume the variables represent positive real numbers.

79. $\dfrac{12}{\sqrt{2}}$

80. $\dfrac{\sqrt{7}}{6x\sqrt{3x}}$

81. $\sqrt[3]{\dfrac{4}{9xy^2}}$

82. $\dfrac{1}{\sqrt[3]{(x + 1)^2}}$

In Exercises 83–86, rationalize the numerator. Assume all variables represent positive real numbers.

83. $\dfrac{7\sqrt{6a}}{5}$

84. $\sqrt{\dfrac{x}{3}}$

85. $\sqrt[3]{\dfrac{(x + 2)^2}{3x}}$

86. $\dfrac{\sqrt[4]{2p^2q^3}}{\sqrt[4]{3}}$

◆ **Critical Thinking**

In Exercises 87–90, rewrite each product using only one radical symbol. [Hint: Use rational exponents to change the given radicals to radicals that have the same index.]

87. $\sqrt{3} \cdot \sqrt[3]{5}$

88. $\sqrt[3]{2} \cdot \sqrt[4]{3}$

89. $\sqrt{x} \cdot \sqrt[3]{y^2}$

90. $\sqrt{2x} \cdot \sqrt[5]{3x^2}$

91. Perform the indicated operations and write the answer in simplified radical form.

(a) $\sqrt[3]{4}(\sqrt[3]{2} - \sqrt[3]{4} + \sqrt[3]{10})$

(b) $\sqrt{abc}(a\sqrt{ab} + b\sqrt{bc} + c\sqrt{ac})$

92. Is $(-1)^{1/n}$ defined when n is an even integer? an odd integer? 0? If it is defined, state its value.

93. The time it takes a pendulum to swing back and forth in completing one cycle is called its *period*. If the length L of a pendulum is known, then its period may be determined by evaluating

$$2\pi\sqrt{\dfrac{L}{32}}.$$

Rationalize the denominator of this expression and then simplify.

94. Which of the following statements is incorrect? Explain.

(i) $[(-2)^6]^{1/2} = (64)^{1/2} = 8$

(ii) $[(-2)^6]^{1/2} = (-2)^{6(1/2)} = (-2)^3 = -8$

95. When the depth of water in a circular above-ground swimming pool is 4 ft, the pool holds 6032 gallons of water and the radius (in feet) of the pool may be found by computing

$$0.1825\left(\dfrac{6032}{\pi}\right)^{1/2}$$

Compute the radius of the pool.

96. To rationalize the denominator of $2/\sqrt{3}$, a student squared both numerator and denominator and obtained $4/3$. Explain why this procedure is wrong.

◆ **1.4** **The Complex Numbers**

◆ **Introductory Comments**

We know that the square roots of -4 and -9 *cannot* be real numbers, since no real number squared is -4 and no real number squared is -9. One of the basic properties of a real number a is that its square is always *nonnegative*:

$$a^2 \geq 0$$

In the 17th century, the French mathematician René Descartes used the word "imaginary" to describe square roots like $\sqrt{-4}$ and $\sqrt{-9}$, and in the 18th century a Swiss mathematician, Leonhard Euler, defined the **imaginary unit i.**

◆ **Imaginary Unit i**

$$i = \sqrt{-1} \qquad \text{where} \qquad i^2 = -1$$

Thus, if a is a real number and $a > 0$, then we can express the *principal square root of $-a$* in term of i:

$$\sqrt{-a} = \sqrt{-1 \cdot a} = \sqrt{-1} \cdot \sqrt{a} = i\sqrt{a}.$$

◆ **Principal Square of a Negative Number**

If a is any real number and $a > 0$, then the **principal square root** of $-a$ is

$$\sqrt{-a} = i\sqrt{a}.$$

Thus, for $\sqrt{-4}$ and $\sqrt{-9}$, we have

$$\sqrt{-4} = i\sqrt{4} = 2i \qquad \text{and} \qquad \sqrt{-9} = i\sqrt{9} = 3i.$$

Note that the squares of $2i$ and $3i$ are -4 and -9, respectively:

$$(2i)^2 = 2^2 i^2 = 4(-1) = -4 \qquad \text{and} \qquad (3i)^2 = 3^2 i^2 = 9(-1) = -9.$$

In the 19th century, the German mathematician Carl Friedrich Gauss extended the real number system to a larger system that includes the real numbers as well as those numbers whose squares are negative. We refer to this larger system as the **complex number system**. Each number in this system is called a **complex number** and may be written in the form $a + bi$, where a and b are real numbers.

◆ **Complex Number in Standard Form**

Any number of the form

$$a + bi,$$

where a and b are real numbers and i is the imaginary unit, is called a **complex number in standard form.**

In the complex number $a + bi$, the real number a is the *real part* and the real number b is the *imaginary part* of the number. If $b = 0$, then $a + bi$

becomes $a + 0i = a$. A complex number of the form $a + 0i$ is a real number. If $b \neq 0$, then $a + bi$ is referred to as an **imaginary number**. Here are some examples of imaginary numbers:

$$2 + 3i \qquad \pi - 6i \qquad 5i \qquad i\sqrt{2}$$

The use of the phrase "imaginary number" indicates the initial uneasiness of mathematicians with nonreal numbers. Today, imaginary numbers often appear in the solutions of quadratic and polynomial equations (see Section 3.4 and Chapter 4) and in applications of electrical circuits and quantum mechanics.

◆ **Operations with Complex Numbers**

We can add, subtract, multiply, and divide complex numbers by applying the same commutative, associative, and distributive properties that we use with real numbers. Here are some illustrations:

$$\sqrt{-4} + \sqrt{-9} = 2i + 3i = (2 + 3)i = 5i$$
$$\sqrt{-4} - \sqrt{-9} = 2i - 3i = (2 - 3)i = -i$$
$$\sqrt{-4} \cdot \sqrt{-9} = (2i)(3i) = (2 \cdot 3)(i \cdot i) = 6i^2 = 6(-1) = -6$$
$$\frac{\sqrt{-4}}{\sqrt{-9}} = \frac{2i}{3i} = \frac{2}{3}$$

CAUTION The property of radicals $\sqrt{a}\sqrt{b} = \sqrt{ab}$ does not apply when both a and b are negative numbers. For example, to write

$$\sqrt{-4} \cdot \sqrt{-9} = \sqrt{(-4)(-9)} = \sqrt{36} = 6 \quad \text{is WRONG!}$$

When working with the principal square root of a negative number, be sure to apply $\sqrt{-a} = i\sqrt{a}$ before using any of the properties of radicals discussed in Section 1.3.

To find the *sum* of two complex numbers $a + bi$ and $c + di$, we add the real and imaginary parts of the numbers separately. Similarly, to find the *difference* of two complex numbers we subtract the real and imaginary parts of the numbers separately.

EXAMPLE 1 Perform the indicated operations and express each answer in standard form.

(a) $(6 - 3i) - (-4 + 7i)$ **(b)** $(3 + 2\sqrt{-3}) + \sqrt{-27}$

◆ **Solution**

(a) $(6 - 3i) - (-4 + 7i) = [6 - (-4)] + (-3 - 7)i$
$$= 10 - 10i$$

(b) $(3 + 2\sqrt{-3}) + \sqrt{-27} = (3 + 2i\sqrt{3}) + i\sqrt{27}$
$$= 3 + (2i\sqrt{3} + 3i\sqrt{3})$$
$$= 3 + 5i\sqrt{3}$$

◆

PROBLEM 1　Find the difference:　$\left(3 + \sqrt{-25}\right) - \left(3 - \sqrt{-36}\right)$ ◆

To find the *product* of two complex numbers $a + bi$ and $c + di$, we multiply each part of the first complex number by each part of the second complex number.

EXAMPLE 2　Perform the indicated operations and express each answer in standard form.

(a) $(3 + 4i)(5 - i)$　　**(b)** $\sqrt{-2}\left(\sqrt{-18} + \sqrt{8}\right)$

◆◆ **Solution**

(a) $(3 + 4i)(5 - i) = 15 - 3i + 20i - 4i^2$

$$(3 + 4i)(5 - i) = 15 - 3i + 20i - 4i^2$$
$$= 15 + 17i - 4(-1)$$
$$= 19 + 17i$$

(b) $\sqrt{-2}(\sqrt{-18} + \sqrt{8}) = i\sqrt{2}(i\sqrt{18} + \sqrt{8})$
$$= i^2\sqrt{36} + i\sqrt{16}$$
$$= 6i^2 + 4i$$
$$= -6 + 4i$$ ◆

PROBLEM 2　Find the product:　$(1 + 3i)(2 + 5i)$ ◆

To find the *quotient* $\dfrac{a + bi}{c + di}$, we multiply numerator and denominator by the **complex conjugate** of the denominator, namely, $c - di$. The product of the complex conjugates is a nonnegative real number:

$$(c + di)(c - di) = c^2 - cdi + cdi - d^2i^2 = c^2 + d^2$$

a nonnegative real number

The fact that the product of complex conjugates is a nonnegative real number enables us to express the quotient $\dfrac{a + bi}{c + di}$ as a complex number in standard form:

$$\frac{a + bi}{c + di} = \frac{(a + bi)(c - di)}{(c + di)(c - di)} = \frac{(ac + bd) + (bc - ad)i}{c^2 + d^2}$$

$$= \frac{ac + bd}{c^2 + d^2} + \frac{bc - ad}{c^2 + d^2}i$$

A complex number in standard form

EXAMPLE 3 Perform the indicated operations and express each answer in standard form.

(a) $\dfrac{3 + 5i}{1 - 5i}$ (b) $\dfrac{8}{-\sqrt{-25}}$

◆◆ **Solution**

a) $\dfrac{3 + 5i}{1 - 5i} = \dfrac{(3 + 5i)(1 + 5i)}{(1 - 5i)(1 + 5i)} = \dfrac{3 + 15i + 5i + 25i^2}{1 - 25i^2}$

The complex conjugate of $1 - 5i$

$= \dfrac{-22 + 20i}{26}$

$= \dfrac{-22}{26} + \dfrac{20}{26}i = -\dfrac{11}{13} + \dfrac{10}{13}i$

b) $\dfrac{8}{-\sqrt{-25}} = \dfrac{8}{-5i} = \dfrac{8(5i)}{-5i(5i)} = \dfrac{40i}{-25i^2} = \dfrac{40i}{25} = \dfrac{8}{5}i$

The complex conjugate of $-5i$

PROBLEM 3 Find the quotient: $\dfrac{1 + \sqrt{-9}}{1 - \sqrt{-9}}$

◆ **Powers of *i***

When we raise i to successive positive integer powers, an interesting pattern develops:

$i^1 = i$ $i^5 = i^4 i = 1(i) = i$ $i^9 = i^8 i = 1(i) = i$

$i^2 = -1$ $i^6 = i^4 i^2 = 1(-1) = -1$ $i^{10} = i^8 i^2 = 1(-1) = -1$

$i^3 = i^2 i = (-1)i = -i$ $i^7 = i^4 i^3 = 1(-i) = -i$ $i^{11} = i^8 i^3 = 1(-i) = -i$

$i^4 = i^2 i^2 = (-1)(-1) = 1$ $i^8 = i^4 i^4 = 1(1) = 1$ $i^{12} = i^8 i^4 = 1(1) = 1$

Notice the values repeat in cycles of four according to the pattern i, -1, $-i$, 1. For higher powers of i, we use the fact that $i^4 = 1$ and apply the laws of exponents from Section 1.2. Here are some illustrations of higher powers:

$$i^{13} = (i^4)^3 i = (1)^3 i = i$$

$$i^{100} = (i^4)^{25} = (1)^{25} = 1$$

$$i^{-22} = \dfrac{1}{i^{22}} = \dfrac{1}{(i^4)^5 i^2} = \dfrac{1}{1^5(-1)} = -1$$

EXAMPLE 4 Find the indicated power and express each answer in standard form.

(a) $\left(\sqrt{-4}\right)^5$ (b) $(1 + 2i)^2$

◆ **Solution**

(a) $\left(\sqrt{-4}\right)^5 = (2i)^5 = 32i^5 = 32i^4 i = 32i$

(b) $(1 + 2i)^2 = (1 + 2i)(1 + 2i)$
$$= 1 + 2i + 2i + 4i^2$$
$$= 1 + 4i + 4(-1)$$
$$= -3 + 4i$$

◆

PROBLEM 4 Find the indicated power: $\left(2 - \sqrt{-9}\right)^2$ ◆

Exercises 1.4

◆ Basic Skills

In Exercises 1–30, perform the indicated operations and express each answer in standard form.

1. $(4 + 3i) - (1 - 5i)$ **2.** $(3 - 2i) + (8 - 3i)$

3. $(9 - 3i) + (6 + 2i) + 4i$

4. $(12 + 6i) - (4 - 3i) + (1 + i)$

5. $8i(2 - 3i)$ **6.** $-4i(5 + 7i)$

7. $(5 + 2i)(9 + 3i)$ **8.** $(8 - 2i)(1 - i)$

9. $\left(\sqrt{2} - 2i\right)\left(\sqrt{2} + 2i\right)$

10. $\left(3 + i\sqrt{5}\right)\left(3 - i\sqrt{5}\right)$

11. $\dfrac{4 + 2i}{i}$ **12.** $\dfrac{1 - 4i}{3i}$

13. $\dfrac{6}{7 - 3i}$ **14.** $\dfrac{2i}{4i - 1}$

15. $\dfrac{1 - 2i}{1 + 2i}$ **16.** $\dfrac{4 - i}{2 - 3i}$

17. $\sqrt{-16} + \sqrt{-25}$ **18.** $\sqrt{-36} - \sqrt{-49}$

19. $\sqrt{3} \cdot \sqrt{-12} \cdot \sqrt{-9}$ **20.** $\dfrac{\sqrt{-125} \cdot \sqrt{36}}{\sqrt{-5}}$

21. $5\sqrt{-5} + \sqrt{-20} - \frac{4}{3}\sqrt{-45}$

22. $2\sqrt{-3} - 2\sqrt{-27} + 6\sqrt{-48}$

23. $\left(3 + 2\sqrt{-36}\right) - \left(5 - 3\sqrt{-49}\right)$

24. $\left(8 - 4\sqrt{-9}\right) + \left(6 + 5\sqrt{-16}\right)$

25. $\sqrt{-3}\left(\sqrt{-12} - \sqrt{-27}\right)$

26. $\sqrt{2}\left(\sqrt{-18} - \sqrt{-50}\right)$

27. $\left(3 - \sqrt{-4}\right)\left(4 + 3\sqrt{-9}\right)$

28. $\left(\sqrt{-25} + 1\right)\left(3\sqrt{-4} - 8\right)$

29. $\dfrac{\sqrt{-4}}{2 - \sqrt{-6}}$ **30.** $\dfrac{9}{1 + 2\sqrt{-3}}$

In Exercises 31–44, find the indicated power and express each answer in standard form.

31. $5i^{10}$ **32.** $13i^{36}$

33. $2i^{-5}$ **34.** $\dfrac{1}{6i^{37}}$

35. $\left(\sqrt{-9}\right)^3$ **36.** $\left(\sqrt{-2}\right)^4$

37. $(3i)^3(2i)^2$ **38.** $(2i)^4(-3i)^3$

39. $(3 - 2i)^2$ **40.** $\left(-1 + 2\sqrt{-25}\right)^2$

41. $\dfrac{1}{(-4 - 5i)^2}$ **42.** $\dfrac{1}{\left(2 + \sqrt{-9}\right)^2}$

43. $i + i^2 + i^3 + i^4$

44. $i^{-1} - i^{-2} + i^{-3} - i^{-4}$

◆ **Critical Thinking**

45. Show that $1 + 0i$ is the *multiplicative identity* for the complex number $a + bi$.

46. Find the *multiplicative inverse*, or *reciprocal*, of the complex number $a + bi$.

47. Evaluate the expression $x^2 - 2x + 2$ when (a) $x = 1 + i$ and (b) $x = 1 - i$.

48. Evaluate the expression $x^2 - 6x + 13$ when (a) $x = 3 - 2i$ and (b) $x = 3 + 2i$.

49. Evaluate the expression $x^2 - 4x + 5$ when (a) $x = 2 - i$ and (b) $x = 2 + i$.

50. Evaluate the expression $x^2 - 2x + 10$ when (a) $x = 1 + 3i$ and (b) $x = 1 - 3i$.

51. What are the two square roots of -9?

52. Show that the two square roots of i are

$$\frac{\sqrt{2} + i\sqrt{2}}{2} \quad \text{and} \quad \frac{-\sqrt{2} - i\sqrt{2}}{2}$$

53. On most calculators, you must enter nonnegative inputs when working with the *square root key*, $\boxed{\sqrt{}}$. (A negative input causes an error message to appear in the display window). Use your calculator in conjunction with the definition of the principal square root of a negative number to evaluate each radical expression to three significant digits. (See Appendix for a review of significant digits.)

(a) $\sqrt{-18}$ (b) $\sqrt{-506}$

(c) $\sqrt{-153.6}$ (d) $\sqrt{-0.8765}$

54. Evaluate the expression

$$\frac{-b \pm \sqrt{b^2 - 4ac}}{2a}$$

for the given values of a, b, and c. Express each answer in standard form.

(a) $a = 1$, $b = -2$, $c = 10$

(b) $a = 2$, $b = 6$, $c = 7$

55. The magnitude of the current (in amperes, A) flowing through the series circuit shown in the sketch may be found by first writing the expression

$$\frac{200}{150 + (188.5 - 76.4)i}$$

in standard form, and then computing $\sqrt{a^2 + b^2}$, where a is the real part and b is the imaginary part of a complex number. Determine the magnitude of the current in this circuit.

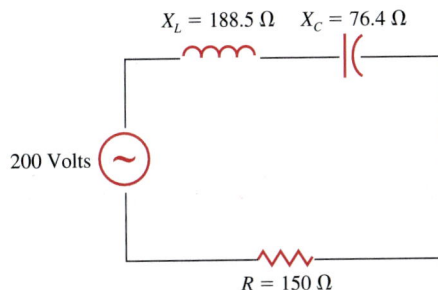

56. The magnitude of the total impedance (in ohms, Ω) for the parallel circuit shown in the sketch may be found by first writing the expression

$$\frac{1}{\dfrac{1}{150} + \dfrac{1}{188.5i} - \dfrac{1}{76.4i}}$$

in standard form, and then computing $\sqrt{a^2 + b^2}$, where a is the real part and b is the imaginary part of a complex number. Determine the magnitude of the total impedance for this circuit.

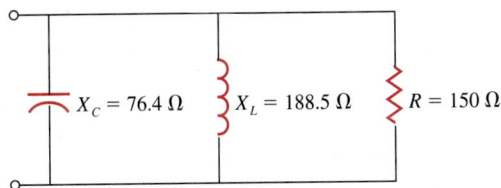

◆

1.5   Polynomials and Special Products

◆ **Introductory Comments**

A collection of variables and constants formed by using addition, subtraction, multiplication, division, powers, or roots is called an **algebraic expression**.

The building block of algebraic expressions is the **term**—a constant, a variable, or a product or quotient of constants and variables. The algebraic expression

$$6x^{-2}y^3 + y - \frac{\pi x^4}{y} - 3$$

| 1st term | 2nd term | 3rd term | 4th term |

consists of four terms. If a term consists of the product of a real number and one or more variables, then the real number is called the **numerical coefficient** (or simply the *coefficient*) of the term. In the preceding algebraic expression the numerical coefficient of the first term is 6; of the second term, 1; of the third term, $-\pi$. The fourth term is called a *constant term* because it contains no variable.

Terms that differ only in their numerical coefficients are called **like terms**. If two or more like terms appear in an algebraic expression, we usually combine them into a single term by applying the distributive property and adding their numerical coefficients. The procedure is called *combining*, or *collecting*, like terms. In each of the following algebraic expressions, we collect like terms:

$$3x^2y + 5x^2y = (3 + 5)x^2y = 8x^2y$$

$$3y^{-1} + 4y^{-1} - y^{-1} = (3 + 4 - 1)y^{-1} = 6y^{-1}$$

$$3\sqrt{5x} + \sqrt{5x} = (3 + 1)\sqrt{5x} = 4\sqrt{5x}$$

A *polynomial* is an algebraic expression in which no variable appears in any denominator or in any radicand, and any variable that does appear is raised to a nonnegative integer power.

◆◇ **Definition of a Polynomial in *x* of Degree *n***

Any algebraic expression of the form

$$a_nx^n + a_{n-1}x^{n-1} + a_{n-2}x^{n-2} + \cdots + a_2x^2 + a_1x + a_0$$

where *n* is a nonnegative integer and $a_0, a_1, a_2, \ldots, a_n$ are real numbers with $a_n \neq 0$, is called a **polynomial in *x* of degree *n***.

If, after like terms are combined, a polynomial consists of only *one* term, it is called a **monomial**; if it consists of *two* terms, it is a **binomial**; and if it consists of *three* terms, it is a **trinomial**. Each of the following algebraic expressions is an example of a polynomial in a single variable:

$$3x^3 \quad \text{(a monomial of degree 3)}$$

$$2y^6 - 7y \quad \text{(a binomial of degree 6)}$$

$$z^2 - 3z + 2 \quad \text{(a trinomial of degree 2)}$$

The **degree of a term** of a polynomial is the sum of the exponents in the term. The algebraic expression

$$5x^2y^4 + xy^3 - 7x^2 + 3$$

is a polynomial in two variables, x and y. The degree of the first term of this polynomial is $2 + 4 = 6$; of the second term, $1 + 3 = 4$; and of the third term, 2. The degree of the fourth term, which may be written as $3x^0$, is regarded as 0. The degree of a polynomial in more than one variable is the degree of the term with the *highest* degree in the polynomial. Hence, the degree of the polynomial $5x^2y^4 + xy^3 - 7x^2 + 3$ is 6.

◆ Addition and Subtraction of Polynomials

To *add* two polynomials, we use the commutative and associative properties to group like terms, and then apply the distributive property to combine the like terms. To *subtract* two polynomials, we use the fact that $a - b = a + (-b)$; that is, we change the sign of each term in the polynomial that comes after the subtraction sign, and then combine like terms.

EXAMPLE 1 Perform the indicated operation.

(a) $(3x + 5y) + (5x^2 - 2x + 7y)$

(b) $(x^3 + 4x^2y + xy^2) - (xy^2 - 2yx^2 + 2y^3)$

◆ **Solution**

(a) $(3x + 5y) + (5x^2 - 2x + 7y)$
$\qquad = 5x^2 + (3x - 2x) + (5y + 7y)$ 　　**Group like terms**

$\qquad = 5x^2 + x + 12y$ 　　**Combine like terms**

(b) $(x^3 + 4x^2y + xy^2) - (xy^2 - 2yx^2 + 2y^3)$
$\qquad = x^3 + 4x^2y + xy^2 \ominus xy^2 \oplus 2yx^2 \ominus 2y^3$ 　　**Change signs**

$\qquad = x^3 + (4x^2y + 2yx^2) + (xy^2 - xy^2) - 2y^3$ 　　**Group like terms**

$\qquad = x^3 + 6x^2y - 2y^3$ 　　**Combine like terms**
◆

PROBLEM 1 Perform the indicated operations:

$(5x^2 - 3x) + (5x - x^2) - (2x + 4x^2)$ 　　◆

◆ Multiplication of Polynomials

To multiply two polynomials, both of which do not contain just one term, we use the distributive property. First we multiply each term of the first polynomial by each term of the second polynomial, and then we collect like terms.

EXAMPLE 2   Perform the indicated operations.

(a) $2x^2(3 - 5x - x^3)$ (b) $(3x - 4y)(2x + 5y)$

(c) $(x + 2)(3x^3 - x^2 + 2x - 4)$

◆ **Solution**

(a) $2x^2(3 - 5x + x^3) = 6x^2 - 10x^3 + 2x^5$

(b) $(3x - 4y)(2x + 5y) = 6x^2 + 15xy - 8xy - 20y^2$

$$= 6x^2 + 7xy - 20y^2$$

(c) $(x + 2)(3x^3 - x^2 + 2x - 4)$

$$= 3x^4 - x^3 + 2x^2 - 4x + 6x^3 - 2x^2 + 4x - 8$$

$$= 3x^4 + 5x^3 - 8$$ ◆

The product in Example 2(c) may also be found by arranging the polynomials *vertically* and multiplying:

$$
\begin{array}{l}
3x^3 - x^2 + 2x - 4 \\
\underline{\hspace{4em} x + 2} \\
3x^4 - x^3 + 2x^2 - 4x \\
\underline{\hspace{3em} 6x^3 - 2x^2 + 4x - 8} \\
3x^4 + 5x^3 \hspace{6em} - 8
\end{array}
$$

Multiply each term of $(3x^3 - x^2 + 2x - 4)$ by x.

Multiply each term of $(3x^3 - x^2 + 2x - 4)$ by 2. Align like terms with the row above.

Add like terms in each row.

The advantage of the vertical method is that like terms are grouped at the same time the multiplication is performed.

PROBLEM 2   Use the vertical method to perform the indicated operation: $(m^3 - m^2 + 3m + 4)(m^2 + 2)$ ◆

◆ **Special Products**

Three *special products* occur frequently when we work with algebraic expressions.

◆ **Special Products**

For algebraic expressions A and B,

1. **Product of a sum and difference:**
   $(A + B)(A - B) = A^2 - B^2$

2. **Square of a binomial:** $(A + B)^2 = A^2 + 2AB + B^2$

3. **Cube of a binomial:** $(A + B)^3 = A^3 + 3A^2B + 3AB^2 + B^3$

We can verify these special products by multiplying in the conventional manner:

1. $(A + B)(A - B) = A^2 - AB + AB + B^2$
$$= A^2 - B^2$$

2. $(A + B)^2 = (A + B)(A + B) = A^2 + AB + AB + B^2$
$$= A^2 + 2AB + B^2$$

3. $(A + B)^3 = (A + B)(A + B)^2$
$$= (A + B)(A^2 + 2AB + B^2)$$
$$= A^3 + 2A^2B + AB^2 + A^2B + 2AB^2 + B^3$$
$$= A^3 + 3A^2B + 3AB^2 + B^3$$

The process of squaring or cubing a binomial is called *expanding* the binomial. To expand a binomial in which the power is greater than 3, the best method to use is the *binomial theorem,* which we will discuss in the last chapter of this text.

EXAMPLE 3 Perform the indicated operations:

(a) $(3x + 2)(3x - 2)$ **(b)** $(4x + 5)^2$ **(c)** $(3x - 2y)^3$

◆ **Solution**

(a) Let $A = 3x$ and $B = 2$. Using

$$(A + B)(A - B) = A^2 - B^2,$$
we have $(3x + 2)(3x - 2) = (3x)^2 - (2)^2$
$$= 9x^2 - 4$$

(b) Let $A = 4x$ and $B = 5$. Using

$$(A + B)^2 = A^2 + 2AB + B^2,$$
we have $(4x + 5)^2 = (4x)^2 + 2(4x)(5) + (5)^2$
$$= 16x^2 + 40x + 25$$

(c) By the definition of subtraction, $(3x - 2y)^3 = [3x + (-2y)]^3$. Letting $A = 3x$ and $B = -2y$, we have

$$(A + B)^3 = A^3 + 3A^2B + 3AB^2 + B^3$$
$$(3x - 2y)^3 = (3x)^3 + 3(3x)^2(-2y) + 3(3x)(-2y)^2 + (-2y)^3$$
$$= 27x^3 - 54x^2y + 36xy^2 - 8y^3$$

◆

PROBLEM 3   Perform the indicated operations.

(a) $(2x - 3)(2x + 3)$ **(b)** $(3x - 5y)^2$ **(c)** $(x^4 + 4)^3$ ◆

◆ **Grouping Symbols and the Order of Operation**

Grouping symbols such as *parentheses* ( ), *brackets* [ ], and *braces* { } are used to show that the terms contained within them represent a single quantity. Parentheses are commonly used for the *innermost* grouping symbols, then brackets, and then braces.

◆ **Order of Operations**

To simplify an algebraic expression that contains more than one set of grouping symbols, we use the following **order of operations** to remove each set of grouping symbols, starting with the innermost set.

1. Raise expressions to powers.

2. Perform the multiplications and divisions in order from left to right.

3. Perform the additions and subtractions in order from left to right.

Repeat the order of operations until all sets of grouping symbols have been removed.

EXAMPLE 4   Perform the indicated operations.

(a) $2x[(8x + 2) - 2(2x + 1)^2]$

(b) $6x^2 + \{-3[x - (2x - 1)(x + 3)]\}$

◆ **Solution**

Start with this power

(a) $2x[(8x + 2) - 2(2x + 1)^2] = 2x[(8x + 2) - 2(4x^2 + 4x + 1)]$
$= 2x[(8x + 2) - 8x^2 - 8x - 2]$
$= 2x[-8x^2]$
$= -16x^3$

Start with this product

(b) $6x^2 + \{-3[x - (2x - 1)(x + 3)]\} = 6x^2 + \{-3[x - (2x^2 + 5x - 3)]\}$

Retain this set of parentheses and subtract *all* terms of this product.

$= 6x^2 + \{-3[x \ominus 2x^2 \ominus 5x \oplus 3]\}$
$= 6x^2 + \{-3x + 6x^2 + 15x - 9\}$
$= 12x^2 + 12x - 9$ ◆

PROBLEM 4 Perform the indicated operations:
$x^2 - [y(x + 2y) - (x + y)(2y - x)]$

◆ **Operations with Nonpolynomial Expressions**

The procedures for performing operations with polynomials can be extended to algebraic expressions that are not polynomials. For example, to find the product of the nonpolynomial expressions $(3 + 2x^{-1})$ and $(2 - 3x^{-1})$, we multiply as we would two binomials:

$$(3 + 2x^{-1})(2 - 3x^{-1}) = 6 - 9x^{-1} + 4x^{-1} - 6x^{-2}$$
$$= 6 - 5x^{-1} - 6x^{-2}$$

EXAMPLE 5 Perform the indicated operations.

(a) $(4\sqrt{xy} - 3y^{-3})(4\sqrt{xy} + 3y^{-3})$ **(b)** $(x^{1/2} + 5x^{-1/2})^2$

◆ **Solution**

(a) Using the special product of a sum and difference with $A = 4\sqrt{xy}$ and $B = 3y^{-3}$, we proceed as follows:

$$(A - B)(A + B) = A^2 - B^2$$
$$(4\sqrt{xy} - 3y^{-3})(4\sqrt{xy} + 3y^{-3}) = (4\sqrt{xy})^2 - (3y^{-3})^2$$
$$= 16xy - 9y^{-6}$$

(b) Using the special product for the square of a binomial with $A = x^{1/2}$ and $B = 5x^{-1/2}$, we proceed as follows:

$$(A + B)^2 = A^2 + 2AB + B^2$$
$$(x^{1/2} + 5x^{-1/2})^2 = (x^{1/2})^2 + 2(x^{1/2})(5x^{-1/2}) + (5x^{-1/2})^2$$
$$= x + 10 + 25x^{-1}$$

PROBLEM 5 Use the special product for the cube of a binomial to find the indicated power: $(x^{-4} + 4)^3$

Exercises 1.5

◆ **Basic Skills**

In Exercises 1–10, write each algebraic expression as a single term.

1. $5x^2 - 3x^2 + 4x^2 - 7x^2$

2. $2y^2 + 4y^2 - 5y^2 - 3y^2$

3. $x^2y + 3x^2y - 5yx^2 - 4yx^2$

4. $3mn^3 - 2mn^3 + n^3m - 3n^3m + mn^3$

5. $\frac{3}{4}x^{-1} + \frac{1}{2}x^{-1} - \frac{7}{2}x^{-1} + \frac{1}{4}x^{-1}$

6. $\frac{2}{3}x^{1/2} + \frac{4}{3}x^{1/2} + \frac{3}{4}x^{1/2} + \frac{1}{4}x^{1/2}$

7. $6\sqrt{3y} - 3\sqrt{3y} + 7\sqrt{3y}$

8. $1.3\sqrt[3]{x} + 4.6\sqrt[3]{x} - 8.6\sqrt[3]{x}$

9. $\sqrt{10x} + 3\sqrt{40x} - 2\sqrt{90x}$

10. $3\sqrt{64p} - \sqrt{100p} - 5\sqrt{49p}$

In Exercises 11–20, determine whether the expression is a polynomial. If it is a polynomial, give its degree. If it is not a polynomial, explain why.

11. $\pi x^2 y^4$

12. $\dfrac{18x^3y^2}{z}$

13. $3x - y^{-1}$

14. $7x^2y^3 - 4z^3$

15. $\sqrt{3}x^2 - 9x + \pi$

16. $5x^{-2}y + z + \sqrt{7}$

17. $9x^3y^4 - 3xy^7 + 9xy - 3$

18. $3x^{1/2}y^3 - 7x + 4y - 8\pi$

19. $3x^2 + y^3 + z^4 + \sqrt{x^2 + y^2}$

20. $\dfrac{7x}{3} + \dfrac{9x^2}{7} + 2x^3 + 8x^4$

In Exercises 21–60, perform the indicated operations.

21. $(3m^2 + 2m - 9) + (4m^2 - 7m)$

22. $(5x^2 - 3x + 4) + (4 - 3x - 5x^2)$

23. $(x^2 + 3xy) - (x^2 + 2yx) - (x^2 - xy)$

24. $(-xy^2 + 4x^2y) - (2yx^2 - 4y^2x)$

25. $-4m^2(8m^3n - 2m^2 + m - 1)$

26. $(4x^2y^3 + xy + y^2)5x^2y^3$

27. $(x^2 - 9y)(3x + 7)$

28. $(2mn + m)(n - m)$

29. $(3x - y)(4x + y)$

30. $(9 - 2x^2)(5 - 3x^2)$

31. $(3x + y)(3x - y)$

32. $(2x + 5)(2x - 5)$

33. $(n - 2)(2n^2 + 5n - 2)$

34. $(p^2 + 4p - 2)(p^2 - p - 3)$

35. $(x + 3)(x^2 - 3x + 9)$

36. $(9x^2 - 6x + 4)(3x + 2)$

37. $(4x^2 + 2x + 1)(2x - 1)$

38. $(2x - 5)(4x^2 + 10x + 25)$

39. $\left(x^2 + x\sqrt{3} + 3\right)\left(x\sqrt{3} - 3\right)$

40. $\left(\sqrt{2} + 3a\right)\left(\sqrt{2} - 5a\right)$

41. $(m + 6)^2$

42. $(y - 8)^2$

43. $(2a - 3b^2)^2$

44. $(3x^3 + 4y)^2$

45. $(3x^3 + 5)^2$

46. $(6a - 5b^2)^2$

47. $(x + 3)^3$

48. $(2 - y)^3$

49. $(4x - 3y^2)^3$

50. $(2m^3 + 5)^3$

51. $n - 2(2n^2 + 5n - 2)$

52. $p^2 + 4p - 2(p^2 - p - 3)$

53. $-m^3 - m[8m - 2m(m + 4)]$

54. $a^2 - [a(2a + b) - (2a - b)(a + b)]$

55. $8x(2x - 1)^2$

56. $1 - 3y(3y + 5)^2$

57. $[(a + 4)(a - 4)]^2$

58. $[(4a - 1)(4a + 1)]^2$

59. $-3x[(x - 2) - 2(3x - 1)^3]$

60. $-x^2 - \{-[x + (3x + 2)(3x - 2)]\}$

In Exercises 61–68, perform the indicated operations.

61. $(3x^{-2} + y) + \left(\dfrac{1}{x^2} - y\right)$

62. $(5t - t^{1/2}) - (2t - \sqrt{t})$

63. $\left(\sqrt{x} + \sqrt{3y}\right)\left(\sqrt{x} - \sqrt{3y}\right)$

64. $\left(\sqrt{x} + 3\right)\left(\sqrt{x} - 5\right)$

65. $(2\sqrt{x} - 7y)^2$

66. $(2x^{-1/3} - x)^3$

67. $7x^{1/3} - (x^{1/3} + 4)(x^{1/3} + 2)$

68. $15x + 2\sqrt{x} - \left(3\sqrt{x} - 2\right)\left(5\sqrt{x} + 4\right)$

◆ **Critical Thinking**

69. Suppose one polynomial P has degree m and another polynomial Q has degree n, where $m > n$. What is the degree of the sum $P + Q$? of the difference $P - Q$? of the product PQ? of the power P^2?

70. Develop a formula for the expansion of $(A + B)^4$. Then use this formula to expand each binomial:

(a) $(2x + 1)^4$ (b) $(x^2 - 2y)^4$

71. The area of a rectangle is the product of its length and width. Find a polynomial that describes the area of the shaded portion of the rectangle shown in the figure.

72. The volume of a rectangular solid is the product of its length, width, and height. Find a polynomial that describes the volume of the rectangular solid shown in the figure.

73. The area of a circle is the product of π and its radius squared. Find a polynomial that describes the area of the shaded portion of the circle shown in the figure.

74. The volume of a sphere is the product of $4\pi/3$ and its radius cubed. Find a polynomial that describes the volume of the sphere in the figure.

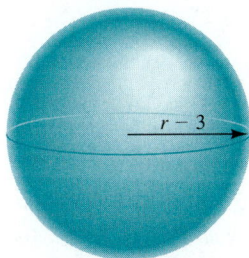

75. Most calculators are programmed to follow the order of operations discussed in this section. Evaluate each expression by following the order of operations, and then compare the answer to the one you obtain by pressing the keys on your calculator in the order written in the expression. Do your answers agree?

(a) $28 + 12 \div 2^2 - 4 \cdot 3$

(b) $(28 + 12) \div (2^2 - 4 \cdot 3)$

76. Use the power key, $\boxed{\wedge}$ or $\boxed{y^x}$, on your calculator to evaluate $(1.01)^3$. Then evaluate $(1 + .01)^3$ by using the special product $(A + B)^3$. Do your answers agree?

77. Each teacher in a school system receives a 6% raise. Find a polynomial that describes a teacher's new salary if x is the previous salary.

78. Suppose a sum of $1000 earns simple interest at the rate of 9% per year, and a sum of $500 earns simple interest at the rate of 6% per year. Find a polynomial that describes the total amount to which these sums will accumulate after t years.

79. A steel beam 36.6 feet long and fixed at both ends carries a uniformly distributed load of 25,000 pounds (lb) as shown in the figure. The deflection of the beam (in inches) at any distance x from one of the fixed ends is given by

$$4.72 \times 10^{-6}x^2(36.6 - x)^2.$$

(a) Write this expression as a trinomial, recording the numerical coefficients in scientific notation rounded to three significant digits. (See Appendix for a review of significant digits.)

(b) The maximum deflection for this beam occurs at midspan (18.3 ft). Determine the maximum deflection to the nearest hundredth of an inch.

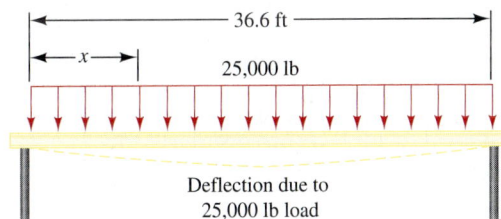

80. A compressor powers the pneumatic tube system used at drive-through bank lanes to carry transactions between a customer and the teller. The horsepower of the compressor designed to carry a tube with a 4-inch diameter a distance of 60 feet is given by

$$0.032p^{3/2}(0.1815 + 0.00348p)$$

where p is the air pressure in the system, in pounds per square inch (psi). Find the product, recording the numerical coefficients in scientific notation rounded to three significant digits.

1.6 Factoring Techniques

◆ **Introductory Comments**

The process of expressing a polynomial as a product of other polynomials is called **factoring** the polynomial. Factoring provides the foundation for simplifying algebraic fractions and for solving several types of equations and inequalities.

In this section our concern is with factoring a polynomial into polynomials that have integer coefficients. This is called *factoring a polynomial over the set of integers*. If a polynomial cannot be written as the product of two polynomials of positive degree that have integer coefficients, it is said to be **prime** over the set of integers. Here are some examples of polynomials that are prime over the set of integers:

$$2x - 3 \qquad x^2 + 1 \qquad x^2 + x - 3 \qquad 3xy^2$$

When a polynomial is written as the product of prime factors, it is said to be **factored completely**. To factor a polynomial completely, we first *factor out* common factors by applying the distributive property from right to left:

$$ab + ac = a(b + c)$$

a is common to each term | *a* is factored out

The process of factoring out common factors is called **common term factoring**. Here are some examples of common term factoring:

$$6x - 9 = 3(2x - 3)$$
$$5x^2(x + 1) + 5(x + 1) = 5(x + 1)(x^2 + 1)$$
$$3x^3y^2 + 3x^2y^2 - 9xy^2 = 3xy^2(x^2 + x - 3)$$

Each of these polynomials is *factored completely over the set of integers*. We can check the factoring process by finding the product of the factors. Monomial factors, such as $3xy^2$, are always considered prime.

◆ **Factoring by Grouping Terms**

From our knowledge of multiplication of polynomials, we know that

$$(a + b)(c + d) = (c + d)(a + b) = ac + ad + bc + bd.$$

Now, consider factoring $ac + ad + bc + bd$. No factor is common to all four terms; however, by grouping the first two terms and the last two terms, we are able to factor by common term factoring:

$$ac + ad + bc + bd = (ac + ad) + (bc + bd) \qquad \text{\textbf{Group terms}}$$

$$= a(c + d) + b(c + d) \qquad \text{\textbf{Factor out common factors from each group}}$$

$$= (a + b)(c + d) \qquad \text{\textbf{Factor out the common binomial } } (c + d)$$

We refer to this process as **factoring by grouping terms**. The key to factoring by grouping terms is to develop a common binomial that can be factored out of each group. As illustrated in Example 1(b), this may require factoring out a negative factor from one of the groups.

EXAMPLE 1   Factor completely over the set of integers.

(a) $x^3 - 3x^2 + 5x - 15$ **(b)** $9 + 2xy - 6x - 3y$

◆ **Solution**

(a) $\begin{aligned} x^3 - 3x^2 + 5x - 15 &= (x^3 - 3x^2) + (5x - 15) \\ &= x^2(x - 3) + 5(x - 3) \\ &= (x - 3)(x^2 + 5) \end{aligned}$

(b) $\begin{aligned} 9 + 2xy - 6x - 3y &= (9 - 6x) + (2xy - 3y) \\ &= 3(3 - 2x) - y(3 - 2x) \end{aligned}$

Factor out $-y$ to obtain the common binomial $(3 - 2x)$.

$$= (3 - 2x)(3 - y)$$ ◆

PROBLEM 1   Terms may usually be grouped in more than one way. Group the first and last terms and the second and third terms of $9 + 2xy - 6x - 3y$, and then factor. You should obtain the same result as in Example 1(b). ◆

◆ **Factoring Trinomials with Binomial Factors**

Binomials such as $(3x + 4)$ and $(x + 2)$ are called **like binomials**, because their first terms are like terms and their second terms are also like terms. The product of two like binomials is a trinomial:

$$(3x + 4)(x + 2) = 3x^2 + 6x + 4x + 8 = 3x^2 + 10x + 8$$

Now, consider factoring the trinomial $3x^2 + 10x + 8$. Because the first term's coefficient is 3, the middle term's coefficient is *positive*, and the last term's coefficient is *positive*, we begin by writing the like binomials as follows:

$$\text{inner}$$
$$(3x + \blacksquare)(x + \blacksquare)$$
$$\text{outer}$$

The two boxes must be filled in with numbers whose product is 8 such that the sum of the inner and outer terms is $10x$. We have four possibilities:

$(3x + 8)(x + 1)$ sum of inner and outer terms is $11x$
$(3x + 1)(x + 8)$ sum of inner and outer terms is $25x$
$(3x + 2)(x + 4)$ sum of inner and outer terms is $14x$
$(3x + 4)(x + 2)$ sum of inner and outer terms is $10x$

Hence, $(3x + 4)(x + 2)$ is the correct choice, and we conclude that

$$3x^2 + 10x + 8 = (3x + 4)(x + 2).$$

CAUTION   Not every trinomial can be factored over the set of integers as the product of two like binomials. For example, the only possible factors of $x^2 + 2x + 3$ are $(x + 1)(x + 3)$. However, since the sum of the inner and outer terms of this product is *not* $2x$, we conclude that $x^2 + 2x + 3$ is prime over the set of integers.

EXAMPLE 2   Factor completely over the set of integers.

(a) $x^2 - 9x + 14$ **(b)** $x^6 + x^3 - 6$

(c) $9x^2 + 24xy + 16y^2$ **(d)** $6x^3y - 3x^2y - 18xy$

◆ **Solution**

(a) For $x^2 - 9x + 14$, we begin with the like binomials

$$(x - \blacksquare)(x - \blacksquare)$$

The two boxes must be filled in with numbers whose product is 14 such that the sum of the inner and outer terms is $-9x$. Checking the possibilities, we find that

$$x^2 - 9x + 14 = (x - 2)(x - 7).$$

(b) For $x^6 + x^3 - 6$, we begin with the like binomials.

$$(x^3 + \blacksquare)(x^3 - \blacksquare)$$

The two boxes must be filled in with numbers whose product is 6 such that the sum of the inner and outer terms is x^3. Checking the possibilities, we find that

$$x^6 + x^3 - 6 = (x^3 + 3)(x^3 - 2).$$

(c) For $9x^2 + 24xy + 16y^2$, we begin with the like binomials

$$(3x + \blacksquare y)(3x + \blacksquare y) \quad \text{or} \quad (9x + \blacksquare y)(x + \blacksquare y)$$

The two boxes must be filled in with numbers whose product is 16 such that the sum of the inner and outer terms is $24xy$. Checking the possibilities, we find that

$$9x^2 + 24xy + 16y^2 = (3x + 4y)(3x + 4y) = (3x + 4y)^2$$

(d) For $6x^3y - 3x^2y - 18xy$, we begin by factoring out the common factor $3xy$:

$$6x^3y - 3x^2y - 18xy = 3xy(2x^2 - x - 6)$$

Now, for $2x^2 - x - 6$, we have the like binomials

$$(2x - \boxed{})(x + \boxed{}) \qquad \text{or} \qquad (2x + \boxed{})(x - \boxed{})$$

The two boxes must be filled in with numbers whose product is 6 such that the sum of the inner and outer terms is $-x$. Checking the possibilities, we find that

$$2x^2 - x - 6 = (2x + 3)(x - 2)$$

Hence,

$$6x^3y - 3x^2y - 18xy = 3xy(2x + 3)(x - 2) \qquad \blacklozenge$$

Referring to Example 2(c), since $9x^2 + 24xy + 16y^2$ may be expressed as the *perfect square* $(3x + 4y)^2$, we refer to $9x^2 + 24xy + 16y^2$ as a **perfect square trinomial**.

PROBLEM 2 Show that $x^2 + 10x + 25$ is a perfect square trinomial. ◆

◆ Factoring Formulas

From our knowledge of multiplication of polynomials, we have the following three products:

1. $(A + B)(A - B) = A^2 - AB + AB - B^2$
$$= A^2 - B^2$$

2. $(A + B)(A^2 - AB + B^2) = A^3 - A^2B + AB^2 + A^2B - AB^2 + B^3$
$$= A^3 + B^3$$

3. $(A - B)(A^2 + AB + B^2) = A^3 + A^2B + AB^2 - A^2B - AB^2 - B^3$
$$= A^3 - B^3$$

These results give us three *factoring formulas*.

◆ **Factoring Formulas**

> **1. Difference of two squares:** $A^2 - B^2 = (A + B)(A - B)$
>
> **2. Sum of two cubes:** $A^3 + B^3 = (A + B)(A^2 - AB + B^2)$
>
> **3. Difference of two cubes:**
> $A^3 - B^3 = (A - B)(A^2 + AB + B^2)$

EXAMPLE 3   Factor completely over the set of integers.

(a) $9x^2 - 4y^2$ (b) $x^6 + 8$

(c) $x^6 + x^3 - 2$ (d) $2x^4 - 2$

◆ **Solution**

(a) Applying the difference of two squares formula, we obtain

$$9x^2 - 4y^2 = (3x)^2 - (2y)^2$$
$$= (3x + 2y)(3x - 2y)$$

(b) Applying the sum of two cubes formula, we obtain

$$x^6 + 8 = (x^2)^3 + (2)^3$$
$$= (x^2 + 2)[(x^2)^2 - (x^2)(2) + (2)^2]$$
$$= (x^2 + 2)(x^4 - 2x^2 + 4)$$

(c) First, we factor the trinomial into two like binomials, and then we factor one binomial as the difference of two cubes:

$$x^6 + x^3 - 2 = (x^3 + 2)(x^3 - 1)$$
$$= (x^3 + 2)(x - 1)(x^2 + x + 1)$$

(d) First, we factor out the common factor 2, and then we factor the difference of two squares twice:

$$2x^4 - 2 = 2(x^4 - 1)$$
$$= 2(x^2 + 1)(x^2 - 1)$$
$$= 2(x^2 + 1)(x + 1)(x - 1)$$

The sum of squares $x^2 + 1$ is prime over the set of integers.   ◆

PROBLEM 3   Factor completely over the set of integers.

(a) $(x + y)^2 - 25y^4$ (b) $64 - x^3$ ◆

Exercises 1.6

◆ Basic Skills

In Exercises 1–74, factor completely over the set of integers.

1. $3a^2b - ab$

2. $48x^2y - 36xy^2$

3. $6x^3y - 9x^2y + 3xy^2$

4. $6x^2y^3 - 18x^3y^4 + 12x^3y^3$

5. $5(x - 1) - x(x - 1)$

6. $5n(2n - 7) + n(2n - 7)$

7. $6m(m + 3)^3 + m(m + 3)^3$

8. $3p(p + 2)^2 + 9p^2(p + 2)^2$

9. $ac + ad + bc + bd$ 10. $xy + 3x + 2y + 6$

11. $x^3 - 3x^2 + 2x - 6$

12. $m^4 - 2m^3 - 3m + 6$

13. $6 - 3x - 2y + xy$

14. $15 + 6y - 20x - 8xy$

15. $15 + 2x^2y - 3x^2 - 10y$

16. $3x^3 - 20y - 15x^2 + 4xy$

17. $a^2 + 3a - 10$

18. $y^2 + 14y + 45$

19. $x^2 - 11xy + 18y^2$

20. $p^2 - pq - 12q^2$

21. $24 + 14x + x^2$

22. $24 - 2m + m^2$

23. $x^4 + 3x^2 - 18$

24. $x^6 + 2x^3 - 63$

25. $2x^2 + 5x + 3$

26. $3x^2 + 8x + 4$

27. $3a^2 + 11ab - 20b^2$

28. $5x^2 + 18xy - 8y^2$

29. $4x^2 - 11x - 3$

30. $6p^2 + 23p - 4$

31. $15 - 16t + 4t^2$

32. $12 - 17t + 6t^2$

33. $12x^2 - 23xy + 10y^2$

34. $18x^2 + 23xy - 6y^2$

35. $2n^4 + 7n^2 + 5$

36. $3y^4 - 13y^2 + 10$

37. $6x^6 - 17x^3y^2 + 10y^4$

38. $10x^8 - x^4y - 3y^2$

39. $8y^{12} + 13y^6 - 6$

40. $12a^8 + 8a^4 - 15$

41. $x^2 - 81$

42. $x^2 - 36$

43. $25x^2 - y^2$

44. $x^2 - 16y^4$

45. $4(x^2 + 5)^2 - 25x^2$

46. $49 - (m + n)^2$

47. $t^3 + 8$

48. $125 + a^3$

49. $27x^3 + 8y^6$

50. $y^3 - 27$

51. $8 - 125n^3$

52. $64n^3 - 1$

53. $t^2 + 8t + 16$

54. $25 - 10x + x^2$

55. $4a^4 - 12a^2b^2 + 9b^4$

56. $9p^6 - 6p^3 + 1$

57. $4x^3y - 28x^2y + 48xy$

58. $3x^4y^2 - 6x^3y^3 - 45x^2y^4$

59. $6x^3 - 8x^2 - 8x$

60. $12x^3 + 3x^2 - 9x$

61. $5a^3 - 45a$

62. $2m^3n - 32mn$

63. $16x^4y + 2xy$

64. $3x^5y^2 - 81x^2y^2$

65. $2n^4 - 32$

66. $81x - 16x^5$

67. $4x^4 - 13x^2 + 9$

68. $9x^4 - 37x^2y^2 + 4y^2$

69. $3a^7 + 6a^4 - 9a$

70. $2x^6y + 26x^3y + 80y$

71. $8x^3y + 8x^2y - 18xy - 18y$

72. $6x^4 + 3x^3 - 24x^2 - 12x$

73. $x^6 - 1$

74. $64 - x^6$

◆ Critical Thinking

In Exercises 75–78, group three terms and factor completely over the integers.

75. $x^2 - 4x + 4 - y^2$

76. $x^2 + 12x + 36 - 9y^2$

77. $4x^2 + 12x - 4y^2 + 9$

78. $9x^2 - 16y^2 - 30x + 25$

In Exercises 79–82, use the factoring formulas given in this section to factor each polynomial over the set of real numbers. [Hint: To factor a polynomial over the set of real numbers, factor the polynomial into polynomials whose coefficients are real numbers.]

79. $x^2 - 10$

80. $3t^2 - 4y^2$

81. $a^3 + 5b^3$

82. $8x^3 - 2$

In Exercises 83–86, use the difference of squares formula to factor each polynomial over the set of complex numbers. [Hint: To factor a polynomial over the set of complex numbers, factor the polynomial into polynomials whose coefficients are complex numbers.]

83. $x^2 + 25$

84. $y^2 + 18$

85. $t^4 - 36$

86. $t^4 - 256$

87. The total amount S of aluminum used to make a beer can is given by the formula

$$S = 2\pi rht + 2\pi r^2t$$

where r is the radius, h is the height, and t is the thick-ness of the beer can. Factor the expression on the right-hand side of this formula.

88. Finding the number of households (in units of 1000) that use a certain product after it has been on the market for n months involves working with the expression

$$32n - \tfrac{4}{3}n^2.$$

Factor this expression so that the coefficient of n in the binomial factor is 1.

89. Finding the response time of mice to a certain stimulus involves working with the expression

$$10.24 + 9.24A^2 - A^4$$

where A is the age (in months) of the subject receiving the stimulus. Given that one factor of this expression is $1 + A^2$, find two other factors with rational coefficients.

90. The velocity V of sound in seawater may be approximated by the formula

$$V = 4800 + 14t - 0.12t^2$$

where t is the temperature of the seawater. Factor the expression on the right-hand side of this formula so that the coefficient of t in one of the binomial factors is 1.

1.7 Algebraic Fractions

◆ **Introductory Comments**

The quotient of two algebraic expressions is called an **algebraic fraction**. If the numerator and denominator of an algebraic fraction are polynomials, then the algebraic fraction is referred to as a **rational expression**. Since division by zero is undefined, we must restrict all the variables in a rational expression to those values that give a nonzero denominator. Thus, when working with a rational expression such as

$$\frac{4y^2 + 5y + 3}{2y - 6},$$

we assume that its denominator $2y - 6 \neq 0$, which implies that $y \neq 3$.

We use the **fundamental property of fractions** to build up an algebraic fraction to higher terms or to reduce an algebraic fraction to lower terms.

◆ **Fundamental Property of Fractions**

For the fraction $\dfrac{a}{b}$ (provided $b \neq 0$), and any real number k (except $k = 0$),

$$\frac{a}{b} = \frac{ak}{bk}.$$

Here are examples of building up a rational expression and reducing a rational expression:

$$\frac{2x}{3} = \frac{2x \cdot 5x}{3 \cdot 5x} = \frac{10x^2}{15x}, x \neq 0 \qquad \frac{10x^2}{15x} = \frac{2x \cdot 5x}{3 \cdot 5x} = \frac{2x}{3}, x \neq 0$$

building up a rational expression to higher terms

reducing a rational expression to lower terms

Two algebraic fractions are said to be **equivalent fractions** if they have the same numerical value for all numbers for which the fractions are defined. The rational expressions $\dfrac{2x}{3}$ and $\dfrac{10x^2}{15x}$ are equivalent fractions, provided $x \neq 0$.

◆ **Reducing a Rational Expression to Lowest Terms**

We say that a rational expression is *reduced to lowest terms* when its numerator and denominator contain no common factor other than ± 1 and we have expressed $-a/(-b)$ as a/b. To reduce a rational expression to lowest terms, we factor its numerator and denominator completely over the set of integers and cancel any common factors. Here is an example of reducing a rational expression to lowest terms:

$$\frac{8x^3y + 12x^2y}{8x^3y} = \frac{4x^2\overset{1}{\cancel{y}}(2x + 3)}{\underset{2}{8x^3\cancel{y}}} = \frac{2x + 3}{2x}, \quad x \neq y \neq 0$$

CAUTION We can cancel common *factors* of the numerator and denominator, but never common *terms*. To reduce the fraction as

$$\frac{\overset{1}{\cancel{8x^3y}} + 12x^2y}{\underset{1}{\cancel{8x^3y}}} \quad \text{is WRONG!}$$

EXAMPLE 1 Reduce each rational expression to lowest terms.

(a) $\dfrac{9 - x^2}{x^2 - 2x - 15}$ **(b)** $\dfrac{1 - x}{x^3 - 1}$

◆ **Solution**

(a) $\dfrac{9 - x^2}{x^2 - 2x - 15} = \dfrac{(3 + x)(3 - x)}{(x + 3)(x - 5)}$ **Factor completely**

$\qquad\qquad\qquad = \dfrac{3 - x}{x - 5}, \quad x \neq -3$ **Reduce to lowest terms**

(b) $\dfrac{1 - x}{x^3 - 1} = \dfrac{1 - x}{(x - 1)(x^2 + x + 1)}$ **Factor completely**

$\qquad\qquad = \dfrac{-(x - 1)}{(x - 1)(x^2 + x + 1)}$ **Use $a - b = -(b - a)$**

$\qquad\qquad = \dfrac{-1}{x^2 + x + 1}, \quad x \neq 1$ **Reduce to lowest terms** ◆

PROBLEM 1 Reduce the rational expression to lowest terms:

$$\frac{16 - x^2}{2x^3 - 5x^2 - 12x}$$ ◆

◆ **Multiplying and Dividing Rational Expressions**

To multiply and divide rational expressions, we use the *multiplication and division properties of fractions.*

◆ **Multiplication and Division Properties of Fractions**

For fractions $\dfrac{a}{b}$ and $\dfrac{c}{d}$ (provided $b \neq 0$ and $d \neq 0$),

Property   **Example**

Multiplication Property

$$\frac{a}{b} \cdot \frac{c}{d} = \frac{ac}{bd} \qquad \frac{2x}{3} \cdot \frac{4x}{5} = \frac{2x \cdot 4x}{3 \cdot 5} = \frac{8x^2}{15}$$

Division Property

$$\frac{a}{b} \div \frac{c}{d} = \frac{a}{b} \cdot \frac{d}{c} = \frac{ad}{bc} \qquad \frac{2x}{3} \div \frac{4x}{5} = \frac{2x}{3} \cdot \frac{5}{4x} = \frac{\overset{1}{2x} \cdot 5}{3 \cdot \underset{2}{4x}}$$

(provided $c \neq 0$)

$$= \frac{5}{6}, \; x \neq 0$$

Before applying these properties to a rational expression, we factor its numerator and denominator completely over the set of integers. In this way, common factors can be cancelled first, yielding a product that is reduced to lowest terms.

EXAMPLE 2   Perform the indicated operations and reduce to lowest terms.

(a) $\dfrac{x-6}{2x^2} \cdot \dfrac{x^2-x}{36-x^2}$

(b) $\dfrac{3x-3}{2x^2+x-6} \div \dfrac{x^2-1}{4x^2-12x+9}$

◆ **Solution**

(a) $\dfrac{x-6}{2x^2} \cdot \dfrac{x^2-x}{36-x^2} = \dfrac{(x-6)\cdot x(x-1)}{2x^2 \cdot (6+x)(6-x)}$ Factor completely and apply the multiplication property

$$= \frac{-(6-x)\cdot x(x-1)}{2x^2 \cdot (6+x)(6-x)}$$ Use $a-b = -(b-a)$

$$= \frac{-(x-1)}{2x(6+x)}, \quad x \neq 0, 6$$ Reduce to lowest terms

(b) $\dfrac{3x-3}{2x^2+x-6} \div \dfrac{x^2-1}{4x^2-12x+9}$

$$= \frac{3x-3}{2x^2+x-6} \cdot \frac{4x^2-12x+9}{x^2-1}$$ Apply the division property

$$= \frac{3(x-1)\cdot(2x-3)^2}{(2x-3)(x+2)\cdot(x+1)(x-1)}$$ Factor completely and apply the multiplication property

$$= \frac{3(2x-3)}{(x+2)(x+1)}, \quad x \neq 1, \tfrac{3}{2}$$ Reduce to lowest terms ◆

PROBLEM 2 Find the quotient: $\dfrac{x^2}{2x^2 - 5x + 2} \div \dfrac{4x}{4 - x^2}$ ◆

◆ **Adding and Subtracting Rational Expressions**

To add and subtract rational expressions, we use the *addition and subtraction properties of fractions.*

◆ **Addition and Subtraction Properties of Fractions**

For fractions $\dfrac{a}{b}$ and $\dfrac{c}{b}$ (provided $b \neq 0$),

Property	*Example*
Addition Property	
$\dfrac{a}{b} + \dfrac{c}{b} = \dfrac{a + c}{b}$	$\dfrac{2x}{3} + \dfrac{5x}{3} = \dfrac{2x + 5x}{3} = \dfrac{7x}{3}$
Subtraction Property	
$\dfrac{a}{b} - \dfrac{c}{b} = \dfrac{a - c}{b}$	$\dfrac{2x}{3} - \dfrac{5x}{3} = \dfrac{2x - 5x}{3}$
	$= \dfrac{-3x}{3} = -x$

After applying these properties, always look for common factors to see if the sum or difference can be reduced to lower terms.

EXAMPLE 3 Perform the indicated operations.

(a) $\dfrac{x + y}{xy} - \dfrac{y - 6x}{xy}$ **(b)** $\dfrac{2x + 4}{x - 5} - \dfrac{x + 3}{5 - x}$

◆ **Solution**

(a) $\dfrac{x + y}{xy} - \dfrac{y - 6x}{xy} = \dfrac{(x + y) - (y - 6x)}{xy}$ Apply the subtraction property. (Be sure to use parentheses.)

$= \dfrac{x + y \ominus y \oplus 6x}{xy}$ Remove parentheses

$= \dfrac{7x}{xy} = \dfrac{7}{y}, \quad x \neq 0$ Combine like terms and reduce to lowest terms

(b) $\dfrac{2x+4}{x-5} - \dfrac{x+3}{5-x} = \dfrac{2x+4}{x-5} \ominus \dfrac{x+3}{\ominus(x-5)}$ Use $a-b = -(b-a)$

$= \dfrac{2x+4}{x-5} + \dfrac{x+3}{x-5}$ Use $a/(-b) = -(a/b)$

$= \dfrac{(2x+4)+(x+3)}{x-5}$ Apply the addition property

$= \dfrac{3x+7}{x-5}$ Combine like terms ◆

PROBLEM 3 Find the difference: $\dfrac{2x^2-9}{2x^2+11x+5} - \dfrac{x^2+16}{2x^2+11x+5}$ ◆

To add or subtract algebraic fractions with *unlike denominators,* we first find their **least common denominator**, which is denoted as **LCD**. The following two-step procedure may be used to find the LCD.

◆ **Procedure for Determining the LCD**

Example: $\dfrac{x-1}{x^2+x} - \dfrac{2x}{x^2+2x+1}$

1. Factor completely each denominator, and use exponential notation to represent repeated prime factors that occur in any one of the denominators.

1. $\dfrac{x-1}{\boxed{x}(x+1)} - \dfrac{2x}{\boxed{(x+1)^2}}$

2. The product of each *different prime factor* to the *highest power* it occurs in any one of the denominators is the LCD.

2. LCD is $x(x+1)^2$.

After determining the LCD, we apply the fundamental property of fractions and build up the fractions to equivalent fractions, each having the LCD as its new denominator. Once the denominators are the same, we use the addition and subtraction properties of fractions to find the sum or difference of the fractions.

EXAMPLE 4 Perform the indicated operations.

(a) $\dfrac{x-1}{x^2+x} - \dfrac{2x}{x^2+2x+1}$

(b) $\dfrac{8}{4-y^2} + \dfrac{y}{2+y} - \dfrac{1}{y-2}$

◆ Solution

(a) $\dfrac{x-1}{x^2+x} - \dfrac{2x}{x^2+2x+1} = \dfrac{x-1}{x(x+1)} - \dfrac{2x}{(x+1)^2}$

> Factor the denominators.
> The LCD is $x(x+1)^2$.

$$= \dfrac{(x-1)\cdot(x+1)}{x(x+1)\cdot(x+1)} - \dfrac{2x\cdot x}{(x+1)^2\cdot x}$$

$$= \dfrac{x^2-1}{x(x+1)^2} - \dfrac{2x^2}{x(x+1)^2}$$

$$= \dfrac{-x^2-1}{x(x+1)^2}$$

(b) $\dfrac{8}{4-y^2} + \dfrac{y}{2+y} - \dfrac{1}{y-2}$

$$= \dfrac{8}{(2+y)(2-y)} + \dfrac{y}{2+y} \oplus \dfrac{1}{2-y}$$

> Factoring and rearranging the last fraction
> yields an LCD of $(2+y)(2-y)$.

$$= \dfrac{8}{(2+y)(2-y)} + \dfrac{y\cdot(2-y)}{(2+y)\cdot(2-y)} + \dfrac{1\cdot(2+y)}{(2-y)\cdot(2+y)}$$

$$= \dfrac{8+(2y-y^2)+(2+y)}{(2+y)(2-y)}$$

$$= \dfrac{10+3y-y^2}{(2+y)(2-y)}$$

We now reduce this answer to lowest terms:

$$\dfrac{10+3y-y^2}{(2+y)(2-y)} = \dfrac{(5-y)(2+y)}{(2+y)(2-y)} = \dfrac{5-y}{2-y}, \quad y \neq -2 \quad ◆$$

PROBLEM 4 Find the difference: $\dfrac{x+1}{x^2-2x+1} - \dfrac{1}{x-1}$ ◆

◆ The Algebra of Calculus

We conclude this section by looking at some of the algebraic support that is needed in calculus. We look at three ideas:

1. Rationalizing the numerator of an algebraic fraction.

2. Simplifying a complex fraction.

3. Writing an expression containing negative exponents as an algebraic fraction.

Note: In calculus, the ratio of the change in the variable y to the change in the variable x is denoted by $\Delta y/\Delta x$ (read "delta y to delta x"). In this text, we use the logo $\boxed{\frac{\Delta y}{\Delta x}}$ in the exercise sets to identify problems that are calculus-related.

To rationalize the numerator of an algebraic fraction that contains the form $\sqrt{a} - \sqrt{b}$, we apply the fundamental property of fractions and multiply both numerator and denominator by $\sqrt{a} + \sqrt{b}$. In computing the product $(\sqrt{a} - \sqrt{b})(\sqrt{a} + \sqrt{b})$, we use the fact that $(A - B)(A + B) = A^2 - B^2$:

$$(\sqrt{a} - \sqrt{b})(\sqrt{a} + \sqrt{b}) = (\sqrt{a})^2 - (\sqrt{b})^2 = a - b$$

Each of the factors $\sqrt{a} - \sqrt{b}$ and $\sqrt{a} + \sqrt{b}$ is called the **conjugate** of the other.

EXAMPLE 5　Rationalize the numerator of $\dfrac{\sqrt{x + h} - \sqrt{x}}{h}$ and simplify the result.

◆ **Solution**　The conjugate of the numerator is $\sqrt{x + h} + \sqrt{x}$. Thus, we multiply both numerator and denominator by this conjugate, and then simplify:

$$\frac{\sqrt{x + h} - \sqrt{x}}{h}$$

$$= \frac{(\sqrt{x + h} - \sqrt{x}) \cdot (\sqrt{x + h} + \sqrt{x})}{h \cdot (\sqrt{x + h} + \sqrt{x})} \qquad \text{\color{red}{\textbf{Apply the fundamental property of fractions}}}$$

$$= \frac{(x + h) - (x)}{h \cdot (\sqrt{x + h} + \sqrt{x})} \qquad \text{\color{red}{\textbf{Multiply}}}$$

$$= \frac{\cancel{h}}{\cancel{h} \cdot (\sqrt{x + h} + \sqrt{x})} \qquad \text{\color{red}{\textbf{Simplify the numerator}}}$$

$$= \frac{1}{\sqrt{x + h} + \sqrt{x}} \qquad \text{\color{red}{\textbf{Reduce to lowest terms}}}$$

◆

PROBLEM 5　Rationalize the numerator and simplify:

$$\frac{\sqrt{(x + h) + 1} - \sqrt{x + 1}}{h}$$

◆

An algebraic fraction that contains fractions in its numerator, denominator, or both its numerator and denominator is called a **complex fraction**. One method of simplifying a complex fraction is to first find the LCD for all fractions appearing in its numerator and denominator. After this is accomplished, we multiply both numerator and denominator of the complex fraction by this LCD and then simplify the resulting expression. The procedure is illustrated in the next example.

EXAMPLE 6　Simplify the complex fraction $\dfrac{\dfrac{3}{x + h} - \dfrac{3}{x}}{h}$.

◆ **Solution**   The LCD for $(x + h)$ and x is $x(x + h)$. Thus, we multiply numerator and denominator of this complex fraction by $x(x + h)$ and then simplify the resulting expression as follows:

$$\frac{\dfrac{3}{x + h} - \dfrac{3}{x}}{h} = \frac{\left(\dfrac{3}{x + h} - \dfrac{3}{x}\right) \cdot [x(x + h)]}{h \cdot [x(x + h)]} \qquad \text{\color{red}\textbf{Apply the fundamental property of fractions}}$$

$$= \frac{3x - 3(x + h)}{h \cdot [x(x + h)]} \qquad \text{\color{red}\textbf{Multiply and reduce}}$$

$$= \frac{-3h}{h \cdot [x(x + h)]} \qquad \text{\color{red}\textbf{Simplify the numerator}}$$

$$= \frac{-3}{x(x + h)} \qquad \text{\color{red}\textbf{Reduce to lowest terms}}$$

◆

PROBLEM 6   As an alternate method of simplifying the complex fraction in Example 6, we can perform the indicated subtraction in the numerator and then divide this result by h. Rework Example 6 using this procedure. Which method to you prefer?   ◆

Algebraic expressions containing negative exponents, such as

$$x[-2(1 - 3x)^{-3}(-3)] + (1 - 3x)^{-2},$$

develop frequently in calculus through a process called differentiation. One method of simplifying such expressions is to factor out the common factor with the *smaller exponent*. The procedure is illustrated in the next example.

EXAMPLE 7   Simplify the expression $x[-2(1 - 3x)^{-3}(-3)] + (1 - 3x)^{-2}$.

◆ **Solution**   First we rewrite $(1 - 3x)^{-2}$ as $(1 - 3x)^{-3}(1 - 3x)^1$, and then we factor out the common factor $(1 - 3x)^{-3}$ from each term:

$$6x(1 - 3x)^{-3} + (1 - 3x)^{-2}$$
$$= 6x(1 - 3x)^{-3} + (1 - 3x)^{-3}(1 - 3x)^1$$
$$= (1 - 3x)^{-3}[6x + (1 - 3x)^1]$$
$$= (1 - 3x)^{-3}(3x + 1)$$
$$= \frac{3x + 1}{(1 - 3x)^3}$$

◆

PROBLEM 7   As an alternate method for simplifying the expression in Example 7, we use $a^{-n} = 1/a^n$ to form a pair of algebraic fractions, and then add the fractions by finding their LCD. Rework Example 7 using this procedure. Which method do you prefer?   ◆

Exercises 1.7

Basic Skills

In Exercises 1–12, reduce the rational expression to lowest terms.

1. $\dfrac{5xy}{5xy - 10x^2y}$

2. $\dfrac{2x - 4x^2}{2x}$

3. $\dfrac{4m - 4n}{12n - 12m}$

4. $\dfrac{9x - 27y}{9y - 3x}$

5. $\dfrac{2x^2 + 11x + 5}{x^2 + 4x - 5}$

6. $\dfrac{16b^2 - a^2}{a^2 + ab - 12b^2}$

7. $\dfrac{n^2 - 5n - 6}{36 - n^2}$

8. $\dfrac{4x^2 - 5x - 6}{8 - x^3}$

9. $\dfrac{x^3 - x}{x^4 + 3x^3 - x^2 - 3x}$

10. $\dfrac{k^4 - 1}{k^8 - 1}$

11. $\dfrac{9 - (x + y)^2}{(x - 3)^2 - y^2}$

12. $\dfrac{x^2 - 2xy + y^2}{x^3 - x^2y - xy^2 + y^3}$

In Exercises 13–38, perform the indicated operations and reduce to lowest terms.

13. $\dfrac{x^2 + 2x + 1}{2xy} \cdot \dfrac{6xy}{x^2 - 1}$

14. $\dfrac{a}{(a - b)^2} \cdot \dfrac{a^2 - b^2}{a^2}$

15. $\dfrac{4 - n^2}{(n - 1)^2} \cdot \dfrac{n^2 + 3n - 4}{n^2 + 2n - 8}$

16. $\dfrac{(x + 2)^3}{x^3 + 8} \cdot \dfrac{x^3 - 2x^2 + 4x}{x^2 + 4x + 4}$

17. $\dfrac{12x^2}{x^2 + 2xy + y^2} \div \dfrac{6x}{x + y}$

18. $\dfrac{m + 3}{5m} \div \dfrac{m^2 - 9}{10m^3}$

19. $\dfrac{3a^3 - 81b^3}{9b^2 - a^2} \div (a^2 + 3ab + 9b^2)$

20. $\dfrac{9 - (x - y)^2}{3xy^2} \div \dfrac{x - y - 3}{3xy}$

21. $\dfrac{3n^2 + 7n + 3}{4n - 1} + \dfrac{5 - n^2}{1 - 4n}$

22. $\dfrac{6x^2 - 12x}{x^2 + 2x + 4} + \dfrac{(x - 2)^3}{x^2 + 2x + 4}$

23. $\dfrac{(x + 1)^2}{3x} - \dfrac{2x + 1}{3x}$

24. $\dfrac{b - a}{2a - b} - \dfrac{a - b}{b - 2a}$

25. $\dfrac{3}{m^2 - 9} + \dfrac{1}{m + 3}$

26. $\dfrac{x + 2}{x - 2} + \dfrac{x - 2}{x + 2}$

27. $\dfrac{1}{(x - 1)^3} + \dfrac{1}{1 - 2x + x^2}$

28. $\dfrac{x + 9}{x^2 - 6x + 9} + \dfrac{2}{x^2 - 9}$

29. $\dfrac{2 - x}{x^2 + 2x} - \dfrac{x}{x^2 + 4x + 4}$

30. $\dfrac{(x - 4)^2}{x + 4} - \dfrac{(x + 4)^2}{x - 4}$

31. $\dfrac{1}{a^2 + a} - \dfrac{1}{a - a^2}$

32. $\dfrac{1}{t - 1} - \dfrac{t^2 + t}{t^3 - 1}$

33. $\dfrac{a + 1}{a^2 + a + 1} + \dfrac{1}{1 - a} + \dfrac{a^2 + 2a + 3}{a^3 - 1}$

34. $\dfrac{x - 3}{3 + x} + \dfrac{x + 3}{3 - x} + \dfrac{36}{x^2 - 9}$

35. $\left(\dfrac{x + y}{x - y} - \dfrac{x - y}{x + y}\right) \div \left(\dfrac{x^2 + y^2}{x^2 - y^2} - \dfrac{x^2 - y^2}{x^2 + y^2}\right)$

36. $\left(x + \dfrac{4}{x}\right)\left(\dfrac{x^2}{4} - 1\right) \div \left(1 - \dfrac{16}{x^4}\right)$

37. $\left(\dfrac{b}{ac} + \dfrac{2}{a} + \dfrac{c}{ab} - \dfrac{a}{bc}\right)\left(1 + \dfrac{b + c}{a - b - c}\right)$

38. $\left(\dfrac{n^2}{25n^2 - 4}\right)\left(5n - 8 - \dfrac{4}{n}\right)\left(5n + 8 - \dfrac{4}{n}\right)$

$\boxed{\frac{\Delta y}{\Delta x}}$ *In Exercises 39–44, rationalize the numerator and reduce to lowest terms.*

39. $\dfrac{3 - \sqrt{t + 3}}{t - 6}$

40. $\dfrac{x - \sqrt{x^2 - 1}}{x + \sqrt{x^2 - 1}}$

41. $\dfrac{\sqrt{2(x + h)} - \sqrt{2x}}{h}$

42. $\dfrac{\sqrt{2 + (x + h)} - \sqrt{2 + x}}{h}$

43. $\dfrac{\sqrt{(x + h)^2 + 1} - \sqrt{x^2 + 1}}{h}$

44. $\dfrac{\sqrt{3(x + h) - 1} - \sqrt{3x - 1}}{h}$

$\frac{\Delta y}{\Delta x}$ *In Exercises 45–50, simplify each complex fraction.*

45. $\dfrac{\dfrac{1}{a} + 1}{\dfrac{1}{a^2} - 1}$

46. $\dfrac{1 - \dfrac{1}{4n^2}}{2 - \dfrac{1}{n}}$

47. $\dfrac{\dfrac{1}{x + h} - \dfrac{1}{x}}{h}$

48. $\dfrac{\dfrac{2}{(x + h) + 1} - \dfrac{2}{x + 1}}{h}$

49. $\dfrac{\dfrac{1}{(x + h)^2} - \dfrac{1}{x^2}}{h}$

50. $\dfrac{\dfrac{5}{\sqrt{x + h}} - \dfrac{5}{\sqrt{x}}}{h}$

$\frac{\Delta y}{\Delta x}$ *In Exercises 51–60, simplify each expression.*

51. $2x[3(x + 3)^2] + (x + 3)^3(2)$

52. $4x^3[5(1 - 2x)^4(-2)] + (1 - 2x)^5(12x^2)$

53. $t[-3(2 - 3t)^{-4}(-3)] + (2 - 3t)^{-3}$

54. $x[-1(3 - 2x)^{-2}(-2)] + (3 - 2x)^{-1}$

55. $\frac{4}{5}n(1 - 4n)^{-6/5} + (1 - 4n)^{-1/5}$

56. $3x[-\frac{1}{3}(3 - x)^{-4/3}(-1)] + 3(3 - x)^{-1/3}$

57. $-t(3 - 4t)^{-3/4} + (3 - 4t)^{1/4}$

58. $\frac{3}{4}y^2(y^2 + 1)^{-2/3} - (y^2 + 1)^{1/3}$

59. $\dfrac{6t^{4/3} - (t^2 + 1)t^{-2/3}}{t^{2/3}}$

60. $\dfrac{(x^2 + 4)^{1/2} - x^2(x^2 + 4)^{-1/2}}{x^2 + 4}$

◆ **Critical Thinking**

61. If we reduce $\dfrac{x - 1}{x^2 - 1}$ to lowest terms, we obtain $\dfrac{1}{x + 1}$. Are these two rational expressions equal for all real values of x? Explain.

$\frac{\Delta y}{\Delta x}$ 62. Rationalize the numerator of

$$\frac{\sqrt[3]{x + h} - \sqrt[3]{x}}{h}$$

by using the rationalizing factor

$$\sqrt[3]{(x + h)^2} + \sqrt[3]{x + h} \cdot \sqrt[3]{x} + \sqrt[3]{x^2}.$$

63. The sum or difference of a polynomial and a rational expression is called a *mixed algebraic expression*. Use the addition and subtraction properties for fractions (reading from right to left) to write each rational expression as a mixed algebraic expression.

(a) $\dfrac{y^2 + 7y - 5}{y}$

(b) $\dfrac{14 - (t + 7)^2}{t + 7}$

64. Use the addition and subtraction properties for fractions (reading from right to left) to write each rational expression as a sum or difference of simpler rational expressions, each reduced to lowest terms.

(a) $\dfrac{12xy^2 - 7x^2y^3 + 6x - 9}{3x^2y}$

(b) $\dfrac{3a(x + y)^2 + (x + y)^3 - 2(x + y)}{2(x + y)^2}$

65. Rationalize the denominator and simplify.

(a) $\dfrac{1}{x - \sqrt{x^2 - 1}}$

(b) $\dfrac{a^2}{\sqrt{a^2 + b^2} + b}$

66. A single lens with focal length 11 cm projects an image of an object onto a screen, as shown in the sketch. The distance (in cm) from the lens to the screen is given by

$$\frac{1}{\dfrac{1}{11} - \dfrac{1}{12}}.$$

Evaluate this expression by (a) simplifying the complex fraction and (b) by using the reciprocal key, $\boxed{1/x}$ or $\boxed{x^{-1}}$, on your calculator.

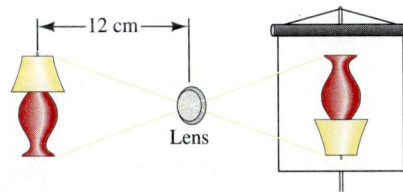

67. If two batteries are connected in parallel, as shown in the figure, and have the same internal resistance r and the same voltage E, then the current i supplied to the external load R is given by the formula

$$i = \frac{E}{R + \dfrac{r}{2}}.$$

(*Exercise 67 continues*)

Simplify the complex fraction that appears on the right-hand side of this formula.

68. If a payment of R dollars is made at the end of each time period for n periods into an annuity that earns interest at the rate of r per period, then the present value A of the annuity is given by the formula

$$A = R\left[\frac{1 - (1 + r)^{-n}}{r}\right].$$

Rewrite the expression on the right-hand side of this formula as a single fraction that contains no negative exponent.

69. Finding the number of crimes committed in a certain part of the city involves working with the expression

$$\frac{60}{2 + \sqrt{n}} - n$$

where n is the number of police that are assigned to the area. Write this expression as a single fraction with a rationalized denominator.

70. The number of items that a machine can produce during an 8-hour day is $t^2 + 16t$, where t is the number of hours that the machine operates. Suppose the total cost to produce these items is $8t + 128$. Find a simplified algebraic expression that represents the *average cost* of production.

1.8　Generating Equivalent Equations and Inequalities

◆ Introductory Comments

An **equation** is a statement declaring that two algebraic expressions are equal. Two examples of equations in one variable are

$$3x + 4x = 7x \quad \text{and} \quad 2x + 1 = 7.$$

The equation $3x + 4x = 7x$ is true no matter what value we choose to replace x. An equation that is true for all permissible values of the variable that it involves is called an **identity**. The equation $2x + 1 = 7$ is true only when we replace x with 3 and is false for all other replacements. An equation that is true for some values of the variable it involves but false for others (or is never true for any value of the variable), is called a **conditional equation**. Any value of the variable that makes an equation a true statement is a **root** of, or **solution** to, the equation. Thus, we say 3 is a root of (or solution to) the equation $2x + 1 = 7$, since 3 is a number that *satisfies* this equation.

An **inequality** is a statement declaring that one algebraic expression is *less than* ($<$), *greater than* ($>$), *less than or equal to* ($\leq$), or *greater than or equal to* ($\geq$) another algebraic expression. Two examples of an inequality in one variable are

$$x + 1 > x \quad \text{and} \quad 2x \geq 6.$$

The inequality $x + 1 > x$ is true no matter what real number we choose to replace x. An inequality that is true for all permissible values of the variable that it involves is called an **absolute inequality**. The inequality $2x \geq 6$ is true only when x is replaced by real numbers greater than or equal to 3. An inequality that is true for some values of the variable it involves but false for others (or is never true for any value of the variable), is called a **conditional inequality**. We refer to values of the variable that make an inequality a true statement as the **solution set** of the inequality. In this text, we use *interval notation* (discussed in Section 1.1) to describe the solution set of an inequality. Hence, we describe the solution set of the inequality $2x \geq 6$ by writing $[3, \infty)$.

◆ **Generating Equivalent Equations**

Equations that have the same solutions are said to be **equivalent equations**. We can generate equivalent equations from a given equation by performing any of the following manipulations.

◆ **Generating Equivalent Equations**

Performing any of the following manipulations on an equation generates an equivalent equation.

Procedure	Example
1. Simplify algebraic expressions that appear on either side.	$9 = 3x - 5x - 1$ is equivalent to $9 = -2x - 1$
2. Add (or subtract) the same quantity to both sides.	$9 = -2x - 1$ is equivalent to $10 = -2x$
3. Multiply (or divide) both sides by the same *nonzero* quantity.	$10 = -2x$ is equivalent to $-5 = x$
4. Interchange the left-hand and right-hand sides.	$-5 = x$ is equivalent to $x = -5$

To **solve** an equation means to find *all* values of the variable that make the equation become a true statement. The usual strategy in solving an equation is to transform the given equation into an equivalent one whose solutions are apparent.

EXAMPLE 1 Solve each equation.

(a) $2y - 11 = 5y - 5$

(b) $\dfrac{2x + 3}{5} - \dfrac{x - 1}{15} = \dfrac{3x - 1}{3}$

(c) $|4t - 3| - 3 = 2$

(d) $5 - 8x^2 = -11$

◆ **Solution**

(a) To solve the equation $2y - 11 = 5y - 5$, we proceed as follows:

$$2y - 11 = 5y - 5$$

$$-11 = 3y - 5 \qquad \text{Subtract } 2y \text{ from both sides}$$

$$-6 = 3y \qquad \text{Add 5 to both sides}$$

$$y = -2 \qquad \text{Divide both sides by 3 and interchange the sides}$$

Since the solution to the last equation is obviously -2, the solution to the original equation must also be -2. By replacing y with -2 in the original equation, we can verify that -2 is a solution.

(b) For equations involving fractions, we eliminate the fractions by multiplying both sides of the equation by the least common denominator (LCD). The LCD for the denominators 5, 15, and 3 is 15. Thus, we multiply both sides of this equation by 15 and cancel the denominators as follows:

$$\frac{2x + 3}{5} - \frac{x - 1}{15} = \frac{3x - 1}{3}$$

$$15\left(\frac{2x + 3}{5} - \frac{x - 1}{15}\right) = 15\left(\frac{3x - 1}{3}\right) \qquad \text{Multiply both sides by 15}$$

$$3(2x + 3) - (x - 1) = 5(3x - 1) \qquad \text{Simplify by canceling denominators}$$

Remove the parentheses

$$6x + 9 - x + 1 = 15x - 5 \qquad \text{Apply the distributive property}$$

$$5x + 10 = 15x - 5 \qquad \text{Collect like terms on the left-hand side}$$

$$10 = 10x - 5 \qquad \text{Subtract } 5x \text{ from both sides}$$

$$15 = 10x \qquad \text{Add 5 to both sides}$$

$$x = \frac{3}{2} \qquad \text{Divide both sides by 10, interchange sides, and reduce}$$

Thus, the solution of the original equation is also $\frac{3}{2}$.

(c) By adding 3 to both sides of the equation $|4t - 3| - 3 = 2$, we obtain the equivalent equation $|4t - 3| = 5$. By the definition of absolute value (Section 1.1), the equation $|4t - 3| = 5$ is equivalent to the following pair of equations:

$$4t - 3 = 5 \qquad \text{or} \qquad -(4t - 3) = 5$$
$$4t = 8 \qquad\qquad\qquad -4t + 3 = 5$$
$$t = 2 \qquad\qquad\qquad\quad -4t = 2$$
$$t = -\frac{1}{2}$$

Thus, the solutions of the original equation are 2 or $-\frac{1}{2}$.

(d) We begin by solving the equation $5 - 8x^2 = -11$ for x^2;

$$5 - 8x^2 = -11$$

$$-8x^2 = -16 \qquad \text{Subtract 5 from both sides}$$

$$x^2 = 2 \qquad \text{Divide both sides by } -8$$

The solutions of $x^2 = 2$ are the numbers whose squares are 2, namely, the square roots of 2. Hence, we *extract the square roots* from this equation and write the solutions $\sqrt{2}$ and $-\sqrt{2}$, or more compactly, $\pm\sqrt{2}$. ◆

Note: In general, the solutions of the equation $x^2 = a$ are $\pm\sqrt{a}$. Each solution is a real number if $a > 0$, an imaginary number if $a < 0$, or zero if $a = 0$.

PROBLEM 1 Try to solve the equation $|4t - 3| + 3 = 2$. Explain why this equation has no solution. ◆

◆ **Generating Equivalent Inequalities**

Inequalities that have the same solution set are said to be **equivalent inequalities**. We can generate equivalent inequalities from a given inequality by performing any of the following manipulations.

◆ **Generating Equivalent Inequalities**

Performing any of the following manipulations on an inequality generates an equivalent inequality.

Procedure	Example
1. Simplify algebraic expressions that appear on either side.	$9 < 3x - 5x - 1$ is equivalent to $9 < -2x - 1$
2. Add (or subtract) the same quantity to both sides.	$9 < -2x - 1$ is equivalent to $10 < -2x$
3. Multiply (or divide) both sides by the same *positive* quantity. Multiply (or divide) both sides by the same *negative* quantity and *reverse the direction of the inequality symbol.*	$10 < -2x$ is equivalent to $-5 > x$
4. Interchange the left-hand and right-hand sides and *reverse the direction of the inequality symbol.*	$-5 > x$ is equivalent to $x < -5$

To **solve** an inequality means to find *all* values of the variable that make the inequality become a true statement. The usual strategy in solving an inequality is to transform the given inequality into an equivalent one whose solution set is apparent.

EXAMPLE 2   Solve each inequality. Use interval notation to describe the solution set.

(a) $5(y + 1) < 2(4 - 3y) + 2y$ **(b)** $\dfrac{2x}{3} - \dfrac{6x - 1}{6} \leq -\dfrac{1}{2}$

(c) $|2x - 3| < 5$ **(d)** $|2 - t| - 3 \geq 1$

◆ **Solution**

(a) To solve this inequality, we proceed as follows:

$$5(y + 1) < 2(4 - 3y) + 2y$$

$$5y + 5 < 8 - 4y \qquad \text{Simplify each side}$$

$$9y + 5 < 8 \qquad \text{Add } 4y \text{ to both sides}$$

$$9y < 3 \qquad \text{Subtract 5 from both sides}$$

$$y < \frac{1}{3} \qquad \begin{array}{l}\text{Divide both sides by 9}\\\text{and reduce}\end{array}$$

Since the solution set of this last inequality is $(-\infty, \frac{1}{3})$, the solution set of the original inequality is also $(-\infty, \frac{1}{3})$.

(b) To solve this inequality, we begin by eliminating fractions as follows:

$$\frac{2x}{3} - \frac{6x - 1}{6} \leq -\frac{1}{2}$$

$$4x - (6x - 1) \leq -3 \qquad \text{Multiply both sides by 6, the LCD}$$

$$-2x + 1 \leq -3 \qquad \text{Simplify the left-hand side}$$

$$-2x \leq -4 \qquad \text{Subtract 1 from both sides}$$

$$x \geq 2 \qquad \begin{array}{l}\text{Divide both sides by } -2 \text{ and reverse}\\\text{the direction of the inequality symbol}\end{array}$$

Since the solution set of this last inequality is $[2, \infty)$, the solution set of the original inequality is also $[2, \infty)$.

(c) From our discussion of absolute value in Section 1.1, we know that the inequality $|2x - 3| < 5$ can be interpreted as the set of all real numbers that are *less than* 5 units from 0. Thus, $|2x - 3| < 5$ is equivalent to the following double inequality:

$$-5 < 2x - 3 < 5$$

$$-2 < \quad 2x \quad < 8 \qquad \text{Add 3 to each member}$$

$$-1 < \quad x \quad < 4 \qquad \text{Divide each member by 2}$$

Hence, the solution set of $|2x - 3| < 5$ is $(-1, 4)$.

(d) Adding 3 to both sides of the inequality $|2 - t| - 3 \geq 1$ gives us the equivalent inequality $|2 - t| \geq 4$. Now, from our discussion of absolute value in Section 1.1, we know that the inequality $|2 - t| \geq 4$ can be interpreted as the set of all real numbers that are *greater than or equal to* 4 units from 0. Thus, $|2 - t| \geq 4$ is equivalent to the following pair of inequalities:

$$\begin{array}{ll} 2 - t \leq -4 & \text{or} \quad 2 - t \geq 4 \\ -t \leq -6 & \qquad -t \geq 2 \\ t \geq 6 & \qquad t \leq -2 \end{array}$$

Hence, the solution set of the original inequality consists of all real numbers belonging to either of the intervals $(-\infty, -2]$ or $[6, \infty)$. Hence, the solution set of $|2 - t| - 3 \geq 1$ is $(-\infty, -2] \cup [6, \infty)$. ◆

PROBLEM 2 Solve the inequality $\dfrac{y + 1}{2} - \dfrac{5(y - 1)}{8} > 1$. ◆

◆ **Literal Equations and Inequalities**

An equation (or inequality) that contains letters other than the variable for which we wish to solve is called a **literal equation** (or **literal inequality**). Here are examples of a literal equation and a literal inequality in which we solve for x;

Literal equation:

$$ax + b = c$$
$$ax = c - b$$
$$x = \frac{c - b}{a}, \quad a \neq 0$$

Literal inequality:

$$ax + b < c$$
$$ax < c - b$$
$$x < \frac{c - b}{a}, \quad a > 0$$
$$\text{or} \quad x > \frac{c - b}{a}, \quad a < 0$$

To solve some literal equations (or inequalities) in which the variable appears *more than once*, we use the rules for generating equivalent equations (or inequalities) and apply the following three steps:

1. *Group* all the variable terms for which we wish to solve on one side of the equation (or inequality).
2. *Factor out* the variable for which we wish to solve.
3. *Divide* each side of the equation (or inequality) by the expression being multiplied by the variable.

The procedure is illustrated in the next example.

EXAMPLE 3 Solve the equation $ax + b = cx + d$ for x.

◆ **Solution** To solve this literal equation for x, we proceed as follows:

$$ax + b = cx + d$$
$$ax - cx = d - b \qquad \textit{Group the x terms on one side}$$
$$x(a - c) = d - b \qquad \textit{Factor out the variable x}$$
$$x = \frac{d - b}{a - c} \qquad \textit{Divide by (a - c),}$$
$$\textit{provided a - c} \neq 0$$
◆

PROBLEM 3   In Example 3, we grouped the x terms on the left-hand side of the equation. Rework Example 3 by grouping the x terms on the right-hand side of the equation and show that the answer is the same as that given in Example 3.   ◆

◆ **Application: Common Formulas**

A **formula** is a mathematical or scientific rule in the form of a literal equation that describes a special relationship between two or more variables. Refer to the formulas listed in the following box as you need them to solve applied problems throughout this text.

◆ **Common Formulas**

Uniform Motion

$$d = rt$$

d is distance
r is rate of speed
t is time

Simple Interest

$$A = P(1 + rt)$$

A is amount accrued
P is amount invested
r is rate of interest
t is time

Temperature

$$F = \tfrac{9}{5}C + 32$$

F is degrees Fahrenheit
C is degrees Celsius

Pythagorean Theorem

$$c^2 = a^2 + b^2$$

a and b are lengths of legs
c is length of hypotenuse

Geometry   (A is area, P is perimeter, C is circumference, V is volume, and S is total surface area)

Rectangle

$$A = lw$$
$$P = 2l + 2w$$

Triangle

$$A = \tfrac{1}{2}bh$$

Circle

$$A = \pi r^2$$
$$C = 2\pi r$$

Trapezoid

$$A = \tfrac{1}{2}h(b_1 + b_2)$$

Rectangular Solid

$$V = lwh$$
$$S = 2lw + 2wh + 2lh$$

Cylinder

$$V = \pi r^2 h$$
$$S = 2\pi r(r + h)$$

Cone

$$V = \tfrac{1}{3}\pi r^2 h$$
$$S = \pi r(r + s)$$

Sphere

$$V = \tfrac{4}{3}\pi r^3$$
$$S = 4\pi r^2$$

EXAMPLE 4

(a) Solve the volume of a sphere formula $V = \dfrac{4}{3}\pi r^3$ for r.

(b) Use the result of part (a) to find the radius of a sphere whose volume is 36π cubic inches.

◆ **Solution**

(a) To solve this formula for r, we proceed as follows:

$$V = \frac{4}{3}\pi r^3$$

$$3V = 4\pi r^3 \qquad \text{\textcolor{red}{\textbf{Multiply both sides by 3}}}$$

$$\frac{3V}{4\pi} = r^3 \qquad \text{\textcolor{red}{\textbf{Divide both sides by } 4\pi}}$$

$$r = \sqrt[3]{\frac{3V}{4\pi}} \qquad \text{\textcolor{red}{\textbf{Extract the real cube root}}}$$

(b) To find the radius of a sphere whose volume is 36π cubic inches, we replace V with 36π:

$$r = \sqrt[3]{\frac{3V}{4\pi}} = \sqrt[3]{\frac{3(36\pi)}{4\pi}} = \sqrt[3]{27} = 3 \text{ inches} \qquad ◆$$

PROBLEM 4 Solve the area of a trapezoid formula $A = \dfrac{1}{2}h(b_1 + b_2)$ for the base b_1. ◆

Exercises 1.8

◆ **Basic Skills**

In Exercises 1–30, find the solutions of each equation, if they exist.

1. $2x - 7 = -21$

2. $14 = 22 + 3y$

3. $26 = 6 - 5t$

4. $12 - 6x = -18$

5. $3n + 4 = n - 4$

6. $12m + 5 = 2m - 5$

7. $8x + 4 + 3x = x - 11$

8. $7x - 3 + 2x = 4x + 3 + x$

9. $9x - 2(4x + 1) = 3$

10. $16 - (3x + 7) = 3x$

11. $\dfrac{5x}{6} - 1 = \dfrac{x}{3}$

12. $4 + \dfrac{n}{6} = \dfrac{n + 4}{2}$

13. $\dfrac{x}{2} + \dfrac{x}{4} = \dfrac{1}{6} + \dfrac{x}{3}$

14. $\dfrac{x}{2} + \dfrac{x}{3} + \dfrac{x}{5} = 1 - \dfrac{x}{6}$

15. $\dfrac{t - 1}{6} = \dfrac{2t + 3}{12} - \dfrac{7t - 4}{10}$

16. $\dfrac{x}{12} - \dfrac{x - 3}{8} = \dfrac{2x - 1}{3}$

17. $|3x| = 12$

18. $3|-2t| = 15$

19. $6 - |5n - 3| = 4$

20. $|4y - 3| - 5 = 4$

21. $5 - 2|9 - 3y| = -15$

22. $2 - |3x - 11| = -4$

23. $x^2 - 3 = 0$

24. $n^2 - 16 = 9$

25. $\frac{1}{2}x^2 + 6 = 14$

26. $2y^2 - 21 = 33$

27. $2x^2 - 5 = 3x^2 - 7$

28. $10 - x^2 = 4 + x^2$

29. $(x - 1)^2 = 49$

30. $3(x + 2)^2 = 24$

In Exercises 31–50, solve each inequality. Use interval notation to describe the solution set.

31. $3x - 2 < 7$

32. $15 - 4x < -3$

33. $2x + 3 \geq 3x + 4$

34. $3m - 4 \leq m - 8$

35. $2(y + 3) > 3 - (4 - y)$

36. $4x - (x - 2) < x - 3$

37. $\dfrac{2p}{3} > \dfrac{p}{9} - \dfrac{5}{6}$

38. $\dfrac{x}{3} + \dfrac{x}{5} < 1 + \dfrac{11x}{15}$

39. $\dfrac{x + 5}{2} - 3 \leq \dfrac{3x}{5} - \dfrac{x - 4}{4}$

40. $\dfrac{h + 1}{2} + \dfrac{2h + 3}{4} \geq 2$

41. $|x| < 3$

42. $|2x| - 2 \geq 8$

43. $\left| \dfrac{2x}{3} \right| - 1 > 0$

44. $5 \left| \dfrac{-x}{10} \right| + 3 < 13$

45. $|5 - 2x| > 3$

46. $|4x + 3| + 2 < 17$

47. $\left| \dfrac{4 - x}{3} \right| \leq \dfrac{1}{2}$

48. $\left| \dfrac{2(4 - 3x)}{3} \right| > 8$

49. $\left| \dfrac{3x}{4} + \dfrac{1 - 3x}{6} \right| \geq 1$

50. $\left| \dfrac{1 - 2x}{4} + \dfrac{1 + 2x}{6} \right| \leq \dfrac{5}{9}$

In Exercises 51–58, solve each literal equation for x.

51. $\dfrac{ax}{b} = c$

52. $\dfrac{a}{bx + c} = 1$

53. $ax = bx + c$

54. $ax - b(c + x) = kx$

55. $\dfrac{a + bx}{k} = 3x + a$

56. $\dfrac{a - x}{b} = \dfrac{c - x}{a}$

57. $|ax + b| = 1$

58. $ax^2 + b = 0$

In Exercises 59–62, solve each literal inequality for x subject to the given conditions.

59. $ax + b \leq cx$, provided $a > c$

60. $ax + b > cx$, provided $a < c$

61. $\dfrac{x}{a} + \dfrac{y}{b} \leq c$, provided $a < 0$ and $b \neq 0$

62. $\dfrac{x}{a} + \dfrac{y}{b} \leq c$, provided $a > 0$ and $b \neq 0$

In Exercises 63–68, use the common formulas given in this section to answer each question.

63. What is the Celsius temperature if the Fahrenheit temperature is 59°F?

64. What time is required for an investment of $10,000 to amount to $20,000 if the interest rate is 5% simple interest?

65. The length of the hypotenuse of an isosceles right triangle is $5\sqrt{2}$ centimeters. What is the length of a leg of this triangle?

66. What is the radius of a circle if its area is 9π square centimeters?

67. What is the height of a cone if its volume is 12π cubic feet and its radius is 2 feet?

68. What is the height of a cylinder if its total surface area is 24π square meters and its radius is 3 meters?

In Exercises 69–78, solve each formula for the indicated variable.

69. Electrical current: $I = \dfrac{E}{R}$

    (a) the voltage E (b) the resistance R

70. Tension in an elevator cable: $T = m(g + a)$

    (a) the mass of the elevator m

    (b) the acceleration a

71. Thermal expansion: $L = L_0(1 + \mu \Delta t)$

    (a) the original length L_0

    (b) the temperature change Δt

72. nth term of an arithmetic sequence:
$a_n = a_1 + (n - 1)d$

    (a) the common difference d

    (b) the number of terms n

73. Amount of money accrued: $A = P + Prt$

    (a) the interest rate r

    (b) the principal invested P

74. Mechanical advantage of a differential hoist:
$$M = \frac{2R}{R - r}$$

    (a) the radius of the small pulley r

    (b) the radius of the large pulley R

75. Thrust of a spaceship's engine: $F = \dfrac{mv - mv_0}{t}$

    (a) the burn time t

    (b) the mass of the spaceship m

76. Sum of a finite geometric series: $S = \dfrac{a_1 - ra_n}{1 - r}$

    (a) the first term a_1

    (b) the common ratio r

77. Total power dissipated by two resistors connected in series: $P = I^2R_1 + I^2R_2$

    (a) the resistance R_1

    (b) the current I

78. Deflection of a cantilever steel beam:
$$D = \frac{8PL^3 + 3WL^3}{24EI}$$

    (a) the concentrated load P

    (b) the length of the beam L

◆ **Critical Thinking**

79. Is the equation $A = B$, where A and B are algebraic expressions, equivalent to the equation $A - B = 0$? to $\dfrac{A}{B} = 1$? to $A^2 = B^2$? Explain.

80. Is the inequality $A < B$, where A and B are algebraic expressions, equivalent to the inequality $A - B < 0$? to $\dfrac{A}{B} < 1$? to $A^2 < B^2$? Explain.

81. Determine the interval for which the expression $\sqrt{3x + 8}$ is undefined in the real number system.

82. Find a value of a such that the given pair of equations are equivalent.

    (a) $a - 4x = 18$ and $2x + 3 = 13$

    (b) $2x - 3a = 5x + 3$ and $3(x + 2) = 6(x + 3)$

83. Given that $x > 5$, find the fallacy in the following argument.

$x > 5$	
$5x > 25$	**Multiply both sides by 5**
$5x - x^2 > 25 - x^2$	**Subtract x^2 from both sides**
$x(5 - x) > (5 + x)(5 - x)$	**Factor each side**
$x > 5 + x$	**Divide both sides by $5 - x$**
$0 > 5$	**Subtract x from both sides**

84. Given that $x = 3$, find the fallacy in the following argument.

$$x = 3$$
$$x - 3 = 0 \qquad \text{Subtract 3 from both sides}$$
$$2x - 6 = 0 \qquad \text{Multiply both sides by 2}$$
$$x^2 + 2x - 6 = x^2 \qquad \text{Add } x^2 \text{ to both sides}$$
$$x^2 + 2x - 15 = x^2 - 9 \qquad \text{Subtract 9 from both sides}$$
$$(x + 5)(x - 3) = (x + 3)(x - 3) \qquad \text{Factor each side}$$
$$x + 5 = x + 3 \qquad \text{Divide both sides by } (x - 3)$$
$$5 = 3 \qquad \text{Subtract } x \text{ from both sides}$$

85. The relationship between the air temperature T_a (in °F), the dew point temperature T_d (in °F), and the height h (in feet) to the base of a cumulus cloud is given by the formula

$$h = \frac{T_a - T_d}{0.0045}.$$

What is the dew point temperature when the base of the cloud is 5200 ft above the ground and the air temperature is 72°F?

86. A pediatrician determines that the weight x in pounds of 95% of the babies born at a certain hospital satisfies the inequality

$$|x - 7.2| \le 2.3.$$

What is the range of the weights of the other 5% of the babies?

Chapter 1 Review

◆ **Questions for Writing or Group Discussion**

1. What is the difference between a *rational number* and an *irrational number*?
2. Is the expression $-x$ always negative? Explain.
3. For what values of a does $\dfrac{|a|}{a} = -1$?
4. What is meant by an *open interval*? a *closed interval*? an *unbounded interval*?
5. List the *laws of exponents*. Illustrate each rule with an example.
6. Look up the meanings of the following prefixes in a dictionary: *pico-, deci-, nano-, giga-,* and *hecto-*. Use these definitions to explain the meanings of picosecond, decimeter, nanofarad, gigawatt, and hectoliter.
7. What is meant by the *principal nth root* of a real number? Is the principal nth root of a real number always a real number? Explain
8. What is a *complex number*? Under what conditions is a complex number a real number? an imaginary number?
9. State the procedures for adding, subtracting, multiplying, and dividing complex numbers.

10. What is a *complex conjugate*? How is it used in finding the quotient of two complex numbers? Illustrate with an example.

11. Does every complex number have a reciprocal? Explain.

12. What is a *polynomial*?

13. What is meant by the *degree of a polynomial*? Illustrate with an example.

14. Explain the procedure for combining *like terms* into a single term.

15. State the procedures for adding, subtracting, and multiplying polynomials.

16. What is meant by *factoring completely* over the set of integers?

17. What are the conditions for a trinomial to be a *perfect square*?

18. From a given fraction, how can an *equivalent fraction* be generated?

19. Describe the process of *reducing* an algebraic fraction to lowest terms.

20. Explain the procedure of determining the *LCD* for a sum or difference of algebraic fractions.

21. What is a *complex fraction*? What is the procedure for simplifying a complex fraction?

22. How are *conjugates* used to rationalize a denominator or numerator?

23. Explain the difference between an *identity* and a *conditional equation*.

24. What is meant by the *solution* of an equation?

25. Does squaring both sides of an equation generate equivalent equations? Explain.

26. When is it necessary to *reverse the direction of the inequality symbol* when generating equivalent inequalities. Illustrate by examples.

◆ **Review Exercises**

In Exercises 1–6, rewrite each statement using an inequality.

1. a is at least 7.
2. b is at most -9.
3. c is negative and greater than -10.
4. d is nonnegative and at most 6.
5. The distance between a and 3 is at most 2.
6. c is more than 5 units from 0.

In Exercises 7 and 8, find the distance from A to B on the real number line.

7. The coordinate of point A is -9 and the coordinate of point B is 6.
8. The coordinate of point A is 3 and the coordinate of point B is π.

In Exercises 9–16, use interval notation to describe the set of all real numbers designated on each real number line.

9.
10.
11.
12.
13.
14.
15.
16.

In Exercises 17–26, simplify each exponential expression. Write the answer without zero or negative exponents. Assume all variables represent positive real numbers.

17. $(-2x^4y^5)(5xy^4)^2$

18. $[-2(x + 3)^2]^3$

19. $(2x + 1)^{-3}(2x + 1)^{-1}(2x + 1)^0$

20. $\left(\dfrac{3x^3}{2}\right)^{-2}$

21. $\dfrac{(x - 1)^4}{(x - 1)^3}$

22. $\dfrac{(2m^2n)^{-4}}{8m^{-2}}$

23. $(x + 1)^{-2/3}(x + 1)^{5/3}$

24. $(36x^4y^{-6})^{3/2}y^8$

25. $\left(\dfrac{16x^2}{y^8}\right)^{-3/4}$

26. $\dfrac{[9(x + 4)^{-3}]^{-1/2}}{(x + 4)^{1/2}}$

In Exercises 27 and 28, perform the indicated operations using scientific notation.

27. $\dfrac{0.00000000018}{60,000,000,000,000}$

28. $(30,000,000,000)(0.0000000002)^3$

In Exercises 29 and 30, rewrite each expression in exponential form.

29. $\sqrt[3]{(x + 2)^2}$

30. $\sqrt{x} \cdot \sqrt[4]{x^3}$

In Exercises 31 and 32, rewrite each expression in radical form.

31. $(x + 4)^{-1/6}$

32. $(2ab)^{3/4}$

In Exercises 33–36, write each radical expression in simplified form. Assume all variables represent positive real numbers.

33. $\sqrt{32(x + y)^3}$

34. $\sqrt{\sqrt[3]{36x^4}}$

35. $\sqrt[4]{32x^2y^2} \cdot \sqrt[4]{2xy^2}$

36. $\sqrt{\dfrac{8m^5}{(n + 2)^6}}$

In Exercises 37 and 38, simplify by assuming the variables represent any real number.

37. $\sqrt{9x^2y^2}$

38. $[(-2)^4(y - 1)^8]^{1/4}$

In Exercises 39–46, perform the indicated operations. Express each answer in standard form.

39. $4\sqrt{-45} + 3\sqrt{-80} - \sqrt{-20}$

40. $\sqrt{-24} \cdot \sqrt{6} \cdot \sqrt{-16}$

41. $(2i)^6$

42. $5i^{-33}$

43. $(9 + i) + (3 - 2i)$

44. $(4 - 3\sqrt{-9}) - (9 + 3\sqrt{-25})$

45. $(3 - 5i)(4 + 7i)$

46. $\dfrac{(1 - i)^2}{3 + 4i}$

In Exercises 47–66, perform the indicated operations with the algebraic expressions.

47. $(a^2 - 3a) + (5a - 4a^2) - (a - 5a^2)$

48. $(x^2 - 2x\sqrt{27x}) - (\sqrt{3x^3} - 3x^2)$

49. $(3\sqrt{40x} - 4\sqrt{90x}) + 2\sqrt{160x}$

50. $3x^{1/2} + (2x^{-1} + 4x^{1/2} - 3x^{-1})$

51. $(x^3 + 2x^2 - 3)(2x^3 - 3x - 1)$

52. $(5n^{-1/2} - 3n^{-3/2})(2n^{3/2} - n^{-1/2})$

53. $(5x^2 + 2y)(6x^2 - 5y)$

54. $(3\sqrt{ab} - 2)(4\sqrt{ab} - 5)$

55. $(2t - 5)(2t + 5)$

56. $(\sqrt{2x} - \sqrt{3y})(\sqrt{2x} + \sqrt{3y})$

57. $(x - 4)(x^2 + 4x + 16)$

58. $(4m^{-2} - 6m^{-1} + 9)(2m^{-1} + 3)$

59. $(3x - 7y)^2$

60. $(\sqrt{2n} + 1)^2$

61. $(2x - 5)^3$

62. $(a^{-2} - 2b)^3$

63. $[(x + 3)(x - 3)]^3$

64. $2(1 - a)^2(-5a) + 10a$

65. $y^2 - [x(3x - y) - (3x - 4y)(x + y)]$

66. $x\{x^2 - [(2x + 3)2x - x(3x + 2)]\}$

In Exercises 67 and 68, find a polynomial that represents the area of the shaded region.

67.

68.

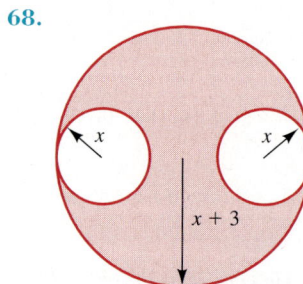

In Exercises 69–88, factor completely over the set of integers.

69. $3y^2(y - 2) + 6y(y - 2)$

70. $x(1 - 3x)^2 + (1 - 3x)^2$

71. $m^3 - 3m^2 + m - 3$ **72.** $12 - 9y - 8x + 6xy$

73. $6x^2 - xy - 12y^2$ **74.** $8t^2 + 6t - 5$

75. $2x^4 - 13x^2 + 15$ **76.** $m^2 + 10mn + 16n^2$

77. $x^2 + 2x - 15$ **78.** $a^6 - 7a^3b + 12b^2$

79. $25x^2 - 9y^4$ **80.** $x^3 + 27y^3$

81. $8x^3 - 1$ **82.** $9m^4 - 12m^2n^2 + 4n^4$

83. $t^3 + 6t^2 + 12t + 8$ **84.** $x^4 - 4y^4$

85. $8x^3y + 2x^2y - 6xy$ **86.** $m^6 - 4m^3 - 32$

87. $2xy + 4x^2y - 18x^3y - 36x^4y$

88. $16t^4 - 1$

In Exercises 89 and 90, reduce each fraction to lowest terms.

89. $\dfrac{n^2 + 5n - 36}{3n^2 - 14n + 8}$ **90.** $\dfrac{16x - x^5}{x^2 - 2x}$

In Exercises 91–106, perform the indicated operations.

91. $\dfrac{-5}{2x} + \dfrac{7}{2x}$ **92.** $\dfrac{2}{\pi} - \dfrac{9}{\pi}$

93. $\dfrac{-9}{5x} \cdot \dfrac{10x^2}{3}$ **94.** $\dfrac{3x}{2} \div \dfrac{15x}{8}$

95. $\dfrac{7y}{8x} + \dfrac{-5x^2}{12y}$ **96.** $\dfrac{5x}{9} - \dfrac{y}{12}$

97. $\dfrac{2x^2 - 3x - 9}{6xy} \cdot \dfrac{2x^2}{9 - x^2}$

98. $\dfrac{2a^2 + ab - 3b^2}{4a^2 - 9b^2} \div \dfrac{b^3 - a^3}{3b - 2a}$

99. $\dfrac{2x + 3}{x^2 - 9} - \dfrac{5x - 6}{x^2 - 9}$

100. $\dfrac{3m(m - n)}{4m - n} + \dfrac{n^2 - m^2}{n - 4m}$

101. $\dfrac{3}{x^2 + 4x} - \dfrac{3}{x^2 + 8x + 16}$

102. $\dfrac{1}{5 + x} + \dfrac{1}{5 - x} - \dfrac{x^2 - 7x}{x^2 - 25}$

103. $\dfrac{1}{x^2 - 4xy + 4y^2} + \dfrac{1}{x^2 - 4y^2}$

104. $\dfrac{y^{-2} - x^{-2}}{x^{-1} - y^{-1}}$

105. $\dfrac{1 + (a - 1)^{-1}}{(a - 1)^{-1}}$ **106.** $\dfrac{1 - \dfrac{1}{25x^2}}{5 - \dfrac{1}{x}}$

In Exercises 107 and 108, rationalize the numerator and simplify.

107. $\sqrt[4]{\dfrac{(x - 5)^3}{3x}}$ **108.** $\dfrac{\sqrt{3x} + \sqrt{3y}}{\sqrt{3xy}}$

In Exercises 109 and 110, rationalize the denominator and simplify.

109. $\dfrac{1}{2x - \sqrt{4x^2 - 1}}$ **110.** $\dfrac{1}{\sqrt[3]{(3 - x)^2}}$

In Exercises 111–118, solve for x.

111. $18 = 24 - 3x$ **112.** $3x + 4 = 7x - 8$

113. $6x - (3 - 2x) = 2(x - 3)$

114. $\dfrac{x}{8} - \dfrac{3x + 1}{10} = \dfrac{3 - x}{20}$

115. $|4 - 3x| = 16$ **116.** $\dfrac{1}{x} - \dfrac{1}{a} = \dfrac{1}{b} - \dfrac{1}{x}$

117. $x^2 - 16 = 0$ **118.** $(2x - 1)^2 = 18$

In Exercises 119 and 120, solve each formula for the indicated variable.

119. Resistance of a conductor: $R = \dfrac{\mu L + \mu Lt}{d^2}$

Solve for the length of the conductor L.

120. Energy output of a battery: $W = I^2R_1t + I^2R_2t$

Solve for the current I.

In Exercises 121–126, solve each inequality. Use interval notation to describe the solution set.

121. $4 - x > 7$ **122.** $3x - 2 < 5x - 4$

123. $\dfrac{x}{4} \le \dfrac{2x}{5} - \dfrac{3}{10}$

124. $\dfrac{1}{6} - \dfrac{t - 2}{12} \ge \dfrac{t - 1}{4}$

125. $|5 - 3n| > 8$ **126.** $\left|\dfrac{2x - 3}{5}\right| < \dfrac{2}{3}$

127. The wavelength λ (the Greek letter lambda) of a radio wave of frequency f is given by the formula

$$\lambda = \dfrac{3 \times 10^8}{f}$$

(*Exercise 127 continues*)

where λ is in meters and f is in hertz. Determine λ if $f = 7.5 \times 10^6$ hertz.

128. The change in length δ (the Greek letter delta) of a concrete slab in a sidewalk due to thermal expansion is given by the formula

$$\delta = 5.5 \times 10^{-6} tL$$

where t is the change in temperature, and L is the original length of the slab. Suppose 30.0-foot concrete slabs are being laid with an expansion gap between the end of one slab and the beginning of the next. What is the minimum expansion gap in *inches* that should be left between the slabs if the temperature of the concrete in the summertime can reach 170°F and the concrete is being laid at a temperature of 44°F?

129. The land area of China is about 3.7×10^6 square miles. In 1980, the population of China was approximately one billion. At that time, what was the amount of land area in *square feet* per person?

130. Presently, the United States natural gas reserve is estimated to be about 180 trillion cubic feet and consumption is averaging approximately 9×10^{11} cubic feet per year. Assuming consumption continues at this rate, how long will the supply last?

131. (a) Solve the volume of a cone formula $V = \frac{1}{3}\pi r^2 h$ for r.

   (b) Use the result of part (a) to find the radius of a cone whose volume is 24π cubic inches and whose height is 2 inches.

132. A pediatrician determines that the length x in inches of 90% of the babies born at a certain hospital satisfies the inequality

$$|x - 20.6| \leq 2.1.$$

What is the range of the lengths of the other 10% of the babies?

The real estate tax T on a property varies directly as its assessed value V.

(a) Express T as a function of V if $T = \$2800$ when $V = \$112,000$.

(b) Generate the graph of the function defined in part (a) for $\$0 \le V \le \$500,000$.

For the solution, see Example 3 in Section 2.7.

Functions and Graphs

2.1 Working in the Cartesian Plane

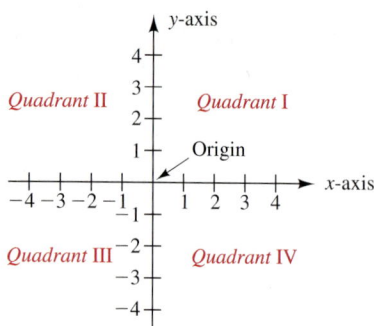

FIGURE 2.1

The coordinate plane, or Cartesian plane

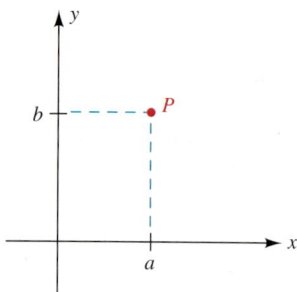

◆ Introductory Comments

Recall from Section 1.1 that for each point on the real number line, there corresponds a real number called its *coordinate*. In this section, we extend this idea by assigning to each point in a plane a pair of real numbers. To do this, we construct horizontal and vertical real number lines that intersect at the zero points of the two lines (see Figure 2.1). The two lines are called *coordinate axes* and the *plane* in which they lie is called the **coordinate plane**, or **Cartesian plane**, after the French mathematician René Descartes (1596–1650). The point where the axes intersect is called the **origin**. The horizontal number line has its positive direction to the right and is usually called the **x-axis**. The vertical number line has its positive direction upward and is usually called the **y-axis**. The coordinate axes divide the plane into four regions, or **quadrants**, which are labeled with Roman numerals, as shown in Figure 2.1.

◆ Plotting Points

FIGURE 2.2

The coordinates of point P are (a, b).

We can assign to each point P in the coordinate plane a unique pair of numbers, called its **rectangular coordinates** or **Cartesian coordinates**. By drawing horizontal and vertical lines through P, we find that the vertical line intersects the x-axis at some point with coordinate a, and the horizontal line intersects the y-axis at some point with coordinate b, as shown in Figure 2.2. Thus, we assign the pair (a, b) to the point P. In the pair (a, b), the first number a is called the **x-coordinate** or **abscissa** of P, and the second number b is called the **y-coordinate** or **ordinate** of P. Since the x-coordinate is always written first, we refer to (a, b) as an **ordered pair** of numbers.

Conversely, we can locate any ordered pair (a, b) in the coordinate plane by constructing a vertical line through a on the x-axis and a horizontal line through b on the y-axis. The intersection of these lines determines a unique point P, which we designate as $P(a, b)$. Thus, for each point in the coordinate plane there corresponds a unique ordered pair of real numbers, and for each ordered pair of real numbers there is a unique point in the coordinate plane. Hence, we have a *one-to-one correspondence* between pairs of real numbers and points in a coordinate plane.

FIGURE 2.3

Plotting points associated with ordered pairs.

EXAMPLE 1   Plot the points $A(2, 3)$, $B(-3, 2)$, $C(-4, -3)$, and $D(\frac{3}{2}, -4)$ on a coordinate plane. Specify the quadrant in which each point lies.

◆ Solution   The point $A(2, 3)$ is in quadrant I, $B(-3, 2)$ is in quadrant II, $C(-4, -3)$ is in quadrant III, and $D(\frac{3}{2}, -4)$ is in quadrant IV, as shown in Figure 2.3.   ◆

A point on the x-axis, y-axis, or at the origin, is not in any of the quadrants.

PROBLEM 1   Plot the points $A(0, 0)$, $B(0, -4)$, and $C(4, 0)$ on a coordinate plane.   ◆

FIGURE 2.4

The length of $\overline{PQ}$, denoted as PQ, may be determined by the Pythagorean theorem.

◆ The Distance Formula

By convention, the *line segment* that joins the points $P(x_1, y_1)$ and $Q(x_2, y_2)$ in the coordinate plane is designated as $\overline{PQ}$, and the *length* of $\overline{PQ}$ is designated as PQ. To determine PQ, we begin by constructing a right triangle, as shown in Figure 2.4. The length of the horizontal leg of the triangle is $|x_2 - x_1|$ and the length of the vertical leg is $|y_2 - y_1|$. By the Pythagorean theorem, the sum of the squares of the lengths of these legs is equal to the square of the length of the hypotenuse PQ:

$$(PQ)^2 = |x_2 - x_1|^2 + |y_2 - y_1|^2$$
$$PQ = \sqrt{|x_2 - x_1|^2 + |y_2 - y_1|^2}$$
$$PQ = \sqrt{(x_2 - x_1)^2 + (y_2 - y_1)^2}$$

We refer to this last equation as the **distance formula**.

◀◆ Distance Formula

> The **distance** between two points $P(x_1, y_1)$ and $Q(x_2, y_2)$ in the coordinate plane is given by
>
> $$PQ = \sqrt{(x_2 - x_1)^2 + (y_2 - y_1)^2}.$$

Note: When we use the distance formula, it does not matter which point is called (x_1, y_1) and which is called (x_2, y_2). This is because $(x_2 - x_1)$ and $(x_1 - x_2)$ are opposites of each other, as are $(y_2 - y_1)$ and $(y_1 - y_2)$, and squaring a number or its opposite gives the same numerical result.

EXAMPLE 2   Find the length of the line segment joining the points $A(-2, 3)$ and $B(4, 1)$.

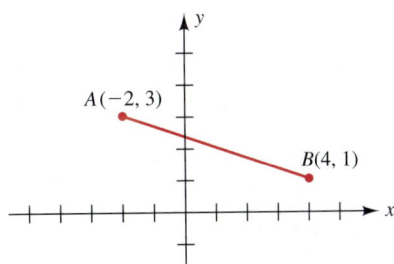

FIGURE 2.5

The length of the line segment joining the points A and B can be found by using the distance formula.

◀◆ Solution   Figure 2.5 shows the points A and B in the coordinate plane. We designate the length of the line segment that joins the points A and B as AB. We use the distance formula with $(x_1, y_1) = (-2, 3)$ and $(x_2, y_2) = (4, 1)$:

$$AB = \sqrt{[4 - (-2)]^2 + (1 - 3)^2}$$
$$= \sqrt{(6)^2 + (-2)^2}$$
$$= \sqrt{40}$$
$$= 2\sqrt{10}$$
$$\approx 6.32$$

PROBLEM 2   Find the length of the line segment joining the points $P(-3, 1)$ and $Q(1, -2)$.

◆ The Midpoint Formula

Given two points $P(x_1, y_1)$ and $Q(x_2, y_2)$ in the coordinate plane, we can find the coordinates of the midpoint M of $\overline{PQ}$ by finding the average value of the *x*-coordinates and the average value of the *y*-coordinates of the endpoints *P* and *Q*. We refer to this fact as the **midpoint formula**.

◆◆ **Midpoint Formula**

The coordinates of the **midpoint** *M* of a line segment joining the points $P(x_1, y_1)$ and $Q(x_2, y_2)$ are

$$\left(\frac{x_1 + x_2}{2}, \frac{y_1 + y_2}{2} \right).$$

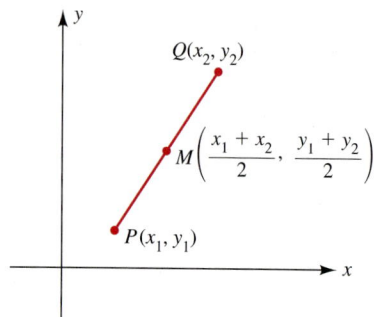

FIGURE 2.6
The midpoint of $\overline{PQ}$ is *M*, provided *PM = MQ* and *PM + MQ = PQ*.

To verify the midpoint formula, we use Figure 2.6 and the distance formula to show that *PM = MQ* and *PM + MQ = PQ*. Note that

$$PM = \sqrt{\left(\frac{x_1 + x_2}{2} - x_1 \right)^2 + \left(\frac{y_1 + y_2}{2} - y_1 \right)^2}$$

$$= \sqrt{\left(\frac{x_2 - x_1}{2} \right)^2 + \left(\frac{y_2 - y_1}{2} \right)^2}$$

$$= \tfrac{1}{2}\sqrt{(x_2 - x_1)^2 + (y_2 - y_1)^2}$$

and

$$MQ = \sqrt{\left(x_2 - \frac{x_1 + x_2}{2} \right)^2 + \left(y_2 - \frac{y_1 + y_2}{2} \right)^2}$$

$$= \sqrt{\left(\frac{x_2 - x_1}{2} \right)^2 + \left(\frac{y_2 - y_1}{2} \right)^2}$$

$$= \tfrac{1}{2}\sqrt{(x_2 - x_1)^2 + (y_2 - y_1)^2}.$$

Thus

$$PM + MQ = \sqrt{(x_2 - x_1)^2 + (y_2 - y_1)^2} = PQ.$$

Since *PM = MQ* and the points *P, M,* and *Q* are *collinear* (*PM + MQ = PQ*), we conclude that *M* is the midpoint of $\overline{PQ}$.

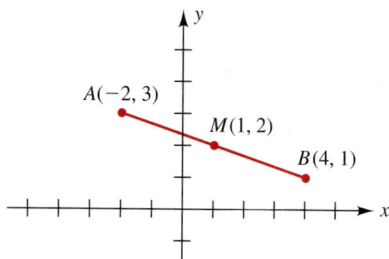

FIGURE 2.7
The coordinates of the midpoint of $\overline{AB}$ are (1, 2).

EXAMPLE 3   Find the coordinates of the midpoint of the line segment joining the points $A(-2, 3)$ and $B(4, 1)$.

◆ **Solution**   Using the midpoint formula with $(x_1, y_1) = (-2, 3)$ and $(x_2, y_2) = (4, 1)$, we have

$$\left(\frac{-2 + 4}{2}, \frac{3 + 1}{2} \right) = (1, 2).$$

Thus, as shown in Figure 2.7, the coordinates of the midpoint *M* are (1, 2). ◆

PROBLEM 3   Find the coordinates of the midpoint of the line segment joining the points $A(-4, -3)$ and $B(5, -1)$. ◆

◆ **Analytic Geometry**

We can place geometric figures on the coordinate plane and solve related problems by using the distance and midpoint formulas. We refer to the branch of mathematics that uses algebra to solve geometric problems as **analytic geometry**.

EXAMPLE 4 Show that the triangle with vertices at the points $A(-2, 2)$, $B(3, 2)$, and $C(1, -2)$ is an isosceles triangle—it has two sides of equal length. Then find the area of triangle ABC.

◆ **Solution** We begin by plotting the points and constructing the triangle, as shown in Figure 2.8. Now, using the distance formula, we can find the length of each side:

$$AB = \sqrt{[3 - (-2)]^2 + (2 - 2)^2} = \sqrt{(5)^2 + (0)^2} = 5$$
$$BC = \sqrt{(1 - 3)^2 + (-2 - 2)^2} = \sqrt{(-2)^2 + (-4)^2} = \sqrt{20} = 2\sqrt{5}$$
$$AC = \sqrt{[1 - (-2)]^2 + (-2 - 2)^2} = \sqrt{(3)^2 + (-4)^2} = \sqrt{25} = 5$$

Since $AB = AC$, we conclude the triangle is isosceles with BC the nonequal side.

To find the area of triangle ABC, recall from geometry that the altitude to the nonequal side of an isosceles triangle bisects that side. Thus, if $\overline{AM}$ is the altitude to the base $\overline{BC}$, then M is the midpoint of $\overline{BC}$, as shown in Figure 2.9. By the midpoint formula, the coordinates of the midpoint M are

$$\left(\frac{1 + 3}{2}, \frac{-2 + 2}{2}\right) = (2, 0)$$

and, by the distance formula, the length of the altitude $\overline{AM}$ is

$$AM = \sqrt{[(2 - (-2)]^2 + (0 - 2)^2} = \sqrt{(4)^2 + (-2)^2}$$
$$= \sqrt{20} = 2\sqrt{5}.$$

Hence, the area of triangle ABC is

$$\text{Area} = \tfrac{1}{2}(\text{base})(\text{height}) = \tfrac{1}{2}(BC)(AM)$$
$$= \tfrac{1}{2}(2\sqrt{5})(2\sqrt{5})$$
$$= 10 \text{ square units.}$$

As an alternative method for finding the area of triangle ABC, let $\overline{CN}$ be the altitude to base $\overline{AB}$. Since $\overline{AB}$ is a horizontal line segment, $CN = 4$. Hence,

$$\text{Area} = \tfrac{1}{2}(AB)(CN) = \tfrac{1}{2}(5)(4) = 10 \text{ square units.} \qquad ◆$$

The *converse* of the Pythagorean theorem is also true. If the square of the length of the longest side of a triangle is equal to the sum of the squares of the lengths of the two other sides, then the triangle is a right triangle, with the right angle opposite the longest side.

PROBLEM 4 Use the distance formula and the converse of the Pythagorean theorem to show that the triangle with vertices at the points $A(2, 5)$, $B(-2, 3)$, and $C(4, 1)$ is an isosceles right triangle. Then find the area of triangle ABC. ◆

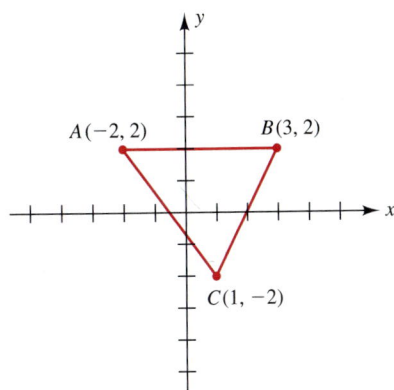

FIGURE 2.8

The triangle formed by the points $A(-2, 2)$, $B(3, 2)$, and $C(1, -2)$

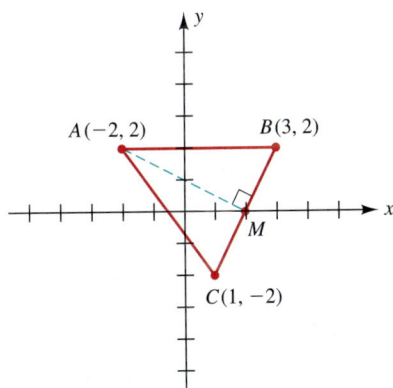

FIGURE 2.9

The altitude to the nonequal side of an isosceles triangle bisects that side. Thus, M is the midpoint of $\overline{BC}$.

Exercises 2.1

Basic Skills

In Exercises 1–6, fill in the blank to complete the statement.

1. If both coordinates are negative, the point is located in quadrant _____.

2. If the *x*-coordinate is _____ and the *y*-coordinate is _____, the point is located in quadrant II.

3. If the *x*-coordinate is _____ and the *y*-coordinate is _____, the point is located in quadrant IV.

4. If the *x*-coordinate is zero, the point is located on the _____.

5. If the *y*-coordinate is zero, the point is located on the _____.

6. If both coordinates are _____, the point is located at the origin.

In Exercises 7 and 8, name the ordered pair associated with each point.

7.

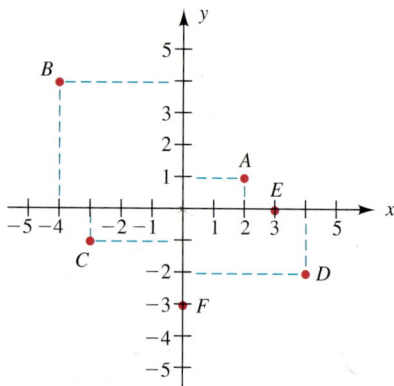

8.

In Exercises 9–18, find

(*a*) the length of the line segment joining the points A and B, and

(*b*) the coordinates of the midpoint M of the line segment joining the points A and B.

9. $A(1, 2)$, $B(4, 6)$ 10. $A(3, -6)$, $B(-2, 6)$

11. $A(-2, 3)$, $B(1, 8)$

12. $A(-1, -2)$, $B(-3, 2)$

13. $A(-5, 2)$, $B(3, -5)$

14. $A(-2, -1)$, $B(-6, -5)$

15. $A\left(-\frac{1}{2}, \frac{2}{3}\right)$, $B\left(-\frac{3}{4}, 1\right)$ 16. $A\left(\frac{1}{2}, \frac{1}{2}\right)$, $B\left(\frac{2}{3}, \frac{5}{8}\right)$

17. $A\left(0, \sqrt{2}\right)$, $B(-4, 0)$

18. $A\left(\sqrt{3}, -1\right)$, $B(0, 2)$

19. Find a formula for the distance *d* between the origin and the point (x, y) in the coordinate plane.

20. Find the coordinates of the midpoint *M* of a line segment joining the origin to the point (x, y) in the coordinate plane.

21. Find the perimeter of a triangle whose vertices are the points $A(-2, 1)$, $B(1, 3)$, and $C(4, -3)$.

22. Find the perimeter of a quadrilateral whose vertices are the points $A(-1, 0)$, $B(2, 4)$, $C(8, -4)$, and $D(4, -12)$.

23. Show that the points $A(1, 4)$, $B(2, -3)$, and $C(-1, -2)$ are the vertices of a right triangle. Then find the area of triangle *ABC*.

24. Show that the triangle whose vertices are the points $A(-5, 14)$, $B(1, 4)$, and $C(11, 10)$ is isosceles. Then find its area.

25. Show that the triangle whose vertices are $A(1, 1)$, $B(-1, -1)$, and $C\left(\sqrt{3}, -\sqrt{3}\right)$ is an equilateral triangle. Then find its area.

26. Show that the triangle whose vertices are $A(0, 6)$, $B(2, 0)$, and $C(8, 2)$ is an isosceles right triangle. Then find its area.

27. A line segment joins the points $A(8, -12)$ and $B(-4, 6)$. Use the midpoint formula to find the coordinates of the three points that divide this line segment into four equal parts.

28. Use the distance formula to determine if the points $A(-2, -3)$, $B(1, 3)$, and $C(2, 5)$ are collinear.

29. A *median* of a triangle is a line segment that joins a vertex of the triangle to the midpoint of the opposite side. Find the lengths of the medians of a triangle whose vertices are $A(-5, 4)$, $B(5, 2)$, and $C(-1, -4)$.

30. The midpoint of a line segment is $M(3, -2)$. One endpoint of the segment has coordinates $(6, 3)$. Find the coordinates of the other endpoint.

━━━◆ **Critical Thinking** ━━━━━━━━━━━━━━━━━━

31. Find the coordinates of a point on the y-axis that are equidistant from the points $A(-2, -4)$ and $B(3, 5)$.

32. Find the coordinates of a point on the x-axis that are equidistant from the points $A(-2, -4)$ and $B(3, 5)$.

For Exercises 33–36, refer to the right triangle with vertices A, O, and B shown in the figure.

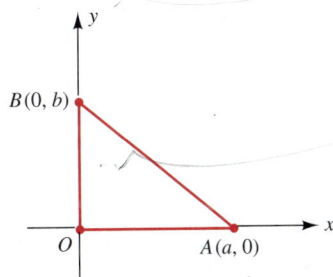

33. Find the coordinates of the midpoint M of the hypotenuse $\overline{AB}$.

34. Find AM and BM.

35. Find OM.

36. What conclusion can you make about the midpoint of the hypotenuse in regard to the three vertices of the right triangle?

For Exercises 37–40, refer to the parallelogram with vertices A, B, C, and O shown in the figure.

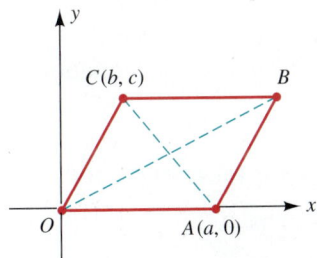

37. Find the coordinates of point B.

38. Find the midpoint of the diagonal $\overline{AC}$.

39. Find the midpoint of the diagonal $\overline{OB}$.

40. What conclusions can you make about the diagonals of a parallelogram?

A point P lies on the line segment joining the points $A(x_1, y_1)$ and $B(x_2, y_2)$ shown in the figure.

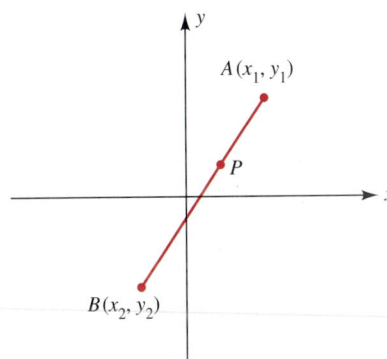

If the ratio AP/AB = r, then the point P has the coordinates

$$P\left(x_1 + r(x_2 - x_1),\ y_1 + r(y_2 - y_1)\right).$$

Use this formula for Exercises 41 and 42.

41. A line segment joins the points $A(6, 4)$ and $B(2, -2)$. Find the coordinates of the two points P_1 and P_2 that divide this line segment into three equal parts.

42. A line segment joins the points $A(5, 6)$ and $B(-3, -4)$. Find the coordinates of the four points P_1, P_2, P_3, and P_4 that divide the line segment into five equal parts.

43. Find the radius and area of a circle that passes through the point $A(2.24, 3.71)$ and has its center at the origin.

44. The line segment that joins the point $A(4.3, 2.5)$ to $B(-2.7, -4.9)$ is the diameter of a circle. Find the center, radius, and area of the circle.

◆ **2.2** **Graphs of Equations and the Graphing Calculator**

◆ **Introductory Comments**

In Section 1.8, we solved various types of equations containing one unknown. We now turn our attention to equations in *two unknowns*. A solution of an equation in two unknowns x and y is an ordered pair (x_1, y_1) such that when x is replaced by x_1 and y by y_1, the equation becomes true. Consider the equation

$$y = 2x + 1$$

with the two unknowns x and y. We say that the ordered pair $(0, 1)$ is a solution of this equation, since

$$y = 2x + 1$$
$$1 = 2(0) + 1$$
$$1 = 1 \quad \text{is true.}$$

To find other ordered pairs that are solutions of this equation, we arbitrarily choose values of x and then determine the corresponding values of y. Under these conditions we say that x is the **independent variable** and y the **dependent variable** in the equation. In order to determine other solutions to this equation, it is convenient to set up a **table of values** such as the one shown in Table 2.1.

TABLE 2.1
Table of values for $y = 2x + 1$

x	-2	-1	0	1	2
$y = 2x + 1$	-3	-1	1	3	5

Thus, along with $(0, 1)$ we have $(-2, -3)$, $(-1, -1)$, $(1, 3)$, and $(2, 5)$ as four other solutions of the equation $y = 2x + 1$. By continuing this process, we can generate many other ordered pairs that satisfy this equation. The **graph** of an equation is the set of *all* ordered pairs (x, y) in the coordinate plane that are solutions of the given equation. The graph of the equation $y = 2x + 1$ is a straight line, as shown in Figure 2.10. The arrowhead at each end of the line indicates that the graph continues indefinitely in that direction. Every point on this line is a solution of the equation $y = 2x + 1$.

◆ **Point-Plotting Method**

We can use the *point-plotting method* to *sketch the graph* of a simple equation.

◆ **Point-Plotting Method**

To sketch the graph of an equation by the **point-plotting method:**

1. Set up a table of values and find a few ordered pairs that satisfy the equation.

2. Plot and label the corresponding points in the coordinate plane.

3. Look for a pattern, and connect the plotted points to form a smooth curve.

Note: Although this method works well for some simple equations in two unknowns, it is inadequate for sketching the graph of more complicated equations. Throughout this chapter, we will introduce graphical aids that are useful for sketching an accurate graph by plotting as few points as possible.

FIGURE 2.10

The graph of the equation $y = 2x + 1$ is a straight line.

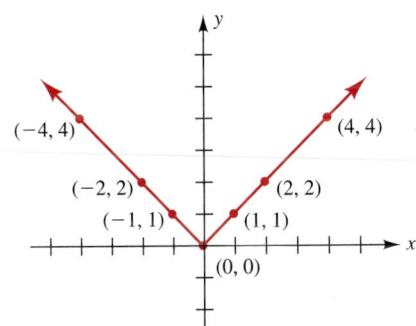

EXAMPLE 1 Sketch the graph of each equation by using the point-plotting method.

(a) $y = |x|$ **(b)** $x = y^2$ **(c)** $2y - x^3 = 0$

◆ **Solution**

(a) We begin by selecting arbitrary values of x and then finding their corresponding values of y. The following table of values organizes our work.

x	-4	-2	-1	0	1	2	4
$y = |x|$	4	2	1	0	1	2	4

We now plot and label the points given by this table and connect them by drawing half-lines through them, according to the suggested pattern. The graph of $y = |x|$ is the **V**-shaped curve shown in Figure 2.11. The arrowheads on the curve indicate that the graph continues upward to the left and upward to the right, according to the pattern suggested by the points we have plotted.

FIGURE 2.11

Graph of $y = |x|$

(b) For the equation $x = y^2$, it is easier to choose arbitrary values of y and then determine the corresponding values of x. In doing so, we are treating y as the independent variable and x as the dependent variable. The following table of values shows several values of y and their corresponding x-values.

y	-3	-2	-1	0	1	2	3
$x = y^2$	9	4	1	0	1	4	9

We plot and label the points given by this table and connect them by drawing a smooth curve, according to the suggested pattern. *Remember that the x-coordinate is always written first in an ordered pair.* The graph of $x = y^2$ is shown in Figure 2.12. It is a cupped-shaped curve called a *parabola*. The arrowheads on the curve indicate that the curve continues indefinitely, according to the pattern suggested by the points we have plotted.

(c) It is best to solve the equation $2y - x^3 = 0$ for either x or y before setting up a table of values. For this equation, solving for y is easier than solving for x, and we obtain

$$y = \tfrac{1}{2}x^3.$$

The following table of values shows several values of x and their corresponding y-values.

FIGURE 2.12

Graph of $x = y^2$

FIGURE 2.13
Graph of $2y - x^3 = 0$

x	-2	-1	0	1	2
$y = \frac{1}{2}x^3$	-4	$-\frac{1}{2}$	0	$\frac{1}{2}$	4

We now plot and label the points given by this table and draw a smooth curve through them. The graph of $2y - x^3 = 0$ is shown in Figure 2.13. The arrowheads on the curve indicate that the graph continues indefinitely, downward to the left and upward to the right. ◆

PROBLEM 1   Sketch the graph of the equation $4x + y = 8$ by using the point-plotting method. ◆

◆ **The Graphing Calculator**

Current technology enables us to generate the graph of an equation by using a personal computer with graphing software or a hand-held graphing calculator. Either of these graphing utilities is capable of plotting several hundred points in a few seconds. Although you may not have access to a personal computer with graphing software, you can purchase a hand-held graphing calculator at a reasonable cost. Throughout this text we use the graphing calculator to explore, investigate, and check many mathematical problems.

The key to success in using a graphing calculator is to select a viewing screen that shows the portion of the graph you wish to display. The viewing screen on a graphing calculator, which is usually called the *viewing rectangle*, represents a portion of the coordinate plane. On most graphing calculators, the WINDOW key or RANGE key is used to select a viewing rectangle. After pressing this key, you enter several values:

Xmin *x*-axis minimum value   **Ymin** *y*-axis minimum value

Xmax *x*-axis maximum value   **Ymax** *y*-axis maximum value

Xscl distance between scale marks on *x*-axis   **Yscl** distance between scale marks on *y*-axis

These values are shown in Figure 2.14.

EXAMPLE 2   Generate the graph of $2y - x^3 = 0$ on a graphing calculator (see Figure 2.13).

◆ **Solution**   We begin by pressing the WINDOW key or RANGE key to choose a viewing rectangle. Although the choice of a viewing rectangle is arbitrary, Figure 2.13 suggests the following viewing rectangle:

FIGURE 2.14
Viewing rectangle on a graphing calculator

Xmin = −5
Xmax = 5
Xscl = 1
Ymin = −5
Ymax = 5
Yscl = 1

After entering these values, we solve the equation $2y - x^3 = 0$ for y to obtain

$$y = \tfrac{1}{2}x^3,$$

and then enter this form of the equation into the calculator. *An equation must be in the form $y = $ ▪ before we can enter the equation into a graphing calculator.* Activating the graphing feature of the calculator, we obtain the graph of $2y - x^3 = 0$, as shown in Figure 2.15. ◆

To generate the graph of $x = y^2$ in the viewing rectangle of a graphing calculator, we first solve the equation for y to obtain $y = \pm\sqrt{x}$. We then enter both equations, $y = \sqrt{x}$ and $y = -\sqrt{x}$, into the calculator.

PROBLEM 2 Generate the graph of $x = y^2$ on a graphing calculator (see Figure 2.12). ◆

Almost every graphing calculator has a ⎣TRACE⎦ key and a ⎣ZOOM⎦ key. The ⎣TRACE⎦ key is used to display the x- and y-coordinates associated with each dot, or *pixel,* along the graph. With the ⎣ZOOM⎦ key, we can magnify the graph n times about any point. These two keys enable us to estimate some special points along the graph, such as the coordinates of the points where the graph intersects the x-axis or y-axis, intersects another graph, or reaches its maximum or minimum value. More advanced graphing calculators have built-in programs for finding these special points. For more detailed explanations about the usage of different types of calculators, consult the manual that came with your calculator or the *Graphing Technology Laboratory Manual* that accompanies this text.

CAUTION Do not use a graphing calculator to obtain a quick solution to an example or exercise with no understanding of the fundamental mathematical ideas being presented. To become dependent upon any machine for solutions can have a disastrous effect on your learning. Instead, use a calculator to explore, investigate, and check the solution of mathematical problems. In this way, you will enhance your learning and understanding of college mathematics.

◆ **Symmetry and Intercepts**

Look at Figure 2.11: If we fold the coordinate plane along the y-axis, then the right-hand and left-hand portions of this graph will coincide. Thus, we say the graph of $y = |x|$ is **symmetric with respect to the y-axis**. This means that for each point (x, y) on the graph there corresponds a point $(-x, y)$ that is also on the graph. Now look at Figure 2.12. If we fold this coordinate plane along the x-axis, the upper and lower portions of the graph will coincide. Thus, we say that the graph of $x = y^2$ is **symmetric with respect to the x-axis**. This means that for each point (x, y) on the graph there corresponds a point $(x, -y)$ that is also on the graph. Finally, consider Figure 2.13. If we fold the coordinate plane along the x-axis, and then along the y-axis, the upper and lower portions of the graph will coincide. Thus, we say the graph of $2y - x^3 = 0$ is **symmetric with respect to the origin**. This means that

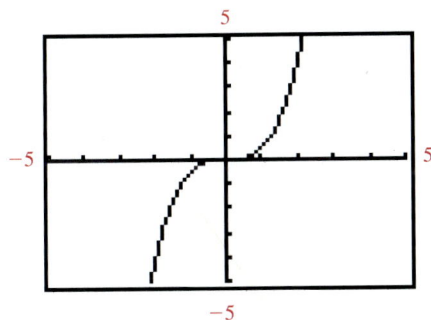

FIGURE 2.15

Graph of $2y - x^3 = 0$ in the viewing rectangle of a graphing calculator for each point (x, y) on the graph, there corresponds a point $(-x, -y)$ that is also on the graph. The following tests can be used to tell whether an equation has any of these three types of symmetry.

◆ **Tests for Symmetry**

1. The graph of an equation in two unknowns x and y is **symmetric with respect to the y-axis** if replacing x with $-x$ yields an equivalent equation.

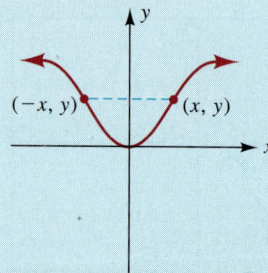

2. The graph of an equation in two unknowns x and y is **symmetric with respect to the x-axis** if replacing y with $-y$ yields an equivalent equation.

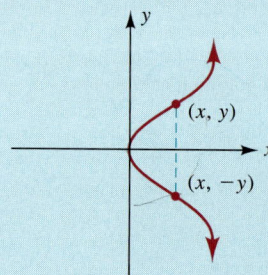

3. The graph of an equation in two unknowns x and y is **symmetric with respect to the origin** if replacing x and y with $-x$ and $-y$, respectively, yields an equivalent equation.

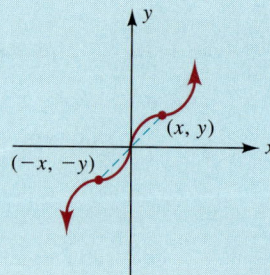

The **intercepts** of a graph are the points where the graph intersects the x-axis and y-axis. An **x-intercept** is the x-coordinate of any point where the graph intersects the x-axis. Since y is zero when the graph intersects the x-axis, the x-intercepts can be found by letting y be zero and solving the equation for x. A **y-intercept** is the y-coordinate of any point where the graph intersects the y-axis. Since x is zero when the graph intersects the y-axis, the y-intercepts can be found by letting x be zero and solving the equation for y.

EXAMPLE 3   Determine if the graph of each equation has symmetry with respect to the x-axis, y-axis, or origin. Then find the x- and y-intercepts, if they exist.

(a) $x^2 + y + 2 = 0$ **(b)** $xy^2 = 1$ **(c)** $x^2 + y^2 = 25$

◆ Solution

(a) *Symmetry:*

x-axis (replace y with −y)	y-axis (replace x with −x)	origin (replace x with −x and y with −y)
$x^2 + y + 2 = 0$	$x^2 + y + 2 = 0$	$x^2 + y + 2 = 0$
$x^2 + (-y) + 2 = 0$	$(-x)^2 + y + 2 = 0$	$(-x)^2 + (-y) + 2 = 0$
$x^2 - y + 2 = 0$	$x^2 + y + 2 = 0$	$x^2 - y + 2 = 0$
not an equivalent equation	an equivalent equation	not an equivalent equation

Since replacing x with $-x$ yields an equivalent equation, we conclude that the graph of $x^2 + y + 2 = 0$ is symmetric with respect to the y-axis.

x-intercepts: Letting $y = 0$ in the equation $x^2 + y + 2 = 0$, we obtain

$$x^2 + (0) + 2 = 0, \quad \text{or} \quad x^2 = -2,$$

which has no real solution. Thus, we conclude that the graph has no x-intercept.

y-intercepts: Letting $x = 0$ in the equation $x^2 + y + 2 = 0$, we obtain

$$(0)^2 + y + 2 = 0, \quad \text{or} \quad y = -2.$$

Hence the y-intercept is -2.

We can use a graphing calculator to verify our work. Solving the equation $x^2 + y + 2 = 0$ for y, we obtain

$$y = -x^2 - 2.$$

Entering this form of the equation into the calculator gives us the graph in Figure 2.16. Note that the graph has symmetry with respect to the y-axis, no x-intercept, and y-intercept -2.

FIGURE 2.16

Graph of $x^2 + y + 2 = 0$

(b) *Symmetry:*

x-axis (replace y with −y)	y-axis (replace x with −x)	origin (replace x with −x and y with −y)
$xy^2 = 1$	$xy^2 = 1$	$xy^2 = 1$
$x(-y)^2 = 1$	$(-x)y^2 = 1$	$(-x)(-y)^2 = 1$
$xy^2 = 1$	$-xy^2 = 1$	$-xy^2 = 1$
an equivalent equation	not an equivalent equation	not an equivalent equation

Since replacing y with $-y$ yields an equivalent equation, we conclude that the graph of $xy^2 = 1$ is symmetric with respect to the x-axis.

x-intercepts: Letting $y = 0$ in the equation $xy^2 = 1$, we obtain

$$x(0)^2 = 1, \quad \text{or} \quad 0x = 1.$$

However, since no number times zero is one, we conclude that the graph has no x-intercept.

y-intercepts: Letting $x = 0$ in the equation $xy^2 = 1$, we obtain

$$(0)y^2 = 1.$$

However, since no number squared times zero is one, we conclude that the graph has no y-intercept.

We can use a graphing calculator to verify our work. Solving the equation $xy^2 = 1$ for y, we obtain

$$y = \pm \sqrt{\frac{1}{x}}.$$

Entering *both* of these equations into the calculator gives us the graph in Figure 2.17. Note that the graph has symmetry with respect to the x-axis, no x-intercept, and no y-intercept.

(c) *Symmetry:*

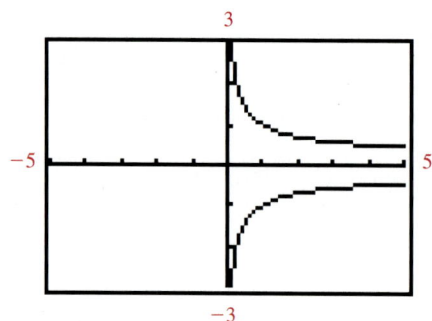

FIGURE 2.17
Graph of $xy^2 = 1$

x-axis (replace y with $-y$)	*y-axis* (replace x with $-x$)	*origin* (replace x with $-x$ and y with $-y$)
$x^2 + y^2 = 25$	$x^2 + y^2 = 25$	$x^2 + y^2 = 25$
$x^2 + (-y)^2 = 25$	$(-x)^2 + y^2 = 25$	$(-x)^2 + (-y)^2 = 25$
$x^2 + y^2 = 25$	$x^2 + y^2 = 25$	$x^2 + y^2 = 25$
an equivalent equation	an equivalent equation	an equivalent equation

Since an equivalent equation is generated in each case, we conclude that the graph of $x^2 + y^2 = 25$ is symmetric with respect to the x-axis, to the y-axis, and to the origin.

x-intercepts: Letting $y = 0$ in the equation $x^2 + y^2 = 25$, we obtain

$$x^2 + (0)^2 = 25, \quad \text{or} \quad x^2 = 25,$$

which implies $x = \pm\sqrt{25}$, or $x = \pm 5$. Hence, the x-intercepts of the graph of $x^2 + y^2 = 25$ are ± 5.

y-intercepts: Letting $x = 0$ in the equation $x^2 + y^2 = 25$, we obtain

$$(0)^2 + y^2 = 25, \quad \text{or} \quad y^2 = 25,$$

which implies $y = \pm\sqrt{25}$ or $y = \pm 5$. Hence, the y-intercepts of the graph of $x^2 + y^2 = 25$ are also ± 5.

We can use a graphing calculator to verify our work. Solving the equation $x^2 + y^2 = 25$ for y, we obtain

$$y = \pm\sqrt{25 - x^2}.$$

Entering *both* of these equations into the calculator gives us the graph in Figure 2.18. To show the true geometric perpective of this graph, which is a circle, we use a *square set viewing rectangle*—one in which the vertical and horizontal tick marks have equal spacing. The graph has symmetry with respect to the x-axis, to the y-axis, and to the origin, and has x-intercepts ± 5, and y-intercepts ± 5. ◆

PROBLEM 3 Test the equation $y^2 + x = 4$ for symmetry. Then find the x- and y-intercepts, if they exist. ◆

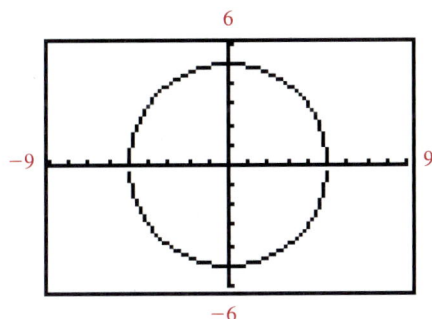

FIGURE 2.18
Graph of $x^2 + y^2 = 25$

To find the x- and y-intercepts for some equations, we factor to form a product and then apply the *zero product property*.

◆ **Zero Product Property**

For real numbers p and q,

if $pq = 0$, then either $p = 0$ or $q = 0$ (or both p and q are zero).

The procedure is illustrated in the next example.

EXAMPLE 4 Find the x- and y-intercepts for the graph of each equation.

(a) $y = x^3 - 3x$ **(b)** $x^2 + 3y^2 - 4x - 12 = 0$

◆ **Solution**

(a) *x-intercepts:* Letting $y = 0$ in the equation $y = x^3 - 3x$, we obtain

$$0 = x^3 - 3x$$
$$0 = x(x^2 - 3) \qquad \text{Factor}$$
$$x = 0 \quad \text{or} \quad x^2 - 3 = 0 \qquad \text{Apply the zero product property}$$
$$x = \pm\sqrt{3} \qquad \text{Solve for } x$$

Hence, we conclude that the x-intercepts are 0, $\sqrt{3} \approx 1.73$, and $-\sqrt{3} \approx -1.73$.

y-intercepts: Letting $x = 0$ in the equation $y = x^3 - 3x$, we obtain

$$y = (0)^3 - 3(0), \quad \text{or} \quad y = 0.$$

Hence, we conclude the y-intercept is 0.

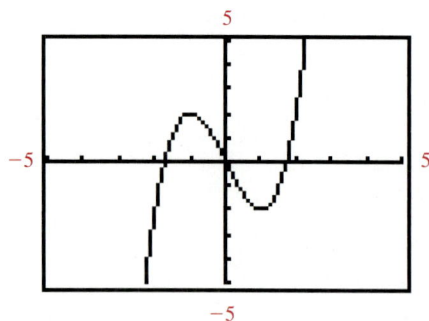

FIGURE 2.19

The graph of $y = x^3 - 3x$ has x-intercepts 0 and $\pm\sqrt{3}$ and y-intercept 0.

Figure 2.19 shows the graph of $y = x^3 - 3x$ in the viewing rectangle of a graphing calculator. We can use the $\boxed{\text{TRACE}}$ key to verify the intercepts.

(b) *x-intercepts:* Letting $y = 0$ in the equation $x^2 + 3y^2 - 4x - 12 = 0$, we obtain

$$x^2 - 4x - 12 = 0$$

$$(x + 2)(x - 6) = 0 \qquad \text{Factor}$$

$$x + 2 = 0 \quad \text{or} \quad x - 6 = 0 \qquad \text{Apply the zero product property}$$

$$x = -2 \qquad\qquad x = 6 \qquad \text{Solve for } x$$

Hence, we conclude that the x-intercepts are -2 and 6.

y-intercepts: Letting $x = 0$ in the equation $x^2 + 3y^2 - 4x - 12 = 0$, we obtain

$$3y^2 - 12 = 0$$

$$y^2 - 4 = 0 \qquad \text{Divide both sides by 3}$$

$$(y + 2)(y - 2) = 0 \qquad \text{Factor}$$

$$y + 2 = 0 \quad \text{or} \quad y - 2 = 0 \qquad \text{Apply zero product property}$$

$$y = -2 \qquad\qquad y = 2 \qquad \text{Solve for } y$$

Hence, we conclude that the y-intercepts are ± 2.

Figure 2.20 shows the graph of $x^2 + 3y^2 - 4x - 12 = 0$, which is equivalent to

$$y = \pm\sqrt{\frac{12 + 4x - x^2}{3}},$$

in the viewing rectangle of a graphing calculator. Note that the x-intercepts are -2 and 6 and the y-intercepts are ± 2. The graph of this equation is called an *ellipse*. ◆

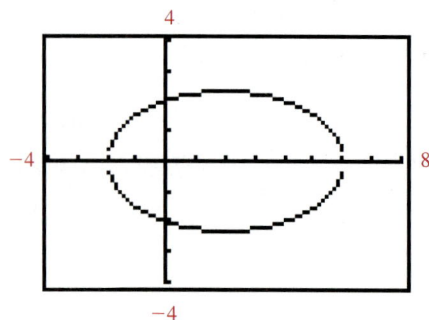

FIGURE 2.20

The graph of $x^2 + 3y^2 - 4x - 12 = 0$ has x-intercepts -2 and 6 and y-intercepts ± 2.

PROBLEM 4 The graph of the equation in Figure 2.19 appears to be symmetric with respect to the origin, and the graph of the equation in Figure 2.20 appears to be symmetric with respect to the x-axis. Algebraically verify these observations. ◆

◆ **Application: Graphs from Experimental Data**

In science and engineering, the relationship between two variables is often given by a table of values obtained from experimentation. Although we may not know a formula that equates the variables in the experiment, we can use the experimental data from the table to draw a graph and to answer questions concerning the relationship between the variables.

EXAMPLE 5 A pizza baked at 400°F was removed from the oven and placed in a room with a constant temperature. The temperature of the pizza was then recorded each minute for the next 6 minutes, as shown in the table of values that follows.

Time t (in minutes)	0	1	2	3	4	5	6
Temperature T (in °F)	400	345	300	262	230	203	181

(a) Plot the graph of this data using t as the horizontal axis and T as the vertical axis.

(b) Estimate the temperature of the pizza for $t = 2.3$ minutes.

◆ **Solution**

(a) We observe from the data that the temperature T decreases as time t increases. Plotting the points (t, T) in the coordinate plane and connecting them to form a smooth curve gives us the graph in Figure 2.21.

(b) We can use the graph in Figure 2.21 to estimate the temperature T when $t = 2.3$ minutes. First we estimate 2.3 in the interval [2, 3] on the t-axis and draw a vertical line from the axis to the graph. Then we draw a horizontal line from this point on the graph to the vertical axis. Reading the value of T on the vertical axis, we find $T \approx 290$ °F when $t = 2.3$ minutes, as shown in Figure 2.21.

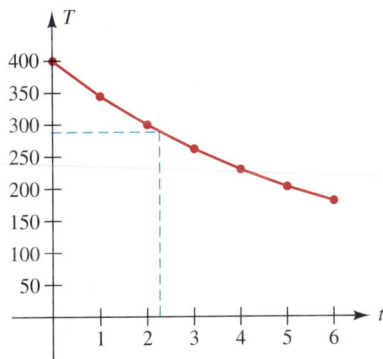

We can also estimate the temperature T by using the **method of linear interpolation**, which assumes that any two consecutive points in the table are connected by a straight line. If values in the table are sufficiently close together, this assumption yields an accurate approximation. Figure 2.22 illustrates that we can set up a direct proportion to find the value of x as follows:

$$\frac{1}{0.3} = \frac{38}{x}$$
$$x = (0.3)(38)$$
$$x = 11.4$$

Therefore, when $t = 2.3$ minutes,

$$T \approx 300 - 11.4 = 288.6 \text{ °F.}$$

FIGURE 2.21

Graph of the experimental data in Example 5

FIGURE 2.22

The value of x may be found by setting up a direct proportion.

PROBLEM 5 Use linear interpolation to estimate the temperature of the pizza for $t = 3.7$ minutes. Use the graph in Figure 2.21 to check your answer.

Exercises 2.2

◆ Basic Skills

In Exercises 1–12, use the point-plotting method to sketch the graph of each equation. Check graphically with a graphing calculator.

1. $y = x$

2. $y = x^2$

3. $y = 2\sqrt{x}$

4. $y = -3\sqrt[3]{x}$

5. $x = y^2 + 2$

6. $x = y^3 - 4$

7. $y - x^3 = 1$

8. $y + |x| = 3$

9. $xy = 1$

10. $y - 3x = 0$

11. $3x + 2y = 12$

12. $4x - 3y = 12$

In Exercises 13–40,

(*a*) *determine algebraically if the graph of each equation has symmetry with respect to the x-axis, y-axis, or origin.*

(*b*) *determine algebraically the x-intercept(s) and y-intercept(s) for the graph of each equation.*

Check each answer graphically with a graphing calculator.

13. $x^2 + 4y = 9$

14. $x - y^4 = 4$

15. $xy = 2$

16. $y = x^{2/3}$

17. $y = 3 - |x|$

18. $x^2 + 4y^2 = 16$

19. $x^2 + 1 = xy$

20. $x^2y + y = 1$

21. $y^2 = \sqrt{x - 2}$

22. $y = |2x - 5|$

23. $6x + 5y = 10$

24. $y^5 - 2x = 32$

25. $x^2 + 3y - 36 = 0$

26. $y = 2x^2 + 5x - 3$

27. $y = 8x - x^4$

28. $y = 12x - x^3$

29. $x^2 + y^2 - 6x - 16 = 0$

30. $4x^2 + 4x - 2y + 1 = 0$

31. $y = \dfrac{\sqrt{x^2 - 2}}{x}$

32. $y = \dfrac{x - 4}{2x - 3}$

33. $3y + 6 = |2x + 9|$

34. $2y - 3x^3 - 2x^2 = 0$

35. $y = 3x - 2x^3 - x^5$

36. $y = x^4 - 4x^2$

37. $x = -\sqrt{4 - y^2}$

38. $3y - \sqrt{x^2 - 9} = 0$

39. $y = \dfrac{1}{x^2 + 1}$

40. $y = \dfrac{x^2 + 1}{x^2}$

41. A bottle of wine at 70 °F is placed in a refrigerator with a constant temperature. The temperature T of the wine is then recorded every 5 minutes for the next 30 minutes, as shown in the table of values that follows.

t (in min)	0	5	10	15	20	25	30
T (in °F)	70	65	61	58	55	52	50

(a) Plot the graph of this data using time t as the horizontal axis and T as the vertical axis.

(b) Use linear interpolation to estimate the temperature of the wine for $t = 12$ minutes. Check your answer by using the graph in part (a).

42. Initially, a culture contains 18 bacteria per microliter. They start to multiply and the total number N present in a microliter is recorded every 3 minutes for the next 15 minutes, as shown in the table of values that follows.

t (in min)	0	3	6	9	12	15
N	18	27	41	61	91	137

(a) Plot the graph of this data using time t as the horizontal axis and N as the vertical axis.

(b) Use linear interpolation to estimate the number of bacteria present in the culture for $t = 7$ minutes. Check your answer by using the graph in part (a).

43. A stone is dropped from the top of a 400-foot building and its height h above the ground is recorded each second for the next 5 seconds, as shown in the table of values that follows.

t (in s)	0	1	2	3	4	5
h (in ft)	400	384	336	256	144	0

(a) Plot the graph of this data using time t as the horizontal axis and h as the vertical axis.

(b) Use linear interpolation to estimate the height of the stone above the ground for $t = 3.4$ seconds. Check your answer by using the graph in part (a).

44. In an electrical circuit with a constant voltage source, the current i (in amperes) is measured through a vari-

able resistor R (in ohms). The data is shown in the table
of values that follows:

Resistance R (in ohms)	10	20	30	40	50
Current i (in amperes)	11.7	5.85	3.90	2.93	2.34

(a) Plot the graph of this data using R as the horizontal
axis and i as the vertical axis.

(b) Use linear interpolation to estimate the current in
the circuit for $R = 38$ ohms. Check your answer by
using the graph in part (a).

◆ Critical Thinking

45. Generate the graph of the equation $y = |x|$ in the viewing rectangle of a graphing calculator. Then, in the same viewing rectangle, generate the graphs of the equations

$$y = |x| + 1, \quad y = |x| + 2, \quad \text{and} \quad y = |x| + 3.$$

Compare each of these graphs to the graph of $y = |x|$.
Describe the graph of $y = |x| + c$ for any constant
$c > 0$.

46. Generate the graph of the equation $y = x^3$ in the viewing rectangle of a graphing calculator. Then, in the same viewing rectangle, generate the graphs of the equations

$$y = x^3 - 1, \quad y = x^3 - 2, \quad \text{and} \quad y = x^3 - 3.$$

Compare each of these graphs to the graph of $y = x^3$.
Describe the graph of $y = x^3 - c$ for any constant $c > 0$.

47. Generate the graph of the equation $y = \sqrt{x}$ in the viewing rectangle of a graphing calculator. Then, in the same viewing rectangle, generate the graphs of the equations

$$y = \sqrt{x - 1}, \, y = \sqrt{x - 2}, \quad \text{and} \quad y = \sqrt{x - 3}.$$

Compare each of these graphs to the graph of $y = \sqrt{x}$.
Describe the graph of $y = \sqrt{x - c}$ for any constant
$c > 0$.

48. Generate the graph of the equation $y = x^2$ in the viewing rectangle of a graphing calculator. Then, in the same viewing rectangle, generate the graphs of the equations

$$y = (x + 1)^2, \quad y = (x + 2)^2, \quad \text{and} \quad y = (x + 3)^2.$$

Compare each of these graphs to the graph of $y = x^2$.
Describe the graph of $y = (x + c)^2$ for any constant
$c > 0$.

49. When a meteorologist gives the outside temperature in degrees Celsius C and in degrees Fahrenheit F, it is observed that $5F - 9C = 160$.

(a) Complete the following table of values for this relationship between C and F. Record each answer to the nearest tenth of a degree.

C (in °C)			12	31	-6
F (in °F)	39	74	-3		

(b) Use the table of values in part (a) to sketch the graph of this relationship between C and F. Use F as the horizontal axis and C as the vertical axis.

(c) Find the F-intercept and C-intercept for the graph in part (b). Interpret their meanings.

(d) Generate the graph of the equation $5F - 9C = 160$ on a graphing calculator. Use the $\boxed{\text{TRACE}}$ key to find the point on the graph where the Celsius and Fahrenheit temperatures are exactly the same. Explain how to find this point algebraically.

50. In a chemistry experiment, it is observed that when the temperature of a confined gas remains constant, the product of its pressure P (in pounds per square inch, psi) and its volume V (in cubic inches, in³) is always equal to 125.

(a) Complete the following table of values for this relationship between P and V. Record each answer to two significant digits.

P (in psi)	3.4	11	41		
V (in in³)				22	6.8	4.0

(Exercise 50 continues)

(b) Use the table of values in part (a) to sketch the graph of this relationship between P and V. Use P as the horizontal axis and V as the vertical axis.

(c) Does the graph in part (b) have axis intercepts? Explain.

2.3   Functions

◆ Introductory Comments

Given an equation in two unknowns x and y, suppose we *choose* values for x and *obtain* corresponding values for y. Hence, we are using x as the independent variable and y as the dependent variable. If for each value of x we choose there corresponds *exactly one* value of y, then we say that the equation defines y as *a function of x.*

◆ Definition of a Function

A **function** from a set X to a set Y is a rule of correspondence that assigns to each element x in X *exactly one* element y in Y. Set X is called the **domain** of the function and set Y contains the **range** of the function.

A function may be represented by a list or table, by a graph, or by a formula or equation. Many functions in this text are specified by equations. For example, consider the equation $y = x^2$. It is useful to think of the x-values as **inputs** and their corresponding y-values as **outputs**. We can think of this equation as a rule that says "Square the input." A *function machine* for this equation is shown in Figure 2.23. If we place $x = 3$ in the input hopper, the machine follows the rule "square the input" and gives us an output of $y = 9$. The rule "square the input" assigns to each input value x one and only one output value y. Thus we say the equation $y = x^2$ *defines y as a function of x.*

Not every equation in two unknowns x and y defines y as a function of x. Consider the equation $y^2 = x$. If we choose the input $x = 9$, the equation becomes $y^2 = 9$ and, consequently, $y = \pm 3$. In this equation we obtain two outputs for the input $x = 9$. Since our definition of a function requires that there be exactly one output for each input, we know that the equation $y^2 = x$ does not define y as a function of x.

◆ Vertical Line Test

Suppose we are given the graph of an equation in two unknowns x and y. If every vertical line that can be drawn in the coordinate plane intersects the graph *at most once,* then we can say the equation has exactly one output y for each input value x that we can assign. Hence, the equation defines y as a function of x. We refer to this graphical method of determining whether y is a function of x as the **vertical line test**.

Input: $x = 3$

Rule: "Square the input."

Output: $y = 9$

FIGURE 2.23

Function machine for $y = x^2$

◆ **Vertical Line Test**

> An equation defines y as a function of x if and only if every vertical line in the coordinate plane intersects the graph of the equation at most once.

The graph of $y = x^2$ (in Figure 2.24) passes the vertical line test, whereas the graph of $y^2 = x$ (in Figure 2.25) fails this test.

FIGURE 2.24

The graph of $y = x^2$ passes the vertical line text.

FIGURE 2.25

The graph of $y^2 = x$ fails the vertical line test.

EXAMPLE 1 Determine if each equation defines y as a function of x.

(a) $x^2 + y - 4 = 0$ **(b)** $x^2 + y^2 - 4 = 0$

◆ **Solution**

(a) If we solve the equation $x^2 + y - 4 = 0$ for y, we obtain

$$y = 4 - x^2.$$

From this form of the equation we can see that for each input value x we choose, one—and only one—output value y corresponds to it. Hence, the equation $x^2 + y - 4 = 0$ defines y as a function of x. Figure 2.26 shows the graph of this equation in the viewing rectangle of a graphing calculator. The equation $x^2 + y - 4 = 0$ passes the vertical line test. Note that every vertical line intersects this graph *at most once*.

(b) If we solve the equation $x^2 + y^2 - 4 = 0$ for y, we obtain

$$y = \pm\sqrt{4 - x^2}.$$

This form of the equation makes it obvious that we have *two* outputs for each input value x in the interval $(-2, 2)$. Hence, the equation

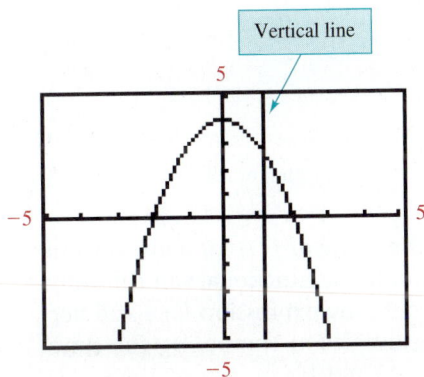

FIGURE 2.26

The graph of $x^2 + y - 4 = 0$ passes the vertical line text. Hence, this equation defines y as a function of x.

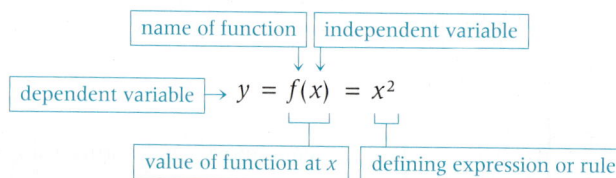

FIGURE 2.27

The graph of $x^2 + y^2 - 4 = 0$ fails the vertical line test. Hence, this equation does not define y as a function of x.

$x^2 + y^2 - 4 = 0$ does *not* define y as a function of x. Figure 2.27 shows the graph of this equation in the viewing rectangle of a graphing calculator. The equation $x^2 + y^2 - 4 = 0$ fails the vertical line test. Note that every vertical line in the interval $(-2, 2)$ intersects the graph twice.   ◆

Note: Taken separately, the equations $y = \sqrt{4 - x^2}$ and $y = -\sqrt{4 - x^2}$ each define y as a function of x, and the graph of each function is a semicircle. Referring to Figure 2.27, we see that the graph of $y = \sqrt{4 - x^2}$ is the semicircle above the x-axis, and the graph of $y = -\sqrt{4 - x^2}$ is the semicircle below the x-axis. Each semicircle passes the vertical line test.

PROBLEM 1   Determine if the equation $2x + y^2 = 4$ defines y as a function of x.   ◆

◆ Functional Notation

The letters f, F, g, G, h, and H are often used to represent functions. We use the notation $f(x)$, read "f of x," to denote the **value of the function f at x.**

◀ **Functional Notation**

> If f is a function and x an element of the domain of f, then the **functional notation**
>
> $$f(x), \quad \text{read "f of x,"}$$
>
> denotes the *value of the function at x.*

CAUTION

The notation $f(x)$ does *not* mean "f times x." Also, the letter f is the *name* of the function, whereas the notation $f(x)$ is the *value* of the function at x. Do not confuse these symbols.

The function f defined by the equation $y = x^2$ may be written using functional notation:

$$\underset{\text{dependent variable}}{y} = \underset{\substack{\text{name of function} \\ \text{value of function at } x}}{f(\underset{\text{independent variable}}{x})} = \underset{\text{defining expression or rule}}{x^2}$$

We refer to the process of determining the value of $f(x)$ for a given value of x as *computing the functional value.* To compute the functional value at some real number a, we replace x with a in the defining expression for f and perform the indicated operations. Here are examples of computing the functional value of $f(x) = x^2$ at two values of x, -3 and 10:

$$\underset{\text{"f of -3"}}{f(-3)} = (-3)^2 = 9 \qquad \underset{\text{"f of 10"}}{f(10)} = (10)^2 = 100$$

Note: The letters we use to represent the function and the independent variable are immaterial. The function g defined by $g(t) = t^2$ describes the same function as the one defined by $f(x) = x^2$.

EXAMPLE 2 Let f be a function defined by $f(x) = x^2 - 4x + 3$. Compute each functional value.

(a) $f(-2)$ (b) $f(3)$ (c) $f(2a)$ (d) $f(x-1)$

◆ **Solution**

(a) The notation $f(-2)$ denotes the value of the function at -2. By replacing x with -2 in the defining expression $x^2 - 4x + 3$, we obtain the functional value at -2:

$$f(-2) = (-2)^2 - 4(-2) + 3$$
$$= 4 + 8 + 3$$
$$= 15$$

(b) The notation $f(3)$ denotes the value of the function at 3. By replacing x with 3 in the defining expression $x^2 - 4x + 3$, we obtain the functional value at 3:

$$f(3) = (3)^2 - 4(3) + 3$$
$$= 9 - 12 + 3$$
$$= 0$$

(c) The notation $f(2a)$ denotes the value of the function at $2a$. By replacing x with $2a$ in the defining expression $x^2 - 4x + 3$, we obtain the functional value at $2a$:

$$f(2a) = (2a)^2 - 4(2a) + 3$$
$$= 4a^2 - 8a + 3$$

(d) The notation $f(x-1)$ denotes the value of the function at $x-1$. By replacing x with $x-1$ in the defining expression $x^2 - 4x + 3$, we obtain the functional value at $x-1$:

$$f(x-1) = (x-1)^2 - 4(x-1) + 3$$
$$= x^2 - 2x + 1 - 4x + 4 + 3$$
$$= x^2 - 6x + 8$$

PROBLEM 2 For the function f defined by $f(x) = x^2 - 4x + 3$, find $f(\pi)$.

In calculus, it is often necessary to evaluate the **difference quotient** for a given function f:

$$\frac{f(x+h) - f(x)}{h}, \quad h \neq 0$$

The notations $f(x + h)$ and $f(x)$ denote the values of the function f at $x + h$
ely.

ind the difference quotient for the function f defined by
$.

o find the difference quotient of a function, we proceed

compute $f(x + h)$:

$+ h) = (x + h)^2 - 4(x + h) + 3$
$\qquad = x^2 + 2xh + h^2 - 4x - 4h + 3$

ubtract $f(x)$ from the result in step 1 and simplify:

$= x^2 + 2xh + h^2 - 4x - 4h + 3 - (x^2 - 4x + 3)$
$= x^2 + 2xh + h^2 - 4x - 4h + 3 \ominus x^2 \oplus 4x \ominus 3$
$= 2xh + h^2 - 4h$

Step 3:   Finally, we divide the result in step 2 by h:

$$\frac{f(x + h) - f(x)}{h} = \frac{2xh + h^2 - 4h}{h}$$

$$= \frac{h(2x + h - 4)}{h} \qquad \text{Factor the numerator}$$

$$= 2x + h - 4 \qquad \text{Reduce to lowest terms} \qquad ◆$$

PROBLEM 3   Find the difference quotient for the function g defined by $g(x) = 4 - 3x$. ◆

As illustrated in the next example, a function can be defined by a multipart rule. We refer to such a function as a **piecewise-defined function**. When we work with a piecewise-defined function, it is important to observe how the rule changes over different subsets of the domain.

EXAMPLE 4   Let h be a function defined by

$$h(x) = \begin{cases} 1 - x^2 & \text{if } x < 1 \\ x - 1 & \text{if } 1 \leq x \leq 3 \\ 4 & \text{if } x > 3. \end{cases}$$

Compute each functional value.

(a) $h(-2)$ **(b)** $h(3)$ **(c)** $h(\pi)$

◆ **Solution**

(a) If $x < 1$, then the function h is defined by $h(x) = 1 - x^2$. Thus, for $x = -2$, we have

$$h(-2) = 1 - (-2)^2 = 1 - 4 = -3.$$

(b) If $1 \le x \le 3$, then the function h is defined by $h(x) = x - 1$. Thus, for $x = 3$, we have

$$h(3) = (3) - 1 = 2.$$

(c) If $x > 3$, then the function h is defined by $h(x) = 4$. Thus, for $x = \pi \approx 3.14$, we have

$$h(\pi) = 4.$$ ◆

PROBLEM 4   For the function h defined in Example 4, find $h(1)$.   ◆

◆ **Finding the Domain of a Function**

We can think of the *domain* of a function as the set of all possible inputs. When a function is defined by an equation, the domain is assumed to be the set of all real numbers that give us real-number outputs. For the function f defined by $f(x) = x^2$, every real number assigned to x leads to a corresponding real number for $f(x)$. Thus, the domain of the function f is the set of all real numbers. In this text, we express the domain of a function by using interval notation (see Section 1.1). Thus, we express the domain of f as $(-\infty, \infty)$.

The *range* of a function is the set of all function values that we obtain from the elements in the domain. Consider the function f defined by $f(x) = x^2$. When real numbers are squared, we obtain *nonnegative* real numbers. Thus, using interval notation, we conclude that the range of f is $[0, \infty)$. For most functions, however, the range is quite difficult to find. The best method for obtaining the range is usually to read it from the graph of the function. We will discuss this method in Section 2.4.

As illustrated in the next example, we must be aware of two properties when determining the domain of a function:

1. Division by zero is undefined.

2. Even roots of negative numbers are *not* real numbers.

EXAMPLE 5   State the domain of each function using interval notation.

(a) $f(x) = \dfrac{1}{x + 4}$

(b) $F(x) = \dfrac{2x + 5}{x^3 - 3x^2 + 2x - 6}$

(c) $G(x) = \sqrt[3]{3 - x}$

(d) $g(x) = \dfrac{\sqrt{2 - x}}{x + 1}$

◆◆ **Solution**

(a) Division by zero is undefined. If $x = -4$, then the denominator becomes zero. Thus, the domain of the function f defined by $f(x) = 1/(x + 4)$ is all real numbers *except* $x = -4$. This domain is shown on the real number line in Figure 2.28. Using interval notation, we express the domain of this function as

$$(-\infty,\ -4) \cup (-4,\ \infty).$$

FIGURE 2.28

The domain of $f(x) = \dfrac{1}{x + 4}$ is $(-\infty, -4) \cup (-4, \infty)$.

(b) Division by zero is undefined. Thus, the domain of the function F defined by

$$F(x) = \frac{2x + 5}{x^3 - 3x^2 + 2x - 6}$$

is all real numbers x except those for which the denominator is zero:

$$
\begin{aligned}
x^3 - 3x^2 + 2x - 6 &= 0 \\
x^2(x - 3) + 2(x - 3) &= 0 && \text{\textbf{Factor by grouping terms}} \\
(x - 3)(x^2 + 2) &= 0 \\
x - 3 = 0 \quad \text{or} \quad x^2 + 2 &= 0 && \text{\textbf{Apply the zero product property}} \\
x &= 3
\end{aligned}
$$

no real solution

Hence, the domain of this function is all real numbers *except* $x = 3$. The domain is shown on the real number line in Figure 2.29. Using interval notation, we can express the domain of this function as

$$(-\infty,\ 3) \cup (3,\ \infty).$$

FIGURE 2.29

The domain of

$F(x) = \dfrac{2x + 5}{x^3 - 3x^2 + 2x - 6}$

is $(-\infty, 3) \cup (3, \infty)$.

(c) Odd roots of real numbers *are* real numbers. Hence, the domain of the function G defined by $G(x) = \sqrt[3]{3 - x}$ is the set of all real numbers. This domain is shown on the real number line in Figure 2.30. Using interval notation, we can express the domain of this function as

$$(-\infty,\ \infty).$$

FIGURE 2.30

The domain of $G(x) = \sqrt[3]{3 - x}$ is $(-\infty, \infty)$

(d) Even roots of negative numbers are *not* real numbers. Thus, for the function

$$g(x) = \frac{\sqrt{2 - x}}{x + 1},$$

the radicand $2 - x$ must be *nonnegative*, that is,

$$2 - x \geq 0$$

$$-x \geq -2 \qquad \text{\textbf{Add} -2 \textbf{to both sides}}$$

$$x \leq 2 \qquad \text{\textbf{Multiply both sides by} -1 \textbf{and reverse the direction of the inequality symbol}}$$

FIGURE 2.31

The domain of $g(x) = \dfrac{\sqrt{2-x}}{x+1}$ is

$(-\infty, -1) \cup (-1, 2]$.

Also, we must exclude $x = -1$ from the domain, since this value of x makes the denominator zero. Hence, the domain of this function is all real numbers less than or equal to 2 except $x = -1$. We can show the domain on the real number line as in Figure 2.31, or write it in interval notation as

$$(-\infty, -1) \cup (-1, 2].$$ ◆

PROBLEM 5 Determine the domain of the function H defined by

$$H(x) = \frac{x+1}{\sqrt{2-x}}.$$ ◆

Exercises 2.3

Basic Skills

In Exercises 1–14, generate the graph of each equation in the viewing rectangle of a graphing calculator. Then use the vertical line test to determine if the equation defines y as a function of x.

1. $x^2 + y - 1 = 0$
2. $x^2 + y - 2x = 0$
3. $x^3 + 7y = 1$
4. $3x + 2y = 6$
5. $y = \sqrt{x^2 - 4}$
6. $y = |x - 5|$
7. $x^2 + y^2 - 4x = 0$
8. $x^2 - y^2 - 4 = 0$
9. $|x|y = 1$
10. $|xy| = 1$
11. $y^3 + x - 1 = 0$
12. $x^3 + y^3 = 1$
13. $xy - x^2 = y$
14. $x^2y + y = 1$

Given the functions f, g, and h, defined by

$$f(x) = 4x^2 - 2x + 1, \; g(x) = \sqrt{x} - 4, \; h(x) = |2x + 3|,$$

compute the functional values given in Exercises 15–34.

15. $f(2)$
16. $g(9)$
17. $h(-4)$
18. $g(\frac{1}{4}) + g(1)$
19. $f(\sqrt{2})$
20. $-h(\frac{3}{2})$
21. $h(ab)$
22. $f(3p)$
23. $g(t^2), t \geq 0$
24. $f(-x)$
25. $f\left(\dfrac{n}{2}\right)$
26. $h(x^2 - 3)$
27. $g(x) - g(0)$
28. $f(x + 2)$
29. $g(1 + x^2)$
30. $f(x + h) - f(x)$
31. $f(\sqrt{x - 2})$
32. $h\left(-\dfrac{x}{2}\right)$
33. $f(\sqrt{x} - 4)$
34. $g(4x^2 - 2x + 1)$

For the piecewise-defined functions in Exercises 35–38, compute each functional value.

35. $f(x) = \begin{cases} 2x - 5 & \text{if } x < 0 \\ x^2 & \text{if } x \geq 0 \end{cases}$
    (a) $f(3)$ (b) $f(-3)$
    (c) $f(2)$ (d) $f(0)$

36. $g(x) = \begin{cases} \sqrt{3 - x} & \text{if } x \leq 3 \\ 2 - x & \text{if } 3 < x < 6 \\ 1 & \text{if } x \geq 6 \end{cases}$
    (a) $g(0)$ (b) $g(4)$
    (c) $g(3)$ (d) $g(7)$

37. $h(x) = \begin{cases} x & \text{if } x > 0 \\ 0 & \text{if } x = 0 \\ -x & \text{if } x < 0 \end{cases}$
    (a) $h(7)$ (b) $h(-3)$
    (c) $h(5) - h(0)$ (d) $h(x + 1)$

38. $f(x) = \begin{cases} 1 & \text{if } x \text{ is a rational number} \\ -1 & \text{if } x \text{ is an irrational number} \end{cases}$
    (a) $f(0)$ (b) $f(6)$
    (c) $f(\pi)$ (d) $f(\sqrt{2})$

In Exercises 39–46, find the difference quotient for the given function f, and simplify the result.

39. $f(x) = 2x + 1$
40. $f(x) = 2 - 3x$
41. $f(x) = x^2$
42. $f(x) = x^3$
43. $f(x) = x^2 + 2x - 3$
44. $f(x) = 2x^2 - 3x$
45. $f(x) = \dfrac{1}{x}$ [Hint: Subtract fractions in the numerator.]
46. $f(x) = \sqrt{x}$ [Hint: Rationalize the numerator.]

In Exercises 47–62, state the domain of each function using interval notation.

47. $f(x) = 3x + 2$

48. $g(x) = 1 - x^2$

49. $f(x) = x^3 + 2$

50. $f(x) = \sqrt[3]{2x - 1}$

51. $H(x) = \sqrt{4 - x}$

52. $h(x) = \sqrt{2x - 3}$

53. $g(x) = -\sqrt{16 - x^2}$

54. $F(x) = \sqrt{x^2 - 4}$

55. $f(x) = \dfrac{x}{x + 2}$

56. $f(x) = \dfrac{-3}{\sqrt{2 - x}}$

57. $G(x) = \dfrac{1}{x^2 - 4}$

58. $f(x) = \dfrac{2x}{x^2 + 3}$

59. $f(x) = \dfrac{-3}{x^2 + 3x - 10}$

60. $f(x) = \dfrac{2x - 3}{x^3 - 4x^2 + x - 4}$

61. $F(x) = \dfrac{\sqrt{2x - 1}}{4 - x}$

62. $H(x) = \sqrt{x^3 - 2x^2 - 8x}$

63. The radius r (in meters) of an oil spill is a function of the time t (in minutes) that the oil has been leaking. This function is given by

$$r(t) = \sqrt{2t - 1}, \quad \text{for } t \geq 1.$$

Find $r(25)$ and describe what this value represents.

64. The bending moment M (in pound-feet, lb-ft) along a simply supported 9-ft steel beam that carries a uniformly distributed load of 800 pounds per foot (lb/ft) is a function of the distance x (in feet) from one end of the beam and is given by

$$\begin{aligned} M(x) &= \tfrac{1}{2}(9)(800)x - \tfrac{1}{2}(800)x^2 \\ &= 3600x - 400x^2, \quad \text{for } 0 \leq x \leq 9. \end{aligned}$$

Find $M(1)$ and describe what this value represents.

65. The cost C (in dollars) for a taxicab fare is a function of the distance x driven (in miles), and is given by

$$C(x) = 1.50 + 1.45x, \quad \text{for } x \geq 0.$$

Find $C(12.4)$ and describe what this value represents.

66. If \$5000 is borrowed at $10\tfrac{1}{2}\%$ simple interest per year, the amount A (in dollars) that must be paid back is a function of the time t (in years) over which it is repaid and is given by

$$A(t) = 5000(1 + 0.105t), \quad \text{for } t \geq 0.$$

Find $A(12)$ and describe what this value represents.

◆ **Critical Thinking**

67. Does the inequality $y < x$ define y as a function of x? Explain.

68. Does the equation $y^n = x$ define y as a function of x when n is an odd integer? an even integer? zero? Explain.

69. Let $X = \{1, 2, 3\}$ and $Y = \{a, b, c, d\}$. Use the definition of a function to determine whether each of the following sets of ordered pairs or figures defines a function from set X to set Y. Explain.

(a) $\{(1, a), (2, b), (3, d)\}$

(b) $\{(1, c), (3, a)\}$

(c)

(d)

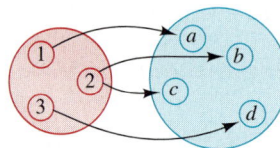

70. Does $f(a + b) = f(a) + f(b)$ for a function f? Use the function f defined by $f(x) = x^2$ to support your claim.

71. For a function f defined as follows, determine the value of A if $f(2) = 12$.

(a) $f(x) = x^2 + Ax$

(b) $f(x) = \sqrt{5x + A}$

72. Give an example of a function f with the property that $f(-x) = f(x)$ for all real numbers x and $-x$ in the domain of f. [Answers are not unique.]

73. The population size P of a certain organism is a function of the temperature t (in degrees Celsius) of the medium in which the organism exists and is given by

$$P(t) = 3t^2, \quad \text{for } 0 \le t \le 40.$$

If t_1 is the present temperature of the medium, compare $P(t_1)$ with $P\left(\frac{1}{2}t_1\right)$, and state the effect of halving the temperature on the population size.

74. In a psychological experiment, a student is given electrical shocks of varying intensities and is asked to rate the intensity of each shock in relation to an initial shock s_0, which is given a rating of 10. It is found that the response number R given by the student is a function of the magnitude (in milliamperes) of the shock s and is defined by

$$R(s) = \frac{s^{3/2}}{40}.$$

Compare $R(s_0)$ with $R(4s_0)$, and state the effect of quadrupling the intensity on the response number.

2.4 Graph of a Function

◆ Introductory Comments

The *graph of a function f* is the same as the graph of the equation $y = f(x)$ and is defined as follows.

◆ **Graph of a Function**

> The **graph of a function f** is the set of all points (x, y) in the coordinate plane such that x is in the domain of f and $y = f(x)$.

The point-plotting method discussed in Section 2.2 may be used to sketch the graph of a simple function. First, we select various values of x from the domain of f and compute the functional values $f(x)$. Next, we plot the points $(x, f(x))$ in the coordinate plane, where x is the directed distance from the y-axis and $f(x)$ is the directed distance from the x-axis. Finally, we connect these points to form a smooth curve, which is the graph of f. The graph of the function f defined by $f(x) = x^2$ is shown in Figure 2.32. The graph of this function is the same as the graph of the equation $y = x^2$ (see Figure 2.24).

To generate the graph of a function f in the viewing rectangle of a graphing calculator, we enter the equation $y = f(x)$. For example, to generate the graph of the function f defined by $f(x) = x^2$ on a graphing calculator, we simply enter the equation $y = x^2$.

◆ Finding the Domain and Range from the Graph of a Function

Observe in Figure 2.32 that the domain and range of the function f defined by $f(x) = x^2$ are evident from its graph. For the domain, we look along the x-axis and note the interval(s) over which the graph is drawn. For the range, we look along the y-axis and note the interval(s) over which the graph is drawn.

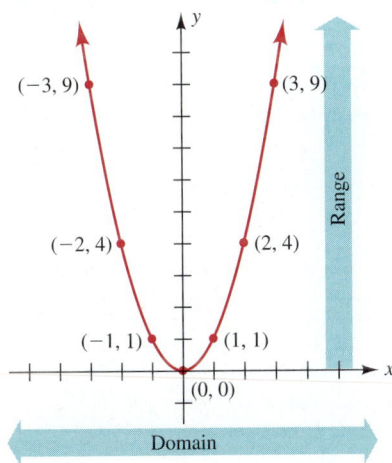

FIGURE 2.32

Graph of $f(x) = x^2$.
Domain: $(-\infty, \infty)$.   Range: $[0, \infty)$.

EXAMPLE 1   The graph of the function F defined by

$$F(x) = -1 + \sqrt{15 + 2x - x^2}$$

FIGURE 2.33

is shown in Figure 2.33. Use this graph to determine the domain and range of F.

◆ **Solution**   As we look along the x-axis, we see that the graph extends from $x = -3$ to $x = 5$. Hence, we conclude that the domain of this function is all real numbers in the closed interval $[-3, 5]$. As we look along the y-axis, we see that the graph extends from $F(-3) = -1$ [or $F(5) = -1$] to $F(1) = 3$. Hence, the range of this function is all real numbers in the closed interval $[-1, 3]$.   ◆

PROBLEM 1   Referring to the function F in Example 1, compute each functional value, if possible. Check each answer with the graph in Figure 2.33.
(a) $F(0)$ **(b)** $F(6)$ ◆

◆   **Symmetry and Real Zeros of Functions**

We say that a function is an *even function* if its graph is symmetric with respect to the y-axis or that it is an *odd function* if its graph is symmetric with respect to the origin. The following tests can be used to determine if a function is even or odd.

◆ **Tests for Even and Odd Functions**

1. A function f is an **even function** if for every x in the domain of f, $-x$ is also in the domain of f, and $f(-x) = f(x)$.

2. A function f is an **odd function** if for every x in the domain of f, $-x$ is also in the domain of f, and $f(-x) = -f(x)$.

The **real zeros** of a function f are the real numbers x for which $f(x) = 0$. When $f(x) = 0$, the graph of the function f intersects the x-axis. Hence, the real zeros of a function f are the x-intercepts of the graph of f.

EXAMPLE 2   Detemine if the function is even, odd, or neither. Then find the real zero(s) of the function.

(a) $g(x) = \sqrt{x^2 - 4}$ **(b)** $h(x) = x - x^3$

◆ **Solution**

(a) *Even, odd, or neither:* The function g is *even* because

$$g(-x) = \sqrt{(-x)^2 - 4} = \sqrt{x^2 - 4} = g(x).$$

Real zeros: The zeros of g are the real roots of the equation $g(x) = \sqrt{x^2 - 4} = 0$. Since the square root of zero is zero, we have

$$x^2 - 4 = 0 \quad \text{or} \quad x = \pm 2.$$

Hence, the real zeros of the function g are ± 2.
    We can use a graphing calculator to verify our work. The graph shown in Figure 2.34 has x-intercepts ± 2 and symmetry with the respect to the y-axis. Thus, the function g is an even function with real zeros ± 2.

(b) *Even, odd, or neither:* The function h is *odd* because

$$h(-x) = (-x) - (-x)^3 = -x + x^3 = -(x - x^3) = -h(x).$$

Real zeros: The zeros of h are the real roots of the equation $h(x) = x - x^3 = 0$. By factoring and then applying the zero product property, we can find the roots:

$$x - x^3 = 0$$
$$x(1 - x^2) = 0$$
$$x = 0 \quad \text{or} \quad 1 - x^2 = 0$$
$$x = \pm 1$$

Hence, the real zeros of the function h are 0 and ± 1.
    We can use a graphing calculator to verify our work. The graph in Figure 2.35 has x-intercepts 0 and ± 1 and symmetry with respect to the origin. Thus, the function h is an odd function with real zeros 0 and ± 1. ◆

PROBLEM 2   Use the graphs of the functions g and h in Figures 2.34 and 2.35 to determine their domains and ranges. ◆

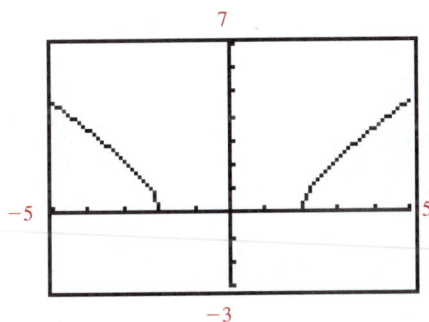

FIGURE 2.34

Graph of $g(x) = \sqrt{x^2 - 4}$, an even function with zeros ± 2

FIGURE 2.35

Graph of $h(x) = x - x^3$, an odd function with zeros 0 and ± 1.

◆ **Solving Equations and Inequalities Graphically**

Example 2 illustrates that the concepts of real zeros, x-intercepts, and real roots of equations are closely related. In general, if f is a function and k is a real number, then the following statements are equivalent:

1. k is an x-intercept for the graph of f.

2. k is a real zero of f.

3. k is a real root of the equation $f(x) = 0$.

In Example 2, we found the real roots of the equation $f(x) = 0$ to get the real zeros of f, which are the x-intercepts of the graph of f. If we reverse the order of these steps, we have a method for *solving an equation graphically*: find the x-intercepts of the graph of f to get the real zeros of f, which are the real roots of the equation $f(x) = 0$. We refer to this procedure as the **x-intercept method for solving an equation**.

Many graphing calculators have a built-in program for finding an x-intercept. If your calculator does not have a built-in program for finding the x-intercept, use the $\boxed{\text{TRACE}}$ key and $\boxed{\text{ZOOM}}$ key on the calculator to estimate the x-intercept.

EXAMPLE 3   Solve each equation by using the x-intercept method.

(a) $5 - 8x^2 = -11$ **(b)** $\sqrt{2x - 1} + 2 = x$

◆ **Solution**

(a) We begin by writing the equation in the form $f(x) = 0$:

$$5 - 8x^2 = -11$$
$$\underbrace{16 - 8x^2}_{f(x)} = 0 \qquad \text{\color{red}{Add 11 to both sides}}$$

Note that for $f(x) = 16 - 8x^2$, we have $f(x) = f(-x)$ for all x in the domain of f. This indicates that f is an even function and its graph is symmetric with respect to the y-axis. We now generate the graph of $y = f(x)$ in the viewing rectangle of a graphing calculator. As shown in Figure 2.36, the x-intercept on the right is approximately 1.414. Because this function is an even function, the x-intercept on the left must be -1.414. Hence, the roots of the original equation are approximately ± 1.414. For the algebraic solution of this equation, see Example 1(d) in Section 1.8.

(b) We begin by writing the equation in the form $f(x) = 0$:

$$\sqrt{2x - 1} + 2 = x$$
$$\underbrace{\sqrt{2x - 1} + 2 - x}_{f(x)} = 0 \qquad \text{\color{red}{Subtract x from both sides}}$$

We now generate the graph of $y = f(x)$ in the viewing rectangle of a graphing calculator. As shown in Figure 2.37, the x-intercept of this graph is 5. Hence, the root of the original equation is also 5. We can check this solution by replacing x with 5 in the original equation:

FIGURE 2.36

The roots of the equation $5 - 8x^2 = -11$ are approximately ± 1.414.

FIGURE 2.37

The root of the equation $\sqrt{2x - 1} + 2 = x$ is 5.

$$\sqrt{2(5) - 1} + 2 \overset{?}{=} (5)$$
$$\sqrt{9} + 2 \overset{?}{=} 5$$
$$3 + 2 \overset{?}{=} 5$$
$$5 = 5$$

◆

Equations that contain an unknown in the radicand, as in Example 3(b), are called **radical equations**. Algebraically, we can eliminate the radical sign in a radical equation that contains one radical expression by isolating the radical on one side of the equation and then raising both sides of the equation to the power that is equal to the index of the radical. But does raising both sides of an equation to a positive integer power always generate equivalent equation? Consider the equation.

$$x = 5$$

Of course, the solution of this equation is 5. If we square both sides of this equation, we obtain

$$x^2 = 25,$$

which implies that $x = \pm 5$. Note that the solution of the original equation, 5, appears as one of the solutions of the equation $x^2 = 25$. However, this equation also has the root -5, which is *not* a solution of the original equation. We refer to a root such as -5 that develops through this algebraic process, but does not satisfy the original equation, as an **extraneous root**.

When raising both sides of an equation to a positive *even* integer power, it is essential to check all apparent solutions which develop in order to make certain that no extraneous root is recorded as a solution to the original equation.

PROBLEM 3 Solve the equation $\sqrt{2x - 1} + 2 = x$ algebraically by isolating the radical expression on one side of the equation and then squaring both sides. What extraneous root develops through this process? ◆

Graphically, we can solve the inequality $f(x) > 0$ by determining the interval(s) on which the graph of f is above the x-axis. To solve $f(x) < 0$, we determine the interval(s) on which the graph of f is below the x-axis. For inequalities containing $\leq$ or $\geq$, we include the real zeros (x-intercepts) of f as part of the solution set. We refer to this procedure as the **x-intercept method for solving an inequality**.

EXAMPLE 4 Solve each inequality by using the x-intercept method.

(a) $|2x - 3| < 5$ **(b)** $\dfrac{x + 1}{2} - \dfrac{5(x - 1)}{8} \geq 1$

◆ **Solution**

(a) We begin by writing the inequality in the form $f(x) < 0$:

$$|2x - 3| < 5$$
$$\underbrace{|2x - 3| - 5}_{f(x)} < 0 \qquad \text{Subtract 5 from both sides}$$

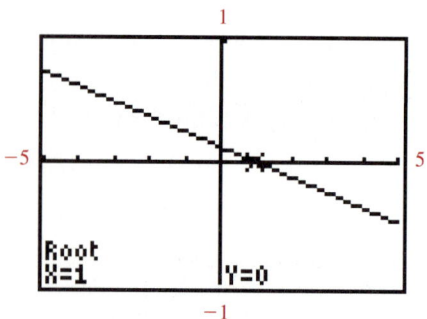

We now generate the graph of $y = f(x)$ in the viewing rectangle of a graphing calculator. Since the inequality symbol is *less than* ($<$), we look for the interval on which the graph is *below* the x-axis. As shown in Figure 2.38, the graph is below the x-axis when $-1 < x < 4$. Hence, the solution set of the original inequality is $(-1, 4)$. For the algebraic solution of this inequality, see Example 2(c) in Section 1.8.

FIGURE 2.38

The solution set of the inequality $|2x - 3| < 5$ is $(-1, 4)$.

(b) We begin by writing the inequality in the form $f(x) \geq 0$:

$$\frac{x + 1}{2} - \frac{5(x - 1)}{8} \geq 1$$

$$\underbrace{\frac{x + 1}{2} - \frac{5(x - 1)}{8} - 1}_{f(x)} \geq 0 \qquad \text{Subtract 1 from both sides}$$

We now generate the graph of $y = f(x)$ in the viewing rectangle of a graphing calculator. Since the inequality symbol is *greater than or equal to* ($\geq$), we look for the interval(s) on which the graph intersects the x-axis or is *above* the x-axis. As shown in Figure 2.39, the graph intersects the x-axis at $x = 1$ and is above the x-axis when $x < 1$. Hence, the solution set of the original inequality is $(-\infty, 1]$.   ◆

FIGURE 2.39

The solution set of the inequality $\frac{x + 1}{2} - \frac{5(x - 1)}{8} \geq 1$ is $(-\infty, 1]$.

PROBLEM 4   Solve the inequality $\frac{x + 1}{2} - \frac{5(x - 1)}{8} \geq 1$ algebraically. Check your answer with the graphical result in Example 4(b).   ◆

◆ **Application: Solving Word Problems**

An essential goal of our study of algebra is to become skillful in reading word problems and changing the words into algebraic equations (or inequalities) that can be solved for the *unknowns*. Unfortunately, no one single procedure can be given to solve all word problems. In general, however, we assign one of the unknowns the variable x and express each of the other unknowns in terms of x. We then develop an equation (or inequality) that relates the known and unknown quantities and solve for x. The procedure is shown in the next example.

EXAMPLE 5   A length of rope hangs from the top of a flagpole, with 6 feet coiled up on the ground at the base of the pole. When the free end of the rope is brought out 18 feet from the base of the pole, it just touches the ground. Assuming the flagpole is on level ground, what is the height of the flagpole?

◆ **Solution**   Referring to Figure 2.40, we let

$$x = \text{height of the flagpole (in feet)}.$$

Since the rope is 6 feet longer than the height of the flagpole, we have

$$x + 6 = \text{length of the rope (in feet)}.$$

From the sketch in Figure 2.40, we can see that x, $x + 6$, and 18 are related by the Pythagorean theorem. Thus,

$$\boxed{(\text{hypotenuse})^2} \;=\; \boxed{(\text{leg})^2} \;+\; \boxed{(\text{leg})^2}$$

$$(x + 6)^2 \;=\; x^2 + 18^2$$

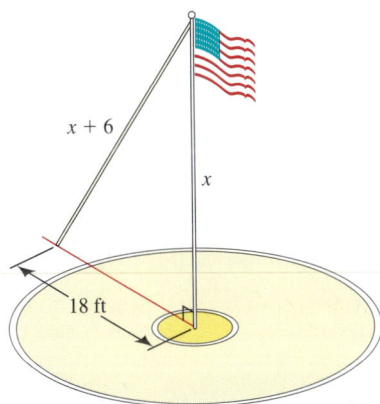

FIGURE 2.40

The quantities x, $x + 6$, and 18 are related by the Pythagorean theorem.

We can solve this equation algebraically or graphically.

Choosing the graphical approach, we write the equation in the form $f(x) = 0$:

$$(x + 6)^2 = x^2 + 18^2$$
$$\underbrace{(x + 6)^2 - x^2 - 18^2}_{f(x)} = 0 \qquad \text{\color{red}{\textbf{Subtract} }} x^2 \text{ \color{red}{and} } 18^2 \text{ \color{red}{\textbf{from both sides}}}$$

We now generate the graph of $y = f(x)$ in the viewing rectangle of a graphing calculator. As shown in Figure 2.41, the x-intercept of this graph is 24. Hence, the root of the original equation is also 24. Therefore, the height of the flagpole is 24 ft.

In Problem 5 you are asked to verify the solution to this problem by solving the equation $(x + 6)^2 = x^2 + 18^2$ algebraically.   ◆

FIGURE 2.41

The root of the equation $(x + 6)^2 - x^2 - 18^2 = 0$ is 24.

PROBLEM 5   Check the result of Example 5 by solving the equation $(x + 6)^2 = x^2 + 18^2$ algebraically.   ◆

Exercises 2.4

◆ Basic Skills

In Exercises 1–10, a function and its graph are given.

(a) *Determine the domain of the function.*

(b) *Determine the range of the function.*

(c) *Find any real zero(s) of the function.*

(d) *State whether the function is even, odd, or neither.*

1. $f(x) = |x| - 2$

2. $g(x) = 3x - 2$

3. $g(x) = x^2 + 2x$

4. $H(x) = 2x^2 - x^4$

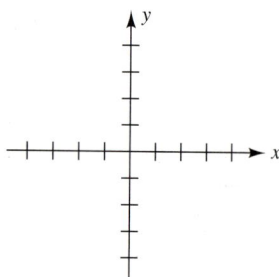

5. $f(x) = \sqrt{5 - x^2}$

6. $g(x) = -\sqrt{x^2 - 3}$

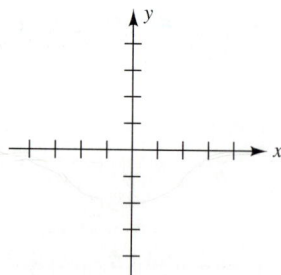

7. $f(x) = \dfrac{-2}{x^2 + 1}$

8. $g(x) = \dfrac{x^2 + 1}{x}$

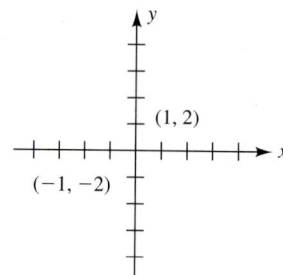

9. $g(x) = x^5 - 5x$,   for $-2 \le x \le 2$

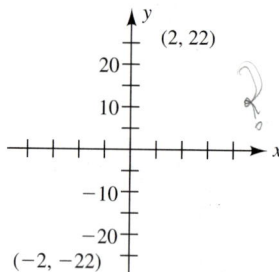

10. $f(x) = \begin{cases} x^2 & \text{if } -2 \le x < 1 \\ 3 - 2x & \text{if } 1 \le x \le 3 \end{cases}$

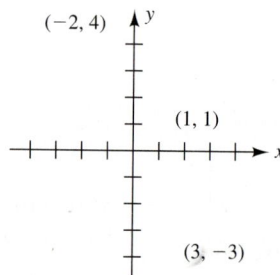

In Exercises 11–20, determine whether the function is even, odd, or neither. Check graphically with a graphing calculator.

11. $F(x) = x$

12. $g(x) = x^3$

13. $f(x) = 3 - |x|$

14. $f(x) = 2x^2 - 4$

15. $f(x) = 4x - x^5$

16. $f(x) = x^4 + 2x^2 + 3$

17. $g(x) = \dfrac{3 - x^2}{2 + x^6}$ **18.** $H(x) = \dfrac{x + x^3}{2|x|}$

19. $h(x) = x^3 - 3x + 1$ **20.** $g(x) = (3x^2 - 2x)^2$

In Exercises 21–30, find the real zero(s) of each function. Check graphically with a graphing calculator.

21. $g(x) = 3x - 5$ **22.** $f(x) = 4 + 5x$

23. $F(x) = x^2 - 3x - 40$ **24.** $g(x) = x^2 - 8x + 12$

25. $h(x) = 2x^2 - 7x + 3$ **26.** $f(x) = 2x^4 - 5x^2 - 12$

27. $g(x) = \dfrac{2x - 3}{x^2 + 4}$ **28.** $G(x) = \dfrac{9 - x^2}{2x + 3}$

29. $H(x) = 5 - \sqrt{x^4 + 9}$

30. $h(x) = x - \sqrt{4x^2 - x - 2}$

In Exercises 31–50, solve each equation or inequality by using the x-intercept method in conjunction with a graphing calculator. Check algebraically.

31. $9x - 2(4x + 1) = 3$ **32.** $\dfrac{x}{6} = \dfrac{x + 4}{2}$

33. $6x^2 = x + 2$

34. $x^3 - 3x^2 - 2x + 6 = 0$

35. $\left|\dfrac{x + 4}{3}\right| = 7$ **36.** $2 - |3x - 11| = -4$

37. $3\sqrt{x + 1} = 4$ **38.** $4\sqrt{3x + 1} = 9$

39. $\sqrt{1 - x} - x = 5$ **40.** $\sqrt{5x - 1} + 3 = x$

41. $\sqrt[3]{x^3 + 8} - x = 2$ **42.** $\sqrt[4]{x^4 - x + 3} = x$

43. $2x + 6 \geq 3x + 4$ **44.** $4x - (x - 2) < x - 3$

45. $\dfrac{x + 5}{2} - 3 \leq \dfrac{3x}{5} - \dfrac{x - 4}{4}$

46. $\dfrac{x + 1}{2} + \dfrac{2x + 3}{4} \geq 2$

47. $|5 - 2x| > 3$ **48.** $|4x + 3| + 2 < 17$

49. $|x^2 - 6| \geq 3$ **50.** $|2 - x^3| < 6$

In Exercises 51–60, set up an equation or inequality and solve by the x-intercept method. Check each result algebraically.

51. The hypotenuse of a right triangle is 1 inch longer than one of its legs. The other leg is 5 inches long. What is the length of the hypotenuse?

52. The length of a rectangle is twice its width. If each side of the rectangle were increased by 3 centimeters, the new perimeter would be 48 centimeters. What are the dimensions of the original rectangle?

53. At a Red Sox baseball game, the price for a box seat is $9 and for grandstand seat, $6. A total of 24,320 seats are sold for a particular game, and the receipts from the box seats equal those from the grandstand seats. How many box seats are sold?

54. A developer purchased 22 acres of land for $900,000. He agreed to pay $50,000 per acre for buildable land and $10,000 per acre for marshland. How many acres of buildable land were purchased?

55. Oak flooring is installed in a square room and a rectangular room of a new home. The square room requires 18 square feet (sq ft) less flooring than is used in the rectangular room. The length of the rectangular room is 6 feet more than a side of the square room, and the width of the rectangular room is 4 feet less than a side of the square room. Find the amount of oak flooring used in each room.

56. How many pounds of cashews, worth $6.40 per pound, must be added to 12 pounds of peanuts, worth $2.40 per pound, to form a mixture that is worth $4.00 per pound?

57. A campsite is set up on the bank of a river that is 40 ft wide. On the other side of the river and 200 ft downstream is another campsite. A newly constructed footbridge and walkway join the two campsites, as shown in the sketch. The walkway, which runs along the edge of the straight river, cost $2 per foot to construct. The footbridge, which is built diagonally over the river, costs $10 per foot. If the total cost of this project is $840, find the length of the walkway.

58. In order to widen a highway, the state takes by eminent domain a strip of land from Lot A, which is shown in the sketch at the top of the next page. What is the largest width for this strip of land that can be taken if Lot A must remain at least one acre after removing the strip? [*Hint:* 1 acre = 43,560 square feet.]

185.6 ft

Lot A

121.5 ft

Land taken by eminent domain

302.4 ft

59. Find the set of three consecutive positive odd integers such that the largest is greater than twice the smallest.

60. Find all sets of four consecutive positive even integers such that the largest is greater than twice the smallest.

Critical Thinking

61. Suppose that f is an even function and one point on its graph is $(-2, 3)$. Find the coordinates of a second point on its graph.

62. Suppose that f is an odd function and one point on its graph is $(-2, 3)$. Find the coordinates of a second point on its graph.

63. Suppose the graph of a function f does not intersect the x-axis. What can we conclude about the solution of the equation $f(x) = 0$?

64. Under what conditions is the function f defined by $f(x) = x^n$ an even function? an odd function?

65. How many y-intercepts can the graph of a function have? How many x-intercepts?

66. Use a graphing calculator to generate the graphs of the functions f and g defined by

$$f(x) = |x| \text{ and } g(x) = \sqrt{x^2}.$$

What can we conclude about $|x|$ and $\sqrt{x^2}$?

67. Can the graph of a nonzero function be symmetric with respect to the x-axis? Explain.

68. Is the function g defined by $g(x) = xf(x)$ even, odd, or neither when the function f is an even function? when the function f is an odd function?

69. Biologists estimate that the population P of the endangered red-spotted turtle is a function of time t (in years) and is given by

$$P(t) = 90 - 10(34t + 22)^{2/5}.$$

In how many years will this species become extinct?

70. A company determines that when it produces q units of a product, its total revenue from sales is $200q^{1/2}$ dollars. Suppose the company can produce no more than 500 units and that the variable cost per unit is \$4 and the fixed cost is \$1600. How many units must the company produce to break even (no profit). [*Hint:* Revenue − cost = profit.]

2.5 Shifting, Reflecting, and Stretching a Graph

◆ Introductory Comments

Figure 2.42 shows the graphs of eight basic functions that occur frequently in mathematics: *constant function, identity function, absolute value function, squaring function, cubing function, reciprocal function, square root function,* and *cube root function.* By plotting points or using a graphing calculator, we can verify the graph of each of these functions.

(a) Constant function
$f(x) = k$
Domain: $(-\infty, \infty)$
Range: $\{k\}$

(b) Identity function
$f(x) = x$
Domain: $(-\infty, \infty)$
Range: $(-\infty, \infty)$

(c) Absolute value function
$f(x) = |x|$
Domain: $(-\infty, \infty)$
Range: $[0, \infty)$

(d) Squaring function
$f(x) = x^2$
Domain: $(-\infty, \infty)$
Range: $[0, \infty)$

(e) Cubing function
$f(x) = x^3$
Domain: $(-\infty, \infty)$
Range: $(-\infty, \infty)$

(f) Reciprocal function
$f(x) = 1/x$
Domain: $(-\infty, 0) \cup (0, \infty)$
Range: $(-\infty, 0) \cup (0, \infty)$

(g) Square root function
$f(x) = \sqrt{x}$
Domain: $[0, \infty)$
Range: $[0, \infty)$

(h) Cube root function
$f(x) = \sqrt[3]{x}$
Domain: $(-\infty, \infty)$
Range: $(-\infty, \infty)$

FIGURE 2.42

Graphs of eight basic functions

FIGURE 2.43

Comparison of the graphs of
$F(x) = |x| + 2$ and $f(x) = |x|$

In this section, we discuss techniques of graphing a function without using a graphing calculator or plotting points. We use the graphs of these eight basic functions to sketch the graphs of many other related functions by using one or more procedures:

1. *Shifting* a graph vertically or horizontally.
2. *Reflecting* a graph about the *x*-axis or *y*-axis.
3. *Stretching* or *compressing* a graph.

◆ Vertical Shift Rule

Figure 2.43 shows the graphs of the absolute value function $f(x) = |x|$ and the function $F(x) = |x| + 2$ in the same viewing rectangle of a graphing calculator. The graph of $F(x) = |x| + 2$ appears to be the same as the graph of $f(x) = |x|$, but *shifted vertically upward* 2 units. When we substitute the same input into the functions F and f, the output $F(x)$ is always 2 more than the output $f(x)$. Without using a graphing calculator or plotting points, we can sketch the graph of $F(x) = |x| + 2$ by shifting the graph of $f(x) = |x|$ vertically upward 2 units. We can apply this rule to any function. In summary, we state this property as the *vertical shift rule*.

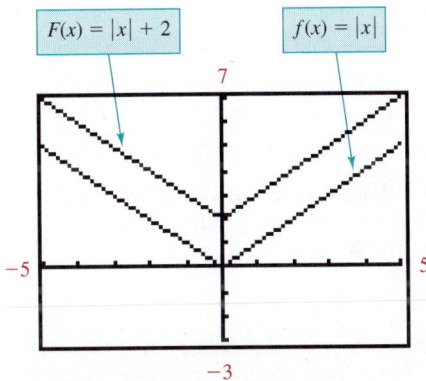

◆ **Vertical Shift Rule**

If f is a function and c is a constant, then the graph of the function F defined by

$$F(x) = f(x) + c$$

is the same as the graph of f shifted *vertically upward* $|c|$ units if $c > 0$ or shifted *vertically downward* $|c|$ units if $c < 0$.

To obtain an accurate sketch, it is good practice to determine and label the *intercepts* of the graph. Recall that the x-intercepts of the graph of a function f are the real zeros of the function and may be found by solving the equation $f(x) = 0$. The y-intercept of the graph of a function f may be found by evaluating $f(0)$. The graph of a function can have at most one y-intercept.

EXAMPLE 1   Sketch the graph of each function and label the intercepts, if they exist.

(a) $F(x) = x^2 + 1$ **(b)** $G(x) = \sqrt{x} - 2$

◆ **Solution**

(a) We can think of this function as $F(x) = f(x) + 1$, where $f(x) = x^2$ is the squaring function in part (d) of Figure 2.42. Thus, by the vertical shift rule, the graph of $F(x) = x^2 + 1$ is the same as the graph of $f(x) = x^2$, but shifted vertically *upward* 1 unit. The graph of the function F is shown in Figure 2.44.

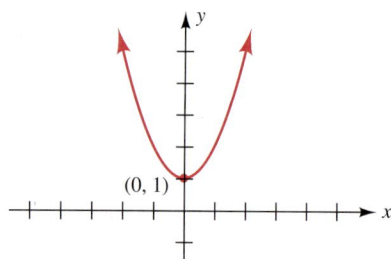

FIGURE 2.44

Graph of $F(x) = x^2 + 1$

x-intercept:

$$F(x) = 0$$
$$x^2 + 1 = 0$$
$$x^2 = -1$$

y-intercept:

$$F(0) = (0)^2 + 1 = 1$$

(b) We can think of this function as $G(x) = f(x) - 2$, where $f(x) = \sqrt{x}$ is the square root function in part (g) of Figure 2.42. Thus, by the vertical shift rule, the graph of $G(x) = \sqrt{x} - 2$ is the same as the graph of $f(x) = \sqrt{x}$, but shifted vertically *downward* 2 units. The graph of the function G is shown in Figure 2.45.

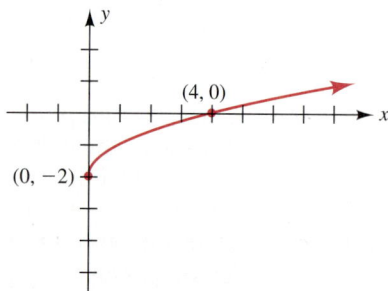

FIGURE 2.45

Graph of $G(x) = \sqrt{x} - 2$

x-intercept:

$$G(x) = 0$$
$$\sqrt{x} - 2 = 0$$
$$\sqrt{x} = 2$$
$$x = 4 \qquad \text{Square both sides}$$

y-intercept:

$$G(0) = \sqrt{0} - 2 = -2$$

$f(x) = |x|$ $F(x) = |x + 2|$

FIGURE 2.46

Comparison of the graphs of
$F(x) = |x + 2|$ and $f(x) = |x|$

PROBLEM 1 Sketch the graph of $h(x) = x^3 - 1$ and label the intercepts. ◆

◆ **Horizontal Shift Rule**

Figure 2.46 shows the graphs of the absolute value function $f(x) = |x|$ and the function $F(x) = |x + 2|$ in the same viewing rectangle of a graphing calculator. The graph of $F(x) = |x + 2|$ appears to be the same as the graph of $f(x) = |x|$, but *shifted horizontally to the left* 2 units. When we substitute an input into F that is 2 *less than* the input we use in f, the values of the outputs $F(x)$ and $f(x)$ are equal. Without using a graphing calculator or plotting points, we can sketch the graph of $F(x) = |x + 2|$ by shifting the graph of $f(x) = |x|$ horizontally to the left 2 units. Like the vertical shift rule, we can apply this property to any function. In summary, we state the *horizontal shift rule*.

◆ **Horizontal Shift Rule**

If f is a function and c is a constant, then the graph of the function F defined by

$$F(x) = f(x + c)$$

is the same as the graph of f shifted *horizontally to the left* $|c|$ units if $c > 0$ or shifted *horizontally to the right* $|c|$ units if $c < 0$.

EXAMPLE 2 Sketch the graph of each function and label the intercepts.

(a) $F(x) = (x + 1)^2$ **(b)** $G(x) = \sqrt[3]{x - 2}$

◆ **Solution**

(a) We can think of this function as $F(x) = f(x + 1)$, where $f(x) = x^2$ is the squaring function in part (d) of Figure 2.42. Thus, by the horizontal shift rule, the graph of $F(x) = (x + 1)^2$ is the same as the graph of $f(x) = x^2$, but shifted horizontally *to the left* 1 unit. The graph of the function F is shown in Figure 2.47.

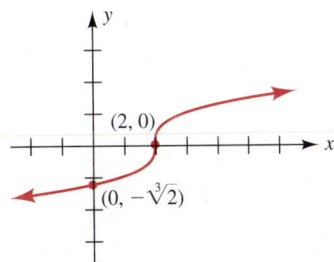

FIGURE 2.47

Graph of $F(x) = (x + 1)^2$

x-intercept:	*y-intercept:*
$F(x) = 0$	$F(0) = (0 + 1)^2 = 1$
$(x + 1)^2 = 0$	
$x + 1 = 0$ **Take the square root of both sides**	
$x = -1$	

(b) We can think of this function as $G(x) = f(x - 2)$, where $f(x) = \sqrt[3]{x}$ is the cube root function in part (h) of Figure 2.42. Thus, by the horizontal shift rule, the graph of $G(x) = \sqrt[3]{x - 2}$ is the same as the graph of $f(x) = \sqrt[3]{x}$, but shifted horizontally *to the right* 2 units. The graph of the function G is shown in Figure 2.48.

FIGURE 2.48

Graph of $G(x) = \sqrt[3]{x - 2}$

x-intercept:

$$G(x) = 0$$
$$\sqrt[3]{x-2} = 0$$
$$x - 2 = 0 \quad \text{Cube both sides}$$
$$x = 2$$

y-intercept:

$$G(0) = \sqrt[3]{0-2} = \sqrt[3]{-2}$$
$$\approx -1.26$$

PROBLEM 2 Sketch the graph of $h(x) = (x-1)^3$ and label the intercepts.

As illustrated in the next example, we can use both the vertical and horizontal shift rules to help sketch the graph of a function.

EXAMPLE 3 Sketch the graph of $F(x) = \dfrac{1}{x-2} + 1$ and label the intercepts.

◆ **Solution**   We can think of this function as $F(x) = f(x-2) + 1$, where $f(x) = 1/x$ is the reciprocal function in part (f) of Figure 2.42. Thus, by the horizontal and vertical shift rules, we obtain the graph of F by shifting the graph of $f(x) = 1/x$ to the right 2 units and upward 1 unit, as shown in Figure 2.49.

FIGURE 2.49

Graph of $F(x) = \dfrac{1}{x-2} + 1$

x-intercept:

$$F(x) = 0$$
$$\frac{1}{x-2} + 1 = 0$$
$$1 + (x-2) = 0 \quad \begin{array}{l}\text{Multiply both sides by}\\ (x-2), \text{ provided}\\ x - 2 \neq 0\end{array}$$
$$x = 1$$

y-intercept:

$$F(0) = \frac{1}{0-2} + 1 = \frac{1}{2}$$

The vertical dashed line that passes through (2, 0) is called a *vertical asymptote* for the graph of F. The horizontal dashed line that passes through (0, 1) is called a *horizontal asymptote* for the graph of F. In Chapter 4 we will discuss asymptotes in more detail.

PROBLEM 3 Sketch the graph of $F(x) = (x+1)^2 - 4$ and label the intercepts.

◆ ***x*-Axis and *y*-Axis Reflection Rules**

Figure 2.50 shows the graphs of the square root function $f(x) = \sqrt{x}$ and the functions $F(x) = -\sqrt{x}$ and $G(x) = \sqrt{-x}$ in the same viewing rectangle of a graphing calculator. It appears the graph of $F(x) = -\sqrt{x}$ is the same as the graph of $f(x) = \sqrt{x}$, but reflected about the *x*-axis. When we substitute the same input into the functions f and F, the output $F(x)$ is always the *opposite* of the output $f(x)$. This property can be applied to any function. In summary, we state the *x-axis reflection rule*.

FIGURE 2.50

Comparison of the graphs of $f(x) = \sqrt{x}$, $F(x) = -\sqrt{x}$, and $G(x) = \sqrt{-x}$

$G(x) = \sqrt{-x}$ $F(x) = -\sqrt{x}$ $f(x) = \sqrt{x}$

◆ **x-Axis Reflection Rule**

If f is a function, then the graph of the function F defined by

$$F(x) = -f(x)$$

is the same as the graph of f, but *reflected about the x-axis*.

Figure 2.50 also shows that the graph of $G(x) = \sqrt{-x}$ is the same as the graph of $f(x) = \sqrt{x}$, but reflected about the y-axis. When we substitute an input into the function G that is the *opposite* of the input we use for f, the values of the outputs $G(x)$ and $f(x)$ are equal. Like the other shift and reflect properties, we can apply this property to any function. In summary, we state the *y-axis reflection rule*.

◆ **y-Axis Reflection Rule**

If f is a function, then the graph of the function F defined by

$$F(x) = f(-x)$$

is the same as the graph of f, but *reflected about the y-axis*.

EXAMPLE 4 Sketch the graph of each function and label the intercepts.

(a) $F(x) = 2 - x^3$ **(b)** $h(x) = \sqrt{2 - x}$

◆ **Solution**

(a) We can think of this function as $F(x) = -f(x) + 2$, where $f(x) = x^3$ is the cubing function in part (e) of Figure 2.42. Thus, by the x-axis reflection rule and the vertical shift rule, the graph of $F(x) = 2 - x^3$ is obtained by reflecting the graph of $f(x) = x^3$ *about the x-axis* and then shifting this graph *upward* 2 units, as shown in Figure 2.51.

FIGURE 2.51

Graph of $F(x) = 2 - x^3$

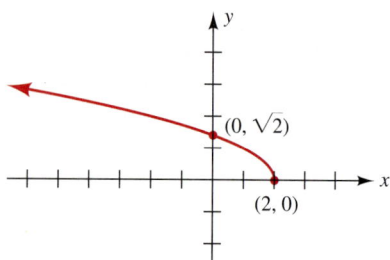

x-intercept:	*y-intercept:*
$F(x) = 0$	$F(0) = 2 - (0)^3 = 2$
$2 - x^3 = 0$	
$x^3 = 2$	
$x = \sqrt[3]{2} \approx 1.26$ Extract the cube root	

(b) The graph of $h(x) = \sqrt{2 - x}$ is the same as the graph of $h(-x) = \sqrt{2 + x}$, but reflected about the *y*-axis. To sketch $h(-x) = \sqrt{2 + x}$, we shift the graph of the square root function $f(x) = \sqrt{x}$ *to the left* 2 units. Now, reflecting this graph *about the y-axis* gives us the graph of $h(x) = \sqrt{2 - x}$, as shown in Figure 2.52.

FIGURE 2.52

Graph of $h(x) = \sqrt{2 - x}$

x-intercept:	*y-intercept:*
$h(x) = 0$	$h(0) = \sqrt{2 + 0} = \sqrt{2}$
$\sqrt{2 - x} = 0$	≈ 1.41
$2 - x = 0$ Square both sides	
$x = 2$	

CAUTION The order in which we apply the shift rules and the axis reflection rules is important. The graph of $F(x) = 2 - x^3$ [in Example 4(a)] is *not* formed by shifting the graph of $f(x) = x^3$ vertically upward 2 units and then reflecting this graph about the *x*-axis. This sequence of steps would produce the graph of the function $G(x) = -(x^3 + 2)$.

PROBLEM 4 Sketch the graph of $F(x) = -|x + 2|$ and label the intercepts.

◆ **Vertical Stretch and Compress Rule**

Figure 2.53 shows the graph of the absolute value function $f(x) = |x|$ and the functions $F(x) = 2|x|$ and $G(x) = \frac{1}{2}|x|$ in the viewing rectangle of

FIGURE 2.53

Comparison of the graphs of $f(x) = |x|$, $F(x) = 2|x|$, and $G(x) = \frac{1}{2}|x|$

a graphing calculator. The graph of $F(x) = 2|x|$ appears to be a steeper **V**-shape than the graph of $f(x) = |x|$. The graph of $G(x) = \frac{1}{2}|x|$ appears to be a flatter **V**-shape than the graph of $f(x) = |x|$. When we substitute the same input into the functions f, F, and G, the output $F(x)$ is *twice* the output $f(x)$, and the output $G(x)$ is *half* the output $f(x)$. The graph of F is *stretched vertically* by a factor of 2, and the graph of G is *compressed vertically* by a factor of 2. In summary, we state the *vertical stretch and compress rules*.

◆ **Vertical Stretch and Compress Rules**

If f is a function and c is a real number with $c > 1$, then the graph of the function F defined by

$$F(x) = cf(x)$$

is similar to the graph of f, but is *stretched vertically* by a factor of c, and the graph of the function G defined by

$$G(x) = \frac{1}{c}f(x)$$

is similar to the graph of f, but is *compressed vertically* by a factor of c.

EXAMPLE 5 Sketch the graph of each function and label the intercepts.

(a) $G(x) = -\dfrac{|x|}{4}$ **(b)** $F(x) = 4(x - 1)$

◆ **Solution**

(a) We can think of this function as $G(x) = -\frac{1}{4}f(x)$, where $f(x) = |x|$. Thus, by the x-axis reflection rule and the vertical compress rule, we obtain the graph of $G(x) = -|x|/4$ by reflecting the graph of the absolute value function about the x-axis and then compressing vertically by a factor of 4, as shown in Figure 2.54.

FIGURE 2.54

The graph of $G(x) = -|x|/4$ is formed from the graph of $y = -|x|$ by compressing vertically by a factor of 4.

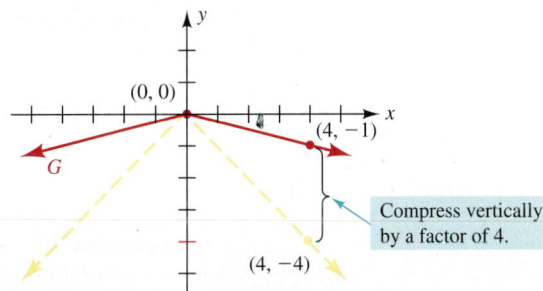

(b) We can think of this function as $F(x) = 4f(x - 1)$, where $f(x) = x$ is the identity function in part (b) of Figure 2.42. By the horizontal shift rule and vertical stretch rule, we obtain the graph of $F(x) = 4(x - 1)$ by shifting the graph of $f(x) = x$ to the right 1 unit and then stretching vertically by a factor of 4, as shown in Figure 2.55.

x-intercept:

$$F(x) = 0$$
$$4(x - 1) = 0$$
$$x - 1 = 0$$
$$x = 1$$

y-intercept:

$$F(0) = 4(0 - 1) = -4$$

FIGURE 2.55

The graph of $F(x) = 4(x - 1)$ is formed from the graph of $y = x - 1$ by stretching vertically by a factor of 4.

PROBLEM 5   Sketch the graph of $F(x) = 4x - 1$ and label the intercepts.

◆ Increasing, Decreasing, and Constant Functions

A function f is said to be an *increasing function* if as x increases, the value of $f(x)$ also *increases*. A function f is said to be a *decreasing function* if as x increases, $f(x)$ *decreases*. A function f is said to be a *constant function* if as x increases, $f(x)$ *remains the same* [see part (a) of Figure 2.42.] More precisely, we state the following definitions.

◆ **Increasing, Decreasing, and Constant Functions**

1. A function f is an **increasing function** if for all a and b in the domain of f, $f(a) < f(b)$ whenever $a < b$.

2. A function f is a **decreasing function** if for all a and b in the domain of f, $f(a) > f(b)$ whenever $a < b$.

3. A function f is a **constant function** if for all a and b in the domain of f, $f(a) = f(b)$.

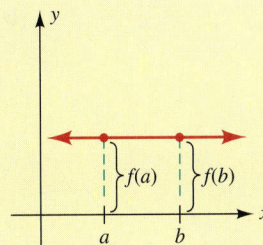

From the graph of a function, we may determine whether the function is increasing or decreasing. The graph of $F(x) = 4(x - 1)$, shown in Figure 2.55, always *rises* as we move from left to right along the x-axis. Thus, as x increases, $F(x)$ also increases, and we conclude the function F is an increasing function. The graph of $h(x) = \sqrt{2 - x}$, shown in Figure 2.52, always *falls* as we move from left to right along the x-axis. Hence, as x increases, $h(x)$ decreases, and we conclude the function h is a decreasing function.

The graph of $G(x) = -|x|/4$, shown in Figure 2.54, rises as we move from left to right until we reach the y-axis. The graph then falls as we continue to move along the x-axis from the y-axis toward the right. Thus, this function is *neither increasing nor decreasing* over its entire domain. This situation is typical of many functions. A function may be increasing over some intervals in its domain and decreasing over other intervals. For $G(x) = -|x|/4$, we say that G is *increasing on the interval* $(-\infty, 0)$ and *decreasing on the interval* $(0, \infty)$.

FIGURE 2.56

Graphs of $y = 1 - x^2$, $y = x - 1$, and $y = 4$, with dashed lines

EXAMPLE 6

(a) Sketch the graph of the piecewise-defined function g:

$$g(x) = \begin{cases} 1 - x^2 & \text{if } x < 1 \\ x - 1 & \text{if } 1 \le x \le 3 \\ 4 & \text{if } x > 3. \end{cases}$$

(b) Use the graph of function g to determine the open intervals on which the function is increasing, decreasing, or constant.

◆◆ **Solution**

(a) First, we apply the graphing techniques from this section to sketch the graphs of $y = 1 - x^2$, $y = x - 1$, and $y = 4$ with dashed lines, as shown in Figure 2.56. To form the graph of g, we darken the graphs of

$$y = 1 - x^2 \quad \text{for } x < 1$$

$$y = x - 1 \quad \text{for } 1 \le x \le 3$$

and $\qquad y = 4 \qquad \text{for } x > 3$

FIGURE 2.57

Graph of

$$g(x) = \begin{cases} 1 - x^2 & \text{if } x < 1 \\ x - 1 & \text{if } 1 \le x \le 3 \\ 4 & \text{if } x > 3 \end{cases}$$

Increasing on intervals $(-\infty, 0)$ and $(1, 3)$
Decreasing on the interval $(0, 1)$
Constant on the interval $(3, \infty)$

as shown in Figure 2.57. We place a *solid dot* at the point $(3, 2)$ to indicate that this point is part of the graph of g, and we place an *open circle* at the point $(3, 4)$ to indicate that this point is *not* part of the graph of g.

(b) As we move along the *x*-axis from left to right, the graph *rises* until we reach the *y*-axis. From the *y*-axis to $x = 1$, the graph *falls*. From $x = 1$ to $x = 3$, the graph again *rises*. Finally, to the right of $x = 3$, the graph remains the same. Thus, we conclude the function is *increasing* on the intervals $(-\infty, 0)$ and $(1, 3)$, *decreasing* on the interval $(0, 1)$, and *constant* on the interval $(3, \infty)$.   ◆

To generate the graph of a piecewise-defined function on a graphing calculator, we enter each part of the function separately and include with each part of the function the interval over which that part is defined. Here is a typical display showing the entries for the piecewise-defined function in Example 6:

$$\boxed{y_1 =} \ (1 - x^2)(x < 1)$$

$$\boxed{y_2 =} \ (x - 1)(x \geq 1)(x \leq 3)$$

$$\boxed{y_3 =} \ (4)(x > 3)$$

Also, when graphing a piecewise-defined function on a graphing calculator it is better to use the "dot" mode instead of the "connected" mode. In connected mode, the calculator may erroneously connect one part of the graph to the next part.

PROBLEM 6   Consult the manual that came with your calculator or the *Graphing Technology Laboratory Manual* that accompanies this text regarding piecewise-defined functions. Then use a graphing calculator to generate the graph of the function *g* in Example 6.   ◆

Exercises 2.5

◆ Basic Skills

In Exercises 1–42, use the shift rules, axis reflection rules, and stretch and compress rules to sketch the graph of each function. Label the axis intercepts, if they exist. Use a graphing calculator to verify each graph.

1. $f(x) = x + 3$

2. $g(x) = x^2 - 1$

3. $G(x) = x^3 - 2$

4. $H(x) = \dfrac{1}{x} + 2$

5. $F(x) = 3 + \sqrt{x}$

6. $f(x) = -2 + |x|$

7. $f(x) = |x - 1|$

8. $g(x) = \sqrt{x + 3}$

9. $H(x) = (x + 2)^2$

10. $F(x) = (x + 1)^3$

11. $F(x) = \dfrac{1}{x + 2}$

12. $f(x) = \sqrt[3]{x - 1}$

13. $h(x) = -\dfrac{1}{x}$

14. $g(x) = -x^3$

15. $f(x) = -x^2$

16. $H(x) = \sqrt[3]{-x}$

17. $g(x) = \sqrt{1 - x}$

18. $f(x) = -\sqrt{3 - x}$

19. $F(x) = (x - 1)^2 + 3$

20. $G(x) = |x - 4| - 1$

21. $f(x) = \sqrt{x + 2} - 1$

22. $h(x) = \dfrac{1}{x + 1} - 2$

23. $h(x) = -|x - 3|$

24. $f(x) = -(x + 1)$

25. $G(x) = 3 - x^2$

26. $H(x) = 2 - \sqrt{x}$

27. $F(x) = 2 - (x + 4)^3$

28. $h(x) = 1 - |x - 1|$

29. $f(x) = \sqrt[3]{x + 2} - 3$

30. $f(x) = 2 - \sqrt{4 - x}$

31. $H(x) = \dfrac{1}{2-x}$

32. $g(x) = 3 + (2-x)^2$

33. $g(x) = 3|x|$

34. $f(x) = -2x$

35. $f(x) = -\frac{1}{3}x^2$

36. $G(x) = \dfrac{x^3}{4}$

37. $G(x) = 2x - 1$

38. $F(x) = 3 - \frac{1}{2}x$

39. $F(x) = 2\sqrt{x-4}$

40. $g(x) = 3 - 2|x|$

41. $F(x) = \dfrac{5}{x+2}$

42. $h(x) = \dfrac{(x-1)^2}{4}$

43. Which of the eight basic functions in Figure 2.42 are increasing functions? decreasing functions?

44. For each of the eight basic functions in Figure 2.42 that is neither an increasing nor decreasing function, determine the open intervals over which the function is increasing, decreasing, or constant.

In Exercises 45–52, determine whether the function is increasing, decreasing, or neither.

45. The function G in Exercise 3

46. The function g in Exercise 8

47. The function g in Exercise 17

48. The function g in Exercise 14

49. The function G in Exercise 25

50. The function G in Exercise 20

51. The function f in Exercise 29

52. The function f in Exercise 34

In Exercises 53–58, use the shift rules, axis reflection rules, and stretch and compress rules to sketch the graph of each piecewise-defined function. Label the axis intercepts, if they exist. Use the graph to determine the open intervals over which the function is increasing, decreasing, or constant. Use a graphing calculator to verify each graph.

53. $g(x) = \begin{cases} x + 1 & \text{if } x < 2 \\ 3 & \text{if } x \ge 2 \end{cases}$

54. $f(x) = \begin{cases} -x & \text{if } x \le 4 \\ \sqrt{x} & \text{if } x > 4 \end{cases}$

55. $f(x) = \begin{cases} x + 2 & \text{if } x < 0 \\ 2 & \text{if } 0 \le x < 3 \\ (x-3)^2 & \text{if } x \ge 3 \end{cases}$

56. $h(x) = \begin{cases} x^3 & \text{if } x < 1 \\ 2 - x^2 & \text{if } 1 \le x < 3 \\ -7 & \text{if } x \ge 3 \end{cases}$

57. $h(x) = \begin{cases} |x| & \text{if } x < 1 \\ -x^2 & \text{if } 1 \le x \le 3 \\ 1/x & \text{if } x > 3 \end{cases}$

58. $g(x) = \begin{cases} x^2 - 1 & \text{if } x \le 1 \\ x - 1 & \text{if } 1 < x < 3 \\ 3 & \text{if } 3 \le x \le 5 \\ 13 - 2x & \text{if } x > 5 \end{cases}$

◆ **Critical Thinking**

The graph of $y = f(x)$ is shown in the figure. Use this graph and the techniques of shifting and reflecting to sketch the graph of each of the equations in Exercises 59–62.

its graph on a graphing calculator to determine if it matches the given graph.

(0, 1)

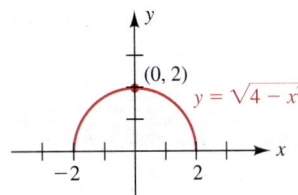

(0, 2) $y = \sqrt{4 - x^2}$

59. $y = -f(x)$

60. $y = f(x) - 3$

61. $y = f(x - 2)$

62. $y = 2 - f(x + 1)$

The graph of $y = \sqrt{4 - x^2}$ is shown in the figure. Use this graph to write an equation for each of the functions whose graphs are shown in Exercises 63–66. Check the equation by generating

63.

(0, 3)

64.

(3, 2)

65.

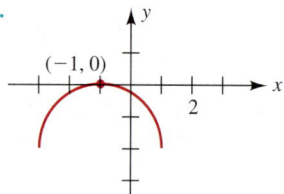

(−1, 0)

2

66.

(2, −1)

4

In Exercises 67–70, use a graphing calculator to generate the graphs of the given equations in the same viewing rectangle. What rule in this section verifies the picture you observe?

67. $y = |x|$, $y = |x| + 1$, $y = |x| + 2$, $y = |x| - 1$, $y = |x| - 2$

68. $y = x^3$, $y = (x + 1)^3$, $y = (x + 2)^3$, $y = (x - 1)^3$, $y = (x - 2)^3$

69. $y = \sqrt{x}$, $y = -\sqrt{x}$, $y = \sqrt{-x}$, $y = -\sqrt{-x}$

70. $y = x^2$, $y = 0.25x^2$, $y = 0.5x^2$, $y = 2x^2$, $y = 4x^2$

71. Use your observations in Exercises 67–70 to fill in the following table:

To graph:	Draw the graph of $y = f(x)$ and then
$y = f(x) + c$ for $c > 0$	
$y = f(x) - c$ for $c > 0$	
$y = f(x + c)$ for $c > 0$	
$y = f(x - c)$ for $c > 0$	
$y = -f(x)$	
$y = f(-x)$	
$y = cf(x)$ for $c > 1$	
$y = cf(x)$ for $0 < c < 1$	

72. In mathematics, the notation $[\![x]\!]$ denotes the largest integer n such that $n \le x$. For example,

$$[\![8.9]\!] = 8, \quad [\![5]\!] = 5, \quad [\![\pi]\!] = 3,$$
$$[\![-8.9]\!] = -9, \quad [\![-\sqrt{3}]\!] = -2, \quad \text{and so on.}$$

The **greatest integer function** f is defined by $f(x) = [\![x]\!]$. Sketch the graph of this function and explain why the function is classified as a *step function*.

73. The cost C (in cents) to mail a letter with first-class postage is a function of its weight w (in ounces) and is given by

$$C(w) = \begin{cases} 32 & \text{if } 0 < w \le 1 \\ 55 & \text{if } 1 < w \le 2 \\ 78 & \text{if } 2 < w \le 3 \\ 101 & \text{if } 3 < w \le 4 \\ 124 & \text{if } 4 < w \le 5 \end{cases}$$

This function is another example of a step function (see Exercise 72).

(a) Sketch the graph of this function.

(b) Find $C(2.4)$ and describe what this value represents.

74. The voltage v (in volts) that is applied to an electrical circuit is a function of time t (in milliseconds) and is given by

$$v(t) = \begin{cases} t^2 & \text{if } 0 \le t \le 2 \\ 4 & \text{if } 2 < t \le 6 \\ (t - 8)^2 & \text{if } 6 < t \le 8 \end{cases}$$

(a) Sketch the graph of this function.

(b) Find $v(7)$ and describe what this value represents.

2.6 ◆ **Composite and Inverse Functions**

◆ **Introductory Comments**

If we select a number, cube it, and then take the cube root of the result, we obtain the number which we selected to cube. Reversing the procedure, that is, taking the cube root first and then cubing the result, also results in the original number. Since one operation "undoes" the other, we say that cubing and taking a cube root are *inverse operations*. In this section, we discuss functions that behave in a similar manner. We begin by discussing a way in which we may combine two functions.

◆ Composition of Functions

One method of combining two functions is called the *composition* of functions. If *f* and *g* are functions, then *f* composed with *g*, denoted *f* ∘ *g*, is formed by using the output of *g* as the input of *f*.

◆ **Composition of Two Functions**

If *f* and *g* are functions, then the **composite function** *f* ∘ *g* is defined by

$$(f \circ g)(x) = f(g(x)).$$

The domain of *f* ∘ *g* is the set of all *x* in the domain of *g* such that *g*(*x*) is in the domain of *f*.

EXAMPLE 1 Use the graphs of *f* and *g* in Figure 2.58 to compute each functional value.

FIGURE 2.58

The graphs of *f* and *g* for Example 1.

(a) $(f \circ g)(4)$ **(b)** $(g \circ f)(0)$

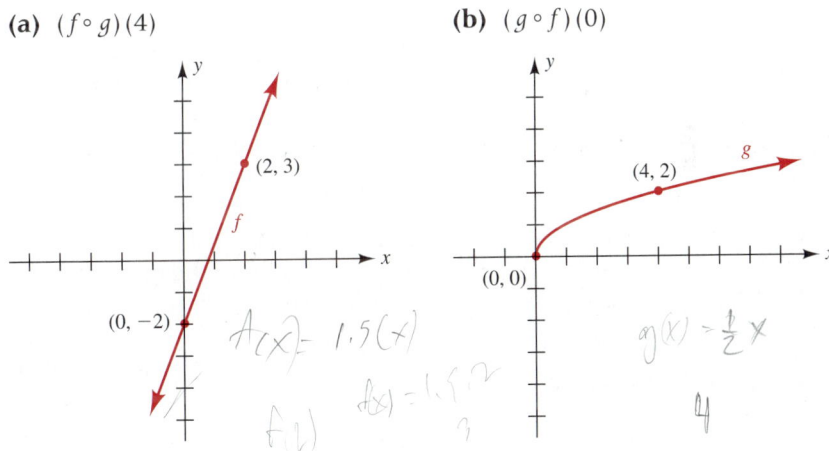

◆ Solution

(a) From the graph of *g* in Figure 2.58, we observe that *g*(4) = 2. Thus,

$$(f \circ g)(4) = f(g(4)) = f(2).$$

Now, from the graph of *f* in Figure 2.58, we observe that *f*(2) = 3. Hence,

$$(f \circ g)(4) = 3.$$

(b) The composite function $g \circ f$ is defined by $(g \circ f)(x) = g(f(x))$. From the graph of f, we observe that $f(0) = -2$. Thus,

$$(g \circ f)(0) = g(f(0)) = g(-2).$$

However, as shown in Figure 2.58, -2 is *not* in the domain of g. Hence, we conclude that

$$(g \circ f)(0) \quad \text{is } undefined.$$

For $g(f(x))$ to make sense, the value of $f(x)$ must be in the domain of the function g. ◆

PROBLEM 1   Using the graphs of f and g in Figure 2.58, find $(f \circ g)(0)$. ◆

As illustrated in the next example, the composite functions $f \circ g$ and $g \circ f$ are *not* necessarily the same function.

EXAMPLE 2   Given the functions f and g defined by $f(x) = \sqrt{x}$ and $g(x) = x^2 - 4$, find each of the following functions. Then determine the domain of each function.

(a) $f \circ g$ **(b)** $g \circ f$

◆ **Solution**

(a) The function $f \circ g$ is defined by

$$(f \circ g)(x) = f(g(x)) = f(x^2 - 4) = \sqrt{x^2 - 4}.$$

The domain of $f \circ g$ is all real numbers in the domain of g such that $g(x)$ is in the domain of f. The domain of g is $(-\infty, \infty)$, and the domain of f is $[0, \infty)$. Thus, from the interval $(-\infty, \infty)$ we can select only inputs x such that the output $g(x)$ is *nonnegative*:

$$x^2 - 4 \geq 0$$

We can solve this inequality by generating the graph of $y = x^2 - 4$ in the viewing rectangle of a graphing calculator. As shown in Figure 2.59, the graph intersects the x-axis at $x = \pm 2$ and is above the x-axis when $x < -2$ or $x > 2$. Hence, the solution set of this inequality is

$$x \leq -2 \quad \text{or} \quad x \geq 2.$$

Therefore, the domain of $f \circ g$ is $(-\infty, -2] \cup [2, \infty)$.

(b) The function $g \circ f$ is defined by

$$(g \circ f)(x) = g(f(x)) = g(\sqrt{x}) = (\sqrt{x})^2 - 4 = x - 4.$$

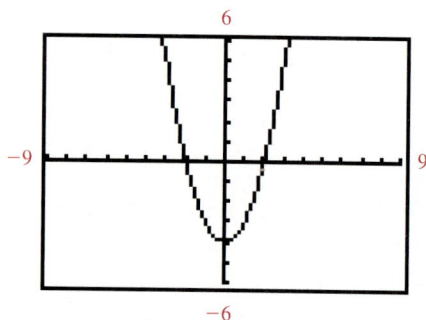

FIGURE 2.59

The solution set of the inequality $x^2 - 4 \geq 0$ is $(-\infty, -2] \cup [2, \infty)$.

The domain of $g \circ f$ is all real numbers in the domain of f such that $f(x)$ is in the domain of g. Since the domain of f is $[0, \infty)$, we can choose inputs only from this interval. Since each *nonnegative* input x gives us an output $f(x)$ that is in the domain of g, the domain of $g \circ f$ is also $[0, \infty)$. Hence, we write

$$(g \circ f)(x) = x - 4, \quad \text{for } x \geq 0.$$

Note that in this example, $(g \circ f)(x) \neq (f \circ g)(x)$. ◆

PROBLEM 2 For the functions f and g given in Example 2, find $(g \circ f)(9)$ by each of the following methods.

(a) First find $f(9)$, and then evaluate $g(f(9))$.

(b) Use the result of Example 2(b), $(g \circ f)(x) = x - 4$ for $x \geq 0$. ◆

◆ **One-to-One Functions**

A function f is said to be *one-to-one* if each element in the range of f is associated with only one element in its domain. More precisely, we state the following definition.

◀▷ **One-to-One Function**

> A function f is a **one-to-one function** if, for all a and b in the domain of f,
>
> $$f(a) = f(b) \quad \text{implies} \quad a = b.$$

A simple method for determining whether a function is one-to-one is to look at its graph. If every horizontal line that can be drawn in the coordinate plane intersects the graph at most once, then each output of the function is associated with only one input. Hence, the function is one-to-one. We refer to this graphical method of determining whether a function is one-to-one as the **horizontal line test**.

◀▷ **Horizontal Line Test**

> A function f is one-to-one if no horizontal line in the coordinate plane intersects the graph of the function in more than one point.

Every function that is either an increasing function or a decreasing function (see Section 2.5) passes the horizontal line test and, therefore, is also a one-to-one function.

EXAMPLE 3 Determine if each function is a one-to-one function.

(a) $f(x) = (x - 1)^3$ **(b)** $h(x) = x^2 - 1$

(0, −1)   (1, 0)

Each horizontal line intersects the graph at most once.

FIGURE 2.60

The graph of $f(x) = (x - 1)^3$ passes the horizontal line test. Hence, f is a one-to-one function.

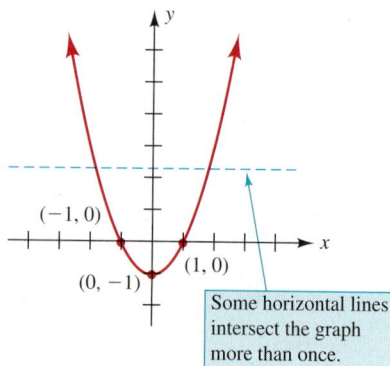

(−1, 0)

(0, −1)   (1, 0)

Some horizontal lines intersect the graph more than once.

FIGURE 2.61

The graph of $h(x) = x^2 - 1$ fails the horizontal line test. Hence, h is *not* a one-to-one function.

◆ **Solution**

(a) Using a graphing calculator or the techniques of graphing a function that we discussed in Section 2.5, we generate the graph of $f(x) = (x - 1)^3$, as shown in Figure 2.60. Since the function f is an increasing function, we know that each horizontal line drawn in the coordinate plane must intersect the graph at most once. Thus, by the horizontal line test, the function f is a one-to-one function. Also, note that

$$f(a) = f(b)$$

implies that $(a - 1)^3 = (b - 1)^3$

$$a - 1 = b - 1 \qquad \text{Take the cube root of both sides}$$

$$a = b \qquad \text{Add 1 to both sides}$$

Since $f(a) = f(b)$ implies $a = b$, we conclude that f is a one-to-one function.

(b) Using a graphing calculator or the techniques of graphing a function that we discussed in Section 2.5, we generate the graph of $h(x) = x^2 - 1$, as shown in Figure 2.61. Notice that some horizontal lines intersect the graph more than once. Thus, by the horizontal line test, the function h is not a one-to-one function. Also, note that

$$h(a) = h(b)$$

implies that $a^2 - 1 = b^2 - 1$

$$a^2 = b^2 \qquad \text{Add 1 to both sides}$$

$$a = \pm b \qquad \text{Take the square root of both sides}$$

Since $h(a) = h(b)$ does not imply $a = b$, we may conclude that h is not a one-to-one function.   ◆

If a function is *not* one-to-one, we can often find a suitable restriction on its domain in order to form a new function that *is* one-to-one. For example, if we restrict the domain of the function h defined in Example 3(b) to *nonnegative* numbers, the new function that is formed is one-to-one.

PROBLEM 3   Show that the function H defined by $H(x) = x^2 - 1$ for $x \geq 0$ is a one-to-one function.   ◆

◆ **Inverse Functions**

Suppose f is a one-to-one function, and for each element x in its domain X there corresponds an element $f(x)$ in its range Y. Because f if a one-to-one function, we can assign to each output $f(x)$ the input x from which it came. Hence, we can define a function g from Y to X in which

$$g(f(x)) = x \quad \text{for every } x \text{ in } X$$

We refer to g as the **inverse function** of f, as shown in Figure 2.62.

FIGURE 2.62
If g is the inverse function of f, then $g(f(x)) = x$ for every x in the domain of f.

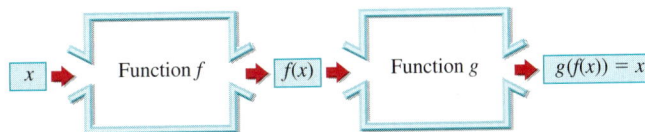

x → Function f → $f(x)$ → Function g → $g(f(x)) = x$

If we let $f(x) = y$, then $g(f(x)) = x$ implies $g(y) = x$. Hence, we also have the following composition:

$$f(g(y)) = y \quad \text{for every } y \text{ in } Y.$$

For notation purposes, we replace y with x and write

$$f(g(x)) = x \quad \text{for every } x \text{ in } Y.$$

◆ **Inverse Function**

Let f be a one-to-one function with domain X and range Y. The **inverse function** of f is a function g with domain Y and range X such that

$$g(f(x)) = x \quad \text{for every } x \text{ in } X$$
and
$$f(g(x)) = x \quad \text{for every } x \text{ in } Y$$

If g is the inverse of f, then f is the inverse of g; that is, inverse functions always come in pairs. Consider the cubing function f defined by $f(x) = x^3$ and the cube root function g defined by $g(x) = \sqrt[3]{x}$. Both of these functions are one-to-one functions and the domain and range of each is the set of all real numbers. [See parts (e) and (h) of Figure 2.42 in Section 2.5.] Note that

$$g(f(x)) = g(x^3) = \sqrt[3]{x^3} = x$$
and
$$f(g(x)) = f(\sqrt[3]{x}) = (\sqrt[3]{x})^3 = x.$$

In summary, the cubing function and the cube root function are *inverses of each other.*

The graphs of the cubing function f and the cube root function g are shown in Figure 2.63. If we fold the coordinate plane along the dotted line $y = x$, the graphs of f and g will coincide. In other words, the graphs of f and g are *reflections of one another in the line $y = x$.* For every point (a, b) on the graph of f there corresponds a point (b, a) on the graph of g. This special relationship between the graphs of a pair of inverse functions gives us a graphical method for verifying whether one function is the inverse of another.

FIGURE 2.63
The graphs of a pair of inverse functions are reflections of one another in the line $y = x$.

EXAMPLE 4   Determine whether the functions f and g are inverses of each other:

$$f(x) = 2x - 8 \qquad g(x) = \tfrac{1}{2}x + 4$$

$g(x) = \frac{1}{2}x + 4$ $f(x) = 2x - 8$

FIGURE 2.64

The functions f and g defined by $f(x) = 2x - 8$ and $g(x) = \frac{1}{2}x + 4$ appear to be inverses of each other.

◆ **Solution**   Figure 2.64 shows the graphs of $f(x) = 2x - 8$ and $g(x) = \frac{1}{2}x + 4$ in a square-set viewing rectangle of a graphing calculator. The functions f and g are one-to-one functions and their domains and ranges are the set of all real number. It appears that the graphs of f and g are reflections of one another in the line $y = x$, which indicates that these functions are inverses of each other. We can verify this conjecture algebraically:

$$f(g(x)) = f\left(\tfrac{1}{2}x + 4\right) \qquad \text{and} \qquad g(f(x)) = g(2x - 8)$$
$$= 2\left(\tfrac{1}{2}x + 4\right) - 8 \qquad\qquad\qquad = \tfrac{1}{2}(2x - 8) + 4$$
$$= x + 8 - 8 \qquad\qquad\qquad\qquad = x - 4 + 4$$
$$= x \qquad\qquad\qquad\qquad\qquad\quad = x$$

Since $f(g(x)) = g(f(x)) = x$, we conclude that the functions f and g are inverses of each other.   ◆

PROBLEM 4   Determine whether $f(x) = (x - 2)^5$ and $g(x) = \sqrt[5]{x} + 2$ are inverses of each other. Check graphically with a graphing calculator.   ◆

◆ **Finding the Inverse of a Function**

We usually denote the inverse function of the function f by using the notation f^{-1} (read "f inverse"). Thus, if f and f^{-1} are inverses of each other, then we write

$$f(f^{-1}(x)) = f^{-1}(f(x)) = x$$

CAUTION   In the notation f^{-1}, the superscript -1 denotes an inverse, *not* an exponent; that is, f^{-1} does not mean $1/f$.

We now discuss a procedure that we can use to find the inverse function of a one-to-one function defined by the equation $y = f(x)$. If we apply f^{-1} to both sides of the equation $y = f(x)$, we obtain

$$f^{-1}(y) = f^{-1}(f(x))$$
$$= x$$

The statement $x = f^{-1}(y)$ suggests that we can find an equation that defines f^{-1} by solving the equation $y = f(x)$ for x. For example, to find the inverse of the function f defined by $f(x) = 2x - 8$, we solve the equation $y = f(x)$ for x:

$$y = 2x - 8$$

$$y + 8 = 2x \qquad \textbf{Add 8 to both sides}$$

$$x = \tfrac{1}{2}y + 4 \qquad \textbf{Divide both sides by 2}$$

Hence, the equation $x = f^{-1}(y) = \frac{1}{2}y + 4$ defines the inverse function of f. The letter we use in the defining expression for f^{-1} is immaterial; however, when we work with functions, it is customary to use the letter x as the independent variable and y as the dependent variable. Thus, replacing x with y, we write

$$f^{-1}(x) = \tfrac{1}{2}x + 4,$$

which agrees with our work in Example 4.

The following procedure may be used to find the inverse of a one-to-one function.

◆ **Procedure for Finding the Inverse of a One-to-One Function**

If f is a one-to-one function, then its inverse function f^{-1} may be found by the following procedure:

1. Set $y = f(x)$.

2. Solve the equation $y = f(x)$ for x in terms of y.

3. Interchange x and y.

4. Replace y with $f^{-1}(x)$ and restrict the domain of f^{-1} (if necessary) so that it equals the range of f.

This procedure is illustrated in the next example.

EXAMPLE 5   Find the inverse of each one-to-one function.

(a) $f(x) = x^3 - 1$ **(b)** $g(x) = \sqrt{x - 1}$

◆ **Solution**

(a) To find the inverse of the function f defined by $f(x) = x^3 - 1$, we proceed as follows:

$$y = x^3 - 1 \qquad \text{Replace } f(x) \text{ with } y$$
$$y + 1 = x^3 \qquad \text{Solve for } x$$
$$x = \sqrt[3]{y + 1}$$
$$y = \sqrt[3]{x + 1} \qquad \text{Interchange } x \text{ and } y$$
$$f^{-1}(x) = \sqrt[3]{x + 1} \qquad \text{Replace } y \text{ with } f^{-1}(x)$$

Since the domains and ranges of both f and f^{-1} are the set of all real numbers, we conclude that the inverse function of f is defined by

$$f^{-1}(x) = \sqrt[3]{x + 1}.$$

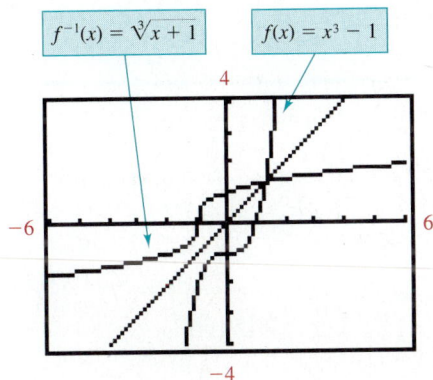

FIGURE 2.65

The graphs of $f(x) = x^3 - 1$ and $f^{-1}(x) = \sqrt[3]{x + 1}$ are reflections of one another in the line $y = x$.

Figure 2.65 shows the graphs of f and f^{-1} in the same viewing rectangle of a graphing calculator. Note that the graphs are reflections of one another in the line $y = x$. This relationship between the graphs verifies our work.

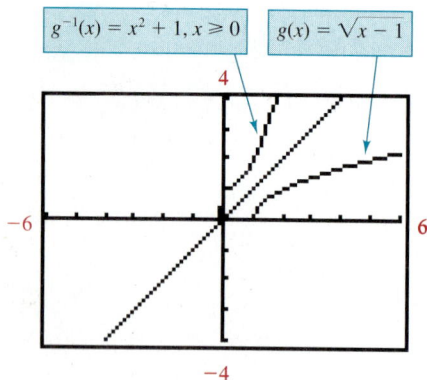

(b) To find the inverse of the function g defined by $g(x) = \sqrt{x - 1}$, we proceed as follows:

$$y = \sqrt{x - 1} \qquad \text{Replace } g(x) \text{ with } y$$

$$y^2 = x - 1 \qquad \text{Solve for } x$$

$$x = y^2 + 1$$

$$y = x^2 + 1 \qquad \text{Interchange } x \text{ and } y$$

$$g^{-1}(x) = x^2 + 1 \qquad \text{Replace } y \text{ with } g^{-1}(x)$$

Since the domain of g^{-1} must be equal to the range of g, we must restrict the domain of g^{-1} to the interval $[0, \infty)$. Hence, we conclude that g^{-1} is defined by

$$g^{-1}(x) = x^2 + 1 \quad \text{for } x \geq 0.$$

Figure 2.66 shows the graphs of g and g^{-1} in the same viewing rectangle of a graphing calculator. The graphs are reflections of one another in the line $y = x$. This relationship between the graphs verifies our work. ◆

FIGURE 2.66

The graphs of $g(x) = \sqrt{x - 1}$ and $g^{-1}(x) = x^2 + 1$, for $x \geq 0$, are reflections of one another in the line $y = x$.

PROBLEM 5 Find the inverse of the one-to-one function f defined by $f(x) = \dfrac{1}{x + 2}$. ◆

Exercises 2.6

◆ Basic Skills

In Exercises 1–8, use the graphs of f, g, and h to compute (if possible) the given functional value.

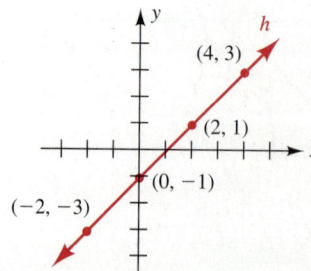

1. $(g \circ f)(0)$

2. $(h \circ f)(-2)$

3. $(f \circ h)(-2)$

4. $(h \circ g)(3)$

5. $(f \circ g)(2)$

6. $(f \circ f)(-2)$

7. $[h \circ (f \circ g)](2)$

8. $[h \circ (g \circ f)](0)$

In Exercises 9–20, find each composite function, given that f, g, h, F, G, and H are defined as follows:

$$f(x) = 2x + 1 \qquad F(x) = \sqrt{x^2 - 4}$$

$$g(x) = x^2 + 1 \qquad G(x) = \frac{1}{x}$$

$$h(x) = \sqrt{x} \qquad H(x) = \frac{2}{x - 1}$$

Then find the domain of the composite function.

9. $f \circ g$

10. $g \circ f$

11. $G \circ f$

12. $G \circ F$

13. $F \circ h$

14. $g \circ h$

15. $G \circ H$

16. $H \circ G$

17. $H \circ f$

18. $G \circ G$

19. $f \circ f$

20. $H \circ H$

21. Use the result of Exercise 15 to evaluate $(G \circ H)(2)$.

22. Use the result of Exercise 16 to evaluate $(H \circ G)(2)$.

In Exercises 23–32,

(a) *determine whether the function is one-to-one.*

(b) *if the function is not one-to-one, find a suitable restriction on its domain in order to form a new function that is one-to-one. (More than one restriction is possible.)*

23. $f(x) = 2x$

24. $f(x) = \frac{|x^3}{3}$

25. $f(x) = x^2$

26. $f(x) = |x| - 3$

27. $f(x) = \sqrt{9 - x}$

28. $f(x) = \frac{1}{x - 3}$

29. $f(x) = |x + 2|$

30. $f(x) = 1 - \sqrt[3]{x}$

31. $f(x) = (x - 2)^2 - 1$

32. $f(x) = (x - 1)^{2/3}$

In Exercises 33–44, use a graphing calculator to determine whether the functions f and g appear to be inverses of each other. If the functions appear to be inverses, verify this fact algebraically.

33. $f(x) = 3x - 1$

$$g(x) = \frac{x + 1}{3}$$

34. $f(x) = 5 - 2x$

$$g(x) = \frac{5 - x}{2}$$

35. $f(x) = \sqrt{x}$

$$g(x) = x^2$$

36. $f(x) = x$

$$g(x) = \frac{1}{x}$$

37. $f(x) = x^3 - 8$

$$g(x) = \sqrt[3]{x + 8}$$

38. $f(x) = (x - 2)^3$

$$g(x) = \sqrt[3]{x} + 2$$

39. $f(x) = \frac{1}{x - 1}$

$$g(x) = \frac{1}{x} + 1$$

40. $f(x) = \frac{x}{x + 3}$

$$g(x) = \frac{-3x}{x - 1}$$

41. $f(x) = \frac{3x + 1}{x}$

$$g(x) = \frac{1}{x - 3}$$

42. $f(x) = \frac{x}{x - 2}$

$$g(x) = \frac{2x}{x - 1}$$

43. $f(x) = (x - 1)^2 \quad \text{for } x \geq 1$

$$g(x) = \sqrt{x} + 1$$

44. $f(x) = x^2 + 2 \quad \text{for } x \geq 0$

$$g(x) = \sqrt{x - 2}$$

Each function f in Exercises 45–62 is a one-to-one function.

(a) *Find the inverse function f^{-1}.*

(b) *Verify the answer to part (a) by showing that the graphs of f and f^{-1} are reflections of one another in the line $y = x$.*

45. $f(x) = 2x$

46. $f(x) = \frac{-x}{4}$

47. $f(x) = 3 - x$

48. $f(x) = x + 1$

49. $f(x) = 2 - x^3$

50. $f(x) = (x + 1)^3$

51. $f(x) = \sqrt[3]{x - 4}$

52. $f(x) = 3 + \sqrt[3]{x}$

53. $f(x) = \frac{1}{x + 1}$

54. $f(x) = \frac{1}{x} - 2$

55. $f(x) = \sqrt{x - 4} + 1$

56. $f(x) = \sqrt{x} + 3$

57. $f(x) = (x - 1)^2, \, x \geq 1$

58. $f(x) = (x - 3)^2 + 2, \, x \geq 3$

59. $f(x) = \sqrt{9 - x^2}, \, 0 \leq x \leq 3$

60. $f(x) = \sqrt{x^2 - 9}, \, x \geq 3$

61. $f(x) = x^2 + 3, \, x \leq 0$

62. $f(x) = 2 - x^2, \, x \leq 0$

◆ **Critical Thinking**

For Exercises 63–68, refer to the following table of values.

x	$f(x)$	$g(x)$
-1	4	3
0	7	2
2	3	-1
3	-1	-5

63. Given that $h(x) = (f \circ g)(x)$, find $h(2)$.

64. Given that $h(x) = (g \circ f)(x)$, find $h(2)$.

65. Given that $g(x) = (f \circ h)(x)$, find $h(2)$.

66. Given that $f(x) = (g \circ h)(x)$, find $h(2)$.

67. Given that $f(x) = (g \circ h)(x)$, find $h(3)$.

68. Given that $g(x) = (f \circ h)(x)$, find $h(-1)$.

In Exercises 69–72, the graph of a function f is shown. Determine whether the function f has an inverse. If so, sketch the graph of f^{-1} and complete the following table of values.

x	-2	0	3
$f^{-1}(x)$			

69.

70.

71.

72.

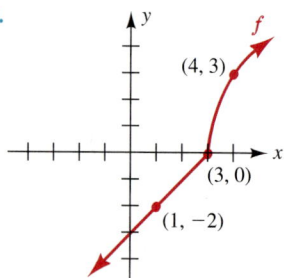

For Exercises 73 and 74, assume the functions f and g are defined as follows:

$$f(x) = \frac{x - 2}{3} \quad \text{and} \quad g(x) = 3x + 1$$

73. Compare $(f \circ g)^{-1}(x)$ with $(g^{-1} \circ f^{-1})(x)$. What can you conclude?

74. Compare $(g \circ f)^{-1}(x)$ with $(f^{-1} \circ g^{-1})(x)$. What can you conclude?

In Exercises 75 and 78, determine the range of the given function by examining the domain of its inverse function.

75. $f(x) = \dfrac{1}{2x - 1}$

76. $f(x) = \dfrac{x}{6 - 3x}$

77. $f(x) = \dfrac{2x + 1}{3x - 2}$

78. $f(x) = \dfrac{9x - 3}{3x + 1}$

79. Find $f(x)$ if $(g \circ f)(x) = 2x - 8$ and $g(x) = 6x + 4$.

80. (a) Determine whether each of the eight basic functions in Figure 2.42 (in Section 2.5) is a one-to-one function.

(b) Which of the one-to-one functions in part (a) is its own inverse?

81. A spherical balloon is being inflated with air. The volume V of the balloon is a function of its radius r and is given by $V(r) = \frac{4}{3}\pi r^3$. In turn, the radius r (in inches) is a function of the time t (in seconds) after the inflation process begins and is given by $r(t) = 3(t + 1)$.

(a) Determine $(V \circ r)(t)$.

(b) What does the composition function $V \circ r$ describe?

(c) Find the volume of the balloon after 3 seconds of inflation.

(d) Find the time at which the volume is $36,000\pi$ cubic inches.

82. In a ski factory, the daily cost C (in dollars) of producing n pairs of skis is given by

$$C(n) = 225n - 0.8n^2 \quad \text{for } 0 \le n \le 40$$

and the number of pairs produced in t hours is given by

$$n(t) = 4.5t \quad \text{for } t \le 10.$$

(a) Determine $(C \circ n)(t)$.

(b) What does the composition function $C \circ n$ describe?

(c) Find the daily production cost (to the nearest dollar) if the factory runs for 8.2 hours per day.

(d) How many hours does the factory operate if the daily production cost is $4228?

2.7 Applied Functions and Variation

◆ **Introductory Comments**

To solve many types of word problems—especially those that require the techniques of calculus—we must begin by setting up an equation that defines a function. We may then analyze the function, draw its graph, and answer questions concerning the functional relationship between the quantities. In this section we practice setting up functions from words.

◆ **Applied Functions**

For many applied problems, we begin with an established formula and then use *substitution* to obtain a functional relationship between the desired variables. The procedure is illustrated in the next two examples.

EXAMPLE 1   Suppose 200 feet of fencing is needed to enclose a rectangular garden.

(a) Express the area A of the rectangular garden as a function of its length x.

(b) State the domain of the function in part (a).

(c) Generate the graph of the function in part (a). Use the graph to find the length x of the garden for which the area A is the *largest* possible.

◆ **Solution**

(a) Drawing a rectangular garden, we let

$$x = \text{the length} \quad \text{and} \quad y = \text{the width}$$

as shown in Figure 2.67. Recall that the area A of a rectangle is the product of its length and width:

$$A = xy$$

For the area A to be a function of its length x, we need to write A in terms of only x. Since 200 feet of fencing are needed to enclose the garden, we have

$$200 = 2x + 2y.$$

Solving this equation for y gives us

$$y = 100 - x.$$

Now, we substitute $100 - x$ for y in the area formula, $A = xy$, and obtain

$$A = x(100 - x), \quad \text{or} \quad A = 100x - x^2.$$

FIGURE 2.67
Rectangular garden of length x and width y

The equation $A = 100x - x^2$ defines the area A of this rectangle as a function of its length x. If we wish to emphasize this functional relationship, we may use functional notation to write

$$A(x) = 100x - x^2.$$

(b) Algebraically, we know that the domain of this function is the set of all real numbers but, geometrically, this does not make sense. This is because x represents the length of the rectangle, and a length can only be *positive*. Thus, we must have $x > 0$. Also, since the perimeter is 200 feet, we must have $x < 100$, otherwise, the width y would be zero or negative. Because of the geometric restrictions required for this function, the domain is (0, 100).

(c) Figure 2.68 shows the graph of $A = 100x - x^2$, with $0 < x < 100$, in the viewing rectangle of a graphing calculator. The x-axis denotes the length x of the rectangle and the y-axis denotes its area A. The coordinates (x, A) at the peak of the curve tell us the length x for which the area A is the *largest* possible. We refer to the value of A at the peak of the curve as the *maximum functional value*. Most graphing calculators have a built-in program for finding the maximum value of a function. If your calculator does not have a built-in program for finding the maximum value, use the $\boxed{\text{TRACE}}$ key and $\boxed{\text{ZOOM}}$ key to estimate the maximum value. Using either method, we find that the maximum area is 2500 square feet and this occurs when $x = 50$ ft. ◆

3500

0   Maximum
    X=50   Y=2500   100

FIGURE 2.68

The coordinates at the peak of the curve tell us the length x of the rectangle for which the area A is the largest possible.

$\boxed{\textbf{PROBLEM 1}}$ Determine the width y of the rectangle in Example 1 for which the area is the *largest* possible. What special type of rectangle gives the maximum area for a fixed perimeter? ◆

$\boxed{\textbf{EXAMPLE 2}}$ Water is flowing into a conical funnel at the rate of 8 cubic inches per minute. The diameter of the base of the funnel is 8 inches and its height is 12 inches.

(a) Express the volume V of water in the funnel as a function of the height h of the water in the funnel.

(b) State the domain and range of the function in part (a). Then generate its graph.

(c) Determine the time t (in minutes) it takes for the water to rise to a height of 6 inches.

◆◆ **Solution**

(a) Drawing a conical funnel as in Figure 2.69, we let

$$h = \text{height of the water in the funnel}$$

and $r = \text{radius of the surface of the water.}$

From elementary geometry, we know that the volume of water in the funnel is given by

$$V = \tfrac{1}{3}\pi r^2 h.$$

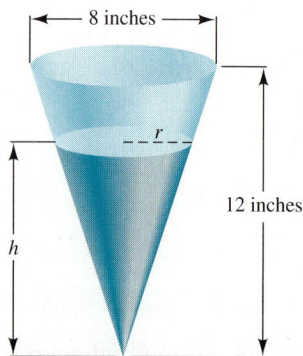

8 inches

r

12 inches

h

FIGURE 2.69

A conical funnel 8 inches wide and 12 inches deep

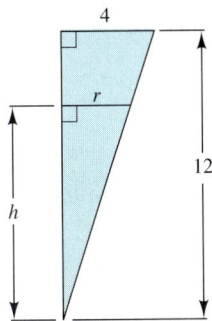

FIGURE 2.70

Similar triangles formed from the cross-section of the funnel

For the volume V to be a function of h, we must write the volume in terms of only h. To do this, we use similar triangles from the cross-section of the funnel shown in Figure 2.70. Since corresponding sides of similar triangles are proportional, we have

$$\frac{4}{12} = \frac{r}{h}, \qquad \text{or} \qquad r = \frac{h}{3}.$$

Now, we substitute $h/3$ for r in the volume formula, $V = \frac{1}{3}\pi r^2 h$, and obtain

$$V = \frac{1}{3}\pi \left(\frac{h}{3}\right)^2 h, \qquad \text{or} \qquad V = \frac{1}{27}\pi h^3.$$

If we wish to emphasize this functional relationship, we use functional notation to write

$$V(h) = \frac{1}{27}\pi h^3.$$

(b) Since the height of the funnel is 12 inches, the water level must be between 0 and 12 inches. Thus, the domain of this function is [0, 12]. The function defined by the equation $V = \frac{1}{27}\pi h^3$ is an increasing function. By the vertical compress rule (Section 2.5), we know that the graph of this function is similar to the graph of the cubing function. For $h = 0$ inch, we have $V = 0$ cubic inch and for $h = 12$ inches, we have $V = 64\pi \approx 201$ cubic inches. Hence, the range of this function is $[0, 64\pi] \approx [0, 201]$. This information suggests that an appropriate viewing rectangle for observing the graph of $V = \frac{1}{27}\pi h^3$ is

> hmin = 0
> hmax = 12
> hscl = 1
> Vmin = 0
> Vmax = 201
> Vscl = 20

Figure 2.71 shows the graph of $V = \frac{1}{27}\pi h^3$ in this viewing rectangle. The x-axis denotes the height of water h, and the y-axis denotes the volume of water V.

(c) When $h = 6$ inches, the volume of water in the funnel is

$$V(6) = \frac{1}{27}\pi(6)^3 = 8\pi \approx 25.1 \text{ cubic inches}$$

Figure 2.71 verifies this fact. Since the water is flowing into the funnel at the rate of 8 cubic inches per minute, it takes

$$\frac{8\pi}{8} = \pi \approx 3.14 \text{ minutes}$$

for the water to reach a height of 6 inches.

FIGURE 2.71

Graph of $V = \frac{1}{27}\pi h^3$
Domain: [0, 12]
Range: $[0, 64\pi] \approx [0, 201]$

PROBLEM 2   Referring to Example 2, express the volume *V* of water in the funnel as a function of the radius *r* of water in the funnel.   ◆

◆ Variation

In business, engineering, and the sciences, the functional relationship between two quantities is often given in terms of *variation*. When we state that **y varies directly as x** or that **y is directly proportional to x**, we mean that the ratio of *y* to *x* is always the same. In other words,

$$\frac{y}{x} = k \qquad \text{or} \qquad y = kx$$

where *k* is called the *variation constant*.

EXAMPLE 3   *Chapter Opening Problem*
The real estate tax *T* on a property varies directly as its assessed value *V*.

(a) Express *T* as a function of *V* if *T* = \$2800 when *V* = \$112,000.

(b) Generate the graph of the function defined in part (a) for \$0 ≤ *V* ≤ \$500,000.

◆ **Solution**

(a) Since *T* varies directly as *V*, we write

$$T = kV$$

where *k* is the variation constant to be determined. To find *k*, we replace *T* with 2800 and *V* with 112,000:

$$T = kV$$
$$2800 = k(112,000)$$
$$k = \frac{2800}{112,000} = \frac{1}{40}$$

Thus, $T = \frac{1}{40}V,$ or $T = \frac{V}{40}$ with $V \geq 0$.

(b) The function defined by the equation *T* = *V*/40 for *V* ≥ 0 is an increasing function. By the vertical compress rule (Section 2.5), we know that the graph of this function is similar to the graph of the identity function. For *V* = \$0, we have *T* = \$0 and for *V* = \$500,000, we have *T* = \$500,000/40, or *T* = \$12,500. This information suggests that an appropriate viewing rectangle for observing the graph of *T* = *V*/40 with 0 ≤ *V* ≤ 500,000 is

```
Vmin = 0
Vmax = 500000
Vscl = 50000
Tmin = 0
Tmax = 12500
Tscl = 1000
```

Figure 2.72 shows the graph of $T = V/40$ in this viewing rectangle. The x-axis denotes the assessed value V, and the y-axis denotes the tax T. Note that $(112000, 2800)$ is a point on this graph.

FIGURE 2.72

Graph of the function defined by the equation $T = V/40$ with $0 \leq V \leq 500,000$

PROBLEM 3 Use the result from Example 3(a) to find the tax on a property with an assessed value of $168,000. Check your answer by using the graph in Figure 2.72. ◆

When we state that *y* **varies inversely as** *x* or that *y* **is inversely proportional to** *x*, we mean that the product of y and x is always the same. In other words,

$$yx = k \qquad \text{or} \qquad y = \frac{k}{x}$$

where k is the *variation constant*.

EXAMPLE 4 The time t it takes a person to travel a fixed distance varies inversely as the rate of speed r at which the person travels.

(a) Express t as a function of r if $t = 4$ hours when $r = 60$ mi/h.

(b) Generate the graph of the function in part (a) for $0 \text{ mi/h} < r \leq 120 \text{ mi/h}$.

◆ **Solution**

(a) Since t varies inversely as r, we write

$$t = \frac{k}{r}$$

where k is the variation constant to be determined. To find k, we replace t with 4 and r with 60 as follows:

$$t = \frac{k}{r}$$

$$4 = \frac{k}{60}$$

$$k = 240$$

Thus, $t = \dfrac{240}{r}$ for $r > 0$.

(b) The function defined by the equation $t = 240/r$ for $r > 0$ is a decreasing function. By the vertical stretch rule (Section 2.5), we know that the graph of this function is similar to the graph of the reciprocal function. Values of r that are close to 0 yield large values of t. For example, if $r = 5$ miles per hour (mi/h), then $t = \frac{240}{5}$, or $t = 48$ h. For $r = 120$ mi/h, we have $t = \frac{240}{120}$, or $t = 2$ h. An appropriate viewing rectangle that shows this information as well as the axes for this graph is

> rmin = 0
> rmax = 120
> rscl = 10
> tmin = 0
> tmax = 50
> tscl = 10

Figure 2.73 shows the graph of $t = 240/r$ in this viewing rectangle. The x-axis denotes the rate of speed r, and the y-axis denotes the time t. Note that (60, 4) is a point on this graph.

FIGURE 2.73

Graph of the function defined by the equation $t = 240/r$ for $r > 0$

PROBLEM 4   Use the results of Example 4(a) to find the time t it takes to travel a fixed distance if the rate of speed is 50 mi/h. Check your answer by using the graph in Figure 2.73. ◆

The following table gives several other types of variation that occur frequently. In each case, k denotes the variation constant.

Statement	Formula
1. y varies directly as the nth power of x.	$y = kx^n$
2. y varies inversely as the nth power of x.	$y = \dfrac{k}{x^n}$
3. y varies directly as x and inversely as z.	$y = \dfrac{kx}{z}$
4. y varies jointly as x and z.	$y = kxz$

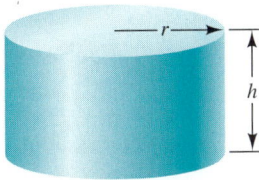

FIGURE 2.74

A right circular cylinder

EXAMPLE 5 The volume V of a right circular cylinder (see Figure 2.74) varies jointly as its height h and the square of its radius r. Describe what happens to the volume of a cylinder if its height and radius are doubled.

◆ **Solution** Since V varies jointly as h and the square of r, we write

$$V = kr^2h$$

where k is the variation constant. If we double both the height and the radius, we obtain

$$V = k(2r)^2(2h) = 8kr^2h.$$

From this equation we can see that the volume of the cylinder is *eight times the original volume*. Thus, when the height and radius of a cylinder are doubled, its volume becomes eight times as large. ◆

PROBLEM 5 The electric resistance R of a wire varies directly as its length L and inversely as the square of its radius r. Describe what happens to the resistance of a wire if its length and radius are doubled. ◆

Exercises 2.7

◆ Basic Skills

1. Express the diameter d of a circle as a function of its circumference C.

2. The length of a rectangle is 5 cm. Express its width w as a function of its perimeter P.

3. Suppose an automobile leasing company charges $40 per day plus $0.20 per mile to rent a car. Express the daily cost C (in dollars) of renting a car as a function of the distance driven n (in miles).

4. Suppose a salesperson earns $200 per week plus 25% commision on all sales. Express the weekly earnings E (in dollars) of a salesperson as a function of the amount A of merchandise (in units) that she sells.

5. A square with side s and diagonal d is shown in the figure.

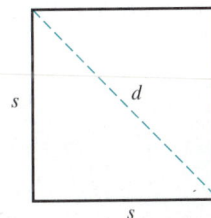

(a) Express s as a function of d.

(b) Express the area A of the square as a function of d.

(*Exercise 5 continues*)

(c) Express the perimeter P of the square as a function of d.

6. An equilateral triangle with side s and height h is shown in the figure.

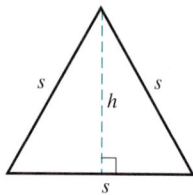

(a) Express h as a function of s.

(b) Express the area A of the triangle as a function of s.

(c) Express the perimeter P of the triangle as a function of h.

7. A small company has a present net worth of $125,000. It is estimated that the future weekly income and expenses for the company will be $30,000 and $26,000, respectively. Express the net worth W of the company at the end of t weeks as a function of time, t.

8. An automobile that was bought for $15,000 depreciates 20% of its original value each year. Express the value V of the automobile at the end of t years as a function of time, t.

9. The radius r of a pile of sand in the shape of a cone is twice its height h.

(a) Express the volume V of sand as a function of h.

(b) Express the volume V of sand as a function of r.

10. The height h of a tin can in the shape of a cylinder is equal to its diameter d.

(a) Express the volume V of the tin can as a function of h.

(b) Express the surface area S of the tin can as a function of d.

11. The point $P(x, y)$ lies on the graph of the circle $x^2 + y^2 = 1$ as shown in the figure.

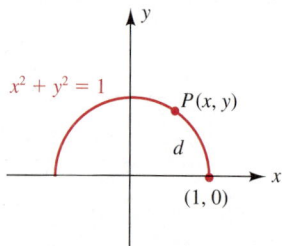

(a) Express the distance d from point P to the point $(1, 0)$ as a function of the x-coordinate of P.

(b) State the domain of the function in part (a).

12. The point $P(x, y)$ lies on the graph of the parabola $2y = x^2$ shown in the figure.

(a) Express the distance d from the point P to the point $(0, 1)$ as a function of the y-coordinate of P.

(b) State the domain of the function in part (a).

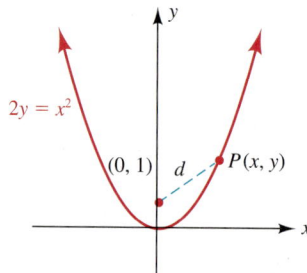

13. The point $Q(x, y)$ lies on the graph of $y = \sqrt{x}$ shown in the figure.

(a) Express the area A of the shaded right triangle as a function of the x-coordinate of Q.

(b) Express the perimeter P of the shaded right triangle as a function of the x-coordinate of Q.

14. The point $Q(x, y)$ lies on the graph of the semicircle $y = \sqrt{25 - x^2}$ shown in the figure.

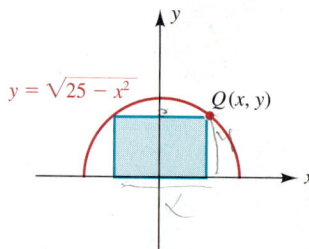

(a) Express the area A of the shaded rectangle as a function of the x-coordinate of Q.

(b) Express the perimeter P of the shaded rectangle as a function of the x-coordinate of Q.

15. The volume of a rectangular box with a square base is 64 cubic inches. Express its surface area S as a function of the width x of its base.

16. The volume of a rectangular box with a square base and an open top is 100 cubic meters. The material used to construct the base costs $4 per square meter, and the material for the sides cost $2.50 per square meter. Express the cost C to construct the rectangular box as a function of the width x of its base.

17. The ends of a water trough are isosceles triangles, as shown in the figure.

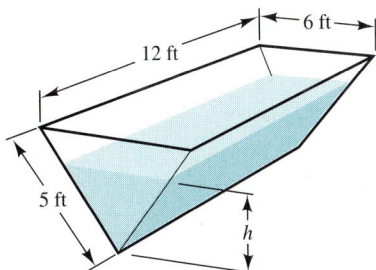

(a) Express the volume V of water in the trough as a function of the height h of the water in the trough.

(b) Give the domain of the function in part (a), and

(c) Sketch the graph of the function in part (a).

18. Suppose the trough described in Exercise 17 is initially empty and then begins to fill with water at the rate of 1 cubic foot per minute (ft^3/min). Express the height h of the water in the trough as a function of the time t (in minutes) that the water flows into the trough.

19. A baseball diamond is a square with 90-foot base paths, as shown in the figure. Suppose a runner on first base runs toward second base at the rate of 30 feet per second (ft/s) as soon as the pitcher throws the ball to home plate.

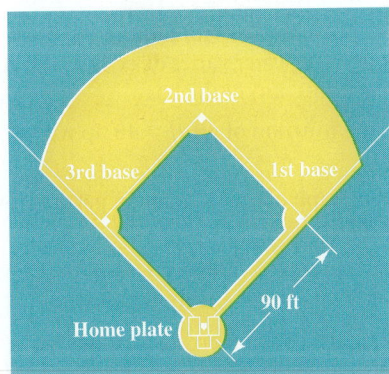

(a) Express the runner's distance d from home plate as a function of the time t (in seconds) after the pitcher throws the ball.

(b) Describe the domain and range of the function in part (a).

20. The fixed cost to run a wood-stove company is $25,000 per month, and the variable cost to produce each stove is $200 per unit. If each stove sells for x dollars, with $200 \le x \le 500$, then the number of stoves sold per month is estimated to be $1000 - 2x$. Express the monthly profit of the company as a function of x. [*Hint:* Profit = total revenue − total cost.]

21. The length of the hypotenuse of a right triangle is 16.2 inches.

(a) Express the area A of the triangle as a function of the length x of one of the legs.

(b) Find the area when $x = 12.4$ inches.

22. Two joggers start from the same place at the same time. One runs due east at 8.2 miles per hour (mi/h) and the other runs due north at 6.3 mi/h.

(a) Determine the distance d between the joggers as a function of the time t (in hours) that they have been running.

(b) Find the distance between the joggers after 1 hour 24 minutes.

In Exercises 23–36, the functional relationship between the quantities is given in terms of variation.

23. *Hooke's law* states that the distance d a spring stretches varies directly as the force F applied to the spring, as long as the elastic limit of the spring is not exceeded. Assume that the elastic limit of a spring occurs at 2400 newtons (N).

(a) Express d as a function of F if $d = 8$ cm when $F = 400$ N.

(b) Generate the graph of the function in part (a) for $0 \text{ N} \le F \le 2400$ N.

(c) Find the distance the spring stretches when $F = 1800$ N. Check your answer by using the graph in part (b).

24. The electrical resistance R of a wire with a constant diameter is directly proportional to the length l of the wire.

(a) Express R as a function of l if $R = 2$ ohms when $l = 30$ meters (m).

(b) Generate the graph of the function in part (a) for $0 \text{ m} \le l \le 100$ m.

(c) Find the resistance of the wire when $l = 75$ m. Check your answer by using the graph in part (b).

25. The weight W (in pounds, lb) of a steel beam with a constant cross-sectional area varies directly as its length l (in feet).

(a) Express W as a function of l if $W = 1500$ lb when $l = 12$ ft.

(b) Generate the graph of the function in part (a) for $0 \text{ ft} \le l \le 48$ ft.

(*Exercise 25 continues*)

(c) Find the weight of the beam when $l = 32$ ft. Check your answer by using the graph in part (b).

26. The weekly payroll P (in dollars) of a company varies directly as the number n of workers assigned to the job.

(a) Express P as a function of n if $P = \$5200$ when $n = 13$ workers.

(b) Generate the graph of the function in part (a) for $0 \leq n \leq 50$.

(c) Find the weekly payroll when $n = 41$. Check your answer by using the graph in part (b).

27. The number N of long-distance phone calls per day between two towns, each with a population of approximately 25,000 people, varies inversely as the distance d (in miles) between the towns.

(a) Express N as a function of d if $N = 20$ calls when $d = 500$ miles.

(b) Generate the graph of the function in part (a) for $0 \text{ mi} < d \leq 2500 \text{ mi}$.

(c) Find the number of calls made per day between two towns when $d = 800$ miles. Check your answer by using the graph in part (b).

28. *Boyle's law* states that when the temperature of a confined gas remains constant, the pressure P (in pounds per square inch, psi) it exerts varies inversely as the volume V (in cubic inches, in³) it occupies.

(a) Express P as a function of V if $P = 4$ psi when $V = 30$ in³.

(b) Generate the graph of the function in part (a) for $0 \text{ in}^3 < V \leq 120 \text{ in}^3$.

(c) Find the pressure exerted when $V = 24$ in³. Check your answer by using the graph in part (b).

29. The wavelength W (in meters, m) of a radio wave varies inversely as its frequency f (in megahertz, MHz).

(a) Express W as a function of f if $W = 40$ m when $f = 7.5$ MHz.

(b) Generate the graph of the function in part (a) for $0 \text{ MHz} < f \leq 50 \text{ MHz}$.

(c) Find the wavelength when $f = 30$ MHz. Check your answer by using the graph in part (b).

30. If the voltage in an electric circuit is constant, the current I (in milliamperes, mA) through a resistor varies inversely as its resistance R (in ohms, Ω).

(a) Express I as a function of R if $I = 2$ mA when $R = 100 \, \Omega$.

(b) Generate the graph of the function in part (a) for $0 \, \Omega < R \leq 200 \, \Omega$.

(c) Find the current generated when $R = 25 \, \Omega$. Check your answer by using the graph in part (b).

31. The gravitational force F (in newtons, N) between two objects varies inversely as the square of the distance d (in meters, m) between them.

(a) Express F as a function of d if $F = 100$ N when $d = 40$ m.

(b) Generate the graph of the function in part (a) for $0 \text{ m} < d \leq 400 \text{ m}$.

(c) Find the force exerted when $d = 125$ m. Check your answer by using the graph in part (b).

32. The lift L (in pounds per square foot, lb/ft²) of an airplane wing varies directly as the square of the speed v (in miles per hour, mi/h) of air flowing over it.

(a) Express L as a function of v if $L = 225$ lb/ft² when $v = 150$ mi/h.

(b) Generate the graph of the function in part (a) for $0 \text{ mi/h} \leq v \leq 500 \text{ mi/h}$.

(c) Find the lift required when $v = 300$ mi/h. Check your answer by using the graph in part (b).

33. The period T (in seconds, s) of a pendulum is directly proportional to the square root of the length l (in meters, m) of the pendulum.

(a) Express T as a function of l if $T = 2.1$ s when $l = 1.095$ m.

(b) Generate the graph of the function in part (a) for $0 \text{ m} \leq l \leq 9 \text{ m}$.

(c) Find the period when $l = 2.405$ m. Check your answer by using the graph in part (b).

34. The power H (in horsepowers, hp) required to propel a motorboat through the water varies directly as the cube of the speed s (in knots, kn) of the boat.

(a) Express H as a function of s if $H = 48$ hp when $s = 11.2$ kn.

(b) Generate the graph of the function in part (a) for $0 \text{ kn} \leq s \leq 20 \text{ kn}$.

(c) Find the power required when $s = 15.6$ kn. Check your answer by using the graph in part (b).

35. The centripetal force acting on an object traveling in a circular path varies directly as the square of its velocity and inversely as the radius of the circle. Describe what happens to the centripetal force acting on the object if the velocity and radius are both tripled.

36. The power produced by an electric generator varies jointly as the load resistance and the square of the current. Describe what happens to the power produced by the generator if the load resistance is doubled and the current is halved.

◆ **Critical Thinking**

37. A rectangular pasture along a river is to be enclosed using 1000 feet of fencing, as shown in the figure. No fencing is needed along the river.

(a) Express the area A of the pasture as a function of the length of a side, x.

(b) State the domain of the function defined in part (a).

(c) Use a graphing calculator to generate the graph of the function defined in part (a). Trace to the peak of this curve, then state the dimensions of the rectangular pasture whose area is the *largest* possible.

(d) Use the result from part (c) to compare the sum of the two lengths x to the length y. What do you observe? Make a conjecture about the dimensions of x and y if a fixed amount of fencing k is used to enclose three sides of a rectangular pasture so that its area is the largest possible.

38. Three adjacent rectangular corrals are to be built with 120 feet of fencing, as shown in the figure.

(a) Express the total enclosed area A as a function of the common length x.

(b) State the domain of the function defined in part (a).

(c) Use a graphing calculator to generate the graph of the function defined in part (a). Trace to the peak of this curve, then state the overall dimensions of the corrals so that the total enclosed area is the *largest* possible.

(d) Use the result from part (c) to compare the sum of the four lengths x to the sum of the two lengths y. What do you observe? Make a conjecture about the dimensions of x and y if a fixed amount of fencing k is used to enclose n adjacent rectangular corrals so that the area is the largest possible.

39. A church window has the shape of a rectangle surmounted by a semicircle, as shown in the figure.

(a) Express the area A of the window as a function of the radius r of the semicircle if the perimeter of the window is 30 ft.

(b) Use a graphing calculator to generate the graph of the function defined in part (a). Trace to the peak of this curve, then state the value of r that makes the area of the window the *largest* possible.

(c) Use the value of r from part (b) to estimate the length y when the area of the window is the largest possible. Make a conjecture about the values of r and y when the area of the window is the largest possible.

40. A football stadium has the shape of a rectangle with semicircular ends as shown in the figure.

(a) Express the area A of the stadium as a function of the radius r of the semicircle if the perimeter of the stadium is 1 mile.

(b) Use a graphing calculator to generate the graph of the function defined in part (a). Trace to the peak of this curve, then estimate (to three significant digits) the value of r that makes the area of the stadium the *largest* possible.

(c) Use the value of r from part (b) to estimate the distance y when the area of the stadium is the largest possible. Make a conjecture about the values of r and y when the area of the stadium is the largest possible. What is the actual shape of the stadium?

41. A baseball player hits the ball from home plate to the outfield and reaches second base safely. Suppose the player runs at the rate of 30 ft/s, directly on the base paths. (See the figure in Exercise 19.) Express the straight-line distance d from home plate to the runner as a function of the time t (in seconds) after he hits the ball. [*Hint:* Use a piecewise-defined function.]

42. An author is paid a royalty of $3.00 per book for the first 1000 books sold. The royalty increases to $3.05 per book for each book sold in excess of 1000. Express the author's royalty R as a function of the number n of books sold. [*Hint:* Use a piecewise-defined function.]

43. The time required for an elevator to lift its passengers varies jointly as the weight of the passengers and the distance they are lifted and inversely as the power of the motor used. Suppose it takes 10 seconds for an elevator to lift 800 pounds to a height of 40 feet with a 20-horsepower motor. Determine the time it takes this elevator to lift 1000 pounds to a height of 80 feet with the same motor.

44. The safe-load capacity of a wooden rectangular beam supported at both ends varies jointly as its width and the square of its depth and inversely as the distance between the supports. Suppose the safe-load capacity of a beam 4 inches wide and 12 inches deep is 3600 pounds when the distance between the supports is 16 feet. Determine the safe-load capacity of a similar beam that is 8 inches wide and 8 inches deep if the distance between the supports remains 16 feet.

◆ Chapter 2 Review

◆ Questions for Writing or Group Discussion

1. Explain the difference between the ordered pair (a, b) and the interval (a, b).

2. Explain in words how to find the *midpoint* of a line segment.

3. Explain in words how to find the *distance* between two points in the *Cartesian plane*.

4. What is the difference between the *independent variable* and the *dependent variable* in an equation?

5. Describe the *point-plotting method* of sketching the graph of a simple equation.

6. What is meant by the *x-intercept* and *y-intercept* of a graph? How can the intercepts be determined from a given equation?

7. How can the *vertical line test* be used to determine if an equation defines y as a function of x?

8. What is a *function*? What is meant by its *domain* and *range*?

9. How can the *zeros* of a function be determined? Graphically, what do the real zeros represent?

10. What is the difference between an *even function* and an *odd function*? Give an example of each.

11. Which of the eight basic functions in Figure 2.42 are *increasing functions*? *decreasing functions*?

12. Does the *squaring function* pass the *horizontal line test*? Explain. What does this tell us about the squaring function?

13. Does the *cubing function* pass the *horizontal line test*? Explain. What does this tell us about the cubing function?

14. How does the graph of a function relate to the graph of its inverse?

15. Explain the algebraic procedure for finding the inverse of a one-to-one function.

16. Can the graph of a function be symmetric with respect to the *x*-axis? Explain.

17. How are the graphs of each of the following functions related to the graph of $y = x^2$?

(a) $y = x^2 + 3$ (b) $y = (x - 2)^2$

(c) $y = -x^2$ (d) $y = (x + 2)^2 - 1$

18. What is the difference between $f \circ g$ and $g \circ f$? Illustrate with examples.

19. How may *composition* be used to show that the functions *f* and *g* are *inverse functions*?

20. If *f* is an even function, what can you conclude about f^{-1}? Explain.

21. Can a function be its own inverse? If so, give an example.

22. Explain how the *distance formula* may be used to determine if three points are *collinear.*

23. Suppose *y varies inversely as x* and *x varies inversely as t*. What can you conclude about *y* and *t*?

24. The volume *V* of helium in a balloon is a function of the radius *r* of the balloon, and the radius *r* of the balloon is a function of time *t*. What does the composite function $V \circ r$ describe?

◆ **Review Exercises**

In Exercises 1–6, find

(a) the length of the line segment joining the points A and B, and

(b) the coordinates of the midpoint M of the line segment joining the points A and B.

1. $A(0, 3);$ $B(4, 6)$

2. $A(2, 0);$ $B(7, 12)$

3. $A(-2, 1);$ $B(2, 3)$

4. $A(4, -1);$ $B(1, -4)$

5. $A(-3, -5);$ $B(6, -1)$

6. $A(7, -3);$ $B(-3, -5)$

7. Show that the quadrilateral that joins the points $A(2, 1)$, $B(4, 3)$, $C(2, 5)$, and $D(0, 3)$ is a square. Then find its perimeter and area.

8. Show that the triangle that joins the points $A(-1, 8)$, $B(5, -2)$, and $C(15, 4)$ is an isosceles triangle. Then find its perimeter and area.

For Exercises 9–12, refer to the triangle with vertices A, O, and B shown in the figure.

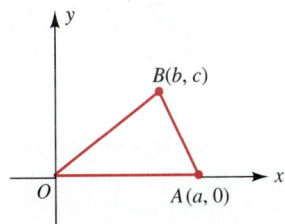

9. Find the length of the line segment joining the midpoints of the sides $\overline{OB}$ and $\overline{AB}$. Compare this length to *OA*, the length of the third side of the triangle.

10. Find the length of the line segment joining the midpoints of the sides $\overline{OA}$ and $\overline{AB}$. Compare this length to *OB*, the length of the third side of the triangle.

11. Find the length of the line segment joining the midpoints of the sides $\overline{OA}$ and $\overline{OB}$. Compare this length to *AB*, the length of the third side of the triangle.

12. What conclusion can you make about the length of a line segment joining the midpoints to any two sides of a triangle with respect to the length of the third side of the triangle?

For each of the equations in Exercises 13–24,

(a) determine any intercepts,

(b) check for symmetry with respect to the x-axis, y-axis, and origin,

(c) generate the graph by plotting points or by using a graphing calculator, and

(d) determine if the equation defines y as a function of x.

13. $y = \frac{1}{2}x^3$

14. $4x = y^2$

15. $3x - |y| = 0$

16. $y - 2\sqrt{x} = 0$

17. $x^2y = 1$

18. $xy = 2$

19. $2x + 3y = 6$

20. $3x - y = 9$

21. $x^2 - 3y = 9$

22. $x + |y| = 3$

23. $|x| + 2|y| = 2$

24. $|x + y| = 1$

Given the function f, g, and h, defined by

$$f(x) = 3x + 4, \ g(x) = 2x^2 - x, \ h(x) = \frac{x + 1}{x - 3}$$

compute the functional values given in Exercises 25–44.

25. $f(-1)$

26. $g(3)$

27. $h(4)$

28. $f(2x)$

29. $g(x - 3)$

30. $h\left(\dfrac{a}{2}\right)$

31. $f(\frac{1}{3}) - f(2)$

32. $g(2) \cdot f(2)$

33. $\dfrac{f(7)}{h(4)}$

34. $(f \circ h)(2)$

35. $(g \circ f)(x^2)$

36. $(f \circ f)(-5)$

37. $h^{-1}(x)$

38. $f(x) + f^{-1}(x)$

39. $f^{-1}(x + 2)$

40. $h^{-1}(-x)$

41. $(h \circ h^{-1})(2)$

42. $(g \circ f^{-1})(x)$

43. $\dfrac{g(x + h) - g(x)}{h}$

44. $\dfrac{f^{-1}(a + h) - f^{-1}(a)}{h}$

Answer questions (a)–(g) for each of the functions in Exercises 45–64.

(*a*) *Find the domain of the function.*

(*b*) *Find any real zeros of the function.*

(*c*) *Determine if the function is even, odd, or neither.*

(*d*) *Sketch the graph of the function (if it is not given) by using the techniques of shifting, reflecting, stretching, and compressing.*

(*e*) *Determine the open intervals where the function is increasing, decreasing, or constant.*

(*f*) *Determine if the function is a one-to-one function.*

(*g*) *Use the graph to find the range of the function.*

45. $f(x) = x + 3$

46. $g(x) = -2x$

47. $h(x) = \sqrt[3]{2x}$

48. $F(x) = \sqrt[3]{x - 1}$

49. $G(x) = x^2 - 9$

50. $H(x) = |x| + 5$

51. $g(x) = (x + 3)^2$

52. $F(x) = \sqrt{3 - x}$

53. $G(x) = x^3 + 8$

54. $h(x) = \dfrac{1}{x - 4}$

55. $F(x) = |x - 3| - 1$

56. $H(x) = 8 - (x + 1)^3$

57. $H(x) = 6x - x^2$

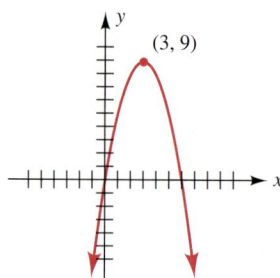

58. $f(x) = |x + 1| + |x - 1|$

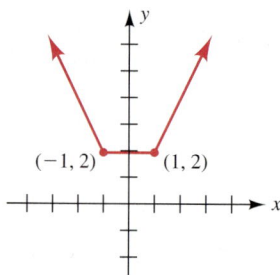

59. $f(x) = \sqrt{100 - x^2}$

60. $g(x) = \sqrt{x^2 - 49}$

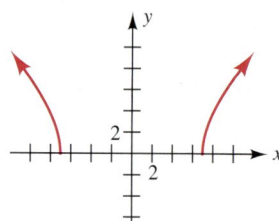

61. $g(x) = 2x^2 - x^4$

62. $H(x) = 2x^3 - 6x$

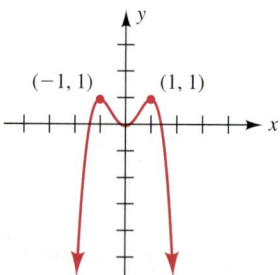

63. $F(x) = \begin{cases} x^2 + 1 & \text{if } x \le 0 \\ 1 & \text{if } 0 < x \le 3 \end{cases}$

64. $h(x) = \begin{cases} \sqrt{x - 1} & \text{if } 1 \le x \le 5 \\ x - 3 & \text{if } x > 5 \end{cases}$

In Exercises 65–70, use the graph of $y = x^{2/3}$ in the figure and the techniques of shifting, reflecting, stretching, and compressing to sketch the graph of the function defined by each equation.

65. $y = x^{2/3} + 2$ **66.** $y = (x - 3)^{2/3}$

67. $y = (x + 1)^{2/3} - 4$ **68.** $y = 3 - x^{2/3}$

69. $y = -3x^{2/3}$ **70.** $y = \dfrac{x^{2/3}}{3}$

In Exercises 71–78, each function f is a one-to-one function.

(a) *Find the inverse function f^{-1}.*

(b) *Check your answer in part (a) by generating the graphs of f and f^{-1} to determine if they are reflections of one another in the line $y = x$.*

71. $f(x) = x^2, x \leq 0$ **72.** $f(x) = 7x - 8$

73. $f(x) = 3 - x^3$ **74.** $f(x) = \dfrac{1}{x} - 3$

75. $f(x) = \sqrt{x - 2}$ **76.** $f(x) = \sqrt{x} + 3$

77. $f(x) = \begin{cases} x^3 & \text{if } x < 1 \\ \dfrac{x + 1}{2} & \text{if } x \geq 1 \end{cases}$

78. $f(x) = \begin{cases} x^{2/3} & \text{if } x \geq 0 \\ -(x^2 + 2) & \text{if } x < 0 \end{cases}$

In Exercises 79–86, refer to the graph of the function f shown in the figure.

79. Specify the domain of f.

80. Specify the range of f.

81. Find $f(0)$. **82.** Find $f(-2)$.

83. Find $f(f(0))$. **84.** Find $f(f(-2))$.

85. Draw the graph of f^{-1}.

86. Compute each functional value.

(a) $f^{-1}(0)$ (b) $f^{-1}(f(4))$ (c) $f(f^{-1}(-6))$

In Exercises 87–92, solve each equation or inequality by using the x-intercept method in conjunction with a graphing calculator. Check algebraically.

87. $6x - (3 - 2x) = 2(x - 3)$

88. $\dfrac{x}{8} - \dfrac{3x + 1}{10} = \dfrac{3 - x}{20}$

89. $\sqrt{2 - 2x} = 3x + 1$ **90.** $4(2x - 1)^2 = 25$

91. $\dfrac{1}{6} - \dfrac{t - 2}{12} \geq \dfrac{t - 1}{4}$ **92.** $\left| \dfrac{2x - 3}{5} \right| < \dfrac{2}{3}$

In Exercises 93 and 94, set up an equation and solve by the x-intercept method. Check algebraically.

93. If \$12,500 is invested, a portion at 8% simple interest and the remainder at 9%, the interest earned yearly is \$41 more than if all the funds had been invested at 8%. How much is invested at each rate?

94. If the radius of a circle is increased by 1 meter, the area of the circle is increased by 19π square meters. Find the radius of the circle.

95. The cost C (in dollars) for a daily truck rental is a function of the distance x driven (in miles) and is given by

$$C(x) = 32 + 0.25x.$$

(a) Evaluate $C(125.7)$, rounding to the nearest cent.

(b) Determine the distance driven when $C(x) = \$138.40$.

96. The volume of a rectangular box with a square base and open top is 9 cubic feet. Express the total surface area S as a function of the width x of its base.

97. The monthly charge for water in a small town is \$0.015 per gallon for the first 1000 gallons used and \$0.02 per gallon for each gallon used in excess of 1000 gallons. Express the charge C for the water as a function of the number n of gallons used.

98. The fixed cost to run a company that manufactures picnic tables is \$10,000 per month, and the variable cost to produce each table is \$100 per unit. If each table sells for x dollars, then the number of tables sold per month is estimated to be $800 - 2x$. Express the monthly profit P of the company as a function of x. [*Hint:* Profit = total revenue − total cost.]

99. The point $P(x, y)$ lies on the right-hand portion of the parabola $y = 3 - x^2$ and is in quadrant I, as shown in the figure.

(a) Express the area A of the shaded isosceles triangle as a function of the x-coordinate of P.

(b) State the domain of the function defined in part (a).

(*Exercise 99 continues*)

(c) Use a graphing calculator to generate the graph of the function defined in part (a). Trace to the peak of this curve, then state the coordinates of point P that make the area of the triangle the *largest* possible.

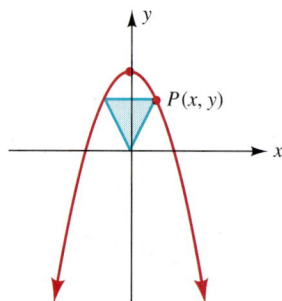

100. A street light 30 ft above level ground casts the shadow of a man 6 ft tall.

(a) Express the shadow length s as a function of the man's distance d from the base of the light pole.

(b) If the man stands 8 ft from the base of the light pole and then walks away from the pole at the rate of 4 ft/s, express the shadow length s (in feet) as a function of the time t (in seconds).

101. The excise tax T (in dollars) on an automobile varies directly as the vehicle's actual value V (in dollars).

(a) Express T as a function of V if $T = \$200$ when $V = \$5000$.

(b) Generate the graph of the function in part (a) for $\$0 \leq V \leq \$20,000$.

(c) Find the tax when $V = \$12,000$. Check your answer by using the graph in part (b).

102. The intensity I (in candelas, cd) of a light source varies inversely as the square of the distance d (in feet) from the light source.

(a) Express I as a function of d if $I = 4$ cd when d is 5 ft.

(b) Generate the graph of the function in part (a) for 0 ft $< d \leq 20$ ft.

(c) Find the intensity when $d = 10$ ft. Check your answer by using the graph in part (b).

103. A large meadow is capable of supporting life for a maximum of 2000 field mice. Suppose the rate of growth G on the mouse population varies jointly as the number n of mice present and the difference between the maximum number supportable by the field and the number present.

(a) Express G as a function of n if $G = 50$ field mice per week when $n = 1200$ mice.

(b) Find the rate of growth when 1500 field mice are living in the field.

104. The time t required for the excavation of a sewer line varies jointly as its length l, width w, and square of its depth d and inversely as the number n of backhoes used. Describe what happens to the time for excavation if the depth is halved and the number of backhoes is doubled.

A four-year-old car has a value of $3000. When one year old, its value was $9000.

(a) Assuming the car's depreciation is linear, express the value V of the car as a function of its age x (in years).

(b) What is the age of the car when its value is fully depreciated?

For the solution, see Example 5 in Section 3.2.

Linear and Quadratic Functions

3.1 ▶ Linear Functions

◆ Introductory Comments

Recall from Section 2.5 that the graph of the identity function $f(x) = x$ is a straight line that passes through the origin and splits quadrants I and III in half [see part (b) of Figure 2.42 in Section 2.5]. By the vertical stretch and compress rule (Section 2.5), we know that the graph of $g(x) = ax$ is also a straight line, but with a *steeper* incline than the graph of $f(x) = x$ if $a > 1$, or a *flatter* incline if $0 < a < 1$. By the x-axis reflection rule (Section 2.5), the line slants *upward* if $a > 0$, or *downward* if $a < 0$, as we move from left to right along the x-axis. In summary, the value of a determines the steepness and direction of the line. In Figure 3.1 we use a graphing calculator to illustrate this fact by generating the graph of $g(x) = ax$ for $a = 2$, -5, and $\frac{1}{4}$.

By the vertical shift rule (Section 2.5), we know that the graph of $F(x) = ax + b$ is the same as the graph of $g(x) = ax$, but shifted vertically *upward* b units if $b > 0$, or *downward* $|b|$ units if $b < 0$. Hence, the value of b is the y-intercept of the graph of F. In Figure 3.2 we use a graphing calculator to illustrate this fact by generating the graph of $F(x) = ax + b$ for several values of a and b.

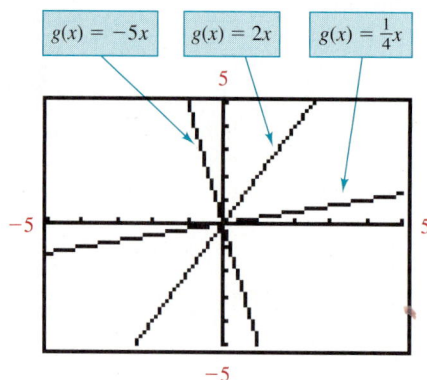

FIGURE 3.1

In the graph of $g(x) = ax$, the value of a determines the steepness and direction of the line.

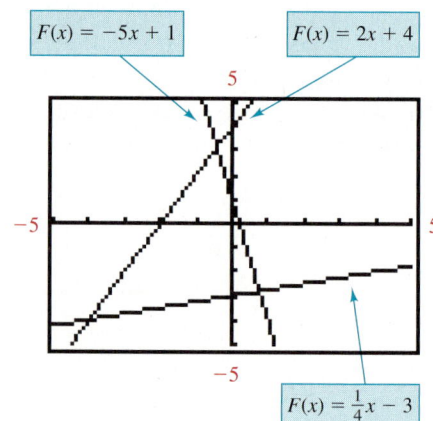

FIGURE 3.2

In the graph of $F(x) = ax + b$, the value of b is the y-intercept of the line.

We refer to the function F defined by $F(x) = ax + b$ as a *linear function*: its graph is a straight line with y-intercept b.

◆ Linear Function

If a and b are any real numbers with $a \neq 0$, then the function F defined by

$$F(x) = ax + b$$

is a **linear function.** The graph of this function is a **straight line** with y-intercept b.

In this section, we study linear functions and related equations whose graphs are straight lines. We begin our discussion by defining the *slope* of a line.

◆ **The Slope of a Line**

The slope of a nonvertical line is a measure of the line's steepness and is defined as the ratio of the vertical rise to the horizontal run between any two distinct points on the line:

$$\text{Slope} = \frac{\text{vertical rise}}{\text{horizontal run}}$$

FIGURE 3.3

The slope of the line through the points P and Q is the ratio of the vertical rise $(y_2 - y_1)$ to the horizontal run $(x_2 - x_1)$.

Consider the straight line through the points $P(x_1, y_1)$ and $Q(x_2, y_2)$ shown in Figure 3.3. The vertical rise between the points P and Q is $y_2 - y_1$, and the horizontal run between these points is $x_2 - x_1$. Thus, the ratio of $(y_2 - y_1)$ to $(x_2 - x_1)$ defines the slope of this line. It is common practice to designate the slope of a line with the lowercase letter m.

◆ **Slope Formula**

The **slope** m of a nonvertical line through the distinct points $P(x_1, y_1)$ and $Q(x_2, y_2)$, provided $x_1 \neq x_2$, is

$$m = \frac{y_2 - y_1}{x_2 - x_1}.$$

If $x_1 = x_2$, the line is vertical and its slope is *undefined*.

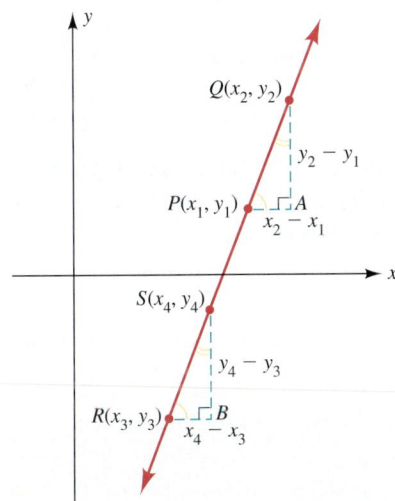

Regardless of which two distinct points we choose on the line, the slope of the line is always the same. To show this, consider the straight line that passes through the points $P(x_1, y_1)$ and $Q(x_2, y_2)$ in Figure 3.3 and also through two other distinct points $R(x_3, y_3)$ and $S(x_4, y_4)$ as shown in Figure 3.4. Notice that the triangles PAQ and RBS are similar, because their corresponding angles are equal. Thus, the corresponding sides of these triangles are proportional, and we have

$$\frac{y_2 - y_1}{x_2 - x_1} = \frac{y_4 - y_3}{x_4 - x_3}.$$

FIGURE 3.4

Since the triangles PAQ and RBS are similar, the slope of the line is the same value, regardless of which two distinct points we choose on the line.

Since these two ratios are equal, and we have already defined the slope of this line as $m = \dfrac{y_2 - y_1}{x_2 - x_1}$, we know that $\dfrac{y_4 - y_3}{x_4 - x_3}$ also defines the slope of this line.

EXAMPLE 1   Determine the slope of the line that passes through the given points.

(a) $(-1, 2)$ and $(3, 4)$ **(b)** $(2, -3)$ and $(-2, 4)$

(c) $(-2, 3)$ and $(3, 3)$ **(d)** $(-1, 1)$ and $(-1, -4)$

◆ **Solution**

(a) Let $(x_1, y_1) = (-1, 2)$ and $(x_2, y_2) = (3, 4)$. Then

$$m = \frac{y_2 - y_1}{x_2 - x_1} = \frac{4 - 2}{3 - (-1)} = \frac{2}{4} = \frac{1}{2} \quad \text{(See Figure 3.5.)}$$

(b) Let $(x_1, y_1) = (2, -3)$ and $(x_2, y_2) = (-2, 4)$. Then

$$m = \frac{y_2 - y_1}{x_2 - x_1} = \frac{4 - (-3)}{-2 - 2} = -\frac{7}{4} \quad \text{(See Figure 3.6.)}$$

(c) Let $(x_1, y_1) = (-2, 3)$ and $(x_2, y_2) = (3, 3)$. Then

$$m = \frac{y_2 - y_1}{x_2 - x_1} = \frac{3 - 3}{3 - (-2)} = \frac{0}{5} = 0 \quad \text{(See Figure 3.7.)}$$

(d) Since the x-coordinates of the points are the same, $x_1 = x_2 = -1$, the slope is undefined (see Figure 3.8).

FIGURE 3.5
A line with a *positive* slope

FIGURE 3.6
A line with a *negative* slope

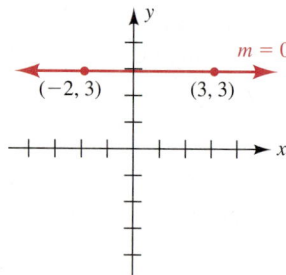

FIGURE 3.7
A line with a *zero* slope

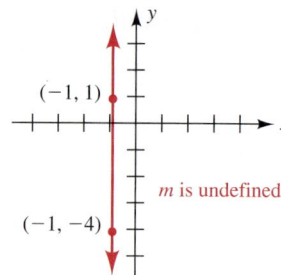

FIGURE 3.8
A line with an *undefined* slope

PROBLEM 1   When determining the slope, does it matter which point you label (x_1, y_1) or (x_2, y_2)? Interchange the labels of the points (x_1, y_1) and (x_2, y_2) in parts (a)–(c) of Example 1, and then determine the slope. What do you conclude?

We can use Example 1 to make some observations about the slope of a line:

1. If the line slants *upward* (increases) as we move from left to right along the x-axis, as in Figure 3.5, then the slope of the line is *positive*.

2. If the line slants *downward* (decreases) as we move from left to right along the x-axis, as in Figure 3.6, then the slope of the line is *negative*.

3. If the line is *horizontal* (parallel to the x-axis), as in Figure 3.7, then $y_1 = y_2$, and therefore the slope of the line is *zero*.

4. If the line is *vertical* (perpendicular to the *x*-axis), as in Figure 3.8, then $x_1 = x_2$, and therefore the slope of the line is *undefined*.

EXAMPLE 2 Sketch the line that passes through the point $(2, -1)$ and has the given slope.

(a) $m = 3$ **(b)** $m = -\frac{2}{3}$

◆ **Solution**

(a) A slope of $3 = \frac{3}{1}$ represents a vertical rise of 3 to a horizontal run of 1 (or a vertical rise of 6 to a horizontal run of 2, and so on). Thus, starting at the point $(2, -1)$, we can go *up 3 units* and run *right 1 unit* to locate another point on the line. Connecting these points with a straightedge gives us a sketch of this straight line, as shown in Figure 3.9.

(b) A slope of $-\frac{2}{3}$ represents a vertical drop of 2 to a horizontal run of 3 (or a vertical drop of 4 to a horizontal run of 6, and so on.) Thus, starting at the point $(2, -1)$, we can go *down 2 units* and run *right 3 units* to locate another point on the line. Connecting these points with a straightedge gives us a sketch of this straight line, as shown in Figure 3.10.

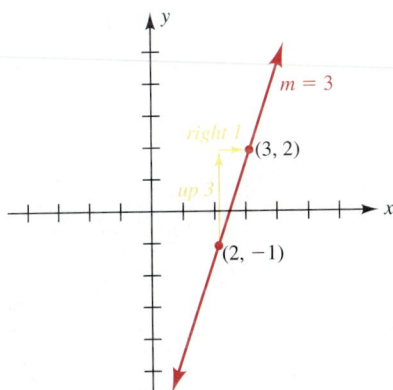

FIGURE 3.9
A line passing through the point $(2, -1)$ with a slope of 3

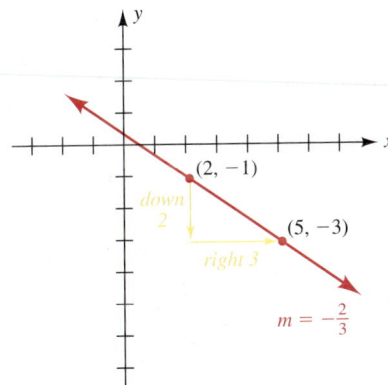

FIGURE 3.10
A line passing through the point $(2, -1)$ with a slope of $-\frac{2}{3}$

PROBLEM 2 Sketch the line that passes through the point $(-3, 0)$ and has slope $\frac{5}{2}$.

◆ **Sketching the Graph of a Linear Function**

The slope *m* of the line with the equation $y = ax + b$ is the number *a*. To show this, we let $P(x_1, y_1)$ and $Q(x_2, y_2)$ be any two distinct points on this line. Since these two points must satisfy the equation $y = ax + b$, we have

$$y_1 = ax_1 + b \quad \text{and} \quad y_2 = ax_2 + b.$$

Thus,

$$m = \frac{y_2 - y_1}{x_2 - x_1} = \frac{(ax_2 + b) - (ax_1 + b)}{x_2 - x_1}$$

$$= \frac{a(x_2 - x_1)}{x_2 - x_1}$$

$$= a$$

When we replace a with m in the equation $y = ax + b$, we obtain

$$y = mx + b$$

which is called the **slope-intercept form** for the equation of a straight line. As illustrated in the next example, we can quickly sketch the graph of a linear function simply by noting its slope and y-intercept.

EXAMPLE 3   Identify the slope and y-intercept, and then sketch the graph of each linear function.

(a) $f(x) = \frac{4}{3}x - 2$ **(b)** $y = 3 - 5x$

◆ **Solution**

(a) The linear function f defined by

$$f(x) = \frac{4}{3}x + (-2),$$

has slope $\frac{4}{3}$ and y-intercept -2. To graph this linear function, we begin by plotting the point $(0, -2)$. Next, remember that the slope is the ratio of the rise to the run. Thus, from the point $(0, -2)$, we go *up 4 units* and run *right 3 units* to locate another point on the line. Connect these two points to form the graph of $f(x) = \frac{4}{3}x - 2$, as shown in Figure 3.11.

(b) The linear function defined by the equation

$$y = 3 - 5x = -5x + 3$$

has slope -5, or $-\frac{5}{1}$, and y-intercept 3. Since the slope is negative, we start at the point $(0, 3)$ and go *down 5 units*, then run *right 1 unit* to locate another point on the line. Connecting these two points, we have the graph of $y = 3 - 5x$, as shown in Figure 3.12.   ◆

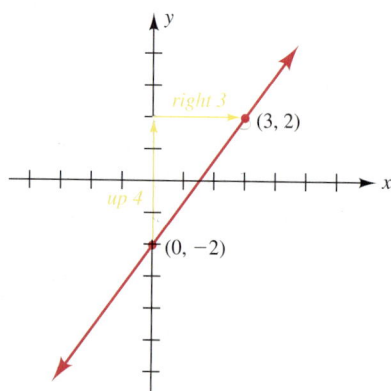

FIGURE 3.11
The graph of $f(x) = \frac{4}{3}x - 2$ has slope $\frac{4}{3}$ and y-intercept -2.

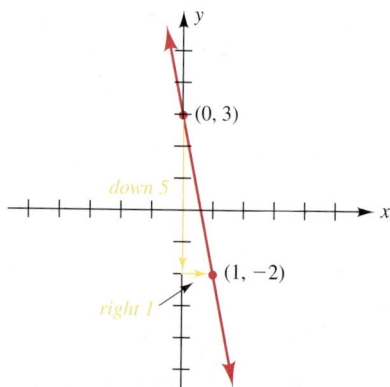

FIGURE 3.12
The graph of $y = 3 - 5x$ has slope -5 and y-intercept 3.

PROBLEM 3   Identify the slope and y-intercept of the function g defined by $g(x) = 3x - \frac{1}{2}$. Then sketch the graph of this linear function.   ◆

◆ **General Form for the Equation of a Straight Line**

Consider the equation $Ax + By + C = 0$ with constants A, B, and C, provided $B \neq 0$. Solving this equation for y, we obtain

$$y = \underbrace{\left(-\frac{A}{B}\right)}_{\text{slope}}x + \underbrace{\left(-\frac{C}{B}\right)}_{\text{y-intercept}}.$$

We know that this is the *slope-intercept form* for the equation of a straight line. Thus, we also know that the graph of the equation $Ax + By + C = 0$ is a straight line with slope $-A/B$ and y-intercept $-C/B$, provided $B \neq 0$. We refer to the equation

$$Ax + By + C = 0$$

as the **general form** for the equation of a straight line.

Suppose $A = 0$ in the equation $Ax + By + C = 0$ with $B \neq 0$. Solving for y, we obtain

$$y = \underbrace{0x}_{\text{slope}} + \underbrace{\left(-\frac{C}{B}\right)}_{\text{y-intercept}}.$$

Since the slope of the line is zero, we conclude that $0x + By + C = 0$, or

$$By + C = 0$$

is the equation of a *horizontal line* with y-intercept $-C/B$. Actually, the equation $By + C = 0$ defines a *constant function*.

If $B = 0$ in the equation $Ax + By + C = 0$ with $A \neq 0$, then we cannot write the equation in the slope-intercept form $y = mx + b$, because division by zero is not defined. However, regardless of what value we choose for y in the equation $Ax + 0y + C = 0$, the quantity $0y$ is always zero. This means that ordered pairs that satisfy $Ax + 0y + C = 0$, or $Ax + C = 0$, can have any y-value as long as the x-value is $-C/A$. Thus, we conclude that

$$Ax + C = 0$$

is the equation of a *vertical line* with x-intercept $-C/A$. Although the graph of $Ax + C = 0$ is a straight line, the equation $Ax + C = 0$ does *not* define a linear function, because this graph fails the vertical line test (see Section 2.3).

FIGURE 3.13
Graph of $2x - 3y = 12$

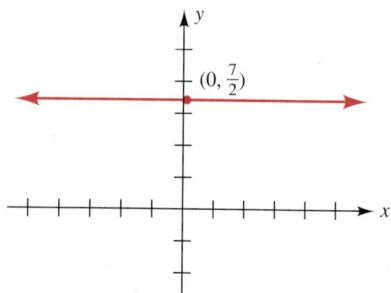

FIGURE 3.14
Graph of $2y - 7 = 0$

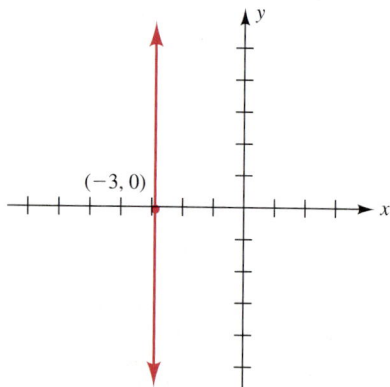

FIGURE 3.15
Graph of $2x + 6 = 0$

EXAMPLE 4 Sketch the graph of each equation.

(a) $2x - 3y = 12$ **(b)** $2y - 7 = 0$ **(c)** $2x + 6 = 0$

◆ **Solution**

(a) Solving for y, we obtain

$$2x - 3y = 12$$
$$-3y = -2x + 12 \qquad \text{Subtract } 2x \text{ from both sides}$$
$$y = \tfrac{2}{3}x - 4 \qquad \text{Divide both sides by } -3$$

This last equation is the slope-intercept form for the equation of a line. The graph of this equation is a straight line with slope $\tfrac{2}{3}$ and y-intercept -4, as shown in Figure 3.13.

(b) The equation $2y - 7 = 0$ can be written in slope-intercept form as

$$y = 0x + \tfrac{7}{2}.$$

slope y-intercept

Since the slope is zero, the graph of this equation is a horizontal line with y-intercept $\tfrac{7}{2}$, as shown in Figure 3.14.

(c) Although the equation $2x + 6 = 0$ cannot be written in slope-intercept form, we can write it in general form as

$$2x + 0y + 6 = 0.$$

Ordered pairs that satisfy this equation can have any y-value provided that $x = -3$. Thus, the graph of this equation is a vertical line with x-intercept -3, as shown in Figure 3.15. ◆

To graph the equation $Ax + By + C = 0$, where A, B, and C are nonzero numbers, we can use an alternate method: find the x-intercept and y-intercept. Connecting these intercepts with a straightedge also gives us the graph of the straight line. For the equation $2x - 3y = 12$ in Example 4(a), we find the intercepts as follows.

Find the x-intercept:	*Find the y-intercept:*
$2x - 3y = 12$ Let $y = 0$: $2x - 3(0) = 12$ $2x = 12$ $x = 6$	$2x - 3y = 12$ Let $x = 0$: $2(0) - 3y = 12$ $-3y = 12$ $y = -4$
Coordinates of x-intercept: $(6, 0)$	Coordinates of y-intercept: $(0, -4)$

Connecting the points $(6, 0)$ and $(0, -4)$ also gives us the graph of the straight line illustrated in Figure 3.13.

To display the graph of a nonvertical line $Ax + By = C$ on a graphing calculator, we must enter the equation in the form $y = f(x)$. To show a true geometric perspective of the slope of the line, we use a square set viewing rectangle—one in which the vertical and horizontal tick marks have equal spacing. Figure 3.16 shows how the graph of the equation $2x - 3y = 12$ (or $y = \frac{2}{3}x - 4$) looks in different viewing rectangles. Only the line in the square set viewing rectangle in part (a) visually appears to have a slope of $\frac{2}{3}$.

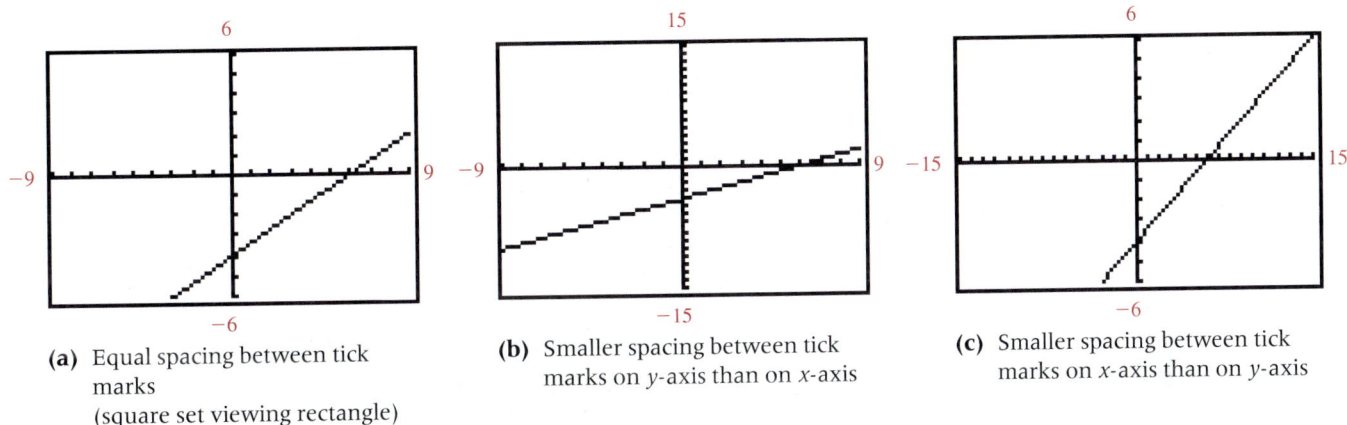

(a) Equal spacing between tick marks
(square set viewing rectangle)

(b) Smaller spacing between tick marks on *y*-axis than on *x*-axis

(c) Smaller spacing between tick marks on *x*-axis than on *y*-axis

FIGURE 3.16

Appearance of $2x - 3y = 12$ in different viewing rectangles.

PROBLEM 4 Find the *x*-intercept and *y*-intercept for the equation $3x + 8y = 12$. Then sketch its graph. Use a graphing calculator to verify the sketch. ◆

◆ **Application: Tax Rate Schedule**

The amount of federal income tax you pay each year is a function of your taxable income for that year. This amount is usually given by the Internal Revenue Service (IRS) in terms of a *tax rate schedule*. Each line in the tax rate schedule can be expressed as a different linear function. If a function is defined by different linear functions on distinct subsets of its domain, it is called a **piecewise linear function**.

EXAMPLE 5 A recent tax rate schedule for a single person with a taxable income up to $93,130 is given by the IRS as follows.

Taxable income	Tax
$0 to $18,550	15% of the amount over $0
$18,550 to $44,900	$2782.50 + 28% of the amount over $18,550
$44,900 to $93,130	$10,160.50 + 33% of the amount over $44,900

(a) Find the piecewise linear function defined by this tax rate schedule.

(b) Use a graphing calculator to generate the graph of the function in part (a).

(c) Use the piecewise linear function in part (a) to determine the amount of tax a single person is required to pay for a taxable income of $39,440: Then use the tax rate schedule to find this tax.

◆ **Solution**

(a) Using this tax rate schedule, we can write three equations:

1. If the taxable income x is $0 to $18,550, then the required tax T (in dollars) is

$$T(x) = 0.15x$$

2. If the taxable income x is more than $18,550 but not more than $44,900, the required tax T (in dollars) is

$$T(x) = 2782.50 + 0.28(x - 18,550)$$
$$= 0.28x - 2411.50$$

3. If the taxable income x is more than $44,900 but not more than $93,130, the required tax T (in dollars) is

$$T(x) = 10,160.50 + 0.33(x - 44,900)$$
$$= 0.33x - 4656.50$$

Thus, the piecewise linear function defined by this tax rate schedule has three parts:

$$T(x) = \begin{cases} 0.15x & \text{if } 0 \le x \le 18550 \\ 0.28x - 2411.50 & \text{if } 18,550 < x \le 44,900 \\ 0.33x - 4656.50 & \text{if } 44,900 < x \le 93,130 \end{cases}$$

(b) To generate the graph of this piecewise linear function on a graphing calculator, we enter each piece of the function separately, specifying the interval over which that piece is defined. Figure 3.17 shows the graph of the function T defined in part (a). The x-axis denotes the taxable income x, and the y-axis denotes the required tax T.

FIGURE 3.17

Graph of the piecewise linear function defined in Example 5(a)

30000

0   93130

X=39440   Y=8631.7

(c) Since a taxable income of $39,440 is more than $18,550 but not more than $44,900, we use the second part of the function to compute the tax:

$$T(x) = 0.28x - 2411.50$$

$$T(39,440) = 0.28(39,440) - 2411.50 = \$8631.70$$

Also, using the second line of the tax rate schedule, we find that the tax is

$$\$2782.50 + 0.28 (\$39,440 - \$18,550) = \$8631.70.$$

The amount over $18,550

Referring to Figure 3.17, we note that (39440, 8631.70) is a point on the graph. ◆

PROBLEM 5 Use the graph in Figure 3.17 to estimate the taxable income of a single person who is required to pay an income tax of $15,000. Then use the piecewise linear function in Example 5(a) to find the taxable income to the nearest cent. ◆

Exercises 3.1

◆ **Basic Skills**

In Exercises 1–12, find the slope (if it exists) of the line that passes through the given points.

1. (2, 0) and (3, 2)

2. (4, 1) and (−1, −1)

3. (−3, 4) and (6, −2)

4. (3, −2) and (4, −3)

5. (−4, 3) and (−4, 1)

6. (−2, 5) and (4, 5)

7. $\left(\frac{3}{2}, -1\right)$ and $\left(-\frac{1}{2}, 7\right)$

8. $\left(\frac{3}{4}, \frac{2}{3}\right)$ and $\left(-1, \frac{1}{6}\right)$

9. (4, 3) and (0, b)

10. (a, 0) and (0, b)

11. (a, b) and (b, a)

12. $\left(a, \frac{a}{b}\right)$ and $\left(b, \frac{b}{a}\right)$

In Exercises 13–18, sketch the line that passes through the given point and has the given slope.

13. Through (3, 1) with slope 3

14. Through (−3, −1) with slope $\frac{4}{3}$

15. Through $\left(0, -\frac{2}{3}\right)$ with slope $-\frac{1}{3}$

16. Through $\left(\frac{1}{2}, \frac{3}{2}\right)$ with slope −2

17. Through (−1, 2) with slope 0

18. Through (2, 3) with undefined slope

In Exercises 19–30, identify the slope and y-intercept, and then sketch the graph of the linear function. Use a graphing calculator to verify each graph.

19. $y = 3x$

20. $y = -2x$

21. $f(x) = -\frac{1}{3}x$

22. $F(x) = \frac{3x}{4}$

23. $y = 2x + 3$

24. $y = 3x - 4$

25. $g(x) = -4x + 1$

26. $G(x) = 5 - 2x$

27. $y = \frac{x}{2} - 1$

28. $y = \frac{2}{3}x - 2$

29. $h(x) = \frac{1 - 3x}{4}$

30. $H(x) = \frac{-x + 3}{2}$

In Exercises 31–42, sketch the graph of each equation, state the slope, and label the x-intercept and y-intercept (if they exist). Use a graphing calculator to verify the graph of each nonvertical line.

31. $x + y = 2$

32. $x - y = -3$

33. $2x + 5y = 10$

34. $3x + y = -6$

35. $21x - 28y = -56$

36. $9x - 18y = 60$

37. $x = 2y$

38. $4x - 3y = 0$

39. $x = -2$

40. $y = 4$

41. $\frac{4}{3}y - 3 = 0$

42. $2.5 - 0.2x = 0$

In Exercises 43–48, sketch the graph of each piecewise linear function. Use a graphing calculator to verify each graph.

43. $f(x) = \begin{cases} x & \text{if } x \le 2 \\ 2x - 2 & \text{if } x > 2 \end{cases}$

44. $F(x) = \begin{cases} 2x & \text{if } x \le 1 \\ 3 - x & \text{if } x > 1 \end{cases}$

45. $h(x) = \begin{cases} 2x + 5 & \text{if } x \le -2 \\ x + 3 & \text{if } x > -2 \end{cases}$

46. $H(x) = \begin{cases} -(3x + 4) & \text{if } x \le -1 \\ \dfrac{x - 2}{3} & \text{if } x > -1 \end{cases}$

47. $g(x) = \begin{cases} 3x - 1 & \text{if } x \le 0 \\ x - 1 & \text{if } 0 < x \le 2 \\ 6 - x & \text{if } x > 2 \end{cases}$

48. $G(x) = \begin{cases} x + 3 & \text{if } x \le 2 \\ 7 - 2x & \text{if } 2 < x \le 4 \\ x - 5 & \text{if } x > 4 \end{cases}$

In Exercises 49–56, use a graphing calculator to generate the graph of each equation in the viewing rectangle. Then estimate the x-intercept and y-intercept by tracing to these points. Round each estimate to three significant digits. Check each estimate algebraically.

49. $7x + 4y = 24$

50. $16y - 18x = -33$

51. $14x - 18y + 57 = 0$

52. $124x = 234y - 167$

53. $6.45x - 8.76y = 34.5$

54. $345.6x - 124.8y + 187.4 = 0$

55. $7.89y - 1.23x = 14.8$

56. $18.4x = 67.3 - 18.7y$

57. The linear equation $F = \frac{9}{5}C + 32$ expresses the Fahrenheit temperature F as a function of the Celsius temperature C.

(a) Identify the slope and F-intercept.

(b) Sketch the graph of the linear equation.

58. The linear equation $C = \frac{5}{9}(F - 32)$ expresses the Celsius temperature C as a function of the Fahrenheit temperature F.

(a) Identify the slope and C-intercept.

(b) Sketch the graph of the linear equation.

59. The linear equation $v = 44 - 3t$ expresses the velocity v (in feet per second, ft/s) of an object as a function of the time t (in seconds) the object travels at a constant deceleration of 3 ft/s². [*Note:* The unit ft/s² is a more compact way of writing *feet per second per second*, or $\frac{\text{ft/s}}{\text{s}}$.]

(a) Identify the slope and v-intercept.

(b) Sketch the graph of the linear equation.

(c) What is the significance of the t-intercept?

60. The linear equation $A = 500 + 3.75t$ expresses the amount A (in dollars) accumulated in a certain savings account after t months ($0 \le t \le 12$) if the money earns 9% simple interest per year.

(a) Identify the slope and A-intercept.

(b) Sketch the graph of the linear equation for $0 \le t \le 12$.

(c) What is the significance of the slope and the A-intercept?

61. The cost C of electricity is a function of the amount x of electricity consumed (in kilowatt-hours, kwh). Suppose the cost of electricity is given by the following rate schedule.

Usage	Cost
First 30 kwh or less	$3.00
Next 70 kwh	$0.06 per kwh
Next 200 kwh	$0.04 per kwh
All kilowatt-hours over 300	$0.01 per kwh

(a) Find the piecewise linear function that describes this rate schedule.

(b) Use a graphing calculator to generate the graph of the function in part (a).

(c) Use the rate table to find the cost of using 240 kwh of electricity. Then use the piecewise linear function defined in part (a) to find the cost.

(d) Use the graph from part (b) to estimate the number of kilowatts-hours consumed if the cost is $14.08. Then use the piecewise linear function from part (a) to find the exact number of kilowatt-hours consumed.

62. The amount of federal income tax T you pay is a function of your taxable income x. A recent tax rate schedule for a single person is given:

Taxable income	Tax
$0 to $1800	11% of the amount over $0
$1800 to $16,800	$198 + 15% of the amount over $1800
$16,800 to $27,000	$2448 + 28% of the amount over $16,800
$27,000 to $54,000	$5304 + 35% of the amount over $27,000
$54,000 and up	$14,754 + 38.5% of the amount over $54,000

(a) Find the piecewise linear function defined by this rate schedule.

(b) Use a graphing calculator to generate the graph of the function in part (a).

(c) Use the tax rate table to determine the amount of tax a single person is required to pay for a taxable

income of \$35,000. Then use the piecewise linear function from part (a) to find the tax.

(d) Use the graph from part (b) to estimate the taxable income of a single person whose income tax is \$7,000. Then use the piecewise linear function defined in part (a) to find this taxable income to the nearest cent.

◆ Critical Thinking

63. The center of a circle of radius 5 has coordinates $(-1, 2)$. Determine the coordinates of the endpoints of a diameter whose slope is $-\frac{3}{4}$.

64. The midpoint of a line segment 20 units long is $(2, -1)$. If the slope of the line segment is $\frac{4}{3}$, what are the coordinates of its endpoints?

65. For what value of a will the line through the points $(3, -2)$ and $(a, 2)$ have the given slope?

(a) $m = -2$ (b) $m = \frac{2}{3}$ (c) $m = -\frac{1}{2}$

66. For what value of a will the points $(-1, 3)$, $(2, a)$, and $(3, -3)$ be collinear (lie on the same line)?

67. Find the inverse function f^{-1} for the linear function $f(x) = mx + b$ with $m \neq 0$.

68. Suppose f and g are linear functions defined by $f(x) = m_1x + b_1$ and $g(x) = m_2x + b_2$, where $m_1 \neq 0$ and $m_2 \neq 0$.

(a) Find $f \circ g$.

(b) Find $g \circ f$.

(c) Is $f \circ g$ a linear function? If so, give its slope and y-intercept.

(d) Is $g \circ f$ a linear function? If so, give its slope and y-intercept.

69. Consider the equation $Ax + By = C$ with $A \neq 0$, $B \neq 0$, and $C \neq 0$. Divide both sides of this equation by C and write the equation in the form

$$\frac{x}{a} + \frac{y}{b} = 1.$$

Explain why this equation is called the *equation of a line in intercept form*.

70. Use the results of Exercise 69 to sketch the graph of each equation. Use a graphing calculator to verify each sketch.

(a) $\dfrac{x}{2} + \dfrac{y}{3} = 1$ (b) $\dfrac{x}{5} - \dfrac{y}{2} = 1$

(c) $\dfrac{y}{4} - \dfrac{x}{6} = 1$ (d) $-\dfrac{x}{\frac{3}{2}} - \dfrac{y}{\frac{8}{3}} = 1$

3.2 ◆ Determining the Equation of a Line

◆ Introductory Comments

In Section 3.1, we were given the equation of a line and then found some information about the line, such as its slope or x- and y-intercepts. In this section, we reverse the procedure—from some given information about a line, we determine its equation. We begin by developing the *point-slope form* for the equation of a straight line. This form is useful for determining the equation of a line when we are given the slope of the line and the coordinates of a fixed point on the line.

◆ Point-Slope Form for the Equation of a Line

Let m be the given slope of a line and $P(x_1, y_1)$ be a given point on the line. Also, let $Q(x, y)$ be *any* point other than P on the given line, as shown in

Figure 3.18. By the slope formula (in Section 3.1), we have

$$m = \frac{y - y_1}{x - x_1}$$

$$y - y_1 = m(x - x_1) \qquad \textcolor{red}{\textbf{Multiply both sides by } (x - x_1)}$$

This last equation is called the **point-slope form** for the equation of a line.

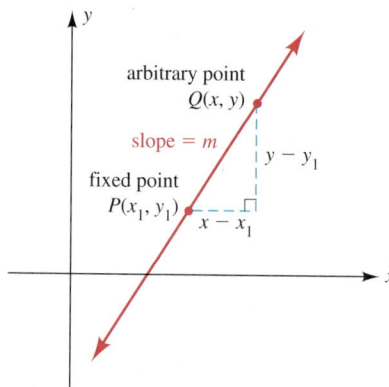

FIGURE 3.18

The slope m of the line through a fixed point $P(x_1, y_1)$ and an arbitrary point $Q(x, y)$ on the line is $(y - y_1)/(x - x_1)$.

◆ Point-Slope Form

The equation of a line having a given slope m and passing through the fixed point $P(x_1, y_1)$ and arbitrary point $Q(x, y)$ is

$$y - y_1 = m(x - x_1).$$

EXAMPLE 1 Find the equation of a line that passes through the point $(2, -1)$ and has a slope of 3. Write this equation in slope-intercept form (see Section 3.1).

◆ Solution   Using the point-slope form for the equation of a line with $(x_1, y_1) = (2, -1)$ and $m = 3$, we obtain the following equation:

$$y - y_1 = m(x - x_1)$$
$$y - (-1) = 3(x - 2) \qquad \textcolor{red}{\textbf{Substitute}}$$
$$y + 1 = 3x - 6 \qquad \textcolor{red}{\textbf{Multiply}}$$
$$y = 3x - 7 \qquad \textcolor{red}{\textbf{Write in slope-intercept form}}$$

Figure 3.19 shows the graph of the equation $y = 3x - 7$ in the viewing rectangle of a graphing calculator. Note that this line passes through the given point $(2, -1)$ and has slope 3.   ◆

FIGURE 3.19

The graph of $y = 3x - 7$ is a line that passes through the point $(2, -1)$ and has slope 3.

PROBLEM 1   Find the equation of a line with slope $\frac{1}{2}$ and x-intercept -3. Write this equation in slope-intercept form.   ◆

If two points on a line are given, then the equation of the line can be determined by using the slope formula (from Section 3.1) together with the point-slope form for the equation of a line.

EXAMPLE 2 Find the equation of the line that passes through the points $(-2, -3)$ and $(2, 1)$. Write the equation in slope-intercept form.

◆ **Solution** The slope of the line that passes through the points $(-2, -3)$ and $(2, 1)$ is

$$m = \frac{1 - (-3)}{2 - (-2)} = \frac{4}{4} = 1.$$

Using $(x_1, y_1) = (-2, -3)$ as the fixed point and $m = 1$ as the slope, we have

$$y - y_1 = m(x - x_1)$$

$$y - (-3) = 1(x - (-2)) \qquad \textbf{Substitute}$$

$$y + 3 = x + 2 \qquad \textbf{Simplify}$$

$$y = x - 1 \qquad \textbf{Write in slope-intercept form}$$

Figure 3.20 shows the graph of the equation $y = x - 1$ in the viewing rectangle of a graphing calculator. Note that this line passes through the given points $(-2, -3)$ and $(2, 1)$. ◆

PROBLEM 2 Use $(x_1, y_1) = (2, 1)$ as a fixed point and $m = 1$ as the slope to determine the equation of a line. Write the equation in slope-intercept form. Your answer should agree with the result in Example 2. ◆

◆ **Parallel Lines**

Suppose two distinct, nonvertical lines $y = m_1 x + b_1$ and $y = m_2 x + b_2$ are parallel, as shown in Figure 3.21.

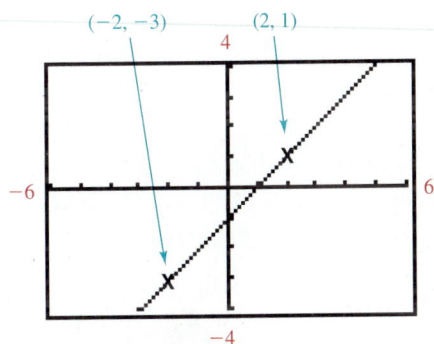

FIGURE 3.20

The graph of $y = x - 1$ is a line that passes through the points $(-2, -3)$ and $(2, 1)$.

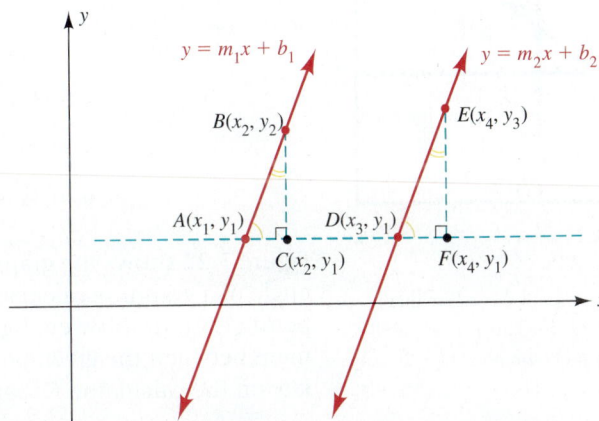

FIGURE 3.21

If the lines $y = m_1 x + b_1$ and $y = m_2 x + b_2$ are *parallel*, then the triangles *ABC* and *DEF* are similar and their corresponding sides are proportional.

The triangles *ABC* and *DEF* are similar, since their corresponding angles are equal. Thus, their corresponding sides are proportional and we have

$$\frac{y_2 - y_1}{x_2 - x_1} = \frac{y_3 - y_1}{x_4 - x_3}.$$

However, $\dfrac{y_2 - y_1}{x_2 - x_1} = m_1$ and $\dfrac{y_3 - y_1}{x_4 - x_3} = m_2.$

Therefore, $m_1 = m_2$. Thus, if two nonvertical lines are parallel, they have the same slope. The converse of this statement is also true: If two distinct nonvertical lines have the same slope, then the lines are parallel.

◆ **Parallel Lines**

Two distinct nonvertical lines with slopes m_1 and m_2 are **parallel** if and only if

$$m_1 = m_2.$$

EXAMPLE 3   Find the equation of the line that passes through the point $(1, 3)$ and is parallel to the line that passes through the points $(-2, -4)$ and $(2, -1)$. Write this equation in slope-intercept form.

◆ **Solution**   The slope of the line that passes through the points $(-2, -4)$ and $(2, -1)$ is

$$m = \frac{-1 - (-4)}{2 - (-2)} = \frac{3}{4}.$$

Since two parallel lines have the same slope, we know that the slope of the line that passes through the point $(1, 3)$ is also $\frac{3}{4}$. Now using the point-slope form for the equation of a line with fixed point $(x_1, y_1) = (1, 3)$ and slope $m = \frac{3}{4}$, we can find the equation of the desired line as follows:

$$y - y_1 = m(x - x_1)$$
$$y - 3 = \tfrac{3}{4}(x - 1) \qquad \text{Substitute}$$
$$y - 3 = \tfrac{3}{4}x - \tfrac{3}{4} \qquad \text{Multiply}$$
$$y = \tfrac{3}{4}x + \tfrac{9}{4} \qquad \text{Write in slope-intercept form}$$

FIGURE 3.22

The graph of $y = \frac{3}{4}x + \frac{9}{4}$ is a line that passes through $(1, 3)$. This line is parallel to the line that joins the points $(-2, -4)$ and $(2, -1)$.

Figure 3.22 shows the graph of the equation $y = \frac{3}{4}x + \frac{9}{4}$ in the viewing rectangle of a graphing calculator. Note that the line passes through the given point $(1, 3)$. In this viewing rectangle, we have also constructed a line segment between the given points $(-2, -4)$ and $(2, -1)$ by using the DRAW key on the calculator. Observe that the lines appear to be parallel. ◆

> **PROBLEM 3** Find the equation of a line that has x-intercept -3 and is parallel to a line whose slope is $\frac{2}{3}$. Write this equation in slope-intercept form.
◆

◆ Perpendicular Lines

Suppose two nonvertical lines $y = m_1x + b_1$ and $y = m_2x + b_2$ are perpendicular, as shown in Figure 3.23.

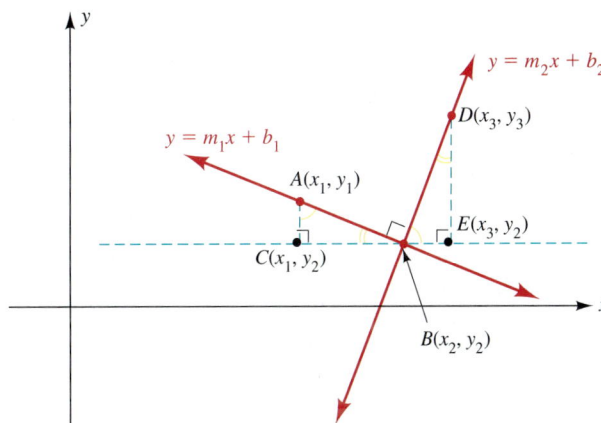

FIGURE 3.23

If the lines $y = m_1x + b_1$ and $y = m_2x + b_2$ are *perpendicular*, then the triangles *BAC* and *DBE* are similar and their corresponding sides are proportional.

Angle *BAC* is complementary to angle *ABC*, and angle *ABC* is complementary to angle *DBE*. Thus, angles *BAC* and *DBE* are equal and so triangles *BAC* and *DBE* are similar. Using the fact that the corresponding sides of these triangles are proportional, we have

$$\frac{y_3 - y_2}{x_3 - x_2} = \frac{x_2 - x_1}{y_1 - y_2}$$

However,

$$\frac{y_3 - y_2}{x_3 - x_2} = m_2,$$

and

$$\frac{x_2 - x_1}{y_1 - y_2} = -\frac{x_1 - x_2}{y_1 - y_2} = -\frac{1}{\dfrac{y_1 - y_2}{x_1 - x_2}} = -\frac{1}{m_1}.$$

Therefore, $m_2 = -\dfrac{1}{m_1}$. Thus, if two nonvertical lines are perpendicular, the slope of one line is the negative reciprocal of the slope of the other. The converse of this statement is also true: If the slope of one line is the negative reciprocal of the slope of another, then the lines are perpendicular.

◆◆ **Perpendicular Lines**

Two nonvertical lines with slopes m_1 and m_2 are **perpendicular** if and only if

$$m_2 = -\frac{1}{m_1}.$$

EXAMPLE 4 Find the equation of a line that passes through the point (2, 1) and is perpendicular to the line $3x - 5y = 10$.

◆ **Solution**   We can determine the slope of the line $3x - 5y = 10$ by writing the equation in slope-intercept form:

$$3x - 5y = 10$$
$$-5y = -3x + 10$$
$$y = \frac{3}{5}x - 2$$

slope

Since the slope of the line $3x - 5y = 10$ is $\frac{3}{5}$, the slope m of the perpendicular line is

$$m = -\frac{1}{\frac{3}{5}} = -\frac{5}{3}.$$

Negative reciprocal of $\frac{3}{5}$

Now, using the point-slope form for the equation of a line with fixed point $(x_1, y_1) = (2, 1)$ and $m = -\frac{5}{3}$, we can find the desired equation:

$$y - y_1 = m(x - x_1)$$
$$y - 1 = -\frac{5}{3}(x - 2) \qquad \text{Substitute}$$
$$y - 1 = -\frac{5}{3}x + \frac{10}{3} \qquad \text{Multiply}$$
$$y = -\frac{5}{3}x + \frac{13}{3} \qquad \text{Write in slope-intercept form}$$

Figure 3.24 shows the graph of the equation $y = -\frac{5}{3}x + \frac{13}{3}$ in the viewing rectangle of a graphing calculator. Note that the line passes through the given point (2, 1). In this viewing rectangle, we also have generated the graph of the equation $3x - 5y = 10$. Observe that the lines appear to be perpendicular. It is important to use a square set viewing rectangle to show the true geometric perspective of the slopes and, hence, the perpendicularity of the lines.   ◆

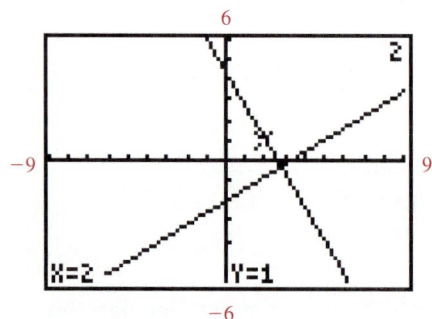

FIGURE 3.24
The graph of $y = -\frac{5}{3}x + \frac{13}{3}$ is a line that passes through (2, 1). This line is perpendicular to the line $3x - 5y = 10$.

PROBLEM 4
Find the equation of a line that passes through the point (3, −4) and is perpendicular to a line that passes through the points (3, 0) and (2, −3). ◆

◆ Application: Linear Depreciation

One type of applied problem that occurs frequently in business is *linear depreciation*. In linear depreciation, the value of an asset *decreases* linearly over time.

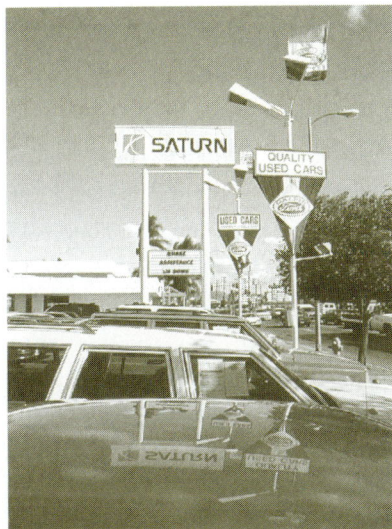

EXAMPLE 5 *Chapter Opening Problem*

A four-year-old car has a value of $3000. When one year old, its value was $9000.

(a) Assuming the car's depreciation is linear, express the value V of the car as a function of its age x (in years).

(b) What is the age of the car when its value is fully depreciated?

◆ Solution

(a) To illustrate the linear depreciation of the car between the ages of 1 and 4 years, we use the ⌈DRAW⌋ key on a graphing calculator to construct a line segment between the points (1, 9000) and (4, 3000), as shown in Figure 3.25. The slope of this line segment is

$$m = \frac{\$9000 - \$3000}{1 \text{ yr} - 4 \text{ yr}} = -\$2000 \text{ per year.}$$

(1, 9000)   (4, 3000)

FIGURE 3.25

A graph of the linear depreciation of the car between the ages of 1 and 4 years

Note that the slope tells us the amount of *depreciation per year*. Now, to express the value V of the car as a function of its age x, we use the point-slope form for the equation of a line with fixed point $(x_1, V_1) = (1, 9000)$ and slope $m = -2000$. Thus,

$$V - V_1 = m(x - x_1)$$

$$V - 9000 = -2000(x - 1) \qquad \text{\textcolor{red}{\textbf{Substitute}}}$$

$$V - 9000 = -2000x + 2000 \qquad \text{\textcolor{red}{\textbf{Multiply}}}$$

$$V = 11,000 - 2000x \qquad \text{\textcolor{red}{\textbf{Write in slope-intercept form}}}$$

(b) The car's value is fully depreciated when $V = 0$. Assuming the car continues to depreciate according to the equation $V = 11,000 - 2000x$, we have

$$0 = 11,000 - 2000x$$
$$2000x = 11,000$$
$$x = 5\tfrac{1}{2}$$

FIGURE 3.26

The x-intercept of the graph of $V = 11,000 - 2000x$ denotes the time when the car's value is fully depreciated.

Thus, the car's value is fully depreciated when it is $5\tfrac{1}{2}$ years old. Figure 3.26 shows that $x = 5\tfrac{1}{2}$ is the x-intercept of the graph of $V = 11,000 - 2000x$, which confirms our answer. ◆

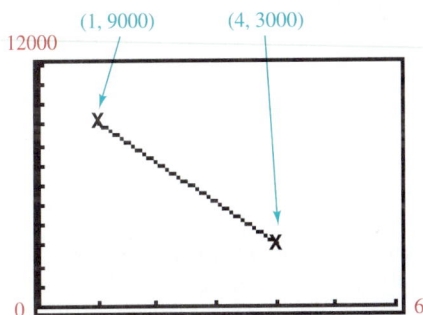

PROBLEM 5 Referring to Example 5, what is the significance of the V-intercept in the equation $V = 11,000 - 2000x$? ◆

Exercises 3.2

◆ Basic Skills

In Exercises 1–38, find the equation of a line that satisfies the given conditions. For nonvertical lines, write the equation in slope-intercept form. For vertical lines, write the equation in general form with integer coeficients.

1. y-intercept 3 and slope $\frac{3}{4}$

2. y-intercept -2 and slope $-\frac{1}{2}$

3. Through $(-1, 2)$ with slope 3

4. Through $(3, 4)$ with slope -2

5. x-intercept 4 and slope $-\frac{7}{3}$

6. x-intercept -3 and slope $\frac{2}{5}$

7. Through the origin with slope -3

8. Through the origin and $(6, 5)$

9. Through $(4, -1)$ and $(2, 5)$

10. Through $(-1, -2)$ and $(4, -3)$

11. Through $\left(-2, \frac{4}{3}\right)$ and $(3, -1)$

12. Through $\left(\frac{6}{5}, -2\right)$ and $\left(-\frac{3}{2}, 1\right)$

13. Through $(-3, 4)$ with y-intercept $\frac{1}{6}$

14. Through $(5, 3)$ with x-intercept -3

15. x-intercept 3 and y-intercept -2

16. x-intercept -4 and y-intercept 6

17. x-intercept $\frac{1}{4}$ and y-intercept 6

18. x-intercept -3 and y-intercept $-\frac{5}{3}$

19. y-intercept $\frac{3}{2}$ and slope 0

20. Through $(4, -3)$ and $(-1, -3)$

21. Through $(3, -2)$ and $(3, 4)$

22. x-intercept $-\frac{2}{3}$ and undefined slope

23. Through $(5, -1)$ with same y-intercept as $3x - 2y = 4$

24. Through $\left(\frac{1}{2}, -1\right)$ with same x-intercept as $4x - y + 4 = 0$

25. Through $(2, 3)$ with x-intercept a and y-intercept a, provided $a \neq 0$

26. Through $\left(\frac{1}{3}, -5\right)$ with x-intercept a and y-intercept $-a$, provided $a \neq 0$

27. x-intercept -2 and parallel to a line with slope $-\frac{1}{2}$

28. y-intercept $\frac{4}{3}$ and parallel to a line with slope -4

29. Through $(-3, -1)$ and parallel to the line that passes through $(1, 0)$ and $(2, 3)$

30. Through $(4, 5)$ and parallel to the line with x-intercept 3 and y-intercept -2

31. Through $(2, -3)$ and parallel to the line $2x - 3y + 6 = 0$

32. Through the origin and parallel to the line $x - 2y - 3 = 0$

33. y-intercept 3 and perpendicular to a line with slope -4

34. x-intercept $\frac{5}{2}$ and perpendicular to a line with slope $\frac{8}{5}$

35. Through $(-1, 0)$ and perpendicular to the line that passes through $(2, 0)$ and $(-1, 4)$

36. Through $(5, -2)$ and perpendicular to the y-axis

37. Through the origin and perpendicular to the line $2x + 4y = 5$

38. Through $(2, -7)$ and perpendicular to the line $3x - 2y - 9 = 0$

39. Use a graphing calculator to generate the graphs of the following equations in the same viewing rectangle:

$$2.795x - 1.625y = 8.325$$

$$6.622x - 3.850y = -5.765$$

Use the graphs to determine if the pair of lines appear to be parallel. Verify your answer algebraically.

40. Use a graphing calculator to generate the graphs of the following equations in the same viewing rectangle:

$$2.740x + 0.8768y = 10.65$$

$$0.6672x - 2.085y = 15.76$$

Use the graphs to determine if the pair of lines appear to be perpendicular. Verify your answer algebraically.

41. At $60°F$ a cricket chirps 50 times per minute and at $80°F$ it chirps 100 times per minute.

    (a) Assuming that the rate of chirping varies linearly with the temperature, express the cricket's chirping rate r as a function of the temperature T.

    (b) Determine the cricket's chirping rate when $T = 40°F$.

42. A salesperson's weekly pay is $450 for weekly sales of $3000 and $500 for weekly sales of $4000.

    (a) Assuming that weekly pay varies linearly with the amount of weekly sales, express the weekly pay P as a function of the amount of weekly sales S.

(b) Determine the base pay and the rate of commission for this salesperson.

43. A new dump truck cost $60,000 and depreciates linearly to a value of $6000 after 6 years.

(a) Express the value V of the truck as a function of its age x (in years).

(b) What is the value of the truck after 4 years?

44. A factory owner buys a new machine for $25,000. The value of the machine depreciates linearly, and the machine has no resale value after 10 years.

(a) Express the value V of the machine as a function of its age x (in years).

(b) What is the resale value of the machine after 6 years?

45. A house cost $50,000 to build 8 years ago. Its value has appreciated linearly and today the house is worth $170,000.

(a) Express the value V of the house as a function of its age x (in years).

(b) What is the projected value of the house 4 years from now?

46. A diamond ring was purchased for $600. Fifteen years later the ring was appraised for $2400. Assume the ring appreciated linearly.

(a) Express the value V of the ring as a function of its age x (in years).

(b) Find the value of the ring 20 years after it was purchased.

47. Trolley tracks are laid with expansion gaps between the steel rails so that the rails can expand without distortion. As the temperature increases, the width of the gap between the rails decreases linearly. Suppose the gap is 2.5 mm wide when the temperature is $51\,°F$ and 1.2 mm wide when the temperature is $83\,°F$.

(a) Express the width W of the expansion gap as a function of the temperature T.

(b) What is the approximate width of the expansion gap when the temperature is $35\,°F$?

(c) At what temperature will the rails just touch?

48. The fuel tank of an automobile is filled with 22.3 gallons of gasoline. When the auto is traveling at a constant rate of speed, the amount of fuel remaining in the tank decreases linearly. After 1 hour 12 minutes of travel, the tank contains 18.1 gallons of fuel.

(a) Express the number N of gallons of gasoline in the tank as a function of the amount of time t traveled.

(b) Approximately how many gallons of gasoline remain after traveling for 2 hours 21 minutes?

(c) If the auto continues to travel at this constant rate, when is the fuel tank empty?

Critical Thinking

49. The *perpendicular bisector* of a line segment is a line that is perpendicular to the segment at its midpoint. Find the equation of the perpendicular bisector of the line segment that joins the given points.

(a) $(4, 0)$ and $(0, -12)$ (b) $(3, -2)$ and $(-1, 1)$

50. Use slopes to show that the triangle with vertices $A(-2, 1)$, $B(4, -8)$, and $C(7, -6)$ is a right triangle. Then find the area of this right triangle.

In Exercises 51–54, find the linear function that satisfies the given conditions.

51. $f(2) = 7$ and $f(-1) = 3$

52. $h(3) = 1$ and $h(0) = 2$

53. $g(2) = -1$ and the graph of g has slope 3

54. $F(-1) = -3$ and the graph of F has slope -2

55. Find $H(3)$ given that H is a linear function with $H(5) = 4$ and $H(0) = 0$

56. Find $G(-4)$ given that G is a linear function with $G(-7) = 3$ and $G(5) = 1$

57. Find the value of k such that the line $kx + 2y + 4 = 0$ satisfies the given condition.

(a) passes through $(3, 1)$

(b) is parallel to the x-axis

(c) is parallel to the line $y = 3x - 4$

(d) is perpendicular to the line $2x - 4y = 5$

58. Find the value of k such that the line $3x + ky = 6$ satisfies the given condition.

(a) passes through $(-2, 4)$

(b) is perpendicular to the x-axis

(c) is parallel to the line $y = 5 - 2x$

(d) is perpendicular to the line $x - y = 0$

59. A *tangent line* to a curve is a line that just touches the curve at a single point and is closer to the curve in the vicinity of the point than any other line. Using calculus, it can be shown that the *slope of the tangent line* (denoted m_{tan}) to the curve $y = 1 - 3x^2$ at any point (x, y) on the curve is given by $m_{tan} = -6x$. Find the equation of the tangent line that touches the point $(1, -2)$, as shown in the sketch.

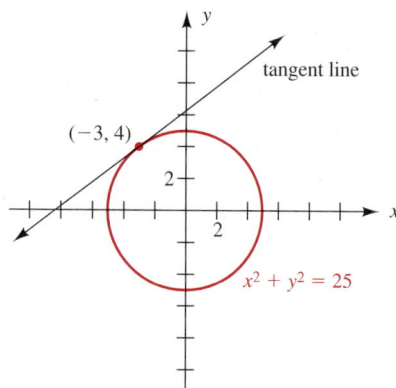

tangent line

$y = 1 - 3x^2$ $(1, -2)$

tangent line

$(-3, 4)$

2

2

$x^2 + y^2 = 25$

60. Refer to Exercise 59. Using calculus, it can be shown that the slope of the tangent line to the circle $x^2 + y^2 = 25$ at any point (x, y) on the circle is given by $m_{\tan} = -x/y$. Find the equation of the tangent line that touches the point $(-3, 4)$ as shown in the sketch.

61. Find the equation of the line that passes through the point $P(x_1, y_1)$ and is parallel to the line $Ax + By + C = 0$. Write the equation in slope-intercept form.

62. Find the equation of the line that passes through the origin and is perpendicular to the line $Ax + By + C = 0$. Write the equation in general form.

3.3 Quadratic Functions

◆ Introductory Comments

axis of symmetry $x = 0$

vertex $(0, 0)$

FIGURE 3.27

The graph of $f(x) = x^2$ is a parabola with vertex $(0, 0)$ and axis of symmetry $x = 0$.

Recall from Section 2.5 that the graph of the squaring function $f(x) = x^2$ is a cupped-shaped curve that is symmetric with respect to the y-axis [see part (d) of Figure 2.42 in Section 2.5]. We refer to this curve as a **parabola** and the vertical line $x = 0$ (the y-axis) as its **axis of symmetry**. The point at which the parabola intersects its axis of symmetry is called the **vertex** of the parabola (see Figure 3.27).

By the vertical stretch and compress rule (in Section 2.5), we know that the graph of $g(x) = ax^2$ is also a parabola, but with a *narrower* opening than the graph of $f(x) = x^2$ if $|a| > 1$, or a *wider* opening if $0 < |a| < 1$. By the x-axis reflection rule (Section 2.5), the parabola opens *upward* if $a > 0$, or *downward* if $a < 0$. In summary, the value of a determines the width and direction of the opening of the parabola. In Figure 3.28, we use a graphing calculator to illustrate this fact by generating the graph of $g(x) = ax^2$ for $a = 2, -5$, and $\frac{1}{4}$.

By the horizontal shift rule and the vertical shift rule (in Section 2.5), we know that the graph of $F(x) = a(x - h)^2 + k$ is the same as the graph of $g(x) = ax^2$, but shifted horizontally to the *right h* units if $h > 0$, or to the *left* $|h|$ units if $h < 0$, and shifted vertically *upward k* units if $k > 0$, or *downward* $|k|$ units if $k < 0$. Hence, the coordinates of the vertex of the parabola are now (h, k) and its axis of symmetry is the vertical line $x = h$. In Figure 3.29, we use a graphing calculator to illustrate this fact by generating the graph of $F(x) = a(x - h)^2 + k$ for several values of a, h, and k.

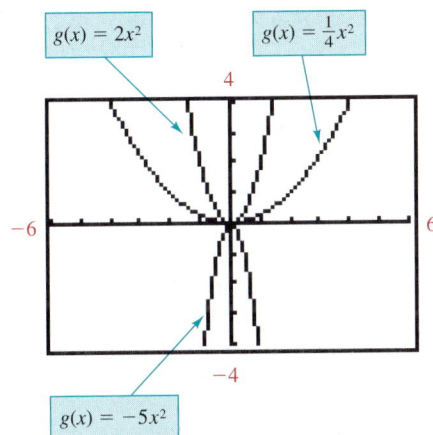

$g(x) = 2x^2$ $g(x) = \frac{1}{4}x^2$

$g(x) = -5x^2$

FIGURE 3.28

In the graph of $g(x) = ax^2$, the value of a determines the width and direction of the opening of the parabola.

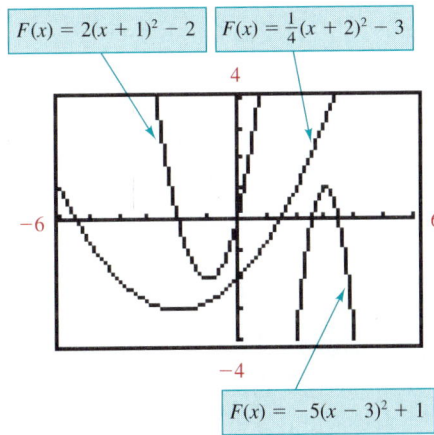

$F(x) = 2(x + 1)^2 - 2$ $F(x) = \frac{1}{4}(x + 2)^2 - 3$

$F(x) = -5(x - 3)^2 + 1$

FIGURE 3.29

In the graph of $F(x) = a(x - h)^2 + k$, the coordinates of the vertex of the parabola are (h, k).

We refer to the function F defined by $F(x) = a(x - h)^2 + k$ as a *quadratic function in standard form*. Its graph is a parabola with vertex (h, k) and axis of symmetry $x = h$.

◆◆ **Quadratic Function in Standard Form**

> If a, h, and k are real numbers with $a \neq 0$, then the function F defined by
>
> $$F(x) = a(x - h)^2 + k$$
>
> is called a **quadratic function in standard form**. The graph of this function is a **parabola** with vertex (h, k) and axis of symmetry $x = h$.

In this section, we study quadratic functions and their applications.

◆ **Working with Quadratic Functions in Standard Form**

To sketch the graph of the quadratic function $F(x) = a(x - h)^2 + k$ without plotting points or using a graphing calculator, we need to locate the vertex (h, k) and determine the y-intercept.

EXAMPLE 1   Determine the vertex and y-intercept for the graph of the quadratic function $f(x) = -2(x + 1)^2 + 8$. Then sketch the graph of this function.

◆ **Solution**   We begin by writing this quadratic function in standard form:

$$f(x) = -2[x - (-1)]^2 + 8$$

a h k

Since $a = -2$ (so $a < 0$), we know the graph is a parabola that opens *downward*. Since $h = -1$ and $k = 8$, its vertex is $(-1, 8)$.

To find the y-intercept, we let $x = 0$ and determine $f(0)$ as follows:

$$f(0) = -2(0 + 1)^2 + 8 = -2 + 8 = 6$$

y-intercept

The graph of $f(x) = -2(x + 1)^2 + 8$ is the parabola shown in Figure 3.30. We draw the right-hand side of the parabola by using the vertex and the y-intercept. We then use the fact that the parabola is symmetric with respect to the vertical line $x = -1$ to draw the left-hand side of this curve.   ◆

PROBLEM 1   Determine the vertex and intercepts for the quadratic function g defined by $g(x) = (x - 3)^2 - 1$.   ◆

As illustrated in the next example, we can determine the equation of a quadratic function if we know the vertex of the parabola and the coordinates of just one other point on the parabola.

EXAMPLE 2   Determine the equation of the quadratic function whose graph passes through the point $(4, 1)$ and has its vertex at $(2, 3)$.

◆ **Solution**   Since the vertex is $(2, 3)$, the quadratic function must have the form

$$y = a(x - 2)^2 + 3.$$

To determine the value of a, we use the fact that the graph passes through the point $(4, 1)$:

$$y = a(x - 2)^2 + 3$$
$$1 = a(4 - 2)^2 + 3 \qquad \text{Substitute } x = 4 \text{ and } y = 1$$
$$1 = 4a + 3 \qquad \text{Solve for } a$$
$$-2 = 4a$$
$$a = -\tfrac{1}{2}$$

Thus, the required quadratic function is $y = -\tfrac{1}{2}(x - 2)^2 + 3$. Figure 3.31 shows the graph of this quadratic function in the viewing rectangle of a graphing calculator. The parabola passes through the point $(4, 1)$ and has its vertex at $(2, 3)$; these facts verify our work.   ◆

PROBLEM 2   Determine the equation of the quadratic function whose graph passes through the point $(2, 3)$ and has its vertex at $(4, 1)$.   ◆

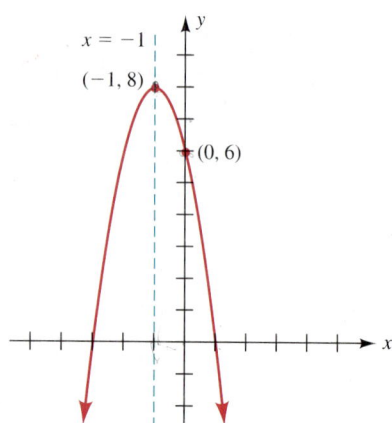

FIGURE 3.30

The graph of $f(x) = -2(x + 1)^2 + 8$ is a parabola that opens *downward* with vertex $(-1, 8)$. The graph is symmetric with respect to the vertical line $x = -1$.

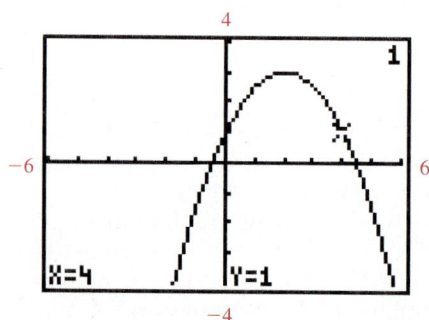

FIGURE 3.31

The graph of the quadratic function $y = -\tfrac{1}{2}(x - 2)^2 + 3$ passes through the point $(4, 1)$ and has its vertex at $(2, 3)$.

◆ Quadratic Functions in General Form

If we square the expression $(x - h)$ in $F(x) = a(x - h)^2 + k$, we obtain

$$F(x) = a(x^2 - 2hx + h^2) + k$$
$$= ax^2 + (-2ah)x + (ah^2 + k)$$

Replace the constant $-2ah$ with b and the constant $ah^2 + k$ with c.

$$= ax^2 + bx + c$$

where a, b, and c are constants and $a \neq 0$. We refer to a function of this form as a *quadratic function in general form.*

◆ Quadratic Function in General Form

If a, b, and c are real numbers and $a \neq 0$, then the function F defined by

$$F(x) = ax^2 + bx + c$$

is a **quadratic function in general form**.

If a quadratic function is written in the general form $F(x) = ax^2 + bx + c$, then we know that its graph is a parabola. We can find the x-coordinate of its vertex by writing the first two terms, $ax^2 + bx$, in the form of $a(x - h)^2$ and noting the value of h. To accomplish this task, we use an algebraic method called **completing the square**. First, we factor a from the first two terms:

$$ax^2 + bx = a\left(x^2 + \frac{b}{a}x\right).$$

Next, we add a special constant to the expression $x^2 + \frac{b}{a}x$ to form a perfect square trinomial that factors to the form $(x - h)^2$. If we add *the square of half the coefficient of x, or* $\left(\frac{b}{2a}\right)^2$, we obtain

$$x^2 + \frac{b}{a}x + \left(\frac{b}{2a}\right)^2 \quad \text{or} \quad \left(x + \frac{b}{2a}\right)^2.$$

The expression $\left(x + \frac{b}{2a}\right)^2$ is of the form $(x - h)^2$ with $h = -\frac{b}{2a}$. Hence, we conclude that $-\frac{b}{2a}$ is the x-coordinate of the vertex of the parabola defined by the function $F(x) = ax^2 + bx + c$. The y-coordinate of the vertex of the parabola may be found by evaluating $F\left(-\frac{b}{2a}\right)$. We refer to this fact as the **vertex formula for a parabola**.

◆ **Vertex Formula for a Parabola**

The graph of the quadratic function $F(x) = ax^2 + bx + c$ is a parabola with

$$\text{axis of symmetry } x = -\frac{b}{2a}$$

$$\text{vertex } \left(-\frac{b}{2a},\ F\left(-\frac{b}{2a}\right)\right).$$

EXAMPLE 3 Determine the vertex and y-intercept for the graph of each quadratic function. Then sketch the graph of the function.

(a) $f(x) = x^2 - 3x - 4$ **(b)** $y = -2x^2 + 4x - 5$

◆ **Solution**

(a) For $f(x) = x^2 - 3x - 4$, we have $a = 1$, $b = -3$, and $c = -4$. Thus, the x-coordinate of the vertex of the parabola is

$$x = -\frac{b}{2a} = -\frac{-3}{2(1)} = \frac{3}{2}$$

and the y-coordinate is

$$y = f\left(-\frac{b}{2a}\right) = f\left(\frac{3}{2}\right) = \left(\frac{3}{2}\right)^2 - 3\left(\frac{3}{2}\right) - 4 = -\frac{25}{4}.$$

Hence, the vertex of the parabola is $\left(\frac{3}{2}, -\frac{25}{4}\right)$. Since $a = 1$ (so $a > 0$), we know the parabola opens upward from the point $\left(\frac{3}{2}, -\frac{25}{4}\right)$.

We find the y-intercept by letting $x = 0$ and evaluating $f(0)$:

$$f(0) = (0)^2 - 3(0) - 4 = -4$$

> y-intercept

The graph of $f(x) = x^2 - 3x - 4$ is shown in Figure 3.32. We draw the left-hand side of the parabola by using the vertex and the y-intercept. We then use the fact that the parabola is symmetric with respect to the line $x = \frac{3}{2}$ to draw the right-hand side of the curve.

(b) For $y = -2x^2 + 4x - 5$, we have $a = -2$, $b = 4$, and $c = -5$. Thus, the x-coordinate for the vertex of the parabola is

$$x = -\frac{b}{2a} = -\frac{4}{2(-2)} = 1.$$

FIGURE 3.32

The graph of $f(x) = x^2 - 3x - 4$ is a parabola that opens upward and has vertex $\left(\frac{3}{2}, -\frac{25}{4}\right)$.

To find the y-coordinate of the vertex, we replace x with 1 in the equation $y = -2x^2 + 4x - 5$ as follows:

$$y = -2(1)^2 + 4(1) - 5 = -3$$

Hence, the vertex has coordinates $(1, -3)$ and, since $a = -2$ (so $a < 0$), the parabola opens downward from the point $(1, -3)$.

We find the y-intercept by letting $x = 0$ and solving for y as follows:

$$y = -2(0)^2 + 4(0) - 5 = -5$$

y-intercept

The graph of $y = -2x^2 + 4x - 5$ is shown in Figure 3.33. We draw the left-hand side of the parabola by using the vertex and the y-intercept. We then use the fact that the parabola is symmetric with respect to the line $x = 1$ to draw the right-hand side of the curve. ◆

FIGURE 3.33

The graph of $y = -2x^2 + 4x - 5$ is a parabola that opens downward and has vertex $(1, -3)$.

PROBLEM 3 Use factoring and the zero product property (in Section 2.2) to find the x-intercepts of the parabola in Example 3(a). Check your answers with the graph in Figure 3.32. ◆

◆ **Maximum and Minimum Values of Quadratic Functions**

If $a > 0$, then the vertex (h, k) is the *lowest point* on the graph of the function $F(x) = a(x - h)^2 + k$. Hence, we say that the function F has a **minimum value** of k when $x = h$. For example, Figure 3.32 indicates that the lowest point on the graph of $f(x) = x^2 - 3x - 4$ is its vertex, $\left(\frac{3}{2}, -\frac{25}{4}\right)$. We refer to $-\frac{25}{4}$ as the *minimum value* for this quadratic function and say that it occurs when $x = \frac{3}{2}$.

If $a < 0$, then the vertex (h, k) is the *highest point* on the graph of the function $F(x) = a(x - h)^2 + k$. Hence, we say that the function F has a **maximum value** of k when $x = h$. For example, Figure 3.33 indicates that the highest point on the graph of $y = -2x^2 + 4x - 5$ is its vertex, $(1, -3)$. We refer to -3 as the *maximum value* for this quadratic function and say that it occurs when $x = 1$.

In general, calculus is needed to find the maximum and minimum values of a function. However, if the function is quadratic, we may use the vertex formula for a parabola to find the maximum or minimum value of the function.

EXAMPLE 4 Determine if the function $g(x) = 3x^2 + 6x - 8$ has a maximum or minimum value. Then find this value.

◆ **Solution** Since the function g is quadratic, we can use the vertex formula for a parabola to determine the maximum or minimum value.

FIGURE 3.34

The quadratic function
$g(x) = 3x^2 + 6x - 8$ has a
minimum value of -11, and this
occurs when $x = -1$.

Since $a = 3$ (so $a > 0$), we know the parabola opens upward. Thus, g has a *minimum value* at

$$x = -\frac{b}{2a} = -\frac{6}{2(3)} = -1.$$

To find the minimum value, we replace x with -1 and evaluate $g(-1)$:

$$g(-1) = 3(-1)^2 + 6(-1) - 8 = -11$$

Thus, the minimum value for the quadratic function $g(x) = 3x^2 + 6x - 8$ is -11, and it occurs when $x = -1$. Figure 3.34 shows the graph of the function g in the viewing rectangle of a graphing calculator. The coordinates of the vertex of the parabola are $(-1, -11)$; this fact confirms our work. ◆

PROBLEM 4   Determine whether the function defined by the equation $y = 3 - 3x^2$ has a minimum or maximum value. Then find this value. ◆

◆ Application: Maximizing the Area of a Garden

The idea of maximum and minimum values occurs in many applied problems. We illustrate one such application in our next example.

EXAMPLE 5   One hundred feet of fencing is to be used to enclose a rectangular garden that abuts a barn. No fencing is needed along the barn. What is the *largest* possible area that can be enclosed?

◆ **Solution**   In this problem we want the *area* to be the *largest* possible. Thus, we express the area as a quadratic function and then find its *maximum value* by using the vertex formula for a parabola. We begin by letting

w = the width of the rectangle   and l = the length of the rectangle

FIGURE 3.35

A rectangular garden of length l and width w with no fencing along the barn

as shown in Figure 3.35. Then the area A of the rectangle is

$$A = lw.$$

Since 100 ft of fencing is available to enclose the rectangular garden, we have

$$100 = l + 2w \quad \text{or} \quad l = 100 - 2w.$$

Substituting $100 - 2w$ for l in the area formula $A = lw$, we obtain

$$A = (100 - 2w)w$$
$$A = 100w - 2w^2$$

A quadratic function with $a = -2$, $b = 100$, and $c = 0$

Since $a = -2$ (so $a < 0$), we know that the graph of $A = 100w - 2w^2$ is a parabola that opens downward. Thus, this quadratic function has a maximum value that occurs at

$$w = -\frac{b}{2a} = -\frac{100}{2(-2)} = 25.$$

Replacing w with 25 in $A = 100w - 2w^2$, we obtain

$$A = 100(25) - 2(25)^2 = \underbrace{1250}_{\text{maximum value}}.$$

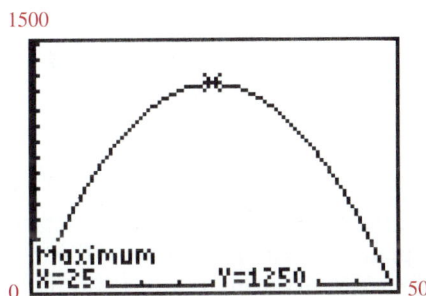

Thus, the largest possible area that can be enclosed is 1250 square feet. This occurs when the width w is 25 ft and the length l is $100 - 2(25)$, or 50 ft. Figure 3.36 shows the graph of $A = 100w - 2w^2$ for $0 < w < 50$ in the viewing rectangle of a graphing calculator. The x-axis denotes the width w of the rectangular garden and the y-axis denotes the area A of the garden. Note that the coordinates of the vertex of the parabola are $(25, 1250)$; this fact confirms our work. ◆

FIGURE 3.36

The quadratic function $A = 100w - 2w^2$ has a maximum value of 1250, and this occurs when $w = 25$.

PROBLEM 5 Suppose 100 ft of fencing is to be used to enclose a rectangular garden on all four sides. What is the *largest* possible area that can be enclosed? ◆

◆ **Application: Law of Motion for Freely Falling Objects**

Nearly four hundred years ago, Galileo discovered the *law of motion for freely falling objects:* If air resistance is not considered for a projectile that is fired straight upward with an initial velocity v_0 from an initial height s_0, then its distance s above the ground t seconds later is given by

$$s = -\tfrac{1}{2}gt^2 + v_0 t + s_0$$

where g is the acceleration due to gravity (approximately 32 ft/s^2, or 9.8 m/s^2). This equation defines a quadratic function, and we can use it to find maximum and minimum values.

EXAMPLE 6 The roof in an enclosed baseball stadium is 150 ft above home plate. A baseball player standing at home plate hits a pop fly straight upward with an initial velocity of 96 ft/s from an initial height of 4 ft. Does the ball hit the roof?

◆ **Solution** Since the measurements are given in feet, we use $g = 32$ ft/s^2 with $v_0 = 96$ ft/s and $s_0 = 4$ ft. Hence, the quadratic function that defines the distance s of the ball above the ground is

$$s = -16t^2 + 96t + 4.$$

Since this equation defines a quadratic function, we can use the vertex formula for a parabola to determine the maximum distance above the ground that the ball travels. Since $a = -16$ (so $a < 0$), we know the parabola opens downward and, therefore, has a maximum value. The maximum value occurs when

$$t = -\frac{b}{2a} = -\frac{96}{2(-16)} = 3 \text{ seconds.}$$

To find the maximum distance above the ground, we replace t with 3 in the equation $s = -16t^2 + 96t + 4$ and evaluate s:

$$s = -16(3)^2 + 96(3) + 4 = 148 \text{ ft}$$

Hence, the maximum distance above the ground that the ball travels is 148 ft. Since the roof is 150 ft above home plate, the ball misses hitting the roof by 2 ft. The graph in Figure 3.37 verifies our work. This graph is not the actual path of the ball (which is hit straight upward), but rather the relationship between time and height. ◆

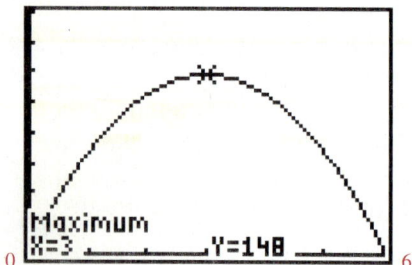

FIGURE 3.37

The quadratic function
$s = -16t^2 + 96t + 4$ has a maximum value of 148, and this occurs when $t = 3$.

PROBLEM 6   Refer to Example 6. Suppose the catcher catches the ball 4 feet above the ground. How long is the ball in flight? ◆

Exercises 3.3

◆ Basic Skills

In Exercises 1–24, determine the vertex and y-intercept for the graph of each quadratic function. Then sketch the graph of the function. Use a graphing calculator to confirm the graph.

1. $y = (x - 1)^2 + 3$

2. $y = (x + 2)^2 - 1$

3. $f(x) = -(x + 3)^2 + 4$

4. $F(x) = -2(x - 1)^2 - 2$

5. $y = 3 - 3(x + 1)^2$

6. $y = \frac{1}{4}(x - 4)^2$

7. $G(x) = \frac{1}{2}(x - 4)^2 - 1$

8. $y = 2\left(x - \frac{1}{2}\right)^2 - \frac{3}{2}$

9. $y = x^2 - 1$

10. $F(x) = 2 - \frac{1}{2}x^2$

11. $f(x) = 4x - x^2$

12. $g(x) = x^2 - 6x$

13. $y = x^2 + 2x - 3$

14. $G(x) = x^2 + 8x + 12$

15. $f(x) = 5 + 4x - x^2$

16. $y = -x^2 - 6x + 7$

17. $y = x^2 + x - 6$

18. $f(x) = 2x^2 - 3x - 5$

19. $h(x) = -2x^2 + 8x - 3$

20. $y = 3x^2 - 6x + 1$

21. $y = 3x^2 - 2x + 4$

22. $y = 3 + 5x - 5x^2$

23. $y = 1 + 4x + 4x^2$

24. $H(x) = 4x^2 - 12x + 9$

In Exercises 25–32, determine the equation of the quadratic function in general form whose graph satisfies the given conditions. Then generate the graph of the equation on a graphing calculator and verify that these conditions have been satisfied.

25. Vertex at $(0, 0)$ and passing through $(2, 1)$

26. Vertex at the origin and passing through $(-1, -3)$

27. Vertex $(0, 2)$ and x-intercepts ± 2

28. Vertex $(0, -4)$ and passing through $(1, 4)$

29. Vertex $(3, 2)$ and passing through the origin

30. Vertex $(2, -1)$ and passing through $(-1, -2)$

31. Vertex $\left(-1, \frac{4}{3}\right)$ and y-intercept 2

32. Vertex $\left(-\frac{3}{4}, -\frac{49}{8}\right)$ and passing through $(-1, 0)$

In Exercises 33–38, determine if the quadratic function has a maximum or minimum value. Then find this value. Use a graphing calculator to verify your answer.

33. $y = 8x - x^2$

34. $f(x) = x^2 + 5x$

35. $G(x) = 4x^2 + 6x + 3$

36. $y = -2x^2 - 4x + 8$

37. $h(x) = -3(x - 7)^2 + 9$

38. $F(x) = 5(x + 4)^2 - 11$

In Exercises 39–42, use a graphing calculator to find the coordinates of the vertex for the graph of each quadratic function. Round each coordinate to three significant digits. Check algebraically by using the vertex formula for a parabola.

39. $f(x) = 12.5x^2 - 18.6x + 10.7$

40. $f(x) = 1.24x^2 + 4.56x - 9.55$

41. $f(x) = -0.024x^2 + 0.126x + 1.344$

42. $f(x) = 107 - 425x - 121x^2$

In Exercises 43 and 44, use the law of motion for freely falling objects given on page 179.

43. A ball is thrown vertically upward with an initial velocity of 64 ft/s from a height of 80 ft above the ground.
  (a) Use the law of motion to express the distance s (in feet) of the ball above the ground after t seconds.
  (b) When does the ball reach its maximum height?
  (c) What is the ball's maximum height?
  (d) How long is the ball in flight?

44. A model rocket (without its parachute) is fired directly upward with an initial velocity of 147 m/s from ground level.
  (a) Use the law of motion to express the distance s (in meters) of the rocket above the ground after t seconds.
  (b) When does the rocket reach its maximum height?
  (c) What is the rocket's maximum height?
  (d) How long is the rocket in flight?

45. The daily manufacturing cost C (in dollars) for a company that makes skis is given by

$$C(x) = x^2 - 100x + 4800,$$

where x is the number of pairs of skis produced per day.
  (a) How many pairs of skis should be produced to minimize the daily manufacturing cost?
  (b) What is the minimum daily manufacturing cost?

46. The daily profit P (in dollars) for a company that makes tennis rackets is given by

$$P(x) = -x^2 + 240x - 5400,$$

where x is the number of rackets produced per day.
  (a) How many tennis rackets should be produced to maximize the daily profit?
  (b) What is the maximum daily profit?

47. Find two numbers such that their sum is 124 and their product is the largest possible.

48. Find two numbers such that their sum is -30 and the sum of their squares is a minimum.

49. Of all rectangles with a perimeter of 44 inches, what are the dimensions of the one with the maximum area?

50. A long piece of sheet metal 12 inches wide is to be made into a rain gutter by bending up two long edges to make straight sides. The gutter will have a rectangular cross-sectional area, as shown in the figure. What depth will give the greatest water carrying capacity?

51. Two hundred feet of fencing is to be used to enclose two identical rectangular animal pens that abut a barn, as shown in the figure. No fencing is needed along the barn. What dimensions for the total enclosure make the area of the pens the largest possible?

52. A football stadium is to be built in the shape of a rectangle with semicircular ends, as shown in the figure at the top of the next page. If the perimeter of the stadium is 1 mile, find the dimensions of x and r so that the area of the rectangular part of the stadium is the largest possible.

53. For the electrical circuit shown in the sketch, the power P (in watts) delivered to the load resistance R_L (in ohms) is given by

$$P = 21.6i - 0.34i^2$$

where i is the current (in amperes) flowing through the circuit. What is the maximum power that can be delivered to the load resistance?

54. A wire 21.6 cm long is cut into two pieces. One of the pieces is bent into a square and the other piece is bent into a circle. Where should the wire be cut if the total area of the square and circle is to be a minimum?

55. A bus service between Plymouth and Boston charges $8 per person for a one-way trip and carries 400 passengers per day. Research shows that for each $1 increase in fare the company loses 25 passengers per day.

(a) What daily rate gives the company the maximum revenue?

(b) What is this maximum revenue?

56. A tutor charges $30 per hour for math tutoring and averages 21 students per week in one-hour sessions. The tutor has found that for each $2 decrease in the hourly rate, 3 more students per week request tutoring services.

(a) What hourly rate should the tutor charge for maximum revenue?

(b) What is this maximum revenue?

◆ **Critical Thinking**

57. If a parabola has two distinct x-intercepts a and b, what is the x-coordinate of its vertex?

58. Find all values of k such that the vertex of the graph of $y = x^2 + kx + 9$ lies on the x-axis.

59. Determine the quadratic function f with zeros -1 and 3 and range $[-2, \infty)$.

60. If the vertex of the graph of $y = 2x^2 + bx + c$ has coordinates $(3, 8)$, find the values of b and c.

61. If the graph of $f(x) = ax^2 + bx + c$ passes through the origin, what is the value of c?

62. Consider the linear function $f(x) = mx + b$ and the quadratic function $g(x) = a(x - h)^2 + k$ with $m \neq 0$ and $a \neq 0$.

(a) Find $f \circ g$.

(b) Find $g \circ f$.

(c) Classify the functions $f \circ g$ and $g \circ f$ as linear, quadratic, or neither.

63. Use a graphing calculator to generate the five graphs of $y = x^2 + 4x + c$ for $c = -2, -1, 0, 1,$ and 2 in the same viewing rectangle. What effect does the constant c have on the axis of symmetry of these parabolas?

64. Use a graphing calculator to generate the four graphs of $y = a(x - 2)^2 + 1$ for $a = -2, -1, 1,$ and 2 in the same viewing rectangle. What effect does the constant a have on the vertex of these parabolas?

3.4 **Quadratic Equations and Inequalities**

◆ **Introductory Comments**

Any equation of the form

$$ax^2 + bx + c = 0$$

where x is the variable and a, b, and c are any real numbers such that $a \neq 0$, is called a **quadratic equation in standard form**. If we replace the equal sign with one of the inequality symbols ($<$, $>$, $\leq$, or $\geq$), we have a **quadratic inequality in standard form**. In this section, we discuss methods for solving quadratic equations and quadratic inequalities.

◆ Solving Quadratic Equations

Some quadratic equations can be solved easily by factoring. This method relies on the zero product property (in Section 2.2). For example, to solve the quadratic equation $x^2 - x = 12$, we may proceed as follows:

$$x^2 - x = 12$$

$$x^2 - x - 12 = 0 \qquad \text{Write in standard form}$$

$$(x - 4)(x + 3) = 0 \qquad \text{Factor the trinomial}$$

$$x - 4 = 0 \quad \text{or} \quad x + 3 = 0 \qquad \text{Apply the zero product property}$$

$$x = 4 \qquad\qquad x = -3 \qquad \text{Solve for } x$$

Thus, the roots of $x^2 - x = 12$ are 4 and -3.

CAUTION The quadratic equation must be in *standard form* before we factor and apply the zero product property. To write

$$x^2 - x = 12$$
$$x(x - 1) = 12$$
$$x = 12 \quad \text{or} \quad x - 1 = 12$$
$$x = 13 \quad \text{is WRONG!}$$

From our preceding work, we know that the solutions are 4 and -3, *not* 12 and 13.

Recall from Section 2.4 that the concepts of real roots of equations, x-intercepts of graphs, and real zeros of functions are all closely related. In the preceding example, we see that the factors of $x^2 - x - 12$, namely $(x - 4)$ and $(x + 3)$, are also closely related to these concepts. The roots of the equation $x^2 - x = 12$ are 4 and -3, the x-intercepts of the graph of $f(x) = x^2 - x - 12$ are 4 and -3 (see Figure 3.38), the real zeros of the function f are 4 and -3, and the factors of $x^2 - x - 12$ are $(x - 4)$ and $(x + 3)$. In general, if f is a quadratic function and k is a real number, then the following statements are equivalent:

1. k is a real root of the equation $f(x) = 0$.
2. k is an x-intercept of the graph of the function f.
3. k is a real zero of f.
4. $x - k$ is a factor of $f(x)$.

FIGURE 3.38

The roots of the equation $x^2 - x = 12$ are the x-intercepts of the graph of $f(x) = x^2 - x - 12$.

EXAMPLE 1 Write each equation in the standard quadratic form $ax^2 + bx + c = 0$. Then factor and solve by using the zero product property.

(a) $(x - 3)(x - 10) = x - 10$ **(b)** $\dfrac{x - 2}{8} + 16 = \dfrac{x^2 - 3}{2}$

◆◆ **Solution**

(a) To solve the equation $(x - 3)(x - 10) = x - 10$, we proceed as follows:

$$(x - 3)(x - 10) = x - 10$$

$$x^2 - 13x + 30 = x - 10 \qquad \textcolor{red}{\textbf{Multiply}}$$

$$x^2 - 14x + 40 = 0 \qquad \textcolor{red}{\textbf{Write in standard form}}$$

$$(x - 4)(x - 10) = 0 \qquad \textcolor{red}{\textbf{Factor the trinomial}}$$

$$x - 4 = 0 \quad \text{or} \quad x - 10 = 0 \qquad \textcolor{red}{\textbf{Apply the zero product property}}$$

$$x = 4 \qquad\qquad x = 10$$

Thus, the roots of the original equation are 4 and 10. Figure 3.39 shows the graph of $f(x) = (x - 3)(x - 10) - x + 10$ in the viewing rectangle of a calculator. Note that the graph is a parabola with x-intercepts 4 and 10.

(b) To solve the equation $\dfrac{x - 2}{8} + 16 = \dfrac{x^2 - 3}{2}$, we multiply both sides of the equation by 8 (the least common denominator), cancel the denominators, and then write the equation in standard form:

$$\frac{x - 2}{8} + 16 = \frac{x^2 - 3}{2}$$

$$8\left(\frac{x - 2}{8} + 16\right) = 8\left(\frac{x^2 - 3}{2}\right) \qquad \textcolor{red}{\textbf{Multiply both sides by 8}}$$

$$(x - 2) + 128 = 4(x^2 - 3) \qquad \textcolor{red}{\textbf{Simplify by canceling denominators}}$$

$$4x^2 - x - 138 = 0 \qquad \textcolor{red}{\textbf{Write in standard form}}$$

To help factor $4x^2 - x - 138$, we can look at the x-intercepts of the graph of $f(x) = 4x^2 - x - 138$. The graph in Figure 3.40 indicates that one of the x-intercepts is 6. If 6 is an x-intercept, then $(x - 6)$ is one of the factors, and we continue as follows:

$$4x^2 - x - 138 = 0$$

$$(x - 6)(4x + 23) = 0 \qquad \textcolor{red}{\textbf{Factor the trinomial}}$$

$$x - 6 = 0 \quad \text{or} \quad 4x + 23 = 0 \qquad \textcolor{red}{\textbf{Apply the zero product property}}$$

$$x = 6 \qquad\qquad x = -\tfrac{23}{4}$$

Thus, the roots of the original equation are 6 and $-\tfrac{23}{4}$.   ◆

PROBLEM 1   Write the equation $x^2 - x - 12 = 3(x + 2)(x - 2)$ in standard quadratic form. Then factor and solve by using the zero product property.   ◆

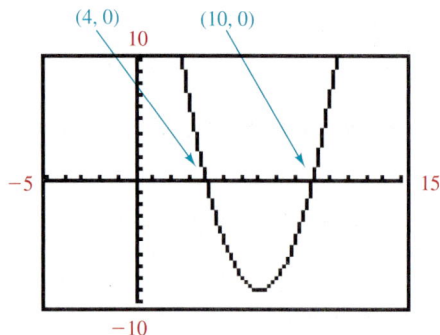

FIGURE 3.39

The roots of the equation $(x - 3)(x - 10) = x - 10$ are the x-intercepts of the graph of $f(x) = (x - 3)(x - 10) - x + 10$.

FIGURE 3.40

If 6 is an x-intercept of the graph of $f(x) = 4x^2 - x - 138$, then $(x - 6)$ is a factor of $4x^2 - x - 138$.

We can solve the quadratic equation $ax^2 + bx + c = 0$ by the zero product property only if we recognize the factors of the trinomial $ax^2 + bx + c$. The method of completing the square (in Section 3.3) enables us to derive a formula that can be used to solve a quadratic equation whether or not we recognize the factors.

Beginning with $ax^2 + bx + c = 0$, we proceed as follows:

$$ax^2 + bx + c = 0$$

$$ax^2 + bx = -c \qquad \text{Subtract } c \text{ from both sides}$$

$$x^2 + \frac{b}{a}x = -\frac{c}{a} \qquad \text{Divide both sides by } a$$

$$x^2 + \frac{b}{a}x + \left(\frac{b}{2a}\right)^2 = \left(\frac{b}{2a}\right)^2 - \frac{c}{a} \qquad \text{Complete the square by adding } (b/2a)^2 \text{ to both sides}$$

$$\boxed{(\text{half of } b/a)^2}$$

$$\left(x + \frac{b}{2a}\right)^2 = \frac{b^2 - 4ac}{4a^2} \qquad \text{Factor the perfect square trinomial and add fractions on the right-hand side}$$

$$x + \frac{b}{2a} = \pm\sqrt{\frac{b^2 - 4ac}{4a^2}} \qquad \text{Extract the square roots}$$

$$x + \frac{b}{2a} = \pm\frac{\sqrt{b^2 - 4ac}}{2a} \qquad \text{Simplify the radical expression}$$

$$x = \frac{-b \pm \sqrt{b^2 - 4ac}}{2a} \qquad \text{Add } -b/2a \text{ to both sides}$$

This last equation represents a formula that can be used to find the roots of *any* quadratic equation that is written in standard form. We refer to this formula as the *quadratic formula*.

◆ **Quadratic Formula**

If $ax^2 + bx + c = 0$, provided $a \neq 0$, then

$$x = \frac{-b \pm \sqrt{b^2 - 4ac}}{2a}.$$

In the quadratic formula, the quantity $b^2 - 4ac$ is called the **discriminant**. It indicates the *nature of the roots* of the quadratic equation $ax^2 + bx + c = 0$ whenever a, b, and c are real numbers. We consider three cases.

Case 1 If $b^2 - 4ac > 0$, then $\sqrt{b^2 - 4ac}$ is a *real number* and the quadratic formula gives two distinct real roots:

$$\frac{-b + \sqrt{b^2 - 4ac}}{2a} \qquad \text{and} \qquad \frac{-b - \sqrt{b^2 - 4ac}}{2a}$$

If *a*, *b*, and *c* are rational numbers and $b^2 - 4ac$ is the square of a rational number, then these two distinct real roots are rational numbers. Otherwise, the roots are irrational numbers.

Case 2   If $b^2 - 4ac < 0$, then $\sqrt{b^2 - 4ac}$ is an *imaginary number* (see Section 1.4) and the quadratic formula gives two complex conjugate roots:

$$-\frac{b}{2a} + \frac{\sqrt{|b^2 - 4ac|}}{2a}i \qquad \text{and} \qquad -\frac{b}{2a} - \frac{\sqrt{|b^2 - 4ac|}}{2a}i$$

Case 3   If $b^2 - 4ac = 0$, then $\sqrt{b^2 - 4ac}$ is *zero* and the quadratic formula gives only one real root:

$$-\frac{b}{2a}$$

The quantity $-\dfrac{b}{2a}$ is called a **repeated root of multiplicity two**.

We now summarize the nature of the roots of the quadratic equation $ax^2 + bx + c = 0$ with real numbers *a*, *b*, and *c*:

Discriminant	Nature of the roots	Typical Graph
$b^2 - 4ac > 0$	Two distinct real roots	
$b^2 - 4ac < 0$	Two complex conjugate roots	
$b^2 - 4ac = 0$	One real root of multiplicity two	

EXAMPLE 2   Solve each quadratic equation by using the quadratic formula.

(a) $x^2 + 6x = 18$ **(b)** $x^2 + 5 = 2x$ **(c)** $2x^2 - 2\sqrt{10}\,x + 5 = 0$

◆ **Solution**

(a) Before identifying *a*, *b*, and *c*, we write $x^2 + 6x = 18$ in standard form:

$$x^2 + 6x - 18 = 0$$

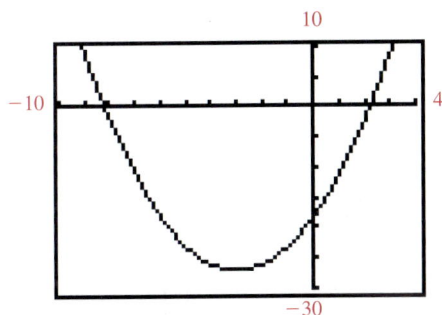

Now, $a = 1$, $b = 6$, and $c = -18$. Using the quadratic formula, we have

$$x = \frac{-6 \pm \sqrt{6^2 - 4(1)(-18)}}{2(1)} = \frac{-6 \pm \sqrt{108}}{2}$$

$$= \frac{-6 \pm 6\sqrt{3}}{2} \qquad \text{Simplify the radical}$$

$$= \frac{2(-3 \pm 3\sqrt{3})}{2} \qquad \text{Factor}$$

$$= -3 \pm 3\sqrt{3} \qquad \text{Reduce}$$

Thus, the two real roots are $-3 + 3\sqrt{3} \approx 2.20$ and $-3 - 3\sqrt{3} \approx -8.20$. Figure 3.41 shows the graph of $f(x) = x^2 + 6x - 18$ in the viewing rectangle of a graphing calculator. Note that the x-intercepts of the parabola are approximately 2.20 and -8.20.

(b) We begin by writing $x^2 + 5 = 2x$ in standard form:

$$x^2 - 2x + 5 = 0$$

Therefore, $a = 1$, $b = -2$, and $c = 5$. Using the quadratic formula, we have

$$x = \frac{-(-2) \pm \sqrt{(-2)^2 - 4(1)(5)}}{2(1)} = \frac{2 \pm \sqrt{-16}}{2}$$

$$= \frac{2 + 4i}{2} \qquad \text{Simplify the radical}$$

$$= \frac{2(1 \pm 2i)}{2} \qquad \text{Factor}$$

$$= 1 \pm 2i \qquad \text{Reduce}$$

Thus, the roots are the complex conjugates $1 + 2i$ and $1 - 2i$. Figure 3.42 shows the graph of $f(x) = x^2 - 2x + 5$ in the viewing rectangle of a graphing calculator. Note that the parabola has no x-intercept. This indicates that the roots of the quadratic equation are not real numbers.

(c) For $2x^2 - 2\sqrt{10}\,x + 5 = 0$, we have $a = 2$, $b = -2\sqrt{10}$, and $c = 5$. Using the quadratic formula, we find

$$x = \frac{-(-2\sqrt{10}) \pm \sqrt{(-2\sqrt{10})^2 - 4(2)(5)}}{2(2)} = \frac{2\sqrt{10} \pm \sqrt{0}}{4}$$

$$= \frac{2\sqrt{10}}{4}$$

$$= \frac{\sqrt{10}}{2} \approx 1.58$$

We refer to $\sqrt{10}/2$ as a repeated root of multiplicity two. Figure 3.43 shows the graph of $f(x) = 2x^2 - 2\sqrt{10}\,x + 5$ in the viewing rectangle

FIGURE 3.41

The x-intercepts of the graph of $f(x) = x^2 + 6x - 18$ are the roots of the equation $x^2 + 6x = 18$.

FIGURE 3.42

The graph of $f(x) = x^2 - 2x + 5$ has no x-intercept. This indicates that the roots of the equation $x^2 + 5 = 2x$ are not real numbers.

FIGURE 3.43

The graph of $f(x) = 2x^2 - 2\sqrt{10}\,x + 5$ is a parabola with its vertex on the x-axis. This indicates that $2x^2 - 2\sqrt{10}\,x + 5 = 0$ has one root of multiplicity two.

of a graphing calculator. Note that the vertex of the parabola is the only x-intercept of the graph. This indicates that the quadratic equation has one real root of multiplicity two.   ◆

To check an imaginary root, like either of the complex conjugates in Example 2(b), we replace x in the original equation with the imaginary root and perform the indicated operations using the fact that $i^2 = -1$.

PROBLEM 2   Check the solutions to the quadratic equation in Example 2(b).   ◆

◆ **Solving Quadratic Inequalities**

The inequalities

$$x^2 - x - 6 < 0 \quad \text{and} \quad x^2 - x - 6 > 0$$

are examples of *quadratic inequalities in standard form.* We can use the graph of the related quadratic function $f(x) = x^2 - x - 6$ to help us find the solution set of each of these inequalities. We know that the graph of the quadratic function f is a parabola. The parabola opens upward and the coordinates of its vertex are

$$\left(-\frac{b}{2a}, f\left(-\frac{b}{2a}\right)\right) = \left(-\frac{1}{2}, -\frac{25}{4}\right).$$

The y-intercept is $f(0) = -6$, and the x-intercepts may be found by factoring and applying the zero product property as follows:

$$x^2 - x - 6 = 0$$
$$(x + 2)(x - 3) = 0$$
$$x + 2 = 0 \quad \text{or} \quad x - 3 = 0$$
$$x = -2 \qquad x = 3$$

The graph of $f(x) = x^2 - x - 6$ is shown in Figure 3.44.

We refer to the x-intercepts, -2 and 3, as the **boundary numbers** of the inequality $x^2 - x - 6 > 0$, since these are the values of x for which $x^2 - x - 6$ is *neither positive nor negative.* As illustrated in Figure 3.44, these two boundary numbers divide the x-axis into three intervals:

$$(-\infty, -2) \qquad (-2, 3) \qquad (3, \infty)$$

The graph of $f(x) = x^2 - x - 6$ is *below* the x-axis in the interval $(-2, 3)$. Hence, $f(x) < 0$ in this interval. Thus, the solution set of the inequality $x^2 - x - 6 < 0$ is

$$(-2, 3)$$

The graph of $f(x) = x^2 - x - 6$ is *above* the x-axis in the interval $(-\infty, -2)$

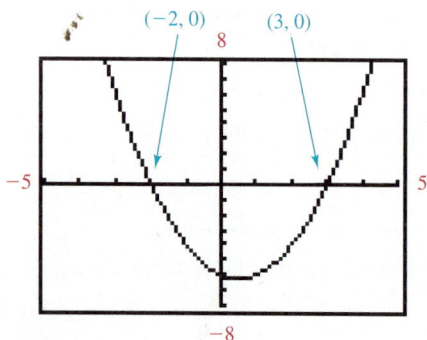

FIGURE 3.44

The x-intercepts of the graph of $f(x) = x^2 - x - 6$ divide the x-axis into three intervals: $(-\infty, -2)$, $(-2, 3)$, and $(3, \infty)$.

or $(3, \infty)$. Hence, $f(x) > 0$ in these intervals. Thus, the solution set of the inequality $x^2 - x - 6 > 0$ is

$$(-\infty, -2) \cup (3, \infty)$$

If an inequality contains the symbols $\leq$ or $\geq$, we include its boundary numbers as part of the solution set. For example, the solution set of $x^2 - x - 6 \leq 0$ is $[-2, 3]$, and the solution set of $x^2 - x - 6 \geq 0$ is $(-\infty, -2] \cup [3, \infty)$.

EXAMPLE 3　Solve each quadratic inequality.

(a) $2x^2 + 3x < 20$　　**(b)** $2x + 9 \leq x^2$

◆ **Solution**

(a) We begin by subtracting 20 from both sides of the inequality and writing it in standard form:

$$2x^2 + 3x - 20 < 0$$

The boundary numbers of this inequality are the x-intercepts of the graph of $f(x) = 2x^2 + 3x - 20$. We can find these boundary numbers by solving the equation $f(x) = 0$:

$$2x^2 + 3x - 20 = 0$$

$$(2x - 5)(x + 4) = 0 \qquad \textbf{Factor}$$

$$2x - 5 = 0 \quad \text{or} \quad x + 4 = 0 \qquad \textbf{Apply the zero product property}$$

$$x = \tfrac{5}{2} \qquad\qquad x = -4$$

boundary numbers

These two boundary numbers divide the x-axis into three intervals:

Interval 1　　　　　Interval 2　　　　　Interval 3
$(-\infty, -4)$　　　　$(-4, \tfrac{5}{2})$　　　　$(\tfrac{5}{2}, \infty)$

-5　-4　-3　-2　-1　0　1　2　$\tfrac{5}{2}$　3　4 → x

We can confirm these intervals by generating the graph of $f(x) = 2x^2 + 3x - 20$ in the viewing rectangle of a graphing calculator. As shown in Figure 3.45, the graph of f is *below* the x-axis in the interval $\left(-4, \tfrac{5}{2}\right)$. Hence, $f(x) < 0$ in this interval. Thus, the solution set of the inequality $2x^2 + 3x - 20 < 0$ is $\left(-4, \tfrac{5}{2}\right)$.

(b) We begin writing the inequality in standard form with a nonnegative x^2 term:

$$2x + 9 \leq x^2$$

$$-x^2 + 2x + 9 \leq 0 \qquad \textbf{Subtract } x^2 \textbf{ from both sides}$$

$$x^2 - 2x - 9 \geq 0 \qquad \textbf{Multiply both sides by } -1 \textbf{ and reverse the inequality}$$

FIGURE 3.45

The graph of $f(x) = 2x^2 + 3x - 20$ is below the x-axis in the interval $\left(-4, \tfrac{5}{2}\right)$.

The boundary numbers of this inequality are the x-intercepts of the graph of $f(x) = x^2 - 2x - 9$. We can find these boundary numbers by solving the equation $x^2 - 2x - 9 = 0$. Since $x^2 - 2x - 9$ is not factorable over the set of integers, we solve this equation by using the quadratic formula:

$$x = \frac{-(-2) \pm \sqrt{(-2)^2 - 4(1)(-9)}}{2(1)} = 1 \pm \sqrt{10}$$

Hence, the two boundary numbers are $1 + \sqrt{10} \approx 4.2$ and $1 - \sqrt{10} \approx -2.2$. These two boundary numbers divide the x-axis into three intervals:

Interval 1 $(-\infty, 1 - \sqrt{10})$ Interval 2 $(1 - \sqrt{10}, 1 + \sqrt{10})$ Interval 3 $(1 + \sqrt{10}, \infty)$

$1 - \sqrt{10} \approx -2.2$ $1 + \sqrt{10} \approx 4.2$

We can confirm these intervals by generating the graph of $f(x) = x^2 - 2x - 9$ in the viewing rectangle of a graphing calculator. As shown in Figure 3.46, the graph of f is *above* the x-axis in the interval $\left(-\infty, 1 - \sqrt{10}\right)$ or $\left(1 + \sqrt{10}, \infty\right)$. Hence, $f(x) \geq 0$ in the interval $\left(-\infty, 1 - \sqrt{10}\right]$ or $\left[1 + \sqrt{10}, \infty\right)$. Thus, the solution set of the inequality $x^2 - 2x - 9 \geq 0$ is

$$\left(-\infty, 1 - \sqrt{10}\right] \cup \left[1 + \sqrt{10}, \infty\right).$$

FIGURE 3.46

The graph of $f(x) = x^2 - 2x - 9$ is above the x-axis in the intervals $\left(-\infty, 1 - \sqrt{10}\right)$ and $\left(1 + \sqrt{10}, \infty\right)$.

Look at the graphs in Figures 3.45 and 3.46: In each of the intervals formed by the boundary numbers, the graph of $f(x) = ax^2 + bx + c$ is entirely above the x-axis or entirely below the x-axis. We can use this fact to help solve a quadratic inequality in standard form without referring to a graph. We simply select an arbitrary *test number* from each interval formed by the boundary numbers, and then determine the algebraic sign of the quadratic expression $ax^2 + bx + c$ at this number. If the quadratic expression is negative for a particular test number in an interval, then it is negative for every value of x in that interval. If the quadratic expression is positive for a particular test number in an interval, then it is positive for every value of x in that interval. We refer to this algebraic procedure of solving a quadratic inequality as the **test number method**.

For example, to solve the quadratic inequality $2x^2 + 3x - 20 < 0$ in Example 3(a) by using the test number method, we select an arbitrary test number from each of the intervals $(-\infty, -4)$, $\left(-4, \frac{5}{2}\right)$, and $\left(\frac{5}{2}, \infty\right)$, and determine the algebraic sign of the quadratic expression $2x^2 + 3x - 20$ at each test number. The following table organizes our work:

Interval	Test Number	Sign of $2x^2 + 3x - 20$
$(-\infty, -4)$	-5	$2(-5)^2 + 3(-5) - 20 = 15 > 0$
$\left(-4, \frac{5}{2}\right)$	1	$2(1)^2 + 3(1) - 20 = -15 < 0$
$\left(\frac{5}{2}, \infty\right)$	3	$2(3)^2 + 3(3) - 20 = 7 > 0$

Since $2x^2 + 3x - 20 < 0$ when $x = 1$, we know that $2x^2 + 3x - 20 < 0$ for every value of x in the interval $\left(-4, \frac{5}{2}\right)$. Hence, we conclude that the solution set of this inequality is $\left(-4, \frac{5}{2}\right)$, which agrees with our work in Example 3(a).

PROBLEM 3 Solve the quadratic inequality $x^2 < 16x$ by using the test number method. ◆

◆ **Application: Parking Lot Addition**

Many applied problems lead to quadratic equations. In the next example, we illustrate how a construction company uses such equations to determine a uniform width that doubles the size of an existing rectangular parking lot.

EXAMPLE 4 An existing rectangular parking lot is 100 ft by 220 ft. A construction company is asked to double the size of this parking lot by adding an area of uniform width, as shown in Figure 3.47.

FIGURE 3.47

(a) What is the uniform width of the additional parking area?

(b) What are the dimensions and area of the new parking lot?

◆ **Solution**

(a) Suppose we let

x = the uniform width (in feet) of the additional parking area.

Then

$220 + 2x$ = the overall length (in feet) of the new parking lot

and

$100 + x$ = the overall width (in feet) of the new parking lot

as shown in Figure 3.48. Since the area of a rectangle is the product of its length and width, we have

FIGURE 3.48

If x is the uniform width, then the dimensions of the new parking lot are $(220 + 2x)$ ft by $(100 + x)$ ft.

| New length | · | New width | = | *Double* the existing area |

$$(220 + 2x) \cdot (100 + x) = 2(220)(100)$$

$$22{,}000 + 420x + 2x^2 = 44{,}000$$

$$2x^2 + 420x - 22{,}000 = 0 \qquad \text{Write in standard form}$$

$$x^2 + 210x - 11{,}000 = 0 \qquad \text{Divide both sides by 2}$$

To solve this quadratic equation, we apply the quadratic formula:

$$x = \frac{-210 \pm \sqrt{(210)^2 - 4(1)(-11{,}000)}}{2(1)} = \frac{-210 \pm \sqrt{88{,}100}}{2}$$

Using a calculator, we find that the approximate roots are

$$\frac{-210 + \sqrt{88{,}100}}{2} \approx 43.4 \qquad \text{and} \qquad \frac{-210 - \sqrt{88{,}100}}{2} \approx -253.4.$$

Since a width cannot be negative, we discard the negative solution, and we conclude that the uniform width of the additional parking area is 43.4 ft. Figure 3.49 shows the graph of $f(x) = x^2 + 210x - 11{,}000$ in the viewing rectangle of a graphing calculator. Note that the positive x-intercept of the parabola is approximately 43.4.

FIGURE 3.49

The positive x-intercept of the graph of $f(x) = x^2 + 210x - 11000$ is the uniform width of the additional parking area, 43.4 ft.

(b) The dimensions of the new parking lot (to the nearest tenth of a foot) are

$$\text{Width:} \quad 100 + x = 100 + 43.4 = 143.4 \text{ ft}$$

$$\text{Length:} \quad 220 + 2x = 220 + 2(43.4) = 306.8 \text{ ft}$$

Hence, the area of the new lot is

$$(143.4 \text{ ft})(306.8 \text{ ft}) = 43{,}995.12 \text{ sq ft} \approx \mathbf{44{,}000 \text{ sq ft}},$$

which is double the original area ($100 \times 220 = \mathbf{22{,}000 \text{ sq ft}}$). ◆

PROBLEM 4 Repeat Example 4 if the construction company is asked to double the size of the existing parking lot by adding an area of uniform width as shown in Figure 3.50.

Additional parking area

100 ft

Existing parking area

220 ft

FIGURE 3.50 ◆

◆ **Application: Hitting a Pop Fly**

Many applied problems lead to quadratic inequalities rather than quadratic equations. In the next example, we use the law of motion for freely falling objects (see Section 3.3) to determine the time interval when a pop fly is more than a given distance above the ground.

EXAMPLE 5 A baseball player hits a pop fly straight upward with an initial velocity of 96 ft/s from an initial height of 4 ft. What is the time interval when the baseball is more than 84 feet above the ground?

◆◆ **Solution** From Example 6 in Section 3.3, the quadratic function that defines the distance s of the ball above the ground is

$$s = -16t^2 + 96t + 4.$$

Figure 3.51 shows the graph of this quadratic function, along with the graph of the horizontal line $s = 84$. The time interval where the parabola is *above* the horizontal line $s = 84$ represents the time interval when the baseball is *more than* 84 ft above the ground. To determine this time interval algebraically, we solve the quadratic inequality

$$-16t^2 + 96t + 4 > 84.$$

Writing this inequality in standard form, we obtain

$$-16t^2 + 96t - 80 > 0 \qquad \text{Subtract 84 from both sides}$$

$$t^2 - 6t + 5 < 0 \qquad \text{Divide both sides by } -16 \text{ and } \textit{reverse the inequality}$$

$$\underbrace{}_{f(t)}$$

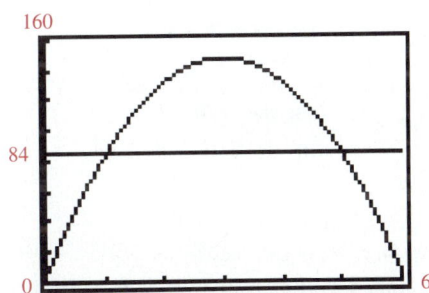

FIGURE 3.51
Graphs of $s = -16t^2 + 96t + 4$ and $s = 84$ in the viewing rectangle of a graphing calculator

We find the boundary numbers of this inequality by solving the equation $f(t) = 0$:

$$t^2 - 6t + 5 = 0$$

$$(t - 1)(t - 5) = 0 \qquad \text{Factor}$$

$$t - 1 = 0 \quad \text{or} \quad t - 5 = 0 \qquad \text{Apply the zero product property}$$

$$t = 1 \qquad\qquad t = 5$$

Hence, the baseball is more than 84 feet above the ground during the time interval from 1 second to 5 seconds.   ◆

PROBLEM 5   Generate the graphs in Figure 3.51 on a graphing calculator and trace to the points of intersection of the parabola and the line. Show that this graphical approach also gives the time interval $(1, 5)$.   ◆

Exercises 3.4

◆ Basic Skills

In Exercises 1–12, write each quadratic equation in the standard form $ax^2 + bx + c = 0$. Then factor and solve by using the zero product property. Check graphically.

1. $x^2 + 2x = 8$

2. $x^2 - 48 = 13x$

3. $6y^2 = 12y$

4. $3t^2 = t$

5. $x^2 - 36 = 0$

6. $100 - 9n^2 = 0$

7. $n(2n - 3) = 5$

8. $(x - 1)(3x + 2) = 2$

9. $(x + 4)(x + 6) = x + 6$

10. $(m - 7)(m + 4) = 2(m - 7)$

11. $y^2 + (y - 2)(y + 8) = 2y$

12. $(x + 3)^2 = x(3x - 1)$

In Exercises 13–24, write each quadratic equation in the standard form $ax^2 + bx + c = 0$. Then solve by using the quadratic formula. Check graphically.

13. $x^2 = 11x + 1$

14. $m^2 - 11 = 2m$

15. $20y^2 + y = 12$

16. $18x^2 + 3x = 10$

17. $5x^2 + 1 = 4x$

18. $2x^2 + 3 = 4x$

19. $(2x - 1)(x + 4) = (x + 1)^2$

20. $y(y - 3) = (2y + 3)(y - 1)$

21. $2(x^2 + 2) = \sqrt{7}\, x$

22. $3x^2 + 2(1 + \sqrt{6}\, x) = 0$

23. $1 - \dfrac{2n^2}{5} = \dfrac{1}{2}$

24. $\dfrac{2x}{3} = x^2 + \dfrac{1}{2}$

In Exercises 25–40, solve each quadratic equation by using any convenient method.

25. $4p^2 - 15 = 0$

26. $3x^2 - 25 = 0$

27. $4x^2 + 49 = 0$

28. $16x^2 + 75 = 0$

29. $(4y + 3)^2 = 32$

30. $(2m - 1)^2 + 24 = 0$

31. $n^2 - 10n + 25 = -18$

32. $4t^2 + 4t + 1 = 28$

33. $x^2 + 12x = 12$

34. $x^2 = 10x + 15$

35. $n^2 - 4n - 12 = 0$

36. $y^2 + 14y + 24 = 0$

37. $x^2 = 3x + 9$

38. $m^2 + 5m = 5$

39. $p^2 - \frac{4}{5}p = \frac{21}{25}$

40. $x^2 + \frac{3}{2}x = \frac{7}{16}$

In Exercises 41–52, solve each quadratic inequality. Write the solution set using interval notation.

41. $x^2 - 9 < 0$

42. $p^2 > 25$

43. $y^2 \geq 12y$

44. $x^2 + 20x \leq 0$

45. $x^2 - 8x \leq 9$

46. $y(y + 10) \geq 24$

47. $4x(x + 3) > 7$

48. $4x^2 + 9 < 20x$

49. $n^2 < 11n + 1$

50. $x^2 + 3x > 8 - 3x$

51. $(x - 1)^2 \geq (2x + 3)^2$

52. $(4n - 1)(2n + 5) \geq (3n + 1)^2$

In Exercises 53–56, use a graphing calculator to solve each equation. Write each solution as a decimal number rounded to three significant digits.

53. $3.81x^2 - 8.94x = 12.6$

54. $0.238x^2 + 0.342x - 1.14 = 0$

55. $872x^2 = 169x + 906$

56. $-968x^2 + 543 = 272x$

In Exercises 57–60, use a graphing calculator to solve each inequality. Describe each solution set using interval notation, and round each decimal number to three significant digits.

57. $27.93x^2 - 92.24 > 0$

58. $2.6x^2 - 5.4x \leq 9.2$

59. $42.6x^2 \leq 18.3x + 92.4$

60. $298x^2 < 987x - 500$

In Exercises 61–78, set up an equation or inequaltity, then solve.

61. Find two integers whose sum is 32 and whose product is 252.

62. Find two consecutive, positive even integers such that the sum of their squares is 340.

63. The area of a triangle is 54 square inches and its base is 3 inches shorter than its height. Find the base and height of the triangle.

64. A rectangle has a perimeter of 84 ft and a diagonal of 30 ft. Find the length and width of the rectangle.

65. After purchasing a coal-burning stove, a homeowner constructs a coal bin with rectangular sides to hold four tons of coal. The height of the coal bin is 4 feet and the length is 2 feet more than the width. Allowing 35 cubic feet for each ton of coal, find the dimensions of the coal bin.

66. A picture is encased in a wood frame of uniform width as shown in the sketch. The dimensions of the picture are 8 inches by 10 inches and the total area of both picture and frame is $131\frac{1}{4}$ square inches. Find the width of the frame.

67. A rectangular flower garden, 16.6 ft by 32.2 ft, is to be surrounded by a brick walkway of uniform width, as shown in the sketch. The contractor has enough brick to cover an area of 400 square feet. How wide should the walk be if all the bricks are to be used?

68. A 4-ft walkway surrounds a circular flower garden, as shown in the sketch. The area of the walk is 44% of the area of the garden. Find the radius of the garden.

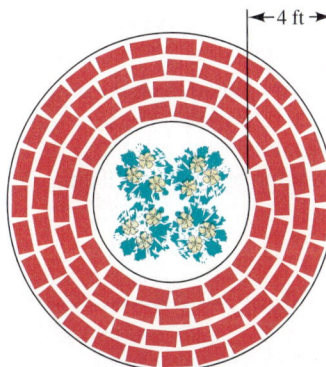

69. When the owners of an apartment complex charge rent of $600 per month for each unit, all 120 apartments are occupied. For each $20 increase in the monthly rent, 3 apartments become vacant. What should the owners of the complex charge for monthly rent if they wish to earn $73,500 per month from rental income?

70. A campus bookstore can order sweatshirts stamped with the university emblem for $12 per sweatshirt when less than 20 shirts are ordered from the supplier. If more than 20 but less than 200 sweatshirts are ordered, the price per shirt is reduced by 3 cents times the number of shirts ordered. How many sweatshirts can the bookstore order for $900?

71. The length of a rectangle is 2 meters more than its width, and its area is less than 99 square meters. What is the range of values for the length of the rectangle?

72. One leg of a right triangle is 3 inches longer than the other leg. If the hypotenuse is less than $\sqrt{39}$ inches, what is the range of values for the shorter leg?

73. Find all sets of three consecutive, even integers such that the square of the largest integer is greater than the sum of the squares of the other two integers.

74. Find all sets of two consecutive integers such that the square of the smaller integer is less than four times the larger integer.

75. When the selling price of a product is p dollars, a manufacturer will *supply* the market with $4p^2 - 5p$ units of their product and consumers will *demand* to buy $30 - p^2$ units. Under what conditions is supply greater than demand?

76. A highway safety group recommends the formula

$$d = 0.045v^2 + v$$

for determining the minimum safe distance d (in feet) that should be maintained between two cars traveling along a highway at a speed of v miles per hour (mi/h).

Exercise 76 continues

If the distance between two cars does not exceed 200 ft, find the interval of safe speeds.

77. If a rocket is fired straight upward from the ground with an initial velocity of 320 ft/s, what is the time interval during which the rocket is more than 1200 ft above the ground? (For the law of motion of freely falling objects, see Section 3.3).

78. If an arrow is shot straight upward from an initial height of 1.2 meters (m) with an initial velocity of 32.0 m/s, what are the time intervals when the arrow is less than 19.2 m above the ground? (For the law of motion of freely falling objects, see Section 3.3).

◆ Critical Thinking

In Exercises 79–84, solve each literal equation for x.

79. $ax^2 + bx = 0$

80. $a^2x^2 - b^2 = 0$

81. $x^2 + 2ax + a^2 = b^2$

82. $a^2x^2 + b^2 = 2abx$

83. $x^2 + 2bx - a^2 = 2ab$

84. $x^2 + x = a^2 + a$

In Exercises 85–90, find the value(s) of k such that each quadratic equation has exactly one root of multiplicity two.

85. $x^2 + 16x + k = 0$

86. $kx^2 - 4x + 2 = 0$

87. $2x^2 - kx + 3 = 0$

88. $5x^2 - (\sqrt{3}k)x + 3 = 0$

89. $kx^2 - kx + 2 = 0$

90. $3x^2 - kx + k = 0$

In Exercises 91 and 92, solve each double inequality. Use interval notation to describe the solution set.

91. $6 < x^2 - x < 20$

92. $10 - x < x^2 - 4x \leq 2x + 16$

93. Find the range of values for x if the area of the region shaded in the sketch is any size from 42 to 127, inclusive.

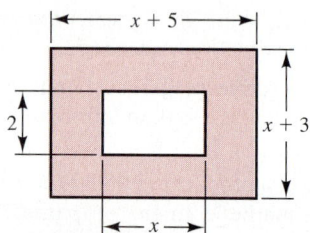

94. Find the range of values for x if the area of the region shaded in the sketch is any size from 21π and 33π, inclusive.

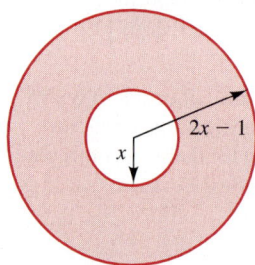

In Exercises 95 and 96, use interval notation to state the domain of each function. Check graphically.

95. $f(x) = \sqrt{x^2 - 2x - 8}$

96. $f(x) = \sqrt[4]{x^2 + 7x + 10}$

In Exercises 97–100, determine the values of k for which each quadratic equation has real roots.

97. $x^2 + kx + 6 = 0$

98. $2x^2 + kx + 3 = 0$

99. $x^2 + kx + k = 0$

100. $kx^2 + x + k = 0$

◆ 3.5 Intersection Points of Two Graphs

◆ Introductory Comments

In Figure 3.52, we show the graphs of the equations

$$y = f(x) \quad \text{and} \quad y = g(x)$$

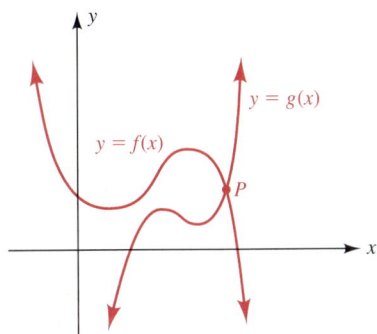

FIGURE 3.52

If the graphs of two equations intersect at point P, then P is called an *intersection point*.

in the same coordinate plane. We refer to the point P as an **intersection point** of the graphs of the two equations. If P is an intersection point of the graphs of the two equations, then the coordinates of point P must satisfy *both* equations. One method of finding the x-coordinates of point P is to substitute $f(x)$ for y in the equation $y = g(x)$ and solve the equation

$$f(x) = g(x)$$

for x. We refer to this procedure of solving two equations simultaneously as the **substitution method**. We begin by discussing the intersection of two lines.

◆ **Intersection of Two Lines**

Figure 3.53 illustrates the three possibilities for the intersection of two lines graphed on the same coordinate plane.

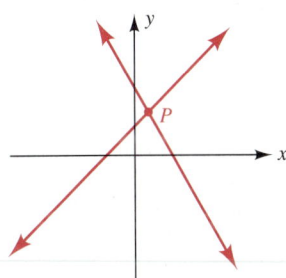

(a) The lines intersect at one point.

(b) The lines are parallel and do not intersect.

(c) The lines coincide and intersect at an infinite number of points.

FIGURE 3.53

The three possibilities for the intersection of two lines graphed on the same coordinate plane.

EXAMPLE 1 Find the coordinates of the intersection point for the lines with the given equations.

(a) $y = 3 - 5x$ and $y = \frac{3}{2}x - 4$ **(b)** $y = 2x + 4$ and $y = 2x - 1$

◆ **Solution**

(a) Figure 3.54 shows the graphs of the lines $y = 3 - 5x$ and $y = \frac{3}{2}x - 4$ in the viewing rectangle of a graphing calculator. The x-coordinate of the intersection point appears to be about 1, and the y-coordinate appears to be between -2 and -3. One method of finding the x-coordinate of the intersection point is to solve the two equations simultaneously by using substitution. We substitute $3 - 5x$ for y in the equation $y = \frac{3}{2}x - 4$ and solve for x as follows:

FIGURE 3.54

The lines with equations $y = 3 - 5x$ and $y = \frac{3}{2}x - 4$ intersect at one point.

$$3 - 5x = \tfrac{3}{2}x - 4$$

$6 - 10x = 3x - 8$ **Multiply both sides by 2**

$14 = 13x$ **Add 10x and 8 to both sides**

$x = \tfrac{14}{13}$ **Divide both sides by 13**

We can find the y-coordinate of the intersection point by substituting $\frac{14}{13}$ for x in one of the original equations. Choosing the equation $y = 3 - 5x$, we find

$$y = 3 - 5\left(\tfrac{14}{13}\right) = 3 - \tfrac{70}{13} = -\tfrac{31}{13}.$$

Thus, the lines intersect at the point $\left(\frac{14}{13}, -\frac{31}{13}\right)$.

(b) The graphs of $y = 2x + 4$ and $y = 2x - 1$ are lines with the same slope, 2, but different y-intercepts (4 and -1). Thus, we conclude these lines are *parallel* and hence have *no intersection point* (see Figure 3.55). If we solve the equations simultaneously by substituting $2x + 4$ for y in the equation $y = 2x - 1$, we find that

FIGURE 3.55
The lines with equations $y = 2x + 4$ and $y = 2x - 1$ are parallel. Hence there is no intersection point.

$$y = 2x - 1$$

$$2x + 4 = 2x - 1 \qquad \text{Substitute } 2x + 4 \text{ for } y$$

$$4 = -1 \qquad \text{Subtract } 2x \text{ on both sides}$$

a false statement

When the variables drop out and the resulting statement is *false*, we conclude that the lines are *parallel* and have *no intersection point.*   ◆

PROBLEM 1   Verify that the point with coordinates $\left(\frac{14}{13}, -\frac{31}{13}\right)$ satisfies the equations of both lines in Example 1(a).   ◆

We now discuss an alternative method for solving two equations simultaneously. Consider the two equations $A = B$ and $C = D$. The equation $C = D$ states that C is the same as D. Thus, if we add C to the left-hand side and D to the right-hand side of the equation $A = B$, we are actually adding the *same* quantity to both sides of the equation $A = B$. Therefore, we can state the following:

$$\begin{array}{ll} \text{If} & A = B \\ \text{and} & C = D \\ \hline \text{then} & A + C = B + D \end{array} \qquad \text{Add the left-hand and right-hand sides}$$

This idea gives us an alternative method for finding the intersection point of two lines. We solve the equations simultaneously by adding the left-hand and right-hand sides of the equations in order to eliminate one of the variables. To accomplish this task, it may be necessary to first multiply one of the equations by a constant so that the coefficient of a variable (either x or y) in one equation is the opposite of the same variable's coefficient in the other equation. In fact, we must sometimes multiply both equations (by different constants) to obtain coefficients of a variable that can be eliminated by adding the equations. We refer to this method as the **elimination method** and illustrate the procedure in the next example.

EXAMPLE 2 Find the coordinates of the intersection point for the lines with the given equations.

(a) $3x - 3y = 10$ and $5x + 6y = 2$
(b) $-6x + 2y = 1$ and $9x + 3y = -\frac{3}{2}$

◆ **Solution**

(a) Figure 3.56 shows the graphs of the equations $3x - 3y = 10$ (or $y = x - \frac{10}{3}$) and $5x + 6y = 2$ (or $y = -\frac{5}{6}x + \frac{1}{3}$) in the viewing rectangle of a graphing calculator. The x-coordinate of the intersection point appears to be about 2 and the y-coordinate appears to be between -1 and -2. One method of finding the x-coordinates of the intersection point is to solve the equations simultaneously by using the elimination method:

$$3x - 3y = 10 \quad \xrightarrow{\substack{\text{Multiply both}\\ \text{sides by 2}}} \quad 6x - 6y = 20$$
$$5x + 6y = 2 \quad \xrightarrow{\text{Leave as is}} \quad 5x + 6y = 2$$

$$\overline{\quad 11x + 0y = 22} \quad \text{Add columns}$$

$$x = 2 \quad \text{Solve for } x$$

To find the y-coordinate of the intersection point, we substitute 2 for x in either of the original equations and solve for y. Replacing x with 2 in the equation $3x - 3y = 10$, we find $y = -\frac{4}{3}$. Thus, the lines intersect at the point $\left(2, -\frac{4}{3}\right)$.

(b) We first write both equations in slope-intercept form, $y = mx + b$:

$$-6x + 2y = 1 \qquad\qquad 9x - 3y = -\frac{3}{2}$$
$$2y = 6x + 1 \qquad\qquad -3y = -9x - \frac{3}{2}$$
$$y = 3x + \tfrac{1}{2} \qquad\qquad y = 3x + \tfrac{1}{2}$$

Since these equations have the same slope and y-intercept, we conclude that the lines *coincide*. Thus, every point on each line satisfies the equation of the other line, and we conclude that the lines have an *infinite number of intersection points*. If we solve the equations simultaneously by using the elimination method, we find that

$$-6x + 2y = 1 \quad \xrightarrow{\text{Multiply both sides by 3}} \quad -18x + 6y = 3$$
$$9x - 3y = -\tfrac{3}{2} \quad \xrightarrow{\text{Multiply both sides by 2}} \quad 18x - 6y = -3$$

$$\overline{\quad 0 = 0}$$

a true statement

When the variables drop out and the resulting statement is true, we conclude that the lines coincide and, therefore, have an infinite number of intersection points. ◆

FIGURE 3.56

The lines with equations $3x - 3y = 10$ and $5x + 6y = 2$ intersect at one point.

PROBLEM 2   Find the coordinates of the intersection point for the lines with the equations $2x - y = 3$ and $y = 4x + 5$ by using each method.

(a) Substitution method   **(b)** Elimination method   ◆

◆ **Intersection of a Line and a Parabola**

Figure 3.57 illustrates the three possibilities for the intersection of a nonvertical line and a parabola graphed on the same coordinate plane.

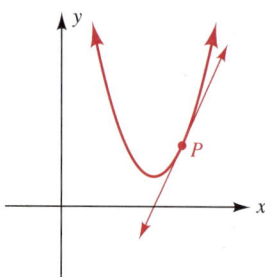

(a) The line and the parabola intersect at two points.

(b) The line and the parabola intersect at one point.

(c) The line and the parabola do not intersect.

FIGURE 3.57
The three possibilities for the intersection of a nonvertical line and a parabola graphed on the same coordinate plane.

Although the notion of *tangency* is studied in detail in calculus, we mention the idea at this time. If a line is not parallel to the axis of a parabola and touches a parabola at only one point, as shown in Figure 3.57(b), then the line is said to be **tangent** to the parabola at that point.

We may find the coordinates of the intersection points of a line and a parabola by solving their equations simultaneously by using the substitution method. The coordinates of the intersection points must satisfy *both* the equation of the line and the equation of the parabola.

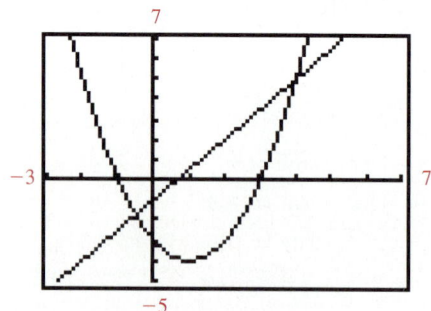

FIGURE 3.58
The line with the equation $3x - 2y = 2$ and the parabola with the equation $y = x^2 - 2x - 3$ intersect at two points.

EXAMPLE 3   Find the coordinates of the intersection points for the line and the parabola with the given equations.

(a) $3x - 2y = 2$ and $y = x^2 - 2x - 3$

(b) $y = x - 1$ and $y = x^2 - 2x + 3$

◆ **Solution**

(a) Figure 3.58 shows that the graphs of $3x - 2y = 2$ $\left(\text{or } y = \frac{3}{2}x - 1\right)$ and $y = x^2 - 2x - 3$ intersect at two points. To find the x-coordinates of these intersection points, we solve the equations simultaneously by using the substitution method:

$$3x - 2y = 2$$

$$3x - 2(x^2 - 2x - 3) = 2 \qquad \text{Substitute } x^2 - 2x - 3 \text{ for } y$$

$$3x - 2x^2 + 4x + 6 = 2 \qquad \text{Simplify}$$

$$2x^2 - 7x - 4 = 0 \qquad \text{Write the quadratic equation in standard form}$$

$$(2x + 1)(x - 4) = 0 \qquad \text{Factor}$$

$$2x + 1 = 0 \quad \text{or} \quad x - 4 = 0 \qquad \text{Apply the zero product property}$$

$$x = -\tfrac{1}{2} \qquad\qquad x = 4 \qquad \text{Solve for } x$$

x-coordinates of the intersection points

To find the y-coordinates of the intersection points that correspond to these x-coordinates, we substitute the x-values of $-\frac{1}{2}$ and 4 in either one of the original equations and solve for y. Substituting $x = -\frac{1}{2}$ and $x = 4$ into the equation $y = x^2 - 2x - 3$, we have

$$y = (-\tfrac{1}{2})^2 - 2(-\tfrac{1}{2}) - 3 = -\tfrac{7}{4} \qquad \text{and} \qquad y = (4)^2 - 2(4) - 3 = 5.$$

y-coordinates of the intersection points

Thus, the coordinates of the intersection points are $\left(-\frac{1}{2}, -\frac{7}{4}\right)$ and $(4, 5)$.

(b) The graphs of $y = x - 1$ and $y = x^2 - 2x + 3$ are shown in Figure 3.59. These graphs appear to have no intersection. If we solve the equations simultaneously by the substitution method, we find

$$x - 1 = x^2 - 2x + 3$$

$$x^2 - 3x + 4 = 0 \qquad \text{Write the quadratic equation in standard form}$$

$$x = \frac{-(-3) \pm \sqrt{(-3)^2 - 4(1)(4)}}{2(1)} \qquad \text{Apply the quadratic formula}$$

$$x = \frac{3 \pm \sqrt{-7}}{2} \qquad \text{Simplify}$$

not a real solution

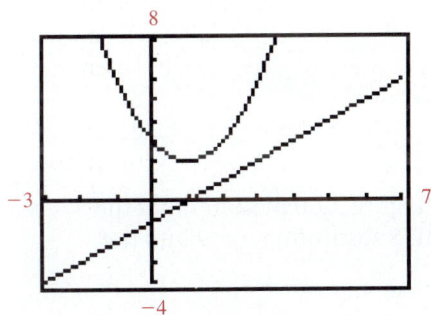

FIGURE 3.59

The line with the equation $y = x - 1$ and the parabola with the equation $y = x^2 - 2x + 3$ do *not* intersect.

If the equations of a line and a parabola have no real solution when solved simultaneously, then the line and the parabola do not intersect. ◆

PROBLEM 3 Find the coordinates of the intersection points for the line and the parabola with equations $4x - y = 6$ and $y = x^2 - 2x + 3$, respectively. ◆

◆ Intersection Points of Other Curves

As illustrated in the next example, the substitution and elimination methods can be used to find the intersection points of other curves as well.

EXAMPLE 4   Find the coordinates of the intersection points of the curves with the given equations.

(a) $y = \sqrt{x}$ and $x + y = 6$ **(b)** $x^2 + y^2 = 4$ and $x^2 - 3y = 0$

◆ Solution

(a) Figure 3.60 shows the graphs of the equations $y = \sqrt{x}$ and $x + y = 6$ (or $y = 6 - x$) in the viewing rectangle of a graphing calculator. To find the x-coordinate of the intersection point, we solve the equations simultaneously by substituting $\sqrt{x}$ for y in the equation $x + y = 6$ as follows:

$$x + y = 6$$

$$x + \sqrt{x} = 6 \qquad \text{Substitute } \sqrt{x} \text{ for } y$$

$$\sqrt{x} = 6 - x \qquad \text{Isolate the radical}$$

$$x = 36 - 12x + x^2 \qquad \text{Square both sides}$$

> Squaring both sides may introduce extraneous roots.

$$0 = 36 - 13x + x^2 \qquad \text{Solve the quadratic equation}$$

$$0 = (9 - x)(4 - x)$$

$$9 - x = 0 \quad \text{or} \quad 4 - x = 0$$

$$x = 9 \quad \text{or} \qquad x = 4$$

We have found two x-coordinates, but Figure 3.60 indicates that the curves have only one intersection point. Substituting $x = 9$ into the equation $y = \sqrt{x}$, we find

$$y = \sqrt{9} = 3.$$

However, the coordinates $(9, 3)$ do not satisfy the equation $x + y = 6$. Thus, $(9, 3)$ is *not* an intersection point.
    Substituting $x = 4$ into the equation $y = \sqrt{x}$, we find

$$y = \sqrt{4} = 2.$$

Since the coordinates $(4, 2)$ also satisfy the equation $x + y = 6$, we conclude that the only intersection point is $(4, 2)$.

(b) Figure 3.61 shows the graphs of the equations $x^2 + y^2 = 4$ (or $y = \pm\sqrt{4 - x^2}$) and $x^2 - 3y = 0$ (or $y = \frac{1}{3}x^2$) in the viewing rect-

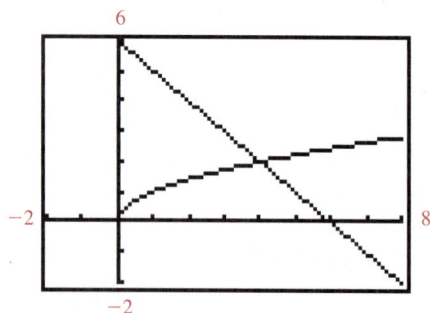

FIGURE 3.60

The graphs of the equations $y = \sqrt{x}$ and $x + y = 6$ intersect at one point.

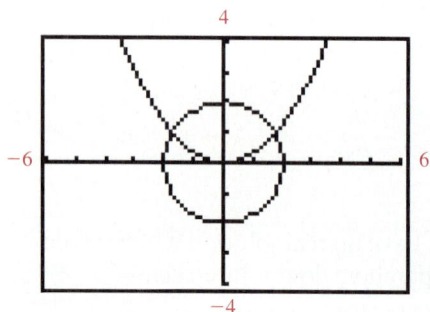

FIGURE 3.61

The graphs of the equations $x^2 + y^2 = 4$ (circle) and $x^2 - 3y = 0$ (parabola) intersect at two points.

angle of a graphing calculator. To find the y-coordinates of the intersection points, we eliminate the x^2 terms by using the elimination method:

$$x^2 + y^2 = 4 \xrightarrow{\text{No change}} x^2 + y^2 \qquad = 4$$

$$x^2 - 3y = 0 \xrightarrow[\text{sides by } -1]{\text{Multiply both}} \underline{\quad -x^2 + \qquad 3y = 0}$$

$$\qquad\qquad\qquad\qquad\qquad y^2 + 3y = 4 \qquad \textbf{\textcolor{brown}{Add and solve the}}$$

quadratic equation for y.

$$y^2 + 3y - 4 = 0$$
$$(y - 1)(y + 4) = 0$$
$$y - 1 = 0 \quad \text{or} \quad y + 4 = 0$$
$$y = 1 \qquad\qquad y = -4$$

Substituting $y = 1$ into the equation $x^2 - 3y = 0$ gives us

$$x^2 = 3 \qquad \text{or} \qquad x = \pm\sqrt{3}.$$

Substituting $y = -4$ into the equation $x^2 - 3y = 0$ gives us

$$x^2 = -12.$$

This equation has no real solution, so the only intersection points are $\left(\sqrt{3},\, 1\right)$ and $\left(-\sqrt{3},\, 1\right)$. ◆

PROBLEM 4 Find the coordinates of the intersection points for the parabolas with equations $y = x^2$ and $y = 4x - x^2$. ◆

◆ Approximating the Intersection Points of Two Curves

The substitution procedure sometimes yields an equation in one unknown that may be challenging to solve. For problems of this nature, we can use a graphing calculator to find the approximate points of intersection.

EXAMPLE 5 Find the approximate coordinates of the intersection points for the curves with equations $y = x^3$ and $4x^2 + y^2 = 16$.

◆ **Solution** To determine the x-coordinates of the intersection points, we substitute x^3 for y in the equation $4x^2 + y^2 = 16$ and obtain

$$4x^2 + (x^3)^2 = 16 \qquad \text{or} \qquad x^6 + 4x^2 - 16 = 0.$$

Note that the equation $x^6 + 4x^2 - 16 = 0$ is *not* of quadratic type and, hence, the chance of solving this equation by any of the algebraic methods that we have discussed in this chapter seems bleak. Instead, we approximate the coordinates of the intersection points by using a graphing calculator.

Figure 3.62 shows the graphs of the equations $y = x^3$ and $4x^2 + y^2 = 16$ $\left(\text{or } y = \pm 2\sqrt{4 - x^2}\,\right)$ in the viewing rectangle of a graphing calculator.

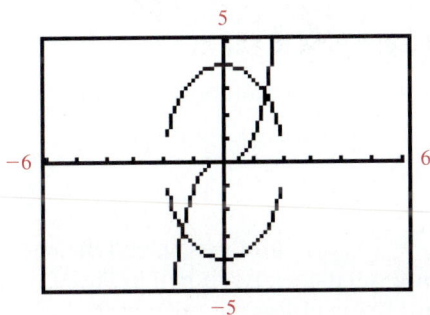

FIGURE 3.62

The graphs of the equations $4x^2 + y^2 = 16$ (ellipse) and $y = x^3$ intersect at two points.

Many graphing calculators have a built-in program for finding an intersection point. If your calculator does not have a built-in program for finding an intersection point, use the $\boxed{\text{TRACE}}$ key and $\boxed{\text{ZOOM}}$ key on the calculator to estimate the coordinates of these points. Either way, we find that the coordinates of the intersection points are $(-1.414, -2.828)$ and $(1.414, 2.828)$ when rounded to the nearest thousandth.   ◆

$\boxed{\text{PROBLEM 5}}$ Referring to Example 5, when we substitute x^3 for y in the equation $4x^2 + y^2 = 16$, we obtain $x^6 + 4x^2 - 16 = 0$. With some *creative* factoring, we can find the *exact* solutions of this equation. First, rewrite this equation as

$$(x^6 - 8) + 4x^2 - 8 = 0.$$

Now, factor $x^6 - 8$ as the difference of cubes (Section 1.6) and then look for a common binomial factor between this expression and $4x^2 - 8$. State the exact coordinates of the intersection points.   ◆

◆ **Solving Equations by Finding Intersection Points**

To determine the intersection points of the graphs of the equations $y = f(x)$ and $y = g(x)$, we find the real roots of the equation $f(x) = g(x)$. Conversely, we can find the real roots of an equation of the form $f(x) = g(x)$ by graphing $y = f(x)$ and $y = g(x)$ in the same coordinate plane and noting the x-coordinates of their intersection points. We refer to this graphical procedure for solving an equation as the **intersection point method**. Hence, we have two graphical methods for solving an equation of the form $f(x) = g(x)$.

1. **x-intercept method**: Write the equation in the form $f(x) - g(x) = 0$, then generate the graph of $y = f(x) - g(x)$. The x-intercepts of this graph are the real solutions of the equation $f(x) = g(x)$. This procedure is outlined in Section 2.4.

2. **Intersection point method**: Generate the graphs of $y = f(x)$ and $y = g(x)$ in the same coordinate plane. The x-coordinates of the intersection points are the real solutions of the equation $f(x) = g(x)$.

$\boxed{\text{EXAMPLE 6}}$ Solve graphically by using the intersection point method.

(a) $2x - 11 = 5x - 5$ **(b)** $(x - 3)(x - 10) = x - 10$

◆ **Solution**

(a) We begin by generating the graphs of $y = 2x - 11$ and $y = 5x - 5$ in the same viewing rectangle of a graphing calculator. As shown in Figure 3.63, the lines intersect at one point. Using either the calculator's built-in program for finding the intersection point or the $\boxed{\text{TRACE}}$ key and $\boxed{\text{ZOOM}}$ key, we find that the x-coordinate of the intersection point is -2. Hence, the real root of the original equation is also -2. For the algebraic solution of this equation, see Example 1(a) in Section 1.8.

(b) We begin by generating the graphs of $y = (x - 3)(x - 10)$ and $y = x - 10$ in the viewing rectangle of a graphing calculator. As shown in Figure 3.64, the parabola and the line intersect at two points. Using either the calculator's built-in program for finding the intersection points or the TRACE key and ZOOM key, we find that the x-coordinates of the intersection points are 4 and 10. Hence, the real roots of the original equation are also 4 and 10. For the algebraic solution of this equation, see Example 1(a) in Section 3.4.

FIGURE 3.63

The real root of the equation $2x - 11 = 5x - 5$ is the x-coordinate of the intersection points of the lines $y = 2x - 11$ and $y = 5x - 5$.

FIGURE 3.64

The real roots of the equation $(x - 3)(x - 10) = x - 10$ are the x-coordinates of the intersection points of the parabola $y = (x - 3)(x - 10)$ and the line $y = x - 10$.

PROBLEM 6 Solve graphically: $x^2 - x - 12 = 3(x + 2)(x - 2)$.
(For the algebraic solution, see Problem 1 in Section 3.4.)

Exercises 3.5

◆ Basic Skills

In Exercises 1–12, use the substitution method or the elimination method to find the coordinates of any intersection point for the lines with the given equations. Check graphically.

1. $3x + 5y = 8$ and $x = 5y$
2. $y = x + 6$ and $3x + 5y = 4$
3. $y = 8x + 3$ and $2x - 3y = 2$
4. $y = 3x - 5$ and $y = 5 - 2x$
5. $y = 4 - 3x$ and $y = 3 - 5x$
6. $y + 2x = 0$ and $3x - 2y = 14$
7. $7x - y = 2$ and $5x - y = 8$
8. $2x + 3y = 7$ and $3x - 2y = 4$

9. $3x - 3y = 10$ and $x - y = 1$
10. $4x - 3y = 8$ and $4x - 2y = 7$
11. $5x - 2y = 10$ and $3x + 6y = 12$
12. $4y = x + 6$ and $3x - 12y = -18$

In Exercises 13–24, use the substitution method to find the coordinates of any intersection point for the line and parabola with the given equations. Check graphically.

13. $y = 2x$ and $y = 3 - x^2$
14. $x + y = 6$ and $y = x^2$
15. $y = x + 3$ and $y = 2x^2 - 3$
16. $6x - 2y = 31$ and $x^2 - 2y - 6x + 5 = 0$

17. $3x - 2y = 5$ and $y = x^2 - 4x$

18. $x + y = -6$ and $y = x^2 - 2x - 8$

19. $y = \frac{5}{2}x + 2$ and $y = 3 - x - 2x^2$

20. $y = \frac{1}{4}x + 2$ and $y = \frac{1}{4}x^2 - x + 3$

21. $2y = x$ and $x^2 - y + 6x + 5 = 0$

22. $5x + 2y = 10$ and $y = 4 - 2x^2$

23. $y = \frac{1}{4}x^2$ and $5x - 2y = 6$

24. $y = x - 2$ and $x^2 - 4x + 3y - 6 = 0$

In Exercises 25–50, use the substitution method or the elimination method to find the coordinates of the intersection points for the graphs with the given equations. Check graphically with a graphing calculator.

25. $y = x^2 - 4x + 3$ and $y = 3 + 2x - x^2$

26. $y = x^2 - 4x$ and $y = 6x - x^2$

27. $x^2 + y^2 = 36$ and $y^2 - 5x = 0$

28. $x^2 + y^2 = 1$ and $y = 2x^2 - 1$

29. $x^2 + y^2 = 25$ and $y = \frac{1}{3}x + 3$

30. $x^2 + y^2 = 4$ and $y = \dfrac{2}{x}$

31. $y = \dfrac{1}{x}$ and $2x - y = 1$

32. $y = \sqrt{x}$ and $y = x - 2$

33. $x - 2y = 4$ and $y = \sqrt{x - 4}$

34. $y = \sqrt{x + 4}$ and $x = -y^2$

35. $y = \frac{1}{3}x + 1$ and $y = \sqrt{x}$

36. $x + y = 3$ and $xy = 3$

37. $9x^2 + y^2 = 36$ and $x^2 + y^2 = 28$

38. $x^2 + 2y^2 = 18$ and $2x^2 + y^2 = 15$

39. $y = x + 1$ and $x^2 + 4y^2 = 4$

40. $y = x^2 - 1$ and $4x^2 + y^2 = 16$

41. $x^2 + 2y^2 = 8$ and $y = \sqrt{x}$

42. $9x^2 + 4y^2 = 25$ and $y = \dfrac{2}{x}$

43. $9y^2 - 4x^2 = 36$ and $2x^2 + y^2 = 4$

44. $4x^2 - y^2 = 16$ and $x^2 + y^2 = 9$

45. $y = x - 2$ and $x^2 - y^2 = 1$

46. $y = x^2 - 1$ and $x^2 - 4y^2 = 1$

47. $x^2 - 5y^2 = 36$ and $x = y^2$

48. $10y^2 - x^2 = 16$ and $y = \sqrt{x}$

49. $4x^2 + y^2 - 8x = 0$ and $y = 6x - 4$

50. $x^2 + 4y^2 - 6x + 5 = 0$ and $y = 2x - 5$

In Exercises 51–58, use a graphing calculator to approximate the coordinates of the intersection points of the graphs with the given equations. Round each coordinate to the nearest thousandth.

51. $y = 3x - 1$ and $y = x^3$

52. $y = 2x^2 - 3x - 5$ and $y = x^3$

53. $x^2 + y^2 = 16$ and $y = x^2 - 2x$

54. $x^2 - y^2 = 4$ and $y = 4x - x^2$

55. $x^2 + 4y^2 - 4x = 0$ and $y = x - 1$

56. $y^2 + 16x = 0$ and $y^2 - 9x^2 - 18x = 18$

57. $y = 4 - 3x - x^2$ and $y = 2x^4$

58. $y = \sqrt[3]{x}$ and $y = 1 - x^2$

In Exercises 59–74, solve each equation graphically by using the intercept method in conjunction with a graphing calculator. Check algebraically.

59. $3n + 4 = n - 4$ **60.** $12m + 5 = 2m - 5$

61. $\dfrac{5x}{6} - 1 = \dfrac{x}{3}$ **62.** $4 + \dfrac{n}{6} = \dfrac{n + 4}{2}$

63. $n(2n - 3) = 5$ **64.** $(x - 1)(3x + 2) = 2$

65. $(x + 4)(x + 6) = x + 6$

66. $(m - 7)(m + 4) = 2(m - 7)$

67. $y^2 + (y - 2)(y + 8) = 2y$

68. $(x + 3)^2 = x(3x - 1)$

69. $5x^2 + 1 = 4x$ **70.** $2x^2 + 3 = 4x$

71. $(2x - 1)(x + 4) = (x + 1)^2$

72. $y(y - 3) = (2y + 3)(y - 1)$

73. $2(x^2 + 2) = \sqrt{7}\,x$ **74.** $\dfrac{2x}{3} = x^2 + \frac{1}{2}$

◆ **Critical Thinking**

75. Find the coordinates of the vertices of a triangle whose sides lie on the lines $x + y = 4$, $3x - y = 4$, and $x - y = 6$.

76. Show that the triangle in Exercise 75 is a right triangle. Then find its area.

77. Two sides of a parallelogram lie on the lines $2x - y = -3$ and $x + 2y = 11$, and one of its vertices is $(7, -3)$. Find the coordinates of the other three vertices of the parallelogram.

78. Show that the parallelogram in Exercise 77 is a rectangle. Then find its area.

79. Find the length of the line segment that joins the intersection points of the line $y = 2x + 3$ and the parabola opening upward with vertex $(-2, -4)$ and y-intercept 0.

80. Find the coordinates of the midpoint of the line segment that joins the intersection points of the parabola $y = 3 - x^2$ and the line with slope 1 and y-intercept 1.

$\frac{\Delta y}{\Delta x}$ *Calculus can be used to find the area of a region bounded by two or more curves. In order to find the area, we must first find the intersection points of the curves. In Exercises 81–86, find the coordinates of the intersection points P, Q, R, and (when applicable) S.*

81.

82.

83.

84.

85.

86.

Chapter 3 Review

◆   **Questions for Writing or Group Discussion**

1. What is meant by the *slope* of a straight line? Does every line have a slope? Explain.

2. How can you find the slope of a straight line from its equation? from its graph?

3. Describe the graph of a *linear function*. Explain the procedure for graphing a linear function.

4. Explain the procedure for finding the equation of a straight line when two points on the line are given.

5. On a coordinate plane, sketch two lines that have the given conditions:
   (a) negative reciprocal slopes and the same y-intercept
   (b) the same slope and different y-intercepts

6. Describe the graph of a *quadratic function*. Explain the procedure for graphing a quadratic function.

7. How can we determine whether a quadratic function has a *maximum* or *minimum value*? Explain how the maximum or minimum value can be found.

8. State the *quadratic formula*. Illustrate its use with an example.

9. Describe the method of *completing the square*.

10. By using the *discriminant*, explain how you can determine whether the roots of a quadratic equation are (a) real, (b) imaginary, or (c) equal.

11. Given that a is a real number, are the roots of the equation $5x^2 + 2ax + a^2 = 0$ real or imaginary? Explain your reasoning.

12. Is it possible to find three consecutive integers whose sum equals the product of the smallest and largest of these integers? Explain.

13. Illustrate with sketches some possibilities of intersection for two lines graphed in the same coordinate plane.

14. Describe two methods that can be used to determine the coordinates of the intersection points of two curves.

15. What can we conclude if the equations of two curves have no real solution when solved simultaneously?

16. Describe two graphical methods that can be used to solve an equation of the form $f(x) = g(x)$. Illustrate with an example.

◆   **Review Exercises**

In Exercises 1–14, identify the graph of each equation as a line or a parabola. If the graph is a line, give its slope and find any x-intercept and y-intercept. If the graph is a parabola, give the coordinates of its vertex and find any x- and y-intercept(s).

1. $y = 2x - 3$

2. $y = 1 - 4x$

3. $y = -\frac{3}{4}x - 1$

4. $y = \frac{2}{3}x + 2$

5. $y = (x - 2)^2 - 3$

6. $y = -2(x + 1)^2 + 2$

7. $y = 15 + 2x - x^2$

8. $y = 2x^2 + x - 10$

9. $6x - 2y = 8$

10. $3x - 4y + 18 = 0$

11. $3y - 18 = 0$

12. $2x + 5 = 0$

13. $x^2 + 2x - 2y + 5 = 0$

14. $2x^2 - 8x - 3y + 5 = 0$

In Exercises 15–24, determine the equation from the given information.

15. Line with slope $\frac{3}{4}$ and y-intercept -2

16. Line with x-intercept 4 and y-intercept -2

17. Line through $(-2, 4)$ with slope 3

18. Line through the origin and parallel to $x - 3y = 4$

19. Line through $(-2, 1)$ and perpendicular to the line through $(1, 1)$ and $(4, 3)$

20. Line through $(-3, 2)$ and perpendicular to the x-axis

21. Parabola with vertex $(0, -4)$ and x-intercepts ± 2

22. Parabola with vertex $(1, 2)$ and x-intercepts -1 and 3

23. Parabola passing through the origin, with vertex $(-2, -3)$ and symmetric with respect to $x = -2$

24. Parabola passing through $(-1, 0)$, with vertex $(2, 3)$ and symmetric with respect to $x = 2$

In Exercises 25–34, solve each quadratic equation.

25. $x^2 - 6x = 16$

26. $x(3x - 4) = 4$

27. $3x^2 + 48 = 0$

28. $16 + 2x^2 = 0$

29. $x^2 = 4x + 10$

30. $2x^2 + 3x = 1$

31. $x^2 + 5 = 2x$

32. $x^2 + 8x + 20 = 0$

33. $2(x^2 - 3) = \sqrt{6}\,x$

34. $2x^2 = 2\sqrt{2}\,x - 3$

In Exercises 35–40, solve each quadratic inequality. Use interval notation to write the solution set.

35. $3x^2 + 7x \le 6$

36. $25 - x^2 > 16$

37. $x^2 + 2x > 4$

38. $x^2 - 5x + 4 < 0$

39. $x + 4 < 3x(x - 1)$

40. $(2y + 1)(y - 3) \ge 2y - 1$

In Exercises 41–46, solve each equation graphically by using the intersection point method and a graphing calculator.

41. $7x - 6 = 4(3x + 1)$

42. $3x + 4 = 7x - 8$

43. $2 - 2x = 9x^2 + 6x + 1$

44. $4x^2 = (2x - 3)^2 + 9$

45. $x(x - 2) = 4(x - 1)$

46. $2(x - 1)^2 = 3x^2 - 5x$

In Exercises 47–56, use the substitution method or the elimination method to find the intersection points of the graphs with the given equations. Check graphically with a graphing calculator.

47. $y = 4x - 5$ and $2x - 3y = 10$

48. $3x - 4y = 6$ and $6x - 3y = 7$

49. $y = x^2 - 4x + 6$ and $y = x + 2$

50. $x = 4y^2 - 8y$ and $x - 8y = -12$

51. $x^2 + y^2 = 36$ and $y = x^2 - 6$

52. $y = 2 - 2x$ and $4x^2 + 9y^2 = 36$

53. $y^2 - x^2 = 1$ and $x^2 + 4y^2 = 9$

54. $y = \sqrt{x + 1}$ and $x^2 + 4y^2 = 9$

55. $x^2 - 4y^2 - 2x - 15 = 0$ and $x - 4y = 0$

56. $y = x^2 - 2x - 3$ and $4x^2 - 12x + 3y = 0$

In Exercises 57 and 58, use a graphing calculator to approximate the coordinates of the intersection points of the given curves. Round each coordinate to the nearest thousandth.

57. $3x + 2y = 6$ and $y = x^3$

58. $4x^2 + y^2 = 16$ and $y = x^2 - 4x$

59. Given the equations $A_1 x + B_1 y = C_1$ and $A_2 x + B_2 y = C_2$, solve the equations simultaneously (a) for x and (b) for y.

60. Given the equations $A_1 x^2 + B_1 y^2 = C_1$ and $A_2 x^2 + B_2 y^2 = C_2$, solve the equations simultaneously (a) for x and (b) for y.

61. The graph of the line $Ax + By + C = 0$ with $A \ne 0$ and $B \ne 0$ is sketched in the figure.

(a) Determine the equation of the line that passes through the origin and is perpendicular to the line $Ax + By + C = 0$.

(b) Determine the coordinates of the intersection point P.

(c) Show that the distance d from the line $Ax + By + C = 0$ to the origin is given by

$$d \quad \frac{|C|}{\sqrt{A^2 + B^2}}$$

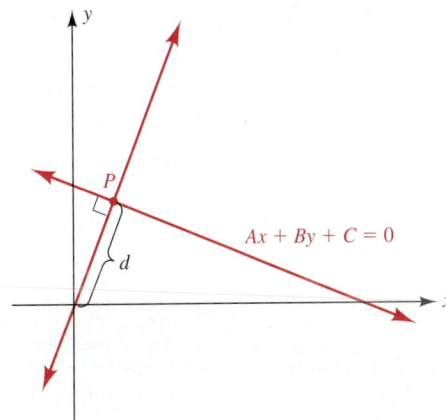

62. Use the formula from Exercise 61(c) to find each distance:

(a) the distance from the origin to the line $3x + 4y - 12 = 0$

Exercise 62 continues

(b) the distance between the parallel lines $3x + y = 4$ and $3x + y = 1$

63. Two sides of a parallelogram lie along the lines $y = x + 4$ and $x + 2y = 2$. One diagonal of the parallelogram is on the line $5x - 2y = -4$. Determine the coordinates of the vertices of the parallelogram.

64. Find the area of the triangle bounded by the lines $2x - y = 6$, $x + 2y = 18$, and $3x + y = 4$.

65. Determine whether the quadratic function $f(x) = 3x^2 - 4x + 5$ has a maximum or minimum value. Then find this value.

66. Determine whether the quadratic function $f(x) = 5x - 2x^2$ has a maximum or minimum value. Then find this value.

67. Determine the value of a such that the quadratic function $f(x) = ax^2 - 4x + 3$ has a maximum value of 5.

68. Let $f(x) = 2x + 3$ and $g(x) = x^2 - 3x + 4$.

    (a) Determine the minimum value of $f \circ g$.

    (b) Determine the minimum value of $g \circ f$.

69. The value of a personal computer depreciates linearly from $2400 at time of purchase to $400 after 5 years.

    (a) Express the value V (in dollars) of the computer as a function of its age x (in years).

    (b) What is the age of the computer when its value is fully depreciated?

70. Each week a salesperson gets paid a base salary plus a commission, which is a fixed percentage of the week's sales. For weekly sales of $1000, total pay is $280, and for weekly sales of $3000, the total pay is $440.

    (a) Express the salesperson's total pay P as a function of weekly sales x.

    (b) What is the base salary?

    (c) What is the significance of the slope of the graph of the function defined in part (a)?

71. In a certain city, the yearly cost C for water that is supplied to a household is a function of the number x of gallons that are used. Suppose the cost of water is given by the following rate schedule.

Usage	Cost
First 5000 gallons or less	$120
Next 10,000 gallons	$0.018 per gallon
Next 15,000 gallons	$0.011 per gallon
All gallons over 30,000	$0.003 per gallon

Find the piecewise linear function that describes this usage schedule. Then sketch its graph.

72. Five hundred feet of fencing is to be used to build three adjacent rectangular corrals (see figure). Find the overall dimensions of the corrals if the total enclosed area is to be the *largest* possible.

73. A triangular piece of land has 400 feet of frontage on a tar road and 250 feet of frontage on a gravel road that is perpendicular to the tar road. What are the dimensions of the *largest* rectangular building lot that can be inscribed in this triangular piece of land if the sides of the rectangular lot lie along the tar and gravel roads?

74. When the owner of an apartment complex charges rent of $600 per month for each unit, all 60 apartments are occupied. Research shows that for each $50 increase in the monthly rent, 3 apartments become vacant. What monthly charge for rent gives the owner maximum revenue?

75. The shape of a cable that supports the weight of a suspension bridge approximates a parabola with its vertex halfway between the supports, as shown in the sketch. Determine the length x of the vertical cable that is 30 meters from the end of a support.

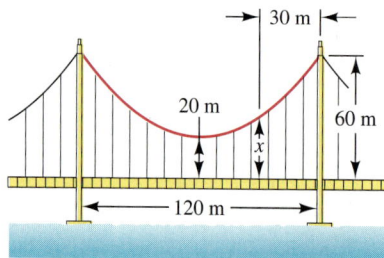

76. The dimensions of a picture are 12 in. by 15 in., as shown in the sketch. Encasing the picture in a frame of uniform width increases its area by 50%. What is the width of the frame?

77. An open box is made from a 6-in. by 9-in. piece of metal by cutting out a square from each corner and then turning up the sides along the dashed lines, as shown in the figure. If the area of the bottom of the box is 18 sq in., what is the height of the box?

78. Find the length and width of a rectangle if its diagonal is 2 centimeters more than the longer side, which in turn is twice the shorter side.

79. Find k such that one root of $25x^2 + kx + 1 = 0$ is two more than the other root.

80. The base of a triangle is 4 ft more than its height, and the area of the triangle is less than 96 sq ft (or ft²). What is the range of values for the base of the triangle?

1. Factor completely over the set of integers.

   (a) $x(x + 1)(2x - 5) - 12(x + 1)$

   (b) $x^2 - 4xy - 4a^2 + 8ay$

2. Simplify each fractional expression.

   (a) $\dfrac{(x - 3)^{1/2} - 4(x - 3)^{-1/2}}{x - 7}$

   (b) $\dfrac{x^{-1} + y^{-1}}{(x + y)^{-1}}$

3. Express $8^0 + 3\frac{1}{2} - 6^{-1} + \dfrac{1}{\sqrt[3]{-27}} + 8^{2/3}$ as an integer.

4. If $\dfrac{\sqrt{19} - \sqrt{10}}{x} = \dfrac{x}{\sqrt{19} + \sqrt{10}}$, find x.

5. For what values of k does the equation $4x^2 - 12x + k = 0$ have the given roots?

   (a) one root that is zero   (b) equal roots

   (c) real roots   (d) imaginary roots

6. For the equation $ax^2 - 5x + 2 = 0$, find the value of a if one root is to be the reciprocal of the other.

7. Expand and simplify $(1 + i)^2$, where $i = \sqrt{-1}$.

8. Arrange the given numbers in order from smallest to largest:

$$4^{-3/2} \qquad \left(\frac{1}{4}\right)^{-2} \qquad (4^{-1/3})^6 \qquad \left(\frac{1}{4}\right)^{-1/2} \qquad \frac{1}{4^0}$$

9. A rectangular field is twice as long as it is wide and is enclosed by x feet of fencing. Express the area A of the field as a function of x.

10. Simplify $(x^2 + 10x + 25)^{1/2} - (x^2 - 10x + 25)^{1/2}$ given the indicated conditions.

    (a) x is any real number   (b) $x \geq 5$

11. Perform the indicated operations and reduce to lowest terms.

    (a) $\dfrac{a}{(a - b)(a - c)} + \dfrac{b}{(b - c)(b - a)} + \dfrac{a}{(c - a)(c - b)}$

    (b) $\dfrac{1 + \dfrac{1}{x^2} + \dfrac{1}{x^4}}{1 + \dfrac{1}{x} + \dfrac{1}{x^2}}$

12. Find the x- and y-intercepts for the graph of the equation $y^2 = x(x - 2)^2$.

13. If the acceleration of an object is constant, its velocity increases linearly with time. Suppose that after 1 second a certain object with a constant acceleration has a velocity of 12 feet/second (ft/s), and after 5 seconds it has a velocity of 20 ft/s.

    (a) Express the velocity v of the object as a function of time t.

    (b) Sketch the graph of the function defined in part (a) and state the significance of the y-intercept.

14. Find the perimeter and the area of the triangle formed by the intersecting lines $x + y = -3$, $3x - 4y = -9$, and $4x - 3y = 2$.

15. For the graph of the equation $y = f(x)$, what is the y-intercept?

16. Sketch the graph of each quadratic function. Label the vertex and the x- and y-intercept(s).

    (a) $f(x) = x^2 - 4x + 3$ (b) $g(x) = 1 - 4x - x^2$

17. A car radiator contains 15 quarts of a 20% antifreeze solution. How many quarts of 20% solution must be drained and replaced with pure antifreeze (100% solution) to achieve a solution of at least 60% antifreeze?

18. Determine k so that the line $x - 3y + k = 0$ passes through the point $(3, -1)$.

19. Find the area of a triangle whose vertices are the points $(8, 8)$, $(-7, -2)$, and $(1, -6)$.

20. Given that $f(x) = \sqrt{x}$, express

$$\frac{f(x + h) - f(x)}{h}$$

    in simplest form with a rationalized numerator. Then evaluate this expression as h approaches 0.

21. Determine whether the graph of each equation has symmetry with respect to the x-axis, y-axis, or origin.

    (a) $x^2y = 9$ (b) $x + y = x^3$

    (c) $x^2 + y^2 - 6x = 16$ (d) $y = \pm\sqrt{9 - x^2}$

22. Determine what two amounts $22,400 can be split into so that the annual interest earned from one amount, which is invested at 8%, is double the interest from the other amount, which is invested at 10%.

23. Find the equation of a line that passes through the point of intersection of the lines $3x - 2y = 7$ and $y = 7 - 2x$ and also satifies the given condition.

    (a) is parallel to the x-axis

    (b) is perpendicular to $x + y = 8$

    (c) passes through $(5, -2)$

    (d) has a slope of -3

24. A business determines that consumers will purchase x units of a product each month when the price (in dollars) of the unit is $(60 - x)/3$. How many units are sold if the total monthly revenue from sales is $225?

25. Sketch the graph of the equation $y = 1/x^2$. Then use the ideas of shifting and reflecting to help sketch the graphs

of the following functions. Label all intercepts and asymptotes.

(a) $f(x) = 2 - \dfrac{1}{x^2}$

(b) $g(x) = \dfrac{1}{(x-1)^2}$

26. A point is equidistant from $(2, 1)$ and $(-4, 3)$, and the slope of the straight line joining this point to the origin is 2. Determine the coordinates of the point.

27. Determine the value of m so that the line $y = mx - 2$ passes through the point of intersection of the lines $y = 3x + 2$ and $y = x - 4$.

28. Determine the quadratic function with real coefficients that has $2 \pm 3i$ as its zeros.

29. A square piece of sheet metal is to be made into a rectangular box with an open top by cutting out a 2-in. by 2-in. square from each corner and bending along the dashed lines, as shown in the sketch. Determine the range of dimensions of the original piece of sheet metal that results in a box volume from 50 to 98 cubic inches.

Cutouts

30. The power P (in watts) dissipated in a certain electrical circuit is directly proportional to the square of the current i (in amperes) flowing through the circuit. Express P as a function of i if $P = 5$ watts when $i = 20$ milliamperes.

31. If a farmer picks his pumpkin crop today, he will have 30,000 pounds of pumpkins worth 20¢ per pound. If he waits, the crop will grow at 2,000 pounds per week, but the price will drop by 1¢ per pound per week.

(a) Express the amount A of money in dollars that the farmer receives at the end of x weeks as a function of x.

(b) When is the most profitable time for the farmer to pick and sell the pumpkin crop?

32. A tank contains 1000 gallons of water. The spigot is opened and water flows out of the tank at the uniform rate of $\frac{2}{3}$ gallon per minute.

(a) Express the amount A of water (in gallons) in the tank as a function of the time t (in minutes) that the spigot is opened.

(b) State the domain and range of the function defined in part (a).

A silo has the shape of a right circular cylinder surmounted by a hemisphere and holds 3000 cubic feet of grain. If the height of the silo is 60 ft, what is the radius of its base?

For the solution, see Example 5 in Section 4.4.

Polynomial and Rational Functions

4.1   Polynomial Functions and Their Graphs

◆ **Introductory Comments**

Any function P defined by

$$P(x) = a_n x^n + a_{n-1} x^{n-1} + a_{n-2} x^{n-2} + \cdots + a_2 x^2 + a_1 x + a_0$$

where n is a nonnegative integer and $a_n \neq 0$, is called a **polynomial function of degree n**. In this text, we restrict our discussion to polynomial functions with *real coefficients*, that is, $a_0, a_1, a_2, \ldots, a_n$ are real numbers with $a_n \neq 0$. The domain of any polynomial function is $(-\infty, \infty)$.

A polynomial function of *degree 0* is a constant function $f(x) = a_0$, of degree 1 is a linear function $g(x) = a_1 x + a_0$, and of degree 2 is a quadratic function $h(x) = a_2 x^2 + a_1 x + a_0$. Since we have discussed constant functions, linear functions, and quadratic functions in previous chapters, we devote this chapter to polynomial functions of degree 3 or greater, such as the following:

1. Cubic function: $F(x) = a_3 x^3 + a_2 x^2 + a_1 x + a_0$
2. Quartic function: $G(x) = a_4 x^4 + a_3 x^3 + a_2 x^2 + a_1 x + a_0$
3. Quintic function: $H(x) = a_5 x^5 + a_4 x^4 + a_3 x^3 + a_2 x^2 + a_1 x + a_0$

The graph of a polynomial function of degree 3 or greater is *continuous* with smooth, rounded turns—it contains no gap, break, jump, or sharp turn. The sketches in Figure 4.1. illustrate the differences between the graphs of polynomial and nonpolynomial functions. In this section, we discuss the important characteristics of the graph of a polynomial function.

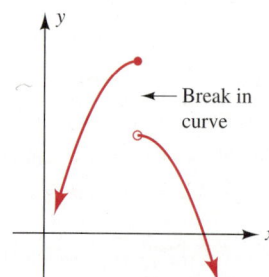

(a) Polynomial function: Smooth, rounded turns; no gap, break, jump, or sharp turn

(b) Nonpolynomial function: Sharp turn

(c) Nonpolynomial function: Break in curve (discontinuous)

FIGURE 4.1

Graphs of polynomial and nonpolynomial functions

◆ Power Functions

The simplest polynomial function contains a single term and has the form

$$P(x) = x^n$$

with n a positive integer. In this text, we refer to this type of polynomial function as a **power function**.

When n is an *even* integer with $n \geq 4$, the graph of $P(x) = x^n$ is similar to the graph of the squaring function $f(x) = x^2$ (Section 2.5). When n is an *odd* integer with $n \geq 5$, the graph of $P(x) = x^n$ is similar to the graph of the cubing function $f(x) = x^3$ (Section 2.5). The greater the value of n (odd or even), the flatter the graph of P in the interval $(-1, 1)$ and the steeper the graph of P in the intervals $(-\infty, -1)$ and $(1, \infty)$. Figure 4.2 shows the graphs of $P(x) = x^4$ and $f(x) = x^2$ in the same viewing rectangle of a graphing calculator, and Figure 4.3 shows the graphs of $P(x) = x^5$ and $f(x) = x^3$ in the same viewing rectangle of a graphing calculator. Note that in both figures, in the interval $(-1, 1)$, the graph of P is flatter than the graph of f, but in the intervals $(-\infty, -1)$ and $(1, \infty)$ the graph of P is steeper than the graph of f.

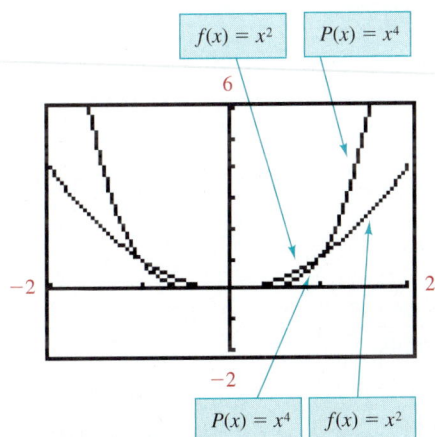

FIGURE 4.2

The graph of $P(x) = x^4$ resembles the graph of $f(x) = x^2$.

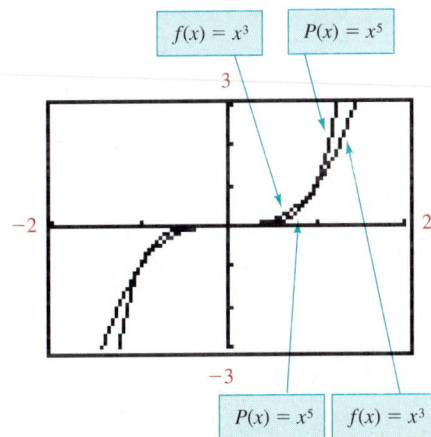

FIGURE 4.3

The graph of $P(x) = x^5$ resembles the graph of $f(x) = x^3$.

We can use the graph of a power function in conjunction with the shift rules, reflection rules, and stretch and compress rules (Section 2.5) to sketch the graphs of some polynomial functions.

EXAMPLE 1 Sketch the graph of each function and label the intercepts.

(a) $P(x) = \frac{1}{2}(x - 1)^5$ **(b)** $P(x) = 16 - x^4$

◆ Solution

(a) We can think of this function as $P(x) = \frac{1}{2}f(x - 1)$ where $f(x) = x^5$. Thus, by the horizontal shift rule and the vertical compress rule, the

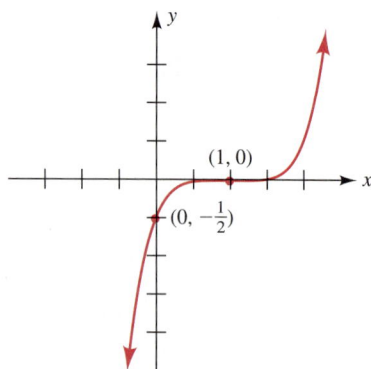

FIGURE 4.4

Graph of $P(x) = \frac{1}{2}(x - 1)^5$

FIGURE 4.5

Graph of $P(x) = 16 - x^4$

graph of $P(x) = \frac{1}{2}(x - 1)^5$ is obtained by shifting the graph of $f(x) = x^5$ right 1 unit, and then compressing this graph by a factor of 2. The graph of the function P is shown in Figure 4.4.

x-intercept:

$$P(x) = 0$$
$$\tfrac{1}{2}(x - 1)^5 = 0$$
$$(x - 1)^5 = 0 \quad \text{\color{red}{Multiply both sides by 2}}$$
$$x - 1 = 0 \quad \text{\color{red}{Take the fifth root of both sides}}$$
$$x = 1$$

y-intercept:

$$P(0) = \tfrac{1}{2}(0 - 1)^5 = -\tfrac{1}{2}$$

(b) We can think of this function as $P(x) = -f(x) + 16$ where $f(x) = x^4$. Thus, by the *x*-axis reflection rule and the vertical shift rule, the graph of $P(x) = 16 - x^4$ is obtained by reflecting the graph of $f(x) = x^4$ about the *x*-axis and then shifting this graph upward 16 units. The graph of the function P is shown in Figure 4.5.

x-intercept:

$$P(x) = 0$$
$$16 - x^4 = 0$$
$$x^4 = 16$$
$$x = \pm\sqrt[4]{16} \quad \text{\color{red}{Extract the two real fourth roots}}$$
$$x = \pm 2$$

y-intercept:

$$P(0) = 16 - (0)^4 = 16$$

◆

PROBLEM 1   Sketch the graph of $P(x) = 1 - (x + 2)^6$ and label the intercepts. ◆

◆ **Left and Right Behavior and Relative Extrema**

When a polynomial function is written in descending powers of x, the coefficient of the term with the highest degree is the **leading coefficient**. In the polynomial function

$$P(x) = a_n x^n + a_{n-1} x^{n-1} + \cdots + a_1 x + a_0$$

a_n is the leading coefficient. For very large values of x (either negative or positive), the highest-degree term of the polynomial function, $a_n x^n$, becomes dominant in relation to the other terms of the polynomial. Because of this, the graph of a polynomial function P eventually rises or falls, depending on

     (i) the sign of its leading coefficient a_n, and

     (ii) whether n is even or odd.

Figure 4.6 illustrates the four possibilities for the left behavior of the graph of a polynomial function P as x decreases without bound ($x \to -\infty$) and for the right behavior as x increases without bound ($x \to \infty$). For convenience, we use quadrants I and IV to illustrate the possible shapes.

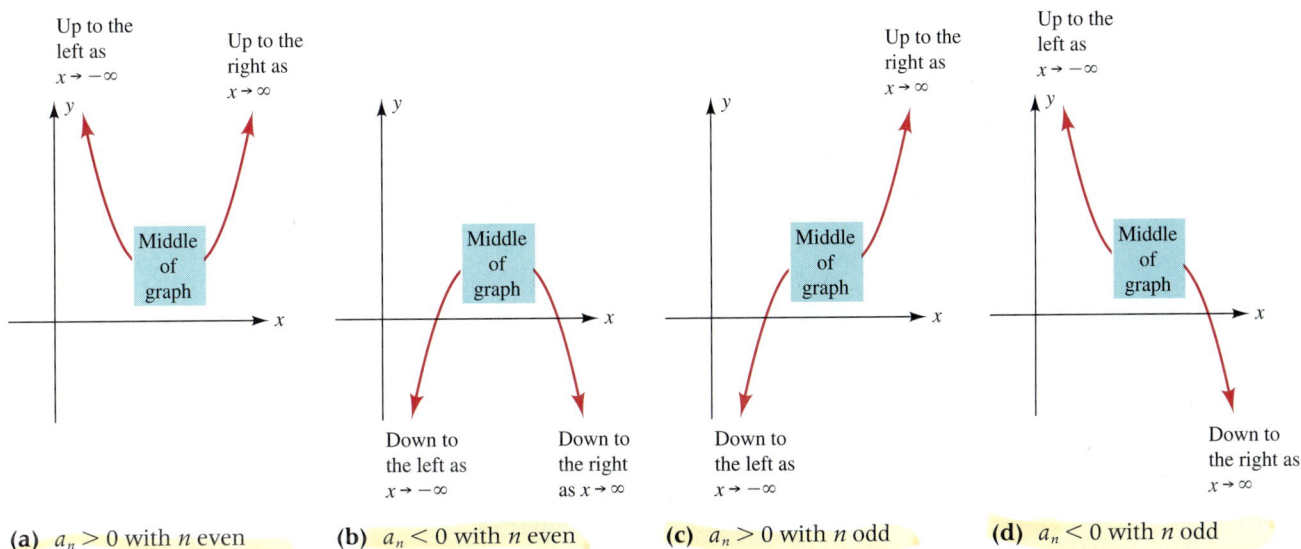

(a) $a_n > 0$ with n even   **(b)** $a_n < 0$ with n even   **(c)** $a_n > 0$ with n odd   **(d)** $a_n < 0$ with n odd

FIGURE 4.6
Left and right behavior of the graph of $P(x) = a_n x^n + a_{n-1} x^{n-1} + \cdots + a_1 x + a_0$

On the graph of a polynomial function, any point at which the function changes from increasing to decreasing or from decreasing to increasing is called a **relative extremum**. The graph of the polynomial function in Figure 4.7 has four such points, or extrema, labeled A, B, C, and D. The points A and C are called **relative maxima** (peaks), and the points B and D are called **relative minima** (valleys) of the function. A single extremum point is called a relative *maximum* or *minimum*.

FIGURE 4.7

The graph of this polynomial function has relative maxima at A and C and relative minima at B and D.

Using calculus, it can be shown that if P is a polynomial function of degree n, then the number of relative extrema either is equal to $n - 1$ or is less than this number ($n - 1$) by an even number. This fact enables us to determine the possible shapes for the shaded portions of the graphs in Figure 4.6.

◆ **Relative Extrema Rule**

If P is a polynomial function of degree n with real coefficients, then the number of relative extrema either is equal to $n - 1$ or is less than this number by an even number.

(a) Up to the left and up to the right, with three relative extrema

(b) Up to the left and up to the right, with one relative extremum

FIGURE 4.8

Sketches of two possible shapes for the graph of a polynomial function of degree 4 with a positive leading coefficient

EXAMPLE 2   For the polynomial function P defined by $P(x) = 3x^4 - 9x^2 - 6x - 1$, determine the left and right behavior of its graph and the possible number of relative extrema.

◆ Solution   For $P(x) = 3x^4 - 9x^2 - 6x - 1$, we have $a_n = 3$ and $n = 4$. Since $a_n > 0$ and n is even, we know that the graph of P goes up to the left as $x \to -\infty$ and up to the right as $x \to \infty$, as shown in Figure 4.6(a). Since the degree of P is 4, its graph has either $4 - 1 = 3$ relative extrema or one relative extremum for the middle portion of the graph of P. Sketches of two possible shapes for the graph of P are shown in Figure 4.8.   ◆

When you use a graphing calculator to generate the graph of a polynomial function, it is important to choose a viewing rectangle that shows the *complete graph*, that is, a viewing rectangle that shows the left and right behavior of the graph as well as any relative extrema in the middle of the graph.

PROBLEM 2   Use a graphing calculator to generate a complete graph of the polynomial function $P(x) = 3x^4 - 9x^2 - 6x - 1$. Compare your graph with the sketches in Figure 4.8.   ◆

◆ Sketching the Graph of a Polynomial Function

The most efficient method of sketching the graph of a polynomial function of degree $n \geq 3$ is to use calculus. The methods of calculus give the relative extrema of a polynomial function, and joining its relative extrema to form a smooth curve gives us its graph. However, if we can write a polynomial function as a product of linear factors $ax + b$, with a and b real numbers, then we can draw a rough sketch of its graph by using the following algebraic procedure.

◆ Algebraic Procedure for Graphing a Polynomial Function

To sketch the graph of a polynomial function $y = P(x)$ of degree $n \geq 3$:

Step 1   Express $P(x)$ as the product of linear factors $ax + b$, with a and b real numbers. Then determine the x-intercept(s) by solving the equation $P(x) = 0$.

Step 2   Determine the left and right behavior of the graph and the shape of the graph near each x-intercept.

Step 3   Sketch the graph of the function using the information in step 2 as a guide.

(*Note:* To sketch a more accurate graph, determine the y-intercept and plot a few additional points.)

EXAMPLE 3   Sketch the graph of each polynomial function. Label the intercepts.

(a) $P(x) = 2x^3 - x^4$ **(b)** $P(x) = x^3 - 2x^2 - 4x + 8$

◆ Solution

(a) Step 1 We use the techniques of factoring from Section 1.6:

$$P(x) = 2x^3 - x^4$$
$$= x^3(2 - x) \qquad \text{\textcolor{red}{\textbf{Factor out }} } x^3$$
$$= (x - 0)^3(2 - x) \qquad \text{\textcolor{red}{\textbf{Write as the product of linear factors}}}$$

To find the x-intercepts of the graph of P, we solve the equation $P(x) = 0$ by applying the zero product property (Section 2.2):

$$(x - 0)^3(2 - x) = 0$$
$$(x - 0)^3 = 0 \quad \text{or} \quad 2 - x = 0$$
$$x = 0 \qquad\qquad x = 2$$

$$\boxed{x - \text{intercepts}}$$

Step 2 For $P(x) = 2x^3 - x^4$, we have $a_n = -1$ and $n = 4$. Since $a_n < 0$ and n is even, we know that the graph of P goes down to the left as $x \to -\infty$ and down to the right as $x \to \infty$. To determine the shape of the graph near an x-intercept, we retain the factor corresponding to the intercept and evaluate the other factor(s) at the x-intercept:

Near $x = 0$:

$$P(x) = (x - 0)^3(2 - x)$$

retain | evaluate at $x = 0$

$$\approx (x - 0)^3(2 - 0)$$
$$\approx 2x^3$$

a cubing function stretched vertically by a factor of 2

Near $x = 2$:

$$P(x) = (x - 0)^3(2 - x)$$

evaluate at $x = 2$ | retain

$$\approx (2 - 0)^3(2 - x)$$
$$\approx 16 - 8x$$

a line with slope -8

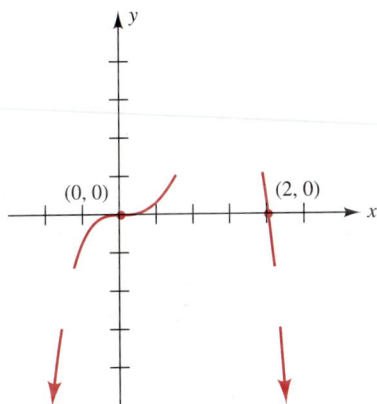

FIGURE 4.9

The left and right behavior of the graph of $P(x) = 2x^3 - x^4$ and its shape near each x-intercept

Figure 4.9 summarizes the information in Step 2.

Step 3 The information in Figure 4.9 enables us to sketch a rough graph of this function. For a more accurate sketch, we also plot the point $(1, 1)$, as shown in Figure 4.10. The graph of this polynomial function has one relative extremum, which satisfies the relative extrema rule for $n = 4$. Using calculus, it can be shown that this relative maximum point has coordinates $\left(\frac{3}{2}, \frac{27}{16}\right)$.

(b) Step 1 We again use the techniques of factoring from Section 1.6:

$$P(x) = x^3 - 2x^2 - 4x + 8$$
$$= x^2(x - 2) - 4(x - 2) \qquad \text{\textcolor{red}{\textbf{Factor by grouping terms}}}$$
$$= (x - 2)(x^2 - 4)$$
$$= (x - 2)^2(x + 2) \qquad \text{\textcolor{red}{\textbf{Factor the difference of squares}}}$$

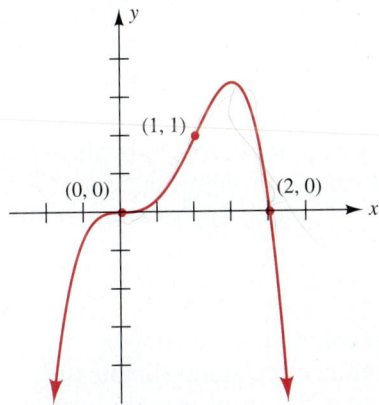

FIGURE 4.10

Graph of $P(x) = 2x^3 - x^4$

To find the x-intercepts of the graph of P, we solve the equation $P(x) = 0$ by applying the zero product property (Section 2.2):

$$(x - 2)^2(x + 2) = 0$$
$$(x - 2)^2 = 0 \quad \text{or} \quad x + 2 = 0$$
$$x = 2 \qquad\qquad x = -2$$

x – intercepts

Step 2　For $P(x) = x^3 - 2x^2 - 4x + 8$, we have $a_n = 1$ and $n = 3$. Since $a_n > 0$ and n is odd, we know that the graph of P goes down to the left as $x \to -\infty$ and up to the right as $x \to \infty$. To determine the shape of the graph near an x-intercept, we retain the factor corresponding to the intercept and evaluate the other factor(s) at the x-intercept:

Near $x = 2$:

$$P(x) = (x - 2)^2(x + 2)$$

retain | evaluate at $x = 2$

$$\approx (x - 2)^2(2 + 2)$$
$$\approx 4(x - 2)^2$$

a parabola stretched vertically by a factor of 4

Near $x = -2$:

$$P(x) = (x - 2)^2(x + 2)$$

evaluate at $x = -2$ | retain

$$\approx (-2 - 2)^2(x + 2)$$
$$\approx 16x + 32$$

a line with slope 16

FIGURE 4.11

The left and right behavior of the graph of $P(x) = x^3 - 2x^2 - 4x + 8$ and its shape near each x-intercept

Figure 4.11 summarizes the information in Step 2.

Step 3　The information in Figure 4.11 enables us to sketch a rough graph of this function. For a more accurate sketch, we also plot the point $(-1, 9)$ and the y-intercept, as shown in Figure 4.12. Note that the graph of this polynomial function has two relative extrema, which satisfies the relative extrema rule for $n = 3$. The relative minimum point has coordinates $(2, 0)$. Using calculus, it can be shown that the relative maximum point has coordinates $\left(-\frac{2}{3}, \frac{256}{27}\right)$.　　◆

Most graphing calculators have built-in programs for determining the relative extrema of a polynomial function. If a calculator does not have these built-in programs, we use the TRACE key and ZOOM key to estimate the coordinates of any relative extrema.

FIGURE 4.12

Graph of $P(x) = x^3 - 2x^2 - 4x + 8$

PROBLEM 3　Generate the graph of each polynomial function in Example 3 in the viewing rectangle of a graphing calculator. Estimate the coordinates of any relative extrema and compare the answers to those given in the solution.　　◆

◆ Application: Bending Moment of a Beam

When a load (or weight) is applied to a steel or wood beam, the force produced by the load creates a *bending moment*. Due to this bending moment, the beam will deform slightly, resulting in a curvature of the beam. Structural engineers must investigate the bending moment at various points along a beam in order to select a beam of proper size to safely carry (support) a particular load. Bending moments are described frequently by polynomial functions of degree $n \geq 3$.

FIGURE 4.13

Simple beam with uniformly increasing load

EXAMPLE 4 For a simply supported beam of length L that carries a uniformly increasing load of weight W, as shown in Figure 4.13, the bending moment M at any distance x from the left end of the beam is given by

$$M(x) = \frac{Wx}{3L^2}(L^2 - x^2).$$

(a) If $W = 300$ lb and $L = 10$ ft, express the function M as a product of linear factors. Then sketch the graph of this function.

(b) Generate the graph of the function in part (a) in the viewing rectangle of a graphing calculator. Then approximate the distance x (from the left end of the beam) at which the bending moment is greatest.

◆ Solution

(a) Replacing W with 300 and L with 10 gives us

$$
\begin{aligned}
M(x) &= \frac{300x}{3(10)^2}(10^2 - x^2) \\
&= 100x - x^3 \\
&= x(10 + x)(10 - x) \qquad \textbf{\textit{\textcolor{red}{Write as the product of linear factors}}}
\end{aligned}
$$

Observe that M is a polynomial function of degree 3. Since the coefficient of x^3 is negative, the graph of M goes up to the left as $x \to -\infty$ and down to the right as $x \to \infty$. Also, by the relative extrema rule, the graph of M has either no relative extremum or two relative extrema.

To find the x-intercepts of the graph of M, we solve the equation

$$x(10 + x)(10 - x) = 0.$$

By the zero product property (Section 2.2), the x-intercepts are

$$x = 0 \qquad x = -10 \qquad x = 10$$

Now we investigate the shape of the graph near each x-intercept.

Near $x = 0$:

$$M(x) = x(10 + x)(10 - x)$$
$$\approx x(10)(10)$$
$$\approx 100x$$

a line with slope 100

Near $x = -10$:

$$M(x) = x(10 + x)(10 - x)$$
$$\approx -10(10 + x)(20)$$
$$\approx -2000 - 200x$$

a line with slope -200

Near $x = 10$:

$$M(x) = x(10 + x)(10 - x)$$
$$\approx 10(20)(10 - x)$$
$$\approx 2000 - 200x$$

Figure 4.14 summarizes the information we have gathered. The large slopes near the x-intercepts indicate that $M(x)$ may be quite large for certain values of x. For example, when $x = 5$, we obtain $M(5) = 375$. Graphically, we can compensate for these large values of $M(x)$ by selecting different scales for the x-axis and M-axis. Although the choice of scales is arbitrary, a reasonable choice is to choose 5 units between tick marks on the x-axis and 100 units between tick marks on the M-axis. The graph of M is shown in Figure 4.15. Since the length of the beam is 10 ft, we must have $0 \leq x \leq 10$. We indicate this restriction on the domain of this function by showing a solid curve in the interval $[0, 10]$ and a dashed curve elsewhere.

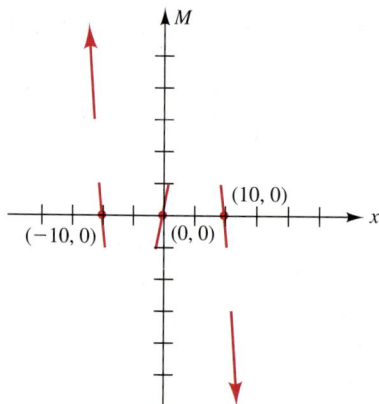

FIGURE 4.14

The left and right behavior of the graph of $M(x) = 100x - x^3$ and its shape near each x-intercept

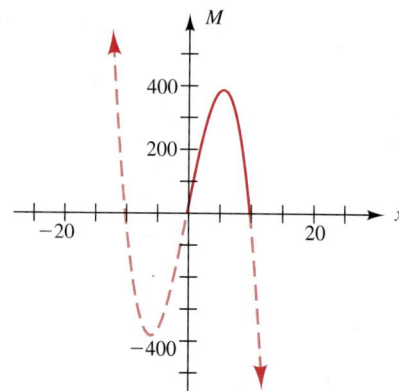

FIGURE 4.15

Graph of $M(x) = 100x - x^3$

(b) To generate the graph of $y = M(x) = 100x - x^3$ for $0 \leq x \leq 10$ on a graphing calculator, we begin by choosing a viewing rectangle. We use our work in part (a) to make a reasonable choice:

Xmin = 0
Xmax = 10
Xscl = 1
Ymin = 0
Ymax = 400
Yscl = 100

400

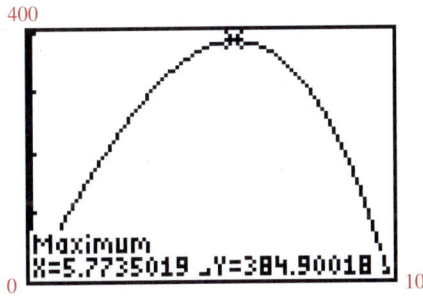

Maximum
X=5.7735019 _Y=384.90018

0   10

FIGURE 4.16
The graph of $M(x) = 100x - x^3$,
$0 \le x \le 10$, in the viewing rectangle of a
graphing calculator

Now, by entering the equation $y = 100x - x^3$, we obtain the desired portion of the graph in the viewing rectangle (see Figure 4.16). The point where the bending moment is greatest is at the peak of this curve. Using either the calculator's built-in program for finding a relative maximum or the TRACE key and ZOOM key, we find that the coordinates of the relative maximum are (5.774, 384.900) to the nearest thousandths. In summary, the maximum bending moment along the beam is approximately 384.900 pound-feet (lb-ft), and this occurs when $x \approx 5.774$ ft. Using calculus, it can be shown that the *exact* maximum bending moment is $\dfrac{2000\sqrt{3}}{9}$ lb-ft, and this occurs when $x = \dfrac{10\sqrt{3}}{3}$ ft. ◆

PROBLEM 4 Is $M(x) = 100x - x^3$ an even function, an odd function, or neither? Explain. ◆

Exercises 4.1

◆ Basic Skills

In Exercises 1–12, use the shift rules, axis reflection rules, and stretch and compress rules (from Section 2.5) to sketch the graph of each polynomial function. Label any intercepts. Use a graphing calculator to verify each graph.

1. $f(x) = -x^5$
2. $f(x) = -x^4$
3. $f(x) = x^4 + 1$
4. $f(x) = x^5 - 1$
5. $f(x) = \frac{1}{4}(x - 1)^7$
6. $f(x) = 2(x + 1)^6$
7. $f(x) = 64 - x^6$
8. $f(x) = 3 - 3x^7$
9. $f(x) = (x - 3)^4 - 16$
10. $f(x) = (x + 1)^5 + 32$
11. $f(x) = 1 - (2 - x)^5$
12. $f(x) = 16 - (1 - x)^4$

For each polynomial function in Exercises 13–20, determine the following features of its graph: (a) the left and right behavior and (b) the possible number of relative extrema.

13. $P(x) = 2x^3 - 7x^2 + 3x - 2$
14. $P(x) = 7 - 5x - 4x^3$
15. $F(x) = 6 - 2x + 8x^2 - 5x^4$
16. $f(x) = 3x^8 - 2x^7 + x^6 - 3x + 1$
17. $P(x) = -x^7 - 2x^5 + 4x - 3$
18. $P(x) = 2x^5 - 20x^4 + 3x - 9$
19. $g(x) = 3x^6 - 2x^4 + 3x^2 + 1$
20. $G(x) = -5x^6 + 3x^5 + 18x^4 - 1$

In Exercises 21–28, generate a complete graph of each polynomial function in the viewing rectangle of a graphing calculator. Then determine the coordinates of any relative extrema. Round each coordinate to three significant digits.

21. The function P in Exercise 13
22. The function P in Exercise 14
23. The function F in Exercise 15
24. The function f in Exercise 16
25. The function P in Exercise 17
26. The function P in Exercise 18
27. The function g in Exercise 19
28. The function G in Exercise 20

In Exercises 29–46, sketch the graph of each polynomial function by using the three-step procedure outlined in this section. Label any intercepts. Use a graphing calculator to verify each graph.

29. $P(x) = x^3 + x^2 - 2x$
30. $P(x) = 12 + 4x - 3x^2 - x^3$
31. $G(x) = x^4 - 10x^2 + 9$
32. $P(x) = 3 - 2x^2 - x^4$
33. $h(x) = 3x^3 - x^4$
34. $f(x) = 4x^3 - 4x^2 + x$
35. $P(x) = x^5 - x^4$
36. $F(x) = 9x^3 - 4x^5$
37. $P(x) = 3x^4 - 12x^3 + 12x^2$
38. $Q(x) = 2x^5 - 12x^4 + 18x^3$
39. $f(x) = x^4 - 2x^2 + 1$

40. $h(x) = 16x^2 - 8x^4 + x^6$

41. $D(x) = x^3 + 2x^2 - 4x - 8$

42. $P(x) = 8x^3 + 4x^2 - 2x - 1$

43. $P(x) = x^2(x - 2)^3$ **44.** $P(x) = 2x(x + 3)^4$

45. $f(x) = 3x^3(x + 1)^2(x - 4)^3$

46. $G(x) = 4x^4(x - \frac{1}{2})^2(x + 6)^2$

47. The response time T (in microseconds) to a certain stimulus is a function of one's age x (in years) and is given by

$$T(x) = 20 + 79x + 76x^2 - 4x^3 \quad \text{for } 10 \le x \le 18.$$

Generate the graph of this function in the viewing rectangle of a graphing calculator. Then use this graph to estimate the age (to the nearest year) when the response time is greatest.

48. The profit P (in thousands of dollars) that a manufacturer makes is a function of the number of units x (in hundreds) that are produced and is given by

$$P(x) = 15 + 4x^2 - x^3 \quad \text{for } 0 \le x \le 6.$$

Generate the graph of this function in the viewing rectangle of a graphing calculator. Then use this graph to estimate the number of units (to the nearest whole number) that should be produced to maximize the manufacturer's profit.

49. A rectangular storage bin, with a square base of dimension x and an open top, is made from 36 square feet of material.

(a) Show that the volume V of the box can be written as the product of linear factors:

$$V = \tfrac{1}{4}x(6 - x)(6 + x)$$

Then state the domain of the function.

(b) Use the three-step procedure outlined in this section to sketch the graph of the polynomial function defined in part (a).

(c) Generate the graph of the function defined in part (a) on a graphing calculator. Then find the dimension x for the storage bin with the largest volume.

50. A rectangle has its base on the x-axis and its two upper vertices touch the parabola $y = 3 - x^2$ at the points $P(x, y)$ and $Q(-x, y)$.

(a) Show that the area A of the rectangle can be written as the product of linear factors:

$$A = 2x(\sqrt{3} - x)(\sqrt{3} + x)$$

Then state the domain of the function.

(b) Use the three-step procedure that is outlined in this section to sketch the graph of the polynomial function defined in part (a).

(c) Generate the graph of the function defined in part (a) on a graphing calculator. Then find the coordinates of points P and Q for the rectangle with the largest area.

A trucking company will accept a package for shipment only if the sum of the length and girth (the distance around the package) does not exceed 180 inches. For the packages shown in Exercises 51 and 52, answer the following questions.

(a) *Express the volume V of the package as a function of x. Then state the domain of the function.*

(b) *Use the three-step procedure outlined in this section to sketch the graph of the polynomial function defined in part (a).*

(c) *Generate the graph of the function defined in part (a) on a graphing calculator. Then find the dimension x for the package with the largest volume that can be accepted for shipping.*

51.

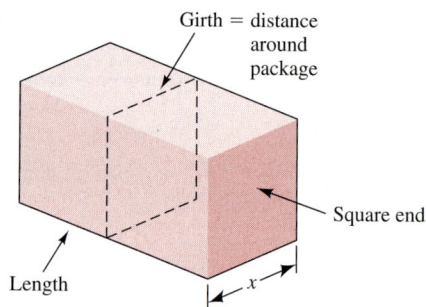

Girth = distance around package

Length

Square end

x

52.

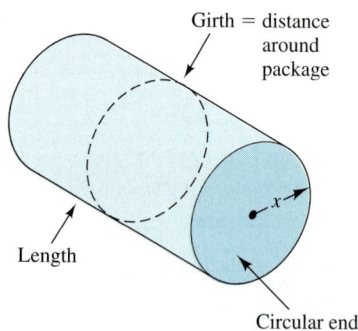

Girth = distance around package

Length

Circular end

x

53. A piece of cardboard, 8 cm by 16 cm, is cut and folded as shown in the sketch to form a rectangular box with a closed top and flap.

(a) Express the volume V of this box as a function of x.

(b) Use the three-step procedure outlined in this section to sketch the graph of the polynomial function defined in part (a).

(c) Generate the graph of the function in part (a) on a graphing calculator. Then find the approximate value of x for which the volume of the box is greatest.

54. A beam of length L is fixed at one end and is simply supported at the other end. It carries a uniformly increasing load of weight W, as shown in the figure. The bending moment M at any distance x from the left end of this beam is given by

$$M(x) = \frac{Wx}{15L^2}(3L^2 - 5x^2).$$

(a) Determine the bending moment function when $W = 1500$ lb and $L = 10$ ft.

(b) Use the three-step procedure outlined in this section to sketch the graph of the polynomial function defined in part (a).

(c) Generate the graph of the function in part (a) on a graphing calculator. Then find the approximate distance x (from the left end of the beam) at which the bending moment is greatest.

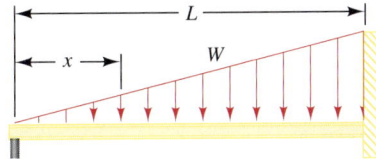

Critical Thinking

In Exercises 55–62,

(a) *sketch the possible shapes of the graph of a polynomial function with the given characteristics.*

(b) *use a graphing calculator to find a particular polynomial function whose graph has the shape given in part (a). Answers are not unique.*

55. Degree 3 with a negative leading coefficient

56. Degree 2 with a positive leading coefficient

57. Degree 4 with a positive leading coefficient

58. Degree 4 with a negative leading coefficient

59. Degree 5 with a positive leading coefficient

60. Degree 7 with a negative leading coefficient

61. Degree 6 with a negative leading coefficient

62. Degree 6 with a positive leading coefficient

63. By solving the equations simultaneously, determine the coordinates of the point(s) where the graphs of the equations intersect. Check graphically.

(a) $y = x^3 - x^2$ and $y = 3x - 3$

(b) $y = x^3 - 3x^2 - 4x$ and $x - y = 15$

64. Determine the maximum number of times the graphs of two polynomial functions P and Q may intersect if the degree of P is m and the degree of Q is n, provided that $m > n$? Explain.

In Exercises 65 and 66, sketch the graph of a polynomial function P with the given characteristics.

65. Degree 3; an odd function; $P(-2) = 0$; $P(4) = -12$

66. Degree 4; an even function; $P(0) = -27$; graph is tangent to the x-axis at $(3, 0)$

67. May a polynomial function of degree 5 be an even function? If so, give an example. If not, explain why.

68. Shown in the sketch are the graphs of several power functions $y = x^n$ with $n \geq 2$ and $0 \leq x \leq 1$. Which graph represents the function with the largest exponent n? Explain.

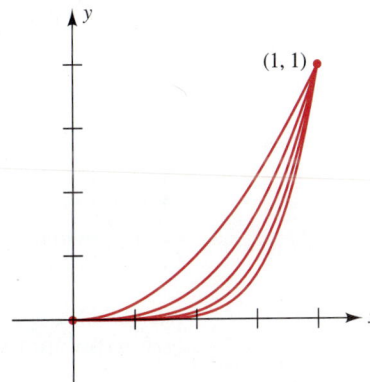

4.2 Polynomial Division

◆ **Introductory Comments**

In order to work with polynomial functions of degree 3 or greater, we must be able to divide one polynomial by another. The procedure that we use is similar to the one for dividing whole numbers. For example, to divide 349 by 15, we can use the long division scheme from arithmetic:

$$
\begin{array}{r}
23 \quad\longleftarrow \text{quotient}\\
\text{divisor} \longrightarrow 15\overline{)349} \quad\longleftarrow \text{dividend}\\
\underline{30}\\
49\\
\underline{45}\\
4 \quad\longleftarrow \text{remainder}
\end{array}
$$

Since 23 is the quotient and 4 is the remainder, we write

$$
\underbrace{\frac{349}{15}}_{\text{divisor}} = \underbrace{23}_{\text{quotient}} + \underbrace{\frac{4}{15}}_{\text{divisor}} \qquad \text{or} \qquad 23\tfrac{4}{15}.
$$

(dividend, quotient, remainder)

In a similar manner, we can divide one polynomial $P(x)$ by another polynomial $D(x)$, provided that $D(x) \neq 0$. This long division scheme, which is called the **division algorithm for polynomials**, yields a quotient polynomial $Q(x)$ and a remainder polynomial $R(x)$ whose degree is less than that of the divisor $D(x)$ such that

$$
\frac{P(x)}{D(x)} = Q(x) + \frac{R(x)}{D(x)}
$$

◆ **Polynomial Long Division**

We refer to the procedure of dividing one polynomial by another polynomial as *polynomial long division*.

EXAMPLE 1   Divide $P(x) = 2x^3 - 5x^2 + 3x - 4$ by $D(x) = x - 2$ using polynomial long division.

◆ **Solution**   We begin by dividing the first term of the divisor, x, into the first term of the dividend, $2x^3$, to obtain the first term of the quotient:

divisor $D(x)$ | first term of quotient $Q(x)$

$$\begin{array}{r} 2x^2 \\ x - 2\overline{)2x^3 - 5x^2 + 3x - 4} \end{array}$$ — dividend $P(x)$

Next, we multiply the first term of the quotient by the divisor and *subtract* this product from the dividend to obtain the first remainder:

divisor $D(x)$ | first term of quotient $Q(x)$

$$\begin{array}{r} 2x^2 \\ x - 2\overline{)2x^3 - 5x^2 + 3x - 4} \\ \underline{2x^3 - 4x^2 } \\ -x^2 + 3x - 4 \end{array}$$

dividend $P(x)$

first remainder

Now, using the first remainder as the *new* dividend, we continue this process until we obtain a remainder that is either zero or a polynomial whose degree is lower than that of the divisor:

divisor $D(x)$ | quotient $Q(x)$

$$\begin{array}{r} 2x^2 - x + 1 \\ x - 2\overline{)2x^3 - 5x^2 + 3x - 4} \\ \underline{2x^3 - 4x^2 } \\ -x^2 + 3x - 4 \\ \underline{-x^2 + 2x } \\ x - 4 \\ \underline{x - 2} \\ -2 \end{array}$$

dividend $P(x)$

remainder $R(x)$

Thus, we write

$$\frac{2x^3 - 5x^2 + 3x - 4}{x - 2} = 2x^2 - x + 1 + \frac{-2}{x - 2}.$$

◆

To check polynomial long division, we use the same method of checking as for arithmetic division, and show that

dividend | quotient

$$P(x) = D(x)Q(x) + R(x).$$

divisor | remainder

PROBLEM 1 Check the polynomial long division in Example 1. ◆

Before beginning the polynomial long division process, we usually write the dividend and divisor in descending powers of x, using 0 as a coefficient of any term that is missing from either the dividend or divisor. This helps us align like terms in the same column when we carry out the division.

EXAMPLE 2   Divide $P(x) = 4x^4 + 3x^5 - 2x^2 - 3x - 2$ by $D(x) = x^2 - 1$ using polynomial long division.

◆ **Solution**   We first arrange the dividend and divisor in descending powers of x, using 0 as a coefficient of any missing term. Then carry out the long division as follows:

$$
\begin{array}{r}
3x^3 + 4x^2 + 3x + 2 \\
x^2 + 0x - 1\overline{)\,3x^5 + 4x^4 + 0x^3 - 2x^2 - 3x - 2} \\
\underline{3x^5 + 0x^4 - 3x^3} \\
4x^4 + 3x^3 - 2x^2 - 3x - 2 \\
\underline{4x^4 + 0x^3 - 4x^2} \\
3x^3 + 2x^2 - 3x - 2 \\
\underline{3x^3 + 0x^2 - 3x} \\
2x^2 + 0x - 2 \\
\underline{2x^2 + 0x - 2} \\
0
\end{array}
$$

Since the remainder is 0, we write

$$\frac{3x^5 + 4x^4 - 2x^2 - 3x - 2}{x^2 - 1} = 3x^3 + 4x^2 + 3x + 2. \quad ◆$$

PROBLEM 2   Divide $P(x) = 6x^3 - 3x^2$ by $D(x) = 3x^2 - 2$ using polynomial long division. ◆

◆ **Synthetic Division**

As we illustrated in Example 1, when a polynomial $P(x)$ of degree $n \geq 1$ is divided by a first-degree polynomial $D(x) = x - r$, the remainder $R(x)$ is a constant k and the degree of the quotient polynomial $Q(x)$ is one less than the degree of $P(x)$. That is,

$$\frac{P(x)}{x - r} = Q(x) + \frac{k}{x - r}.$$

A shortcut called *synthetic division* may be used to divide a polynomial of degree $n \geq 1$ by a first-degree polynomial $x - r$. In **synthetic division**, we use only the essential data from the long division process by eliminating any power of x or coefficient that is a duplicate of that directly above it (in the same column). Also, we replace the divisor $x - r$ by the *synthetic divisor r* so that we may *add* rather than subtract columns. In the next example, we re-work Example 1 using synthetic division.

EXAMPLE 3 Divide $P(x) = 2x^3 - 5x^2 + 3x - 4$ by $D(x) = x - 2$ using synthetic division.

◆ **Solution**   We begin by replacing the divisor $x - 2$ with the synthetic divisor 2 and writing the coefficients of the dividend:

synthetic divisor ——— $2\overline{)2 \quad -5 \quad 3 \quad -4}$ ——— coefficients of $P(x)$

We bring down the leading coefficient two rows, multiply the synthetic divisor by the leading coefficient, and record this product in the middle row under the second coefficient of $P(x)$. Then we add to obtain the sum of the second column:

synthetic divisor ——— $2\overline{)2 \quad -5 \quad 3 \quad -4}$ ——— coefficients of $P(x)$
$ 4$
leading coefficient ——— $2 \quad -1$
 sum of second column

Now, we multiply the synthetic divisor by the sum of the second column; record this product in the middle row under the third coefficient of $P(x)$; then add to obtain the sum of the third column. We continue in this manner until all columns have a sum:

synthetic divisor ——— $2\overline{)2 \quad -5 \quad 3 \quad -4}$ ——— coefficients of $P(x)$
$ 4 \quad -2 \quad 2$
leading coefficient ——— $2 \quad -1 \quad 1 \quad -2$
sum of second column
sum of third column
sum of last column

The sum of the last column is the constant remainder k. The preceding sums, to the left of the last sum, are the coefficients of the successive terms in the quotient polynomial $Q(x)$, whose degree is one less than that of $P(x)$. Thus, from the synthetic division

$$2\overline{)2 \quad -5 \quad 3 \quad -4}$$
$$4 \quad -2 \quad 2$$
$$\boxed{2} \quad \boxed{-1} \quad \boxed{1} \quad \boxed{-2}$$

we can state

$$\frac{2x^3 - 5x^2 + 3x - 4}{x - 2} = 2x^2 - x + 1 + \frac{-2}{x - 2}$$

This result agrees with the quotient and remainder polynomials we obtained in Example 1.   ◆

PROBLEM 3 Divide $P(x) = x^3 - 5x^2 + 7x - 3$ by $D(x) = x - 1$ using synthetic division. ◆

CAUTION The shortcut of synthetic division works only when the divisor $D(x)$ is a first-degree polynomial of the form $x - r$. For division problems such as Example 2, where the divisor $D(x)$ is quadratic, synthetic division does *not* apply. See Exercises 71–76 for the procedure to use when the divisor $D(x)$ is a first-degree polynomial of the form $ax + b$.

In synthetic division, we must record the coefficients of the dividend $P(x)$ in the order that they appear when $P(x)$ is written in descending powers of x, and if a power of x is missing, we must record 0 as its coefficient.

EXAMPLE 4 Divide $P(x) = 2x^4 - 15x^2 + 9x + 1$ by $D(x) = x + 3$ using synthetic division.

◆ Solution Since the divisor $D(x) = x + 3 = x - (-3)$, we use -3 as the synthetic divisor. Now, after inserting 0 for the coefficient of the x^3 term in the dividend, we carry out the synthetic division as follows:

synthetic divisor —— $-3\overline{)2 \quad 0 \quad -15 \quad 9 \quad 1}$ —— coefficients of $P(x)$

$$\quad\quad\quad\quad -6 \quad 18 \quad -9 \quad 0$$

$$2 \quad -6 \quad 3 \quad 0 \quad \boxed{1}$$ —— constant remainder k

coefficients of $Q(x)$, whose degree is 1 less than $P(x)$

Thus,

$$\frac{2x^4 - 15x^2 + 9x + 1}{x + 3} = 2x^3 - 6x^2 + 3x + \frac{1}{x + 3}.$$ ◆

We can check the synthetic division process by showing that

$$P(x) = (x - r)Q(x) + k.$$

PROBLEM 4 Check the synthetic division in Example 4. ◆

◆ The Remainder Theorem

If $Q(x)$ is the quotient polynomial and k is the constant remainder obtained when a polynomial $P(x)$ is divided by a polynomial $x - r$, then we have

$$\frac{P(x)}{x - r} = Q(x) + \frac{k}{x - r}.$$

Multiplying both sides of this equation by $x - r$, we obtain

$$P(x) = (x - r)Q(x) + k.$$

Now, replacing x with r, we find

$$P(r) = (r - r)Q(r) + k$$
$$= 0 \cdot Q(r) + k$$
$$= k$$

In other words, $P(r)$ is the same number as the remainder k obtained by dividing $P(x)$ by $x - r$. We refer to this fact about polynomial division as the **remainder theorem**.

◆ **Remainder Theorem**

> When a polynomial $P(x)$ is divided by $x - r$, the remainder is $P(r)$.

EXAMPLE 5 Use the remainder theorem to determine the remainder when

$$P(x) = x^4 - 3x^3 - 4x^2 + 13x - 8 \qquad \text{is divided by} \qquad x - 3.$$

◆ **Solution** The remainder theorem tells us that when $P(x)$ is divided by $x - 3$, the remainder is $P(3)$. Evaluating $P(3)$, we find

$$P(3) = (3)^4 - 3(3)^3 - 4(3)^2 + 13(3) - 8$$
$$= 81 - 81 - 36 + 39 - 8$$
$$= -5$$

Thus, when $P(x)$ is divided by $x - 3$, the remainder is -5. ◆

PROBLEM 5 Verify the remainder in Example 5 by dividing $P(x)$ by $x - 3$ using synthetic division. ◆

As illustrated in the following example, we can evaluate $P(r)$ by using the remainder theorem and synthetically dividing $P(x)$ by $x - r$. This approach is often easier than finding $P(r)$ by direct substitution.

EXAMPLE 6 Given that $P(x) = 4x^4 - 6x^3 + 4x - 3$, find $P\left(-\frac{1}{2}\right)$ by using synthetic division and the remainder theorem.

◆ **Solution** According to the remainder theorem, $P\left(-\frac{1}{2}\right)$ is the remainder obtained when $P(x)$ is divided by $x - \left(-\frac{1}{2}\right) = x + \frac{1}{2}$. Using synthetic division, we find the remainder:

$$
\begin{array}{r|rrrrr}
-\frac{1}{2} & 4 & -6 & 0 & 4 & -3 \\
 & & -2 & 4 & -2 & -1 \\
\hline
 & 4 & -8 & 4 & 2 & \boxed{-4} \quad \leftarrow \text{remainder}
\end{array}
$$

FIGURE 4.17
For the polynomial function
$P(x) = 4x^4 - 6x^3 + 4x - 3$, the graph confirms that $P\left(-\frac{1}{2}\right) = -4$ is a functional value.

Thus, $P\left(-\frac{1}{2}\right) = -4$. Figure 4.17 shows the graph of this polynomial function in the viewing rectangle of a graphing calculator. Note that $\left(-\frac{1}{2}, -4\right)$ is a point on this graph, which confirms our answer. ◆

PROBLEM 6   Given that $P(x) = 4x^4 - 6x^3 + 4x - 3$, find $P(\frac{1}{2})$ by using synthetic division and the remainder theorem. Check graphically, and compare your result with Figure 4.17.   ◆

If $P(x)$ is a polynomial and r is a complex number (see Section 1.4 for a discussion of complex numbers) such that $P(r) = 0$, we say that r is a **zero** of the polynomial function P or a **root** of the polynomial equation $P(x) = 0$. Also, recall from Section 2.4 that any *real* zero of a function P represents an x-intercept of the graph of P.

EXAMPLE 7   Given the function P defined by $P(x) = 3x^3 - 8x^2 + 10x - 4$, use synthetic division and the remainder theorem to determine whether each number is a zero of P:

(a) $\frac{2}{3}$ **(b)** $1 + i$

◆ **Solution**

(a) If $\frac{2}{3}$ is a zero of the function P, then $P(\frac{2}{3}) = 0$. Using synthetic division, we find

$$\frac{2}{3}\overline{)\begin{array}{rrrr} 3 & -8 & 10 & -4 \\ & 2 & -4 & 4 \\ \hline 3 & -6 & 6 & \boxed{0} \end{array}} \quad\text{remainder}$$

Since the remainder is 0, we know that $P(\frac{2}{3}) = 0$, and so $\frac{2}{3}$ is a zero of the polynomial function P defined by $P(x) = 3x^3 - 8x^2 + 10x - 4$. Figure 4.18 shows the graph of P in the viewing rectangle of a graphing calculator. Note that $\frac{2}{3}$ is an x-intercept of this graph.

(b) If $1 + i$ is a zero of the function P, then $P(1 + i) = 0$. Using synthetic division and recalling from Section 1.4 that $i^2 = -1$, we find

$$1 + i\,\overline{)\begin{array}{rrrr} 3 & -8 & 10 & -4 \\ & 3 + 3i & -8 - 2i & 4 \\ \hline 3 & -5 + 3i & 2 - 2i & \boxed{0} \end{array}} \quad\text{remainder}$$

Since the remainder is 0, we know that $P(1 + i) = 0$, and so $1 + i$ is a zero of the polynomial function P defined by $P(x) = 3x^3 - 8x^2 + 10x - 4$. Although the imaginary number $1 + i$ is a zero of P, it is not an x-intercept of the graph of P. Remember, only real zeros of a function P represent x-intercepts of the graph of P.   ◆

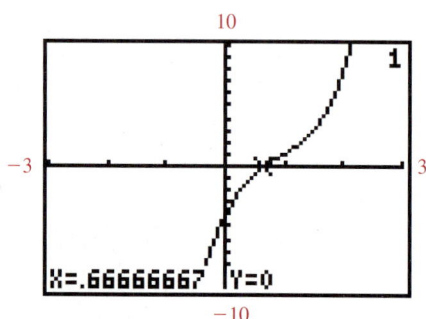

FIGURE 4.18

If $\frac{2}{3}$ is a real zero of $P(x) = 3x^3 - 8x^2 + 10x - 4$, then $\frac{2}{3}$ is an x-intercept of the graph of P.

PROBLEM 7   Use synthetic division and the remainder theorem to determine if $1 - i$ is a zero of the polynomial function P defined in Example 7.   ◆

Exercises 4.2

◆ **Basic Skills**

In Exercises 1–16, divide P(x) by D(x) using polynomial long division.

1. $P(x) = 3x^3 - 2x^2 + x - 3$
$D(x) = x - 1$

2. $P(x) = 2x^4 + 3x^3 - 4x^2 - 2x + 1$
$D(x) = x + 2$

3. $P(x) = 3x^4 - 10x^3 + 16x^2 - 14x + 3$
$D(x) = x^2 - 2x + 3$

4. $P(x) = 2x^3 - 7x^2 + 11x - 4$
$D(x) = x^2 - 3x + 4$

5. $P(x) = 4x^3 + 3x - 3$
$D(x) = 2x - 1$

6. $P(x) = 9x^4 + 6x^3 + 3x - 2$
$D(x) = 3x + 2$

7. $P(x) = 2x^5 + 2x^3 - 2x^2 + 5$
$D(x) = 2x^3 + 1$

8. $P(x) = x^6 - 2x^2 - 3x + 4$
$D(x) = x^3 - 3x$

9. $P(x) = x^3$
$D(x) = x + 1$

10. $P(x) = x^4$
$D(x) = x^2 + 1$

11. $P(x) = 6x^5 - x^3 + 2x^2 - x - 1$
$D(x) = 1 - 2x^2$

12. $P(x) = 3x^4 + 4x^3 - 15x^2 - 4x - 12$
$D(x) = 6 - x - x^2$

13. $P(x) = 3 - 6x$
$D(x) = 2x + 1$

14. $P(x) = 8 - 2x - x^2$
$D(x) = x^2 + 1$

15. $P(x) = 8x^3$
$D(x) = (2x - 1)^2$

16. $P(x) = (x + 1)^4$
$D(x) = x^2 + 1$

In Exercises 17–34, divide P(x) by D(x) using synthetic division. In each case, verify the remainder by using the remainder theorem.

17. $P(x) = 2x^2 - 3x + 4$
$D(x) = x - 1$

18. $P(x) = 3x^2 - 6x - 5$
$D(x) = x + 3$

19. $P(x) = x^3 + 5x^2 - 2x + 4$
$D(x) = x + 2$

20. $P(x) = x^4 - 6x^3 + 4x^2 + 15x + 2$
$D(x) = x - 4$

21. $P(x) = 2x^4 - 7x^3 - 3x^2 - 16$
$D(x) = x - 4$

22. $P(x) = 3x^3 - 70x + 25$
$D(x) = x + 5$

23. $P(x) = 4x^4 + 25x^3 - 32x$
$D(x) = x + 6$

24. $P(x) = x^5 - 6x^3 + 7x^2 - 3$
$D(x) = x - 2$

25. $P(x) = 3 - x^5$
$D(x) = x - 2$

26. $P(x) = 7 + x - 2x^3$
$D(x) = x + 1$

27. $P(x) = 3x^3 - 8x^2 + 19x - 10$
$D(x) = x - \frac{2}{3}$

28. $P(x) = 4x^3 - 8x^2 + x + 3$
$D(x) = x + \frac{1}{2}$

29. $P(x) = 2x^4 + x^3 - 19x + 8$
$D(x) = x + \frac{5}{2}$

30. $P(x) = 4x^4 + 3x^2 - 27$
$D(x) = x - \frac{3}{2}$

31. $P(x) = 14.5x^3 + 266.4x^2 - 828.1$
$D(x) = x + 18.2$

32. $P(x) = 2.5x^3 - 6.0x^2 - 22.3x + 41.4$
$D(x) = x - 3.6$

33. $P(x) = 275x^4 - 126x^2 - 627x - 1350$
$D(x) = x - 1.8$

34. $P(x) = 125x^4 + 2680x^3 - 1625x^2 - 1824x + 1012$
$D(x) = x + 22$

In Exercises 35–50, use synthetic division and the remainder theorem to find the indicated functional value. Check graphically whenever possible.

35. $P(x) = 2x^3 - 3x^2 + 4x + 5$; $P(1)$

36. $Q(x) = 3x^3 - 4x^2 + x - 9$; $Q(-1)$

37. $F(x) = x^4 + 2x^3 - 5x - 3$; $F(-2)$

38. $g(x) = x^3 - 2x^2 - 5x + 6$; $g(3)$

39. $P(x) = 2x^3 - 7x^2 + 2x + 1$; $P(4)$

40. $Q(x) = -x^3 + 4x^2 + 12x + 2$; $Q(6)$

41. $Q(x) = 4x^3 + 2x^2 - 1$; $Q(\frac{1}{2})$

42. $P(x) = 3x^4 - x^3 + x - 1$; $P(-\frac{2}{3})$

43. $f(x) = 3x^4 - 5x^3 + 7x^2 + 9x + 2$; $f(-\frac{1}{3})$

44. $H(x) = 5x^4 - 7x^2 + 12x - 4$; $H(\frac{2}{5})$

45. $h(x) = 2x^3 - 2x^2 + 4x + 7$; $h(1 - i)$

46. $G(x) = 2x^4 + x^3 - 8x^2 - 1$; $G(-2i)$

47. $P(x) = x^4 - x^3 + 3x^2 + 2x + 2$; $P(\sqrt{2})$

48. $P(x) = 3x^3 - 21x - 4$; $P(2 + \sqrt{3})$

49. $P(x) = 2.6x^3 - 9.8x^2 + 3.7x + 4.2$; $P(1.7)$

50. $P(x) = 0.035x^4 + 1.025x^3 - 2134$; $P(-32.2)$

In Exercises 51–58, use synthetic division to determine whether the given value of x is a zero of the polynomial function P. Check graphically whenever possible.

51. $P(x) = 2x^3 + x^2 - 14x - 3$; $x = -3$

52. $P(x) = 4x^3 + 2x^2 - 11x - 18$; $x = 2$

53. $P(x) = -2x^4 + 3x^3 + 8x^2 - 18$; $x = \frac{3}{2}$

54. $P(x) = -6x^4 + x^3 + 2$; $x = \frac{2}{3}$

55. $P(x) = x^4 - 3x^3 + 13x^2 - 27x + 36$; $x = 3i$

56. $P(x) = 4x^3 - 17x^2 + 24x - 5$; $x = 2 - i$

57. $P(x) = 3x^3 - 10x^2 + 5x + 4$; $x = 1 - \sqrt{2}$

58. $P(x) = x^4 - 2x^3 + x^2 + 6x - 12$; $x = \sqrt{3}$

In Exercises 59–66, use synthetic division to determine whether the given value of x is a root of the polynomial equation. Check graphically whenever possible.

59. $3x^4 - 5x^3 - 12x^2 + 19x + 2 = 0$; $x = -2$

60. $-6x^3 + 2x^2 + 5x + 21 = 0$; $x = 3$

61. $6x^3 + 2x^2 + 3x + 1 = 0$; $x = -\frac{1}{3}$

62. $25x^5 + x^3 - 2x^2 + 5x + 2 = 0$; $x = \frac{2}{5}$

63. $2x^3 - 3x^2 + 8x - 12 = 0$; $x = -2i$

64. $x^4 + 2x^3 + x^2 - 2x - 2 = 0$; $x = -1 + i$

65. $3x^4 - 2x^3 - 8x^2 + 4x + 4 = 0$; $x = \sqrt{2}$

66. $x^4 - 2x^3 - 6x^2 + 8x + 8 = 0$; $x = 1 + \sqrt{3}$

67. When a Patriot missile is fired from its launchpad, the height h (in meters) of the missile above the ground is a function of the horizontal distance x (in kilometers) from the launchpad and is given by

$$h(x) = 224x + 35x^2 - 24x^3 \quad \text{for } 0 \le x \le 4$$

Use synthetic division and the remainder theorem to find $h(2.45)$. Round your answer to three significant digits.

68. The current i (in milliamperes) flowing through an electric circuit is a function of time t (in seconds) and is given by

$$i(t) = 545 + 8.2t^2 - 1.2t^4 \quad \text{for } 0 \le t \le 5$$

Use synthetic division and the remainder theorem to find $i(3.8)$. Round your answer to three significant digits.

◆ Critical Thinking

69. When the polynomial $P(x)$ is divided by $x + 6$, the quotient is $x^3 + x^2 + x + 5$ and the remainder is -12. Find $P(x)$.

70. When $P(x) = 2x^3 - 5x^2 + 8x - 11$ is divided by the polynomial $D(x)$, the quotient is $x^2 - 2x + 3$ and the remainder is -8. Find $D(x)$.

Synthetic division may be used to divide a polynomial P(x) of degree n ≥ 1 by a first-degree polynomial D(x) = ax + b. Since

$$\frac{P(x)}{ax + b} = \frac{P(x)}{a\left(x + \dfrac{b}{a}\right)} = \frac{\dfrac{1}{a}P(x)}{x + \dfrac{b}{a}},$$

we can obtain the quotient and remainder by using a synthetic divisor $-b/a$ with dividend $(1/a)P(x)$. In Exercises 71–76, divide P(x) by D(x) using synthetic division.

71. $P(x) = 6x^3 + 7x^2 - 18x + 16$
     $D(x) = 3x + 8$

72. $P(x) = 8x^4 + 2x^3 - 3x^2 + 2x - 1$
     $D(x) = 2x - 1$

73. $P(x) = 32x^4 - 4x^3 + 7x^2 - 10x + 2$
     $D(x) = 4x - 1$

74. $P(x) = 3x^3 + x^2 + 2x + 8$
     $D(x) = 3x + 4$

75. $P(x) = x^4 - 2x^3 + 2x^2 - 7x + 6$
     $D(x) = 2 - x$

76. $P(x) = 24 - 20x - 12x^2 + 28x^3 - 15x^4$
     $D(x) = 6 - 5x$

In Exercises 77 and 78, use synthetic division to determine a value of k such that −2 is a zero of the given polynomial function. Check graphically.

77. $P(x) = x^4 + 2x^3 + 6x^2 + kx + 8$

78. $P(x) = kx^3 + 3x^2 + 6x - 16$

In Exercises 79 and 80, use synthetic division to determine a value of k such that 4 is a root of the given polynomial equation. Check graphically.

79. $kx^3 - 10x^2 + kx + 24 = 0$

80. $2x^3 + k^2x^2 - 2x + 88k = 0$

4.3 Factors and Zeros of Polynomial Functions

◆ Introductory Comments

Finding the zeros of a polynomial function of degree 2 is simple: apply the quadratic formula. Finding the zeros of a polynomial function of degree 3 or greater is usually much more difficult, but mathematicians have developed formulas for this purpose. Unlike the quadratic formula, however, formulas for finding the zeros of a polynomial function of degree 3 or 4 are extremely complicated. Mathematicians have also shown that no such general formulas exist for finding the zeros of polynomial functions of degree 5 or greater.

In this section, we discuss some theorems that we can use to help find the zeros of a polynomial function. We begin by proving the factor theorem, which shows a relationship between the zeros and factors of a polynomial function.

◆ The Factor Theorem

If $Q(x)$ is the quotient polynomial and k the constant remainder obtained when a polynomial $P(x)$ is divided by $x - r$, then we have

$$\frac{P(x)}{x - r} = Q(x) + \frac{k}{x - r}.$$

Multiplying both sides of this equation by $x - r$, we obtain

$$P(x) = (x - r)Q(x) + k.$$

Now by the remainder theorem (Section 4.2), we know that $P(r) = k$. So we can also write

$$P(x) = (x - r)Q(x) + P(r).$$

However, if the remainder $P(r) = 0$, then

$$P(x) = (x - r)Q(x)$$

and we conclude that $(x - r)$ is a factor of $P(x)$. Conversely, if $x - r$ is a factor of $P(x)$, then

$$P(x) = (x - r)Q(x)$$

and it follows that

$$P(r) = (r - r)Q(r)$$
$$= 0 \cdot Q(r)$$
$$= 0.$$

We refer to this consequence of the remainder theorem as the **factor theorem**.

◆ **Factor Theorem**

A first-degree polynomial $x - r$ is a factor of a polynomial $P(x)$ if and only if $P(r) = 0$ or, equivalently, if and only if r is a zero of P.

It is important to recognize that the concepts of factors, roots, and zeros are closely related. In general, if P is a polynomial function and r is a complex number, then the following statements are equivalent:

1. $x - r$ is a factor of the polynomial $P(x)$.

2. r is a zero of the function P.

3. r is a root of the equation $P(x) = 0$.

Also, if r is a real number, then

4. r is an x-intercept of the graph of P.

EXAMPLE 1   Given that $P(x) = 2x^3 - 3x^2 + 8x - 12$, determine whether each expression is a factor of $P(x)$.

(a) $x - 2$ **(b)** $x - \frac{3}{2}$ **(c)** $x + 2i$

◆ **Solution**

(a) According to the factor theorem, $x - 2$ is a factor of $P(x)$ if $P(2) = 0$. To determine $P(2)$, we can use either direct substitution or synthetic division. Choosing synthetic division, we find

$$
\begin{array}{r|rrrr}
2 & 2 & -3 & 8 & -12 \\
  &   & 4 & 2 & 20 \\
\hline
  & 2 & 1 & 10 & \boxed{8} \longleftarrow P(2)
\end{array}
$$

Since $P(2) = 8 \neq 0$, we conclude that $x - 2$ is not a factor of $P(x)$.

(b) According to the factor theorem, $x - \frac{3}{2}$ is a factor of $P(x)$ if $P(\frac{3}{2}) = 0$. To determine $P(\frac{3}{2})$, we can use either direct substitution or synthetic division. Choosing synthetic division, we find

$$
\begin{array}{r|rrrr}
\frac{3}{2} & 2 & -3 & 8 & -12 \\
            &   & 3 & 0 & 12 \\
\hline
            & 2 & 0 & 8 & \boxed{0} \longleftarrow P(\tfrac{3}{2})
\end{array}
$$

Since $P(\frac{3}{2}) = 0$, we conclude that $x - \frac{3}{2}$ is a factor of $P(x)$. The graph in Figure 4.19 also illustrates this fact.

(c) According to the factor theorem, $x + 2i$ is a factor of $P(x)$ if $P(-2i) = 0$. To determine $P(-2i)$, we can use direct substitution or

FIGURE 4.19

Since $\frac{3}{2}$ is an x-intercept of the graph of $P(x) = 2x^3 - 3x^2 + 8x - 12$, we know that $x - \frac{3}{2}$ is a factor of $P(x)$.

synthetic division. Choosing synthetic division and recalling from Section 1.4 that $i^2 = -1$, we find

$$
\begin{array}{r|rrrr}
-2i & 2 & -3 & 8 & -12 \\
& & -4i & 6i-8 & 12 \\
\hline
& 2 & -3-4i & 6i & \boxed{0}
\end{array}
$$
$\qquad\qquad\qquad\qquad\qquad\qquad P(-2i)$

Since $P(-2i) = 0$, we conclude that $x + 2i$ is a factor of $P(x)$. ◆

PROBLEM 1   Determine whether $x - 2i$ is a factor of
$P(x) = 2x^3 - 3x^2 + 8x - 12$. ◆

◆ Fundamental Theorem of Algebra

A linear function $f(x) = ax + b$, with $a \neq 0$, has just one zero, $-b/a$. A quadratic function $g(x) = ax^2 + bx + c$, with $a \neq 0$, has two zeros:

$$
\frac{-b + \sqrt{b^2 - 4ac}}{2a} \quad \text{and} \quad \frac{-b - \sqrt{b^2 - 4ac}}{2a}
$$

How many zeros for a cubic function? a quartic function? a quintic function? The **fundamental theorem of algebra** begins to answer these questions.

◆ **Fundamental Theorem of Algebra**

> Every polynomial function P of degree $n \geq 1$ has at least one zero in the set of complex numbers.

The proof of this theorem was provided by Carl Friedrich Gauss in 1799, but is beyond the scope of this book. Thus, we shall simply accept its validity and use its results.

Consider the polynomial function P of degree $n \geq 1$ defined by

$$
P(x) = a_n x^n + a_{n-1} x^{n-1} + \cdots + a_1 x + a_0 \quad \text{for } a_n \neq 0.
$$

According to the fundamental theorem of algebra, the function P has at least one zero. If this zero is r_1, then, by the factor theorem, $x - r_1$ is a factor of the polynomial $P(x)$. Thus, we can write

$$
P(x) = (x - r_1)Q_1(x),
$$

where the polynomial $Q_1(x)$ is of degree $n - 1$. Now if $n - 1 \geq 1$, then, by the fundamental theorem of algebra, the function Q_1 has at least one zero. If this zero is r_2, then, by the factor theorem, $x - r_2$ is a factor of the polynomial $Q_1(x)$ and we have

$$
Q_1(x) = (x - r_2)Q_2(x),
$$

where the polynomial $Q_2(x)$ is of degree $n - 2$. Thus, using substitution, we obtain

$$P(x) = (x - r_1)(x - r_2)Q_2(x).$$

If $n - 2 \geq 1$, we continue this procedure until we obtain a quotient polynomial $Q_n(x)$ of degree 0, that is, until $Q_n(x)$ is the constant a_n. We can then write

$$P(x) = \underbrace{(x - r_1)(x - r_2)\cdots(x - r_n)}_{n \text{ linear factors}} a_n,$$

and conclude that P has exactly n linear factors. Hence, P has exactly n zeros, namely $r_1, r_2, \ldots, r_n$.

◆ Extension of the Fundamental Theorem

> Every polynomial function P of degree $n \geq 1$ can be expressed as the product of n linear factors and, consequently, P has exactly n zeros in the set of complex numbers.

Thus, a cubic function can be expressed as the product of three linear factors, and it has three zeros; a quartic function can be expressed as the product of four linear factors, and it has four zeros; a quintic function can be expressed as the product of five linear factors, and it has five zeros; and so on.

The n zeros of a polynomial function of degree $n \geq 2$ are not necessarily distinct. For example, consider the polynomial function P defined by

$$P(x) = 2x^5 - 4x^4 + 2x^3.$$

Using the factoring techniques from Section 1.6, we can write $P(x)$ as the product of five linear factors:

$$P(x) = 2x^5 - 4x^4 + 2x^3 = 2x^3(x^2 - 2x + 1) \qquad \textbf{Factor out } 2x^3$$

$$= 2x^3(x - 1)^2 \qquad \textbf{Factor the trinomial}$$

$$= 2\underbrace{(x - 0)^3(x - 1)^2}_{\text{five linear factors}} \qquad \textbf{Rewrite}$$

Since the linear factor $(x - 0)$ appears three times and the linear factor $(x - 1)$ appears twice, we say that 0 is a zero of *multiplicity three* and 1 is a zero of *multiplicity two*. In general, if a linear factor $(x - r)$ appears k times, then r is a zero of multiplicity k.

Figure 4.20 shows the graph of

$$P(x) = 2x^5 - 4x^4 + 2x^3 = 2(x - 0)^3(x - 1)^2$$

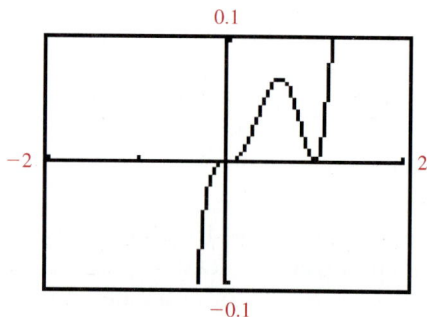

FIGURE 4.20

The graph of $P(x) = 2x^5 - 4x^4 + 2x^3$ is tangent to the x-axis at $x = 0$ and $x = 1$.

in the viewing rectangle of a graphing calculator. Note the appearance of the graph near the zeros of the function. In general, if r is a zero of multiplicity k, provided $k \geq 2$ for a polynomial function P, then the graph of P is *tangent* to the x-axis at $x = r$. If k is even, then the tangency looks similar to either one of the sketches shown in Figure 4.21, that is, the graph "bounces off" the x-axis. If k is odd, then the tangency looks similar to either one of the sketches shown in Figure 4.22, that is, the graph goes through the x-axis in an S-turn.

FIGURE 4.21

Appearance of tangency when r is a zero of multiplicity k and k is even

FIGURE 4.22

Appearance of tangency when r is a zero of multiplicity k and k is odd

If we know the n zeros of a polynomial function P of degree n and the coordinates of just one other point on its graph, then we can find the equation that defines the polynomial function. This procedure is illustrated in the next example.

EXAMPLE 2 Find the equation of a polynomial function P of degree 4 if the graph passes through the point $(2, 27)$ and the zeros of P are 1 of multiplicity two and -1 of multiplicity two.

◆ **Solution** By the factor theorem, since the zeros of P are 1 of multiplicity two and -1 of multiplicity two, the general form of the polynomial function is

$$P(x) = a(x - 1)^2(x + 1)^2.$$

To determine the value of a, we use the fact that the graph passes through the point $(2, 27)$:

$$P(x) = a(x - 1)^2(x + 1)^2$$

$$27 = a(2 - 1)^2(2 + 1)^2 \quad \text{Substitute } x = 2 \text{ and } P(x) = 27$$

$$27 = 9a \quad \text{Solve for } a$$

$$a = 3$$

Hence, the polynomial function with the given characteristics is

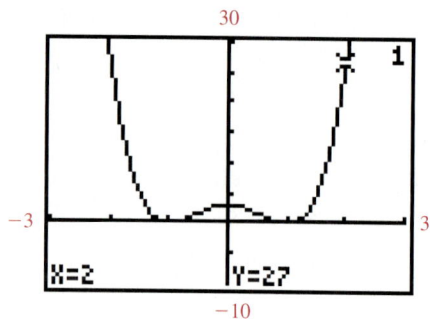

FIGURE 4.23

The graph of $P(x) = 3x^4 - 6x^2 + 3$ is tangent to the x-axis at $x = \pm 1$ and passes through the point $(2, 27)$.

$$
\begin{aligned}
P(x) &= 3(x - 1)^2(x + 1)^2 \\
&= 3[(x - 1)(x + 1)]^2 \\
&= 3(x^2 - 1)^2 \\
&= 3(x^4 - 2x^2 + 1) \\
&= 3x^4 - 6x^2 + 3
\end{aligned}
$$

Figure 4.23 shows the graph of $P(x) = 3x^4 - 6x^2 + 3$ in the viewing rectangle of a graphing calculator. Note that the graph satisfies the given conditions. ◆

PROBLEM 2   Find the equation of a polynomial function P of degree 4 if its graph passes through the point $(1, 8)$ and the zeros of P are 0 of multiplicity three and -3. ◆

◆ **Rational Zero Theorem**

Unfortunately, the zeros of a polynomial function are not always as easily attainable as those found for

$$P(x) = 2x^5 - 4x^4 + 2x^3 = 2(x - 0)^3(x - 1)^2$$

because not every polynomial can be factored by the methods discussed in Section 1.6. For example, consider finding the zeros of the polynomial function P defined by

$$P(x) = 2x^3 - 11x^2 + 13x - 4.$$

We can't factor out any common term, and factoring by grouping terms doesn't work for this polynomial. Then how do we factor this polynomial? To help us find some of the linear factors of a polynomial function, we use the **rational zero theorem**.

◆ **Rational Zero Theorem**

Let

$$P(x) = a_nx^n + a_{n-1}x^{n-1} + \cdots + a_1x + a_0$$

be a polynomial function with integer coefficients and with $a_n \neq 0$ and $a_0 \neq 0$. If the rational number b/c, in lowest terms, is a zero of P, then b must be a factor of the constant term a_0 and c must be a factor of the coefficient of the highest-degree term a_n.

CAUTION   The rational zero theorem does not say that a polynomial function with integer coefficients *has* rational zeros. It simply enables us to develop a list of all *possible* rational zeros of a polynomial function P with integer coefficients. By using direct substitution or synthetic division, we can then determine if any of the rational numbers in this list *are* zeros of P.

In the next example, we illustrate that if we can express a polynomial function in terms of linear and quadratic factors, we can then find *all* zeros of the function.

EXAMPLE 3   Find all zeros of $P(x) = 2x^3 - 11x^2 + 13x - 4$.

◆ **Solution**   For $P(x) = 2x^3 - 11x^2 + 13x - 4$, the constant term is -4 and the coefficient of the highest-degree term is 2. Now the factors of -4 are ± 1, ± 2, and ± 4, and the factors of 2 are ± 1 and ± 2. Thus, according to the rational zero theorem, the possible rational zeros of P are

$$\frac{\pm 1}{\pm 1}, \quad \frac{\pm 2}{\pm 1}, \quad \frac{\pm 4}{\pm 1} \quad \text{and} \quad \frac{\pm 1}{\pm 2}, \quad \frac{\pm 2}{\pm 2}, \quad \frac{\pm 4}{\pm 2},$$

which simplify to ± 1, ± 2, ± 4, and $\pm \frac{1}{2}$. Using this list of numbers and the graph of P in Figure 4.24, we note that 4, 1, and $\frac{1}{2}$ may be rational zeros of P. Testing 4 using synthetic division, we find

$$
\begin{array}{r|rrrr}
4) & 2 & -11 & 13 & -4 \\
   &   & 8 & -12 & 4 \\
\hline
   & 2 & -3 & 1 & \boxed{0} \quad \text{remainder}
\end{array}
$$

coefficients of $Q(x)$, whose degree is 1 less than $P(x)$

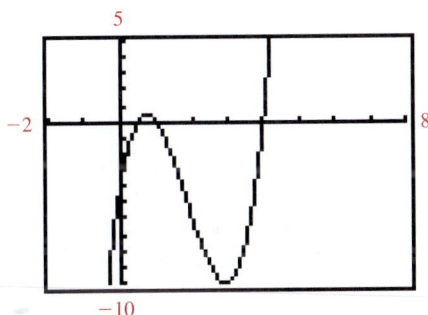

FIGURE 4.24

The polynomial function $P(x) = 2x^3 - 11x^2 + 13x - 4$ appears to have rational zeros of $x = 4$, 1, and $\frac{1}{2}$.

Since the remainder is 0 when we divide by the synthetic divisor 4, we know that 4 is a rational zero of P, and hence $x - 4$ is a factor of $P(x)$. Now we can write $P(x)$ as the product of linear factors:

$$
\begin{aligned}
P(x) &= (x - 4)Q(x) = (x - 4)(2x^2 - 3x + 1) \\
     &= (x - 4)(x - 1)(2x - 1)
\end{aligned}
$$

Hence, by the factor theorem, the three zeros of P are 4, 1, and $\frac{1}{2}$.   ◆

PROBLEM 3   Find all zeros of $P(x) = 3x^3 + 22x^2 + 25x + 6$.   ◆

We can find repeated rational zeros of a polynomial function by inspecting the tangency of the graph to the x-axis.

EXAMPLE 4   Find all zeros of $P(x) = x^4 - 8x^3 + 26x^2 - 48x + 45$.

◆ **Solution**   For $P(x) = x^4 - 8x^3 + 26x^2 - 48x + 45$, the constant term is 45 and the coefficient of the highest-degree term is 1. When the coefficient of the highest-degree term is 1, the possible rational zeros of P are simply the factors of the constant term:

$$\pm 1, \quad \pm 3, \quad \pm 5, \quad \pm 9, \quad \pm 15, \quad \text{and} \quad \pm 45.$$

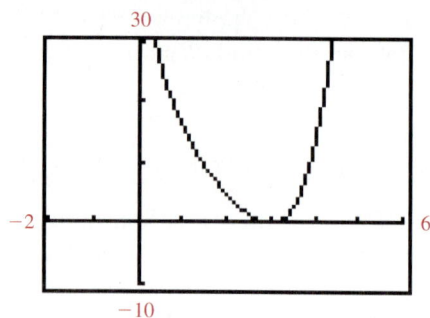

FIGURE 4.25

The polynomial function
$P(x) = x^4 - 8x^3 + 26x^2 - 48x + 45$
appears to have a rational zero of
multiplicity two at $x = 3$.

Using this list of numbers and the graph of P in Figure 4.25, we note that 3 may be a rational zero of even multiplicity. Testing 3 twice using synthetic division, we find the coefficients of $Q(x)$:

$$
\begin{array}{r|rrrrr}
3) & 1 & -8 & 26 & -48 & 45 \\
  &   & 3 & -15 & 33 & -45 \\
\hline
3) & 1 & -5 & 11 & -15 & \boxed{0} \quad \text{— remainder} \\
  &   & 3 & -6 & 15 & \\
\hline
  & 1 & -2 & 5 & \boxed{0} \quad \text{— remainder}
\end{array}
$$

coefficients of $Q(x)$ whose degree is 2 less than $P(x)$

Since the remainder is 0 each time we divide by the synthetic divisor 3, we know that 3 is a rational zero of multiplicity two, and hence $(x - 3)^2$ is a factor of $P(x)$. Therefore,

$$P(x) = (x - 3)^2 Q(x) = (x - 3)^2(x^2 - 2x + 5).$$

We can find the zeros of the quadratic function $Q(x) = x^2 - 2x + 5$ by applying the quadratic formula:

$$x = \frac{-(-2) \pm \sqrt{(-2)^2 - 4(1)(5)}}{2(1)} = \frac{2 \pm \sqrt{-16}}{2} = \frac{2 \pm 4i}{2} = 1 \pm 2i.$$

Thus, the four zeros of the function P are 3 of multiplicity two, $1 + 2i$, and $1 - 2i$.   ◆

Using the factor theorem, we can express the polynomial function in Example 4 as the product of four linear factors:

$$P(x) = (x - 3)(x - 3)[x - (1 + 2i)][x - (1 - 2i)].$$

PROBLEM 4   Find the product $(x - 3)^2[x - (1 + 2i)][x - (1 - 2i)]$. You should obtain the original form of the polynomial given in Example 4.   ◆

As illustrated in the next example, we can use the rational zero theorem to factor completely a polynomial over the set of integers.

EXAMPLE 5   Factor completely
$P(x) = 2x^5 + x^4 + x^3 + 13x^2 + 5x - 6$ over the set of integers.

◆ **Solution**   For $P(x) = 2x^5 + x^4 + x^3 + 13x^2 + 5x - 6$, the constant term is -6 and the coefficient of the highest-degree term is 2. The factors of -6 are ± 1, ± 2, ± 3, and ± 6, and the factors of 2 are ± 1, ± 2. According to the rational zero theorem, the possible rational zeros of P are

$$\frac{\pm 1}{\pm 1}, \quad \frac{\pm 2}{\pm 1}, \quad \frac{\pm 3}{\pm 1}, \quad \frac{\pm 6}{\pm 1} \quad \text{and} \quad \frac{\pm 1}{\pm 2}, \quad \frac{\pm 2}{\pm 2}, \quad \frac{\pm 3}{\pm 2}, \quad \frac{\pm 6}{\pm 2},$$

which simplify to ± 1, ± 2, ± 3, ± 6, $\pm \frac{1}{2}$, and $\pm \frac{3}{2}$. Using this list of numbers and the graph of P in Figure 4.26, we note that $-\frac{3}{2}$, -1, and $\frac{1}{2}$ may be rational zeros of P. Testing $-\frac{3}{2}$ using synthetic division, we find

$$
\begin{array}{r|rrrrrr}
-\frac{3}{2}) & 2 & 1 & 1 & 13 & 5 & -6 \\
 & & -3 & 3 & -6 & -\frac{21}{2} & \frac{33}{4} \\
\hline
 & 2 & -2 & 4 & 7 & -\frac{11}{2} & \boxed{\frac{9}{4}}
\end{array}
$$

$\longleftarrow$ remainder

Since the remainder is *not* 0 when we divide by the synthetic divisor $-\frac{3}{2}$, we know that $-\frac{3}{2}$ is *not* a rational zero of P. Since the real zero between -2 and -1 isn't a rational number, we conclude that it must be an irrational number.

We now test our other possible rational zeros, -1, and $\frac{1}{2}$. The order in which we test these numbers is immaterial:

$$
\begin{array}{r|rrrrrr}
-1) & 2 & 1 & 1 & 13 & 5 & -6 \\
 & & -2 & 1 & -2 & -11 & 6 \\
\frac{1}{2}) & 2 & -1 & 2 & 11 & -6 & \boxed{0} \\
 & & 1 & 0 & 1 & 6 & \\
\hline
 & 2 & 0 & 2 & 12 & \boxed{0} &
\end{array}
$$

$\longleftarrow$ remainder

$\longleftarrow$ remainder

coefficients of $Q(x)$ whose degree is 2 less than $P(x)$

Since the remainder is 0 when we divide by the synthetic divisors -1 and $\frac{1}{2}$, we conclude that -1 and $\frac{1}{2}$ are both rational zeros of P. Therefore,

$$
\begin{aligned}
P(x) &= (x + 1)(x - \tfrac{1}{2})Q(x) \\
&= (x + 1)(x - \tfrac{1}{2})(2x^3 + 2x + 12).
\end{aligned}
$$

Thus, we can factor completely the polynomial P over the set of integers:

$$
P(x) = (x + 1)(2x - 1)(x^3 + x + 6) \qquad \blacklozenge
$$

We know that the polynomial function in Example 5 has an irrational zero in the interval $(-2, -1)$ and two rational zeros, -1 and $\frac{1}{2}$. Since the graph of this function intersects the x-axis only three times, the other two zeros of this fifth-degree polynomial function must be imaginary numbers. Although we cannot find these imaginary zeros by the methods discussed in this section, we can estimate the irrational zero by using the [TRACE] key and [ZOOM] key on a graphing calculator or the calculator's built-in program for finding an x-intercept.

PROBLEM 5 Use a graphing calculator to estimate (to the nearest thousandth) the irrational zero in the interval $(-2, -1)$ for the polynomial P defined in Example 5. ◆

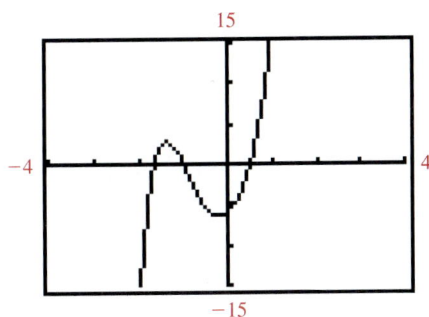

FIGURE 4.26

The polynomial function
$P(x) = 2x^5 + x^4 + x^3 + 13x^2 + 5x - 6$
might have rational zeros at $x = -\frac{3}{2}$, -1, and $\frac{1}{2}$.

◆ Conjugate Pair Theorem

In Example 4, we found that two zeros of the polynomial function

$$P(x) = x^4 - 8x^3 + 26x^2 - 48x + 45$$

are the imaginary numbers $1 + 2i$ and $1 - 2i$. Together, these imaginary numbers are called a *conjugate pair*. If a polynomial function with real coefficients has imaginary zeros, then these zeros always occur in conjugate pairs. We refer to this fact as the **conjugate pair theorem**.

◆ **Conjugate Pair Theorem**

> If P is a polynomial function of degree $n \geq 1$ with real coefficients, and if the imaginary number $a + bi$ is a zero of P, then its conjugate $a - bi$ is also a zero of P.

We can state two consequences of the conjugate pair theorem:

1. A polynomial function with real coefficients of *odd* degree must have at least one real zero. In fact, such polynomial functions must have an odd number of real zeros (see Examples 3 and 5).

2. A polynomial function with real coefficients of *even* degree must have either an even number of real zeros or no real zero at all (see Examples 4 and 6).

EXAMPLE 6   Find all zeros of $P(x) = x^4 - 6x^3 + 15x^2 - 18x + 10$, given that $2 - i$ is a zero of P.

◆ **Solution**   This fourth-degree polynomial function must have four zeros in the set of complex numbers. By the conjugate pair theorem, since $2 - i$ is a zero of P and P has real coefficients, we know that $2 + i$ is also a zero of P. Thus, $x - (2 - i)$ and $x - (2 + i)$ are factors of the polynomial $P(x)$. Hence, we can write

$$P(x) = [x - (2 - i)][x - (2 + i)]Q(x).$$

Now we can find $Q(x)$ by dividing $P(x)$ by $x - (2 - i)$ using synthetic division, and then dividing the result of this division by $x - (2 + i)$, using synthetic division again:

Thus, $Q(x) = x^2 - 2x + 2$. We can determine the zeros of this quadratic function by applying the quadratic formula:

$$x = \frac{-(-2) \pm \sqrt{(-2)^2 - 4(1)(2)}}{2(1)} = \frac{2 \pm \sqrt{-4}}{2} = 1 \pm i.$$

Thus, the four zeros of P are the (nonreal) conjugate pairs $2 \pm i$ and $1 \pm i$.

◆

PROBLEM 6   Use a graphing calculator to generate the graph of the polynomial function in Example 6. How can we tell from the graph that this function has no real zero?

◆

Exercises 4.3

◆ Basic Skills

In Exercises 1–12, use the factor theorem to determine if the first polynomial is a factor of the second polynomial.

1. $x - 1$; $x^3 - 4x^2 + x - 4$
2. $x + 2$; $x^3 - 4x^2 + x - 4$
3. $x + 3$; $x^4 - 7x^3 + 10x^2 + 3x - 15$
4. $x - 5$; $x^4 - 7x^3 + 10x^2 + 3x - 15$
5. $x - 4$; $x^4 - 3x^3 - 12x - 16$
6. $x + 1$; $x^4 - 3x^3 - 12x - 16$
7. $x + \frac{2}{3}$; $6x^3 + x^2 + 7x + 6$
8. $x - \frac{1}{3}$; $6x^3 + x^2 + 7x + 6$
9. $x - i$; $x^4 - 2x^2 + 1$ 10. $x + i$; $x^4 - 2x^2 + 1$
11. $x - (1 + i)$; $x^3 - 2x + 4$
12. $x - (1 - i)$; $x^3 - 2x + 4$

In Exercises 13–20, determine the polynomial function P with the given characteristics.

13. Degree 3; zeros are 1, -1, and 2; graph passes through the point (3, 16)
14. Degree 3; zeros are 0, -2, and 4; graph passes through the point $(-1, -15)$
15. Degree 4; zeros are -1, 1, and 2 of multiplicity two; graph passes through the point (0, 16)
16. Degree 3; zeros are 2 of multiplicity three; graph passes through the point (1, 7)

17. Degree 4; zeros are ± 1 and $\pm i$; graph passes through the point (3, 80)
18. Degree 4; zeros are $1 \pm i$ of multiplicity two; graph passes through the point $(0, -12)$
19. Degree 5; zeros are 0 of multiplicity three and $\pm\sqrt{2}$; graph passes through the point (1, 5)
20. Degree 6; zeros are 0 of multiplicity two, 1 of multiplicity two, and $1 \pm \sqrt{3}$; graph passes through the point $(2, -24)$

For the polynomial function P defined in Exercises 21–32, (a) write P(x) as the product of linear factors, and (b) find all the zeros of P.

21. $P(x) = x^3 + x^2 - 5x + 3$
22. $P(x) = 2x^3 + 9x^2 + x - 12$
23. $P(x) = 6x^3 + x^2 - 9x + 4$
24. $P(x) = 18x^3 - 33x^2 + 20x - 4$
25. $P(x) = x^4 - 4x^3 + 5x^2 - 4x + 4$
26. $P(x) = x^5 - 5x^4 + 10x^3 - 10x^2 + 5x - 1$
27. $P(x) = 2x^4 - 17x^3 + 53x^2 - 72x + 36$
28. $P(x) = 8x^4 - 14x^3 - 71x^2 - 10x + 24$
29. $P(x) = x^4 - 3x^3 - 12x - 16$
30. $P(x) = 2x^4 - 3x^3 - 3x - 2$
31. $P(x) = x^4 - 4x^3 - 7x^2 + 50x - 50$; $3 - i$ is a zero

32. $P(x) = x^4 - 6x^3 + 18x^2 - 30x + 25;\ [x - (1 + 2i)]$ is a factor

In Exercises 33–50, (a) find all rational zeros of each polynomial function and (b) factor the polynomial completely over the set of integers.

33. $P(x) = x^3 - 6x^2 + 11x - 6$

34. $P(x) = x^3 + 7x^2 + 16x + 12$

35. $f(x) = 2x^3 - 5x^2 - 14x + 8$

36. $F(x) = 6x^3 + 35x^2 - 8x - 12$

37. $H(x) = 4x^3 + 9x^2 + 26x + 6$

38. $G(x) = 2x^3 - 9x^2 + 14x - 10$

39. $P(x) = 12x^3 + 4x^2 - 17x + 6$

40. $Q(x) = 24x^3 - 70x^2 + 63x - 18$

41. $f(x) = 2x^4 + x^3 + 3x - 18$

42. $F(x) = x^4 - 11x^3 - 14x - 26$

43. $g(x) = 3x^4 - 14x^3 + 16x^2 - 3x + 8$

44. $h(x) = 4x^4 - 3x^3 - 12x^2 + 25x - 12$

45. $P(x) = 2x^4 + 9x^3 + 14x^2 + 9x + 2$

46. $f(x) = 4x^4 - 28x^3 + 67x^2 - 63x + 18$

47. $f(x) = x^5 + x^4 - 6x^3 + 8x - 16$

48. $P(x) = 6x^5 - 8x^4 + 5x^3 - 16x^2 - 25x + 10$

49. $F(x) = 2x^5 + 5x^4 - 4x^3 - 11x^2 + 4x + 4$

50. $D(x) = 16x^5 - 80x^4 + 127x^3 - 59x^2 - 8x + 4$

◆ Critical Thinking

51. Is $x + a$, with $a > 0$, a factor of $x^n - a^n$ when n is even? when n is odd? Explain.

52. Is $x + a$, with $a > 0$, a factor of $x^n + a^n$ when n is even? when n is odd? Explain.

53. Determine the value of k if $x - 2$ is a factor of $x^4 + 3x^3 + kx^2 - 5x - 2$.

54. Determine the value of k if $x + 3$ is a factor of $2x^4 + 4x^3 - 5x^2 + 2k + k^2$.

55. Find all integers k for which the polynomial function $P(x) = x^3 - 2x^2 + kx - 3$ has one rational zero and two imaginary zeros.

56. What are the possible rational zeros of

$$P(x) = x^n + a_{n-1}x^{n-1} + a_{n-2}x^{n-2} + \cdots + a_1 x + a_0$$

if each coefficient is an integer and a_0 is a prime number?

In Exercises 57–62, find all rational zeros of each polynomial function.
[Hint: To extend the rational zero theorem to include a polynomial function with rational coefficients, multiply both sides of the function by the LCD to eliminate all fractions. The new function that is formed has the same zeros as the original function.]

57. $P(x) = \frac{1}{2}x^3 + \frac{5}{6}x^2 + \frac{8}{3}x - 1$

58. $f(x) = \frac{1}{2}x^3 + x^2 - \frac{1}{8}x - \frac{1}{4}$

59. $g(x) = \frac{3}{4}x^4 + x^3 - \frac{1}{4}x^2 + x - 1$

60. $G(x) = \frac{1}{10}x^4 - \frac{1}{2}x^2 - x - \frac{3}{5}$

61. $f(x) = 0.2x^3 - 2.3x^2 + 3.1x + 1.4$

62. $P(x) = 0.6x^4 - 1.1x^3 + 0.7x^2 - 2.2x + 2.4$

The graphs of cubic functions are shown in Exercises 63 and 64. Determine the equation of each function.

63.

64.

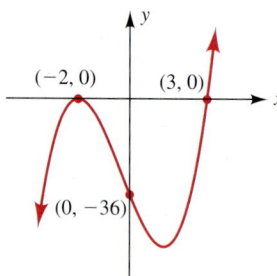

The graphs of quartic functions are shown in Exercises 65 and 66. Determine the equation of each function.

65.

66.

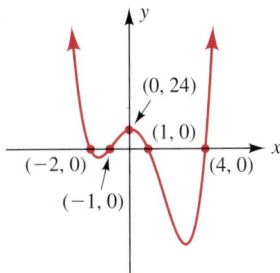

67. The velocity v (in feet per second, or ft/s) of the current at the mouth of a tidal river is a function of the time t (in hours) after high tide and is given by

$$v(t) = -0.240t^3 + 0.064t^2 + 8.527t - 2.982$$
$$\text{for } 0 \leq t \leq 6$$

Find, to the nearest minute, the time(s) t when the velocity of the current is 0 ft/s.

68. The deflection d (in millimeters, mm) of a 34.0-meter steel beam is a function of the distance x (in meters, m) from the left end of the beam and is given by

$$d(x) = 0.002x^4 - 0.129x^3 + 2.480x^2 - 13.804x$$
$$\text{for } 0 \leq x \leq 34$$

Find the distance(s) x for which the deflection of the beam is 0 mm.

4.4 Polynomial Equations and Inequalities

◆ Introductory Comments

Any equation of the form

$$a_n x^n + a_{n-1}x^{n-1} + a_{n-2}x^{n-2} + \cdots + a_2 x^2 + a_1 x + a_0 = 0$$

where n is a nonnegative integer and $a_n, a_{n-1}, a_{n-2}, \ldots, a_2, a_1,$ and a_0 are real numbers with $a_n \neq 0$, is called a **polynomial equation of degree n in standard form**. If we replace the equal sign with one of the inequality symbols ($<$, $>$, $\leq$, or $\geq$) we have a **polynomial inequality of degree n in standard form**. In this section, we discuss methods for solving polynomial equations and polynomial inequalities.

◆ Solving Polynomial Equations

We can solve some polynomial equations by using the factoring techniques introduced in Section 1.6.

EXAMPLE 1 Find all roots of each polynomial equation.

(a) $(x - 2)^2(x + 1) = 2x - 2$ 　　　**(b)** $\dfrac{4x^2(x - 3)^2}{5} + \dfrac{x(x - 3)^3}{3} = 0$

◆ Solution

(a) We begin by writing the equation in standard form:

$$(x - 2)^2(x + 1) = 2x - 2$$

$$(x^2 - 4x + 4)(x + 1) = 2x - 2 \qquad \textcolor{red}{\text{Expand } (x - 2)^2}$$

$$x^3 - 3x^2 + 4 = 2x - 2 \qquad \textcolor{red}{\text{Multiply polynomials}}$$

$$x^3 - 3x^2 - 2x + 6 = 0 \qquad \textcolor{red}{\text{Write in standard form}}$$

This third-degree polynomial equation must have three roots in the set of complex numbers. We can find these three roots as follows:

$$x^3 - 3x^2 - 2x + 6 = 0$$
$$x^2(x - 3) - 2(x - 3) = 0 \qquad \text{Factor by grouping terms}$$
$$(x - 3)(x^2 - 2) = 0$$

$$x = 3, \quad x = \pm\sqrt{2}$$

Hence, the three roots of the original equation are 3 and $\pm\sqrt{2} \approx \pm1.414$.

Figure 4.27 shows the graph of $P(x) = (x - 2)^2(x + 1) - 2x + 2$ in the viewing rectangle of a graphing calculator. The graph indicates that the x-intercepts are 3 and approximately ±1.414; these facts confirm our answer.

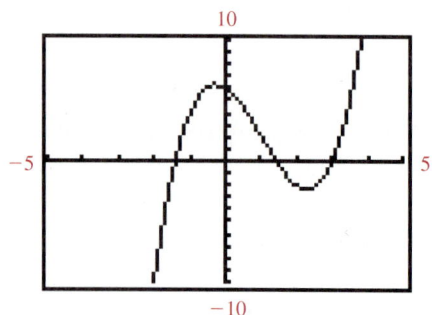

(b) We begin by eliminating the fractions:

$$\frac{4x^2(x - 3)^2}{5} + \frac{x(x - 3)^3}{3} = 0$$

$$15\left[\frac{4x^2(x - 3)^2}{5} + \frac{x(x - 3)^3}{3}\right] = 15(0) \qquad \text{Multiply both sides by the LCD}$$

$$12x^2(x - 3)^2 + 5x(x - 3)^3 = 0 \qquad \text{Simplify}$$

It is best *not* to write this equation in standard form. Instead, we factor out the common factors from each term:

$$12x^2(x - 3)^2 + 5x(x - 3)^3 = 0$$

$$x(x - 3)^2[12x + 5(x - 3)] = 0 \qquad \text{Factor out } x(x-3)^2$$

$$x(x - 3)^2(17x - 15) = 0 \qquad \text{Simplify the third factor}$$

$$x = 0, \quad x = 3, \quad x = \tfrac{15}{17}$$

Thus, the four roots of the original fourth-degree polynomial equation are 0, 3 of multiplicity two, and $\frac{15}{17} \approx 0.882$. The graph in Figure 4.28 supports our work. ◆

PROBLEM 1 Find all roots of the equation $(x^2 + 4)(x^2 - 3x - 4) = 0$. ◆

If a polynomial equation is not factorable by the techniques discussed in Section 1.6, then we apply the rational zero theorem (Section 4.3) and write the equation as the product of linear and quadratic factors.

EXAMPLE 2 Find all roots of each polynomial equation.

(a) $\dfrac{x^3 + 5x}{3} - \dfrac{x - 3x^2}{2} = 1$ **(b)** $x^2(x - 1)^2 = 4(3x - 5)$

FIGURE 4.27

The x-intercepts of the graph of $P(x) = (x - 2)^2(x + 1) - 2x + 2$ are 3 and $\pm\sqrt{2} \approx \pm1.414$.

FIGURE 4.28

The x-intercepts of the graph $P(x) = \dfrac{4x^2(x - 3)^2}{5} + \dfrac{x(x - 3)^3}{3}$ are 0, 3, and $\frac{15}{17} \approx 0.882$.

◆ **Solution**

(a) We begin by eliminating the fractions:

$$\frac{x^3 + 5x}{3} - \frac{x - 3x^2}{2} = 1$$

$$6\left[\frac{x^3 + 5x}{3} - \frac{x - 3x^2}{2}\right] = 6(1) \qquad \text{Multiply both sides by the LCD}$$

$$2(x^3 + 5x) - 3(x - 3x^2) = 6 \qquad \text{Simplify}$$

$$2x^3 + 9x^2 + 7x - 6 = 0 \qquad \text{Write in standard form}$$

This third-degree polynomial equation must have three roots in the set of complex numbers. By the rational zero theorem (Section 4.3), the possible rational roots of this equation are

$$\pm 1, \quad \pm 2, \quad \pm 3, \quad \pm 6, \quad \pm\frac{1}{2}, \quad \text{and} \quad \pm\frac{3}{2}.$$

The graph in Figure 4.29 indicates that -3, -2, and $\frac{1}{2}$ may be possible roots. Testing -3, we find

$$
\begin{array}{r|rrrr}
-3 & 2 & 9 & 7 & -6 \\
   &   & -6 & -9 & 6 \\
\hline
   & 2 & 3 & -2 & \boxed{0} \\
\end{array}
\quad \longleftarrow \text{remainder}
$$

coefficients of $Q(x)$ whose degree is 2

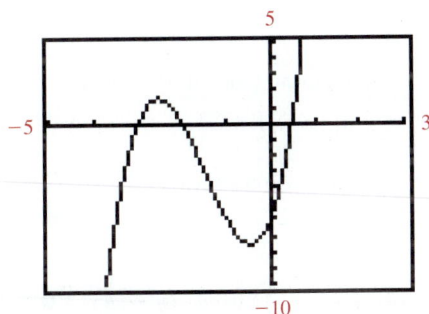

FIGURE 4.29

The graph of
$P(x) = 2x^3 + 9x^2 + 7x - 6$ indicates
that -3, -2, and $\frac{1}{2}$ may be rational roots of
$2x^3 + 9x^2 + 7x - 6 = 0$.

Since the remainder is zero, we know that -3 is a root of the equation. Hence, $x + 3$ is a factor of the polynomial $2x^3 + 9x^2 + 7x - 6$. We can now write the equation as the product of three linear factors:

$$2x^3 + 9x^2 + 7x - 6 = 0$$
$$(x + 3)(2x^2 + 3x - 2) = 0$$
$$(x + 3)(x + 2)(2x - 1) = 0$$

$$x = -3, \; x = -2, \; x = \tfrac{1}{2}$$

Therefore, the three roots of the original equation are -3, -2, and $\frac{1}{2}$.

(b) We begin by writing the equation in standard form:

$$x^2(x - 1)^2 = 4(3x - 5)$$
$$x^2(x^2 - 2x + 1) = 12x - 20$$
$$x^4 - 2x^3 + x^2 - 12x + 20 = 0$$

This fourth-degree polynomial equation must have four roots in the set of complex numbers. By the rational zero theorem (Section 4.3), the possible rational roots of this equation are

$$\pm 1, \quad \pm 2, \quad \pm 4, \quad \pm 5, \quad \pm 10, \quad \text{and} \quad \pm 20$$

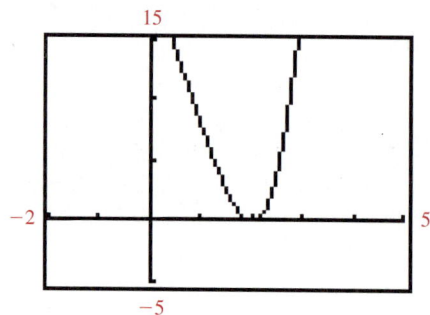

FIGURE 4.30

The graph of
$P(x) = x^4 - 2x^3 + x^2 - 12x + 20$
indicates that 2 may be a rational root of
even multiplicity of
$x^4 - 2x^3 + x^2 - 12x + 20 = 0$.

The graph in Figure 4.30 indicates that 2 may be a root with even multiplicity. Checking 2, we find

$$\begin{array}{r}
2\overline{)}1 \quad -2 \quad\ \ 1 \quad -12 \quad\ \ 20 \\
2 \quad\ \ 0 \quad\ \ 2 \quad -20 \\
\hline
2\overline{)}1 \quad\ \ 0 \quad\ \ 1 \quad -10 \quad\ \ \textcircled{0} \\
2 \quad\ \ 4 \quad\ \ 10 \\
\hline
1 \quad\ \ 2 \quad\ \ 5 \quad\ \ \textcircled{0}
\end{array}$$

remainder

remainder

coefficients of $Q(x)$ whose degree is 2

Since the remainder is 0 each time we divide by the synthetic divisor 2, we know that 2 is a rational root of multiplicity two and, hence, $(x - 2)^2$ is a factor of $x^4 - 2x^3 + x^2 - 12x + 20$. We can now write the equation as the product of linear and quadratic factors:

$$x^4 - 2x^3 + x^2 - 12x + 20 = 0$$
$$(x - 2)^2(x^2 + 2x + 5) = 0$$

Apply the quadratic formula.

$$x = 2 \qquad x = 1 \pm 2i$$

Therefore, the four roots of the original equation are 2 of multiplicity two and the complex conjugates $1 + 2i$ and $1 - 2i$. ◆

PROBLEM 2　Find all roots of the equation $x^2(x - 1) = 2(5x + 4)$. ◆

Although we may not be able to find all complex roots of a polynomial equation $P(x) = 0$ by using the techniques discussed in this chapter, we can find all its real roots by locating the x-intercepts of the graph of P.

EXAMPLE 3　Find all real roots of each polynomial equation. Record any rational root as a reduced fraction and any irrational root as a decimal number rounded to the nearest thousandth.

(a) $x^3 = x - 3$ 　　　**(b)** $4(x^2 - 1)(x^2 + 1) = 2x^3 - 2x^2 - 7x$

◆ **Solution**

(a) Writing the equation in standard form, we obtain

$$x^3 - x + 3 = 0.$$

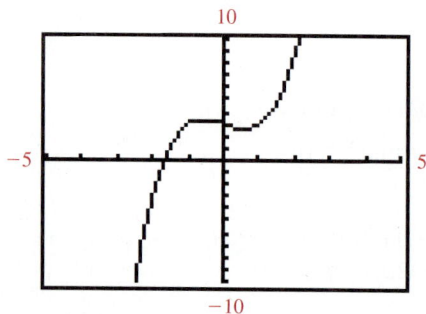

FIGURE 4.31

The equation $x^3 - x + 3 = 0$ has one real root in the interval $(-2, -1)$.

As a consequence of the conjugate pair theorem (Section 4.3), this third-degree polynomial equation has either 1 or 3 real roots. Figure 4.31 shows a complete graph of $P(x) = x^3 - x + 3$ in the viewing rectangle of a graphing calculator. Note that the graph of P intersects the *x*-axis once. Hence, the equation $x^3 - x + 3 = 0$ has only one real root, and this root is in the interval $(-2, -1)$.

By the rational zero theorem (Section 4.3), the only possible rational roots of the equation $x^3 - x + 3 = 0$ are ± 1 and ± 3. Therefore, the root in the interval $(-2, -1)$ must be irrational. To find this irrational root, we use a graphing calculator to estimate the *x*-intercept of the graph. Using the calculator's built-in program for finding an *x*-intercept or the [TRACE] key and [ZOOM] key, we find that $x \approx -1.672$, rounded to the nearest thousandth. Hence, the only real root of the original equation is approximately -1.672.

(b) We begin by writing the equation in standard form:

$$4(x^2 - 1)(x^2 + 1) = 2x^3 - 2x^2 - 7x$$
$$4(x^4 - 1) = 2x^3 - 2x^2 - 7x$$
$$4x^4 - 2x^3 + 2x^2 + 7x - 4 = 0$$

As a consequence of the conjugate pair theorem (Section 4.3), this fourth-degree polynomial equation has either 0, 2, or 4 real roots. Figure 4.32 shows a complete graph of $P(x) = 4x^4 - 2x^3 + 2x^2 + 7x - 4$ in the viewing rectangle of a graphing calculator. Since the graph of P intersects the *x*-axis twice, the equation $4x^4 - 2x^3 + 2x^2 + 7x - 4 = 0$ must have two real roots.

By the rational zero theorem (Section 4.3), the possible rational roots of the equation are ± 1, ± 2, ± 4, $\pm \frac{1}{2}$, and $\pm \frac{1}{4}$. Using this list of possibilities and the graph of P in Figure 4.32, we note that $\frac{1}{2}$ and -1 may be rational roots. Testing $\frac{1}{2}$, we find

FIGURE 4.32

The equation
$4x^4 - 2x^3 + 2x^2 + 7x - 4 = 0$ has two
real roots.

$$
\begin{array}{r|rrrrr}
\tfrac{1}{2}) & 4 & -2 & 2 & 7 & -4 \\
  &   & 2 & 0 & 1 & 4 \\
\hline
  & 4 & 0 & 2 & 8 & \boxed{0} \quad \text{— remainder}
\end{array}
$$

Since the remainder is zero, we conclude that $\frac{1}{2}$ is a rational root of the original equation. Testing -1 as a possible rational root, we find

$$
\begin{array}{r|rrrrr}
-1) & 4 & -2 & 2 & 7 & -4 \\
  &   & -4 & 6 & -8 & 1 \\
\hline
  & 4 & -6 & 8 & -1 & \boxed{-3} \quad \text{— remainder}
\end{array}
$$

Since the remainder is *not* zero, we conclude that -1 is *not* a rational root. Hence, the root near $x = -1$ must be irrational. Using the calculator's built-in program for finding an *x*-intercept or the [TRACE] key and [ZOOM] key, we find that $x \approx -1.128$, rounded to the nearest thousandth. In summary, the real roots of the original equation are $\frac{1}{2}$ and approximately -1.128. ◆

PROBLEM 3 Find all roots of the equation $x^2(x + 1)^2 = 2x + 8$. ◆

◆ **Solving Polynomial Inequalities**

We can use the procedure for solving a quadratic inequality (Section 3.4) to help solve a polynomial inequality. To solve the polynomial inequality $P(x) > 0$, we find the *boundary numbers* of the inequality—the real values of x for which the inequality is neither positive nor negative. These boundary numbers are the x-intercepts of the graph of P, and they divide the x-axis into intervals. The solution set of the inequality $P(x) > 0$ includes all the intervals on which the graph of P is *above* the x-axis. For the inequality $P(x) < 0$, the solution set of the inequality is all the intervals on which the graph of P is *below* the x-axis. If the inequality symbol is $\leq$ or $\geq$, we include the boundary numbers as part of the solution set.

> **EXAMPLE 4**   Solve each polynomial inequality.
>
> **(a)** $9 > x^4 - 8x^2$ **(b)** $x^3 - 2x^2 + 8 \geq 10x$

◆ **Solution**

(a) We begin by writing the inequality in standard form:

$$9 > x^4 - 8x^2$$
$$0 > x^4 - 8x^2 - 9$$
$$x^4 - 8x^2 - 9 < 0 \qquad \text{\color{red}{Interchange sides and reverse the direction of the inequality}}$$

The boundary numbers of this inequality are the x-intercepts of the graph of $P(x) = x^4 - 8x^2 - 9$. We can find these boundary numbers by solving the equation $P(x) = 0$:

$$x^4 - 8x^2 - 9 = 0$$

$$(x^2 - 9)(x^2 + 1) = 0 \qquad \text{\color{red}{Factor the trinomial}}$$

$$(x + 3)(x - 3)(x^2 + 1) = 0 \qquad \text{\color{red}{Factor the difference of squares}}$$

$$x = -3 \quad x = 3 \quad \text{no real solution}$$

boundary numbers

These two boundary numbers divide the x-axis into three intervals, as shown:

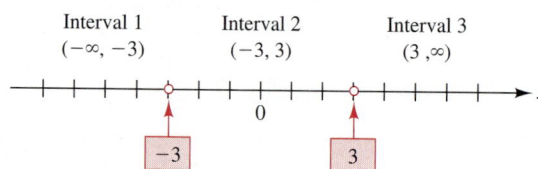

FIGURE 4.33

The graph of $P(x) = x^4 - 8x^2 - 9$ is below the x-axis in the interval $(-3, 3)$.

FIGURE 4.34

The graph of $P(x) = x^3 - 2x^2 - 10x + 8$ indicates that 4 may be a rational root of the equation $x^3 - 2x^2 - 10x + 8 = 0$.

We can confirm these intervals by generating the graph of $P(x) = x^4 - 8x^2 - 9$ in the viewing rectangle of a graphing calculator. As shown in Figure 4.33, the graph of $P(x) = x^4 - 8x^2 - 9$ is *below* the x-axis in the interval $(-3, 3)$. Hence, $P(x) < 0$ in this interval. Thus, the solution set of the original inequality is $(-3, 3)$.

(b) We begin by subtracting $10x$ from both sides of the inequality and then writing the inequality in standard form:

$$x^3 - 2x^2 - 10x + 8 \geq 0$$

The boundary numbers of this inequality are the x-intercepts of the graph of $P(x) = x^3 - 2x^2 - 10x + 8$. We can find these boundary numbers by solving the equation $x^3 - 2x^2 - 10x + 8 = 0$. By the rational zero theorem (Section 4.3), the possible rational roots of this equation are ± 1, ± 2, ± 4, and ± 8. The graph in Figure 4.34 indicates that 4 may be a possible root, so we test the number:

$$
\begin{array}{r}
4\overline{)\begin{array}{rrrr} 1 & -2 & -10 & 8 \\ & 4 & 8 & -8 \\ \hline 1 & 2 & -2 & \enclose{circle}{0} \end{array}}
\end{array}
\;\;\text{remainder}
$$

coefficients of $Q(x)$ whose degree is 2

If 4 is a root of the equation, then we have

$$x^3 - 2x^2 - 10x + 8 = 0$$
$$(x - 4)(x^2 + 2x - 2) = 0 \qquad \text{Apply the quadratic formula.}$$

$$x = 4 \qquad x = -1 \pm \sqrt{3}$$

boundary numbers

These three boundary numbers divide the x-axis into four intervals:

Interval 2
$(-1 - \sqrt{3}, -1 + \sqrt{3})$

Interval 1
$(-\infty, -1 - \sqrt{3})$

Interval 3
$(-1 + \sqrt{3}, 4)$

Interval 4
$(4, \infty)$

$-1 - \sqrt{3} \approx -2.73$

$-1 + \sqrt{3} \approx 0.73$

4

We can confirm these intervals by generating the graph of $P(x) = x^3 - 2x^2 - 10x + 8$ in the viewing rectangle of a graphing calculator. As shown in Figure 4.34, the graph of P is *above* the x-axis in the interval $\left(-1 - \sqrt{3}, -1 + \sqrt{3}\right)$ or $(4, \infty)$. Hence, $P(x) \geq 0$ in the interval $\left[-1 - \sqrt{3}, -1 + \sqrt{3}\right]$ or $[4, \infty]$. Thus, the solution set of the original inequality is $\left[-1 - \sqrt{3}, -1 + \sqrt{3}\right] \cup [4, \infty)$.   ◆

Observe in Figures 4.33 and 4.34 that in each of the intervals formed by the boundary numbers, the graph of the polynomial function is entirely above the x-axis or entirely below the x-axis. Hence, we can use the test number method (Section 3.4) to help solve a polynomial inequality in standard form without referring to a graph. We simply select an arbitrary test number from each interval formed by the boundary numbers and then determine the algebraic sign of the polynomial. If the polynomial is negative for a particular test number in an interval, then it is negative for every value of x in that interval. If the polynomial is positive for a particular test number in an interval, then it is positive for every value of x in that interval.

For example, to solve the polynomial inequality $x^4 - 8x^2 - 9 < 0$ in Example 4(a) by using the test number method, we select an arbitrary test number from each of the intervals $(-\infty, -3)$, $(-3, 3)$, and $(3, \infty)$ and then determine the algebraic sign of the polynomial $x^4 - 8x^2 - 9$ for each test number. The following table organizes our work:

Interval	Test Number	Sign of $x^4 - 8x^2 - 9$
$(-\infty, -3)$	-4	$(-4)^4 - 8(-4)^2 - 9 = 119 > 0$
$(-3, 3)$	1	$(1)^4 - 8(1)^2 - 9 = -16 < 0$
$(3, \infty)$	4	$(4)^4 - 8(4)^2 - 9 = 119 > 0$

Since $x^4 - 8x^2 - 9 < 0$ when $x = 1$, then we know that $x^4 - 8x^2 - 9 < 0$ for every value of x in the interval $(-3, 3)$. Hence, we conclude that the solution set of this inequality is $(-3, 3)$, which agrees with our work in Example 4(a).

PROBLEM 4   Solve the polynomial inequality $x^3 + 6 \geq 7x$ by using the test number method.   ◆

◆ **Application: Silo for Storing Grain**

Many applied problems involve the solution of a polynomial equation of the form

$$a_n x^n + a_{n-1} x^{n-1} + \cdots + a_1 x + a_0 = 0,$$

where $n \geq 3$. Such equations often have solutions that are irrational numbers. We illustrate one such application in the next example.

EXAMPLE 5   *Chapter Opening Problem*

A silo has the shape of a right circular cylinder surmounted by a hemisphere and holds 3000 cubic feet of grain. If the height of the silo is 60 ft, what is the radius of its base?

FIGURE 4.35

A silo with height 60 ft and radius x

FIGURE 4.36

The equation
$\pi x^3 - 180\pi x^2 + 9000 = 0$ has only one real root in the interval (0, 60).

◆ **Solution**   Let

$$x = \text{the radius of the cylindrical part of the silo.}$$

Since the radius of the hemisphere is the same as the radius of the cylinder, the height of the cylindrical part of the silo is $60 - x$, as shown in Figure 4.35. Now the volume V of the silo is the sum of the volumes of the cylinder and hemisphere:

cylinder   hemisphere

$$V = \pi r^2 h \qquad + \frac{2}{3}\pi r^3$$

$$3000 = \pi x^2(60 - x) \ + \frac{2}{3}\pi x^3$$

$$9000 = 3\pi x^2(60 - x) + 2\pi x^3 \qquad \textbf{Multiply both sides by 3}$$

$$\pi x^3 - 180\pi x^2 + 9000 = 0 \qquad \textbf{Write in standard form}$$

Because the irrational number π is a coefficient of only some terms of this polynomial equation, the equation cannot have rational roots. Also, since the height of the cylindrical part of the silo is $60 - x$, we must have $x < 60$. As shown in Figure 4.36, the graph of $P(x) = \pi x^3 - 180\pi x^2 + 9000$ intersects the x-axis only once in the interval (0, 60). Hence, the equation $\pi x^3 - 180\pi x^2 + 9000 = 0$ has only one real root in this interval. Using the calculator's built-in program for finding an x-intercept or the $\boxed{\text{TRACE}}$ key and $\boxed{\text{ZOOM}}$ key, we find that $x \approx 4.035$, rounded to the nearest thousandth. In summary, the radius of the base of the silo is approximately 4.035 ft.   ◆

PROBLEM 5   If the silo described in Example 5 holds 4000 cubic feet of grain, what is the radius of its base?   ◆

◆ **Application: Ash Pan for a Wood Stove**

For some applied problems, it is necessary to solve a polynomial inequality of degree $n \geq 3$. We illustrate one such application in the next example.

EXAMPLE 6   An ash pan for a wood stove is made from a rectangular piece of sheet metal, 8 inches by 15 inches, by cutting out a square from each corner and turning up the resulting sides. For safety reasons, the volume of the pan that is formed must be greater than 90 cubic inches. Find an interval for the acceptable dimensions of the square cut-out.

◆ **Solution**   Recall from elementary geometry that the volume V of a rectangular box is the product of its length, width, and height. If we let

$$x = \text{the length of the side of the square cut-out,}$$

then the dimensions of the ash pan are

$$\text{height} = x \qquad \text{length} = 15 - 2x \qquad \text{width} = 8 - 2x$$

as shown in Figure 4.37. Thus,

$$V = (15 - 2x)(8 - 2x)x.$$

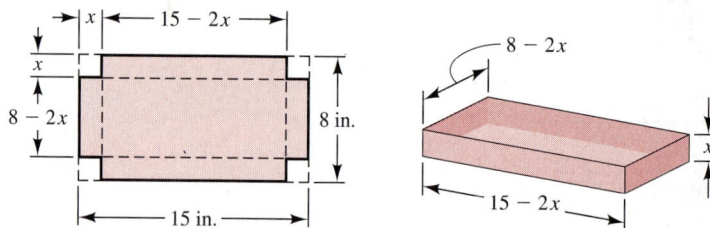

FIGURE 4.37

The dimensions of the ash pan described in Example 6

Now, to find the interval when $V > 90$ cubic inches we solve the corresponding inequality:

$$(15 - 2x)(8 - 2x)x > 90$$

$$(120 - 46x + 4x^2)x > 90 \qquad \text{Multiply the binomials}$$

$$4x^3 - 46x^2 + 120x - 90 > 0 \qquad \text{Write in standard form}$$

$$2x^3 - 23x^2 + 60x - 45 > 0 \qquad \text{Divide both sides by 2}$$

By the rational zero theorem, the possible rational boundary numbers of this inequality are:

$$\pm 1, \quad \pm 3, \quad \pm 5, \quad \pm 9, \quad \pm 15, \quad \pm 45, \quad \pm \tfrac{1}{2}, \quad \pm \tfrac{3}{2}, \quad \pm \tfrac{5}{2}, \quad \pm \tfrac{9}{2}, \quad \pm \tfrac{15}{2}, \quad \pm \tfrac{45}{2}$$

However, the length of the side of the cut-out cannot be a negative number, nor can it be greater than 4 inches (since the piece of metal is 8 inches wide). Since $0 \leq x \leq 4$, we check only 1, 3, $\tfrac{1}{2}$, $\tfrac{3}{2}$, and $\tfrac{5}{2}$ as possible rational boundary numbers. The graph in Figure 4.38 indicates that $\tfrac{3}{2}$ may be a boundary number of the inequality. Testing $\tfrac{3}{2}$, we find

$$
\begin{array}{r|rrrr}
\tfrac{3}{2}) & 2 & -23 & 60 & -45 \\
  &   & 3 & -30 & 45 \\
\hline
  & 2 & -20 & 30 & 0
\end{array}
\quad \text{remainder}
$$

FIGURE 4.38

The graph of
$P(x) = 2x^3 - 23x^2 + 60x - 45$
indicates that $\tfrac{3}{2}$ may be a critical value of the inequality
$2x^3 - 23x^2 + 60x - 45 > 0$.

Since the remainder is zero, we know that $x = \tfrac{3}{2}$ is a boundary number of the inequality. The other boundary number in the interval $(1, 2)$ must be irrational. Using the calculator's built-in program for finding an x-intercept or the $\boxed{\text{TRACE}}$ key and $\boxed{\text{ZOOM}}$ key, we find that $x \approx 1.838$, rounded to the nearest thousandth. In summary, the volume of the pan is greater than 90 cubic inches when the dimensions of the square cut-out are in the interval between 1.5 inches and approximately 1.838 inches. ◆

PROBLEM 6   Factor the polynomial in Example 6,
$2x^3 - 23x^2 + 60x - 45$, completely over the set of integers. Then show
that the exact value of the irrational boundary number in the interval $(1, 2)$
is $5 - \sqrt{10}$.   ◆

Exercises 4.4

◆ Basic Skills

In Exercises 1–20, find all roots of each equation and give the multiplicity of any repeated root.

1. $y^3 - y = 0$

2. $2x^4 = 16x^2$

3. $x^4 - 3x^3 - 2x^2 = 0$

4. $10x^2 + 3x = 8x^2$

5. $x^4(2x - 3) = x^3$

6. $x^2(5x - 8) = 2x$

7. $\dfrac{x(x^2 + 2)}{3} = \dfrac{x^2 + 2}{2}$

8. $\dfrac{n^5(n - 2)}{6} - \dfrac{n(n - 2)}{2} = 0$

9. $3x^2(x + 5)^2 + 2x(x + 5)^3 = 0$

10. $4x^2(x + 1)^3 + 2x(x + 1)^4 = 0$

11. $6x^3 + x^2 - 31x + 24 = 0$

12. $6x^3 + x^2 - 29x = 14$

13. $x^2(9x^2 + 10) - 6x(x^2 + 1) = -1$

14. $12(4x - 3) - 25x^2 = 4x^3(x - 3)$

15. $x^4 + 27x = 0$ **16.** $y^3 = 8y^6$

17. $x^2(x + 1)^2 = 4(2x + 5)$

18. $x^2(2x^2 - 17x + 57) = 97x + 65$

19. $\dfrac{x(43x^2 + x - 2)}{6} + 4 = x^2(x^2 + 13)$

20. $\dfrac{x^6 - x^2}{10} - \dfrac{2x^4 - 2}{5} = \dfrac{x^5 - 5x^3 + 4x}{3}$

In Exercises 21–40, find the real roots of each equation. Record any rational root as a reduced fraction and any irrational root as a decimal number rounded to the nearest hundredth.

21. $x^3 + x = 4$

22. $2x^3 + x + 9 = 0$

23. $x^2(2x + 1) = -5$

24. $x^2(x - 6) = 5$

25. $\dfrac{x^2(x - 3)}{6} = x - 1$

26. $\dfrac{2x^3 - 5}{4} = x - x^2$

27. $x(3x + 4)(x - 2) = 2$

28. $(3 - 5x)(1 + x) = x^3$

29. $\dfrac{x(x^3 - 4)}{3} - 1 = 0$

30. $\dfrac{x^3(x + 3)}{2} + 1 = 0$

31. $2x^4 + 2x^3 + x^2 - 3x = 4$

32. $4x^4 - 4x^2 + x = 1$

33. $x[x^2(2x + 3) + 10x + 17] + 5 = 2$

34. $x[x^2(2x - 7) + 3x - 4] + 8 = 6$

35. $x[x^2(3x^2 - 5) - 2x(2x^2 + 1) + 4] = -8$

36. $x[2x^2(x + 1)(x - 1) - x(7x^2 + 5) - 11] = 4$

37. $\dfrac{x^4(x + 3)}{5} - \dfrac{x^2(2x + 5)}{2} + \dfrac{4(2x + 1)}{10} = 0$

38. $\dfrac{x^3(x^2 + 1)}{11} + \dfrac{x^2(x^2 - 2)}{3} = \dfrac{4(3x - 2)}{33}$

39. $\dfrac{x(x^5 - x^4 + 2)}{2} + \dfrac{x^2(x - 1)}{3} = 1$

40. $x^2(x^4 - 3x^3 - 2) - \dfrac{x^3(x - 3)}{4} + \dfrac{1}{2} = 0$

In Exercises 41–56, solve each inequality.

41. $x^3 \le 4x^2 + 8x$

42. $2x^4 + 3x^3 > 2x^2$

43. $t^4 + 3t^3 > 2t^2 + 6t$

44. $\dfrac{2h^3 - 9}{3} \le h(h - 2)$

45. $\dfrac{n^2(n^2 + 4)}{5} < 9$

46. $x^6 \ge x^3 + 2$

47. $2x^2(3x - 2)^3 + (3x - 2)^4 \ge 0$

48. $t(2t + 1)^2(t - 3)^3(t + 2) < 0$

49. $2x^4 - 3x^3 - 3x^2 - 3x \le 5$

50. $3x^4 - 4x^3 + 8x^2 - 16x > 16$

51. $3 - 7x > x^2(x - 5)$

52. $x^2(2x^2 - 11x + 18) \le 4(x + 2)$

53. $6x^4 - 41x^3 \le 19(4x - 1) - 5(23x^2 - 1)$

54. $3x^2(x + 2)(x - 3) + 11x(x^2 - 3) < 2(x - 15)$

55. $\dfrac{(x^2 + 5)(x^2 - 3)}{6} \geq x^3 - 3x$

56. $\dfrac{x^5 - x^2 + 8}{4} - \dfrac{x^4 - x}{2} > 2x^3$

For the applied problems in Exercises 57–68, express any rational solution as a reduced fraction and any irrational solution as a decimal number rounded to the nearest thousandth.

57. As a result of poaching, the number of elephants in a certain herd is declining rapidly. It is estimated that the population P of the herd is a function of time t (in years) and is given by

$$P(t) = -t^3 - t^2 + t + 150.$$

In approximately how many years will the herd be extinct?

58. The charge q (in coulombs) on a certain electrical capacitor is a function of the amount of time t (in seconds) that the capacitor has been charging and is given by

$$q(t) = 3t^3 + 4t.$$

After how many seconds of charging does the capacitor reach a charge of 10 coulombs?

59. The total profit P (in dollars) a company makes is a function of the number of units x it sells and is given by

$$P(x) = x^3 - x^2 - 12x - 210 \quad \text{for } x \geq 0.$$

How many units must the company sell to make a profit?

60. Under certain conditions, the velocity v (in feet per second, or ft/s) of an object as a function of time t (in seconds) is given by

$$v(t) = 4t^3 - 11t^2 - 6t + 9 \quad \text{for } 0 \leq t \leq 4.$$

Determine the time interval(s) in which $v(t) \leq 0$ ft/s.

61. The rectangular jewelry box shown in the figure has a square base, and its height is 2 inches less than the length of its base. Find the dimensions of the box if the volume of the box is 48 cubic inches.

62. The allergy medication capsule shown in the figure has the shape of a cylinder surmounted by a hemisphere on each end. The total length of the capsule is 10 millimeters and it holds 25π cubic millimeters of medicine. Find the diameter d of the capsule.

63. The width of a rectangular toy box is twice its height and the length of the box is 1 foot more than its width, as shown in the figure. The volume of the box is 18 cubic feet. What are the dimensions of the box?

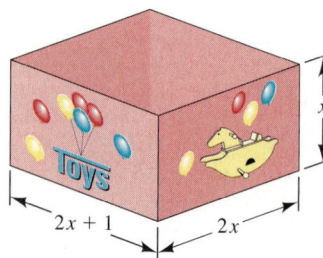

64. Recall from elementary geometry that the volume V of a cone with radius r and height h is $V = \frac{1}{3}\pi r^2 h$. If a conical sandpile has a slant height of 10 meters, as shown in the sketch, for what value(s) of h is $V = 128\pi$ cubic meters?

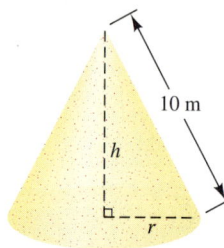

65. When the length of one side of a cube is reduced by 2 inches, the volume of the remaining rectangular solid is 75 cubic inches. What is the length of the side of the original cube?

66. A rectangular box has length 2 inches, width 2 inches, and height 3 inches. Each of these dimensions is increased uniformly until a new box is formed whose volume is triple that of the original box. What are the dimensions of the new box?

67. A piece of cardboard, 4 cm by 5 cm, is made into an open rectangular box by cutting out a square from each corner and turning up the resulting sides, as shown in the figure. The volume of the box that is formed is less than 3 cubic centimeters. Find an interval(s) for the acceptable dimensions of the square cut-out.

Square
Cutout

4 cm

←— 5 cm —→

68. Recall from elementary geometry that the volume V of a cylinder with radius r and height h is $V = \pi r^2 h$. Sup-

pose the slant distance from the top of an oil drum to the bottom is 12 feet, as shown in the figure. Find an interval for the height h of the drum if its volume is greater than 160π cubic feet.

r

12 ft

h

◆ **Critical Thinking**

A *cubic formula* may be used to solve cubic equations of the form $x^3 + ax + b = 0$, where a and b are real numbers. The formula states that the three roots of this equation are

$$A + B, \qquad -\frac{A + B}{2} + \frac{A - B}{2}\sqrt{-3},$$

and

$$-\frac{A + B}{2} - \frac{A - B}{2}\sqrt{-3}$$

where

$$A = \sqrt[3]{-\frac{b}{2} + \sqrt{\frac{b^2}{4} + \frac{a^3}{27}}}$$

and

$$B = \sqrt[3]{-\frac{b}{2} - \sqrt{\frac{b^2}{4} + \frac{a^3}{27}}}.$$

In Exercises 69 and 70, use the cubic formula to approximate the three roots of each equation.

69. $x^3 + x - 4 = 0$ **70.** $2x^3 + x + 9 = 0$

The cubic equation $y^3 + py^2 + qy + r = 0$ can be written in the form $x^3 + ax + b = 0$ by replacing y with $x - \frac{p}{3}$. In Exercises 71 and 72,

(a) use this substitution to write the given equation in the form $x^3 + ax + b = 0$.

(b) use the cubic formula (see Exercises 69 and 70) to approximate the three roots of the given equation.

71. $y^3 - 3y^2 - 9y + 27 = 0$

72. $y^3 - 6y^2 - 5 = 0$

In Exercises 73 and 74, find the coordinates of the intersection point(s) of the graphs of the two equations by solving the two equations simultaneously. Check graphically with a graphing calculator.

73. $y = x^3$ and $21x - 4y = 10$

74. $y = x^4$ and $20x + 27y = 7$

75. Use a graphing calculator to generate the graph of $P(x) = x^3 + ax^2 + b$ for several positive values of a and b. Then discuss the nature of the roots of the equation $x^3 + ax^2 + b = 0$.

76. Use a graphing calculator to generate the graph of $P(x) = x^3 + ax^2 + b$ for several negative values of a and b. Then discuss the nature of the roots of the equation $x^3 + ax^2 + b = 0$.

77. Use a graphing calculator to generate the graph of $P(x) = x^4 + ax + b$ for several negative values of a and b. Then discuss the nature of the roots of the equation $x^4 + ax + b = 0$.

78. Use a graphing calculator to generate the graph of $P(x) = x^4 + ax + b$ for several positive values of a and negative values b. Then discuss the nature of the roots of the equation $x^4 + ax + b = 0$.

4.5   Rational Functions and Their Graphs

◆ Introductory Comments

Any function f of the form

$$f(x) = \frac{P(x)}{D(x)},$$

where $P(x)$ and $D(x)$ are polynomials and $D(x) \neq 0$, is called a **rational function**. Since division by zero is undefined, the domain of a rational function f is the set of all real numbers other than those numbers that make its denominator zero, that is, all real numbers except the real zeros of D.

When working with rational functions, we often need to evaluate expressions of the form $1/c$, where $|c|$ is a very large number or a very small number. If $|c|$ is a very large number, then $1/c$ is close to 0, and if $|c|$ is a very small number, then $1/c$ is very large in absolute value. Here are some illustrations:

$$\underbrace{\frac{1}{1000} = 0.001 \qquad \frac{1}{-1000} = -0.001}_{\text{close to zero}} \qquad \underbrace{\frac{1}{0.001} = 1000 \qquad \frac{1}{-0.001} = -1000}_{\text{very large in absolute value}}$$

◆ Vertical Asymptotes

The reciprocal function $f(x) = 1/x$ (Section 2.5) is one of the simplest rational functions. Its domain is the set of all real numbers except 0. Using interval notation, we write the domain as $(-\infty, 0) \cup (0, \infty)$. Figure 4.39 shows the graph of the reciprocal function. We say that the y-axis (the line $x = 0$) is a *vertical asymptote* of the graph of f because the graph approaches the y-axis as x gets closer and closer to zero.

As x approaches zero from the *left* $(x \to 0^-)$, the value of $f(x)$ *decreases* without bound $[f(x) \to -\infty]$:

$$\text{As } x \to 0^-, \quad f(x) = \frac{1}{x} \to \frac{1}{-0} \to -\infty$$

$$\underbrace{}_{\text{a negative number close to 0}}$$

As x approaches zero from the *right* $(x \to 0^+)$, the value of $f(x)$ *increases* without bound $[f(x) \to +\infty]$:

$$\text{As } x \to 0^+, \quad f(x) = \frac{1}{x} \to \frac{1}{+0} \to +\infty$$

$$\underbrace{}_{\text{a positive number close to 0}}$$

We define a vertical asymptote as follows.

$y = 0$

$x = 0$

FIGURE 4.39

Graph of $f(x) = \dfrac{1}{x}$

◆ **Vertical Asymptotes**

The line $x = a$ is a **vertical asymptote** of the graph of a rational function f if at least one of the following statements is true.

(a) As $x \to a^-$, $f(x) \to +\infty$.

(b) As $x \to a^+$, $f(x) \to +\infty$.

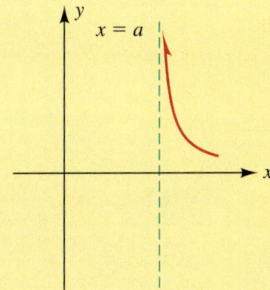

(c) As $x \to a^-$, $f(x) \to -\infty$.

(d) As $x \to a^+$, $f(x) \to -\infty$.

To determine the vertical asymptote(s) of a rational function f defined by $f(x) = P(x)/D(x)$, where $P(x)/D(x)$ is reduced to lowest terms, we find the real zeros of D. If a is a real zero of D, then the vertical line $x = a$ is a vertical asymptote of the graph of f.

EXAMPLE 1 Find the vertical asymptote(s) of the graph of each function, and then determine the appearance of its graph near each vertical asymptote.

(a) $F(x) = \dfrac{3x - 2}{x - 1}$

(b) $f(x) = \dfrac{x + 3}{x^2 - 2x - 8}$

◆ **Solution**

(a) The zero of the denominator $D(x) = x - 1$ is 1. Therefore, since $P(x)/D(x)$ is reduced to lowest terms, we know that $x = 1$ is the only vertical asymptote of the graph of F.

To determine whether $F(x) \to +\infty$ or $F(x) \to -\infty$ as x approaches the vertical asymptote $x = 1$, we check the algebraic sign of $F(x)$ near the asymptote as follows:

As $x \to 1^-$, $F(x) = \dfrac{3x - 2}{x - 1} \to \dfrac{1}{-0} \to -\infty$

a negative number close to 0

As $x \to 1^+$, $F(x) = \dfrac{3x - 2}{x - 1} \to \dfrac{1}{+0} \to +\infty$

a positive number close to 0

Thus, we conclude that $F(x) \to -\infty$ as $x \to 1^-$ and that $F(x) \to +\infty$ as $x \to 1^+$. Figure 4.40 summarizes this information.

(b) Factoring the denominator, we have

$$f(x) = \frac{x + 3}{x^2 - 2x - 8} = \frac{x + 3}{(x + 2)(x - 4)}.$$

Thus, the zeros of $D(x) = x^2 - 2x - 8$ are -2 and 4. Since $P(x)/D(x)$ is in reduced form, we conclude that $x = -2$ and $x = 4$ are vertical asymptotes of the graph of f.

To determine whether $f(x) \to +\infty$ or $f(x) \to -\infty$ as x approaches each vertical asymptote, we check the algebraic sign of $f(x)$ near each asymptote as follows:

As $x \to -2^-$, $f(x) = \dfrac{x + 3}{(x + 2)(x - 4)} \to \dfrac{1}{(-0)(-6)} \to +\infty$

a negative number close to 0

As $x \to -2^+$, $f(x) = \dfrac{x + 3}{(x + 2)(x - 4)} \to \dfrac{1}{(+0)(-6)} \to -\infty$

a positive number close to 0

As $x \to 4^-$, $f(x) = \dfrac{x + 3}{(x + 2)(x - 4)} \to \dfrac{7}{(6)(-0)} \to -\infty$

a negative number close to 0

As $x \to 4^+$, $f(x) = \dfrac{x + 3}{(x + 2)(x - 4)} \to \dfrac{7}{(6)(+0)} \to +\infty$

a positive number close to 0

Thus, we conclude that $f(x) \to +\infty$ as $x \to -2^-$, $f(x) \to -\infty$ as $x \to -2^+$, $f(x) \to -\infty$ as $x \to 4^-$, and $f(x) \to +\infty$ as $x \to 4^+$. Figure 4.41 summarizes this information.   ◆

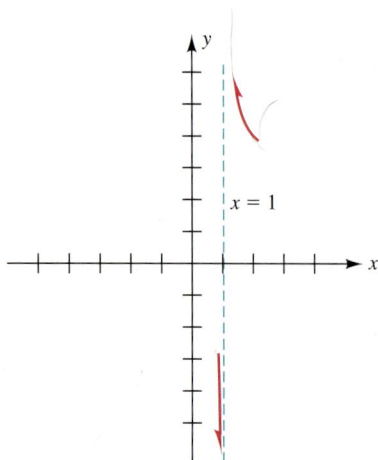

FIGURE 4.40

Appearance of the graph of $F(x) = \dfrac{3x - 2}{x - 1}$ as x approaches the vertical asymptote $x = 1$

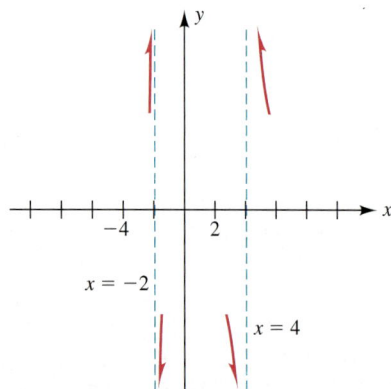

FIGURE 4.41

Appearance of the graph of $f(x) = \dfrac{x + 3}{x^2 - 2x - 8}$ as x approaches each vertical asymptote

PROBLEM 1   State the domain of each of the rational functions defined in Example 1.   ◆

◆ Horizontal Asymptotes

Recall the graph of the reciprocal function, $f(x) = 1/x$, sketched in Figure 4.39. We say that the x-axis (the line $y = 0$) is a *horizontal asymptote* of the graph of f because the graph approaches the x-axis for very large values of x (either positive or negative). As x increases without bound ($x \to +\infty$), $f(x)$ approaches zero through values greater than zero [$f(x) \to 0^+$], and as x decreases without bound ($x \to -\infty$), $f(x)$ approaches zero through values less than zero [$f(x) \to 0^-$]. We define a horizontal asymptote as follows.

◆ **Horizontal Asymptotes**

If $y = b$ is a **horizontal asymptote** of the graph of a rational function f, then at least one of the following statements is true.

(a) As $x \to +\infty$, $f(x) \to b^+$.   **(b)** As $x \to +\infty$, $f(x) \to b^-$.

(c) As $x \to -\infty$, $f(x) \to b^+$.   **(d)** As $x \to -\infty$, $f(x) \to b^-$.

To determine whether a rational function has a horizontal asymptote, we begin by dividing each term of the numerator and denominator by the highest power of x that appears in the rational expression. In doing so, we develop fractions of the form k/x^n, where k is a constant and n is a positive integer. Now, for very large values of x (either positive or negative), x^n become very large and, consequently, expressions of the form k/x^n become very small. We indicate this by writing

$$\frac{k}{x^n} \to 0 \quad \text{as} \quad |x| \to \infty.$$

Although a rational function may have many vertical asymptotes, it can have at most one horizontal asymptote. By using the method described above, it can be shown that the rational function $f(x) = P(x)/D(x)$, where $P(x)/D(x)$ is reduced to lowest terms, follows these rules for the existence of a horizontal asymptote:

1. one horizontal asymptote if the degree of $P(x)$ is less than or equal to the degree of $D(x)$;

2. no horizontal asymptote if the degree of $P(x)$ is greater than the degree of $D(x)$.

EXAMPLE 2   Find the horizontal asymptote of the graph of each function. Then determine the appearance of its graph as $|x| \to \infty$.

(a) $F(x) = \dfrac{3x - 2}{x - 1}$ **(b)** $f(x) = \dfrac{x + 3}{x^2 - 2x - 8}$

◆ **Solution**

(a) Dividing each term of the numerator and denominator by x, we obtain

$$F(x) = \frac{3x - 2}{x - 1} = \frac{3 - \dfrac{2}{x}}{1 - \dfrac{1}{x}}.$$

Now, as $|x| \to \infty$, each of the fractions $\dfrac{2}{x}$ and $\dfrac{1}{x}$ approaches zero. Thus,

$$\text{as } |x| \to \infty, \qquad F(x) \to \frac{3 - 0}{1 - 0} = 3.$$

Hence, the line $y = 3$ is a horizontal asymptote of the graph of F.

To determine if $F(x)$ approaches this horizontal asymptote through values greater than three $[F(x) \to 3^+]$ or through values less than three $[F(x) \to 3^-]$ as $|x| \to \infty$, we choose an arbitrary large value of $|x|$ and determine if the approach is from above or from below the horizontal line $y = 3$. Here we choose $|x| = 100$:

$$F(100) = \frac{3(100) - 2}{(100) - 1} = \frac{298}{99} \approx 3.01$$

> approaching $y = 3$ through values greater than 3

$$F(-100) = \frac{3(-100) - 2}{(-100) - 1} = \frac{-302}{-101} \approx 2.99$$

> approaching $y = 3$ through values less than 3

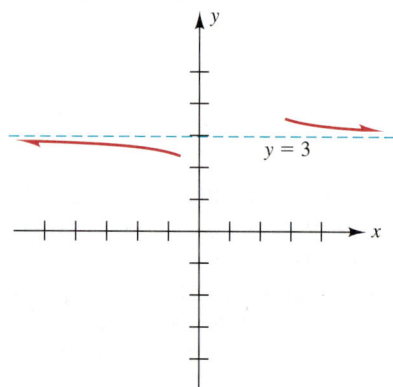

FIGURE 4.42

Appearance of the graph of

$F(x) = \dfrac{3x - 2}{x - 1}$ as $|x| \to \infty$

Thus, we conclude $F(x) \to 3^+$ as $x \to +\infty$ and $F(x) \to 3^-$ as $x \to -\infty$. The appearance of the graph of F as $|x| \to \infty$ is shown in Figure 4.42.

(b) The highest power of x that appears in the rational expression is x^2. Dividing each term of the numerator and denominator by x^2, we obtain

$$f(x) = \frac{x + 3}{x^2 - 2x - 8} = \frac{\dfrac{1}{x} + \dfrac{3}{x^2}}{1 - \dfrac{2}{x} + \dfrac{8}{x^2}}.$$

Now, as $|x| \to \infty$, each of the fractions $\dfrac{1}{x}$, $\dfrac{3}{x^2}$, $\dfrac{2}{x}$, and $\dfrac{8}{x^2}$ approaches zero. Thus,

$$\text{as } |x| \to \infty, \qquad f(x) \to \frac{0 + 0}{1 - 0 - 0} = 0.$$

Hence, the line $y = 0$ (the x-axis) is a horizontal asymptote of the graph of f. To determine whether

$$f(x) = \frac{x + 3}{x^2 - 2x - 8} = \frac{x + 3}{(x + 2)(x - 4)}$$

approaches the x-axis through values greater than zero $[\,f(x) \to 0^+\,]$ or through values less than zero $[\,f(x) \to 0^-\,]$ as $|x| \to \infty$, we choose an arbitrary large value of $|x|$ and determine if the approach is from above or from below the x-axis. We choose $|x| = 100$:

$$f(100) = \frac{103}{(102)(96)} \approx +0.01$$

$$f(-100) = \frac{-97}{(-98)(-104)} \approx -0.01$$

Thus, we conclude $f(x) \to 0^+$ as $x \to +\infty$ and $f(x) \to 0^-$ as $x \to -\infty$. The appearance of the graph of f as $|x| \to \infty$ is shown in Figure 4.43. ◆

Although the graph of a rational function can not intersect any of its vertical asymptotes, it can intersect its horizontal asymptote. To determine if the rational function

$$f(x) = \frac{x + 3}{x^2 - 2x - 8}$$

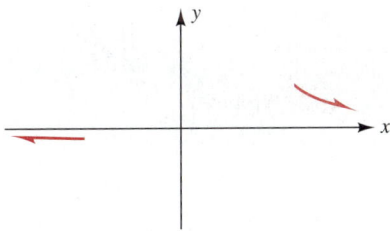

FIGURE 4.43

Appearance of the graph of
$f(x) = \dfrac{x + 3}{x^2 - 2x - 8}$ as $|x| \to \infty$

in Example 2(b) intersects its horizontal asymptote $y = 0$, we solve the equation $f(x) = 0$ as follows:

$$\frac{x + 3}{x^2 - 2x - 8} = 0$$

$$x + 3 = 0 \qquad \textcolor{red}{\textbf{Multiply both sides by } x^2 - 2x - 8 \neq 0}$$

$$x = -3$$

Hence, we conclude that the graph of this rational function intersects its horizontal asymptote at $x = -3$.

PROBLEM 2   Show that the rational function defined in Example 2(a) does *not* intersect its horizontal asymptote $y = 3$ by showing that the equation $F(x) = 3$ leads to a contradiction.   ◆

◆ **Oblique Asymptotes**

If the degree of the numerator $P(x)$ of a rational function is one more than the degree of the denominator $D(x)$ and $P(x)/D(x)$ is reduced to lowest terms, then the graph of the rational function $f(x) = P(x)/D(x)$ has an **oblique asymptote**—one that is neither vertical nor horizontal. To determine the oblique asymptote of the rational function $f(x) = P(x)/D(x)$, where the degree of $P(x)$ is one more than the degree of $D(x)$, we use polynomial long division to write the rational function in the form

$$f(x) = (mx + b) + \frac{R(x)}{D(x)}$$

where the degree of $R(x)$ is less than the degree of $D(x)$. Now,

$$\text{as } |x| \to \infty, \qquad \frac{R(x)}{D(x)} \to 0.$$

Hence, $f(x) \to mx + b$, which we recognize as the equation of a straight line with slope m and y-intercept b.

EXAMPLE 3   Find the oblique asymptote for the graph of the rational function g defined by $g(x) = \dfrac{2x^3}{x^2 + 3}$. Then determine the appearance of the graph of g as $|x| \to \infty$.

◆ **Solution**   Using polynomial long division, we divide $2x^3$ by $x^2 + 3$ as follows:

$$
\begin{array}{r}
2x \\
x^2 + 3 \overline{\smash{)}2x^3 } \\
\underline{2x^3 + 6x} \\
-6x
\end{array}
$$

Hence, we can write

$$g(x) = 2x + \frac{-6x}{x^2 + 3} = 2x + \frac{\dfrac{-6}{x}}{1 + \dfrac{3}{x^2}}$$

As $|x| \to \infty$, each of the fractions $\dfrac{-6}{x}$ and $\dfrac{3}{x^2}$ approaches zero. Thus,

$$\text{as } |x| \to \infty, \qquad g(x) \to 2x + \frac{0}{1+0} = 2x.$$

In summary, the line $y = 2x$ is an oblique asymptote of the graph of g.

To determine the appearance of the graph of g as $|x| \to \infty$, we choose an arbitrary large value of $|x|$ and determine if the approach is from above the line $y = 2x$ or from below this line. Choosing $x = 100$, we have $y = 2(100) = 200$, and

$$g(100) = \frac{2(100)^3}{(100)^2 + 3} \approx 199.94$$

approaching $y = 2x$ from below the line

Choosing $x = -100$, we have $y = 2(-100) = -200$, and

$$g(-100) = \frac{2(-100)^3}{(-100)^2 + 3} \approx -199.94$$

approaching $y = 2x$ from above the line

The appearance of the graph of g as $|x| \to \infty$ is shown in Figure 4.44. ◆

Although the graph of a rational function cannot intersect its vertical asymptotes, the graph of a rational function *may* intersect its oblique asymptote. To determine whether the graph of a rational function f intersects its oblique asymptote $y = mx + b$, we solve the equation $f(x) = mx + b$ for x.

PROBLEM 3 Show that the rational function defined in Example 3 intersects its oblique asymptote $y = 2x$ at the origin. ◆

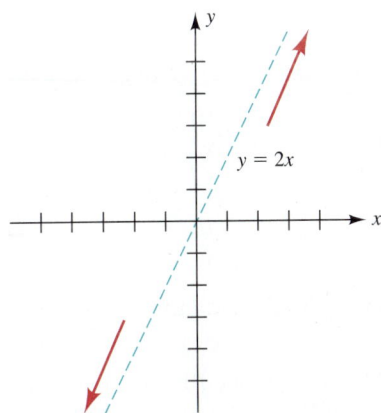

FIGURE 4.44
Appearance of the graph of
$g(x) = \dfrac{2x^3}{x^2 + 3}$ as $|x| \to \infty$

◆ Sketching the Graph of a Rational Function

We can usually determine the basic shape of the graph of a rational function by noting the manner in which the graph approaches its asymptotes. To complete the graph of the function, we determine the intercepts, observe any symmetry and, if necessary, plot a few additional points. We now state a general procedure for graphing a rational function

$$f(x) = \frac{P(x)}{D(x)},$$

where $P(x)/D(x)$ is reduced to lowest terms.

◆ **General Procedure for Graphing a Rational Function**

1. Find the vertical asymptotes of the graph of f, and determine the appearance of the graph as x approaches each vertical asymptote from the left and from the right.

2. Find the horizontal or oblique asymptote of the graph of f, and determine the appearance of the graph of f as $|x| \to \infty$.

3. (a) Determine whether the function f is an even function, an odd function, or neither.

   (b) Find the x-intercepts for the graph of f. Since $f(x) = 0$ only when its numerator $P(x) = 0$, the x-intercepts of the graph of f are the real zeros of P.

   (c) Find the y-intercept for the graph of f. This is obtained by evaluating $f(0)$.

   (d) Determine whether the graph crosses its horizontal or oblique asymptote.

4. Plot a few additional points, if necessary, and sketch the graph.

EXAMPLE 4   Sketch the graph of each rational function. Label the asymptotes and the intercepts.

(a) $F(x) = \dfrac{3x - 2}{x - 1}$ (b) $f(x) = \dfrac{x + 3}{x^2 - 2x - 8}$ (c) $g(x) = \dfrac{2x^3}{x^2 + 3}$

◆ **Solution**

(a) By combining the information in Figures 4.40 and 4.42, we begin to see the shape of the graph of $F(x) = \dfrac{3x - 2}{x - 1}$. To complete the graph of this function, we determine the x- and y-intercepts.

$$x\text{-intercept:} \quad 3x - 2 = 0 \qquad y\text{-intercept:} \quad F(0) = \frac{3(0) - 2}{(0) - 1} = 2$$

$$x = \frac{2}{3}$$

Plotting the intercepts and noting the manner in which the graph approaches its asymptotes (Figures 4.40 and 4.42), we sketch the graph of this function, as shown in Figure 4.45.

(b) By combining the information in Figures 4.41 and 4.43, we begin to see the shape of the graph of $f(x) = \dfrac{x + 3}{x^2 - 2x - 8}$. To complete the graph of this function, we determine the x- and y-intercepts:

$$x\text{-intercept:} \quad x + 3 = 0 \qquad y\text{-intercept:} \quad f(0) = \frac{(0) + 3}{(0)^2 - 2(0) - 8}$$

$$x = -3 \qquad\qquad = -\frac{3}{8}$$

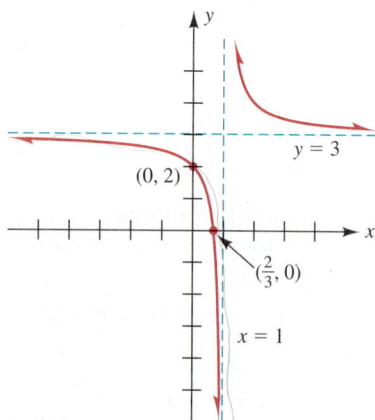

FIGURE 4.45

Graph of $F(x) = \dfrac{3x - 2}{x - 1}$

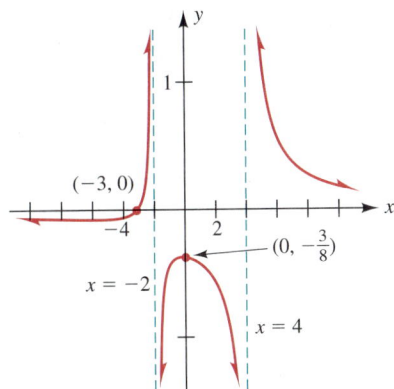

FIGURE 4.46

Graph of $f(x) = \dfrac{x + 3}{x^2 - 2x - 8}$

FIGURE 4.47

Graph of $g(x) = \dfrac{2x^3}{x^2 + 3}$

FIGURE 4.48

Graph of $f(x) = \dfrac{x + 3}{x^2 - 2x - 8}$ as shown

in two viewing rectangles

(a) Distorted graph

(b) Realistic graph

Plotting the intercepts and noting the manner in which the graph approaches its asymptotes (see Figures 4.41 and 4.43), we sketch the graph of $f(x) = \dfrac{x + 3}{x^2 - 2x - 8}$, as shown in Figure 4.46.

(c) Using the information in Figure 4.44 and the fact that the graph of g has no vertical asymptote (why?), we begin to see the shape of the graph of $g(x) = \dfrac{2x^3}{x^2 + 3}$. Also, we note that $g(-x) = -g(x)$ for all x, so we know that g is an odd function. Hence, its graph is symmetric with respect to the origin. To complete the graph of this function, we determine the x- and y-intercepts.

$$x\text{-intercept:} \quad 2x^3 = 0 \qquad y\text{-intercept:} \quad g(0) = \dfrac{2(0)^3}{(0)^2 + 3}$$

$$x = 0 \qquad\qquad\qquad\qquad = 0$$

Using the intercepts, the manner in which the graph approaches its oblique asymptote, and the fact that the graph is symmetric with respect to the origin, we sketch the graph of this function, as shown in Figure 4.47. ◆

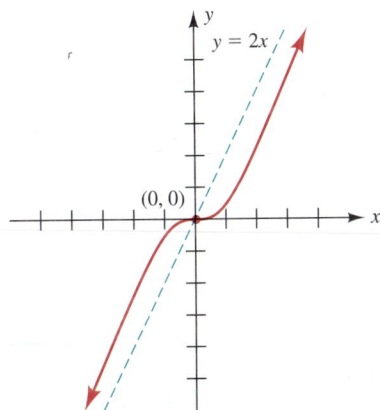

Note that the graph of the rational function f in Figure 4.46 has a relative maximum point in the interval $(-2, 4)$. Although the methods of calculus are needed to determine the exact coordinates of this relative maximum point, we can estimate its coordinates by using a graphing calculator. When generating the graph of a rational function on a graphing calculator, we may obtain a distorted picture in the viewing rectangle. Figure 4.48 shows how the graph of $f(x) = \dfrac{x + 3}{x^2 - 2x - 8}$ looks in two different viewing rectangles.

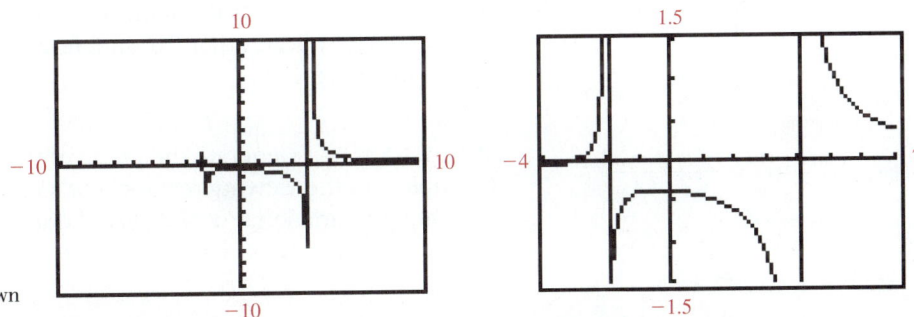

Note that the graph in part (a) of Figure 4.48 contains spikes and seems distorted, whereas the graph in part (b) is a more realistic picture. The vertical line segments through $x = -2$ and $x = 4$ approximate the vertical asymptotes of the graph and are not actually part of the graph. They occur in the viewing rectangle because the calculator evaluates the function only for a finite number of values, plots these values, and then attempts to connect these values to make a "continuous" curve. To generate a graph without these vertical lines, use the "dot" mode instead of the "connected" mode.

PROBLEM 4   Use a graphing calculator to estimate the coordinates of the relative maximum point for $f(x) = \dfrac{x + 3}{x^2 - 2x - 8}$. Record the coordinates to three significant digits.   ◆

◆ Rational Functions with Common Factors

To sketch the graph of a rational function $f(x) = P(x)/D(x)$, where $P(x)$ and $D(x)$ have common factors, we begin by reducing $P(x)/D(x)$ to lowest terms. The graph of f is the same as the graph of the new function that is formed by the reducing process, except it may have one or more "holes," which indicates that f is not a continuous function.

EXAMPLE 5   Sketch the graph of $f(x) = \dfrac{x^2 + x - 2}{x - 1}$.

◆ **Solution**   Factoring the numerator, we obtain

$$f(x) = \frac{(x + 2)(x - 1)}{x - 1}$$

and reducing to lowest terms gives us the new function

$$F(x) = x + 2.$$

Note that the functions f and F are *not* the same functions. The domain of f is the set of all real numbers except 1, whereas the domain of F is the set of all real numbers. The graph of F is a straight line with slope 1 and y-intercept 2. The graph of f is the same as the graph of F, except for a hole at (1, 3), which indicates that f is undefined at this point. The graph of f is shown in Figure 4.49.   ◆

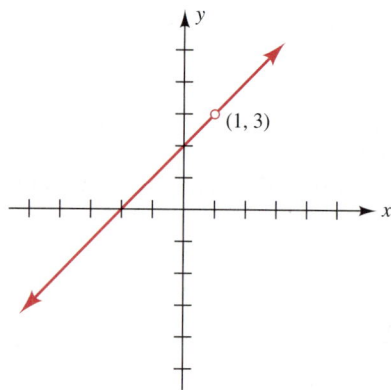

FIGURE 4.49

The graph of $f(x) = \dfrac{x^2 + x - 2}{x - 1}$ is discontinuous at (1, 3).

PROBLEM 5   Generate the graph of the rational function defined in Example 5 in the viewing rectangle of a graphing calculator. Then try to evaluate the function at $x = 1$. What does the calculator display suggest?   ◆

◆ Application: Average Cost Function

The *total cost* to run a business is the sum of the *fixed cost* (costs that do not depend on the level of production, such as mortgage payment, insurance, utilities, and so on) and *variable cost* (costs that vary with the level of production, such as maintenance of equipment, cost of materials, labor, and so on). If $T(x)$ is the total cost of producing x units of a product, then the **average cost function** C is defined by

$$C(x) = \frac{T(x)}{x}$$

EXAMPLE 6   A certain manufacturer produces a kit for building a log home. The fixed cost for the company is $141,120 per year and the variable cost (in dollars) of producing x log-home kits per year is $12,000x + 20x^2$ for $0 \le x \le 300$.

(a) Define a rational function that represents the average cost per kit.

(b) Generate the graph of the function in part (a) in the viewing rectangle of a graphing calculator. Then find the approximate number of kits that should be produced in order to minimize the average cost per kit.

◆ **Solution**

(a) If $T(x)$ is the total cost of producing x log-home kits per year, then

$$T(x) = \underbrace{141,120}_{\text{fixed cost}} + \underbrace{(12,000x + 20x^2)}_{\text{variable cost}} \quad \text{for } 0 \le x \le 300.$$

Hence, the average cost per kit is

$$C(x) = \frac{141,120 + 12,000x + 20x^2}{x} \quad \text{for } 0 < x \le 300.$$

(b) Observe the following characteristics of the graph of this function:

Vertical asymptote: $x = 0$

$$\text{As } x \to 0^+, \quad C(x) = \frac{141,120 + 12,000x + 20x^2}{x} \to +\infty$$

Oblique asymptote: Dividing numerator and denominator by x, we obtain

$$C(x) = \frac{141,120}{x} + 12,000 + 20x.$$

Now,

$$\text{as } |x| \to \infty, \quad \frac{141,120}{x} \to 0.$$

Hence, $C(x) \to 12,000 + 20x$ as $|x| \to \infty$. Thus, the line $y = 20x + 12,000$ is an oblique asymptote of the graph of C. This information suggests that an appropriate viewing rectangle for observing the graph of this average cost function is as follows:

Xmin = 0
Xmax = 300
Xscl = 100
Ymin = 0
Ymax = 30000
Yscl = 10000

30000

Minimum
X=84.000029   Y=15360

0   300

FIGURE 4.50

Graph of

$$C(x) = \frac{141{,}120 + 12{,}000x + 20x^2}{x}$$

for $0 < x \le 300$ and its oblique asymptote $y = 20x + 12{,}000$

Figure 4.50 shows the graphs of the average cost function and its oblique asymptote in this viewing rectangle. The x-coordinate of the lowest point on this curve represents the number of kits that should be produced per year in order to minimize the average cost per kit. Using either the calculator's built-in program for finding a relative minimum point or the TRACE key and ZOOM key, we find that the coordinates of the relative minimum point are (84, 15360), rounded to the nearest whole number. In summary, the minimum average cost per kit is $15,360 and this occurs when 84 kits are produced per year.   ◆

PROBLEM 6   For the problem described in Example 6, determine how many kits are produced per year if the average cost per kit is $16,640.   ◆

Exercises 4.5

◆ Basic Skills

In Exercises 1–10,

(a) *find the vertical asymptote(s) of the graph of f, if any exist.*

(b) *if the graph of f has vertical asymptotes, determine the appearance of the graph near each vertical asymptote. Verify the appearance by using a graphing calculator.*

1. $f(x) = \dfrac{x}{x + 2}$

2. $f(x) = \dfrac{x + 1}{2x - 3}$

3. $f(x) = \dfrac{1}{x^2 + 4}$

4. $f(x) = \dfrac{x + 3}{x^2 - 4}$

5. $f(x) = \dfrac{x^3 - 1}{2x^2 + 5x - 3}$

6. $f(x) = \dfrac{-1}{x^2 + x + 4}$

7. $f(x) = \dfrac{x + 4}{x^3 - 2x - 4}$

8. $f(x) = \dfrac{x^2 + 2}{x^3 + x^2 - 2}$

9. $f(x) = \dfrac{2 - 3x}{x^3 - 5x^2 + 2x + 8}$

10. $f(x) = \dfrac{2x + 1}{2x^4 - x^3 - 18x^2 + 9x}$

In Exercises 11–20,

(a) *find the horizontal asymptote of the graph of f, if one exists.*

(b) *if the graph of f has a horizontal asymptote, determine the appearance of the graph as $|x| \to \infty$. Verify the appearance by using a graphing calculator.*

(c) *if the graph of f has a horizontal asymptote, determine whether the graph crosses the horizontal asymptote and, if so, where. Verify the coordinates by using a graphing calculator.*

11. $f(x) = \dfrac{x + 2}{x - 3}$

12. $f(x) = \dfrac{x^2 - 4}{2x + 1}$

13. $f(x) = \dfrac{x^2 - 1}{2x^2 + 7x - 15}$

14. $f(x) = \dfrac{2x^2 - 3x + 4}{x^3 + x - 1}$

15. $f(x) = \dfrac{x^3 + 3}{1 - 3x + 2x^2}$

16. $f(x) = \dfrac{x^3 + 2x - 1}{x^3 + 8}$

17. $f(x) = \dfrac{3x^2 - 2x + 1}{4x - x^2}$

18. $f(x) = \dfrac{1 - 4x^2}{x^2}$

19. $f(x) = \dfrac{3x - 1}{x^3 - 27}$

20. $f(x) = \dfrac{2x^3 - 3x + 4}{1 + x - 3x^3}$

In Exercises 21–36, sketch the graph of each rational function. Label the x- and y-intercepts and the vertical and horizontal asymptote(s), if they exist. Use a graphing calculator to verify each sketch.

21. $f(x) = \dfrac{3x}{x - 3}$

22. $f(x) = \dfrac{2x + 1}{x + 2}$

23. $F(x) = \dfrac{x + 6}{x^2 - 9}$

24. $F(x) = \dfrac{2}{4 - x^2}$

25. $g(x) = \dfrac{-1}{x^2 + 1}$

26. $g(x) = \dfrac{3x}{x^2 + 9}$

27. $G(x) = \dfrac{x}{x^2 - 4x - 12}$

28. $G(x) = \dfrac{10 - x}{x^2 + 3x - 10}$

29. $h(x) = \dfrac{x - 2}{2x^2 - x - 3}$

30. $h(x) = \dfrac{2x - 1}{x^2 + 2x + 1}$

31. $H(x) = \dfrac{2x^2 + 2x + 2}{x^2 - 4x}$

32. $H(x) = \dfrac{x^2 + 6x + 9}{2x^2 - 6x}$

33. $f(x) = \dfrac{6}{2x^3 + 9x^2 + 3x - 4}$

34. $F(x) = \dfrac{x^2}{x^3 + 3x^2 - x - 3}$

35. $G(x) = \dfrac{x + 2}{x^4 - 2x^3 + 2x - 1}$

36. $g(x) = \dfrac{-x^4}{x^4 + 2x^3 - 7x^2 + 2x - 8}$

In Exercises 37–44,

(a) *find the oblique asymptote of the graph of f.*

(b) *determine whether the graph of f crosses the oblique asymptote and, if so, where.*

(c) *sketch the graph of f and label the x- and y-intercepts and any asymptote(s). Use a graphing calculator to verify each sketch.*

37. $f(x) = \dfrac{x^2 - 4}{x}$

38. $f(x) = \dfrac{x^3 + 1}{x^2}$

39. $f(x) = \dfrac{x^2}{x - 3}$

40. $f(x) = \dfrac{x^2 + 1}{x + 1}$

41. $f(x) = \dfrac{x^2 - 3x - 4}{x + 3}$

42. $f(x) = \dfrac{x^2 - x}{x - 2}$

43. $f(x) = \dfrac{x^3 + 8}{x^2 - 3x - 4}$

44. $f(x) = \dfrac{x^3 - 2x - 4}{x^2 - 1}$

Each of the rational functions P(x)/D(x) in Exercises 45–50 is not reduced to lowest terms. Sketch the graph of the function and label the x- and y-intercepts and any asymptote or "hole." Use a graphing calculator to verify each sketch and to confirm any "hole" in the graph.

45. $F(x) = \dfrac{x^2 + 3x - 10}{x - 2}$

46. $f(x) = \dfrac{x^3 + 2x^2}{x}$

47. $g(x) = \dfrac{x^2 - x - 6}{x^2 + 4x + 4}$

48. $R(x) = \dfrac{2x^2 - 5x - 3}{x^2 - 4x + 3}$

49. $f(x) = \dfrac{x^4 - 2x^3 - x + 2}{x^3 - 2x^2}$

50. $h(x) = \dfrac{2x^4 + x^3 - 3x^2 + 5x - 5}{x^3 + x^2 + x - 3}$

51. The margins at the top and bottom of a poster are 2 inches each, and the margins at the sides are 4 inches each, as shown in the figure. The poster contains 50 square inches of printed matter.

(a) Show that the area A of the entire poster is given by $A = \dfrac{4x^2 + 18x}{x - 8}$.

(b) Generate the graph of the rational function defined in part (a) on a graphing calculator. Then find the overall dimensions of the poster whose area is the smallest possible.

52. A rectangular box with an open top has a square base, as sketched in the figure. The volume of the box is 64 cubic inches.

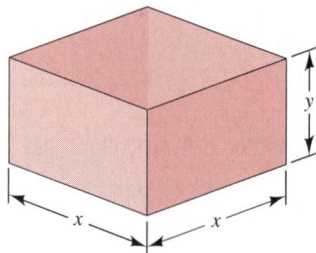

(a) Show that the surface area S of the box is given by
$$S = \frac{x^3 + 256}{x}.$$

(b) Generate the graph of the rational function defined in part (a) on a graphing calculator. Then find the

dimensions of the box whose surface area is the smallest possible.

53. A certain manufacturer produces wood-burning stoves. The fixed cost for the company is $81,000 per year. The variable cost (in dollars) of producing x stoves per year is $400x + 0.1x^2$ for $0 \le x \le 2000$.

(a) Define a function that represents the average cost per stove.

(b) Generate the graph of the function in part (a) on a graphing calculator. Then find the number of stoves that should be produced per year in order to minimize the average cost per stove.

54. A trucking company determines that the cost of operating one of its trucks is $(7 + 0.008x)$ dollars per mile when the truck is driven at x miles per hour $(x \le 75)$. In addition, the company pays the driver $20 per hour.

(a) Express the total cost C per mile as a function of x.

(b) Generate the graph of the function in part (a) on a graphing calculator. Then find the average speed at which the trucks should be driven in order to minimize the total cost per mile.

◆ **Critical Thinking**

55. Can the graph of a rational function have two or more horizontal asymptotes? Explain.

56. Can the graph of a rational function have both a horizontal and oblique asymptote? Explain.

57. Give an example of a rational function whose graph crosses its horizontal asymptote more than once.

58. Give an example of a rational function whose graph crosses its oblique asymptote more than once.

59. Is $x = 1$ always a vertical asymptote of the graph of the rational function

$$f(x) = \frac{P(x)}{x - 1}$$

Explain.

60. If the degree of the numerator of a rational function is 2 more than the degree of its denominator, then the rational function has a *parabolic asymptote*. Find the parabolic asymptote of each function. Then use the parabolic asymptote and any vertical asymptotes and intercepts to help sketch the graph of each function. Use a graphing calculator to verify each sketch.

(a) $f(x) = \dfrac{x^3 + 1}{x}$

(b) $f(x) = \dfrac{x^4 + 6x^2 + 9}{x^2 - 1}$

61. Sketch the graph of each equation.

(a) $x + x^2y - y = 0$

(b) $2x^2 - xy + 3y - 1 = 0$

62. Sketch the graphs of the two given equations on the same coordinate plane. Then, by solving the equations simultaneously, determine the coordinates of the points where their graphs intersect. Use a graphing calculator to verify the coordinates.

(a) $y = \dfrac{1}{x^2 - 4}$ and $3x - 5y = 8$

(b) $y = \dfrac{x^3 - 1}{x^2}$ and $y = -2x - 4$

63. It is estimated that the cost C (in millions of dollars) to remove x percent of the pollution from Boston Harbor is given by $C(x) = \dfrac{500x}{100 - x}$. What can be said about the cost of cleaning up the harbor as $x \to 100^-$?

64. When a 100-ohm resistor is connected in parallel with a variable resistor x, the total resistance R is given by $R(x) = \dfrac{100x}{x + 100}$. What is the significance of the horizontal asymptote for the graph of R?

Rational Equations and Inequalities

◆ Introductory Comments

Recall from Section 1.7 that if the numerator and denominator of a fraction are polynomials of degree $n \geq 0$, then the fraction is called a *rational expression*. Since division by zero is undefined, the variables in a rational expression must be restricted to those values that give a nonzero denominator. For example, when working with the rational expression

$$\frac{4x^2 + 5x + 3}{2x - 6},$$

we assume that its denominator $2x - 6 \neq 0$, which implies that $x \neq 3$.

If both sides of an equation are composed of rational expressions, the equation is called a **rational equation**, and if both sides of an inequality are composed of rational expressions, the inequality is called a **rational inequality**. In this section, we discuss methods of solving rational equations and rational inequalities in one unknown.

◆ Solving Rational Equations

A rational equation is reducible to a polynomial equation (Section 4.4) after the denominators in the rational equation have been eliminated. To eliminate the denominators, we multiply both sides of the rational equation by the least common denominator (LCD) of all the fractions in the equation. When we multiply both sides of a rational equation by an LCD that contains a variable, we may introduce an *extraneous root*—a root that develops through the algebraic process but does not satisfy the original equation. This will occur when we multiply both sides of an equation by an LCD whose value is zero for certain values of the variable. Remember, multiplying both sides of an equation by zero does *not* generate an equivalent equation.

EXAMPLE 1   Solve each rational equation.

(a) $\dfrac{9}{x - 1} - 4 = \dfrac{1}{2}$ **(b)** $\dfrac{3}{2} - \dfrac{1}{m + 2} = \dfrac{m}{2m + 4}$

◆ **Solution**

(a) The LCD for the denominators $(x - 1)$ and 2 is $2(x - 1)$. We multiply both sides of this rational equation by $2(x - 1)$, assuming that $2(x - 1) \neq 0$:

$$2(x - 1)\left[\frac{9}{x - 1} - 4\right] = 2(x - 1)\left[\frac{1}{2}\right] \qquad \text{**Multiply both sides by }2(x - 1)\text{**}$$

$$18 - 8(x - 1) = x - 1 \qquad \text{**Simplify**}$$

$$26 - 8x = x - 1 \qquad \begin{array}{l}\text{**Collect like terms on the**}\\\text{**left-hand side**}\end{array}$$

$$27 = 9x \qquad \text{**Add }(8x + 1)\text{ to both sides**}$$

$$x = 3 \qquad \begin{array}{l}\text{**Divide both sides by 9 and**}\\\text{**interchange sides**}\end{array}$$

Note that if $x = 3$, then the LCD $2(x - 1) \neq 0$. Hence, we have generated equivalent equations and, therefore, the solution of the original equation must also be 3. The graph in Figure 4.51 confirms our work.

(b) The LCD for the denominators 2, $m + 2$, and $2m + 4$ [which factors to $2(m + 2)$] is $2(m + 2)$. We multiply both sides of this rational equation by $2(m + 2)$, assuming that $2(m + 2) \neq 0$:

FIGURE 4.51

The x-intercept of the graph of $f(x) = \dfrac{9}{x - 1} - \dfrac{9}{2}$ is the solution of the equation $\dfrac{9}{x - 1} - 4 = \dfrac{1}{2}$.

$$2(m + 2)\left[\frac{3}{2} - \frac{1}{m + 2}\right] = 2(m + 2)\left[\frac{m}{2(m + 2)}\right] \qquad \begin{array}{l}\text{**Multiply both sides**}\\\text{**by }2(m + 2)\text{**}\end{array}$$

$$3(m + 2) - 2 = m \qquad \text{**Simplify**}$$

$$3m + 4 = m \qquad \begin{array}{l}\text{**Collect like terms**}\\\text{**on left-hand side**}\end{array}$$

$$4 = -2m \qquad \begin{array}{l}\text{**Subtract }3m\text{ from**}\\\text{**both sides**}\end{array}$$

$$m = -2 \qquad \begin{array}{l}\text{**Divide both sides by }-2\text{**}\\\text{**and interchange sides**}\end{array}$$

However, if $m = -2$, then the LCD $2(m + 2) = 0$. Since multiplying both sides of an equation by zero does *not* generate an equivalent equation, the apparent solution, -2, must be discarded. Therefore, we conclude this rational equation has *no solution*. Also, note that the graph in Figure 4.52 does not intersect the x-axis, which indicates that the original equation has no solution. ◆

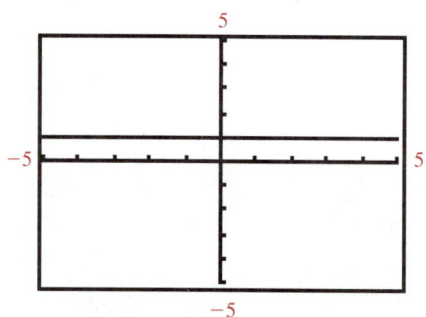

FIGURE 4.52

The graph of $f(m) = \dfrac{3}{2} - \dfrac{1}{m + 2} - \dfrac{m}{2m + 4}$ does not intersect the m-axis.

PROBLEM 1   Rewrite the function

$$f(m) = \frac{3}{2} - \frac{1}{m + 2} - \frac{m}{2m + 4}$$

with a single fraction on the right-hand side. Then show that the new function that is formed is $F(m) = 1$ provided $m = -2$. Compare your graph with Figure 4.52. ◆

The rational equations in Example 1 are reducible to polynomial equations of degree 1 (linear equations). In the next example we solve rational equations that are reducible to polynomial equations of degree $n \geq 2$.

EXAMPLE 2 Solve each rational equation.

(a) $\dfrac{1}{n-3} - \dfrac{6}{n^2-9} = 2$ **(b)** $x + \dfrac{3x+1}{x-1} + \dfrac{6}{x} = 12$

◆ **Solution**

(a) The LCD for $n-3$ and n^2-9 [which factors to $(n-3)(n+3)$] is $(n-3)(n+3)$. We can eliminate fractions by multiplying both sides of this rational equation by the LCD, provided that $(n-3)(n+3) \neq 0$:

$$\frac{1}{n-3} - \frac{6}{n^2-9} = 2$$

$$(n+3) - 6 = 2(n-3)(n+3) \qquad \text{\textcolor{red}{\textbf{Multiply both sides by } $(n-3)(n+3)$}}$$

$$n - 3 = 2n^2 - 18 \qquad \text{\textcolor{red}{\textbf{Simplify both sides}}}$$

$$2n^2 - n - 15 = 0 \qquad \text{\textcolor{red}{\textbf{Write in standard form}}}$$

$$(2n+5)(n-3) = 0 \qquad \text{\textcolor{red}{\textbf{Factor}}}$$

$$\text{\textcolor{red}{\textbf{Solve for } n}}$$

$$n = -\tfrac{5}{2} \quad n = 3$$

Note that if $n = -\tfrac{5}{2}$, then the LCD $(n-3)(n+3) \neq 0$. Hence, we have generated equivalent equations and, therefore, a solution of the original equation must also be $-\tfrac{5}{2}$. However, if $n = 3$, then the LCD $(n-3)(n+3) = 0$. Since multiplying both sides of an equation by zero does not generate an equivalent equation, the apparent solution, 3, must be discarded. Therefore, we conclude this rational equation has only one solution, namely, $-\tfrac{5}{2}$, as illustrated in Figure 4.53.

(b) The LCD for $x-1$ and x is $x(x-1)$. We can eliminate fractions by multiplying both sides of this rational equation by the LCD, provided that $x(x-1) \neq 0$:

$$x + \frac{3x+1}{x-1} + \frac{6}{x} = 12$$

$$x[x(x-1)] + (3x+1)x + 6(x-1) = 12[x(x-1)] \qquad \text{\textcolor{red}{\textbf{Multiply both sides by } $x(x-1)$}}$$

$$x^3 - 10x^2 + 19x - 6 = 0 \qquad \text{\textcolor{red}{\textbf{Write in standard form}}}$$

By the rational zero theorem (Section 4.3), the possible rational roots of this polynomial equation are ± 1, ± 2, ± 3, and ± 6. The graph in

FIGURE 4.53

The n-intercept of the graph of
$$f(n) = \frac{1}{n-3} - \frac{6}{n^2-9} - 2$$
is the solution of the equation
$$\frac{1}{n-3} - \frac{6}{n^2-9} = 2.$$

FIGURE 4.54

The graph of
$P(x) = x^3 - 10x^2 + 19x - 6$ indicates
that 2 might be a rational root of the
equation $x^3 - 10x^2 + 19x - 6 = 0$.

Figure 4.54 indicates that 2 may be a rational root, so we test 2 using
synthetic division:

$$
\begin{array}{r|rrrr}
2 & 1 & -10 & 19 & -6 \\
  &   & 2 & -16 & 6 \\
\hline
  & 1 & -8 & 3 & \boxed{0}
\end{array}
$$
remainder

Since the remainder is zero, we continue as follows:

$$x^3 - 10x^2 + 19x - 6 = 0$$
$$(x - 2)(x^2 - 8x + 3) = 0$$

Apply the quadratic formula.

$$x = 2 \qquad x = 4 \pm \sqrt{13}$$

Note that if $x = 2$ or $x = 4 \pm \sqrt{13}$, then the LCD $x(x - 1) \neq 0$.
Hence, we have generated an equivalent equation and, therefore, the
solution of the original equation must also be

$$x = 2, \qquad 4 + \sqrt{13} \approx 7.61, \quad \text{and} \quad 4 - \sqrt{13} \approx 0.394.\qquad \blacklozenge$$

PROBLEM 2   Generate the graph of

$$f(x) = x + \frac{3x + 1}{x - 1} + \frac{6}{x} - 12$$

in the viewing rectangle of a graphing calculator, and confirm that the
roots are $x = 2$, $4 + \sqrt{13} \approx 7.61$, and $4 - \sqrt{13} \approx 0.394$, as we found
in Example 2(b).   ◆

◆ Solving Rational Inequalities

An inequality that can be written with a rational expression on the left-hand
side and zero on the right-hand side is called a **rational inequality in standard form**. The rational inequality

$$\frac{5x - 2}{x - 1} \leq 4,$$

can be written in standard form by proceeding as follows:

$$\frac{5x - 2}{x - 1} - 4 \leq 0 \qquad \textbf{Subtract 4 from both sides}$$

$$\frac{(5x - 2) - 4(x - 1)}{x - 1} \leq 0 \qquad \textbf{Write as a single fraction}$$

$$\frac{x + 2}{x - 1} \leq 0 \qquad \textbf{Simplify}$$

CAUTION

When writing a rational inequality in standard form, we cannot multiply both sides of the inequality by an expression that contains the unknown. Since we don't know whether the unknown quantity is positive or negative, we don't know whether to change the direction of the inequality symbol. For example, to eliminate the fraction by writing

$$\frac{5x - 2}{x - 1} \leq 4 \quad \text{as} \quad 5x - 2 \leq 4(x - 1) \quad \text{is WRONG!}$$

This procedure assumes that $x - 1 > 0$, which may not be true.

To solve the rational inequality in standard form $f(x) > 0$, we begin by finding the values of the unknown that make the numerator or denominator of the rational expression zero. We refer to these values of the unknown as the *boundary numbers* of the rational inequality. These boundary numbers are the x-intercepts and vertical asymptotes of the graph of f, and they divide the x-axis into intervals. The solution set of the rational inequality $f(x) > 0$ is all the intervals on which the graph of f is *above* the x-axis. For the rational inequality $f(x) < 0$, the solution set is all the intervals on which the graph of f is *below* the x-axis. If the inequality symbol is $\leq$ or $\geq$, we include the boundary numbers that represent the x-intercepts. Since division by zero is undefined, the solution set of a rational inequality cannot include any boundary numbers that represents a vertical asymptote.

EXAMPLE 3 Solve each rational inequality.

(a) $\dfrac{5x - 2}{x - 1} \leq 4$ **(b)** $\dfrac{2x + 3}{x - 3} > \dfrac{x + 1}{x - 1}$

◆ **Solution**

(a) As we have found in the preceding discussion, the rational inequality $\dfrac{5x - 2}{x - 1} \leq 4$ is equivalent to the inequality

$$\frac{x + 2}{x - 1} \leq 0.$$

We find the boundary numbers of a rational inequality in standard form by setting its numerator and denominator equal to zero.

For numerator: $x + 2 = 0$ *For denominator:* $x - 1 = 0$
$x = -2$ $x = 1$

These two boundary numbers divide the real number line into three intervals:

Interval 1 $(-\infty, -2)$ Interval 2 $(-2, 1)$ Interval 3 $(1, \infty)$

We can confirm these intervals by generating the graph of
$f(x) = \dfrac{x + 2}{x - 1}$ in the viewing rectangle of a graphing calculator. As
shown in Figure 4.55, the graph of f is *below* the x-axis in the interval
$(-2, 1)$. Hence, $f(x) \leq 0$ in the interval $[-2, 1)$. Thus, the solution set
of the original inequality is $[-2, 1)$.

(b) We begin by writing the inequality in standard form:

$$\frac{2x + 3}{x - 3} > \frac{x + 1}{x - 1}$$

$$\frac{2x + 3}{x - 3} - \frac{x + 1}{x - 1} > 0 \qquad \text{Subtract } \frac{x + 1}{x - 1} \text{ from both sides}$$

$$\frac{(2x + 3)(x - 1) - (x + 1)(x - 3)}{(x - 3)(x - 1)} > 0 \qquad \text{Subtract fractions}$$

$$\frac{x^2 + 3x}{(x - 3)(x - 1)} > 0 \qquad \text{Simplify the numerator}$$

FIGURE 4.55

The graph of $f(x) = \dfrac{x + 2}{x - 1}$ is below the
x-axis in the interval $(-2, 1)$.

We then determine the boundary numbers:

For numerator:

$$x^2 + 3x = 0$$
$$x(x + 3) = 0$$

$$x = 0 \quad x = -3$$

For denominator:

$$(x - 3)(x - 1) = 0$$

$$x = 3 \quad x = 1$$

Thus, the four boundary numbers are -3, 0, 1, and 3. These four
boundary numbers divide the real number line into five intervals:

Interval 1	Interval 2	Interval 3	Interval 4	Interval 5
$(-\infty, -3)$	$(-3, 0)$	$(0, 1)$	$(1, 3)$	$(3, \infty)$

We can confirm these intervals by generating the graph of
$f(x) = \dfrac{x^2 + 3x}{(x - 3)(x - 1)}$ in the viewing rectangle of a graphing calculator. As shown in Figure 4.56, the graph of f is above the x-axis in
the interval $(-\infty, -3)$, $(0, 1)$, or $(3, \infty)$. Hence, $f(x) > 0$ in the interval $(-\infty, -3)$, $(0, 1)$, or $(3, \infty)$. Thus, the solution set of the original
inequality is

$$(-\infty, -3) \cup (0, 1) \cup (3, \infty).$$

FIGURE 4.56

The graph of $f(x) = \dfrac{x^2 + 3x}{(x - 3)(x - 1)}$ is
above the x-axis in the interval
$(-\infty, -3)$, $(0, 1)$, or $(3, \infty)$.

Observe in Figures 4.55 and 4.56 that in each of the intervals formed by
the boundary numbers, the graph of the rational function is entirely above
the x-axis or entirely below the x-axis. Hence, we can use the test number
method (Section 3.4) to help solve a rational inequality in standard form

without referring to a graph. We simply select an arbitrary test number from each interval formed by the boundary numbers, and then determine the algebraic sign of the rational expression. If the rational expression is negative for a particular test number in an interval, then it is negative for every value of x in that interval. If the rational expression is positive for a particular test number in an interval, then it is positive for every value of x in that interval.

For example, to solve the rational inequality

$$\frac{x + 2}{x - 1} \leq 0$$

in Example 3(a) by using the test number method, we select an arbitrary test number from each of the intervals $(-\infty, -2)$, $(-2, 1)$, and $(1, \infty)$, and then determine the algebraic sign of the rational expression $\frac{x + 2}{x - 1}$ for each of the test numbers. The following table organizes our work:

Interval	*Test Number*	*Sign of* $\frac{x + 2}{x - 1}$
$(-\infty, -2)$	-3	$\frac{-3 + 2}{-3 - 1} = \frac{1}{4} > 0$
$(-2, 1)$	-1	$\frac{-1 + 2}{-1 - 1} = -\frac{1}{2} < 0$
$(1, \infty)$	2	$\frac{2 + 2}{2 - 1} = 4 > 0$

Since $\frac{x + 2}{x - 1} < 0$ when $x = -1$, we know that $\frac{x + 2}{x - 1} < 0$ for every value of x in the interval $(-2, 1)$. Hence, we conclude that the solution set of $\frac{x + 2}{x - 1} \leq 0$ is

$$[-2, 1)$$

Never include the value of the asymptote in a solution set.

This answer agrees with our work in Example 3(a).

PROBLEM 3   Solve the rational inequality $\frac{x^2 + 2x - 4}{x + 2} \geq 1$ by using the test number method.   ◆

◆ **Application: Uniform Motion**

Objects that move at a constant rate of speed are in *uniform motion*. The formula

$$d = rt$$

is used to find the distance d traveled when given the constant rate of speed r over a given time t. We may also write this formula as follows:

$$r = \frac{d}{t} \quad \text{and} \quad t = \frac{d}{r}$$

As illustrated in the next example, uniform motion problems often involve the solution of a rational equation.

EXAMPLE 4   A college rowing team practices on the Charles River in Boston. In order to qualify for the NCAA championship trials, the crew must be able to row at a constant rate of 18 mi/h in still water. Their coach observes that while rowing at a constant rate of speed, the crew travels 3 miles downstream in the same time it takes them to row 2 miles upstream. He knows that the Charles River flows at a constant rate of 3 mi/h. Will his crew qualify for the NCAA championship trials?

◆ **Solution**   Let

x = the crew's rate of rowing (in mi/h) in still water.

Since the river flows at a constant rate of 3 mi/h, we can state that

$x + 3$ = the crew's rate of rowing (in mi/h) *downstream*

and

$x - 3$ = the crew's rate of rowing (in mi/h) *upstream*.

We can arrange these facts as a table:

	Rate, r	Distance, d	Time, $t = \dfrac{d}{r}$
Downstream	$x + 3$	3	$\dfrac{3}{x + 3}$
Upstream	$x - 3$	2	$\dfrac{2}{x - 3}$

Since the problem states that the time to travel 3 miles downstream *is the same as* the time to travel 2 miles upstream, we can write the following equation:

FIGURE 4.57

The x-intercept of the graph of

$$f(x) = \frac{3}{x+3} - \frac{2}{x-3} \text{ is 15.}$$

| time downstream | = | time upstream |

$$\frac{3}{x+3} = \frac{2}{x-3}$$

$$3(x-3) = 2(x+3) \qquad \text{Multiply both sides by the LCD, } (x+3)(x-3)$$

$$3x - 9 = 2x + 6$$

$$x = 15$$

Thus, the crew's rate in still water is 15 mi/h. At this rate, the team will not qualify for the NCAA championship trials. The graph in Figure 4.57 supports our work. ◆

PROBLEM 4 Refer to Example 4: If the speed of the river is 3 mi/h and the crew can row 4 miles downstream in the same time it takes them to row 3 miles upstream, will the team qualify for the NCAA trials? ◆

◆ Application: Rate of Work

Another type of applied problem that involves solving a rational equation is the *rate of work problem*. Suppose it takes t hours to complete a certain job when working at a constant rate of speed. Then, $1/t$ of the job is done in 1 hour. For example, if it takes 3 hours to mow a lawn, then $\frac{1}{3}$ of the job is done in 1 hour. We use this idea to solve the next example.

EXAMPLE 5 Two spillways from a reservoir can be used to flood a cranberry bog. If both spillways are used together, the bog floods in 24 hours. If only the larger spillway is used, the bog will flood in 10 hours less time than if only the smaller spillway is used. What time is required using each spillway alone to flood the cranberry bog?

◆ **Solution**   Let

$$x = \text{amount of time (in hours) to flood the bog} \atop \text{using only the smaller spillway}$$

and

$$x - 10 = \text{amount of time (in hours) to flood the} \atop \text{bog using only the larger spillway.}$$

Hence, we have the following equation:

| part of job done by smaller spill-way in 1 hour | + | part of job done by larger spill-way in 1 hour | = | part of job done by both spillways together in 1 hour |

$$\frac{1}{x} + \frac{1}{x-10} = \frac{1}{24}$$

$$24(x-10) + 24x = x(x-10) \qquad \text{Multiply both sides by the LCD}$$

$$x^2 - 58x + 240 = 0 \qquad \text{Write in standard form}$$

FIGURE 4.58

An x-intercept of the graph of
$$f(x) = \frac{1}{x} + \frac{1}{x - 10} - \frac{1}{24}$$ is
approximately 53.5.

Now, to solve this quadratic equation in standard form, we apply the quadratic formula:

$$x = \frac{-(-58) \pm \sqrt{(-58)^2 - 4(1)(240)}}{2(1)} \approx 4.5 \quad \text{or} \quad 53.5$$

We disregard the solution of 4.5, since this value yields a negative time for the larger spillway to flood the bog. Hence, the only acceptable solution for this quadratic equation is 53.5. In summary, the bog floods in 53.5 hours using only the smaller spillway and in 43.5 hours using only the larger spillway. The graph in Figure 4.58 supports our work. ◆

PROBLEM 5 Suppose that a cranberry bog floods in 48 hours when a smaller spillway is used alone and it floods in 36 hours when a larger spillway is used alone. How long does it take to flood the cranberry bog if both spillways operate simultaneously? ◆

Exercises 4.6

◆ Basic Skills

In Exercises 1–28, solve each rational equation. Check graphically with a graphing calculator.

1. $\dfrac{3}{x + 2} - \dfrac{1}{x - 2} = \dfrac{x}{x^2 - 4}$

2. $\dfrac{1}{x + 1} - \dfrac{1}{x - 1} = \dfrac{2x}{x^2 - 1}$

3. $\dfrac{2x - 1}{4x + 1} - \dfrac{3x + 2}{6x + 1} = 0$

4. $\dfrac{2x + 3}{3x} + \dfrac{1}{3} = \dfrac{x - 3}{x + 3}$

5. $\dfrac{3}{x - 2} - \dfrac{1}{2 - 3x} = \dfrac{2}{3x^2 - 8x + 4}$

6. $\dfrac{2}{x^2 + 5x + 4} + \dfrac{1}{x^2 + x} = \dfrac{2}{x^2 + 4x}$

7. $\dfrac{3}{x^3 + 1} + \dfrac{x}{x^2 - x + 1} = \dfrac{1}{x + 1}$

8. $\dfrac{1}{8x^3 - 1} = \dfrac{2x}{4x^2 + 2x + 1} - \dfrac{1}{2x - 1}$

9. $\dfrac{n - 1}{5} = \dfrac{4}{n}$

10. $\dfrac{x}{27} - \dfrac{1}{x} = \dfrac{2}{9}$

11. $\dfrac{1}{x} - \dfrac{1}{6} = \dfrac{x - 1}{x + 1}$

12. $\dfrac{1}{p} + \dfrac{3}{4} = \dfrac{2p + 3}{3p - 1}$

13. $\dfrac{5 - x}{x - 3} = \dfrac{3}{x^2 - 4x + 3}$

14. $\dfrac{7}{x^2 - 4} - \dfrac{2}{x + 2} = 1$

15. $\dfrac{x + 2}{x + 1} - \dfrac{x - 3}{x - 1} = \dfrac{x^2 + 2x - 1}{x^2 - 1}$

16. $\dfrac{x}{2x + 1} - \dfrac{3}{3 - x} = \dfrac{13}{2x^2 - 5x - 3}$

17. $\dfrac{n - 1}{n - 3} - 2 = \dfrac{1}{2} - \dfrac{n - 5}{n^2 - 5n + 6}$

18. $\dfrac{2}{m - 1} + \dfrac{1}{m^2 - 4m + 3} + 2 = \dfrac{3}{m - 3}$

19. $\dfrac{1}{n} - \dfrac{2n}{5} = \dfrac{1}{2n}$

20. $\dfrac{2x}{3} = x^2 + \dfrac{1}{2}$

21. $\dfrac{1}{p} - \dfrac{1}{3} = \dfrac{4}{p - 3}$

22. $\dfrac{x}{x + 1} - \dfrac{x + 1}{x} = \dfrac{1}{2}$

23. $\dfrac{x}{x^2 - 4} + \dfrac{1}{x + 2} = 1$

24. $\dfrac{2}{3t + 2} + 1 = \dfrac{t}{3t - 2} + \dfrac{8}{4 - 9t^2}$

25. $\dfrac{4}{x} + \dfrac{2}{x-1} = \dfrac{6x}{x+1}$

26. $\dfrac{5}{x+2} - \dfrac{2x}{3} = \dfrac{x-6}{x}$

27. $\dfrac{x}{x+3} + \dfrac{16x}{x^2-9} = \dfrac{3x^2}{2}$

28. $\dfrac{x+1}{x-2} + \dfrac{x-1}{x+2} = \dfrac{3x}{x^2-4} + \dfrac{3}{x}$

In Exercises 29–48, solve each rational inequality. Write each solution set using interval notation.

29. $\dfrac{1}{x} > 2$

30. $\dfrac{2}{3x} < 1$

31. $\dfrac{x+3}{x-1} < 0$

32. $\dfrac{3-2x}{5x} > 0$

33. $\dfrac{x-4}{2x+1} \geq 3$

34. $\dfrac{3x+2}{x-1} \leq 1$

35. $\dfrac{x^2+3x-4}{x-2} \leq 0$

36. $\dfrac{x^2+5x-14}{x^2+5x-24} \geq 0$

37. $\dfrac{x^2+6x+2}{x^2-25} < 0$

38. $\dfrac{3x-4}{x^2+4x+1} < 0$

39. $\dfrac{1}{x} > \dfrac{2x+1}{x+8}$

40. $\dfrac{2x-3}{x+4} < \dfrac{x-1}{x+3}$

41. $\dfrac{1}{x-2} + \dfrac{x}{x+3} < \dfrac{1}{2}$

42. $x + \dfrac{x-3}{2x+5} < 3$

43. $\dfrac{7}{x^2+7x+12} \geq \dfrac{x-3}{x+4}$

44. $\dfrac{x}{x^2+2x-15} \geq \dfrac{2}{x^2-25}$

45. $\dfrac{x-2}{x^3+2x^2+5x+10} < 0$

46. $\dfrac{x^3-3x+2}{x^2+1} > 0$

47. $\dfrac{5}{x^2-4} \geq \dfrac{4x}{3x+3}$

48. $\dfrac{5x+3}{x^2+2x+5} \leq \dfrac{3x}{2x^2-2}$

In Exercises 49–60, set up a rational equation, and then solve it.

49. One number is three times another, and the sum of their reciprocals is $\frac{1}{3}$. Find the two numbers.

50. The denominator of a fraction exceeds its numerator by 18. The reduced value of the fraction is $\frac{4}{7}$. What is the fraction?

51. A girl, starting from her home, leaves on a bicycle trip. After riding 2 miles at the rate of 12 mi/h, the chain on her bicycle becomes disengaged. She immediately jogs back home along the same route. Find the rate of speed (in mi/h) at which she jogs if she returns home 30 minutes after she left.

52. A canoeist paddles 8 miles upstream in twice the time it takes to paddle the same distance downstream. Find the rate of the current in the river if the canoeist can paddle 3 mi/h in still water.

53. Two ferries leave Woods Hole for Nantucket. One ferry travels 2.2 knots faster than the other and requires 24 minutes less time to make the trip. Find the speed (in knots) of each ferry if the distance between ports is 24 nautical miles.

54. A college rowing team rows 6 miles down river and back again in 1 hour 15 minutes. If the river flows at 2 mi/h, find the crew's rate of speed in still water.

55. A tank can be filled by an inlet pipe in 15 hours, and emptied by an outlet pipe in 25 hours. The tank is initially empty, and then both pipes are opened. How long will it take to fill the tank?

56. A man can paint his house in 8 days. His son alone could paint it in 12 days. How long will it take to paint the house if they work together?

57. Using a large sump pump and small sump pump together, a homeowner can drain all the water from a cellar in 24 minutes. Using only one sump pump to drain the water requires 20 minutes longer with the smaller pump than with the larger one. Find the time required to drain this amount of water using each pump alone.

58. A swimming pool can be filled with two hoses in 8 hours. Using the smaller hose alone, the pool can be filled in 20 hours. How long would it take to fill the pool using only the larger hose?

59. A group of college students rents a bus for a ski trip. The cost for the bus is $480, and the students divide this expense evenly. Because of sickness, four students are unable to travel. This increases the share that each remaining person must pay by $10. How many students go on the ski trip?

60. A real estate developer bought two parcels of land for $600,000 each. The larger parcel contains 10 acres more than the smaller parcel yet costs $10,000 per acre less. How many acres are in each parcel?

◆ **Critical Thinking**

In Exercises 61–68, solve each literal equation for x.

61. $\dfrac{1}{m} - \dfrac{1}{x} = \dfrac{1}{n} + \dfrac{1}{x}$

62. $\dfrac{1}{a} - \dfrac{1}{x} = \dfrac{1}{b}$

63. $\dfrac{2x + k}{x - k} - \dfrac{x - k}{x + k} = 1$

64. $\dfrac{a + x}{a - x} = \dfrac{x^2}{a^2 - x^2}$

65. $\dfrac{a}{x - 1} + a = bx$

66. $\dfrac{a}{x} + x = a + \dfrac{1}{x}$

67. $x + \dfrac{1}{a} = a + \dfrac{1}{x}$

68. $\dfrac{x}{a} + \dfrac{a}{x} = \dfrac{a}{b} + \dfrac{b}{a}$

In Exercises 69 and 70, solve each double inequality. Use interval notation to describe the solution set

69. $-3 \le \dfrac{2x - 1}{x + 2} \le 7$

70. $\dfrac{1}{x} \le \dfrac{2 - 6x}{x - 5} < 10$

In Exercises 71 and 72, use interval notation to describe the domain of each function.

71. $f(x) = \sqrt{\dfrac{2x + 3}{x - 1}}$

72. $f(x) = \sqrt{\dfrac{x^2 - 4}{x + 1}}$

73. On a math test, a student multiplied both sides of the inequality $\dfrac{x}{x + 1} < 0$ by $x + 1$ to obtain $x < 0$, and wrote $(-\infty, 0)$ as the solution set. Explain why this procedure is incorrect, and then show the correct method of solving this inequality.

74. Find a rational inequality whose solution set is $(-\infty, -2) \cup [0, \infty)$. Several answers are correct.

75. If two resistances, R_1 and R_2, are connected in parallel in an electric circuit, then the total effective resistance R can be found by using the formula

$$R = \dfrac{R_1 R_2}{R_1 + R_2}$$

If R_1 is 10.6 ohms (Ω), what is the range of values for R_2 if the total effective resistance is positive but not more than 8.2 Ω?

76. In an ant colony, the relationship between the number of adult ants A and the number of larvae L that survive to become adults is given by

$$L = \dfrac{6000A}{A + 1200}.$$

Under what conditions is the number of larvae that survive to become adults less than the number of adult ants in the colony?

Chapter 4 Review

◆ **Questions for Writing or Group Discussion**

1. What are *relative extrema* of a polynomial function? How are extrema used as an aid in graphing a polynomial function?

2. What can be said about the zeros of a polynomial function P of degree n if the graph of P
    (a) does not cross or touch the *x*-axis.
    (b) crosses the *x*-axis n times.

3. Under what conditions is the graph of a polynomial function *tangent* to the x-axis? Discuss the types of tangency that can occur for the graph of a polynomial function.

4. Does each of the following sketches represent the graph of a *polynomial function*? Explain.

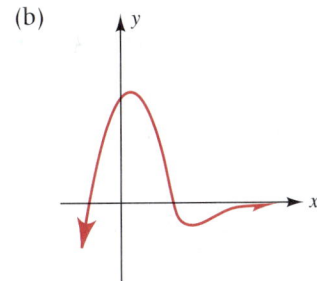

(a)

(b)

5. Explain the method for checking *polynomial long division*. Illustrate with an example.

6. Under what conditions is it possible to divide one polynomial by another using *synthetic division*?

7. Explain how synthetic division and the *remainder theorem* can be used to determine if r is a zero of a polynomial function P.

8. Explain how synthetic division and the *factor theorem* can be used to determine if $x - r$ is a factor of a polynomial $P(x)$.

9. Determine the possible combinations of real and imaginary zeros of a polynomial function P if
   (a) P is a *quartic function* with real coefficients.
   (b) P is a *quintic function* with real coefficients.

10. If P is a *cubic function* with real coefficients, find
    (a) the maximum number of real zeros of P.
    (b) the minimum number of real zeros of P.

11. Explain the procedure for determining the possible *rational zeros* of a polynomial function.

12. Does the *fundamental theorem of algebra* tell us how to find a zero of a polynomial function? If not, what does it tell us?

13. Is it possible for a *polynomial equation* with real coefficients to have three imaginary roots? Explain.

14. How is the domain of a *rational function* determined? Give an example of a rational function whose domain is (a) $(-\infty, a) \cup (a, \infty)$ or (b) all real numbers.

15. Is it possible for the graph of a rational function to cross its *vertical asymptote*? Explain.

16. Discuss the procedure for finding the *horizontal asymptote* of a rational function.

17. Discuss the procedure for finding the *oblique asymptote* of a rational function.

18. Does each of the following sketches represent the graph of a rational function? Explain.

19. Explain how an extraneous root may develop when solving a *rational equation*. Illustrate with an example.

20. What is meant by the *boundary numbers* of a polynomial inequality? of a rational inequality?

◆ **Review Exercises**

In Exercises 1–8, divide the first polynomial by the second polynomial. Use synthetic division whenever possible.

1. $3x^3 + 2x^2 - 3x + 4$; $x^2 + 1$

2. $4x^3 + 3x^2 - x + 6$; $x - 1$

3. $2x^4 + 2x^3 + 3x - 1$; $x + 2$

4. $3x^4 + 8x^2 + 3x - 1$; $x^2 + 2x - 2$

5. $6x^4 - 2x^3 + 9x^2 - 4$; $x - \frac{1}{3}$

6. $4x^4 - x^3 + 5x^2 - 2x + 1$; $x + \frac{3}{4}$

7. $x^6 + 1$; $x^3 + x + 1$

8. $x^5 - 8x^3 - 7x + 6$; $(x + 3)^2$

In Exercises 9–12, use synthetic division and the remainder theorem to find the indicated functional value. Check graphically with a graphing calculator.

9. $P(x) = 2x^3 - 3x^2 + 5x - 8$; $P(2)$

10. $P(x) = 4x^3 + 7x^2 - 8x - 6$; $P(-3)$

11. $f(x) = 3x^4 - x^3 + 4x^2 - 5x - 6$; $f\left(-\frac{2}{3}\right)$

12. $F(x) = 6x^4 - x^3 - 10x^2 - 5x + 3$; $F\left(\frac{3}{2}\right)$

In Exercises 13–18, use the factor theorem to determine whether the first polynomial is a factor of the second polynomial.

13. $x - 2$; $4x^3 - 7x^2 - 4$

14. $x + 3$; $2x^3 + 3x^2 - 7x - 9$

15. $x + \frac{4}{3}$; $3x^4 + x^3 + 2x^2 + 8x + 4$

16. $x - \frac{1}{4}$; $4x^4 + 3x^3 - x^2 - 16x + 4$

17. $x - 2i$; $x^5 + 3x^3 + 2x^2 - 4x + 8$

18. $x - (2 + i)$; $2x^4 + 7x^3 + 5x^2 - 9x - 5$

In Exercises 19–30, (a) write P(x) as the product of linear factors, and (b) find all zeros of P.

19. $P(x) = x^3 - 5x^2 + 2x + 8$

20. $P(x) = 2x^3 + x^2 - 13x + 6$

21. $P(x) = 3x^3 - 10x^2 - 7x + 20$

22. $P(x) = x^3 - 10x^2 + 34x - 40$

23. $P(x) = 2x^4 + 5x^3 + 5x^2 + 20x - 12$

24. $P(x) = 2x^4 - 9x^3 + 8x^2 + 19x - 30$

25. $P(x) = 4x^5 - 4x^4 - 15x^3 + 10x^2 + 11x - 6$

26. $P(x) = 4x^5 + 16x^4 - 23x^3 - 13x^2 + 11x + 5$

27. $P(x) = x^4 - 4x^3 + 9x^2 - 4x + 8$;
$x + i$ is a factor

28. $P(x) = x^4 - 6x^3 + 11x^2 - 10x + 2$;
$x - (1 - i)$ is a factor

29. $P(x) = 2x^5 - 9x^4 + 22x^3 - 41x^2 + 56x - 20$;
$2 + i$ is a zero

30. $P(x) = x^6 + 2x^5 + 5x^4 - 12x^3 - 29x^2 - 70x - 25$;
$2i - 1$ is a zero of multiplicity two

In Exercises 31–44, find all real roots of each equation. Record irrational roots to the nearest hundredth.

31. $x^3 + 2x - 8 = 0$ **32.** $2x^3 + 5x^2 + 3 = 0$

33. $2x^4 - 5x^3 + 6x^2 - 5x = 25$

34. $x^3 + 8x^2 + 2x = 3$ **35.** $x^4 + 6x^3 + 4 = 0$

36. $x^5 - 3x^4 - 2x^2 + x + 15 = 0$

37. $x^5 - 8x^4 + 20x^3 - 19x^2 + 12x = 12$

38. $3x^6 + 9x^5 + 10x^4 - 15x^2 - 17x = 6$

39. $\dfrac{x}{2x-1} + \dfrac{1}{2} = \dfrac{3x-1}{6x-3}$

40. $\dfrac{3}{x+3} - \dfrac{1}{x-3} = \dfrac{x}{x^2-9}$

41. $\dfrac{x}{x+4} - \dfrac{2}{x-1} = \dfrac{20}{x^2+3x-4}$

42. $\dfrac{x-1}{x} = \dfrac{5}{2} - \dfrac{x}{x-1}$

43. $\dfrac{1}{2} - \dfrac{1}{x} = \dfrac{1}{x-2}$

44. $x + \dfrac{1}{x} = \dfrac{1}{3} + 3$

In Exercises 45–52, solve each inequality. Use interval notation to describe the solution set.

45. $x^3 + 2x^2 \geq 4x$

46. $x^4 - 5x^2 + 4 < 0$

47. $x^5 - 6x^4 + 16x^3 - 32x^2 + 48x > 32$

48. $x^6 - 2x^5 - 7x^4 + 16x^3 - 17x^2 + 18x \leq 9$

49. $\dfrac{p-4}{p+2} \geq 0$

50. $\dfrac{x^2-2x+3}{x+1} \leq 0$

51. $\dfrac{1}{x} < \dfrac{3(x-1)}{x+4}$

52. $m + \dfrac{3m}{2-m} > 4$

In Exercises 53–66, sketch the graph of each function. Label the x- and y-intercepts and any asymptote(s). Check each sketch with a graphing calculator.

53. $g(x) = 2x^2 - x^4$

54. $h(x) = 4x^3 - 3x^4$

55. $f(x) = x^3 + 3x^2 - 6x - 8$

56. $F(x) = 10 - 11x - 4x^2 - 3x^3$

57. $G(x) = 4x^4 - 12x^3 - 39x^2 - 3x - 10$

58. $P(x) = x^4 - x^3 - 9x^2 + 9x$

59. $f(x) = \dfrac{3x-2}{x+1}$

60. $F(x) = \dfrac{1}{x^2+4}$

61. $G(x) = \dfrac{x}{x^2-3x-10}$

62. $g(x) = \dfrac{x+4}{x^3-4x^2+4x}$

63. $F(x) = \dfrac{x^2-1}{2x}$

64. $f(x) = \dfrac{x^2+4}{x-1}$

65. $H(x) = \dfrac{x^2-8x+15}{x-5}$

66. $R(x) = \dfrac{x^2-x-12}{x^2-9}$

67. Determine a polynomial function P of degree 3 with zeros 1, −2, and 3 such that
(a) $P(0) = 3$ (b) $P(2) = -8$

68. Determine a polynomial function P of degree 4 with zeros −1, −1, $\frac{1}{2}$, and $-\frac{3}{2}$ such that
(a) $P(0) = -3$ (b) $P(1) = 24$

69. Sketch the graphs of the equations $y = x^3$ and $7x - 4y = -3$ on the same coordinate plane. Then determine the coordinates of the points where their graphs intersect.

70. Sketch the graphs of the functions

$$f(x) = \frac{4}{x} \quad \text{and} \quad g(x) = x^2 - 2x - 8$$

on the same coordinate plane. Then determine to the nearest hundredth the coordinates of the points where their graphs intersect.

71. Determine the horizontal asymptote for

$$f(x) = \frac{2x^2+5x-7}{3x^2+2}.$$

Does the graph of this function cross the horizontal asymptote? If so, find the coordinates of the point where it does cross.

72. Determine the oblique asymptote for

$$f(x) = \frac{x^3-2x^2+3x}{x^2-4x+3}.$$

Does the graph of this function cross the oblique asymptote? If so, find the coordinates of the point where it does cross.

73. A forester estimates that the number N of 2 in. by 3 in. by 8 ft construction-grade studs that can be cut from a fir tree is given by $N(x) = 4x^3 - 8x^2 + 5x - 1$, where x is the circumference (in feet) of the base of the tree trunk and $x \geq 1$. Determine the circumference of a tree that can yield 24 studs.

74. When a 20-microfarad capacitor is connected in series with a variable capacitor x, the total capacitance C is given by $C(x) = \dfrac{20x}{x+20}$.
(a) Sketch the graph of this function for $x \geq 0$.
(b) What is the significance of the horizontal asymptote?

75. The height of a tin can is 3 cm more than its radius. Determine the radius of the can if its volume is 112π cubic cm.

76. An open rectangular box has four sides and a bottom of uniform thickness. The outside dimensions of the box are 3 ft by 3 ft by 2 ft, as shown in the sketch at the top of the next page. Determine the uniform thickness (to the nearest hundredth) if the box holds 12 cubic feet of water when filled to the top.

77. A piece of cardboard, 15 cm by 20 cm, is made into an open rectangular box by cutting out a square from each corner and turning up the sides, as shown in the figure.

(a) Express the volume V of this box as a function of x.

(b) Find *two* possible lengths of the side of the square cut-out if the volume of the box is 250 cubic centimeters.

(c) Generate the graph of the function in part (a) on a graphing calculator. Then approximate the value of x for which the volume of the box is greatest.

78. A beam of length L is simply supported at both ends and carries a load of weight W that decreases uniformly from the center, as shown in the figure. The bending moment M at any distance x ($x \leq L/2$) from the left end of this beam is given by

$$M(x) = \frac{Wx}{6L^2}(3L^2 - 4x^2).$$

(a) Determine the bending moment function when $W = 2400$ lb and $L = 20$ ft. Then generate the graph of this function on a graphing calculator.

(b) At what distance x from the left end of the beam is the bending moment 6336 lb-ft?

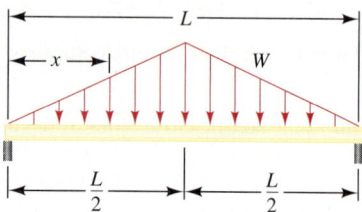

79. A rectangular box has a square base and open top, as shown in the sketch. The volume of the box is 100 cubic meters. The material used to construct the base costs $4 per square meter and the material for the sides cost $2.50 per square meter.

(a) Show that the cost C of materials to construct the box is given by

$$C = \frac{4x^3 + 1000}{x} \quad \text{for } x > 0.$$

(b) Generate the graph of the rational function defined in part (a) on a graphing calculator. Then approximate the dimensions of the box for which the cost is the smallest possible.

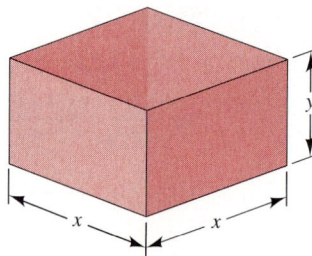

80. A certain manufacturer produces wood-burning stoves. The fixed cost for the company is $81,000 per year and the variable cost (in dollars) of producing x stoves is $400x + 0.1x^2$.

(a) Define a function that represents the average cost per stove.

(b) Generate the graph of the function in part (a) on a graphing calculator. Then approximate the number of stoves that should be produced per year in order to minimize the average cost per stove.

81. A triathlete in training runs 2 miles at the rate of 8 mi/h and returns home, along the same route, riding a bicycle. The entire trip takes 20 minutes. Find the rate of speed at which the triathlete pedals.

82. A gas tank is filled in $3\frac{3}{4}$ hours when two pipelines are used. The time required to fill the tank is four hours less when using the larger pipeline than when using the smaller one. What time is required by each pipeline alone to fill the tank?

S eismologists estimate that the San Francisco earthquake of 1906 measured 8.3 on the Richter scale. How many times more intense was this earthquake than the Loma Prieta quake, which occurred during the 1989 World Series and measured 7.1 on the Richter scale?

For the solution, see Example 5 in Section 5.3.

Exponential and Logarithmic Functions

5.1 Exponential Functions and Their Graphs

◆ Introductory Comments

Any function that can be expressed as sums, differences, products, quotients, powers, or roots of polynomials is classified as an **algebraic function**. All of the functions that we have discussed in Chapters 2, 3, and 4, are algebraic functions. Any function that goes beyond the limits of, or *transcends*, an algebraic function is called a **transcendental function**. In this section we study a class of transcendental functions in which the independent variable appears as an exponent.

◆ **Exponential Function**

> If b is a real number such that $b > 0$ and $b \neq 1$, then the function f defined by
>
> $$f(x) = b^x$$
>
> is called an **exponential function** with base b.

Note: We exclude 1 as a base for exponential functions because $f(x) = 1^x = 1$ is a constant function. We exclude zero as a base because 0^x is undefined when $x \leq 0$. We exclude negative numbers as bases because b^x for $b < 0$ is an imaginary number for infinitely many values of x, such as $x = \frac{1}{2}, \frac{1}{4}, \frac{1}{6}$, and so on.

In Sections 1.2 and 1.3, we evaluated expressions of the form b^x, where b is a *positive real number* and x is a *rational number*. Whether x is a positive integer, zero, a negative integer, a common fraction, or a decimal fraction, we can evaluate b^x as a positive real number, as illustrated in the following examples.

$$4^2 = 16$$

$$4^0 = 1$$

$$4^{-2} = \frac{1}{4^2} = \frac{1}{16}$$

$$4^{3/2} = (4^{1/2})^3 = (\sqrt{4})^3 = (2)^3 = 8$$

$$4^{1.4} = 4^{14/10} = 4^{7/5} \approx 6.9644 \quad \leftarrow$$

We can use a calculator to express the fifth root of 4 to the seventh power as an approximate decimal number.

If b is a positive real number and x is an *irrational* number, then we define b^x as the number approached by b^r as r takes on rational values that get closer and closer to x. For example, we define $4^{\sqrt{2}}$ as the real number approached by 4^r as r takes on rational values that get closer and closer to $\sqrt{2}$:

r	1	1.4	1.41	1.414	1.4142	1.41421
4^r	4	6.964405 …	7.061624 …	7.100891 …	7.102860 …	7.102958 …

From these values we see that $4^{\sqrt{2}} \approx 7.103$, rounded to four significant digits. *For each real number x, b^x is a unique positive real number.* Hence, we conclude that the domain of an exponential function is $(-\infty, \infty)$ and its range is $(0, \infty)$.

◆ Graph of an Exponential Function

Figure 5.1 shows the graph of $g(x) = 4^x$ in the viewing rectangle of a graphing calculator. Notice from the graph that this function is always increasing and its y-intercept is 1. Since $g(x)$ approaches zero [$g(x) \to 0$] as x decreases without bound ($x \to -\infty$), the x-axis is a horizontal asymptote. The graph of an exponential function $f(x) = b^x$ with $b > 4$ rises *more* rapidly than the graph of $g(x) = 4^x$, and the graph of $f(x) = b^x$ with $1 < b < 4$ rises *less* rapidly than the graph of g.

The basic characteristics of the graph of $f(x) = b^x$ for $b > 1$ are as follows.

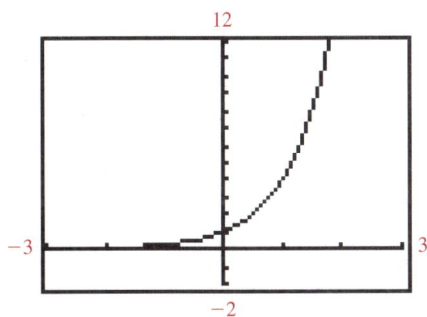

FIGURE 5.1

The graph of $g(x) = 4^x$ rises from left to right. Its y-intercept is 1 and the x-axis is a horizontal asymptote.

◆ Characteristics of the Graph of $f(x) = b^x$ for $b > 1$

1. The y-intercept is 1, and the graph has no x-intercept.
2. The x-axis is a horizontal asymptote [$f(x) \to 0$ as $x \to -\infty$].
3. The graph is always rising.

If the base of an exponential function is a fraction $\frac{1}{b}$ between 0 and 1, we usually write the function in terms of a negative exponent:

$$f(x) = \left(\frac{1}{b}\right)^x = (b^{-1})^x = b^{-x}$$

Figure 5.2 shows the graph of $h(x) = (\frac{1}{4})^x = 4^{-x}$ in the viewing rectangle of a graphing calculator. By the y-axis reflection rule (Section 2.5), the graph of h is the same as the graph of $g(x) = 4^x$ (see Figure 5.1) reflected about the y-axis. The graph of $h(x) = (\frac{1}{4})^x = 4^{-x}$ is always falling and its y-intercept is 1. Since $h(x)$ approaches zero [$h(x) \to 0$] as x increases without bound ($x \to +\infty$), the x-axis is a horizontal asymptote.

The basic characteristics of the graph of $f(x) = b^{-x}$ for $b > 1$ are as follows.

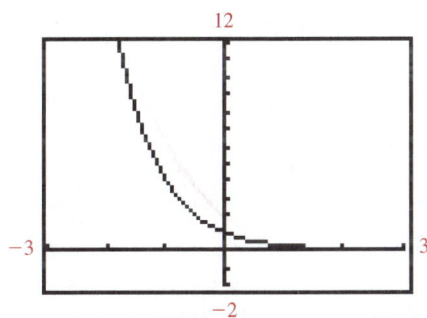

FIGURE 5.2

The graph of $h(x) = (\frac{1}{4})^x = 4^{-x}$ falls from left to right. Its y-intercept is 1 and the x-axis is a horizontal asymptote.

◆ Characteristics of the Graph of $f(x) = b^{-x}$ for $b > 1$

1. The y-intercept is 1, and the graph has no x-intercept.
2. The x-axis is a horizontal asymptote [$f(x) \to 0$ as $x \to +\infty$].
3. The graph is always falling.

Knowing the characteristics of the graph of an exponential function enables us to graph several related functions by applying the shift rules and axis reflection rules (see Section 2.5).

EXAMPLE 1 Sketch the graph of each function. Label the horizontal asymptote and the y-intercept.

(a) $G(x) = 4^{x+2}$ **(b)** $H(x) = 3 - 4^{-x}$

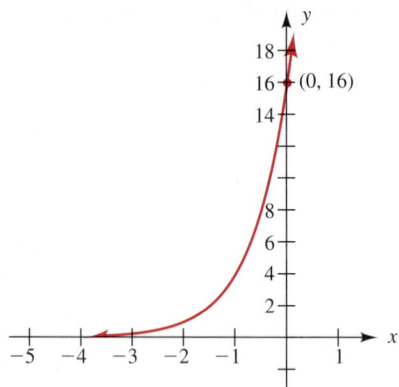

Graph of $G(x) = 4^{x+2}$

FIGURE 5.4
Graph of $H(x) = 3 - 4^{-x}$

◆ **Solution**

(a) By the horizontal shift rule (Section 2.5), the graph of $G(x) = 4^{x+2}$ is the same as the graph of $g(x) = 4^x$ (see Figure 5.1) shifted horizontally to the left 2 units. The y-intercept is

$$G(0) = 4^{0+2} = 4^2 = 16$$

and the horizontal asymptote remains the x-axis. The graph of $G(x) = 4^{x+2}$ is shown in Figure 5.3.

(b) By the x-axis reflection rule and the vertical shift rule (Section 2.5), the graph of $H(x) = 3 - 4^{-x}$ is the same as the graph of $h(x) = 4^{-x}$ (see Figure 5.2) reflected about the x-axis and then shifted vertically upward 3 units. The y-intercept is

$$H(0) = 3 - 4^0 = 3 - 1 = 2.$$

Note that $4^{-x} \to 0$ as $x \to +\infty$. Hence, $H(x) \to 3$ as $x \to +\infty$. Thus, $y = 3$ is the horizontal asymptote. The graph of $H(x) = 3 - 4^{-x}$ is shown in Figure 5.4. ◆

Note: Observe that the graph of H in Figure 5.4 has an x-intercept between 0 and -1. To find the x-intercept, we solve the equation

$$3 - 4^{-x} = 0 \quad \text{or} \quad 4^{-x} = 3$$

for x. However, none of our previous methods can be used to solve an equation in which the unknown appears as an exponent. In Section 5.2, we introduce the *logarithmic function*, which we can use to help solve this type of equation.

PROBLEM 1 State the domain and range of the function H defined in Example 1(b). ◆

◆ **Base e**

In applications of exponential functions, one particular irrational number occurs frequently as the base. This irrational number is denoted by the letter e, and its value is

$$e \approx 2.71828 \ldots$$

The function f defined by

$$f(x) = e^x$$

is called the **exponential function with base e**. With a calculator, we can evaluate $f(x) = e^x$ for $x = 0, \pm 1, \pm 2, \pm 3, \ldots$ by using the $\boxed{e^x}$ key. These values are displayed in the following table.

x	-3	-2	-1	0	1	2	3
e^x	$e^{-3} \approx 0.05$	$e^{-2} \approx 0.14$	$e^{-1} \approx 0.37$	1	$e \approx 2.72$	$e^2 \approx 7.39$	$e^3 \approx 20.1$

Figure 5.5 shows the graphs of $f(x) = e^x$ and $g(x) = e^{-x}$ in the same viewing rectangle of a graphing calculator. The graph of $f(x) = e^x$ rises more rapidly than the graph of $y = 2^x$, but less rapidly than the graph of $y = 3^x$.

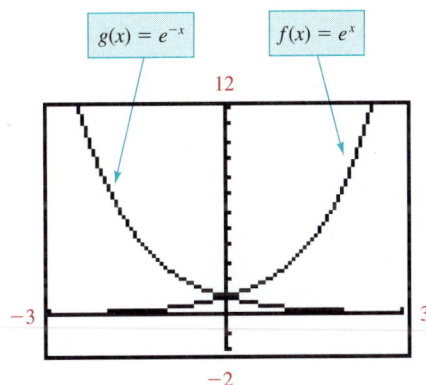

FIGURE 5.5

Graphs of $f(x) = e^x$ and $g(x) = e^{-x}$, as shown in the viewing rectangle of a graphing calculator

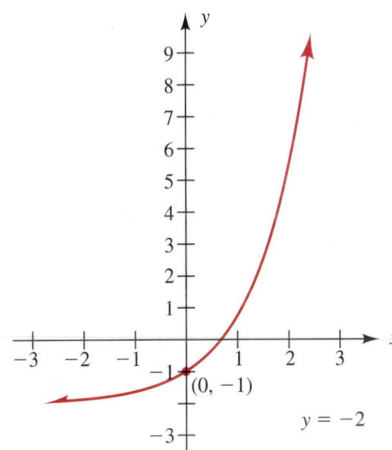

FIGURE 5.6

Graph of $F(x) = e^x - 2$

EXAMPLE 2 Sketch the graph of the function F defined by $F(x) = e^x - 2$. Label the horizontal asymptote and the y-intercept.

◆ **Solution** By the vertical shift rule (Section 2.5), the graph of $F(x) = e^x - 2$ is the same as the graph of $f(x) = e^x$ (see Figure 5.5) shifted vertically downward 2 units, as shown in Figure 5.6. The y-intercept is

$$F(0) = e^0 - 2 = 1 - 2 = -1.$$

Note that $F(x) \to -2$ as $x \to -\infty$. Thus, $y = -2$ is the horizontal asymptote. ◆

PROBLEM 2 Use a graphing calculator to verify the graph of the function shown in Figure 5.6. ◆

◆ Properties of Real Exponents

The laws of exponents given in Section 1.2 apply to all real exponents. We restate them here and refer to them as the **properties of real exponents**.

◆◆ **Properties of Real Exponents**

> If the bases a and b are positive real numbers and the exponents x and y represent real numbers, then
>
> 1. $b^0 = 1$ 2. $b^{-x} = \dfrac{1}{b^x}$
>
> 3. $b^x b^y = b^{x+y}$ 4. $(b^x)^y = b^{xy}$
>
> 5. $\dfrac{b^x}{b^y} = b^{x-y}$ 6. $(ab)^x = a^x b^x$
>
> 7. $\left(\dfrac{a}{b}\right)^x = \dfrac{a^x}{b^x}$ 8. $\dfrac{a^{-x}}{b^{-y}} = \dfrac{b^y}{a^x}$

EXAMPLE 3   Use the properties of real exponents to write each expression as a constant or as an expression in the form b^x or b^{-x} with $b > 1$.

(a) $\dfrac{5^{\pi}}{5^{2x+\pi}}$ **(b)** $4^{-3x} \cdot 8^{2x+1}$

◆◆ **Solution**

(a) Since the bases are the same, we begin by applying property 5:

$$\frac{5^{\pi}}{5^{2x+\pi}} = 5^{\pi-(2x+\pi)} \qquad \text{\color{red}{Subtract exponents}}$$

$$= 5^{-2x} \qquad \text{\color{red}{Simplify}}$$

$$= (5^2)^{-x} \text{ or } 25^{-x} \qquad \text{\color{red}{Rewrite using property 4 and simplify}}$$

Figure 5.7 shows that the graphs of

$$f(x) = \frac{5^{\pi}}{5^{2x+\pi}} \qquad \text{and} \qquad g(x) = 25^{-x}$$

are identical. This fact confirms our work.

(b) First, we write each factor with the same base so that we can apply the properties of real exponents:

$$4^{-3x} \cdot 8^{2x+1} = (2^2)^{-3x} \cdot (2^3)^{2x+1}$$

Now we apply property 4, then property 3:

$$4^{-3x} \cdot 8^{2x+1} = (2^2)^{-3x} \cdot (2^3)^{2x+1}$$

$$= 2^{-6x} \cdot 2^{6x+3} \qquad \text{\color{red}{Multiply the exponents}}$$

$$= 2^3 \text{ or } 8 \qquad \text{\color{red}{Add the exponents and simplify}}$$

Figure 5.8 confirms our work.   ◆

FIGURE 5.7

The graphs of $f(x) = \dfrac{5^{\pi}}{5^{2x+\pi}}$ and $g(x) = 25^{-x}$ are identical.

FIGURE 5.8

The graphs of $f(x) = 4^{-3x} \cdot 8^{2x+1}$ and $g(x) = 8$ are identical.

PROBLEM 3   Use the properties of real exponents to write the expression $\dfrac{125^{x-2}}{25^{2x-3}}$ as an expression in the form b^x or b^{-x}.   ◆

◆ Application: Compound Interest

Exponential functions occur in several types of applied problems. One of the most familiar types involves the interest that we earn on a savings account. The amount of simple interest i earned on a principal P invested at a certain rate of interest r per year over a time of t years is given by the formula

$$i = Prt.$$

Thus, the amount A in the account after t years is

$$A = P + i = P + Prt = P(1 + rt).$$

Compound interest is interest paid both on the principal and on any interest earned previously. If a certain principal P is deposited in a savings account at an interest rate r per year and the interest is *compounded n times per year,* then interest is paid on both principal and any interest earned previously every $(1/n)$th of a year. The interest i earned after the first compounding period is

$$i = Prt = Pr\,\frac{1}{n}.$$

Hence the amount A_1 in the account after the first compounding period is

$$A_1 = P + i$$
$$= P + Pr\,\frac{1}{n}$$
$$= P\left(1 + \frac{r}{n}\right) \qquad \textbf{Factor out } \textbf{\textit{P}}$$

The interest i earned after the second compounding period is

$$i = A_1 r\,\frac{1}{n},$$

and the amount A_2 in the account after the second compounding period is

$$A_2 = A_1 + i = A_1 + A_1 r\,\frac{1}{n}.$$

Now, replacing A_1 with $P\left(1 + \dfrac{r}{n}\right)$, we have

$$A_2 = P\left(1 + \frac{r}{n}\right) + P\left(1 + \frac{r}{n}\right)r\frac{1}{n}$$

$$= P\left(1 + \frac{r}{n}\right)\left[1 + \frac{r}{n}\right] \qquad \text{Factor out } P\left(1 + \frac{r}{n}\right) \text{ from each term}$$

$$= P\left(1 + \frac{r}{n}\right)^2 \qquad \text{Simplify}$$

Continuing in this manner, we can show that after one year (with n compounding periods in a year) the amount A in the account is

$$A = P\left(1 + \frac{r}{n}\right)^n,$$

and after t years the amount in the account is

$$A = P\left(1 + \frac{r}{n}\right)^{nt}.$$

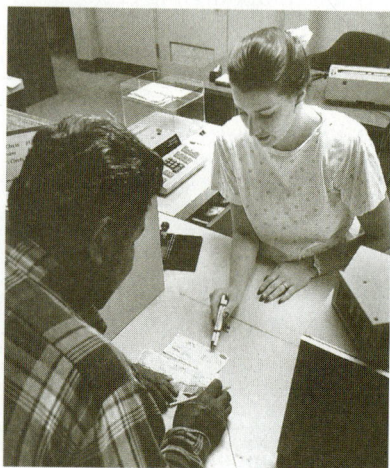

◆ **Compound Interest Formula for n Compoundings Per Year**

If a certain principal P is deposited in a savings account at an interest rate r per year and interest is compounded n times per year, then the amount A in the account after t years is given by the formula

$$A = P\left(1 + \frac{r}{n}\right)^{nt}.$$

EXAMPLE 4   Suppose that $1000 is deposited in a bank account at an interest rate of 8% per year compounded quarterly. Assume the depositor makes no subsequent deposit or withdrawal.

(a) Express the amount in the account as a function of time t.

(b) Find the amount in the account after five years.

◆ **Solution**

(a) If the interest is compounded quarterly, then it is compounded four times per year. Using the formula $A = P\left(1 + \frac{r}{n}\right)^{nt}$ with $P = 1000$, $r = 0.08$, and $n = 4$, we have

$$A = 1000\left(1 + \frac{0.08}{4}\right)^{4t} = 1000(1.02)^{4t}.$$

(b) To find the amount in the account after five years, we substitute $t = 5$ in the formula from part (a), and then use a calculator to evaluate this expression:

$$A = 1000(1.02)^{4\cdot5} \approx 1485.95$$

FIGURE 5.9
The graph of $A = 1000(1.02)^{4t}$ for $t \geq 0$ indicates that for $t = 5$ years, $A \approx \$1485.95$.

Thus, the amount in the account after 5 years is $1485.95. The graph in Figure 5.9 confirms our work. ◆

PROBLEM 4 Use the graph in Figure 5.9 to estimate the time it takes for the amount in the account to reach $2000. ◆

If the number of compoundings per year increases *without bond* ($n \to +\infty$), then we have **continuous compounding**. In the formula

$$A = P\left(1 + \frac{r}{n}\right)^{nt}$$

suppose we let $n/r = m$. Then $n = mr$, and $m \to +\infty$ implies that $n \to +\infty$. Now, by direct substitution and the properties of real exponents, we have

$$A = P\left(1 + \frac{r}{n}\right)^{nt} = P\left(1 + \frac{1}{m}\right)^{mrt} = P\left[\left(1 + \frac{1}{m}\right)^{m}\right]^{rt}$$

Using a calculator, we can show that as $m \to +\infty$, the expression $\left(1 + \frac{1}{m}\right)^{m}$ seems to approach the number $e \approx 2.71828\ldots$. The following table displays some calculated values for this expression.

m	1	10	100	1000	10,000	100,000
$\left(1 + \frac{1}{m}\right)^{m}$	2	2.59374...	2.70481...	2.71692...	2.71814...	2.71826...

Hence,

$$\text{as} \quad m \to +\infty, \quad A = P\left[\left(1 + \frac{1}{m}\right)^{m}\right]^{rt} \to Pe^{rt}$$

◆ **Compound Interest Formula for Continuous Compounding**

If a certain principal P is deposited in a savings account at an interest rate r per year and the interest is compounded continuously, then the amount A in the account after t years is given by the formula

$$A = Pe^{rt}.$$

EXAMPLE 5 Suppose that $1000 is deposited in a bank account at an interest rate of 8% per year compounded continuously. Assume the depositor makes no subsequent deposit or withdrawal.

(a) Express the amount in the account as a function of time t.

(b) Find the amount in the account after five years

◆ **Solution**

(a) If the interest is compounded continuously, then we use the formula $A = Pe^{rt}$ with $P = 1000$ and $r = 0.08$ to obtain

$$A = 1000e^{0.08t}.$$

(b) To find the amount in the account after five years, we substitute $t = 5$ in the formula from part (a), and then use a calculator to evaluate this expression:

$$A = 1000e^{(0.08)(5)} \approx 1491.82$$

Thus, if the interest is compounded continuously, the amount in the account after five years is $1491.82. The graph in Figure 5.10 confirms our work.   ◆

FIGURE 5.10

The graph of $A = 1000e^{0.08t}$ for $t \geq 0$ indicates that for $t = 5$ years, $A \approx \$1491.82$.

PROBLEM 5   Compare the answers in Example 4(b) and Example 5(b). Which type of compounding is better for the investor?   ◆

Exercises 5.1

◆ **Basic Skills**

1. Generate the graphs of the exponential functions defined by the equations $y = 2^x$, $y = 3^x$, and $y = e^x$ in the same viewing rectangle of a graphing calculator. Then answer the following questions.

(a) What is the point of intersection of these graphs?

(b) For an x-value on the interval $(-\infty, 0)$, which equation has the largest y-value? which has the smallest y-value?

(c) For an x-value on the interval $(0, \infty)$, which equation has the largest y-value? which has the smallest y-value?

2. Generate the graphs of the exponential functions defined by the equations $y = 2^{-x}$, $y = 3^{-x}$, and $y = e^{-x}$ in the same viewing rectangle of a graphing calculator. Then answer the following questions.

(a) What is the point of intersection of these graphs?

(b) For an x-value on the interval $(-\infty, 0)$, which equation has the largest y-value? which has the smallest y-value?

(c) For an x-value on the interval $(0, \infty)$, which equation has the largest y-value? which has the smallest y-value?

In Exercises 3–16, use the shift rules and axis reflection rules (Section 2.5) to sketch the graph of each function. For each graph, label the y-intercept and horizontal asymptote. Use a graphing calculator to verify each sketch.

3. $f(x) = -2^x$

4. $f(x) = -3^{-x}$

5. $h(x) = 2^{x-2}$

6. $h(x) = e^{x+1}$

7. $F(x) = -(3^x + 2)$

8. $F(x) = 2^x + 1$

9. $G(x) = 2(\frac{1}{2})^x$

10. $G(x) = 9 \cdot 3^{-x}$

11. $H(x) = 1 - e^{-x}$

12. $H(x) = 2 - (\frac{1}{3})^{-x}$

13. $A(x) = 2 + 3^{2-x}$

14. $A(x) = 2^{x+1} - 3$

15. $h(x) = \dfrac{e}{e^x}$

16. $h(x) = 2 - e^{1-x}$

In Exercises 17–30, use the properties of real exponents to write each expression as a constant or as an expression in the form b^x or b^{-x} with $b > 1$. Check graphically with a graphing calculator.

17. 1^{5x}

18. $1^{x-\pi}$

19. $(6^x)^{-2}$

20. $(\frac{1}{3})^{-2x}$

21. $(4^{x-1})^0$

22. $(2^{2x})^{-2}$

23. $e(e^{x-1})$

24. $e^x \cdot e^{-x}$

25. $(3)(3^{2x-1})$

26. $\dfrac{36}{6^{2-x}}$

27. $(9^{3x-2})(27^{1-2x})$

28. $\dfrac{8^{2-x}}{4^{x+3}}$

29. $\left(\dfrac{5^{2x}}{25^{x-1}}\right)^{1/2}$ **30.** $(2^{-x} \cdot 4^x)^2$

In Exercises 31–38,

(*a*) *express the amount A in the account as a function of time t.*

(*b*) *find the amount in the account after the indicated number of years. Check graphically with a graphing calculator.*

31. $5000 invested at 9% per year compounded annually; 4 years

32. $3000 invested at 7% per year compounded semiannually;   8 years

33. $10,000 invested at $8\frac{1}{2}$% per year compounded quarterly;   12 years

34. $800 invested at $9\frac{3}{4}$% per year compounded daily; 10 years

35. $2000 invested at 6% compounded continuously; 9 years

36. $15,000 invested at 10% compounded continuously; 6 years

37. $120,000 invested at 8% compounded continuously; 6 months

38. $250,000 invested at $12\frac{3}{4}$% compounded continuously; 8 months

39. Suppose P is the air pressure (in pounds per square inch, or psi) at a certain height h (in feet) above sea level, and $P = ke^{-0.00004h}$ where k is a constant.

(a) Determine k if the air pressure at sea level is 15 psi.

(b) What is the pressure outside the cabin of an airplane whose altitude is 20,000 feet?

40. Suppose A is the number of moose living in a protected forest in Maine after a certain time t (in years since 1968), and $A = ke^{0.15t}$ where k is a constant.

(a) Determine k if 201 moose were present in 1988.

(b) What is the expected moose population in 1998?

41. The number A of bacteria in a certain culture is a function of time t (in minutes) and is described by $A(t) = 18e^{0.2t}$.

(a) Find the number of bacteria initially present in the culture.

(b) Find the number of bacteria present in the culture after 10 minutes. Use a graphing calculator to verify the answer.

42. The voltage V (in volts) in an electrical circuit is a function of time t (in milliseconds, or ms) and is given by

$$V(t) = 400e^{-t/10}.$$

(a) Find the initial voltage in the circuit.

(b) Find the voltage in the circuit after 30 ms.

(c) Use a graphing calculator to generate the graph of the function V for $t \geq 0$. Then discuss the voltage in the circuit as t increases without bound $(t \to +\infty)$.

43. The value V (in dollars) of a certain piece of machinery is a function of its age t (in years) and is given by

$$V(t) = 2000 + 10{,}000e^{-0.35t}.$$

(a) What is the value of the machine when it is purchased?

(b) What is the value of the machine 4 years after it is purchased?

(c) Use a graphing calculator to generate the graph of the function V for $t \geq 0$. Then discuss the machine's scrap value, that is, the value of the machinery as t increases without bound $(t \to +\infty)$.

44. For a certain secretarial student, the number N of words typed (correctly) per minute on a new word processor is a function of the time t (in hours) that the student has practiced on the machine and is given by

$$N(t) = 60(1 - e^{-0.05t}).$$

(a) What is the student's rate of typing after 10 hours of practice?

(b) Use a graphing calculator to generate the graph of the function N for $t \geq 0$. What is the best rate of typing this student can hope to achieve on this word processor?

◆ **Critical Thinking**

45. Perform the indicated operations and simplify.

(a) $(2^x + 2^{-x})(2^x - 2^{-x})$

(b) $(3^x + 3^{-x})^2 - (3^x - 3^{-x})^2$

46. Given the function f defined by $f(x) = e^x$, determine whether f is a one-to-one function and whether there exists an inverse function f^{-1}. If f^{-1} exists, sketch its graph.

In Exercises 47–52, find the real zeros of each function. Check graphically with a graphing calculator.

47. $F(x) = 2e^{-2x} - xe^{-2x}$

48. $G(x) = 2x^2e^{5x} - xe^{5x} - e^{5x}$

49. $g(x) = \dfrac{(x^2 - 4x + 4)e^x}{x + 1}$

50. $h(x) = \dfrac{x^2 e^x + 2xe^x}{x - 3}$

51. $H(x) = x^2 3^{2x} - 9^x$

52. $f(x) = x^2 (\tfrac{1}{2})^{3x} - x8^{-x}$

53. The graph of the exponential function $y = a^x$ passes through the point $(-2, 16)$. Determine the value of the base a.

54. Given that f is an exponential function defined by $f(x) = e^x$, show that each of the following statements is true.

(a) $f(a)f(b) = f(a + b)$

(b) $\dfrac{f(a)}{f(b)} = f(a - b)$

(c) $[f(a)]^n = f(na)$

55. Complete the table, given that the function f is defined by $f(x) = (-2)^x$.

x	0	1	2	3	4	-1	-2	-3	-4
$f(x)$									

Explain why we cannot sketch the graph of f by plotting the points in this table and connecting them with a smooth curve.

56. The *hyperbolic sine function f* and the *hyperbolic cosine function g* are defined by

$$f(x) = \frac{e^x - e^{-x}}{2} \quad \text{and} \quad g(x) = \frac{e^x + e^{-x}}{2}.$$

(a) Determine whether f and g are even or odd functions.

(b) Generate the graphs of f and g in the viewing rectangle of a graphing calculator.

57. Over a period of 10 years, which interest rate is the best investment?

(i) 8.1% compounded semiannually

(ii) 8.0% compounded monthly

(iii) 7.9% compounded continuously

58. In calculus, it is shown that if $|x| \le 1$, then

$$e^x \approx 1 + x + \frac{x^2}{2} + \frac{x^3}{6} + \frac{x^4}{24} + \frac{x^5}{120}.$$

Use this formula and a calculator to evaluate each expression.

(a) $e^{0.5}$

(b) $e^{-0.4}$

(c) e

(d) e^{-1}

59. A function is defined as $f(x) = e^x$.

(a) Show that the difference quotient

$$\frac{f(x + h) - f(x)}{h} = e^x \left(\frac{e^h - 1}{h} \right).$$

(b) Use a calculator to evaluate $\dfrac{e^h - 1}{h}$ for the values of h given in the following table.

h	1	0.1	0.01	0.001	0.0001
$\dfrac{e^h - 1}{h}$					

(c) Use your table entries from part (b) to determine the number that $\dfrac{e^h - 1}{h}$ seems to approach as $h \to 0^+$.

60. In statistics, the *normal probability distribution function* is given by

$$f(x) = \frac{1}{\sigma \sqrt{2\pi}} e^{-(x-\mu)^2/2\sigma^2}$$

where μ is the mean and σ the standard deviation of the distribution. The graph of this function is usually referred to as a "bell-shaped" curve. Use a graphing calculator to generate the graph of the bell-shaped curve for $\mu = 0$ and $\sigma = 1$.

5.2 Logarithmic Functions and Their Graphs

◆ Introductory Comments

As we discussed in Section 5.1, the exponential function $f(x) = b^x$ is always increasing when $b > 1$ and is always decreasing when $0 < b < 1$. Recall from Section 2.6 that every function f that is either an increasing function or a decreasing function is also a one-to-one function and has an inverse function f^{-1}. Using the method suggested in Section 2.6, we can attempt to find the inverse function of $f(x) = b^x$ by solving the equation

$$y = b^x$$

for x. However, we have no algebraic procedure that we can use to solve this last equation for x. By convention, we solve this equation for x by writing

$$x = \log_b y \quad \boxed{\text{read "log base } b \text{ of } y\text{"}}$$

The expression $\log_b y$ represents *the power to which b must be raised in order to obtain y.*

If we interchange x and y in the expression $x = \log_b y$, and then replace y with $f^{-1}(x)$, we obtain the inverse of the exponential function $f(x) = b^x$:

$$f^{-1}(x) = \log_b x$$

The function f^{-1} is called the *logarithmic function with base b*. Since the exponential function $f(x) = b^x$ and logarithmic function $f^{-1}(x) = \log_b x$ are inverses of each other, the domain of the logarithmic function must be the range of the exponential function, namely, $(0, \infty)$. In other words, $\log_b x$ is defined only when x is positive.

◀◆ **Logarithmic Function with Base b**

> For $b > 0$ and $b \neq 1$, the **logarithmic function with base b** is defined as
>
> $$y = \log_b x \quad \text{if and only if} \quad b^y = x \quad \text{for } x > 0.$$

From our definition of the logarithmic function, we can state that the equation

$$\log_b u = v \quad \text{is equivalent to} \quad b^v = u.$$

The equation $\log_b u = v$ is said to be in **logarithmic form** and the equation $b^v = u$ is said to be in **exponential form**. If we change an equation from logarithmic form to exponential form, or vice versa, the base b remains the same in each case; that is, in logarithmic form the base is the subscript b, and in exponential form the base is the number b being raised to a power v. Here are some illustrations of equivalent equations:

Logarithmic form		*Exponential form*
$\log_4 16 = 2$	is equivalent to	$4^2 = 16$
$\log_{16} 4 = \frac{1}{2}$	is equivalent to	$16^{1/2} = 4$
$\log_2 \frac{1}{16} = -4$	is equivalent to	$2^{-4} = \frac{1}{16}$

Two logarithmic bases occur in most applied problems—base 10 and base *e*. We refer to a base 10 logarithm ($\log_{10}$) as a **common logarithm** and a base *e* logarithm ($\log_e$) as a **natural logarithm**, or Napierian logarithm (named after the Scottish mathematician John Napier, 1550–1617). The common logarithm of a positive number *a* is usually written **log *a*** (read "log of *a*"), and the natural logarithm of a positive number *a* is usually written **ln *a*** (read "el-en of *a*").

$$\log a \quad \text{means} \quad \log_{10} a \qquad \text{and} \qquad \ln a \quad \text{means} \quad \log_e a$$

The function *F* defined by

$$F(x) = \log x \quad \text{for } x > 0$$

is called the **common logarithmic function**, and the function *G* defined by

$$G(x) = \ln x \quad \text{for } x > 0$$

is called the **natural logarithmic function**.

◆ **Evaluating Logarithms**

To evaluate the logarithmic expression $\log_b a$, we find the power to which *b* can be raised to obtain *a*. The procedure is illustrated in the next example.

EXAMPLE 1   Find the value of each logarithm, if it is defined.

(a) $\log_5 25$ **(b)** $\log \frac{1}{100}$
(c) $\log_8 16$ **(d)** $\ln(-1)$

◆ **Solution**

(a) The logarithm $\log_5 25$ represents the power *y* such that $5^y = 25$. Since $5^2 = 25$, we conclude that $y = 2$. Hence, $\log_5 25 = 2$.

(b) The common logarithm $\log \frac{1}{100} = \log_{10} \frac{1}{100}$ represents the power *y* such that $10^y = \frac{1}{100}$. Since $10^{-2} = \frac{1}{100}$, we conclude that $\log \frac{1}{100} = -2$.

(c) The logarithm $\log_8 16$ represents the power *y* such that $8^y = 16$. We know that $8^1 = 8$ and $8^2 = 64$. Since 16 is between 8 and 64 and the base 8 is a perfect cube, we try a fractional exponent between 1 and 2 whose denominator is 3. Note that $8^{4/3} = 16$. Hence, $\log_8 16 = \frac{4}{3}$.

(d) The domain of the natural logarithmic function $y = \ln x$ is $(0, \infty)$. Thus, for $x \leq 0$ the natural logarithmic function is undefined. Hence, we conclude

$$\ln(-1) = \log_e(-1) \quad \text{is undefined.}$$

In general, $\log_b a$ is undefined whenever $a \leq 0$.

As a direct consequence of the definition of the logarithmic function, we have two identities:

$$\log_b 1 = 0 \qquad \text{and} \qquad \log_b b = 1$$

In each identity, we assume that $b > 0$ and $b \neq 1$.

PROBLEM 1   Find the value of each logarithm: **(a)** $\log 1$ **(b)** $\ln e$ ◆

Most calculators have $\boxed{\text{LOG}}$ and $\boxed{\text{LN}}$ keys, which we can use to evaluate common logarithms and natural logarithms, respectively. Also, we can use the $\boxed{\text{LN}}$ key to evaluate $\log_b a$ by applying the **base e conversion formula**. This formula states that we can express all logarithms in terms of the natural logarithm.

◆ **Base e Conversion Formula**

If $\log_b a$ is defined, then

$$\log_b a = \frac{\ln a}{\ln b}$$

Note: The base e conversion formula is a consequence of the more general *change of base formula*, which we develop in Section 5.3. For now, we shall accept the validity of this formula and use its results.

EXAMPLE 2   Use a calculator to find the approximate value of each logarithm.

(a) $\log 25$ **(b)** $\ln \frac{1}{2}$ **(c)** $\log_4 24$

◆ **Solution**

(a) The common logarithm $\log 25$ represents the power y such that $10^y = 25$. We know that $10^1 = 10$ and $10^2 = 100$. Since 25 is between 10 and 100, we know that $\log 25$ must be some real number between 1 and 2. Using the $\boxed{\text{LOG}}$ key on a calculator, we approximate the value to four significant digits:

$$\log 25 \approx 1.398.$$

In other words, $25 \approx 10^{1.398}$.

(b) The natural logarithm $\ln \frac{1}{2}$ represents the power y such that $e^y = \frac{1}{2}$, or 0.5. Since $e^0 = 1$ and $e^{-1} = 1/e \approx 0.37$, we know that $\ln \frac{1}{2}$ must be some real number between 0 and -1. Using the $\boxed{\text{LN}}$ key on a calculator, we approximate the value to four significant digits:

$$\ln \frac{1}{2} \approx -0.6931.$$

(c) The expression $\log_4 24$ represents the power to which 4 must be raised to get 24. We know that $4^2 = 16$ and $4^3 = 64$. Since 24 is between 16 and 64, $\log_4 24$ must be a real number between 2 and 3. Using the base *e* conversion formula, we write

$$\log_4 24 = \frac{\ln 24}{\ln 4}.$$

Now, using the $\boxed{\text{LN}}$ key on a calculator and rounding to four significant digits, we obtain

$$\log_4 24 = \frac{\ln 24}{\ln 4} \approx 2.292.$$ ◆

PROBLEM 2   Use a calculator to verify the results obtained in Problem 1.
◆

◆   **Graph of a Logarithmic Function**

Recall from Section 2.6 that the graph of a function and its inverse are symmetric with respect to the line $y = x$. Hence, the graph of $f(x) = \log_b x$ may be obtained from the graph of $g(x) = b^x$ by reflecting the graph of *g* in the line $y = x$, as shown in Figure 5.11.

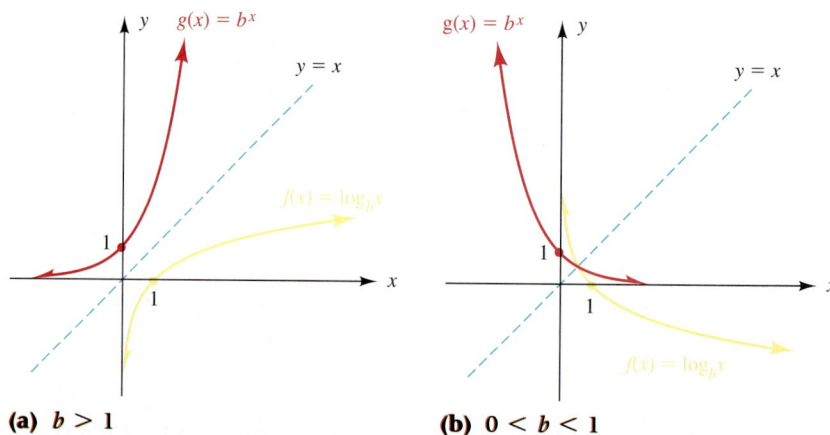

FIGURE 5.11

Like the graphs of any pair of inverse functions, the graphs of $f(x) = \log_b x$ and $g(x) = b^x$ are symmetric with respect to the line $y = x$.

(a) $b > 1$ **(b)** $0 < b < 1$

   In Figure 5.11(a) the graph of $f(x) = \log_b x$ for $b > 1$ is always rising and its *x*-intercept is 1. Also, for $b > 1$, observe that $f(x)$ decreases without bound $[f(x) \to -\infty]$ as *x* approaches 0 from the right $(x \to 0^+)$. Hence, the *y*-axis is a vertical asymptote.
   In Figure 5.11(b) the graph of $f(x) = \log_b x$ for $0 < b < 1$ is always falling, although its *x*-intercept remains 1. Also, for $0 < b < 1$, observe that $f(x)$ increases without bound $[f(x) \to +\infty]$ as *x* approaches 0 from the right $(x \to 0^+)$. Again, the *y*-axis is a vertical asymptote. We can summarize these features of the graph of a logarithmic function as follows.

◆◆ **Characteristics of the Graph of $f(x) = \log_b x$**

1. The x-intercept is 1, and the graph has no y-intercept.

2. The y-axis is a vertical asymptote.

3. If $b > 1$, the graph of $f(x) = \log_b x$ is always rising.

4. If $0 < b < 1$, the graph of $f(x) = \log_b x$ is always falling.

To sketch the graph of a logarithmic function, we may simply use the basic characteristics of the general graph and plot a few points.

EXAMPLE 3   Sketch the graph of each logarithmic function.

(a) $f(x) = \log_4 x$ **(b)** $g(x) = \log_{1/4} x$

◆ **Solution**

(a) Since the base b is 4 (and $4 > 1$), the graph of $f(x) = \log_4 x$ is always rising. The x-intercept is 1 and, since $f(x) \to -\infty$ as $x \to 0^+$, the y-axis is a vertical asymptote. Selecting a few convenient inputs for x, we find their corresponding outputs $f(x)$, as shown in the table:

x	$\frac{1}{4}$	1	4
$f(x)$	-1	0	1

Plotting the points associated with these ordered pairs and connecting them to form a smooth curve gives us the graph of $f(x) = \log_4 x$, as shown in Figure 5.12.

(b) Since the base b is $\frac{1}{4}$ (and $0 < \frac{1}{4} < 1$), the graph of $g(x) = \log_{1/4} x$ is always falling. The x-intercept is 1 and, since $g(x) \to +\infty$ as $x \to 0^+$, the y-axis is a vertical asymptote. Selecting a few convenient inputs for x, we find their corresponding outputs $g(x)$ as follows:

x	$\frac{1}{4}$	1	4
$g(x)$	1	0	-1

Plotting the points associated with these ordered pairs and connecting them to form a smooth curve gives us the graph of $g(x) = \log_{1/4} x$, as shown in Figure 5.13.   ◆

Observe that the graph of $g(x) = \log_{1/4} x$ is the same as the graph of $f(x) = \log_4 x$ reflected about the x-axis. In general, if b is a real number such that $b > 0$ and $b \neq 1$, then the graph of $g(x) = \log_{1/b} x$ is the same as the graph of $f(x) = \log_b x$ reflected about the x-axis.

FIGURE 5.12

The function $f(x) = \log_4 x$ is an increasing function.

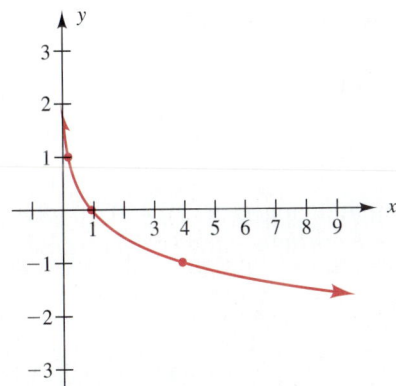

FIGURE 5.13

The function $g(x) = \log_{1/4} x$ is a decreasing function.

To generate the graph of the common logarithmic function or the natural logarithmic function on a graphing calculator, we use the LOG key or LN key, respectively. For bases other than 10 or e, we apply the base e conversion formula:

$$f(x) = \log_b x = \frac{\ln x}{\ln b}$$

For example, to generate the graph of $f(x) = \log_4 x$ in the viewing rectangle of a graphing calculator, we enter the function

$$f(x) = \frac{\ln x}{\ln 4}.$$

The graph is shown in Figure 5.14.

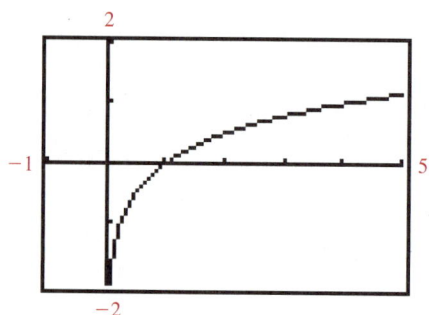

FIGURE 5.14

Graph of $f(x) = \log_4 x$ shown in the viewing rectangle of a graphing calculator

PROBLEM 3   Generate the graph of $g(x) = \log_{1/4} x$ in the viewing rectangle of a graphing calculator. Compare this graph with the sketch in Figure 5.13.   ◆

Knowing the basic shape of the graph of the logarithmic function $f(x) = \log_b x$, enables us to graph related functions by applying shift rules and axis reflection rules, which we discussed in Section 2.5.

EXAMPLE 4   Sketch the graph of each function. Label the vertical asymptote and any x- and y-intercepts.

(a) $F(x) = \log_4(x + 2)$ **(b)** $G(x) = 1 + \log_4(-x)$

◆ Solution

(a) By the horizontal shift rule (Section 2.5), the graph of $F(x) = \log_4(x + 2)$ is the same as the graph of $f(x) = \log_4 x$ (see Figure 5.12) shifted to the left 2 units, as shown in Figure 5.15. We find the y-intercept by evaluating $F(0)$:

$$F(0) = \log_4(0 + 2) = \log_4 2 = \tfrac{1}{2}$$

We find the x-intercept by solving the equation $\log_4(x + 2) = 0$. To solve this equation, we change from logarithmic form to exponential form:

$$\log_4(x + 2) = 0 \quad \text{is equivalent to} \quad 4^0 = x + 2$$
$$1 = x + 2$$
$$x = -1$$

Hence, the x-intercept of the graph of F is -1. Note that $F(x) \to -\infty$ as x approaches -2 from the right ($x \to -2^+$). Thus, the vertical asymptote is the line $x = -2$.

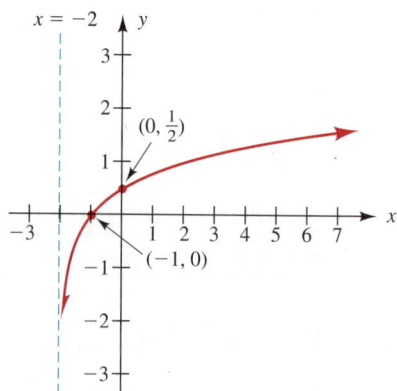

FIGURE 5.15

Graph of $F(x) = \log_4(x + 2)$

(b) By the *y*-axis reflection rule and the vertical shift rule (Section 2.5), the graph of $G(x) = 1 + \log_4(-x)$ is the same as the graph of $f(x) = \log_4 x$ (see Figure 5.12) reflected about the *y*-axis and then shifted vertically upward 1 unit, as shown in Figure 5.16. Since $G(0)$ is undefined, the graph has no *y*-intercept. The *x*-intercept is found by solving the equation $1 + \log_4(-x) = 0$. To solve this equation, we write it in logarithmic form and then change to exponential form:

$$1 + \log_4(-x) = 0$$

$$\log_4(-x) = -1 \qquad \text{is equivalent to} \qquad 4^{-1} = -x$$
$$\tfrac{1}{4} = -x$$
$$x = -\tfrac{1}{4}$$

Thus, the *x*-intercept of the graph of *G* is $-\tfrac{1}{4}$. Note that $G(x) \to -\infty$ as *x* approaches 0 from the left $(x \to 0^-)$. Hence, the *y*-axis is a vertical asymptote. ◆

FIGURE 5.16

Graph of $G(x) = 1 + \log_4(-x)$

PROBLEM 4 Use a graphing calculator to confirm the sketches in Figures 5.15 and 5.16. ◆

◆ **Simplifying Logarithmic Expressions**

We know that the logarithmic function with base *b* and the exponential function with base *b* are inverses of each other. If

$$f(u) = b^u \qquad \text{and} \qquad g(u) = \log_b u \quad \text{for } u > 0$$

then composing *f* with *g* in either order must yield the identity function:

$$f(g(u)) = b^{(\log_b u)} = u \quad \text{for } u > 0$$

and

$$g(f(u)) = \log_b(b^u) = u.$$

In summary, we have

$$b^{\log_b u} = u \quad \text{for } u > 0 \qquad\qquad \text{and} \qquad\qquad \log_b b^u = u$$

We can illustrate each of these identities:

$$e^{\ln 6} = e^{\log_e 6} = 6 \qquad \text{and} \qquad \log\sqrt{10} = \log_{10} 10^{1/2} = \tfrac{1}{2}$$

In the next example, we apply these logarithmic identities to simplify expressions containing logarithms. This procedure is often employed in calculus.

EXAMPLE 5 Simplify each expression.

(a) $e^{\ln(x+1)}$ **(b)** $\log_2 16^{x-2}$

FIGURE 5.17

The graph of $f(x) = e^{\ln(x+1)}$ is identical to the graph of $y = x + 1$ for $x > -1$.

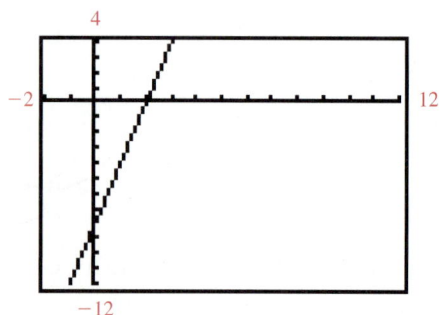

FIGURE 5.18

The graph of $f(x) = \log_2 16^{x-2}$ is identical to the graph of $y = 4x - 8$.

◆ **Solution**

(a) Since $b^{\log_b u} = u$ for $u > 0$, we have

$$e^{\ln(x+1)} = e^{\log_e(x+1)} = x + 1$$

provided that $x + 1 > 0$, which implies that $x > -1$. Figure 5.17 shows the graph of $f(x) = e^{\ln(x+1)}$ in the viewing rectangle of a graphing calculator. The graph of this function is the same as the graph of the line $y = x + 1$ for $x > -1$, which confirms our answer.

(b) We begin by applying the properties of real exponents (Section 5.1) and matching the base of the exponential expression to the base of the logarithm:

$$\log_2 16^{x-2} = \log_2(2^4)^{x-2} = \log_2 2^{4x-8}$$

Since $\log_b b^u = u$ for any permissible base b and any real number u, we have

$$\log_2 16^{x-2} = \log_2 2^{4x-8} = 4x - 8.$$

Figure 5.18 shows the graph of $f(x) = \log_2 16^{x-2}$ in the viewing rectangle of a graphing calculator. The graph of this function is the same as the graph of the line $y = 4x - 8$, which confirms our answer. ◆

PROBLEM 5   Simplify $\log_3 9^{2x-3}$. Use a graphing calculator to verify your answer. ◆

◆ **Application: The Retention Function**

In psychological experiments it has been shown that most of what we learn we forget rapidly and the remainder of the learned material recedes slowly from our memory. The function R that describes the percentage of the learned material we retain over a period of time t is called the **retention function**. The retention function is often defined in terms of logarithms, and its graph is called the **retention curve**.

EXAMPLE 6   In a psychological experiment, a student is asked to memorize a list of nonsense syllables by studying the list until one perfect repetition is performed from memory. At various times during the following week, the student is asked to recall the list of syllables. It is found that the retention function R associated with this experiment is given by

$$R(t) = 100 - 29 \log(27t + 1) \quad \text{for } 0 \le t \le 7,$$

where t is time in days.

(a) What percentage of the learned material does the student retain initially (at $t = 0$)?

(b) What percentage of the learned material does the student retain after 8 hours?

◆ **Solution**

(a) The percentage of the learned material that is retained initially is

$$
\begin{aligned}
R(0) &= 100 - 29 \log[27(0) + 1] \\
&= 100 - 29 \log 1 \\
&= 100 - 29(0) \\
&= 100\%
\end{aligned}
$$

(b) The percentage of the learned material retained after 8 hours ($t = \frac{1}{3}$ day) is

$$
\begin{aligned}
R(\tfrac{1}{3}) &= 100 - 29 \log[27\,(\tfrac{1}{3}) + 1] \\
&= 100 - 29 \log 10 \\
&= 100 - 29(1) \\
&= 71\%
\end{aligned}
$$

Figure 5.19 shows the retention curve associated with this experiment in the viewing rectangle of a graphing calculator. The point $(\frac{1}{3}, 71)$ lies on this curve, which confirms our work. ◆

FIGURE 5.19

Graph of $R(t) = 100 - 29 \log(27t + 1)$ with $0 \le t \le 7$

PROBLEM 6 For the retention function given in Example 6, determine the percentage of the learned material that is retained after 5 days. Check this value with the retention curve in Figure 5.19. ◆

Exercises 5.2

◆ **Basic Skills**

In Exercises 1–30, find the value of each expression, if it is defined.

1. $\log_6 36$

2. $\log_2 8$

3. $\log_5 125$

4. $\log_3 81$

5. $\log_3 3$

6. $\log_7(-7)$

7. $\log_2 \frac{1}{16}$

8. $\log_3 \frac{1}{27}$

9. $\log_{1/3} 9$

10. $\log_{1/2} 16$

11. $\log 100$

12. $\log \frac{1}{1000}$

13. $\ln 1$

14. $\log 1$

15. $\ln e^2$

16. $\ln \sqrt{e}$

17. $\log_{64}(-4)$

18. $\log_9 3$

19. $\log_4 8$

20. $\log_8 32$

21. $\log_{1/8} 16$

22. $\log_{27} \frac{1}{9}$

23. $4^{\log_4 3}$

24. $3^{\log_3 10}$

25. $e^{\ln 5}$

26. $10^{\log 2}$

27. $10^{\ln e}$

28. $e^{\log 10}$

29. $\log(\ln e)$

30. $\ln(\log 10)$

In Exercises 31–42, use a calculator to find the approximate value of each logarithm. Round each answer to four significant digits.

31. $\log 28$

32. $\log 750$

33. $\ln 8$

34. $\ln 56$

35. $\log_2 12$

36. $\log_4 9$

37. $\log_{12} 945$

38. $\log_{20} 1250$

39. $\log_3 \frac{2}{3}$

40. $\log_5 \frac{17}{6}$

41. $\log_{2/3} 34$

42. $\log_{1/2} \frac{3}{4}$

43. Generate the graphs of the logarithmic functions defined by the equations $y = \log_2 x$, $y = \log_3 x$,

$y = \log x$, and $y = \ln x$ in the same viewing rectangle of a graphing calculator. Then answer the questions that follow:

(a) What is the point of intersection of these graphs?

(b) For an x-value on the interval $(0, 1)$, which equation has the largest y-value? which has the smallest y-value?

(c) For an x-value on the interval $(1, \infty)$, which equation has the largest y-value? which has the smallest y-value?

44. Generate the graphs of the logarithmic functions defined by the equations $y = \log_{1/2} x$, $y = \log_{1/3} x$, $y = \log_{1/10} x$, and $y = \log_{1/e} x$ in the same viewing rectangle of a graphing calculator. Then answer the questions that follow:

(a) What is the point of intersection of these graphs?

(b) For an x-value on the interval $(0, 1)$, which equation has the largest y-value? which has the smallest y-value?

(c) For an x-value on the interval $(1, \infty)$, which equation has the largest y-value? which has the smallest y-value?

In Exercises 45–58, use the shift rules and axis reflection rules (Section 2.5) to sketch the graph of each function. In each case, label the vertical asymptote and any x- and y-intercepts. Use a graphing calculator to verify each sketch.

45. $F(x) = -\log_2 x$

46. $g(x) = -\log_{1/3} x$

47. $h(x) = \log_{1/10}(-x)$

48. $f(x) = -\ln(-x)$

49. $g(x) = \log_3 x + 2$

50. $G(x) = \log_2 x - 1$

51. $G(x) = \log_3(x + 2)$

52. $H(x) = \log_2(x - 1)$

53. $f(x) = -3 + \ln(x - 1)$

54. $f(x) = 2 + \log(x + 2)$

55. $H(x) = 2 - \log_2(x + 3)$

56. $F(x) = 1 - \log(x - 4)$

57. $f(x) = \ln(3 - x) + 2$

58. $f(x) = 2 - \log_3(2 - x)$

In Exercises 59–74, simplify each expression. Use a graphing calculator to confirm your answer. Assume any variable is restricted to those real numbers that allow the logarithmic expression to be defined.

59. $\log_4 4^{x-3}$

60. $\log_3 3^{x+1}$

61. $6^{\log_6(x+2)}$

62. $5^{\log_5(x-4)}$

63. $e^{\ln(x^2+4)}$

64. $10^{\log(1/x)}$

65. $\ln e^{x^2+2x}$

66. $(\ln e)^{x^2+2x}$

67. $(\log 10)^{2-x^2}$

68. $\log 10^{2-x^2}$

69. $\log_2 8^{x+2}$

70. $\log_9 3^{2x-2}$

71. $e^{x+\ln x}$ [*Hint*: $a^{x+y} = a^x a^y$]

72. $e^{x-\ln x}$ [*Hint*: $a^{x-y} = a^x/a^y$]

73. $e^{2\ln x}$ [*Hint*: $a^{xy} = (a^x)^y$]

74. $e^{-\ln x}$ [*Hint*: $a^{xy} = (a^x)^y$]

Chemists describe the acidity or alkalinity of a liquid by denoting its pH. An acid has pH < 7 and an alkaline has pH > 7. By definition,

$$\mathrm{pH} = -\log[\mathrm{H}^+],$$

where [H⁺] is the liquid's concentration of hydrogen ions, measured in moles per liter (m/L). In Exercises 75 and 76, determine the pH of each liquid.

75. Milk: $[\mathrm{H}^+] \approx 4.1 \times 10^{-7}$ m/L

76. Apple juice: $[\mathrm{H}^+] \approx 6.2 \times 10^{-4}$ m/L

77. In a psychological experiment, a student is asked to memorize a list of telephone numbers by studying the list until one perfect repetition is performed from memory. At various times during the following week, the student is asked to recall the list of numbers. It is found that the retention function R associated with this experiment is given by

$$R(t) = 100 - 31 \log(36t + 1) \quad \text{for } 0 \le t \le 7$$

where t is time in days.

(a) What percentage of the learned material is retained initially (at $t = 0$)?

(b) What percentage of the learned material is retained after 6 hours?

(c) Use a graphing calculator to generate the graph of the retention curve associated with this function. From this graph, estimate how much time has elapsed when only 33% of the learned material is still retained.

78. Students taking their last calculus course are given a final exam, and the average score of the group is recorded. Each month thereafter the students are given an equivalent exam, and for each exam their average score is recorded. It is found that the average score S is a function of time t (in months) and is defined by

$$S(t) = 72 - 22 \log(t + 1) \quad \text{for } 0 \le t \le 12.$$

(a) What is the average score on the original final exam?

(b) What is the average score on an exam given six months later?

(c) Use a graphing calculator to generate the graph of the retention curve associated with this function. From this graph, estimate how much time has elapsed when the average score is 50.

79. The number N of computers sold by a certain company in a month is a function of the amount x (in thousand of dollars) that is spent on advertising and is given by

$$N(x) = 1000[1 + \ln(x + 1)].$$

(a) If no money is spent on advertising, how many computers are sold in a month?

(b) If \$100,000 is spent on advertising, approximately how many computers are sold in a month?

(c) Use a graphing calculator to generate the graph of the curve associated with this function. From this graph, estimate how much money on advertising must be spent to sell 4000 computers in a month.

80. The time t (in minutes) required for a cup of hot tea to cool to a temperature of T degrees Fahrenheit when placed in a room whose temperature is maintained at $70°$F is given by

$$t = -8 \ln \frac{T - 70}{100} \quad \text{for } 70 < T \le 170.$$

(a) How long does it take the tea to cool to $140°$F?

(b) How long does it take the tea to cool to $100°$F?

(c) Use a graphing calculator to generate the graph of the curve associated with this function. From this graph, estimate the temperature of the tea when $t = 4$ minutes.

◆ Critical Thinking

81. Find the base b of the logarithmic function $f(x) = \log_b x$ if the graph of this function passes through the given point.

(a) $(81, 4)$ (b) $(\frac{1}{8}, -\frac{3}{2})$

82. Without using a calculator, determine between which two consecutive integers the value of each logarithm lies.

(a) $\log_3 99$ (b) $\log_5 150$
(c) $\log_8 7$ (d) $\log_{1/2} 40$

83. Use interval notation to describe the domain of each function.

(a) $f(x) = \dfrac{1}{1 - \ln x}$ (b) $F(x) = \sqrt{10 - \log x}$

84. Given that $f(x) = e^x$ and $g(x) = \ln x$, find $f(g(x))$ and $g(f(x))$. What do these results tell us about the functions f and g?

In Exercises 85–88,

(a) find the inverse of each function.

(b) sketch the graphs of the function and its inverse on the same coordinate plane.

85. $f(x) = 6^x$ **86.** $g(x) = (\frac{1}{6})^x$

87. $f(x) = \log_8 x$ **88.** $g(x) = \log_{1/8} x$

89. Use the $\boxed{\text{LN}}$ key on your calculator to help complete the following table. Then answer the questions that follow.

(a) Compare the values in the table for $\ln(xy)$ and $\ln x + \ln y$. What do you discover?

(b) Compare the values in the table for $\ln \dfrac{x}{y}$ and $\ln x - \ln y$. What do you discover?

(c) Compare the values in the table for $\ln x^y$ and $y \ln x$. What do you discover?

90. In calculus, it is shown that if $|x| < 1$, then

$$\ln(x + 1) \approx x - \frac{x^2}{2} + \frac{x^3}{3} - \frac{x^4}{4} + \frac{x^5}{5} - \frac{x^5}{6}.$$

Use this formula and a calculator to evaluate each logarithm.

(a) $\ln 1.8$ (b) $\ln 0.2$
(c) $\ln 1.25$ (d) $\ln 0.75$

Given the functions f and g defined by

$$f(x) = \ln(x^2 - 2x + 3) \quad \text{and} \quad g(x) = \log \frac{1}{x^2 + 2},$$

compute the functional values given in Exercises 91 and 92. Round each value to three significant digits.

91. $(f \circ g)(4.4)$ **92.** $(g \circ f)(21.8)$

x	y	$\ln(xy)$	$\ln x + \ln y$	$\ln\frac{x}{y}$	$\ln x - \ln y$	$\ln x^y$	$y \ln x$
2	3						
18	4						
24	15						
π	10						

93. The function f is defined as $f(x) = \dfrac{\ln x}{x}$.

  (a) Determine the domain of the function f.

  (b) Determine any zeros of f.

  (c) Explain the behavior of $f(x)$ as $x \to +\infty$.

  (d) Explain the behavior of $f(x)$ as $x \to 0^+$.

  (e) Generate the graph of the function f in the viewing rectangle of a graphing calculator. Then estimate the value of x at which the maximum value of this function seems to occur.

94. The function f is defined as $f(x) = \dfrac{1}{x \ln x}$.

  (a) Determine the domain of the function f.

  (b) Determine any zeros of f.

  (c) Explain the behavior of $f(x)$ as $x \to +\infty$ and as $x \to 1^+$.

  (d) Explain the behavior of $f(x)$ as $x \to 0^+$ and as $x \to 1^-$.

  (e) Generate the graph of this function in the viewing rectangle of a graphing calculator. Then estimate the maximum value of the function f in the interval $(0, 1)$.

5.3   Properties of Logarithms

◆ **Introductory Comments**

Recall from Section 5.1 three properties of real exponents:

  1. $b^x b^y = b^{x+y}$ **2.** $\dfrac{b^x}{b^y} = b^{x-y}$ **3.** $(b^x)^y = b^{xy}$

By using the log identity $b^{\log_b u} = u$ for $u > 0$ and these three properties of real exponents, we can obtain three corresponding **properties of logarithms**:

  1. For $x > 0$ and $y > 0$,

$$b^{\log_b xy} = xy$$

$$= b^{\log_b x} \cdot b^{\log_b y} \qquad \textcolor{red}{\textbf{Rewrite } x \textbf{ and } y \textbf{ using } u = b^{\log_b u}}$$

$$= b^{\log_b x \,+\, \log_b y} \qquad \textcolor{red}{\textbf{Add exponents}}$$

Since the exponential function is a one-to-one function, we conclude that the exponents $\log_b xy$ and $\log_b x + \log_b y$ must be equal, that is,

$$\boxed{\log_b xy = \log_b x + \log_b y}$$

The logarithm of a product of two positive real numbers is the sum of the logarithms of its factors.

  2. For $x > 0$ and $y > 0$,

$$b^{\log_b (x/y)} = \frac{x}{y}$$

$$= \frac{b^{\log_b x}}{b^{\log_b y}} \qquad \textcolor{red}{\textbf{Rewrite } x \textbf{ and } y \textbf{ using } u = b^{\log_b u}}$$

$$= b^{\log_b x \,-\, \log_b y} \qquad \textcolor{red}{\textbf{Subtract exponents}}$$

Since the exponential function is a one-to-one function, we conclude that the exponents $\log_b(x/y)$ and $\log_b x - \log_b y$ must be equal, that is,

$$\log_b \frac{x}{y} = \log_b x - \log_b y$$

The logarithm of a quotient of two positive real numbers is the logarithm of its numerator minus the logarithm of its denominator.

3. For $x > 0$ and any real number n,

$$b^{\log_b x^n} = x^n$$

$$= (b^{\log_b x})^n \qquad \text{\textcolor{red}{Rewrite } } x \text{ \textcolor{red}{using } } u = b^{\log_b u}$$

$$= b^{n \, \log_b x} \qquad \text{\textcolor{red}{Multiply exponents}}$$

Since the exponential function is a one-to-one function, we conclude that the exponents $\log_b x^n$ and $n \log_b x$ must be equal, that is,

$$\log_b x^n = n \log_b x$$

The logarithm of a positive real number raised to a power is the product of the power and the logarithm of that real number.

We now summarize these properties.

◆ **Properties of Logarithms**

If $b > 0$ and $b \neq 1$, then for any positive real numbers x and y,

1. $\log_b xy = \log_b x + \log_b y$

2. $\log_b \dfrac{x}{y} = \log_b x - \log_b y$

3. $\log_b x^n = n \log_b x$ where n is a real number

These properties of logarithms enable us to rewrite certain functions that contain logarithms so that we may sketch the graphs of these functions. Also, these properties enable us to rewrite certain logarithmic equations so that we may solve these equations, as we will show in Section 5.4.

◆ **Expanding Logarithmic Expressions**

We can use the properties of logarithms to write a single logarithmic expression as the sum and difference of simpler logarithmic expressions without logarithms of products, quotients, and powers. This procedure, which is used frequently in calculus, is called *expanding* a logarithmic expression.

EXAMPLE 1   Expand each logarithmic expression.

(a) $\log \sqrt[3]{10x^2}$ **(b)** $\ln \dfrac{\sqrt{x^2 + 4}}{xe^{3x}}$

◆ **Solution**

(a) Assuming that $x > 0$, we have

$$\log \sqrt[3]{10x^2} = \log(10x^2)^{1/3} \qquad \text{Change radical to a rational exponent}$$

$$= \tfrac{1}{3}\log(10x^2) \qquad \text{Apply log property 3}$$

$$= \tfrac{1}{3}(\log 10 + \log x^2) \qquad \text{Apply log property 1}$$

$$= \tfrac{1}{3}(\log 10 + 2 \log x) \qquad \text{Apply log property 3}$$

$$= \tfrac{1}{3}(1 + 2 \log x) \qquad \text{Evaluate log 10}$$

Figure 5.20 shows that the graphs of

$$f(x) = \log \sqrt[3]{10x^2} \quad \text{for } x > 0 \qquad \text{and} \qquad g(x) = \tfrac{1}{3}(1 + 2 \log x)$$

are identical. This fact confirms our work.

(b) Assuming that $x > 0$, we have

$$\ln \frac{\sqrt{x^2 + 4}}{xe^{3x}} = \ln \frac{(x^2 + 4)^{1/2}}{xe^{3x}} \qquad \text{Change radical to a rational exponent}$$

$$= \ln(x^2 + 4)^{1/2} - \ln xe^{3x} \qquad \text{Apply log property 2}$$

$$= \ln(x^2 + 4)^{1/2} - (\ln x + \ln e^{3x}) \qquad \text{Apply log property 1}$$

Keep the parentheses.

$$= \tfrac{1}{2}\ln(x^2 + 4) - (\ln x + 3x \ln e) \qquad \text{Apply log property 3}$$

$$= \tfrac{1}{2}\ln(x^2 + 4) - \ln x - 3x \qquad \text{Evaluate ln } e = 1$$

Figure 5.21 confirms our work.   ◆

FIGURE 5.20

The graphs of $f(x) = \log \sqrt[3]{10x^2}$ for $x > 0$ and $g(x) = \tfrac{1}{3}(1 + 2 \log x)$ are identical.

FIGURE 5.21

The graphs of $f(x) = \ln \dfrac{\sqrt{x^2 + 4}}{xe^{3x}}$ and $g(x) = \tfrac{1}{2}\ln(x^2 + 4) - \ln x - 3x$ are identical.

Note: The logarithmic expression in Example 1(a) is defined for *all* real numbers except 0. We make the assumption that $x > 0$ so that the properties of logarithms may be applied. If we allow that $x < 0$, then we use the absolute value of x to write

$$\log \sqrt[3]{10x^2} = \tfrac{1}{3}(1 + 2 \log |x|).$$

CAUTION   Do not misapply the three properties of logarithms. In Example 1(a), to apply log property 3 and write

$$\ln xe^{3x} \qquad \text{as} \qquad 3x \ln xe \quad \text{is WRONG!}$$

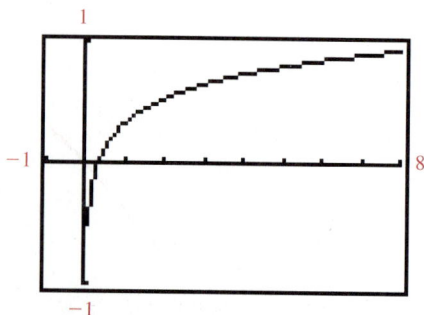

The expression $3x \ln xe$ is the simplification of $\ln(xe)^{3x}$, not of $\ln xe^{3x}$. Several other errors that are often made when working with the properties of logarithms are illustrated in Exercises 1–4.

PROBLEM 1 Show, by using a specific example, that $(\log_b x)^n$ and $n \log_b x$ are *not* equivalent logarithmic expressions. ◆

We often expand a logarithmic function in order to obtain an equivalent function whose graph is easily plotted. When applying the properties of logarithms to a logarithmic function, we must always preserve the domain of the original function. For example, consider the functions F and G defined by

$$F(x) = \log_4 x^2 \quad \text{and} \quad G(x) = 2\log_4 x$$

The domain of the function F is the set of all real numbers x with $x \neq 0$. However, the domain of the function G is the set of all positive real numbers x. Since the functions F and G have different domains, they are not equivalent functions. To apply logarithmic property 3 to function F and preserve the same domain, we must use the absolute value of x and write

$$F(x) = \log_4 x^2 = \log_4 |x|^2 = 2\log_4 |x|.$$

EXAMPLE 2 Sketch the graph of $H(x) = \log_4 \dfrac{1}{x-1}$. Label the vertical asymptote and any x- and y-intercepts.

◆ **Solution** We begin by expanding this logarithmic function:

$$H(x) = \log_4 \frac{1}{x-1} = \log_4 1 - \log_4(x-1) \quad \text{\textcolor{red}{\textbf{Apply log property 2}}}$$

$$= 0 - \log_4(x-1) \quad \text{\textcolor{red}{\textbf{Evaluate} } \log_4 1 = 0}$$

$$= -\log_4(x-1) \quad \text{\textcolor{red}{\textbf{Simplify}}}$$

By the horizontal shift rule and x-axis reflection rule (Section 2.5), the graph of $H(x) = -\log_4(x-1)$ is the same as the graph of $f(x) = \log_4 x$ (see Figure 5.12 in Section 5.2) shifted right 1 unit and then reflected about the x-axis, as shown in Figure 5.22. We find the x-intercept by solving the equation $H(x) = 0$:

$$\log_4 \frac{1}{x-1} = 0 \quad \text{is equivalent to} \quad 4^0 = \frac{1}{x-1}$$

$$1 = \frac{1}{x-1}$$

$$x - 1 = 1$$

$$x = 2$$

FIGURE 5.22
Graph of $H(x) = \log_4 \dfrac{1}{x-1}$

Hence, the x-intercept of the graph of H is 2. Note that $H(x) \to +\infty$ as $x \to 1^+$. Hence $x = 1$ is a vertical asymptote. ◆

PROBLEM 2   Use a graphing calculator to confirm the sketch in Figure 5.22.

◆

◆ Condensing Logarithmic Expressions

The properties of logarithms can also be used to write sums and differences of logarithmic expressions as a single logarithmic expression with a coefficient of 1. (This is a reversal of the *expansion* process.) For problems of this type, we begin by applying logarithmic property 3 and then changing any numerical coefficient to an exponent. Once we have removed numerical coefficients, we can apply properties 1 and 2. This procedure, which is used frequently to solve logarithmic equations (see Section 5.4), is called *condensing* a logarithmic expression.

EXAMPLE 3   Condense each logarithmic expression.

(a) $\log 50 + 2 \log 4 - 3 \log 2$ **(b)** $\ln(x^2 - 1) - 2 \ln(x + 1)$

◆ Solution

(a)

$$\log 50 + 2 \log 4 - 3 \log 2 = \log 50 + \log 4^2 - \log 2^3 \qquad \text{Apply log property 3}$$

$$= \log 50 + \log 16 - \log 8 \qquad \text{Simplify}$$

$$= \log(50 \cdot 16) - \log 8 \qquad \text{Apply log property 1}$$

$$= \log \frac{50 \cdot 16}{8} \qquad \text{Apply log property 2}$$

$$= \log 100 = 2 \qquad \text{Reduce and evaluate}$$

(b) The domain of $\ln(x^2 - 1) - 2 \ln(x + 1)$ is $(1, \infty)$. Hence,

$$\ln(x^2 - 1) - 2 \ln(x + 1) = \ln(x^2 - 1) - \ln(x + 1)^2 \qquad \text{Apply log property 3}$$

$$= \ln \frac{x^2 - 1}{(x + 1)^2} \qquad \text{Apply log property 2}$$

$$= \ln \frac{(x + 1)(x - 1)}{(x + 1)^2} \qquad \text{Factor the numerator}$$

$$= \ln \frac{x - 1}{x + 1} \quad \text{for } x > 1 \qquad \text{Reduce}$$

Figure 5.23 shows that the graphs of $f(x) = \ln(x^2 - 1) - 2 \ln(x + 1)$ and $g(x) = \ln \dfrac{x - 1}{x + 1}$ for $x > 1$ are identical. This fact confirms our work.

◆

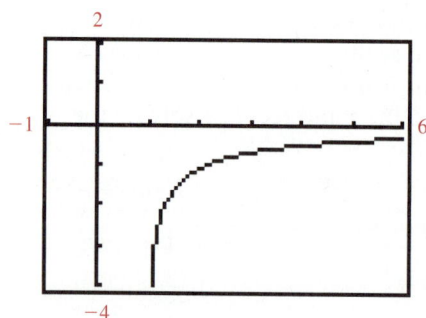

FIGURE 5.23
The graphs of
$f(x) = \ln(x^2 - 1) - 2 \ln(x + 1)$
and $g(x) = \ln \dfrac{x - 1}{x + 1}$ for $x > 1$ are identical.

PROBLEM 3   Use the ⎡LOG⎤ key on a calculator to evaluate $\log 50 + 2 \log 4 - 3 \log 2$. Your answer should agree with the result in Example 3(a).

◆

◆ Change of Base Formula

In Section 5.2 we used the base e conversion formula to evaluate $\log_b a$ and to generate the graph of $y = \log_b x$ in the viewing rectangle of a graphing calculator. The base e conversion formula is a consequence of the more general *change of base formula*, which we now develop.

By the definition of the logarithmic function with base b, we know that

$$y = \log_b x \quad \text{is equivalent to} \quad b^y = x.$$

Taking the base a logarithm of both sides of the equation $b^y = x$, we obtain

$$\log_a b^y = \log_a x$$
$$y \log_a b = \log_a x \qquad \text{\color{red}{Apply log property 3}}$$
$$y = \frac{\log_a x}{\log_a b} \qquad \text{\color{red}{Divide both sides by } \log_a b}$$

However, since $y = \log_b x$, we substitute and obtain

$$\log_b x = \frac{\log_a x}{\log_a b},$$

which is the **change of base formula**.

◆◆ **Change of Base Formula**

> If $\log_b x$ is defined, then
>
> $$\log_b x = \frac{\log_a x}{\log_a b} \quad \text{for } a > 0 \text{ and } a \neq 1.$$

If we let $a = e$ in this formula, we obtain the base e conversion formula:

$$\log_b a = \frac{\ln a}{\ln b}$$

If we let $a = 10$ in this formula, we obtain the *base 10 conversion formula*:

$$\log_b a = \frac{\log a}{\log b}$$

Hence, to evaluate $\log_4 24$ on a calculator, we can use the $\boxed{\text{LN}}$ key (see Example 2(c) in Section 5.2) or the $\boxed{\text{LOG}}$ key:

$$\log_4 24 = \frac{\log 24}{\log 4} \approx 2.292.$$

If we replace x with a in the change of base formula, we obtain

> $$\log_b a = \frac{1}{\log_a b}$$

or, equivalently,

$$(\log_b a)(\log_a b) = 1$$

EXAMPLE 4 Evaluate each logarithmic expression.

(a) $\dfrac{1}{\log_2 12} + \dfrac{1}{\log_6 12}$ (b) $(\log_3 16)(\log_2 27)$

◆ **Solution**

(a) $\dfrac{1}{\log_2 12} + \dfrac{1}{\log_6 12} = \log_{12} 2 + \log_{12} 6$ **Apply** $\dfrac{1}{\log_a b} = \log_b a$

$\qquad\qquad\qquad\qquad\quad = \log_{12}(2 \cdot 6)$ **Apply log property 1**

$\qquad\qquad\qquad\qquad\quad = \log_{12} 12$ **Simplify**

$\qquad\qquad\qquad\qquad\quad = 1$ **Apply** $\log_b b = 1$

(b)

$(\log_3 16)(\log_2 27) = (\log_3 2^4)(\log_2 3^3)$ **Rewrite**

$\qquad\qquad\qquad\quad = (4 \log_3 2)(3 \log_2 3)$ **Apply log property 3**

$\qquad\qquad\qquad\quad = (4 \cdot 3)[(\log_3 2)(\log_2 3)]$ **Rearrange the factors**

$\qquad\qquad\qquad\quad = 12[1]$ **Apply** $(\log_b a)(\log_a b) = 1$

$\qquad\qquad\qquad\quad = 12$ **Simplify**   ◆

PROBLEM 4 Use the ⌐LN⌐ key or the ⌐LOG⌐ key on a calculator to evaluate each logarithmic expression in Example 4. Your answers should agree with the results obtained in the example.   ◆

◆ **Application: Richter Scale**

When a physical quantity varies over a large range of values, it is convenient to work with a *logarithmic scale* in order to obtain a more manageable set of numbers. Some examples of logarithmic scales are the **brightness scale**, used for measuring the magnitude of a star (see Exercises 49 and 50); the **decibel scale**, used to measure the magnitude of sound (see Exercises 51 and 52), and the **Richter scale** (named after the seismologist Charles F. Richter), used for measuring the magnitude of an earthquake.

On the Richter scale, the magnitude R of an earthquake is given by

$$R = \log \dfrac{I}{I_0}$$

where I is the intensity of the earthquake and I_0 is the intensity of a zero-level earthquake having magnitude $R = 0$.

EXAMPLE 5 *Chapter Opening Problem*

Seismologists estimate that the San Francisco earthquake of 1906 measured 8.3 on the Richter scale. How many times more intense was this earthquake than the Loma Prieta quake, which occurred during the 1989 World Series and measured 7.1 on the Richter scale?

◆ **Solution** Let

$$I_a = \text{the intensity of the 1906 earthquake}$$

and $\qquad I_b = \text{the intensity of the 1989 earthquake.}$

We use the formula $R = \log(I/I_0)$ and log property 2:

$$8.3 = \log \frac{I_a}{I_0} \qquad \text{and} \qquad 7.1 = \log \frac{I_b}{I_0}$$

$$8.3 = \log I_a - \log I_0 \qquad\qquad 7.1 = \log I_b - \log I_0$$

Solving these equations simultaneously by using the elimination method (Section 3.5), we find

$$
\begin{array}{ll}
8.3 = \log I_a - \log I_0 & \\
7.1 = \log I_b - \log I_0 & \\
\hline
1.2 = \log I_a - \log I_b & \text{Subtract} \\
1.2 = \log \dfrac{I_a}{I_b} & \text{Apply log property 2}
\end{array}
$$

Now, to find the ratio I_a/I_b, we change logarithmic form to exponential form:

$$1.2 = \log \frac{I_a}{I_b} \quad \text{is equivalent to} \quad \frac{I_a}{I_b} = 10^{1.2} \approx 16$$

Hence, the 1906 earthquake was about 16 times as intense as the 1989 quake. ◆

PROBLEM 5 One of the strongest earthquakes ever recorded occurred in Japan in 1933 and measured 8.9 on the Richter scale. How many times more intense was this quake than the 1989 Loma Prieta quake in California? ◆

Exercises 5.3

◆ Basic Skills

In Exercises 1–4, show by using specific examples that the given logarithmic expressions are not equivalent.

1. $\log_b(x + y)$ and $\log_b x + \log_b y$

2. $\log_b(x - y)$ and $\log_b x - \log_b y$

3. $\dfrac{\log_b x}{\log_b y}$ and $\log_b \dfrac{x}{y}$

4. $(\log_b x)(\log_b y)$ and $\log_b xy$

In Exercises 5–18, expand each logarithmic expression to eliminate logarithms of products, quotients, and powers. Assume that each variable is restricted only to real numbers for which the properties of logarithms can be applied.

5. $\log_3(27^2 \cdot 81)$

6. $\log 100x^2$

7. $\log_2[8x(x + 2)]$

8. $\log_5 \sqrt[4]{25x}$

9. $\ln \dfrac{xe^2}{10}$

10. $\log \dfrac{(x + 4)^2}{2x}$

11. $\log_3 \sqrt[3]{\dfrac{x^2}{27}}$

12. $\log_b\left(\dfrac{x^2 + 2}{x + 3}\right)^3$

13. $\log_b \dfrac{x^2}{\sqrt[3]{(x + 1)^2}}$

14. $\ln \dfrac{1}{2\sqrt{3x^2 + 2}}$

15. $\log_3 \dfrac{1}{9\sqrt{x^3 y^{2/3}}}$

16. $\log \dfrac{\sqrt[3]{1 + y^2}}{(x - 2)(x + 2)^{1/2}}$

17. $\ln\left(\dfrac{e^{x^2}}{e^x + 1}\right)^2$

18. $\ln \dfrac{xe^{-2x}}{\sqrt{2e^x - 1}}$

In Exercises 19–30, condense each expression to a single logarithm with a coefficient of 1. Assume that each variable is restricted only to real numbers for which the properties of logarithms can be applied.

19. $\ln 2 + \ln 3 + \ln 4$

20. $\ln 6 - \ln 3 + \ln 5$

21. $\log 40 - (3 \log 2 - \log 20)$

22. $2 \log 6 - (2 \log 3 + \log 4)$

23. $-2 \log_3 6 + 4 \log_3 2 - \frac{1}{2} \log_3 16$

24. $5 \log_6 2 + \frac{2}{3} \log_6 27 + \frac{3}{2} \log_6 4$

25. $\log_5(x - 1) - 2 \log_5 x + \log_5(x + 3)$

26. $3 \ln x + \frac{2}{3} \ln(x - 1) + \ln 2$

27. $2 \ln(x - 2) - \ln(x^2 - 4)$

28. $2 \ln(x^3 + 1) - \ln(x^2 - x + 1)$

29. $\frac{1}{2}[\ln(x - 3) + \ln(x + 3)] - 2(\ln x - \ln 3)$

30. $3\left[\log\left(x + \sqrt{x^2 - 1}\right) + \log\left(x - \sqrt{x^2 - 1}\right)\right] - \log 2$

In Exercises 31–38, use the properties of logarithms to help sketch the graph of each function. Label any x- and y-intercept(s) and the vertical asymptote.

31. $f(x) = \dfrac{1}{2} \log x^4$

32. $f(x) = \ln x^3$

33. $f(x) = \ln \dfrac{1}{x + 3}$

34. $f(x) = \log_3 \dfrac{9}{x}$

35. $f(x) = \log_{1/2}[8(x - 1)]$

36. $f(x) = \log_2(16x + 32)$

37. $f(x) = \log_3 \dfrac{1}{18 - 9x}$

38. $f(x) = \log \dfrac{10}{(x - 1)^2}$

In Exercises 39–46, use $\log_b a = \dfrac{1}{\log_a b}$ to help express each logarithmic expression as a whole number.

39. $(\log_4 5)(\log_5 4)$

40. $(\log 7)(\log_7 10)$

41. $(\log_3 4)(\log_4 81)$

42. $(\log_5 16)(\log_2 125)$

43. $\log 20 + \dfrac{1}{\log_5 10}$

44. $\dfrac{1}{\log_4 10} + \dfrac{1}{\log_{25} 10}$

45. $\dfrac{1}{\log_a ab} + \dfrac{1}{\log_b ab}$

46. $\dfrac{\log_b x}{\log_{ab} x} - \dfrac{\log_b x}{\log_a x}$

47. The strongest earthquake ever recorded in the United States occurred in Alaska on Good Friday, March 27, 1964, and measured 8.6 on the Richter scale. How many times more intensive was this earthquake than the 1989 Loma Prieta quake in California, which measured 7.1?

48. One of the aftershocks from the 1989 Loma Prieta earthquake was 60 times less intense than the major earthquake (magnitude 7.1). What was the magnitude of the aftershock?

The magnitude m of a star is given by

$$m = -2.5 \log \dfrac{B}{B_0},$$

where B is the brightness of the star and B_0 is the brightness of a zero-level star having magnitude m = 0. Use this formula to answer Exercises 49 and 50.

49. The magnitude of our Sun is -26.8 and the magnitude of Sirius, the brightest star in the heavens, is -1.5. How many times brighter is the sun than Sirius?

50. What is the difference in magnitude between two stars if the brightness of one is 50 times the brightness of the other?

The loudness L (in decibels, db) of a sound is given by

$$L = 10 \log \frac{I}{I_0}$$

where I is the intensity of the sound and I_0 is the intensity of the faintest sound that can be heard. Use this formula to answer Exercises 51 and 52.

51. The loudness of a whisper is 30 db and the loudness of a rock concert is 120 db. How many times more intense is the rock concert than the whisper?

52. If the loudness of two sounds differs by 20 db, how many times more intense is the more audible sound than the other sound?

◀ ◆ **Critical Thinking**

53. Suppose the functions f and g are defined by

$$f(x) = \ln x^2 \quad \text{and} \quad g(x) = 2 \ln x.$$

Are f and g the same function? Explain.

54. Show that if $f(x) = \ln x$, then $f\left(\frac{1}{x}\right) = -f(x)$.

55. Given that $\log_b A = 2$ and $\log_b B = 3$, find the value of each logarithm.

(a) $\log_A b$ (b) $\log_B b$

(c) $\log_A b^2$ (d) $\log_B \sqrt{b}$

(e) $\log_{AB} b$ (f) $\log_{AB}(1/b)$

(g) $\log_{A/B} b$ (h) $\log_{B/A} b$

56. Explain the fallacy in the following argument:

$$\begin{aligned} 3 &= \log_2 8 \\ &= \log_2(4 + 4) \\ &= \log_2 4 + \log_2 4 \\ &= 2 + 2 \\ &= 4 \end{aligned}$$

57. Use the LOG key on your calculator to find the approximate value of each of the following logarithms:

log 6.2   log 62   log 620   log 6200

Compare these values. Do you observe a pattern? Show why this pattern develops by expressing 62, 620, and 6200 in scientific notation and then finding the common logarithms of these expressions by using the properties of logarithms.

58. Use the LOG key on your calculator to find the approximate value of each of the following logarithms:

log 3.5   log 0.35   log 0.035   log 0.0035

Compare these values. Do you observe a pattern? Show why this pattern develops by expressing 0.35, 0.035, and 0.0035 in scientific notation and then finding the common logarithms of these expressions by using the properties of logarithms.

59. Evaluate the following expressions by each of the given procedures:

$$\ln \pi + \ln\left(\frac{\sqrt{2}}{\pi}\right) + \frac{1}{2} \ln\left(\frac{3}{2}\right) - \ln\left(\frac{\sqrt{3}}{e}\right)$$

(a) Use the LN key on your calculator.

(b) Use the properties of logarithms.

60. Use the change of base formula and a calculator to complete the following table.

a	b	$\log_b \dfrac{1}{a}$	$\log_{1/b} a$
2	3		
18	4		
24	15		
π	10		

Compare the values of $\log_b(1/a)$ and $\log_{1/b} a$. Prove the relationship that you observe.

◆ **5.4** ◆ **Exponential and Logarithmic Equations**

◆ **Introductory Comments**

An equation in which the variable appears in an exponent is referred to as an **exponential equation**, and an equation that contains logarithmic expressions is referred to as a **logarithmic equation**. In this section, we discuss procedures for solving these two types of equations. We then look at some applied problems that involve the solution techniques introduced in this section.

◆ **Solving Exponential Equations**

If both sides of an exponential equation can be written as powers of the same base, then we can solve the equation by equating the powers; that is,

$$\text{if} \quad b^x = b^y, \quad \text{then} \quad x = y.$$

For example, we can solve the exponential equation $27^{x-2} = 9$ by writing both sides of the equation in terms of the same base, 3:

$$27^{x-2} = 9$$

$$(3^3)^{x-2} = 3^2 \qquad \textbf{Write both sides with the same base}$$

$$3^{3(x-2)} = 3^2 \qquad \textbf{Multiply exponents}$$

$$3(x - 2) = 2 \qquad \textbf{Equate exponents and solve for } x$$

$$3x = 8$$

$$x = \tfrac{8}{3}$$

Check: $27^{[(8/3)-2]} = 27^{2/3} = 9$

An alternative method for solving this exponential equation is to take the natural (or common) logarithm of both sides of the equation:

$$27^{x-2} = 9$$

$$\ln 27^{x-2} = \ln 9 \qquad \textbf{Take the natural logarithm of both sides}$$

$$(x - 2) \ln 27 = \ln 9 \qquad \textbf{Apply log property 3}$$

$$x - 2 = \frac{\ln 9}{\ln 27} \qquad \textbf{Divide both sides by ln 27}$$

$$x = 2 + \frac{\ln 9}{\ln 27} \qquad \textbf{Add 2 to both sides}$$

Now, since $9 = 3^2$ and $27 = 3^3$, we apply logarithmic property 3 and write

$$x = 2 + \frac{\ln 9}{\ln 27} = 2 + \frac{2 \ln 3}{3 \ln 3} = 2 + \frac{2}{3} = \frac{8}{3},$$

This result agrees with the answer we obtained previously. This alternative procedure is particularly useful when both sides of an exponential equation *cannot* be written as powers of the same base.

◆ **Procedure for Solving an Exponential Equation**

> 1. Isolate the exponential expression on one side of the equation.
>
> 2. Take the natural (or common) logarithm of both sides of this equation, apply the properties of logarithms, and solve for the unknown.

EXAMPLE 1 Solve each exponential equation.

(a) $6^x = 50$ **(b)** $4e^{2x-1} = 100$

◆ **Solution**

(a) Since both sides of this equation cannot be written as powers of the same base, we apply the given procedure:

$$6^x = 50$$

$$\ln 6^x = \ln 50 \qquad \text{Take the natural logarithm of both sides}$$

$$x \ln 6 = \ln 50 \qquad \text{Apply log property 3}$$

$$x = \frac{\ln 50}{\ln 6} \approx 2.183 \qquad \text{Solve for } x$$

Figure 5.24 shows the graph of $f(x) = 6^x - 50$ in the viewing rectangle of a graphing calculator. The x-intercept is approximately 2.183, which confirms our answer.

(b) Since both sides of this equation cannot be written as powers of the same base, we apply the given procedure:

$$4e^{2x-1} = 100$$

$$e^{2x-1} = 25 \qquad \text{Isolate the exponential expression}$$

$$\ln e^{2x-1} = \ln 25 \qquad \text{Take the natural logarithm of both sides}$$

$$2x - 1 = \ln 25 \qquad \text{Simplify the left-hand side}$$

$$x = \frac{\ln 25 + 1}{2} \approx 2.109 \qquad \text{Solve for } x$$

Figure 5.25 shows the graph of $f(x) = 4e^{2x-1} - 100$ in the viewing rectangle of a graphing calculator. The x-intercept is approximately 2.109, which confirms our answer. ◆

Note: To solve the equation in Example 1(a) we may also change the exponential form $6^x = 50$ to its equivalent logarithmic form:

$$x = \log_6 50$$

Now to evaluate $\log_6 50$ we use the base e conversion formula (Section 5.2):

$$x = \log_6 50 = \frac{\ln 50}{\ln 6} \approx 2.183$$

FIGURE 5.24

The x-intercept of the graph of $f(x) = 6^x - 50$ is the root of the equation $6^x = 50$.

FIGURE 5.25

The x-intercept of the graph of $f(x) = 4e^{2x-1} - 100$ is the root of the equation $4e^{2x-1} = 100$.

PROBLEM 1   Solve the equation $5^{x-1} = 325$.   ◆

As illustrated in the next example, we can solve an exponential equation that involves more than one exponential expression by using a similar procedure.

EXAMPLE 2   Solve the equation $3^{x-2} = 2^{-x}$.

◆◆ **Solution**   To solve this equation, we proceed as follows:

$$3^{x-2} = 2^{-x}$$

$$\ln 3^{x-2} = \ln 2^{-x} \qquad \text{\color{red}{Take the natural logarithm of both sides}}$$

$$(x - 2)\ln 3 = -x\ln 2 \qquad \text{\color{red}{Apply log property 3}}$$

> Keep the parentheses.

$$x\ln 3 - 2\ln 3 = -x\ln 2 \qquad \text{\color{red}{Multiply}}$$

$$x\ln 3 + x\ln 2 = 2\ln 3 \qquad \text{\color{red}{Group the x terms on one side}}$$

$$x(\ln 3 + \ln 2) = 2\ln 3 \qquad \text{\color{red}{Factor out x}}$$

$$x\ln 6 = \ln 9 \qquad \text{\color{red}{Apply log properties 1 and 3}}$$

$$x = \frac{\ln 9}{\ln 6} \approx 1.226 \qquad \text{\color{red}{Divide both sides by $\ln 6$}}$$

Figure 5.26 shows the graph of $f(x) = 3^{x-2} - 2^{-x}$ in the viewing rectangle of a graphing calculator. The x-intercept is approximately 1.226, which confirms our answer.   ◆

PROBLEM 2   Solve the equation $e^{x-1} = 3^x$.   ◆

FIGURE 5.26

The x-intercept of the graph of $f(x) = 3^{x-2} - 2^{-x}$ is the root of the equation $3^{x-2} = 2^{-x}$.

◆ Solving Logarithmic Equations

We can solve a logarithmic equation in logarithmic form by changing to exponential form. For example,

> logarithmic form

> exponential form

$$\log_3(x - 12) = 2 \qquad \text{is equivalent to} \qquad 3^2 = x - 12$$
$$x = 3^2 + 12$$
$$x = 21$$

Check: $\log_3[(21) - 12] = \log_3 9 = 2$

To solve many logarithmic equations that are not initially in logarithmic form, we use the following procedure:

◆ **Procedure for Solving Logarithmic Equations**

1. Isolate the logarithmic expressions on one side of the equation.

2. Apply the properties of logarithms, and write the equation in logarithmic form.

3. Change to exponential form, and solve for the unknown.

Note: This procedure may produce extraneous roots.

EXAMPLE 3   Solve each logarithmic equation.

(a) $2 - \log x = \log 3$ **(b)** $\log_4(x - 2) + 2 \log_4 x = 1 + \log_4 2x$

◆ **Solution**

(a) Using the given procedure, we have

$$2 - \log x = \log 3$$

$$2 = \log 3 + \log x \qquad \text{Isolate the logarithms on one side}$$

$$2 = \log 3x \qquad \text{Apply log property 1}$$

$$10^2 = 3x \qquad \text{Change to exponential form}$$

$$x = \frac{100}{3} \qquad \text{Solve for } x$$

Figure 5.27 shows the graph of $f(x) = 2 - \log x - \log 3$ in the viewing rectangle of a graphing calculator. The x-intercept is $\frac{100}{3}$, which confirms our answer.

(b) Using the given procedure, we have

$$\log_4(x - 2) + 2 \log_4 x = 1 + \log_4 2x$$

$$\log_4(x - 2) + 2 \log_4 x - \log_4 2x = 1 \qquad \text{Isolate the logarithmic expressions on one side}$$

$$\log_4 \frac{x^2(x - 2)}{2x} = 1 \qquad \text{Apply the log properties and write in logarithmic form}$$

$$\log_4 \frac{x(x - 2)}{2} = 1 \qquad \text{Reduce, assuming } x \neq 0$$

$$4^1 = \frac{x(x - 2)}{2} \qquad \text{Change to exponential form}$$

$$x^2 - 2x - 8 = 0 \qquad \text{Write in quadratic form and solve for } x$$

$$(x - 4)(x + 2) = 0$$

$$x = 4 \quad x = -2$$

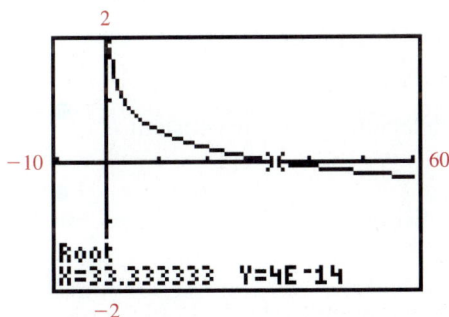

FIGURE 5.27

The x-intercept of the graph of $f(x) = 2 - \log x - \log 3$ is the root of the equation $2 - \log x = \log 3$.

Root
X=33.333333   Y=4E⁻¹⁴

FIGURE 5.28

The graph of $f(x) = \log_4 (x - 2) + 2 \log_4 x - 1 - \log_4 2x$ has only one *x*-intercept.

Figure 5.28 shows the graph of

$$f(x) = \log_4(x - 2) + 2 \log_4 x - 1 - \log_4 2x$$

in the viewing rectangle of a graphing calculator. Note that 4 is the only *x*-intercept. This indicates that the other value, -2, is an extraneous root. Observe that $f(-2)$ is undefined. Hence, $x = 4$ is the only solution.   ◆

PROBLEM 3   Solve the logarithmic equation $2 \ln x - \ln 9 = 4$.   ◆

Many literal equations and formulas that contain logarithmic expressions may be solved for one of the unknowns by using this procedure.

EXAMPLE 4   Solve the literal equation $\ln(y + a) - 2 \ln(x + b) = c$ for *y*. Assume that *a*, *b*, and *c* are constants.

◆ **Solution**   We solve for *y* as follows:

$$\ln(y + a) - 2 \ln(x + b) = c$$

$$\ln \frac{y + a}{(x + b)^2} = c \qquad \text{Apply the log properties and write in logarithmic form}$$

$$e^c = \frac{y + a}{(x + b)^2} \qquad \text{Change to exponential form}$$

$$y = e^c(x + b)^2 - a \qquad \text{Solve for y}$$

Since *c* is a constant, e^c is also a constant. Relabeling e^c as the constant *k* gives us

$$y = k(x + b)^2 - a \qquad \text{where } k = e^c.$$ ◆

PROBLEM 4   Solve the following literal equation for *y*:

$$x + \ln y = \ln c \quad \text{where } c \text{ is a constant}$$ ◆

◆ **Solving Factorable Equations**

We can solve other exponential and logarithmic equations by factoring and applying the zero product property (Section 2.2).

EXAMPLE 5   Solve each equation.

(a) $(\log x)^2 = \log x^2$ **(b)** $e^x + 3e^{-x} = 4$

◆◆ **Solution**

(a) We begin by rewriting the equation to obtain an equivalent equation that we can solve by the zero product property:

$$(\log x)^2 = \log x^2$$

$$(\log x)^2 = 2\log x \qquad \text{Apply log property 3}$$

$$(\log x)^2 - 2\log x = 0 \qquad \text{Subtract 2 log } x \text{ from both sides}$$

Now we can factor out log x from this last equation and solve for x by applying the zero product property:

$$(\log x)^2 - 2\log x = 0$$

$$\log x(\log x - 2) = 0 \qquad \text{Factor out log } x$$

$$\log x = 0 \quad \text{or} \quad \log x - 2 = 0 \qquad \text{Apply the zero product property}$$

$$x = 10^0 \qquad\qquad x = 10^2 \qquad \text{Change to exponential form}$$

$$x = 1 \qquad\qquad\quad x = 100 \qquad \text{Simplify}$$

The graph in Figure 5.29 confirms our answers.

(b) We begin by rewriting the equation to obtain an equivalent equation that we can solve by the zero product property:

$$e^x + 3e^{-x} = 4$$

$$e^{2x} + 3 = 4e^x \qquad \text{Multiply both sides by } e^x$$

$$e^{2x} - 4e^x + 3 = 0 \qquad \text{Subtract } 4e^x \text{ from both sides}$$

This last equation is of the quadratic form $u^2 - 4u + 3 = 0$ with $u = e^x$:

$$u^2 - 4u + 3 = 0$$

$$(u - 1)(u - 3) = 0 \qquad \text{Factor}$$

$$u = 1 \quad \text{or} \quad u = 3 \qquad \text{Solve for } u$$

$$e^x = 1 \qquad e^x = 3 \qquad \text{Replace } u \text{ with } e^x$$

$$x = \ln 1 \qquad x = \ln 3 \qquad \text{Take the natural logarithm of both sides}$$

$$x = 0 \qquad x \approx 1.099 \qquad \text{Evaluate}$$

Figure 5.30 shows the graph of $f(x) = e^x + 3e^{-x} - 4$ in the viewing rectangle of a graphing calculator. The x-intercepts are 0 and approximately 1.099; these values confirm our answer. ◆

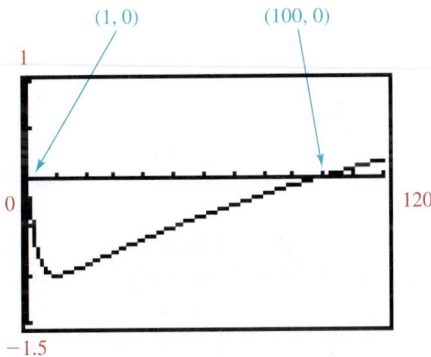

FIGURE 5.29

The graph of $f(x) = (\log x)^2 - \log x^2$ has two x-intercepts.

FIGURE 5.30

The x-intercepts of the graph of $f(x) = e^x + 3e^{-x} - 4$ are $x = 0$ and $x \approx 1.099$.

PROBLEM 5   Solve the equation $\log \sqrt{x} = \sqrt{\log x}$ by squaring both sides.   ◆

◆ **Graphical Methods of Solutions**

If the techniques that we have discussed in this section are insufficient to solve a particular logarithmic or exponential equation, we may use a calculator or computer with graphing capabilities to find the approximate solutions. For example, we cannot solve the equation $2^x + 3^x = 9$ by taking the natural logarithm of both sides. Hence, we use the graphical approach illustrated in the next example.

EXAMPLE 6   Use a graphing calculator to find the approximate solution(s) of the equation $2^x + 3^x = 9$.

◆ **Solution**   One method of solving this equation graphically is to generate the graphs of the equations

$$y = 2^x + 3^x \quad \text{and} \quad y = 9$$

in the same viewing rectangle of a graphing calculator and then determine the x-coordinate of the intersection point(s). As shown in Figure 5.31, the graphs intersect once, and the x-coordinate of the intersection point appears to be between $x = 1$ and $x = 2$.

   Most graphing calculators have a built-in program for finding this intersection point. If a calculator does not have this feature, we can use the TRACE key and ZOOM key to approximate the x-coordinate of the intersection point. By either method, we find that the x-coordinate of the intersection point is $x \approx 1.620$, to the nearest thousandth. We conclude that the solution to the equation $2^x + 3^x = 9$ is approximately 1.620.   ◆

FIGURE 5.31

The graphs of $y = 2^x + 3^x$ and $y = 9$ intersect in one point.

PROBLEM 6   Solve the exponential equation in Example 6 by generating the graph of $f(x) = 2^x + 3^x - 9$ in the viewing rectangle of a graphing calculator and noting the x-intercept. You should obtain the same answer as in Example 6.   ◆

◆ **Application: Exponential Growth and Decay**

Functions that represent an amount of a substance which changes by a fixed multiple over the same increment of time are of the form

$$A(t) = A_0 e^{kt}$$

where $A(t)$ is the amount present at time t, A_0 is the original amount present at time 0, and k is a constant related to the rate of growth or decay. If $k > 0$, then $A(t) = A_0 e^{kt}$ is called an **exponential growth function** and if $k < 0$, then $A(t) = A_0 e^{kt}$ is called an **exponential decay function**.

The compound interest formula

$$A = Pe^{rt}$$

is an example of an exponential growth function. In Section 5.1 we used this formula to find the amount A in a savings account when a certain principal P is invested at an interest rate r per year and interest is compounded continuously. Now that we have defined logarithms, we can also use this formula to find the time t or the interest rate r required for a given amount to reach a particular value.

EXAMPLE 7   An investor deposits $10,000 into a savings account paying 8% interest per year compounded continuously. Assuming the investor makes no subsequent deposit or withdrawal, find the time required for this investment to accumulate to $50,000.

◆ **Solution**   If $10,000 is deposited in a savings account at an interest rate of 8% per year and interest is compounded continuously, then the amount A in the account after t years is given by the formula

$$A = 10,000e^{0.08t}$$

We want to find the time t when $A = \$50,000$. Thus, we replace A with 50,000 and solve for t as follows:

$$50,000 = 10,000e^{0.08t}$$

$$5 = e^{0.08t} \qquad \text{Divide both sides by 10,000}$$

$$\ln 5 = 0.08t \qquad \text{Take the natural logarithm of both sides}$$

$$t = \frac{\ln 5}{0.08} \approx 20.1 \text{ years} \qquad \text{Divide by 0.08}$$

Thus, it takes approximately 20 years for the original investment to accumulate to $50,000. The graph in Figure 5.32 supports our work.   ◆

PROBLEM 7   At what interest rate, compounded continuously, must an amount of money be invested if the amount is to double in 5 years 6 months?   ◆

Under controlled laboratory conditions, a population P of living organisms increases exponentially as a function of time t, and the growth pattern is described by the **Malthusian model**

$$P(t) = P_0 e^{kt}$$

where P_0 is the initial population and k is a positive constant. The Malthusian model (named after the Englishman Thomas Malthus, 1766–1834) is also an exponential growth function.

70000

0   30

X=20.117974 Y=50000

FIGURE 5.32

The graph of $A = 10,000e^{0.08t}$ for $t \geq 0$ shows that $A = \$50,000$ when $t \approx 20.1$ years.

EXAMPLE 8   Suppose 18 bacteria per microliter are present initially in a culture, and 20 minutes later 270 bacteria per microliter are present.

(a) Determine the Malthusian model that describes this growth pattern.

(b) Determine the time t (to the nearest minute) when 360 bacteria per microliter are present in the culture.

◆ **Solution**

(a) Since 18 bacteria are present initially, we know that $P_0 = 18$. Thus, starting with the Malthusian model $P(t) = P_0 e^{kt}$, we can write

$$P(t) = 18e^{kt}$$

We must now determine the constant k. Using the fact that if $t = 20$, then $P(t) = 270$, we have

$$270 = 18e^{20k}$$
$$15 = e^{20k} \qquad \text{Divide both sides by 18}$$
$$\ln 15 = 20k \qquad \text{Take the natural logarithm of both sides}$$
$$k = \frac{\ln 15}{20} \approx 0.1354 \qquad \text{Solve for } k$$

Replacing k with 0.1354 gives us the Malthusian model that describes this growth pattern. Thus, we have

$$P(t) = 18e^{0.1354t}$$

(b) Using the exponential function developed in part (a), we replace $P(t)$ with 360 and solve for t:

$$360 = 18e^{0.1354t}$$
$$20 = e^{0.1354t} \qquad \text{Divide both sides by 18}$$
$$\ln 20 = 0.1354t \qquad \text{Take the natural logarithm of both sides}$$
$$t = \frac{\ln 20}{0.1354} \approx 22 \text{ minutes} \qquad \text{Solve for } t$$

The graph in Figure 5.33 supports our work.   ◆

PROBLEM 8   For the growth model described in Example 8, determine the time t (to the nearest minute) when 540 bacteria per microliter are present in the culture.   ◆

FIGURE 5.33

The graph of $P(t) = 18e^{0.1354t}$ for $t \geq 0$ shows that when 360 bacteria per microliter are present in the culture, $t \approx 22$ minutes.

Radioactive materials disintegrate exponentially over time, and their decay patterns can be described by exponential decay functions. To describe the decay of radioactive materials, the term *half-life* is often used. **Half-life** is defined as the time required for a given mass of a radioactive material to disintegrate to half its original mass.

EXAMPLE 9   Strontium-90, a waste product from nuclear reactors, has a half-life of 28 years. It is estimated that a certain quantity of this material will be safe to handle when its mass is $\frac{1}{1000}$ of its original amount. Determine the time required to store strontium-90 until it disintegrates to this level of safety.

◆ Solution   If the half-life of strontium-90 is 28 years, then half of the original amount is present after 28 years. Thus, starting with the exponential decay function $A(t) = A_0 e^{kt}$, we replace $A(t)$ with $\frac{1}{2}A_0$ and t with 28, and then solve for k as follows:

$$\frac{1}{2}A_0 = A_0 e^{k(28)}$$

$$\frac{1}{2} = e^{k(28)} \qquad \text{Divide both sides by } A_0$$

$$\ln \frac{1}{2} = 28k \qquad \text{Take the natural logarithm of both sides}$$

$$k = \frac{\ln \frac{1}{2}}{28} \approx -0.02476 \qquad \text{Solve for } k$$

Thus, we have

$$A(t) = A_0 e^{-0.02476t}$$

To determine the time t required to store strontium-90 until its mass reaches $\frac{1}{1000}$ of its original amount, we replace $A(t)$ with $0.001A_0$, and solve for t as follows:

$$0.001A_0 = A_0 e^{-0.02476t}$$

$$0.001 = e^{-0.02476t} \qquad \text{Divide both sides by } A_0$$

$$\ln 0.001 = -0.02476t \qquad \text{Take the natural logarithm of both sides}$$

$$t = \frac{\ln 0.001}{-0.02476} \approx 279 \text{ years} \qquad \text{Solve for } t \qquad ◆$$

PROBLEM 9   According to the decay function developed in Example 9, how long does it take 30 grams of strontium-90 to decay to 27 grams? ◆

Exercises 5.4

◆ Basic Skills

In Exercises 1–24, solve each exponential equation. Check graphically with a graphing calculator.

1. $9^x = 27$

2. $16^x = \frac{1}{8}$

3. $2^{x-1} = 16^x$

4. $5^{x+3} = 25^{x-2}$

5. $7^x = 35$

6. $3^x = 36$

7. $e^{3x} = \frac{2}{3}$

8. $10^{-2x} = 60$

9. $10^{3-2x} = 28$

10. $e^{2x-1} = 2$

11. $3^{2x-1} = 40$

12. $6^{x+5} = 75$

13. $4e^{-7x} = 15$

14. $12(10^{1-x}) = 60$

15. $3 - 2(10^{-x}) = -117$

16. $1 + 2e^x = 9$

17. $10^{2x-1} = 4^{-x}$

18. $5^{x-2} = 4^{2x+1}$

19. $e^{x/2} = 2^{x-1}$

20. $4^{-2x} = e^{x+1}$

21. $3 \cdot 2^{x-2} = 6^x$

22. $5 \cdot 3^{-x} = 4 \cdot 2^{x+2}$

23. $2^x - 6 \cdot 8^{3-x} = 0$

24. $\dfrac{10^{x-1}}{e^{2x}} - 4 = 0$

Solve each logarithmic equation in Exercises 25–44. Check graphically with a graphing calculator.

25. $\log x = -2$

26. $\ln x = 3$

27. $2 \log_4 x = 3$

28. $3 \log_{1/8} x = -4$

29. $\ln(x - e) = 1$

30. $-3 \log_8(x + 9) = 1$

31. $\log_2(2x^2 + 5x + 5) = 3$

32. $\ln(x^2 - 3x - 3) = 0$

33. $\log x + \log 5 = 1$

34. $\ln x = 1 + \ln 5$

35. $3 \ln(x + 1) - \ln 27 = 3$

36. $2 \log_6(2 - x) = 2 - \log_6 3$

37. $\log_4(x + 12) - \log_4(x - 3) = 2$

38. $\log_3 x + \log_3(x - 6) = 3$

39. $\log_5 x + \log_5(x + 2) = \log_5(x + 6)$

40. $\log 2 + \log(2x - 3) = \log 3x - \log 3$

41. $\log(x + 6) + 1 = 2 \log(3x - 2)$

42. $\log_2(x - 1) + 2 \log_2 x = 2 + \log_2 3x$

43. $\frac{1}{2} \ln(3 - 2x) - \ln x = 0$

44. $\frac{1}{2} \log(x + 3) + \log 2 = 1$

In Exercises 45–52, solve each literal equation for y. Assume a, b, and c are constants.

45. $\ln y + 2x = \ln c$

46. $2(x + \ln x) = \ln y - \ln c$

47. $2 \log(x + a) = 1 - \log y$

48. $\log y = \log(x + y) + 2$

49. $\ln(x + y) - \ln(x - y) = c$

50. $\ln(x + 2y) + 2 \ln x = \ln y + c$

51. $\dfrac{1}{a} \ln y - \dfrac{1}{a} \ln(a - by) = c$

52. $-\frac{1}{4} \ln (y + 2) + \frac{1}{4} \ln(y - 2) = x + c$

In Exercises 53–58, use factoring and the zero product property to help solve each equation. Check graphically with a graphing calculator.

53. $(\ln x)^2 = \ln x^2$

54. $2[\log(x + 1)]^2 - \log(x + 1)^3 = 5$

55. $e^{4x} - 3e^{2x} = 4$

56. $2^{3x} + 4 \cdot 2^{-3x} = 5$

57. $3 \cdot 10^x - 10^{-x} = 2$

58. $\frac{1}{2}(e^x - 9e^{-x}) = 4$

In Exercises 59–64, use a graphing calculator to find the approximate solution(s) of the equation. Round each answer to four significant digits.

59. $2x + \ln x = 5$

60. $e^{-x} - x^3 = 0$

61. $xe^x = 4$

62. $2^{x+1} + 5^x = 1$

63. $2^x = 3 - x^2$

64. $\ln(x + 1) = x^2 - 4x$

65. An investor deposits $1000 into a savings account paying 7% interest per year compounded continuously. Assuming the investor makes no subsequent deposit or withdrawal, find the time required (in years) for the investment to accumulate to the given amount:

    (a) $1500   (b) $6000   (c) $100,000

66. A certain amount of money is deposited into a savings account paying 10% interest per year compounded continuously. Assuming no subsequent deposit or withdrawal is made, find the time required for the amount to double.

67. When a child is born, his grandfather invests $10,000 for the child's college education. When the child is 18 years old, the balance in the account is $73,000. If interest is compounded continuously, find the interest rate (to the nearest percent) at which the investment is made. Assume no subsequent deposit or withdrawal is made and the interest rate does not change.

68. At what interest rate, compounded continuously, must money be invested if the amount is to quadruple in 18 years?

69. In 1982 foresters found 24 eagles living in a certain national park. In 1992 they counted 144 eagles living in this area.

    (a) Determine the Malthusian model that describes this growth pattern.

    (b) In what year is the eagle population expected to reach 300?

70. Approximately 10^4 bacteria per milliliter are present initially in a culture. Five hours later, approximately 10^7 bacteria per milliliter are present.

    (a) Determine the Malthusian model that describes this growth pattern.

    (b) Approximately how many minutes does it take for this population to double in size?

71. When a plant or animal dies, the amount of carbon-14 in the organism decreases exponentially according to the exponential decay function. Suppose that in 1986 an archaeologist discovers a human skull at an ancient burial site and determines that 60% of the original carbon-14 is still present. If the half-life of carbon-14 is 5600 years, approximate the ancient's year of death.

72. The radioactive substance iodine-131 decays from 30 grams to 25 grams in 50.5 hours.

(a) Determine the exponential decay function that describes this decay pattern.

(b) How many hours will it take for this radioactive substance to decay to 20 grams?

(c) Find the half-life (in days) of this radioactive substance.

73. Due to the presence of acid rain, the population P of fish in a pond in New Hampshire is decreasing according to the equation

$$\log_2 P = -\tfrac{1}{3}t + \log_2 P_0$$

where t is time in years after 1985 and P_0 is the original population in 1985.

(a) Solve the equation for P.

(b) In the year 1991, what percent of the original fish population remained?

74. A student infected with influenza returns to a college campus of 1000 students. If no student leaves the campus, then the time t (in days) it takes for N students to become infected with the virus is given by

$$t = \ln 999N - \ln(1000 - N).$$

(a) In how many days will 20 students be infected?

(b) Determine the number of infected students after 6 days.

◆ Critical Thinking

75. In Section 5.1 we developed the compound interest formula

$$A = P\left(1 + \frac{r}{n}\right)^{nt}.$$

Solve this literal equation for t and express the answer in terms of natural logarithms.

76. Explain the fallacy in the following argument:

$1 < 2$

$\tfrac{1}{4} < \tfrac{1}{2}$ **Divide both sides by 4**

$\ln \tfrac{1}{4} < \ln \tfrac{1}{2}$ **Take the natural logarithm of both sides**

$\ln \left(\tfrac{1}{2}\right)^2 < \ln \tfrac{1}{2}$ **Rewrite $\tfrac{1}{4}$**

$2 \ln \tfrac{1}{2} < \ln \tfrac{1}{2}$ **Apply log property 3**

$2 < 1$ **Divide both sides by $\ln \tfrac{1}{2}$**

77. If y varies directly as the mth power of x, then

$$y = kx^m$$

where k is the variation constant.

(a) Take the natural logarithm of both sides of this equation and show that

$$\ln y = m \ln x + \ln k.$$

(b) Construct the graph of $\ln y = m \ln x + \ln k$ in an XY-plane where $X = \ln x$ and $Y = \ln y$. Describe the graph.

78. Solve for x in terms of b.

(a) $\log_b(x - 3) = -1 + \log_b 5$

(b) $2 + \log_b(4x + 1) = \log_b(2 - x)$

Find the inverse of each function given in Exercises 79–86. Use a graphing calculator to verify each answer by showing that the graph of the function and the graph of the inverse are reflections of one another in the line $y = x$.

79. $h(x) = e^{x/2}$

80. $F(x) = 10^{2x/3}$

81. $G(x) = 1 - 10^{2x}$

82. $H(x) = 3e^{2x+1}$

83. $h(x) = -\ln \dfrac{x}{2}$

84. $F(x) = 2 \log \dfrac{3x}{2}$

85. $G(x) = 1 + 2 \log 3x$

86. $H(x) = 3 \ln(2x - 1)$

In Exercises 87–90, solve each equation for x.

87. $\dfrac{e^x + e^{-x}}{2} = 10$

88. $10^x + 10^{-x} = 6$

89. $2 \log_x 3 + \log_3 x = 3$

90. $2 \log_4 x - 3 \log_x 4 = 5$

91. When a cable or rope is suspended between two points at the same height and allowed to hang under its own weight, it forms a curve called a **catenary** (after the Latin word for chain). The power line shown in the figure is an example of a catenary, and its equation is given by

$$y = 25(e^{x/50} + e^{-x/50})$$

where x and y are measured in feet.

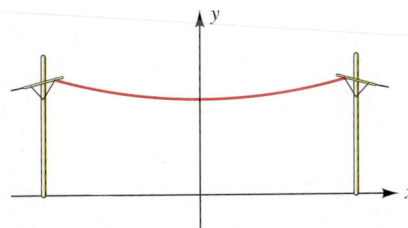

Exercise 91 continues

(a) Find the height of the power line at $x = 0$.

(b) Find x (to the nearest tenth of a foot) when $y = 75$ ft.

92. In a series circuit containing a capacitor C, a resistance R, and a battery source E, the instantaneous current i at any time t is given by

$$\ln i = -\frac{t}{RC} + \ln E - \ln R.$$

(a) Solve the equation for i.

(b) What happens to the current i as t increases without bound ($t \to +\infty$)?

93. In a particular electric circuit, the current i (in amperes) is a function of the time t (in seconds) after the switch is closed and is given by $i(t) = 20(1 - e^{-4.5t})$.

(a) Discuss the behavior of $i(t)$ as t increases without bound ($t \to +\infty$).

(b) Find the time (in milliseconds) at which $i(t) = 6$ amperes.

94. When limited resources such as food supply and space are taken into account, the population P of a living organism as a function of time t is more accurately described by the **logistic law** than by the Malthusian model. The logistic law is given by the function

$$P(t) = \frac{cP_0}{P_0 + (c - P_0)e^{-kt}}$$

where P_0 is the initial population and c and k are positive constants. Suppose the population of wolves in a certain forest follows the logistic law with $P_0 = 200$, $c = 1500$, $k = 0.2$, and time t in years.

(a) Discuss the behavior of $P(t)$ as t increases without bound ($t \to +\infty$).

(b) Find the time (in years) when $P(t) = 600$ wolves.

If a heated object of temperature T_0 is placed in a cooler medium that has a constant temperature of T_1, then the temperature T of the object at any time t is given by the formula

$$T = T_1 + (T_0 - T_1)e^{-kt}$$

where k is a positive constant. This formula is **Newton's law of cooling***, named after Sir Isaac Newton (1642–1727). Use this formula to answer Exercises 95 and 96.*

95. A pizza baked at 400°F is removed from the oven and placed in a room with a constant temperature of 70°F. The pizza cools to 300°F in two minutes.

(a) State the formula that describes this cooling process.

(b) Find the temperature of the pizza after it cools for three minutes.

(c) How soon after it is removed from the oven does the pizza cool to 150°F?

96. A bottle of white zinfandel wine is stored in a room with a temperature of 70°F. The distributor recommends that the wine is best when served slightly chilled to 40°F. At 4:00 P.M. the bottle is placed in a refrigerator having a constant temperature of 35°F, and after 30 minutes the temperature of the wine is 50°F. At what time should the wine be removed from the refrigerator if it is to be served when chilled to 40°F?

Chapter 5 Review

◆ **Questions for Writing or Group Discussion**

1. If $b > 0$ but $b \neq 1$, how is b^r defined when r is a rational number? an irrational number?

2. Under what conditions is the *exponential function $f(x) = b^x$ an increasing function? a decreasing function?*

3. How is the graph of $f(x) = b^x$ related to the graph of $g(x) = b^{-x}$? Discuss the behavior of $f(x)$ and $g(x)$ as $x \to +\infty$.

4. Are the functions $f(x) = e^{x+1}$ and $g(x) = e \cdot e^x$ identical? Explain.

5. Why is an interest rate that is *compounded continuously* better than one that is compounded a fixed number of times per year?

6. What is the meaning of the *base b logarithm* $\log_b x$? How can a calculator be used to evaluate $\log_b x$? When is $\log_b x$ undefined?

7. Complete each *logarithmic identity*.
   (a) $\log_b 1 =$ ____
   (b) $\log_b b =$ ____
   (c) $\log_b b^u =$ ____
   (d) $b^{\log_b u} =$ ____

8. What is the base for *common logarithms*? for *natural logarithms*?

9. State the change of base formula for logarithms. How are the expressions $\log_b a$ and $\log_a b$ related?

10. List some characteristics of the graph of the *logarithmic function* $f(x) = \log_b x$.

11. How can we rewrite the logarithm of a product? of a quotient? of a quantity raised to a power?

12. Are the functions $f(x) = \ln(x+1)^2$ and $g(x) = 2\ln(x+1)$ identical? Explain.

13. How is the graph of $f(x) = \log_b x$ related to the graph of $g(x) = \log_b(1/x)$?

14. What is the domain and range of the *natural logarithmic function* $f(x) = \ln x$? What is the inverse function of $f(x) = \ln x$?

15. What *logarithmic scale* is used to describe the magnitude of an earthquake? the loudness of a sound?

16. Explain the general procedure for solving an *exponential equation*. Illustrate with an example.

17. Explain the general procedure for solving a *logarithmic equation*. Illustrate with an example.

18. Give an example of an *exponential growth function* and an *exponential decay function*.

19. Why is 1 excluded as a base for exponential functions?

20. What is the value of $(1 + 1/n)^n$ as n increases without bound?

◆ **Review Exercises**

In Exercises 1–6, use the properties of real exponents to write each expression as a constant or as an expression in the form b^x or b^{-x} for $b > 1$.

1. 1^{3x-2}
2. $(3^{x+\pi})^0$
3. $e^{2x} \cdot e^{1-2x}$
4. $(2^{-3x})^2$
5. $\dfrac{4^{x+3}}{8^{x+2}}$
6. $(3^{-x} \cdot 9^x)^2$

In Exercises 7–24, find the value of each expression.

7. $\log_7 49$
8. $\log_2 32$
9. $\log \sqrt[3]{100}$
10. $\ln e^3$
11. $\log_{1/2} 8$
12. $\log_{1/3} 9$
13. $\log_8 2$
14. $\log_{25} \frac{1}{5}$
15. $\log_8 \frac{1}{4}$
16. $\log_{16} 64$
17. $\log_3 45$
18. $\log_6 30$
19. $\log_{1/2} 100$
20. $\log_{2/3} \frac{1}{2}$
21. $e^{\ln 4}$
22. $10^{\log 8}$
23. $3^{\log_7 7}$
24. $8^{\log_2 1}$

In Exercises 25–34, simplify each expression. Assume that each variable is restricted only to real numbers that allow the logarithmic expression to be defined.

25. $\log_7 7^{2x}$
26. $5^{\log_5(2-x)}$
27. $e^{\ln(x^2+1)}$
28. $\ln e^{2-3x}$
29. $\log_3 27^{1-x}$
30. $\log_4 2^{4x-4}$
31. $e^{2x+\ln x}$
32. $e^{-\ln(x-1)}$
33. $(\log_3 x)(\log_x 3)$
34. $\dfrac{1}{\log_x(x/2)} - \dfrac{1}{\log_2(x/2)}$

In Exercises 35–40, expand each expression to eliminate logarithms of products, quotients, and powers. Assume that each variable is restricted only to real numbers for which the properties of logarithms can be applied.

35. $\log_3[9x^2(x-1)]$

36. $\log_4\left[4x\sqrt{x+2}\,\right]$

37. $\log \dfrac{1}{x\sqrt{2x-3}}$

38. $\log \dfrac{1000}{xy^{1/3}}$

39. $\ln \dfrac{xe^{-x^2}}{e^x - 1}$

40. $\ln \dfrac{4e^{2x}}{\sqrt{e^x + 2}}$

In Exercises 41–46, condense each expression to a single logarithm with a coefficient of 1. Assume that each variable is restricted only to real numbers for which the properties of logarithms can be applied.

41. $\ln 12 - \ln 3 + \ln 2$

42. $-2\log_2 6 + 3\log_2 3 - \frac{1}{2}\log_2 9$

43. $3\log_3(x+1) - \log_3(x^2 - 1) + \log_3(x-1)$

44. $2\ln x + \frac{1}{2}\ln(x+1) + \ln 2$

45. $\log \dfrac{1}{\sqrt{x} - \sqrt{x-1}} - \log\left(\sqrt{x} + \sqrt{x-1}\right)$

46. $\log\left(\sqrt{x+1} + 1\right) + \log\left(\sqrt{x+1} - 1\right)$

In Exercises 47–58, sketch the graph of each function. Label any x- and y-intercept(s) and horizontal or vertical asymptote.

47. $f(x) = 5^x$

48. $g(x) = 6^{-x}$

49. $h(x) = e^{x-2}$

50. $F(x) = 2 + e^x$

51. $G(x) = -(2 + e^{x+1})$

52. $H(x) = 3^x + 3^{-x}$

53. $h(x) = \log_{1/5} x$

54. $F(x) = \log_6(-x)$

55. $G(x) = -(2 + \log_4 x)$

56. $H(x) = \log_4(x+2)$

57. $f(x) = \ln \dfrac{1}{x+1}$

58. $g(x) = \ln\sqrt{1-x}$

In Exercises 59–80, solve each equation.

59. $10^{2x} = 80$

60. $e^{2x-1} = 9$

61. $16^x = \frac{1}{8}$

62. $3^{x+3} = 27^{1-x}$

63. $3^{2x} = 4^{1-x}$

64. $e^{2x} = 2^{x+1}$

65. $2e^{-x} = 3^{2x-1}$

66. $5 \cdot 2^{-x} = 3 \cdot 5^{x+2}$

67. $\log_5 x = 1$

68. $\ln x = 10$

69. $\ln|x+2| - 3 = 0$

70. $\log_3(x^2 - 3x + 5) = 2$

71. $\log_3(x-7) - \log_3(x+1) = 2$

72. $\log 3 - \log(2-x) = \log 2 - \log 2x$

73. $\frac{1}{2}\log(12-x) - \log x = 0$

74. $\log_2(3x-2) + 2 = 2\log_2(x+2)$

75. $\ln y = c + 2\ln|x|$, where c is a constant; solve for y

76. $3(x - \ln x) = \ln(2y - 3) + \ln c$, where c is a constant; solve for y

77. $x^2 e^{-x} - e^{-x} = 0$

78. $x\ln x - 3\ln x = 0$

79. $\log_x 3 + 2\log_3 x = 3$

80. $e^x + 1 = 2e^{-x}$

In Exercises 81–86, find the inverse of each function.

81. $f(x) = 8^x$

82. $g(x) = e^{-2x}$

83. $h(x) = 2e^{3x-2}$

84. $F(x) = 1 - 8^{1-x}$

85. $G(x) = \log_{1/2}(2x)$

86. $H(x) = -\ln(x-1)$

87. Suppose that \$5000 is invested at an interest rate of 9% per year. Assuming the investor makes no subsequent deposit or withdrawal, determine the amount in the account after 10 years if the interest is compounded (a) semiannually, (b) monthly, or (c) continuously.

88. Suppose that \$100,000 is invested at an interest rate of $7\frac{1}{2}$% per year. Assuming the investor makes no subsequent deposit or withdrawal, determine the amount in the account after 9 months if the interest is compounded (a) quarterly, (b) daily, or (c) continuously.

89. An investor deposits \$500 into a savings account paying $6\frac{3}{4}$% interest per year compounded continuously. Assuming the investor makes no subsequent deposit or withdrawal, find the time it takes for the investment to accumulate to \$800.

90. At what interest rate compounded continuously must money be invested if the amount is to triple in 12 years?

91. Initially, approximately 10^3 bacteria per milliliter are present in a culture. Three hours later approximately 10^5 bacteria per milliliter are present.

(a) Determine the Malthusian model that describes this growth pattern.

(b) Approximately how many minutes does it take for this population to double in size?

92. A radioactive substance disintegrates from 60 grams to 50 grams in four years.

(a) Determine the exponential decay function that describes this disintegration.

(b) How many years will it take for this radioactive substance to decay to 10 grams?

(c) Find the half-life of this radioactive substance.

93. One of the strongest earthquakes ever recorded occurred in Japan in 1993 and measured 8.9 on the Richter scale. How many times more intense was this earthquake than one that measures 6.0 on the Richter scale?

94. Students taking their last French course are given a final exam, and the average score of the group is recorded. Each month thereafter the students are given an equivalent exam, and for each exam their average score is recorded. It is found that the average score S is a function of the time t (in months) and is given by

$$S(t) = 77 - 17 \ln(t + 1) \quad \text{for } 0 \le t \le 12.$$

   (a) What is the average score on the original final exam?

   (b) What is the average score on an exam given three months later?

   (c) How much time has elapsed when the average score is 40?

95. When a hot metal object with a temperature of 400°F is placed in a room whose temperature is 70°F, the object cools at a rate such that its temperature T is a function of the time t (in minutes) that the object has been in the room and is given by Newton's cooling law

$$T(t) = 70 + (400 - 70)e^{-kt}$$

where k is a constant.

   (a) Determine the value of the constant k if the temperature of the object after 10 minutes is 290°F.

   (b) Find $T(60)$, the temperature of the object after 1 hour.

   (c) How much time has elapsed when $T(t) = 150°$F?

96. In a certain forest, the population P of foxes after t years is described by the logistic law

$$P(t) = \frac{25000}{100 + 150e^{-0.1t}}.$$

   (a) Find $P(10)$, the number of foxes in the forest after 10 years.

   (b) Discuss the behavior of $P(t)$ as t increases without bound $(t \to +\infty)$.

   (c) Find the time in years when $P(t) = 210$ foxes.

The parabolic reflector of a searchlight has a diameter of 3 feet at the open end and a depth of 1 foot. How far from the vertex of the parabola should the filament of the bulb be located so that all the light rays emanating from the bulb are parallel to the axis of the parabola?

For the solution, see Example 5 in Section 6.2.

Conic Sections

6.1   The Circle

◆ Introductory Comments

Early Greek mathematicians noted that when a double cone is sliced with a plane, a special family of curves is formed. As illustrated in Figure 6.1, this family of curves has four main members: *circle, parabola, ellipse,* and *hyperbola.* Collectively, this family of curves is called the **conic sections**.

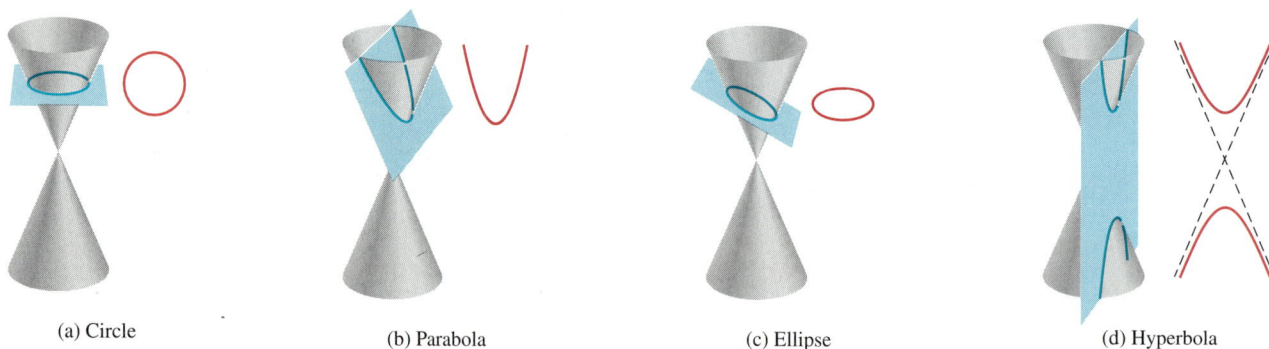

(a) Circle   (b) Parabola   (c) Ellipse   (d) Hyperbola

FIGURE 6.1
The conic sections

When the plane intersects the double cone, it is also possible to obtain a single point, one line, or a pair of lines. Do you see how? These special cases are called the **degenerate conic sections**.

An equation of the form

$$Ax^2 + Bxy + Cy^2 + Dx + Ey + F = 0$$

in which the coefficients A, B, and C are not all zero, is called a **general quadratic equation in two unknowns**. As we will see, if the graph of this type of equation exists, then it is either a conic section or a degenerate conic section.

In this chapter we study the geometric properties of each conic section and the distinguishing characteristics of their equations. We limit our discussion to general quadratic equations in two unknowns that contain no xy-term. By letting $B = 0$, we keep at least one of the coordinate axes parallel to an axis of symmetry of the conic section. If $B \neq 0$, then neither coordinate axis is parallel to an axis of symmetry of the conic section. For example, the graph of the general quadratic equation $xy = 1$ [or $y = 1/x$, the reciprocal function in Figure 2.42(f)] is a hyperbola whose axis of symmetry $y = x$ makes an angle of 45° with the coordinate axes.

◆ Determining the Equation of a Circle

We begin our discussion with the circle. Geometrically, we define a circle as follows.

◆ **Geometric Definition of a Circle**

A **circle** is the set of all points in a plane that lie a fixed distance from a given point. We call the fixed distance the **radius** and the given point the **center** of the circle.

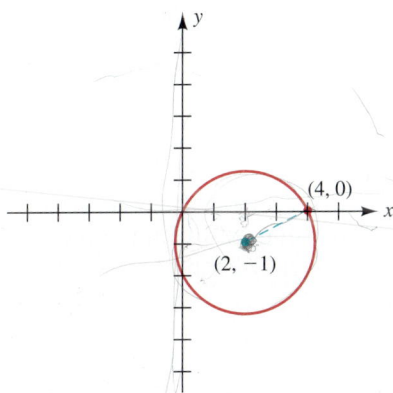

FIGURE 6.2

A circle with center (h, k) and radius r

Shown in Figure 6.2 is a circle with radius r and center (h, k). The point (x, y) lies on this circle if and only if the distance from the center (h, k) to (x, y) is r. By using the distance formula (Section 2.1), we have

$$r = \sqrt{(x - h)^2 + (y - k)^2}.$$

If we square both sides of this equation and then interchange the sides, we obtain

$$(x - h)^2 + (y - k)^2 = r^2.$$

We refer to this last equation as the *algebraic definition of a circle,* or the *equation of a circle in standard form.*

◆ **Equation of a Circle in Standard Form**

The **equation of a circle in standard form** with center (h, k) and radius r is

$$(x - h)^2 + (y - k)^2 = r^2.$$

We can determine the equation of a circle if we know its center (h, k) and the value of its radius r.

EXAMPLE 1 Determine the equation of a circle that has center $(2, -1)$ and passes through the point $(4, 0)$.

◆ **Solution** The circle with the given characteristics is shown in Figure 6.3. Since the center is $(2, -1)$, the equation of this circle must have the form

$$(x - 2)^2 + (y + 1)^2 = r^2$$

Now, the radius r of this circle is the distance from the center $(2, -1)$ to the point $(4, 0)$. By the distance formula, we have

$$r = \sqrt{(4 - 2)^2 + [0 - (-1)]^2} = \sqrt{5}$$

Hence, $r^2 = (\sqrt{5})^2 = 5$. Therefore, the equation of a circle that has center $(2, -1)$ and passes through the point $(4, 0)$ is

$$(x - 2)^2 + (y + 1)^2 = 5.$$

FIGURE 6.3

A circle that has center $(2, -1)$ and passes through the point $(4, 0)$

To generate the graph of $(x - h)^2 + (y - k)^2 = r^2$ in the viewing rectangle of a graphing calculator, we solve the equation for y:

$$(x - h)^2 + (y - k)^2 = r^2$$

$$(y - k)^2 = r^2 - (x - h)^2 \qquad \text{Subtract } (x - h)^2 \text{ from both sides}$$

$$y - k = \pm\sqrt{r^2 - (x - h)^2} \qquad \text{Take the square root of both sides}$$

$$y = k \pm \sqrt{r^2 - (x - h)^2} \qquad \text{Add } k \text{ to both sides}$$

We then graph both equations:

$$y = k + \sqrt{r^2 - (x - h)^2} \qquad \text{and} \qquad y = k - \sqrt{r^2 - (x - h)^2}$$

To obtain a true geometric perspective of the graph, we use a square set viewing rectangle.

PROBLEM 1   Generate the graph of the circle $(x - 2)^2 + (y + 1)^2 = 5$ in the viewing rectangle of a graphing calculator. Verify that this circle passes through the point (4, 0) and that the coordinates of its center are (2, −1). ◆

A **tangent** to a circle is a line that intersects the circle in one and only one point. Referring to Figure 6.4, we say that the line and the circle are *tangent* at point *P*. Point *P* is called the **point of tangency** of the line and circle. *Every tangent to a circle is perpendicular to a radius drawn to the point of tangency.*

EXAMPLE 2   Determine the equation of a circle that has center (−2, 1) and is tangent to the line $x + 2y = 5$.

◆ **Solution**   The circle with the given characteristics is shown in Figure 6.5. The slope of the line $x + 2y = 5$ (or $y = -\frac{1}{2}x + \frac{5}{2}$) is $-\frac{1}{2}$. Since perpendicular lines have negative reciprocal slopes, the slope of the radius drawn to the point of tangency is 2. We can find the equation of the radius drawn to the point of tangency by using the point-slope form of the equation of a straight line (Section 3.2):

$$y - y_1 = m(x - x_1)$$
$$y - 1 = 2(x + 2)$$
$$y = 2x + 5$$

To find the coordinates of the point of tangency *P*, we solve the equations $x + 2y = 5$ and $y = 2x + 5$ simultaneously by using substitution (Section 3.5):

$$x + 2y = 5$$

$$x + 2(2x + 5) = 5 \qquad \text{Substitute } y = 2x + 5$$

$$5x + 10 = 5 \qquad \text{Solve for } x$$

$$x = -1$$

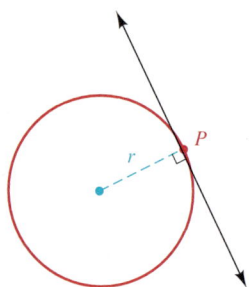

FIGURE 6.4

The line and the circle are *tangent* at point *P*.

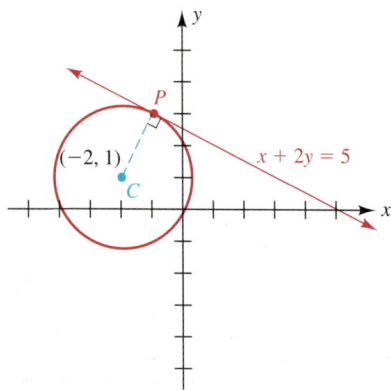

FIGURE 6.5

A circle that has center (−2, 1) and is tangent to the line $x + 2y = 5$

Substituting $x = -1$ into the equation $y = 2x + 5$ gives us $y = 3$. Hence, the coordinates of the point of tangency P are $(-1, 3)$. Now, by the distance formula, the radius of the circle is

$$CP = \sqrt{(3 - 1)^2 + [-1 - (-2)]^2} = \sqrt{5}$$

Therefore, the equation of a circle that has center $(-2, 1)$ and is tangent to the line $x + 2y = 5$ is

$$(x + 2)^2 + (y - 1)^2 = 5.$$

Figure 6.6 shows the graphs of $(x + 2)^2 + (y - 1)^2 = 5$ and $x + 2y = 5$ in the viewing rectangle of a graphing calculator. This display confirms our work. ◆

PROBLEM 2 Determine the x- and y-intercepts for the equation of the circle $(x + 2)^2 + (y - 1)^2 = 5$ obtained in Example 2. ◆

FIGURE 6.6

The circle $(x + 2)^2 + (y - 1)^2 = 5$ is tangent to the line $x + 2y = 5$ at $(-1, 3)$.

◆ **General Form of a Circle**

Squaring the expressions $(x - h)$ and $(y - k)$ in the equation

$$(x - h)^2 + (y - k)^2 = r^2$$

gives us

$$x^2 - 2hx + h^2 + y^2 - 2ky + k^2 = r^2$$
$$x^2 + y^2 - 2hx - 2ky + (h^2 + k^2 - r^2) = 0.$$

Replacing the constants $-2h$ with D, $-2k$ with E, and $(h^2 + k^2 - r^2)$ with F, we have

$$x^2 + y^2 + Dx + Ey + F = 0$$

This general quadratic equation in two unknowns is called the **equation of a circle in general form** and is characterized by the presence of x^2 and y^2 terms, each having a coefficient of 1.

If the equation of a circle is given in general form, then we can find the radius of the circle and the coordinates of its center by converting the equation to standard form and observing the values of h, k, and r. For example, the equation

$$x^2 + y^2 = 25$$

is of the form $x^2 + y^2 + Dx + Ey + F = 0$ with $D = E = 0$ and $F = -25$. By thinking of this equation as

$$(x - 0)^2 + (y - 0)^2 = 5^2$$

standard form of a circle

we know that its graph is a circle with center (0, 0) and radius 5. Figure 6.7 shows the graph of $x^2 + y^2 = 25$ $\left(\text{or } y = \pm\sqrt{25 - x^2}\right)$ in the viewing rectangle of a graphing calculator. The graph confirms the center and radius.

In general, the graph of any equation of the form

$$x^2 + y^2 = r^2$$

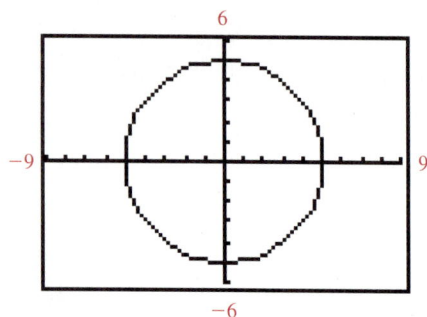

FIGURE 6.7
The graph of $x^2 + y^2 = 25$ is a circle with center (0, 0) and radius 5.

with $r > 0$ is a circle with center at the origin and radius r.

To find the center and radius of a circle defined by the equation

$$x^2 + y^2 + Dx + Ey + F = 0 \qquad \text{with } D \neq 0 \text{ or } E \neq 0,$$

we use the algebraic process of completing the square (Section 3.3) to write the equation in standard form. We can use a graphing calculator to check our work by solving the equation $x^2 + y^2 + Dx + Ey + F = 0$ for y and graphing *both* equations that result. To generate the graph of this equation when $E \neq 0$, we rewrite the equation as a quadratic equation in y:

$$y^2 + Ey + (x^2 + Dx + F) = 0$$

We apply the quadratic formula to this equation:

$$y = \frac{-E \pm \sqrt{E^2 - 4(x^2 + Dx + F)}}{2}$$

EXAMPLE 3   Find the radius and the coordinates of the center of the circle given by each equation.

(a) $x^2 + y^2 + 6x = 0$ **(b)** $4x^2 + 4y^2 - 8x + 20y + 13 = 0$

◆ **Solution**

(a) The equation $x^2 + y^2 + 6x = 0$ appears to be the equation of a circle in general form with $D = 6$, $E = 0$, and $F = 0$. Using the process of completing the square, we have

$(x^2 + 6x \quad) + y^2 = 0$ **Regroup**

$(x^2 + 6x + 9) + y^2 = 9$ **Complete the square by adding 9 to both sides**

$(x + 3)^2 + y^2 = 9$ **Factor the perfect square trinomial**

$[x - (-3)]^2 + (y - 0)^2 = 3^2$ **Write in standard form**

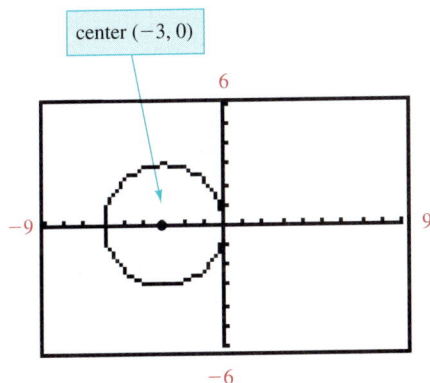

Thus, the center of the circle is $(-3, 0)$ and the radius is 3. Figure 6.8 shows the graph of $x^2 + y^2 + 6x = 0$ $\left(\text{or } y = \pm\sqrt{-x^2 - 6x}\right)$ in the viewing rectangle of a graphing calculator. This graph confirms our work.

FIGURE 6.8
The graph of $x^2 + y^2 + 6x = 0$ is a circle with center $(-3, 0)$ and radius 3.

(b) Dividing both sides of the equation by 4, we obtain

$$x^2 + y^2 - 2x + 5y + \tfrac{13}{4} = 0.$$

This equation appears to be the equation of a circle in general form with $D = -2$, $E = 5$, and $F = \tfrac{13}{4}$. Using the process of completing the square, we have

$$(x^2 - 2x\quad) + (y^2 + 5y +\quad) = -\tfrac{13}{4} \qquad \text{Regroup}$$

Add 1 to both sides.

$$(x^2 - 2x + 1) + (y^2 + 5y + \tfrac{25}{4}) = -\tfrac{13}{4} + (1 + \tfrac{25}{4}) \qquad \text{Complete the squares}$$

Add 25/4 to both sides.

$$(x - 1)^2 + (y + \tfrac{5}{2})^2 = 4 \qquad \text{Factor}$$

$$(x - 1)^2 + [y - (-\tfrac{5}{2})]^2 = 2^2 \qquad \text{Write in standard form}$$

Thus, the center of this circle is $(1, -\tfrac{5}{2})$ and the radius is 2.

To generate the graph of the equation $4x^2 + 4y^2 - 8x + 20y + 13 = 0$ in the viewing rectangle of a graphing calculator, we rewrite the equation as a quadratic equation in y,

$$4y^2 + 20y + (4x^2 - 8x + 13) = 0,$$

and apply the quadratic formula:

$$y = \frac{-(20) \pm \sqrt{(20)^2 - 4(4)(4x^2 - 8x + 13)}}{2(4)} = -\tfrac{5}{2} \pm \sqrt{3 + 2x - x^2}$$

Entering both of these equations into the calculator, we graph the circle shown in Figure 6.9. This graph confirms our work. ◆

FIGURE 6.9
The graph of $4x^2 + 4y^2 - 8x + 20y + 13 = 0$ is a circle with center $(1, -\tfrac{5}{2})$ and radius 2.

CAUTION Not every equation of the form $x^2 + y^2 + Dx + Ey + F = 0$ is the equation of a circle. For example, if the equation is written in the standard form

$$(x - h)^2 + (y - k)^2 = N \qquad \text{and} \qquad N = 0,$$

then the graph of this equation is the single point (h, k). We refer to this single point as a *degenerate circle*. Also, if

$$(x - h)^2 + (y - k)^2 = N \qquad \text{and} \qquad N < 0,$$

then this equation has no graph, since no point (x, y) with real coordinates satisfies this equation.

PROBLEM 3   Is $x^2 + y^2 + 4x + 6y + 13 = 0$ the equation of a circle? Explain.   ◆

◆ Application: Layout of a Circular Curve

Two straight roadways are often connected by a circular curve that has a *central angle* θ (read "theta") and radius r, as shown in Figure 6.10. The point where roadway A is perpendicular to the radius of the circular curve is called the *point of curvature* (PC), and the point where roadway B is perpendicular to the radius of the circular curve is called the *point of tangency* (PT).

A surveyor lays out a circular curve by placing stakes at various points along the curve between PC and PT. We can use the Pythagorean theorem and the ideas discussed in this section to lay out a circular curve in which $\theta = 120°$, $90°$, or $60°$. The methods of trigonometry are needed to lay out circular curves with other central angles.

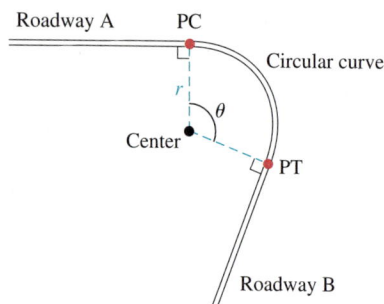

FIGURE 6.10

Two straight roadways connected by a circular curve

EXAMPLE 4   Two straight roadways are connected by a circular curve that has central angle $\theta = 90°$ and radius $r = 80$ ft, as shown in the Figure 6.11. A surveyor starts to lay out this circular curve by placing stakes at PC and PT and also at several other points along the curve, such as point A. Take the coordinate axes as shown in the figure.

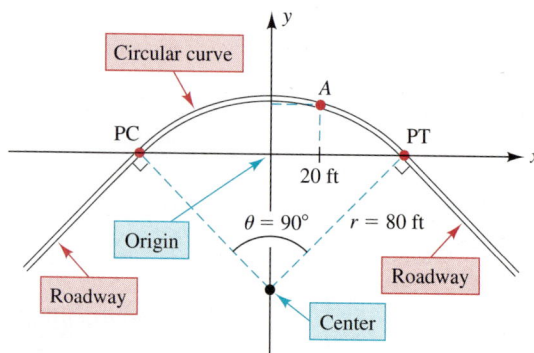

FIGURE 6.11

(a) Find the coordinates of PC and PT and the equation of the circular curve.

(b) Determine the coordinates of point A.

◆ Solution

(a) If $\theta = 90°$, then the triangle with vertices at the center, PT, and PC is an isosceles right triangle (45°–45°–90°). This implies that the triangle with vertices at the origin, center, and PT (or PC) is also an isosceles right triangle. Hence, the distance from the origin to the center is equal to the distance from the origin to PT. If we call this common distance a, then by the Pythagorean theorem we have

$$a^2 + a^2 = 80^2$$
$$2a^2 = 6400$$
$$a^2 = 3200$$
$$a = \sqrt{3200} = 40\sqrt{2} \approx 56.57$$

Hence, the coordinates of the center of the circular curve are $(0, -56.57)$, the coordinates of PT are $(56.57, 0)$ and, by symmetry, the coordinates of PC are $(-56.57, 0)$, with each coordinate measured to the nearest hundredth of a foot. The equation of the circular curve is given by

$$(x - h)^2 + (y - k)^2 = r^2$$
$$x^2 + (y + 56.57)^2 = 80^2, \text{ for } -56.57 < x < 56.57$$

(b) The x-coordinate of point A on the circular curve is 20 ft. We find the y-coordinate of this point by substitution:

$$x^2 + (y + 56.57)^2 = 80^2$$
$$20^2 + (y + 56.57)^2 = 80^2$$
$$(y + 56.57)^2 = 80^2 - 20^2$$
$$y + 56.57 = \sqrt{80^2 - 20^2}$$
$$y = \sqrt{80^2 - 20^2} - 56.57 \approx 20.89$$

So the coordinates of A are $(20, 20.89)$.

Figure 6.12 shows the graph of $x^2 + (y + 56.57)^2 = 80^2$ $\left(\text{or } y = -56.57 \pm \sqrt{80^2 - x^2}\right)$ in the viewing rectangle of a graphing calculator between PC and PT. Note that $(20, 20.89)$ is a point on this graph, which supports our work. ◆

FIGURE 6.12
Graph of $x^2 + (y + 56.57)^2 = 80^2$ between PC and PT

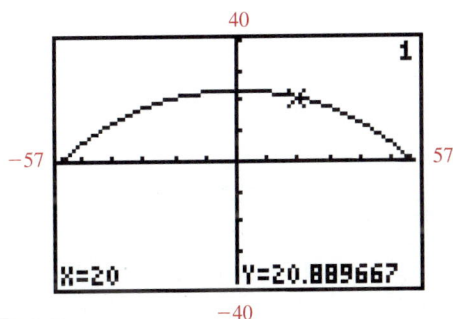

$\boxed{\text{PROBLEM 4}}$ Referring to Example 4, show that the distance from the center of the circular curve to point A is 80 ft, the radius of the circle. ◆

Exercises 6.1

◆ **Basic Skills**

In Exercises 1–16, determine the radius and the coordinates of the center of the circle given by each equation of a circle. Check graphically with a graphing calculator.

1. $x^2 + (y - 2)^2 = 16$

2. $(x + 5)^2 + (y + 2)^2 = 4$

3. $x^2 + y^2 = 9$ **4.** $x^2 + y^2 - 36 = 0$

5. $x^2 + y^2 - 2y = 3$ **6.** $x^2 + y^2 + 5x = 0$

7. $x^2 + y^2 - 6x + 8y + 9 = 0$

8. $x^2 + y^2 - x - 4y = 2$

9. $3x^2 + 3y^2 + 8x = 0$ **10.** $5x^2 + 5y^2 - 10y = 1$

11. $4x^2 + 4y^2 - 8x + 12y = 3$

12. $2x^2 + 2y^2 - x + 20y + 40 = 0$

13. $x^2 + y^2 + 10x - 2y + 26 = 0$

14. $4x^2 + 4y^2 + 20x - 16y + 41 = 0$

15. $3x^2 + 3y^2 - 4x - 8y + 24 = 0$

16. $x^2 + y^2 - 12x + 45 = 0$

Find the x- and y-intercepts of the circle whose equation is given in the indicated exercise.

17. Exercise 1   **18.** Exercise 2

19. Exercise 7   **20.** Exercise 8

In Exercises 21–32, determine the equation of a circle that has the given characteristics. Record each equation in general form.

21. Center at $(0, 0)$; radius 2

22. Center at $(-3, 0)$; radius 3

23. Center at $(2, -3)$; radius $\sqrt{2}$

24. Center at $(-1, 2)$; radius $\frac{2}{3}$

25. Center at the origin; passes through the point $(-3, 4)$

26. Center at $(-1, 3)$; passes through the point $(1, -1)$

27. Line segment from (2, 3) to (−2, 5) is the diameter

28. Line segment from (−1, −1) to (3, 1) is the radius
[*Hint:* Two answers are possible.]

29. Center at (−2, 5); tangent to the *y*-axis

30. Center at (3, −1); tangent to the line $x - 2y = -5$

31. Center at (4, 2); tangent to the line $4x + 3y = -3$

32. Tangent to both the *x*- and *y*-axes; diameter of 10
[*Hint:* Four answers are possible.]

In Exercises 33–36, use a graphing calculator to generate the graph of each circle. Then estimate its center and radius, rounding each answer to three significant digits. Check algebraically.

33. $32x^2 + 32y^2 + 343x - 876 = 0$

34. $x^2 + y^2 + 526x - 365y = 0$

35. $x^2 + y^2 - 7.22x - 4.84y + 12.34 = 0$

36. $4.1x^2 + 4.1y^2 - 24.1x - 16.3y + 25.2 = 0$

Two straight roadways are connected by a circular curve that has central angle θ and radius r, as shown in the sketch. A surveyor wants to lay out this circular curve by placing stakes at PC and PT and also at several other points along the curve, such as point A. Use this information for Exercises 37–40:

(*a*) *Find the coordinates of PC and PT and the equation of the circular curve for the indicated values of θ and r.*

(*b*) *Determine the coordinates of point A for the indicated value of x.*

Round all coordinates to the nearest hundredth of a foot.

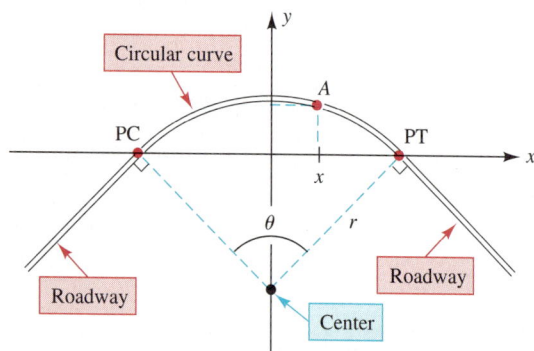

37. $\theta = 90°,\quad r = 125$ ft,$\quad x = 50$ ft

38. $\theta = 90°,\quad r = 75$ ft,$\quad x = 10$ ft

39. $\theta = 60°,\quad r = 54$ ft,$\quad x = 20$ ft

40. $\theta = 120°,\quad r = 210$ ft,$\quad x = 80$ ft

41. A castle entrance has the form of a Gothic arch, as shown in the figure.

(a) Taking the axes as shown in the figure, find the equation of the circle that forms the left-hand side of the arch.

(b) Use the equation in part (a) to find the height *h* of the arch, rounded to three significant digits.

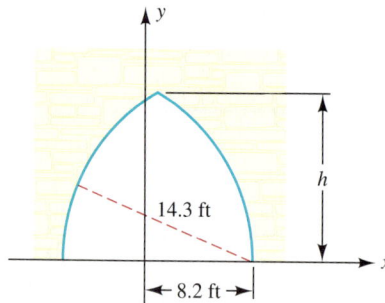

42. The radius of a Ferris wheel is 10.1 meters, as shown in the figure.

(a) Taking the axes as shown, find the equation of the circle that forms the Ferris wheel.

(b) Use the equation in part (a) to find the distance *x* from a chair to the vertical axis of the wheel when the height of the chair is 4.2 meters above the ground. Round the answer to three significant digits.

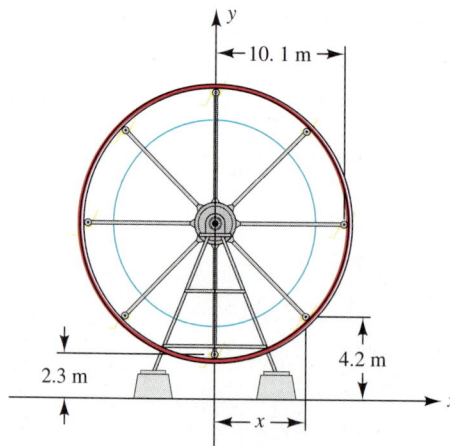

─────── ◄ **Critical Thinking** ───────

43. Two circles having the same center but different radii are called **concentric circles**.

(a) If the ratio of the radii of two concentric circles is 2 : 1 and the equation of the circle with the larger radii is $x^2 + y^2 - 10x + 6y = 2$, determine the equation of the other circle.

(b) Find the area *between* the two circles described in part (a).

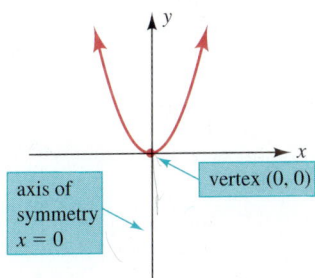

44. Two circles have equations $x^2 + y^2 + 2x + 4y + 1 = 0$ and $x^2 + y^2 - 8y = 0$.

  (a) Find the distance between their centers.

  (b) Find the coordinates of the midpoint of the line segment joining their centers.

45. Find the points of intersection of the circle $x^2 + y^2 - 2x + 4y - 20 = 0$ and the line $y = x - 2$.

46. Show that the line $3x + 4y = -25$ is tangent to the circle $x^2 + y^2 = 25$, and find the coordinates of the point of tangency.

47. Find the equation of the tangent to the circle at the indicated point.

  (a) $x^2 + y^2 = 25$; at $(4, 3)$

  (b) $x^2 + y^2 - 6x + 4y = 4$; at $(2, 2)$

48. For the circle $x^2 + y^2 + Dx + Ey + F = 0$, use the method of completing the square to determine a formula in terms of D, E, and F for the radius of the circle.

49. Find the value(s) of h in the equation $(x - h)^2 + (y + 1)^2 = 25$ so that the circle passes through the point $(4, 2)$.

50. Find the value of D so that the circle with the equation

$x^2 + y^2 + Dx + 2y - 1 = 0$ has its center on the line $x + 3y = 2$.

51. Find the value of m for which the line $y = mx + 5$ is tangent to the circle $x^2 + y^2 = 9$.

52. Find the value of r^2 for which the circle $x^2 + y^2 = r^2$ is tangent to the line $x + y = 4$.

53. Show that $x_1 x + y_1 y = r^2$ is the equation of the tangent to $x^2 + y^2 = r^2$ at the point (x_1, y_1) on the circle.

54. Angle P is inscribed in a semicircle of radius r, as shown in the sketch.

  (a) Find the slopes of $\overline{AP}$ and $\overline{BP}$. Then show that the product of these slopes is -1.

  (b) Using the result of part (a), what conclusion can you make about any angle inscribed in a semicircle?

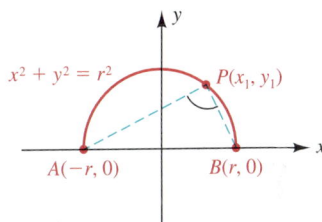

6.2 The Parabola

FIGURE 6.13

The graph of $y = x^2$ is a parabola with vertex $(0, 0)$ and axis of symmetry $x = 0$.

◆ **Introductory Comments**

Recall from Section 3.3 that the graph of the equation $y = x^2$ is a parabola with the vertical line $x = 0$ (the y-axis) as its *axis of symmetry*. The point at which the parabola intersects its axis of symmetry is called the *vertex* of the parabola (see Figure 6.13).

By the vertical stretch and compress rules (Section 2.5), we know that the graph of $y = ax^2$ is also a parabola, but with a *wider* opening than $y = x^2$ if $0 < a < 1$, or a *narrower* opening if $a > 1$, as shown in Figure 6.14. By the x-axis reflection rule (Section 2.5), the parabola opens *upward* if $a > 0$ or *downward* if $a < 0$ (see Figure 6.15). In summary, the value of a in the equation $y = ax^2$ determines the width and direction of the opening of the parabola.

FIGURE 6.14

Graph of $y = ax^2$ for $0 < a < 1$, $a = 1$, and $a > 1$

FIGURE 6.15

Graph of $y = ax^2$ for $a > 0$ and $a < 0$

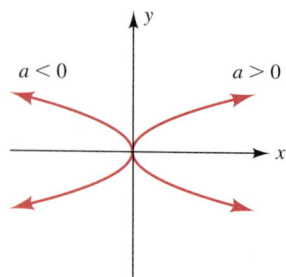

Interchanging x and y in the equation $y = ax^2$ gives us $x = ay^2$, whose graph is also a parabola with its vertex at the origin. However, its axis of symmetry is the horizontal line $y = 0$ (the x-axis) instead of vertical line $x = 0$. If $a > 0$ in the equation $x = ay^2$, the parabola opens to the *right* and if $a < 0$, the parabola opens to the *left*, as shown in Figure 6.16. In this section, we study the geometric properties of a parabola and the distinguishing characteristics of its equation.

FIGURE 6.16

Graph of $x = ay^2$ for $a > 0$ and $a < 0$

◆ Standard Form of a Parabola

If h and k are real numbers and we replace x with $(x - h)$ and y with $(y - k)$ in the equations $y = ax^2$ and $x = ay^2$, we obtain two additional equations:

$$y = ax^2 \qquad\qquad \text{and} \qquad\qquad x = ay^2$$
$$y - k = a(x - h)^2 + k \qquad\qquad x - h = a(y - k)^2$$
$$y = a(x - h)^2 + k \qquad\qquad x = a(y - k)^2 + h$$

By the vertical and horizontal shift rules (Section 2.5), we can state that the graph of $y = a(x - h)^2 + k$ is the same as the graph of $y = ax^2$ with the vertex shifted to the point (h, k) and axis of symmetry the line $x = h$ (see Figure 6.17). The graph of $x = a(y - k)^2 + h$ is also a parabola with vertex (h, k), but with axis of symmetry $y = k$ (see Figure 6.18). We refer to $y = a(x - h)^2 + k$ and $x = a(y - k)^2 + h$ as the *algebraic definitions of a parabola*, or the *equations of a parabola in standard form*.

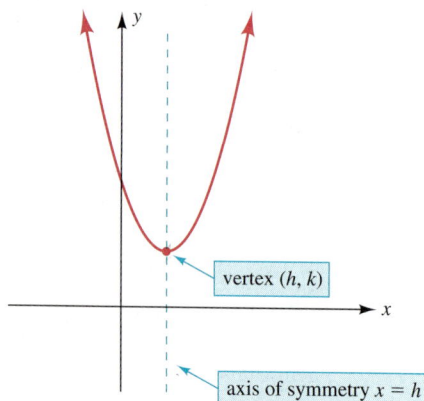

FIGURE 6.17

Graph of $y = a(x - h)^2 + k$ with $a > 0$, $h > 0$, and $k > 0$

FIGURE 6.18

Graph of $x = a(y - k)^2 + h$ with $a < 0$, $h > 0$, and $k > 0$

◆ Equations of a Parabola in Standard Form

The **equation of a parabola in standard form** with vertex (h, k) is

$$y = a(x - h)^2 + k$$

if the axis of symmetry of the parabola is *vertical*, or

$$x = a(y - k)^2 + h$$

if the axis of symmetry of the parabola is *horizontal*.

In each case, a is a nonzero constant that determines the width and direction of the opening of the parabola.

EXAMPLE 1 Determine the equation of a parabola in standard form with vertex (2, 3) and horizontal axis of symmetry if the parabola passes through the point (4, 1).

◆ **Solution** The parabola with the given characteristics is shown in Figure 6.19. Since the vertex is (2, 3) and the axis of symmetry is the *horizontal* line $y = 3$, the equation of this parabola must have the form

$$x = a(y - 3)^2 + 2.$$

FIGURE 6.19

A parabola passing through (4, 1) with vertex (2, 3) and horizontal axis of symmetry

Since the parabola opens to the *right*, we know that $a > 0$. To determine the value of a, we use the fact that the parabola passes through the point (4, 1).

$$x = a(y - 3)^2 + 2$$

$$4 = a(1 - 3)^2 + 2 \qquad \text{Replace } x \text{ with 4 and } y \text{ with 1}$$

$$4 = 4a + 2 \qquad \text{Solve for } a$$

$$a = \tfrac{1}{2}$$

Thus, the required equation is $x = \tfrac{1}{2}(y - 3)^2 + 2$. ◆

To generate the graph of $x = a(y - k)^2 + h$ in the viewing rectangle of a graphing calculator, we solve the equation for y:

$$x = a(y - k)^2 + h$$

$$x - h = a(y - k)^2 \qquad \text{Subtract } h \text{ from both sides}$$

$$\frac{x - h}{a} = (y - k)^2 \qquad \text{Divide both sides by } a$$

$$y - k = \pm \sqrt{\frac{x - h}{a}} \qquad \text{Take the square root of both sides}$$

$$y = k \pm \sqrt{\frac{x - h}{a}} \qquad \text{Add } k \text{ to both sides}$$

We then enter both equations into the graphing calculator:

$$y = k + \sqrt{\frac{x - h}{a}} \qquad \text{and} \qquad y = k - \sqrt{\frac{x - h}{a}}$$

PROBLEM 1   Generate the graph of $x = \frac{1}{2}(y - 3)^2 + 2$ in the viewing rectangle of a graphing calculator. Verify that this parabola passes through the point $(4, 1)$ and that the coordinates of its vertex are $(2, 3)$.   ♦

♦   General Form of a Parabola

If we expand $(x - h)^2$ in the equation $y = a(x - h)^2 + k$, distribute a, and collect all terms on one side of the equation, we obtain a general quadratic equation in two unknowns of the form

$$Ax^2 + Dx + Ey + F = 0$$

where $A, D, E,$ and F are real numbers with $A \neq 0$ and $E \neq 0$. Similarly, if we expand $(y - k)^2$ in the equation $x = a(y - k)^2 + h$, distribute a, and collect all terms on one side of the equation, we obtain a general quadratic equation in two unknowns of the form

$$Cy^2 + Dx + Ey + F = 0$$

where $C, D, E,$ and F are real numbers with $C \neq 0$ and $D \neq 0$. We refer to

$$Ax^2 + Dx + Ey + F = 0 \qquad \text{and} \qquad Cy^2 + Dx + Ey + F = 0$$

as the **equations of a parabola in general form** with vertical and horizontal axes of symmetry, respectively. These general quadratic equations in two unknowns are characterized by the presence of either an x^2 term or y^2 term, but never both.

To generate the graph of the equation $Ax^2 + Dx + Ey + F = 0$ in the viewing rectangle of a graphing calculator, we simply solve the equation for y and enter the equation that results:

$$y = \frac{-Ax^2 - Dx - F}{E}$$

To generate the graph of the equation $Cy^2 + Dx + Ey + F = 0$ in the viewing rectangle of a graphing calculator, we solve the equation for y and enter *both* equations that result. If $E \neq 0$, we rewrite the equation as a quadratic equation in y, namely, $Cy^2 + Ey + (Dx + F) = 0$, and apply the quadratic formula:

$$y = \frac{-E \pm \sqrt{E^2 - 4C(Dx + F)}}{2C}$$

If the equation of a parabola is given in general form, we can find the coordinates of the vertex of the parabola by converting the equation to standard form and observing the values of h and k. To accomplish this task, it may be necessary to use the algebraic process of completing the square (Section 3.3).

EXAMPLE 2   Find the coordinates of the vertex and the direction of the opening of the parabola given by each equation.

(a) $2x^2 + 3y = 6$ **(b)** $y^2 - 2x - 4y + 10 = 0$

◆ **Solution**

(a) The x^2 term indicates that this is the equation of a parabola in general form with a vertical axis of symmetry. To determine the coordinates of the vertex, we solve for y and write the equation in standard form:

$$2x^2 + 3y = 6$$

$$3y = -2x^2 + 6 \qquad \text{Subtract } 2x^2 \text{ from both sides}$$

$$y = -\tfrac{2}{3}x^2 + 2 \qquad \text{Divide both sides by 3}$$

$$y = -\tfrac{2}{3}(x - 0)^2 + 2 \qquad \text{Write in standard form}$$

Thus, the vertex has coordinates $(0, 2)$. Since $a = -\tfrac{2}{3}$ (so $a < 0$), the parabola opens downward from its vertex $(0, 2)$. Figure 6.20 shows the graph of $2x^2 + 3y = 6$ (or $y = -\tfrac{2}{3}x^2 + 2$) in the viewing rectangle of a graphing calculator. This graph confirms our work.

(b) The y^2 term indicates that this is the equation of a parabola in general form with a horizontal axis of symmetry. To determine the coordinates of the vertex, we solve for x and then complete the square as follows:

$$y^2 - 2x - 4y + 10 = 0$$

$$x = \tfrac{1}{2}y^2 - 2y + 5 \qquad \text{Solve for } x$$

$$x = (\tfrac{1}{2}y^2 - 2y \quad) + 5 \qquad \text{Regroup}$$

$$x = \tfrac{1}{2}(y^2 - 4y \quad) + 5 \qquad \text{Factor out } 1/2$$

Add and subtract 2.

$$x = \tfrac{1}{2}(y^2 - 4y + 4) + 5 - 2 \qquad \text{Complete the square}$$

$$x = \tfrac{1}{2}(y - 2)^2 + 3 \qquad \text{Write in standard form}$$

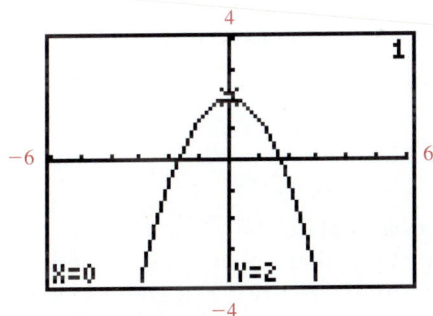

FIGURE 6.20

The graph of the equation $2x^2 + 3y = 6$ is a parabola. It opens downward from its vertex $(0, 2)$.

Thus, the vertex has coordinates $(3, 2)$. Since $a = \tfrac{1}{2}$ (so $a > 0$), the parabola opens to the right from its vertex $(3, 2)$.

To generate the graph of the equation $y^2 - 2x - 4y + 10 = 0$ in the viewing rectangle of a graphing calculator, we rewrite the equation as a quadratic equation in y,

$$y^2 - 4y + (-2x + 10) = 0,$$

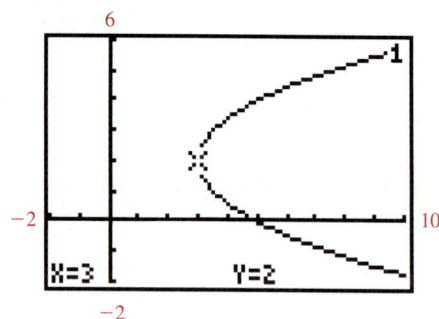

6

−2   10

X=3   Y=2

−2

FIGURE 6.21

The graph of the equation
$y^2 - 2x - 4y + 10 = 0$ is a
parabola. It opens to the right from its
vertex (3, 2).

◆ **Geometric Definition
of a Parabola**

and apply the quadratic formula:

$$y = \frac{-(-4) \pm \sqrt{(-4)^2 - 4(1)(-2x + 10)}}{2(1)} = 2 \pm \sqrt{2x - 6}$$

Entering both of these equations into the calculator, we obtain the parabola that is shown in Figure 6.21. This graph confirms our work.   ◆

PROBLEM 2   Find the *x*-intercepts for the parabolas defined by the equations in Example 2. Check graphically with a graphing calculator.   ◆

◆ **Geometric Definition of a Parabola**

Geometrically, we define a parabola as follows.

A **parabola** is the set of all points in a plane that are equidistant from a fixed line and a fixed point not on the line. The fixed point is called the **focus** of the parabola and the fixed line is called the **directrix** of the parabola.

We now show that our geometric definition of a parabola is consistent with our algebraic definition. Applying our geometric definition, a parabola is the set of all points *P* such that

$$PQ = PF$$

as shown in Figure 6.22. For convenience, let's place the parabola that is shown in Figure 6.22 in the *xy*-coordinate plane so that it opens upward with vertex at the origin and axis of symmetry $x = 0$ (see Figure 6.23). The vertex,

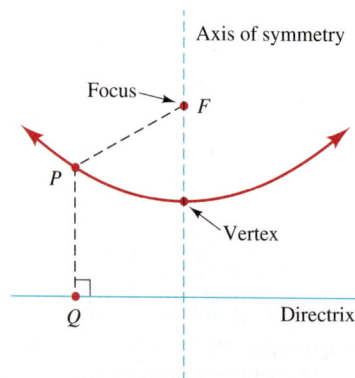

FIGURE 6.22

A parabola is the set of all points *P* such that $PQ = PF$.

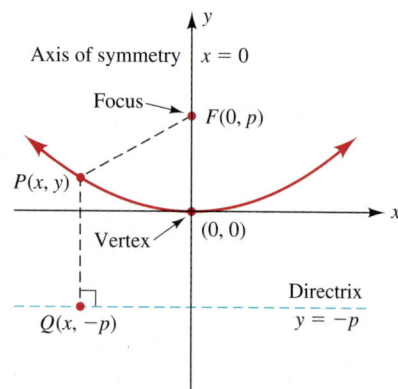

FIGURE 6.23

The vertex of a parabola is the halfway point between the focus and directrix.

like every other point on the parabola, is equidistant from the focus and the directrix. Hence, if the focus F has coordinates $(0, p)$, then the equation of the directrix must be $y = -p$. Referring to Figure 6.23, we have

$$PQ = PF$$

$\sqrt{(x - x)^2 + (y + p)^2} = \sqrt{(x - 0)^2 + (y - p)^2}$	**Apply distance formula**
$(x - x)^2 + (y + p)^2 = (x - 0)^2 + (y - p)^2$	**Square both sides**
$y^2 + 2py + p^2 = x^2 + y^2 - 2py + p^2$	**Expand**
$4py = x^2$	**Solve for y**

$$y = \frac{1}{4p}x^2$$

From our algebraic definition of a parabola, we recognize

$$y = \frac{1}{4p}x^2 = \frac{1}{4p}(x - 0)^2 + 0$$

as the equation of a parabola in standard form with vertex $(0, 0)$ and vertical axis of symmetry $x = 0$. In this equation, $a = 1/(4p)$ is the nonzero constant that determines the width and direction of the opening of the parabola. Hence, we have shown that our geometric and algebraic definitions of a parabola are consistent.

The equation of a parabola in standard form with a vertical axis of symmetry may be written as

$$y = \frac{1}{4p}(x - h)^2 + k$$

where (h, k) is the vertex and $|p|$ is the distance from the focus to the vertex. If $p > 0$, then the focus is *above* the vertex and the parabola opens upward, and if $p < 0$ then the focus is *below* the vertex and the parabola opens downward.

In a similar manner, we can show that if a parabola is placed in the xy-coordinate plane so that it opens to the right with vertex at the origin, then the geometric definition of a parabola yields

$$x = \frac{1}{4p}y^2.$$

Hence, we may rewrite the equation of a parabola in standard form with horizontal axis of symmetry as

$$x = \frac{1}{4p}(y - k)^2 + h$$

where (h, k) is the vertex and $|p|$ is the distance from the focus to the vertex. If $p > 0$, then the focus is to the *right* of the vertex, and the parabola opens toward the right, and if $p < 0$, then the focus is to the *left* of the vertex and the parabola opens toward the left.

EXAMPLE 3 Find the focus and directrix of each parabola.

(a) $y^2 + 12x = 0$ **(b)** $x^2 + 3x - 6y = 0$

◆ **Solution**

(a) Solving for x, we obtain

$$x = -\tfrac{1}{12}y^2 = -\tfrac{1}{12}(y - 0)^2 + 0,$$

which is the equation of a parabola in standard form with horizontal axis of symmetry, vertex at $(0, 0)$, and

$$\frac{1}{4p} = -\frac{1}{12}$$
$$-4p = 12$$
$$p = -3.$$

Since $p < 0$, the focus is 3 units to the *left* of the vertex, and the directrix is 3 units to the right of the vertex. Hence, the coordinates of the focus are $(-3, 0)$ and the directrix is the vertical line $x = 3$, as shown in Figure 6.24.

(b) First, we write the equation in standard form by completing the square as follows:

$$x^2 + 3x - 6y = 0$$
$$y = \tfrac{1}{6}(x^2 + 3x \quad) \qquad \text{Solve for } y$$

Add and subtract 3/8.

$$y = \tfrac{1}{6}(x^2 + 3x + \tfrac{9}{4}) - \tfrac{3}{8} \qquad \text{Complete the square}$$
$$y = \tfrac{1}{6}(x + \tfrac{3}{2})^2 - \tfrac{3}{8} \qquad \text{Factor}$$

We recognize this last equation as the equation of a parabola in standard form with vertical axis of symmetry, vertex at $(-\tfrac{3}{2}, -\tfrac{3}{8})$, and

$$\frac{1}{4p} = \frac{1}{6}$$
$$4p = 6$$
$$p = \tfrac{3}{2}$$

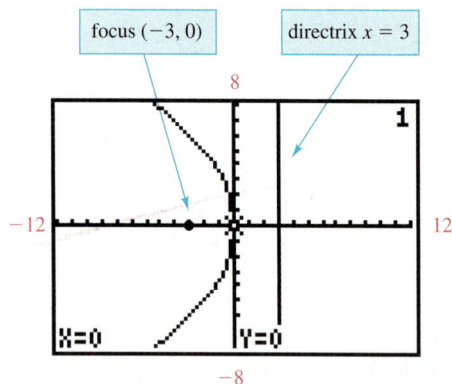

focus $(-3, 0)$ directrix $x = 3$

X=0   Y=0

FIGURE 6.24
The focus of the parabola $y^2 + 12x = 0$ is $(-3, 0)$ and its directrix is $x = 3$.

Since $p > 0$, the focus is $\frac{3}{2}$ units *above* the vertex, and the directrix is $\frac{3}{2}$ units below the vertex. Hence, the coordinates of the focus are $\left(-\frac{3}{2}, \frac{9}{8}\right)$ and the directrix is the horizontal line $y = -\frac{15}{8}$, as shown in Figure 6.25.

PROBLEM 3 Find the focus and directrix of the parabola given by $3y^2 - 8x = 0$.

If we know the focus and directrix of a parabola, then we can determine its equation.

EXAMPLE 4 Find the equation of a parabola with focus $F(-5, 1)$ and directrix $x = 2$.

◆ **Solution** Since the directrix is the vertical line $x = 2$, the parabola must have a horizontal axis of symmetry that passes through the focus $F(-5, 1)$, as shown in Figure 6.26. Thus, the equation of this parabola has the form

$$x = \frac{1}{4p}(y - k)^2 + h,$$

where (h, k) is the vertex and $|p|$ is the distance from the focus to the vertex.

The vertex of a parabola is always located halfway between its focus and directrix. Hence, by the midpoint formula (Section 2.1), the coordinates of the vertex are

$$\left(\frac{-5 + 2}{2}, 1\right) = \left(-\frac{3}{2}, 1\right).$$

The distance from the focus to the vertex is

$$|p| = \left|-5 - \left(-\frac{3}{2}\right)\right| = \frac{7}{2}.$$

Since the focus is to the left of the vertex, the parabola opens to the left. Hence, we choose $p = -\frac{7}{2}$. Therefore, the equation of this parabola is

$$x = \frac{1}{4p}(y - k)^2 + h$$

$$x = \frac{1}{4\left(-\frac{7}{2}\right)}(y - 1)^2 + \left(-\frac{3}{2}\right)$$

$$x = -\frac{1}{14}(y - 1)^2 - \frac{3}{2}$$

PROBLEM 4 Referring to Example 4, express the equation of the parabola in general form with integer coefficients.

FIGURE 6.25

The focus of the parabola $x^2 + 3x - 6y = 0$ is $\left(-\frac{3}{2}, \frac{9}{8}\right)$ and the directrix is $y = -\frac{15}{8}$.

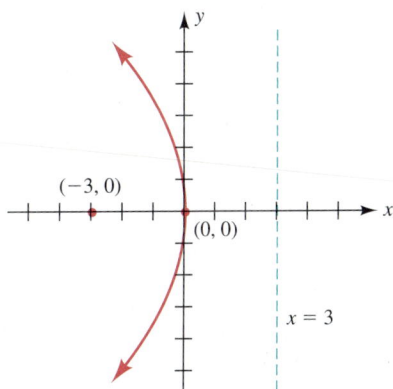

FIGURE 6.26

A parabola with a vertical directrix must have a horizontal axis of symmetry.

◆ **Application: Reflection Property of a Parabola**

A law of physics states that the angle of incidence equals the angle of reflection; that is, the angle at which a light or sound wave hits a surface is equal to the angle at which it reflects from the surface, as shown in Figure 6.27. When this physical law is applied to a parabolic surface, an interesting phenomenon occurs: all light or sound waves that move parallel to the axis of the parabola bounce off the parabola and then pass through the focus of the parabola (see Figure 6.28). We refer to this phenomenon as the **reflection property of a parabola**. Engineers and scientists use this reflection property to design telescopes, radar antennas, satellite dishes, solar heating devices, microphones, and other communications equipment.

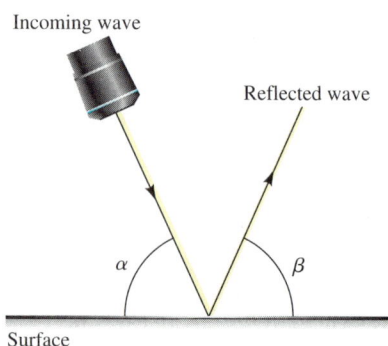

FIGURE 6.27

The angle of incidence α equals the angle of reflection β.

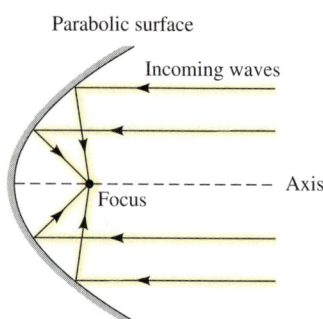

FIGURE 6.28

Light or sound waves entering parallel to the axis reflect to the focus.

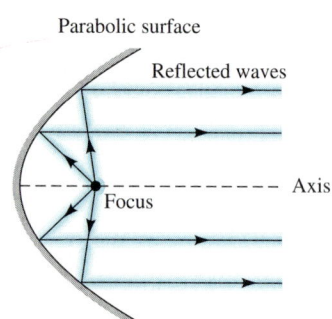

FIGURE 6.29

Light or sound waves emanating from the focus reflect parallel to the axis.

The reflection property of a parabola is also used in the design of searchlights and flashlights: The light rays emanating from a bulb, located at the focus of a parabolic surface, bounce off the surface and shine parallel to the axis of the parabola, thus keeping the light beam at maximum intensity (see Figure 6.29).

EXAMPLE 5   *Chapter Opening Problem*

The parabolic reflector of a searchlight has a diameter of 3 feet at the open end and a depth of 1 foot. How far from the vertex of the parabola should the filament of the bulb be located so that all the light rays emanating from the bulb are parallel to the axis of the parabola?

◆ **Solution**   We place the parabolic reflector in the *xy*-coordinate plane so that its vertex is at the origin and it opens to the right, as shown in Figure 6.30. Now, if the diameter of the opening of the parabolic reflector is 3 feet and the depth is 1 foot, then the coordinates of point Q in Figure 6.30 are $(1, \frac{3}{2})$. Hence,

$$x = \frac{1}{4p}y^2$$

$$1 = \frac{1}{4p}\left(\frac{3}{2}\right)^2 \qquad \text{Substitute } x = 1 \text{ and } y = \tfrac{3}{2}$$

$$\frac{4}{9} = \frac{1}{4p} \qquad \text{Multiply both sides by } \tfrac{4}{9}$$

$$p = \frac{9}{16} \qquad \text{Solve for } p$$

Since p represents the distance from the vertex to the focus and light rays emanating from the focus are all parallel to the axis of the parabola, we conclude that the filament of the bulb should be placed $\frac{9}{16}$ foot (or $6\frac{3}{4}$ inches) from the vertex of the parabolic reflector. ◆

FIGURE 6.30

If the diameter of the parabolic reflector is 3 feet and the depth is 1 foot, then point Q has coordinates $(1, \frac{3}{2})$.

PROBLEM 5 Taking the axes as shown in Figure 6.30, determine the equation of the parabola that may be rotated about the x-axis to generate this parabolic reflector. ◆

Exercises 6.2

◆ Basic Skills

In Exercises 1–16, find the coordinates of the vertex and the direction of the opening of the parabola with the given equation. Check graphically with a graphing calculator.

1. $x = y^2$
2. $x = -3y^2$
3. $y = 2x^2 - 8$
4. $x = 1 - 4y^2$
5. $x = -(y - 1)^2 + 4$
6. $y = 2(x + 1)^2 - 2$
7. $y = 4x - x^2$
8. $4x = y^2 - 4$
9. $x = y^2 - 3y - 4$
10. $y = 6 - x - \frac{1}{3}x^2$
11. $y^2 - 2x - 4y + 2 = 0$
12. $y^2 + 3x + 6y + 3 = 0$
13. $x^2 + 8x - 2y + 16 = 0$
14. $x^2 - 3y - 6x + 15 = 0$
15. $2y^2 + 3x + 16y + 26 = 0$
16. $3y^2 + 4x - 6y + 3 = 0$

In Exercises 17–22, find the focus and directrix of the parabola whose equation is given in the indicated exercise.

17. Exercise 3
18. Exercise 6
19. Exercise 9
20. Exercise 10
21. Exercise 15
22. Exercise 16

In Exercises 23–32, determine the equation of the parabola with the given characteristics.

23. Vertex at the origin, passing through the point (2, 4), and symmetric with respect to the x-axis
24. Vertex at the origin, passing through the point $(-1, 3)$, and symmetric with respect to the y-axis
25. Vertex at $(-1, -1)$, passing through the point (0, 1), and symmetric with respect to the line $x = -1$
26. Vertex at (2, 1), passing through the origin, and symmetric with respect to the line $y = 1$
27. Focus at (0, 1) and directrix $x = 2$
28. Focus at $(-4, -2)$ and directrix $y = 2$
29. Vertex at $(-3, 2)$ and focus at $(-3, -3)$
30. Vertex at $(9, -12)$ and focus at $(1, -12)$

31. Vertex at $(3, -1)$ and directrix $x = -1$

32. Vertex at $(-8, -4)$ and directrix $y = 2$

In Exercises 33–36, use a graphing calculator to generate the graph of each parabola. Then estimate the coordinates of the vertex, rounding each coordinate to three significant digits.

33. $12.5x^2 - 18.6x + 2.4y = 10.7$

34. $1.24y^2 + 4.56x = 9.55$

35. $23.4y^2 - 45.8y - 12.6x = 0$

36. $y^2 - 2.34x - 5.74y = -6.87$

37. The cable that supports the weight of a suspension bridge approximates a parabola with vertex located halfway between the supports, as shown in the sketch. Determine the length x of the vertical cable that is 30 meters from the end of the support.

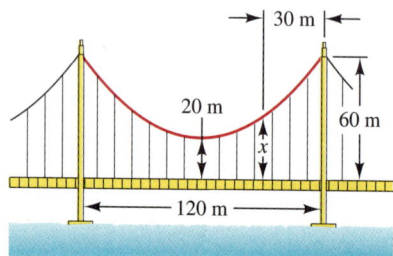

38. The main entrance into a new shopping mall is a parabolic arch with span 12 ft and height 18 ft, as shown in the sketch. Calculate the height h of a security camera that is located 4 feet from the center of the arch.

39. A parabolic satellite dish for receiving television signals has its receiver located at the focus, as shown in the sketch. Find the depth d of the satellite dish.

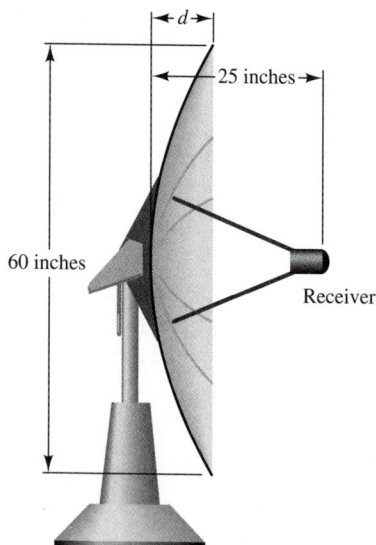

40. A solar collector for a domestic hot water heater is parabolic in shape, as shown in the sketch. At what distance d from the vertex should the water pipe be located in order to take advantage of the reflection property of a parabola?

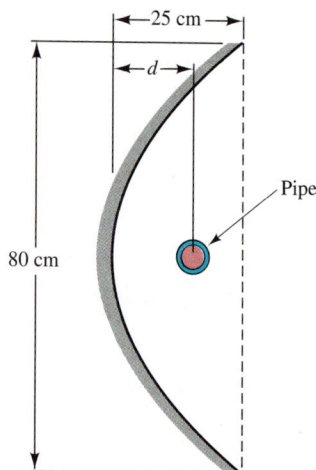

41. In an engineer's new design for an automobile headlight, a parabola is rotated about the x-axis, as shown in the sketch. Determine the equation of the parabola that is used to generate the shape of the headlight. Write the equation in the form $x = ay^2$, with a rounded to three significant digits.

Headlight

16.4 cm

10.2 cm

42. A simply supported beam with a concentrated load at midspan deflects in the shape of a parabola, as shown in the sketch. Determine the equation of the parabola, which is called the *elastic curve*. Write the equation in the form $y = ax^2 + bx$ with a and b each rounded to three significant digits.

392 in.

Load

1.25 in.

Beam

◆◆ **Critical Thinking** ─────────────────────────────

43. If a parabola has two distinct x-intercepts a and b, what is the x-coordinate of its vertex?

44. Find all values of k such that the vertex of the graph of $y = x^2 + kx + 9$ lies on the x-axis.

45. If the graph of $x = ay^2 + by + c$ passes through the origin, what is the value of c?

46. Determine the value of a such that the parabola $x = ay^2 - 4y + 3$ has its axis of symmetry along the line $y = 5$.

47. What positive values of p make the opening of the parabola $y = \dfrac{1}{4p}x^2$ narrower than the opening of the parabola $y = x^2$? Wider than $y = x^2$?

48. The chord of a parabola that passes through its focus and is perpendicular to its axis of symmetry is called the *latus rectum* of the parabola. The latus rectum of the parabola $x = \dfrac{1}{4p}y^2$ is the line segment $\overline{AB}$, as shown in the sketch. What is the length of the latus rectum?

Directrix

A

Focus

B

49. The equation of a parabola is $Ax^2 + Ey = 0$. Find the coordinates of its focus in terms of A and E.

50. What point on a parabola is always closest to its focus?

6.3 The Ellipse

◆ **Introductory Comments**

In this section, we study the geometric properties of the ellipse and the distinguishing characteristics of its equation. First, we consider the equation

$$\frac{x^2}{a^2} + \frac{y^2}{b^2} = 1$$

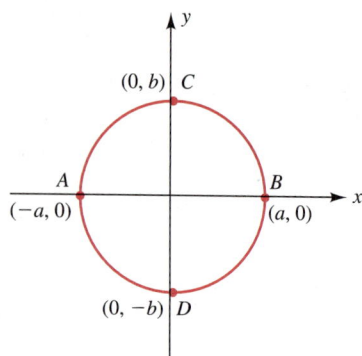

(a) If $a = b$, the graph is a circle with center at the origin.

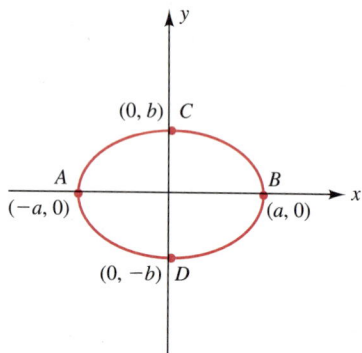

(b) If $a > b$, the graph is an ellipse, elongated horizontally, with center at the origin.

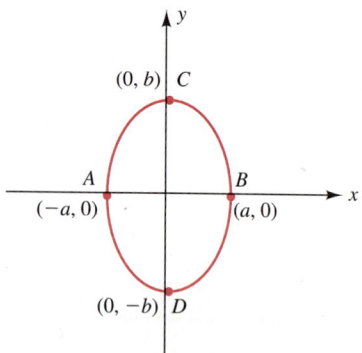

(c) If $a < b$, the graph is an ellipse, elongated vertically, with center at the origin.

FIGURE 6.31

The three possibilities for the graph of
$$\frac{x^2}{a^2} + \frac{y^2}{b^2} = 1$$

with $a > 0$ and $b > 0$. This is a general quadratic equation in two unknowns with $A = 1/a^2$, $C = 1/b^2$, $F = -1$, and $B = D = E = 0$. Using the tests for symmetry (Section 2.2), we find that the graph of this equation is symmetric with respect to the x-axis, the y-axis, and the origin. To find the x-intercepts of the graph we let $y = 0$ and obtain $x = \pm a$. Similarly, letting $x = 0$ gives us the y-intercepts $\pm b$. If we solve the original equation for y, we obtain

$$\frac{x^2}{a^2} + \frac{y^2}{b^2} = 1$$

$$\frac{y^2}{b^2} = 1 - \frac{x^2}{a^2} = \frac{a^2 - x^2}{a^2} \qquad \text{\color{red}{\textbf{Subtract }} } x^2/a^2 \text{ \textbf{from both sides}}$$
$$\text{\color{red}{\textbf{and write as a single fraction}}}$$

$$y^2 = \frac{b^2}{a^2}(a^2 - x^2) \qquad \text{\color{red}{\textbf{Multiply both sides by }} } b^2$$

$$y = \pm \frac{b}{a}\sqrt{a^2 - x^2} \qquad \text{\color{red}{\textbf{Take the square root of both sides}}}$$

Now, if y is to be a real number, then $(a^2 - x^2)$ must be either positive or zero, that is

$$a^2 - x^2 \geq 0 \qquad \text{or} \qquad -a \leq x \leq a.$$

This means that the graph of $(x^2/a^2) + (y^2/b^2) = 1$ exists only between the x-intercepts $\pm a$. Also, from this form of the equation we observe that as x increases from 0 to a, y decreases from b to 0. Figure 6.31 shows the three possibilities for the graph of $(x^2/a^2) + (y^2/b^2) = 1$.

If $a = b$, as shown in Figure 6.31(a), then we have

$$\frac{x^2}{a^2} + \frac{y^2}{a^2} = 1 \qquad \text{or} \qquad x^2 + y^2 = a^2.$$

This is the equation of a circle with center at the origin and radius a (see Section 6.1). If a and b have different values, as shown in parts (b) and (c) of the figure, then the circle is flattened or stretched to form an egg-shaped curve called an **ellipse**.

In each of the graphs shown in part (b) or (c) of Figure 6.31, the horizontal and vertical line segments $\overline{AB}$ and $\overline{CD}$ are the **axes** of the ellipse. We refer to the longer line segment as the **major axis** and the shorter line segment as the **minor axis**. If $a > b$, as in Figure 6.31(b), then $\overline{AB}$ is the major axis and $\overline{CD}$ is the minor axis. However, if $a < b$, as in Figure 6.31(c), then $\overline{CD}$ is the major axis and $\overline{AB}$ is the minor axis. We refer to the endpoints of the major axis as the **vertices** of the ellipse and the midpoint of the major axis and minor axis as its **center**. Half of an axis of an ellipse is called a **semiaxis**. Note that the lengths of the semiaxes are a and b.

◆ **Standard Form of an Ellipse**

If we replace x with $(x - h)$ and y with $(y - k)$ in the equation $(x^2/a^2) + (y^2/b^2) = 1$, we obtain

$$\frac{(x - h)^2}{a^2} + \frac{(y - k)^2}{b^2} = 1.$$

If a and b are positive real numbers such that $a = b$, we have

$$\frac{(x - h)^2}{a^2} + \frac{(y - k)^2}{a^2} = 1 \qquad \text{or} \qquad (x - h)^2 + (y - k)^2 = a^2$$

the equation of a circle in standard form with center (h, k) and radius a (see Section 6.1). If a and b have different positive values, we refer to $(x - h)^2/a^2 + (y - k)^2/b^2 = 1$ as the *algebraic definition of an ellipse*, or the *equation of an ellipse in standard form.*

◆◆ **Equation of an Ellipse in Standard Form**

> The **equation of an ellipse in standard form** with center (h, k), horizontal axis of length $2a$, and vertical axis of length $2b$ is
>
> $$\frac{(x - h)^2}{a^2} + \frac{(y - k)^2}{b^2} = 1.$$

We can determine the equation of an ellipse if we know its center (h, k) and the lengths of its semiaxes a and b.

EXAMPLE 1 Determine the equation of an ellipse with vertices $(3, 3)$ and $(-5, 3)$ and minor axis of length 4.

◆◆ **Solution** The ellipse with the given characteristics is shown in Figure 6.32. Since the vertices of an ellipse are the endpoints of its major axis, we know the major axis is horizontal. Thus, the minor axis is vertical. Now, if the minor axis is vertical and has a length of 4, we have

$$2b = 4 \qquad \text{or} \qquad b = 2.$$

The length of the major axis is the distance between the x-coordinates of the vertices, or $|3 - (-5)| = 8$. Thus,

$$2a = 8 \qquad \text{or} \qquad a = 4.$$

The midpoint of the major axis is the center of the ellipse. Using the midpoint formula (Section 2.1), we find the coordinates of the center:

$$(h, k) = \left(\frac{3 + (-5)}{2}, \frac{3 + 3}{2} \right) = (-1, 3).$$

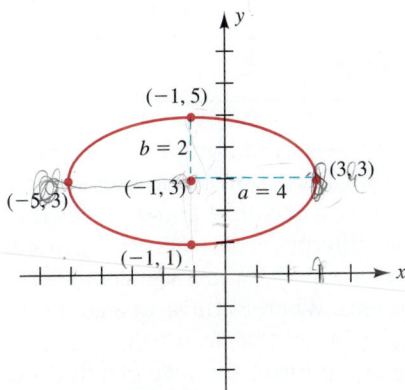

FIGURE 6.32
An ellipse with vertices $(3, 3)$ and $(-5, 3)$ and minor axis of length 4

Thus, the equation of the ellipse is

$$\frac{(x - h)^2}{a^2} + \frac{(y - k)^2}{b^2} = 1$$

$$\frac{[x - (-1)]^2}{4^2} + \frac{(y - 3)^2}{2^2} = 1 \qquad \textcolor{red}{\textbf{Substitute}}$$

$$\frac{(x + 1)^2}{16} + \frac{(y - 3)^2}{4} = 1 \qquad \textcolor{red}{\textbf{Simplify}} \qquad ◆$$

To generate the graph of $(x - h)^2/a^2 + (y - k)^2/b^2 = 1$ in the viewing rectangle of a graphing calculator, we must solve the equation for y and enter both equations that result. To obtain a true geometric perspective of the graph, we use a square set viewing rectangle.

PROBLEM 1   Generate the graph of

$$\frac{(x + 1)^2}{16} + \frac{(y - 3)^2}{4} = 1$$

in the viewing rectangle of a graphing calculator. Verify that this ellipse has vertices $(3, 3)$ and $(-5, 3)$ and minor axis of length 4.   ◆

◆ **General Form of an Ellipse**

If we take the equation of an ellipse in standard form and multiply both sides of the equation by a^2b^2, we obtain an equation of the form

$$A(x - h)^2 + C(y - k)^2 = N$$

with $A = b^2$, $C = a^2$, and $N = a^2b^2$. If we expand $(x - h)^2$ and $(y - k)^2$, distribute A and C, and collect all terms on one side of the equation, we obtain an equation of the form

$$Ax^2 + Cy^2 + Dx + Ey + F = 0$$

We refer to this general quadratic equation in two unknowns as the **equation of an ellipse in general form** with vertical and horizontal axes of symmetry. This equation is characterized by the presence of x^2 and y^2 terms that have *different coefficients* but *like signs*. The difference between the general form of the equation of an ellipse and that of a circle is that the coefficients of the x^2 and y^2 terms of an ellipse are different, whereas those of a circle are the same.

If the equation of an ellipse is given in general form, then we can find the coordinates of its center and vertices by converting the equation to standard form and observing the values of h, k, a, and b. We can use a graphing calculator to check our work by solving the equation

$$Ax^2 + Cy^2 + Dx + Ey + F = 0$$

for y and graphing both equations that result. To generate the graph of this equation when $E \neq 0$, we rewrite the equation as a quadratic equation in y,

$$Cy^2 + Ey + (Ax^2 + Dx + F) = 0,$$

and apply the quadratic formula:

$$y = \frac{-E \pm \sqrt{E^2 - 4C(Ax^2 + Dx + F)}}{2C}$$

EXAMPLE 2   Determine the coordinates of the center and vertices of the ellipse given by each equation.

(a) $4x^2 + y^2 = 9$ **(b)** $x^2 + 9y^2 + 8x - 36y + 43 = 0$

◆ **Solution**

(a) We begin by writing the equation in standard form:

$$4x^2 + y^2 = 9$$

$$\frac{4x^2}{9} + \frac{y^2}{9} = 1 \qquad \text{Divide both sides by 9}$$

$$\frac{x^2}{\frac{9}{4}} + \frac{y^2}{9} = 1 \qquad \text{Invert 4/9 and divide}$$

$$\frac{(x - 0)^2}{\left(\frac{3}{2}\right)^2} + \frac{(y - 0)^2}{3^2} = 1 \qquad \text{Write in standard form}$$

This last equation tells us that the graph of $4x^2 + y^2 = 9$ is an ellipse with $a = \frac{3}{2}$, $b = 3$, and center at the origin. Since $b > a$, the ellipse is elongated vertically with a major axis of length $2b = 6$. Hence, the co-ordinates of the vertices are $(0, 3)$ and $(0, -3)$. Figure 6.33 shows the graph of $4x^2 + y^2 = 9$ $\left(\text{or } y = \pm \sqrt{9 - 4x^2}\right)$ in the viewing rectangle of a graphing calculator, which confirms our work.

(b) We begin by writing the equation in standard form. To do this, we use the process of completing the square (Section 3.3) and proceed as follows:

$$x^2 + 9y^2 + 8x - 36y + 43 = 0$$

$$(x^2 + 8x \quad) + (9y^2 - 36y \quad) = -43 \qquad \text{Regroup}$$

$$(x^2 + 8x \quad) + 9(y^2 - 4y \quad) = -43 \qquad \text{Factor out 9}$$

Add 16 to both sides.

$$(x^2 + 8x + 16) + 9(y^2 - 4y + 4) = -43 + (16 + 36) \qquad \text{Complete the squares}$$

Add 36 to both sides.

$$(x + 4)^2 + 9(y - 2)^2 = 9 \qquad \text{Factor}$$

$$\frac{(x + 4)^2}{9} + \frac{(y - 2)^2}{1} = 1 \qquad \text{Divide both sides by 9}$$

$$\frac{[x - (-4)]^2}{3^2} + \frac{(y - 2)^2}{1^2} = 1 \qquad \text{Write in standard form}$$

FIGURE 6.33

The graph of $4x^2 + y^2 = 9$ is an ellipse with center $(0, 0)$ and vertices $(0, 3)$ and $(0, -3)$.

This last equation tells us that the graph of $x^2 + 9y^2 + 8x - 36y + 43 = 0$ is an ellipse with $a = 3$, $b = 1$, and center $(h, k) = (-4, 2)$. Since $a > b$, the ellipse is elongated horizontally with major axis of length $2a = 6$. Hence, the coordinates of the vertices are $(-7, 2)$ and $(-1, 2)$.

To generate the graph of the equation $x^2 + 9y^2 + 8x - 36y + 43 = 0$ in the viewing rectangle of a graphing calculator, we rewrite the equation as a quadratic equation in y,

$$9y^2 - 36y + (x^2 + 8x + 43) = 0,$$

and apply the quadratic formula:

$$y = \frac{-(-36) \pm \sqrt{(-36)^2 - 4(9)(x^2 + 8x + 43)}}{2(9)} = 2 \pm \frac{\sqrt{-7 - 8x - x^2}}{3}$$

Entering both of these equations into the calculator, we obtain the ellipse that is shown in Figure 6.34. This graph confirms our work.

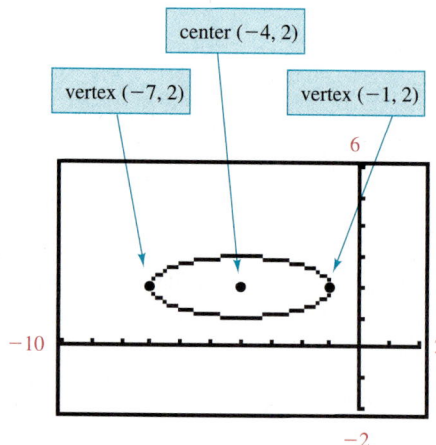

FIGURE 6.34

The graph of $x^2 + 9y^2 + 8x - 36y + 43 = 0$ is an ellipse with center $(-4, 2)$ and vertices $(-7, 2)$ and $(-1, 2)$.

CAUTION Not every equation of the form

$$Ax^2 + Cy^2 + Dx + Ey + F = 0,$$

where the x^2 and y^2 terms have different coefficients but like signs, represents an ellipse. For example, if this equation is written in the form

$$A(x - h)^2 + C(y - k)^2 = N \quad \text{and} \quad N = 0,$$

then the graph of the equation is the single point (h, k). We refer to this single point as a *degenerate ellipse*. Also, if

$$A(x - h)^2 + C(y - k)^2 = N \quad \text{and} \quad N < 0,$$

then this equation has no graph, since no point (x, y) with real coordinates satisfies this equation.

PROBLEM 2 Is $3x^2 + 2y^2 - 6x + 8y + 15 = 0$ the equation of an ellipse? Explain.

◆ **Geometric Definition of an Ellipse**

Geometrically, we define an ellipse as follows.

◆ **Geometric Definition of an Ellipse**

> An **ellipse** is the set of all points in a plane, the sum of whose distances from two fixed points is a constant, which is equal to the length of the major axis of the ellipse. The two fixed points are called the **foci** (plural of focus).

We now show that our geometric definition of an ellipse is consistent with our algebraic definition. Applying our geometric definition, an ellipse is the set of all points P such that

$$PF_1 + PF_2 = \text{length of major axis}$$

as shown in Figure 6.35. For convenience, we can place the ellipse that is shown in Figure 6.35 in the xy-coordinate plane so that its center is at the origin, its major axis is horizontal with length $2a$, and its minor axis is vertical with length $2b$ (see Figure 6.36). We have labeled the coordinates of the foci $(-c, 0)$ and $(c, 0)$, with $a > c$.

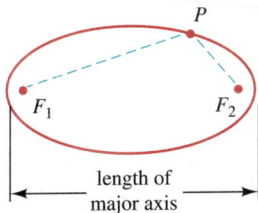

FIGURE 6.35

An ellipse is the set of all points P such that $PF_1 + PF_2 = $ length of major axis.

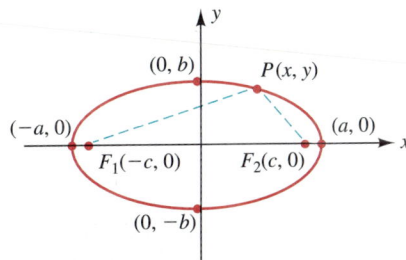

FIGURE 6.36

An ellipse in the xy-coordinate plane with center at the origin, major axis of length $2a$, minor axis of length $2b$, and foci with coordinates $(-c, 0)$ and $(c, 0)$

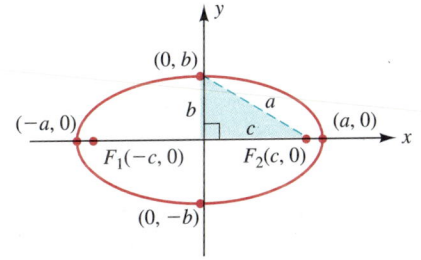

FIGURE 6.37

The constants a, b, and c are related by the Pythagorean theorem.

The constants a, b, and c have a special relationship. Referring to Figure 6.37, the point $(0, b)$ is on the ellipse and is equidistant from each of the foci. Since the sum of the distances from the foci to the point $(0, b)$ must be $2a$, the distance from each focus to $(0, b)$ is a. Thus, the shaded right triangle in Figure 6.37 has legs b and c and hypotenuse a. Hence, by the Pythagorean theorem, we have the relation

$$c^2 = a^2 - b^2$$

in an ellipse with horizontal semimajor axis of length a, vertical semiminor axis of length b, and distance c from the center to each focus. For an ellipse with vertical semimajor axis of length b and horizontal semiminor axis of length a, we have the relation

$$c^2 = b^2 - a^2$$

Now referring to Figure 6.36, we have

$$PF_1 + PF_2 = 2a$$

$$\sqrt{(x+c)^2 + y^2} + \sqrt{(x-c)^2 + y^2} = 2a \qquad \text{Apply the distance formula}$$

$$\sqrt{(x+c)^2 + y^2} = 2a - \sqrt{(x-c)^2 + y^2} \qquad \text{Isolate one of the radicals}$$

$$(x+c)^2 + y^2 = 4a^2 - 4a\sqrt{(x-c)^2 + y^2} + (x-c)^2 + y^2 \qquad \text{Square both sides}$$

$$4a\sqrt{(x-c)^2 + y^2} = 4a^2 + (x-c)^2 + y^2 - (x+c)^2 - y^2 \qquad \text{Isolate the remaining radical}$$

$$a\sqrt{(x-c)^2 + y^2} = a^2 - cx \qquad \begin{array}{l}\text{Simplify the right side}\\ \text{then divide both sides by 4}\end{array}$$

$$a^2[(x-c)^2 + y^2] = a^4 - 2a^2cx + c^2x^2 \qquad \text{Square both sides}$$

$$(a^2 - c^2)x^2 + a^2y^2 = a^2(a^2 - c^2) \qquad \text{Collect like terms}$$

$$\frac{x^2}{a^2} + \frac{y^2}{a^2 - c^2} = 1 \qquad \text{Divide both sides by } a^2(a^2 - c^2)$$

$$\frac{x^2}{a^2} + \frac{y^2}{b^2} = 1 \qquad \text{Replace } c^2 \text{ with } a^2 - b^2$$

From our algebraic definition of an ellipse, we recognize this last equation as the equation of an ellipse in standard form with center $(0, 0)$, horizontal axis of length $2a$, and vertical axis of length $2b$. We have shown that our geometric and algebraic definitions of an ellipse are consistent.

The **eccentricity** e of an ellipse is a measure of its "ovalness" and is defined as

$$e = \frac{\text{distance from center to focus}}{\text{distance from center to vertex}}$$

(Do not confuse the eccentricity e with base e in exponential and logarithmic expressions.) The eccentricity e of an ellipse is a number between 0 and 1. If e is close to 1, then the ellipse is very long and narrow and if e is close to 0, then the ellipse is nearly circular.

EXAMPLE 3   Find the foci and eccentricity of the ellipse defined by each equation.

(a) $16x^2 + 9y^2 = 144$ **(b)** $x^2 + 4y^2 - 6x + 5 = 0$

◆ **Solution**

(a) Dividing both sides by 144, we obtain

$$\frac{x^2}{9} + \frac{y^2}{16} = 1 \qquad \text{or} \qquad \frac{(x-0)^2}{3^2} + \frac{(y-0)^2}{4^2} = 1,$$

the equation of an ellipse in standard form with $a = 3$, $b = 4$, and center at the origin. Since $b > a$, the ellipse is elongated vertically with a semimajor axis of length 4 and semiminor axis of length 3. To find the coordinates of the foci, we apply the relation

$$c^2 = 4^2 - 3^2,$$

which implies $|c| = \sqrt{7} \approx 2.646$. Hence, the coordinates of the foci are

$$\left(0, \sqrt{7}\right) \approx (0, 2.646) \qquad \text{and} \qquad \left(0, -\sqrt{7}\right) \approx (0, -2.646)$$

as shown in Figure 6.38. The eccentricity e of this ellipse is

$$e = \frac{\text{distance from center to focus}}{\text{distance from center to vertex}} = \frac{\sqrt{7}}{4} \approx 0.66.$$

(b) First, we write the equation in standard form by completing the square:

$$x^2 + 4y^2 - 6x + 5 = 0$$

$$(x^2 - 6x \qquad) + 4y^2 = -5 \qquad \text{Regroup}$$

Add 9 to both sides.

$$(x^2 - 6x + 9) + 4y^2 = -5 + 9 \qquad \text{Complete the square}$$

$$(x - 3)^2 + 4y^2 = 4 \qquad \text{Factor}$$

$$\frac{(x-3)^2}{2^2} + \frac{(y-0)^2}{1^2} = 1 \qquad \begin{array}{l}\text{Divide both sides by 4 and}\\\text{write in standard form}\end{array}$$

This is the equation of an ellipse in standard form with $a = 2$, $b = 1$, and center $(3, 0)$. Since $a > b$, the ellipse is elongated horizontally with a semimajor axis of length 2 and semiminor axis of length 1. To find the coordinates of the foci, we apply the relation

$$c^2 = 2^2 - 1^2,$$

which implies that $c = \sqrt{3} \approx 1.732$. Hence, the coordinates of the foci are

$$\left(3 - \sqrt{3}, 0\right) \approx (1.268, 0) \qquad \text{and} \qquad \left(3 + \sqrt{3}, 0\right) \approx (4.732, 0)$$

focus $(0, \sqrt{7})$ focus $(0, -\sqrt{7})$

FIGURE 6.38

The foci of the ellipse $16x^2 + 9y^2 = 144$ are $(0, \sqrt{7})$ and $(0, -\sqrt{7})$.

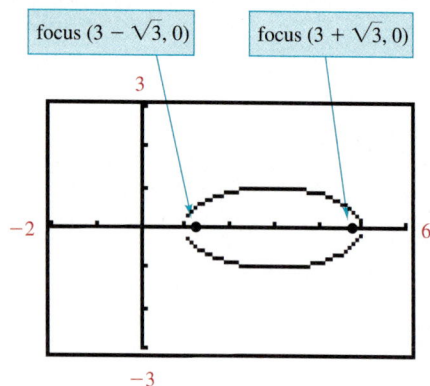

FIGURE 6.39

The foci of the ellipse
$x^2 + 4y^2 - 6x + 5 = 0$ are
$(3 - \sqrt{3}, 0)$ and $(3 + \sqrt{3}, 0)$.

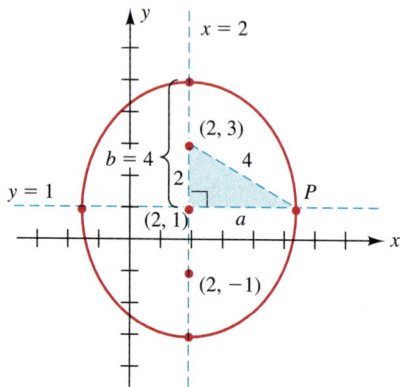

FIGURE 6.40

An ellipse with foci $(2, 3)$ and $(2, -1)$ has center $(2, 1)$.

as shown in Figure 6.39. The eccentricity e of this ellipse is

$$e = \frac{\text{distance from center to focus}}{\text{distance from center to vertex}} = \frac{\sqrt{3}}{2} \approx 0.87.$$ ◆

PROBLEM 3   Find the foci and eccentricity for the ellipse defined by $9x^2 + y^2 - 9 = 0$.   ◆

If we know the foci and eccentricity of an ellipse, then we can determine its equation.

EXAMPLE 4   Find the equation of an ellipse in standard form with foci $(2, 3)$ and $(2, -1)$ and eccentricity $\frac{1}{2}$.

◆ **Solution**   The center of an ellipse is located halfway between its foci. Hence, the center of this ellipse is $(h, k) = (2, 1)$, as shown in Figure 6.40. Therefore, the equation of this ellipse must have the form

$$\frac{(x - 2)^2}{a^2} + \frac{(y - 1)^2}{b^2} = 1.$$

We now proceed to find the lengths of the semimajor axis b and the semiminor axis a. Since the distance from the center to a focus is $c = 2$ and the eccentricity of this ellipse is given to be $\frac{1}{2}$, we have

$$\frac{\text{distance from center to focus}}{\text{distance from center to vertex}} = \frac{2}{b} = \frac{1}{2},$$

which implies that $b = 4$. We find the value of a by applying the relation $c^2 = b^2 - a^2$:

$$2^2 = 4^2 - a^2 \quad \text{Substitute } c = 2 \text{ and } b = 4$$

$$a^2 = 4^2 - 2^2 \quad \text{Solve for } a$$
$$a^2 = 12$$
$$a = 2\sqrt{3}$$

Since $(h, k) = (2, 1)$, $a^2 = (2\sqrt{3})^2 = 12$, and $b^2 = (4)^2 = 16$, the equation of this ellipse in standard form is

$$\frac{(x - 2)^2}{12} + \frac{(y - 1)^2}{16} = 1.$$ ◆

PROBLEM 4   Referring to Example 4, express the equation of the ellipse in general form with integer coefficients.   ◆

◆ **Application: Reflection Property of an Ellipse**

Like the parabola, the ellipse has a *reflection property*. This property explains the "whispering gallery effect" in elliptically shaped dome rooms. If a person standing at a special point on one side of the room whispers, all the sound waves bounce off the sides of the dome and return to another special point on the other side of the room, allowing a person standing at that point to hear the whisper (see Figure 6.41). The two special points in the room where this phenomenon occurs are the *foci* of the ellipse. The whispering gallery effect may be observed in the National Statuary Hall in the Capitol building in Washington, D.C. (Exercises 41 and 42 are applied problems that use the reflection property of an ellipse.)

FIGURE 6.41
In an elliptically shaped dome room, the whisper of one person can be heard by another person on the other side of the room.

Exercises 6.3

◆ **Basic Skills**

In Exercises 1–16, determine the coordinates of the center and vertices of the ellipse defined by each equation. Check graphically with a graphing calculator.

1. $\dfrac{x^2}{16} + \dfrac{y^2}{4} = 1$

2. $\dfrac{x^2}{4} + \dfrac{y^2}{9} = 1$

3. $25x^2 + 4y^2 = 100$

4. $16x^2 + 25y^2 = 400$

5. $x^2 + 16y^2 = 9$

6. $4x^2 + 9y^2 = 16$

7. $3x^2 + 4y^2 - 9 = 0$

8. $x^2 + 5y^2 - 10 = 0$

9. $\dfrac{(x - 2)^2}{16} + \dfrac{(y + 1)^2}{4} = 1$

10. $\dfrac{x^2}{9} + \dfrac{(y - 3)^2}{25} = 1$

11. $9x^2 + 4(y + 2)^2 = 36$

12. $16(x - 4)^2 + (y - 5)^2 = 16$

13. $25x^2 + 4y^2 - 50x + 24y + 45 = 0$

14. $9x^2 + 25y^2 - 36x + 50y = 39$

15. $4x^2 + y^2 + 40x - 4y + 103 = 0$

16. $2x^2 + y^2 + 8x + 8y + 23 = 0$

In Exercises 17–22, find the foci and eccentricity of the ellipse whose equation is given in the indicated exercise.

17. Exercise 3

18. Exercise 4

19. Exercise 9

20. Exercise 12

21. Exercise 13

22. Exercise 16

In Exercises 23–32, determine the equation of the ellipse with the given characteristics.

23. Center at the origin, minor axis of length 2, and horizontal major axis of length 5

24. Vertices $(0, \pm 3)$ and minor axis of length 2

25. Vertices $(2, -6)$ and $(2, 2)$, and minor axis of length 3

26. Vertices $(-1, 2)$ and $(5, 2)$, and endpoints of minor axis $(2, 4)$ and $(2, 0)$

27. Vertices $(-2, -3)$ and $(-2, 3)$, and one of the foci at $(-2, 1)$

28. Foci $(-4, -1)$ and $(2, -1)$, and one of the vertices at $(4, -1)$

29. Foci $(\pm 2, 0)$ and major axis of length 6

30. Foci $(1, \pm 4)$ and minor axis of length 6

31. Foci $(0, 0)$ and $(3, 0)$, and eccentricity $\frac{1}{2}$

32. Foci $(1, -3)$ and $(1, 5)$, and eccentricity $\frac{2}{3}$

In Exercises 33–36, use a graphing calculator to generate the graph of each ellipse. Then estimate the coordinates of the vertices, rounding each coordinate to three significant digits. Check algebraically.

33. $23.4x^2 + 19.6y^2 = 87.3$

34. $8.67x^2 + 3.67y^2 = 9.87y$

35. $0.25x^2 + 0.14y^2 - 1.24y - 8.12 = 0$

36. $5.4x^2 + 3.4y^2 - 2.2x - 4.8y - 11.6 = 0$

37. A road passes through a tunnel whose cross-section is a semiellipse, which is 12 feet high at the center and 30 feet wide, as shown in the sketch. What is the tallest tractor-trailer rig that can fit through the tunnel if the trailer is 10 feet wide?

38. A window positioned above an 8-foot-wide glass door is a semiellipse, as shown in the sketch. What is the length of each brace?

39. A stream 2.3 meters deep passes through an elliptical culvert, as shown in the sketch. What is the width w of the stream?

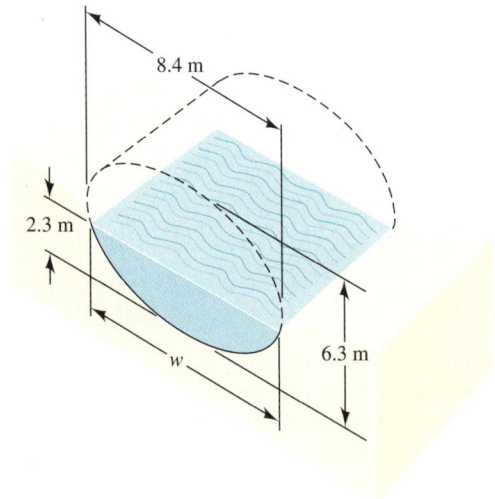

40. In astronomy, the average distance from the Earth to the Sun is called 1 astronomical unit (AU), and $1 \text{ AU} \approx 9.26 \times 10^7$ miles. It is known that Halley's comet travels around the Sun in an elliptical orbit with the Sun at one focus. The major axis of the ellipse is 36.18 AU and the minor axis is 9.12 AU.

(a) What is the distance between the foci of the ellipse?

(b) What is the eccentricity of Halley's comet?

(c) What is the closest that Halley's comet comes to the sun?

(d) What is the furthest that Halley's comet gets from the sun?

41. An architect's design for an elliptically shaped ceiling calls for rotating the upper portion of the ellipse $x^2 + 4y^2 = 400$, 180° about its major axis, as shown in the sketch.

(a) Find the length and width of this ceiling if x and y are measured in feet.

(b) By the reflection property of an ellipse, a whisper spoken at one focus can be heard at the other focus. Where are the whispering and listening positions located?

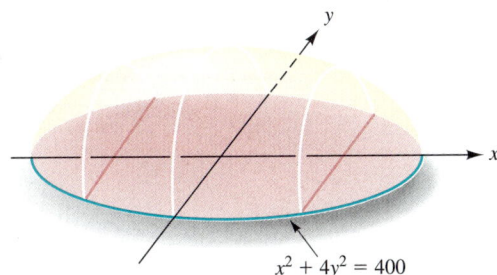

$$x^2 + 4y^2 = 400$$

42. A *lithotripter* is a medical device that uses the reflection property of an ellipse. Ultrahigh-frequency (UHF) sound waves emanating from one focus of a lithotripter are reflected to the other focus of the device to break up a

kidney stone. Suppose a lithotripter is formed by rotating the portion of an ellipse below its minor axis about its major axis, as shown in the sketch. How far from the kidney stone should the opening of the lithotripter be placed in order to break up the stone?

Critical Thinking

43. Suppose you are given two thumbtacks, a piece of string, and a pencil. Using this material, explain how to construct an ellipse.

44. The eccentricity of one ellipse is the constant k_1 and the eccentricity of another ellipse is the constant k_2 such that $k_1 > k_2$. Which ellipse is more circular? Explain.

45. Find the equation of the largest circle that can fit inside the ellipse

$$\frac{x^2}{a^2} + \frac{y^2}{b^2} = 1 \qquad \text{if } a < b.$$

46. The equation of an ellipse with a horizontal major axis is $Ax^2 + Cy^2 = E$. Find the coordinates of its foci in terms of A, C, and E.

47. The chord of an ellipse that passes through a focus and is perpendicular to its major axis is called the *latus rectum* of the ellipse. For the ellipse $(x^2/a^2) + (y^2/b^2) = 1$ with $a > b$, the latus rectum is the line segment $\overline{AB}$ shown in the sketch. Show that the length of the latus rectum is $2b^2/a$.

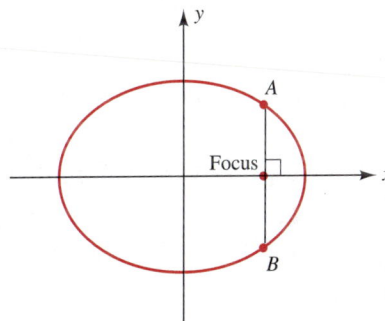

48. The sketch shows a square inscribed in the ellipse $9x^2 + 16y^2 = 144$. What is the length of the side of the square?

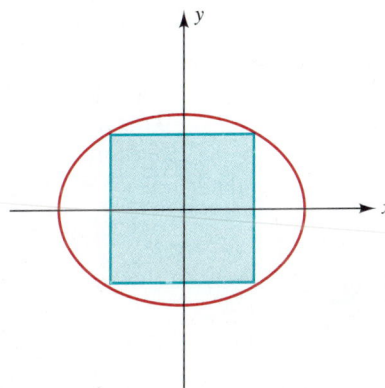

If a and b are positive real numbers, then solving
$(x^2/a^2) + (y^2/b^2) = 1$ *for y yields* $y = \pm(b/a)\sqrt{a^2 - x^2}$.
Taken separately, the equations

$$y = \frac{b}{a}\sqrt{a^2 - x^2} \quad \text{and} \quad y = -\frac{b}{a}\sqrt{a^2 - x^2}$$

define y as a function of x, and the graph of each function is a semi-ellipse (half an ellipse) whenever $a \neq b$. *In Exercises 49 and 50, state the domain and range of each function, and sketch its graph.*

49. $y = \frac{2}{3}\sqrt{9 - x^2}$ **50.** $y = -\sqrt{25 - 9x^2}$

6.4 The Hyperbola

◆ Introductory Comments

In this section, we study the geometric properties of the hyperbola and the distinguishing characteristics of its equation. First, we consider the equation

$$\frac{x^2}{a^2} - \frac{y^2}{b^2} = 1$$

with $a > 0$ and $b > 0$. This is a general quadratic equation in two unknowns with $A = 1/a^2$, $C = -1/b^2$, $F = -1$, and $B = D = E = 0$. Using the tests for symmetry (Section 2.2), we find that the graph of this equation is symmetric with respect to the x-axis, the y-axis, and the origin. To find the x-intercept of the graph we let $y = 0$ and obtain $x = \pm a$. However, if we let $x = 0$, we find that the graph has no y-intercept because the equation

$$-\frac{y^2}{b^2} = 1 \quad \text{or} \quad y^2 = -b^2$$

has no real solution. If we solve the original equation $\dfrac{x^2}{a^2} - \dfrac{y^2}{b^2} = 1$ for y, we obtain

$$y = \pm\frac{b}{a}\sqrt{x^2 - a^2}.$$

Now if y is to be a real number, then $x^2 - a^2$ must be either positive or zero, that is,

$$x^2 - a^2 \geq 0,$$

so $x \geq a \quad \text{or} \quad x \leq -a.$

This means that the graph exists only to the right of the x-intercept a and to the left of the x-intercept $-a$. Intuitively, we can see that for very large values of x (either negative or positive),

$$y = \pm\frac{b}{a}\sqrt{x^2 - a^2} \approx \pm\frac{b}{a}\sqrt{x^2} = \pm\frac{b}{a}x.$$

In other words, the graph of $(x^2/a^2) - (y^2/b^2) = 1$ approaches the lines $y = \pm\dfrac{b}{a}x$ as $|x|$ increases without bound. We refer to the lines

$$y = \pm \frac{b}{a}x$$

as the **asymptotes** of the curve.

The diagonals of a rectangle with vertices (a, b), $(-a, b)$, $(-a, -b)$, and $(a, -b)$ have slopes $\pm b/a$. Extending these diagonals gives us the asymptotes of the curve, as shown in Figure 6.42. Using the x-intercepts and the asymptotes, we can sketch the graph of $(x^2/a^2) - (y^2/b^2) = 1$, which is the **hyperbola** shown in Figure 6.43. The two distinct curves that make up the graph are the **branches** of the hyperbola.

Figure 6.43 shows the characteristics of a hyperbola. The points $(-a, 0)$ and $(a, 0)$ are the **vertices** of the hyperbola, and the horizontal and vertical line segments $\overline{AB}$ and $\overline{CD}$ are called the **axes** of the hyperbola. Half of an axis of a hyperbola is called a **semiaxis**. Note that the lengths of the semiaxes are a and b. The line segment that connects the vertices of the hyperbola is called the **transverse axis** and the other axis of the hyperbola is the **conjugate axis**. In this figure the horizontal line segment $\overline{AB}$ is the transverse axis and the vertical line segment $\overline{CD}$ is the conjugate axis. The midpoint of both the transverse axis and the conjugate axis is the **center** of the hyperbola. The hyperbola in Figure 6.43 has its center at the origin.

The graph of the equation

$$\frac{y^2}{b^2} - \frac{x^2}{a^2} = 1$$

with $a > 0$ and $b > 0$ is also a hyperbola. However, the y-intercepts are now $\pm b$ and the graph has no x-intercept. Thus, the vertices of this hyperbola are $(0, b)$ and $(0, -b)$. The transverse axis is vertical with length $2b$ and the conjugate axis is horizontal with length $2a$. The asymptotes are still the two lines $y = \pm(b/a)x$, and the center of the hyperbola remains at the origin. The graph of $(y^2/b^2) - (x^2/a^2) = 1$ is shown in Figure 6.44.

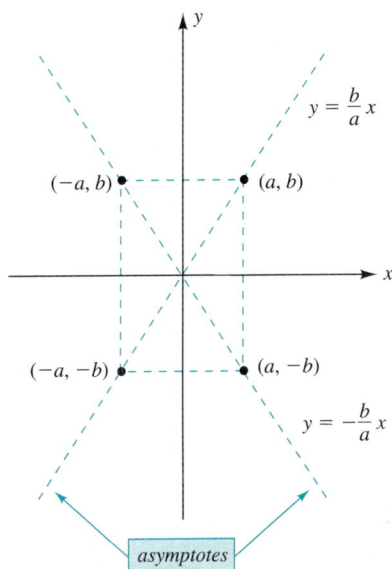

FIGURE 6.42

The asymptotes $y = \pm(b/a)x$ can be constructed from the diagonals of the rectangle with vertices (a, b), $(-a, b)$, $(-a, -b)$, and $(a, -b)$.

FIGURE 6.43

The graph of $\dfrac{x^2}{a^2} - \dfrac{y^2}{b^2} = 1$ is called a

hyperbola. The x-intercepts are $\pm a$ and the asymptotes are $y = \pm(b/a)x$.

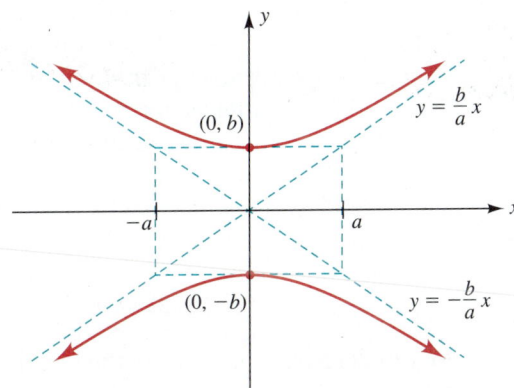

FIGURE 6.44

The graph of $\dfrac{y^2}{b^2} - \dfrac{x^2}{a^2} = 1$ is a hyperbola with y-intercepts $\pm b$ and asymptotes $y = \pm(b/a)x$.

◆   **Standard Form of a Hyperbola**

The graphs of the equations

$$\frac{x^2}{a^2} - \frac{y^2}{b^2} = 1 \quad \text{and} \quad \frac{y^2}{b^2} - \frac{x^2}{a^2} = 1$$

are the same as the graphs of

$$\frac{(x-h)^2}{a^2} - \frac{(y-k)^2}{b^2} = 1 \quad \text{and} \quad \frac{(y-k)^2}{b^2} - \frac{(x-h)^2}{a^2} = 1,$$

respectively, but the center of the hyperbola is now (h, k). We refer to these equations with $a > 0$ and $b > 0$ as the *algebraic definitions of a hyperbola*, or the *equations of a hyperbola in standard form*.

◆ **Equations of a Hyperbola in Standard Form**

The **equation of a hyperbola in standard form** with center (h, k), *horizontal* transverse axis of length $2a$, and vertical conjugate axis of length $2b$ is

$$\frac{(x-h)^2}{a^2} - \frac{(y-k)^2}{b^2} = 1.$$

The **equation of a hyperbola in standard form** with center (h, k), *vertical* transverse axis of length $2b$, and horizontal conjugate axis of length $2a$ is

$$\frac{(y-k)^2}{b^2} - \frac{(x-h)^2}{a^2} = 1.$$

We can determine the equation of a hyperbola if we know its center (h, k) and the lengths of its semiaxes a and b.

EXAMPLE 1   Determine the equation of a hyperbola if its asymptotes are $y = \frac{1}{2}x + 1$ and $y = -\frac{1}{2}x + 1$ and its horizontal transverse axis has length 6.

◆ **Solution**   The hyperbola with the given characteristics is shown in Figure 6.45. The intersection point of the asymptotes is the center of the hyperbola. Hence, the center of this hyperbola is

$$(h, k) = (0, 1).$$

Since we know that the horizontal transverse axis has length 6, we have

$$2a = 6 \quad \text{or} \quad a = 3.$$

We can determine b by using the fact that the slopes of the asymptotes are $\pm b/a = \pm 1/2$. Substituting $a = 3$, we have

$$\pm \frac{b}{3} = \pm \frac{1}{2},$$

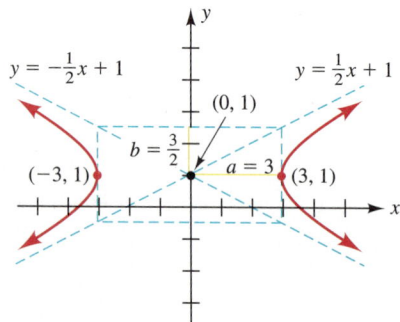

FIGURE 6.45

The hyperbola with asymptotes $y = \pm\frac{1}{2}x + 1$ and horizontal transverse axis of length 6 which implies that $b = \frac{3}{2}$. Therefore, the equation of the hyperbola is

$$\frac{(x - h)^2}{a^2} - \frac{(y - k)^2}{b^2} = 1$$

$$\frac{(x - 0)^2}{3^2} - \frac{(y - 1)^2}{(3/2)^2} = 1 \qquad \textbf{Substitute}$$

$$\frac{x^2}{9} - \frac{4(y - 1)^2}{9} = 1 \qquad \textbf{Simplify}$$ ◆

To generate the graph of

$$\frac{(x - h)^2}{a^2} - \frac{(y - k)^2}{b^2} = 1 \qquad \text{or} \qquad \frac{(y - k)^2}{b^2} - \frac{(x - h)^2}{a^2} = 1$$

in the viewing rectangle of a graphing calculator, we must solve the equation for y and enter both equations that result.

PROBLEM 1 Generate the graph of $\dfrac{x^2}{9} - \dfrac{4(y - 1)^2}{9} = 1$ in the viewing rectangle of a graphing calculator. Verify that this hyperbola has horizontal transverse axis of length 6 and asymptotes $y = \pm\frac{1}{2}x + 1$. ◆

◆ General Form of a Hyperbola

If we take the equations of a hyperbola in standard form and multiply both sides of each of these equations by a^2b^2, we obtain equations of the forms

$$A(x - h)^2 - C(y - k)^2 = N \qquad \text{and} \qquad C(y - k)^2 - A(x - h)^2 = N$$

with $A = b^2$, $C = a^2$, and $N = a^2b^2$. Now, if we expand $(x - h)^2$ and $(y - k)^2$ in these equations, distribute A and C, and collect all terms on one side, we obtain general quadratic equations in two unknowns of the forms

$$Ax^2 - Cy^2 + Dx + Ey + F = 0$$

and

$$Cy^2 - Ax^2 + Dx + Ey + F = 0$$

We refer to these two equations as the **equations of a hyperbola in general form** with vertical and horizontal axes of symmetry, respectively. These general quadratic equations in two unknowns are characterized by the presence of x^2 and y^2 terms that have *different* signs.

If the equation of a hyperbola is given in general form, then we can find the coordinates of its center and vertices by converting the equation to standard form and observing the values of h, k, a, and b. We can use a graphing calculator to check our work by solving the equation in general form for y and graphing both equations that result.

EXAMPLE 2   Determine the coordinates of the center and vertices of the hyperbola given by each equation.

(a) $4x^2 - 5y^2 = 20$ **(b)** $4x^2 - 9y^2 + 8x + 36y + 4 = 0$

◆ **Solution**

(a) We begin by writing the equation in standard form:

$$4x^2 - 5y^2 = 20$$

$$\frac{x^2}{5} - \frac{y^2}{4} = 1 \qquad \text{Divide both sides by 20}$$

$$\frac{(x-0)^2}{(\sqrt{5})^2} - \frac{(y-0)^2}{2^2} = 1 \qquad \text{Write in standard form}$$

This last equation tells us that the graph of $4x^2 - 5y^2 = 20$ is a hyperbola with horizontal semitransverse axis of length $a = \sqrt{5}$, vertical semiconjugate axis of length $b = 2$, and center at the origin. Hence, the coordinates of the vertices are

$$\left(\sqrt{5}, 0\right) \approx (2.236, 0) \qquad \text{and} \qquad \left(-\sqrt{5}, 0\right) \approx (-2.236, 0).$$

Figure 6.46 shows the graph of $4x^2 - 5y^2 = 20$ $\left(\text{or } y = \pm 2\sqrt{\frac{x^2 - 5}{5}}\right)$ in the viewing rectangle of a graphing calculator. This graph confirms our work.

(b) We begin by writing the equation in standard form. To do this, we use the process of completing the square (Section 3.3):

$$4x^2 - 9y^2 + 8x + 36y + 4 = 0$$

$$(4x^2 + 8x) + (-9y^2 + 36y) = -4 \qquad \text{Regroup}$$

$$4(x^2 + 2x) - 9(y^2 - 4y) = -4 \qquad \text{Factor out 4 and } -9$$

Add 4 to both sides.

$$4(x^2 + 2x + 1) - 9(y^2 - 4y + 4) = -4 + (4 - 36) \qquad \text{Complete the squares}$$

Add −36 to both sides.

$$4(x + 1)^2 - 9(y - 2)^2 = -36 \qquad \text{Factor}$$

$$\frac{(y-2)^2}{4} - \frac{(x+1)^2}{9} = 1 \qquad \text{Divide both sides by } -36$$

$$\frac{(y-2)^2}{2^2} - \frac{(x+1)^2}{3^2} = 1 \qquad \text{Write in standard form}$$

The last equation tells us that the graph of $4x^2 - 9y^2 + 8x + 36y + 4 = 0$ is a hyperbola with vertical semitransverse axis of length $b = 2$, horizontal

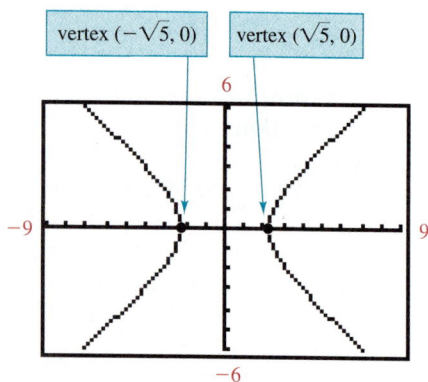

FIGURE 6.46
The graph of $4x^2 - 5y^2 = 20$ is a hyperbola with center at the origin and vertices $(\pm\sqrt{5}, 0)$.

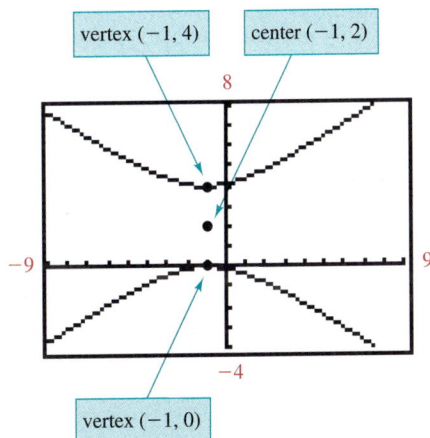

vertex $(-1, 4)$ center $(-1, 2)$

vertex $(-1, 0)$

FIGURE 6.47

The graph of
$4x^2 - 9y^2 + 8x + 36y + 4 = 0$
is a hyperbola with center $(-1, 2)$ and
vertices $(-1, 4)$ and $(-1, 0)$.

semiconjugate axis of length $a = 3$, and center $(h, k) = (-1, 2)$. Hence the coordinates of the vertices are $(-1, 4)$ and $(-1, 0)$.

To generate the graph of the equation $4x^2 - 9y^2 + 8x + 36y + 4 = 0$ in the viewing rectangle of a graphing calculator, we rewrite the equation as a quadratic equation in y,

$$-9y^2 + 36y + (4x^2 + 8x + 4) = 0,$$

and apply the quadratic formula:

$$y = \frac{-36 \pm \sqrt{(36)^2 - 4(-9)(4x^2 + 8x + 4)}}{2(-9)}$$

$$= 2 \pm \frac{2\sqrt{x^2 + 2x + 10}}{3}$$

Entering both of these equations into the calculator, we obtain the hyperbola that is shown in Figure 6.47. This graph confirms our work. ◆

CAUTION   Not every equation of the form

$$Ax^2 - Cy^2 + Dx + Ey + F = 0 \quad \text{or} \quad Cy^2 - Ax^2 + Dx + Ey + F = 0,$$

where the x^2 and y^2 terms have different signs, represents a hyperbola. For example, if $Ax^2 - Cy^2 + Dx + Ey + F = 0$ is written in the form

$$A(x - h)^2 - C(y - k)^2 = N \quad \text{and} \quad N = 0,$$

then the graph of the equation is a pair of lines that intersect at (h, k). We refer to these two lines as a *degenerate hyperbola*.

PROBLEM 2   Is $x^2 - 4y^2 - 6x + 8y + 5 = 0$ the equation of a hyperbola? Explain. ◆

◆ **Geometric Definition of a Hyperbola**

Geometrically, we define a hyperbola as follows.

◆ **Geometric Definition of a Hyperbola**

A **hyperbola** is the set of all points in a plane, the difference of whose distances from two fixed points is a constant, which is equal to the length of the transverse axis of the hyperbola. The two fixed points are called the **foci** (plural of focus).

For convenience, we place a hyperbola in the xy-coordinate plane so that its center is at the origin, its transverse axis is horizontal with length $2a$, and

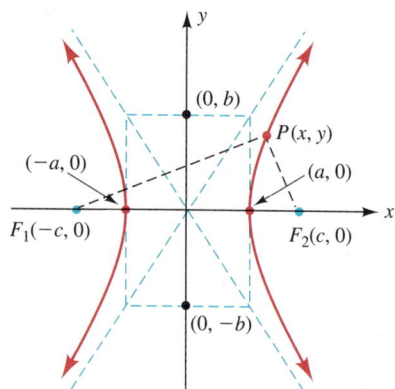

its conjugate axis is vertical with length $2b$ (see Figure 6.48). We have labeled the coordinates of the foci $(-c, 0)$ and $(c, 0)$, with $a < c$.

The relationship between the constants a, b, and c in any hyperbola is

$$c^2 = a^2 + b^2$$

FIGURE 6.48

A hyperbola in the *xy*-coordinate plane with center at the origin, transverse axis of length $2a$, conjugate axis of length $2b$, and foci with coordinates $(-c, 0)$ and $(c, 0)$

Figure 6.49 illustrates this relationship. If c is the length of the line segment from the center to the focus, then c is also the length of the diagonal of the rectangle that is formed by a and b. Now, referring to Figure 6.48 and applying our geometric definition of a hyperbola, we have

$$PF_1 - PF_2 = 2a$$

$$\sqrt{(x + c)^2 + y^2} - \sqrt{(x - c)^2 + y^2} = 2a$$

Using the same algebraic procedure as we used for an ellipse in Section 6.3— isolating the radical, squaring both sides, and simplifying—and using the fact that $c^2 = a^2 + b^2$ in any hyperbola, we obtain the algebraic definition

$$\frac{x^2}{a^2} - \frac{y^2}{b^2} = 1.$$

Hence, we conclude that our algebraic definition of a hyperbola is consistent with our geometric definition.

The **eccentricity** e of a hyperbola is a measure of the "spread of its wings" and is defined as

$$e = \frac{\text{distance from center to focus}}{\text{distance from center to vertex}}$$

FIGURE 6.49

The constants a, b, and c are related by the Pythagorean theorem.

The eccentricity e of a hyperbola is always a number greater than 1. The larger the number, the wider the opening of each branch of the hyperbola.

EXAMPLE 3   Find the foci and eccentricity of the hyperbola defined by each equation.

(a) $16x^2 - 9y^2 = 144$ **(b)** $y^2 - 4x^2 - 6y + 5 = 0$

◆ **Solution**

(a) Dividing both sides by 144, we obtain

$$\frac{x^2}{9} - \frac{y^2}{16} = 1 \quad \text{or} \quad \frac{(x - 0)^2}{3^2} - \frac{(y - 0)^2}{4^2} = 1,$$

the equation of a hyperbola in standard form with center at the origin, horizontal semitransverse axis of length $a = 3$ and vertical semiconjugate axis of length $b = 4$. To find the coordinates of the foci, we apply the relation

$$c^2 = 3^2 + 4^2,$$

which implies that $|c| = 5$. Hence, the coordinates of the foci are $(-5, 0)$ and $(5, 0)$, as shown in Figure 6.50.

The eccentricity e of this hyperbola is

$$e = \frac{\text{distance from center to focus}}{\text{distance from center to vertex}} = \frac{5}{3} \approx 1.67.$$

(b) First, we write the equation in standard form by completing the square:

$$y^2 - 4x^2 - 6y + 5 = 0$$

$$(y^2 - 6y \quad) - 4x^2 = -5 \qquad \text{Regroup}$$

Add 9 to both sides.

$$(y^2 - 6y + 9) - 4x^2 = -5 + 9 \qquad \text{Complete the square}$$

$$(y - 3)^2 - 4x^2 = 4 \qquad \text{Factor}$$

$$\frac{(y - 3)^2}{2^2} - \frac{(x - 0)^2}{1^2} = 1 \qquad \text{Divide both sides by 4 and write in standard form}$$

This is the equation of a hyperbola in standard form with center $(h, k) = (0, 3)$, vertical semitransverse axis of length $b = 2$, and horizontal semiconjugate axis of length $a = 1$. To find the coordinates of the foci, we apply the relation

$$c^2 = 1^2 + 2^2,$$

which implies that $|c| = \sqrt{5} \approx 2.236$. Therefore, the coordinates of the foci are

$$\left(0, 3 + \sqrt{5}\right) \approx (0, 5.236) \quad \text{and} \quad \left(0, 3 - \sqrt{5}\right) \approx (0, 0.764)$$

as shown in Figure 6.51.

The eccentricity e of this hyperbola is

$$e = \frac{\text{distance from center to focus}}{\text{distance from center to vertex}} = \frac{\sqrt{5}}{2} \approx 1.12. \qquad ◆$$

PROBLEM 3 Find the foci and eccentricity for the hyperbola defined by $9x^2 - y^2 = 9 = 0$. ◆

If we know the foci and eccentricity of a hyperbola, then we can determine its equation.

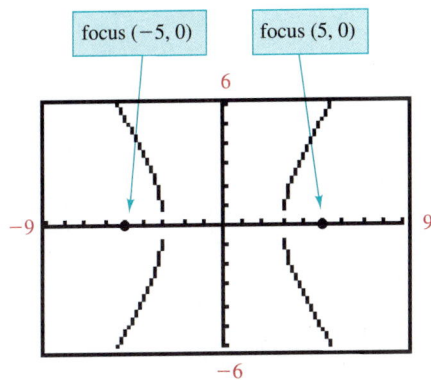

FIGURE 6.50
The foci of the hyperbola
$16x^2 - 9y^2 = 144$ are $(-5, 0)$
and $(5, 0)$.

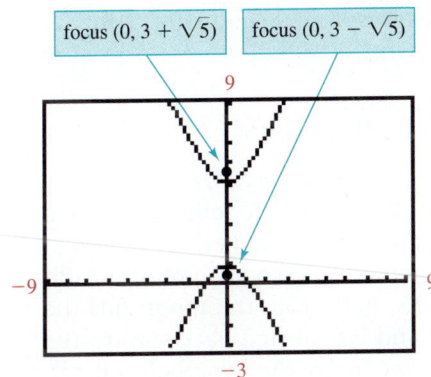

FIGURE 6.51
The foci of the hyperbola
$y^2 - 4x^2 - 6y + 5 = 0$ are
$(0, 3 + \sqrt{5})$ and $(0, 3 - \sqrt{5})$.

EXAMPLE 4   Find the equation of the hyperbola in standard form with foci (2, 5) and (2, −3) and eccentricity 2.

◆ **Solution**   The center of a hyperbola is located halfway between its foci. Hence, the center of this hyperbola is $(h, k) = (2, 1)$, as shown in Figure 6.52. Therefore, since the transverse axis is vertical, the equation of this hyperbola must have the form

$$\frac{(y-1)^2}{b^2} - \frac{(x-2)^2}{a^2} = 1$$

We now proceed to find the lengths of the semitransverse axis b and the semiconjugate axis a. Since the distance from the center to a focus is $c = 4$ and the eccentricity of his hyperbola is given to be 2, we have

$$\frac{\text{distance from center to focus}}{\text{distance from center to vertex}} = \frac{4}{b} = 2,$$

which implies that $b = 2$. We find the value of a by applying the relation $c^2 = a^2 + b^2$:

$$4^2 = a^2 + 2^2 \quad \text{Substitute } c = 4 \text{ and } b = 2$$
$$a^2 = 4^2 - 2^2 \quad \text{Solve for } a$$
$$a^2 = 12$$
$$a = 2\sqrt{3}$$

Since $(h, k) = (2, 1)$, $a^2 = (2\sqrt{3})^2 = 12$, and $b^2 = (2)^2 = 4$, the equation of this hyperbola in standard form is

$$\frac{(y-1)^2}{4} - \frac{(x-2)^2}{12} = 1.$$

PROBLEM 4   Referring to Example 4, express the equation of the hyperbola in general form with integer coefficients.

◆ **Application: Reflection Property of a Hyperbola**

Like the parabola and ellipse, the hyperbola also has a *reflection property*. A ray of light directed at one focus of a hyperbolic mirror will be reflected so as to pass through its other focus. The reflection property of a hyperbola is used in the design of reflecting telescopes such as the Hale telescope on Mount Palomar, California. As illustrated in Figure 6.53, light from the moon hits the primary parabolic mirror in the telescope and is reflected back toward the parabola's focus F_1, which is also one of the two foci of the hyperbolic mirror. The light hits the hyperbolic mirror and is reflected toward the hyperbola's second focus F_2, which is also one of the two foci of an elliptical mirror. The light then hits the elliptical mirror and reflects to the ellipses's second focus F_3, which is the eyepiece of the telescope (see Exercises 37).

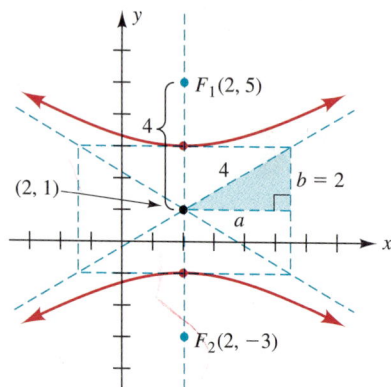

FIGURE 6.52
A hyperbola with foci (2, 5) and (2, −3) has center (2, 1).

FIGURE 6.53
A reflecting telescope uses the reflection properties of the parabola, ellipse, and hyperbola to direct an image to the eyepiece.

Exercises 6.4

◆ **Basic Skills**

In Exercises 1–16, determine the coordinates of the center and vertices of the hyperbola given by each equation. Check graphically with a graphing calculator.

1. $\dfrac{x^2}{4} - \dfrac{y^2}{16} = 1$ **2.** $\dfrac{y^2}{4} - \dfrac{x^2}{4} = 1$

3. $9y^2 - 4x^2 = 36$ **4.** $9x^2 - 16y^2 = 144$

5. $16x^2 - y^2 = 9$ **6.** $4y^2 - 16x^2 = 25$

7. $9x^2 - 25y^2 = -4$ **8.** $4y^2 - x^2 + 12 = 0$

9. $\dfrac{x^2}{9} - \dfrac{(y + 2)^2}{4} = 1$

10. $\dfrac{(y - 1)^2}{16} - (x + 3)^2 = 1$

11. $(y - 3)^2 - (x - 1)^2 = 36$

12. $9(x + 1)^2 - 25(y - 4)^2 = 225$

13. $36x^2 - y^2 - 144x - 6y + 126 = 0$

14. $16y^2 - 9x^2 + 90x = 261$

15. $9y^2 - 3x^2 - 24x + 18y = 120$

16. $2x^2 - 4y^2 + 12x + 16y + 1 = 0$

In Exercises 17–22, find the foci and eccentricity of the hyperbola whose equation is given in the indicated exercise.

17. Exercise 3   **18.** Exercise 4

19. Exercise 9   **20.** Exercise 12

21. Exercise 13   **22.** Exercise 16

In Exercises 23–32, determine the equation of the hyperbola with the given characteristics.

23. Vertices $(0, \pm 4)$ and asymptotes $4y - x = 0$ and $4y + x = 0$

24. Vertices $(2, -2)$ and $(-2, -2)$, and endpoints of conjugate axis $(0, 0)$ and $(0, -4)$

25. Asymptotes $y = \pm x$ and horizontal transverse axis of length 4

26. Asymptotes $y = -x$ and $y = x + 2$, and vertical transverse axis of length 4

27. Vertices $(-2, -3)$ and $(-2, 3)$, and one of the foci at $(-2, 5)$

28. Foci $(-4, -1)$ and $(2, -1)$, and one of the vertices at $(1, -1)$

29. Foci $(\pm 2, 0)$ and transverse axis of length 2

30. Foci $(1, \pm 4)$ and conjugate axis of length 4

31. Foci $(0, 0)$ and $(3, 0)$, and eccentricity 3

32. Foci $(1, -3)$ and $(1, 5)$, and eccentricity $\frac{4}{3}$

In Exercises 33–36, use a graphing calculator to generate the graph of each hyperbola. Then estimate the coordinates of the vertices, rounding each coordinate to three significant digits. Check algebraically.

33. $356x^2 - 129y^2 = 926$

34. $1.25x^2 - 2.70y^2 + 8.55y = 2.35$

35. $0.35x^2 - 0.12y^2 - 2.34y - 7.89 = 0$

36. $3.4x^2 - 6.4y^2 - 1.4x - 4.8y - 18.6 = 0$

37. A Cassegrain telescope contains a primary parabolic mirror and a smaller hyperbolic mirror, as shown in the sketch. A ray of light from the moon strikes the parabolic mirror and is reflected toward the parabola's focus F_1, which is also a focus of the hyperbolic mirror. The ray of light is then reflected off the hyperbolic mirror toward the hyperbola's other focus F_2, which is the eyepiece of the telescope. If the equation of the hyperbolic mirror is $x^2/441 - y^2/400 = 1$ and x and y are measured in centimeters, determine the distance from the hyperbolic mirror to the eyepiece.

38. The figure shows the hyperbolic path of a UFO. This path moves closer toward Earth along the line $y = x/4$ and comes within 6000 miles of the Earth's surface before moving away. Assuming the radius of the Earth

CHAPTER 6 ◆ *Conic Sections*

is 4000 miles, determine the equation of the path of the UFO.

$$y = \frac{x}{4}$$

Earth

6000 mi

39. Allied tanks are located at fixed points *A* and *B*, as shown in the sketch. An enemy tank located at point *P* fires its gun and 1.3 seconds after the shot is heard at point *B*, the shot is heard at point *A*.

 (a) Assuming that the speed of sound is 1100 ft/s, find *PA* − *PB* and explain why point *P* must lie on a branch of a hyperbola.

 (b) Determine the equation of the branch of the hyperbola described in part (a) if the coordinates (in feet) of points *A* and *B* are (−1500, 0) and (1500, 0), respectively.

 (c) An allied patrol unit at point *C* (0, 0) calculates that the distance from them to the enemy tank is 950 ft. Determine the distances *PA* and *PB* from the allied tanks to the enemy tank.

P
Enemy tank
Allied tank
Allied tank
A   C   B
Patrol unit

40. In an engineer's design for the horn on a megaphone, a hyperbola is rotated about the *x*-axis, as shown in the figure. If the distance from *A* to *B* is 10.26 cm, determine the radius of each end of the megaphone.

$$y = 0.42x$$

25.60 cm
5.08 cm

◆ **Critical Thinking**

41. Describe the general shape of a hyperbola whose eccentricity is (a) close to 1 or (b) very large.

42. What is the eccentricity of a hyperbola with asymptotes $y = \pm x$?

43. Suppose the transverse axis and conjugate axis of a hyperbola are both doubled. What effect does this have on the hyperbola's eccentricity and asymptotes?

44. The equation of a hyperbola with a horizontal transverse axis is $Ax^2 - Cy^2 = E$. Find the coordinates of its foci in terms of *A*, *C*, and *E*.

45. Describe the graph of each of the following degenerate conic sections.

 (a) $x^2 + y^2 - 6x - 8y + 25 = 0$

 (b) $9x^2 + 4y^2 + 54x - 16y + 97 = 0$

 (c) $4x^2 - y^2 - 16x - 4y + 12 = 0$

 (d) $y^2 - 9x^2 + 36x - 10y - 11 = 0$

46. When a supersonic jet breaks the sound barrier, it produces a conical shock wave, as shown in the sketch. Describe the intersection of the conical shock wave with the ground if the plane's path is parallel to the ground.

Conical shock wave
Jet
Intersection

If a and b are positive real numbers, then solving

$$\frac{x^2}{a^2} - \frac{y^2}{b^2} = 1 \quad \text{and} \quad \frac{y^2}{b^2} - \frac{x^2}{a^2} = 1$$

for y yields

$$y = \pm \frac{b}{a}\sqrt{x^2 - a^2} \quad \text{and} \quad y = \pm \frac{b}{a}\sqrt{x^2 + a^2},$$

respectively. Taken separately, the equations

$$y = \frac{b}{a}\sqrt{x^2 - a^2} \qquad y = -\frac{b}{a}\sqrt{x^2 - a^2}$$

$$y = \frac{b}{a}\sqrt{x^2 + a^2} \qquad y = -\frac{b}{a}\sqrt{x^2 + a^2}$$

define y as a function of x, and the graph of each function is one-half (not necessarily one branch) of a hyperbola. In Exercises 47–50, state the domain and range of each function, and sketch its graph.

47. $y = \frac{1}{2}\sqrt{x^2 - 4}$ **48.** $y = -3\sqrt{x^2 - 25}$

49. $y = -\sqrt{x^2 + 9}$ **50.** $y = \frac{3}{4}\sqrt{4x^2 + 1}$

Chapter 6 Review

◆ Questions for Writing or Group Discussion

1. Name the four main *conic sections.* Show how each is formed by slicing a double cone with a plane.

2. What is meant by a *degenerate conic section*?

3. Give the *standard form* of the equation of each conic section if its axes of symmetry are parallel to the *xy*-axes.

4. Give the *general form* of the equation of each conic section if its axes of symmetry are parallel to the *xy*-axes.

5. State the geometric definition of a *circle.* Illustrate with a sketch.

6. What is the slope of the *tangent* to the circle $x^2 + y^2 = r^2$ at the point (x_1, y_1) on the circle?

7. State the geometric definition of a *parabola.* Illustrate with a sketch.

8. Under what conditions does the equation of a parabola define *y* as a function of *x*?

9. How can you tell from the equation of a parabola whether the curve opens upward, downward, to the left, or to the right?

10. State the geometric definition of an *ellipse.* Illustrate with a sketch.

11. What is meant by the *major axis* of an ellipse? How can you determine from the equation of an ellipse in standard form whether the major axis is horizontal or vertical?

12. Why is a *circle* a special case of an ellipse?

13. Explain the procedure for finding the coordinates of the *foci* of an ellipse from its equation. Illustrate with an example.

14. What is meant by the *eccentricity* of an ellipse? Describe the shape of an ellipse with eccentricity close to 1 or close to 0.

15. State the geometric definition of a *hyperbola.* Illustrate with a sketch.

16. What is meant by the *transverse axis* of a hyperbola? How can you determine from the equation of a hyperbola in standard form whether the transverse axis is horizontal or vertical?

17. What are the equations of the *asymptotes* of the hyperbola $(x^2/a^2) - (y^2/b^2) = 1$? How are they used as an aid in graphing a hyperbola?

18. What is meant by the *eccentricity* of a hyperbola? Describe the shape of a hyperbola with eccentricity close to 1 or very large.

19. Explain the procedure for finding the coordinates of the *foci* of a hyperbola from its equation. Illustrate with an example.

20. Discuss the *reflection properties* of a parabola, an ellipse, and a hyperbola.

◆ **Review Exercises**

In Exercises 1–28, identify the graph of each equation as a circle, parabola, ellipse, hyperbola, or a degenerate conic section. If the graph is a circle, give its radius and the coordinates of its center. If the graph is a parabola, give the direction of its opening and the coordinates of its vertex. If the graph is an ellipse or a hyperbola, give the coordinates of its center and vertices.

1. $x^2 + y^2 = 16$

2. $9x^2 + 9y^2 = 1$

3. $y^2 - 4x = 0$

4. $2x^2 + y = 0$

5. $x^2 + 9y^2 = 16$

6. $4x^2 + 3y^2 = 9$

7. $x^2 - y^2 = 25$

8. $4y^2 - x^2 = 49$

9. $4(x + 2)^2 + 4(y + 1)^2 = 1$

10. $(x - 1)^2 + y^2 = 5$

11. $y = (x - 2)^2 - 3$

12. $x = -2(y + 1)^2 + 2$

13. $9x^2 + (y - 3)^2 = 9$

14. $4(x + 2)^2 + 9(y - 1)^2 = 36$

15. $9x^2 - 16(y - 1)^2 = 144$

16. $(x + 2)^2 - (y - 1)^2 = -4$

17. $x^2 + y^2 - 10x - 8y + 16 = 0$

18. $x^2 + y^2 - 6y - 16 = 0$

19. $y^2 + x - 2y - 15 = 0$

20. $x^2 + 2x - 2y + 5 = 0$

21. $4x^2 + y^2 - 8x + 4y - 8 = 0$

22. $x^2 + 16y^2 - 32y - 9 = 0$

23. $4y^2 - 9x^2 - 8y + 5 = 0$

24. $16x^2 - 9y^2 - 32x - 36y - 24 = 0$

25. $y^2 - 4x^2 + 8x - 4 = 0$

26. $x^2 + y^2 - 16y + 64 = 0$

27. $2x^2 + 3y^2 - 4x + 12y + 14 = 0$

28. $x^2 - y^2 - 4x - 2y + 3 = 0$

In Exercises 29–32, find the focus and directrix of the conic section whose equation is given in the indicated exercise.

29. Exercise 3

30. Exercise 12

31. Exercise 19

32. Exercise 20

In Exercises 33–36, find the foci and eccentricity of the conic section whose equation is given in the indicated exercise.

33. Exercise 7

34. Exercise 6

35. Exercise 21

36. Exercise 24

In Exercises 37–56, determine the equation of the conic section from the given information.

37. Circle with center $(0, 0)$ and radius 7

38. Circle with center $(0, -2)$ and passing through $(1, 3)$

39. Circle whose diameter is the line segment joining the points $(0, 3)$ and $(4, 6)$

40. Circle with center $(4, -2)$ and tangent to the x-axis

41. Circle with center $(2, 3)$ and tangent to $4x - 3y = 24$

42. Circle with radius $\sqrt{5}$ and passing through $(2, 1)$ and $(2, 5)$

43. Parabola with vertex $(0, -4)$ and x-intercepts ± 2

44. Parabola with vertex $(2, 1)$ and y-intercepts -1 and 3

45. Parabola passing through the origin, vertex $(-2, -3)$, and symmetric with respect to $y = -3$

46. Parabola passing through $(-1, 0)$, vertex $(2, 3)$, and symmetric with respect to $x = 2$

47. Parabola with focus $(1, 0)$ and directrix $y = -1$

48. Parabola with vertex $(3, -2)$ and focus $(4, -2)$

49. Ellipse with vertices $(0, 0)$ and $(6, 0)$, and minor axis of length 4

50. Ellipse with center $(2, 5)$, minor axis of length 2, and horizontal major axis of length 8

51. Ellipse with foci $(0, 0)$ and $(4, 0)$, and eccentricity $\frac{1}{2}$

52. Ellipse with vertices $(-1, -3)$ and $(-1, 3)$, and one of the foci at $(-1, 2)$

53. Hyperbola with asymptotes $y = \pm \frac{2}{3}x$ and vertical transverse axis of length 6

54. Hyperbola with center $(2, -1)$, horizontal transverse axis of length 10, and conjugate axis of length 4

55. Hyperbola with vertices $(-3, -1)$ and $(3, -1)$, and one of the foci at $(4, -1)$

56. Hyperbola with foci $(\pm 3, 0)$ and eccentricity 2

57. Discuss the position of the circle $(x - h)^2 + (y - k)^2 = r^2$ relative to the coordinate axes for the indicated condition.

(a) $h = 0$ (b) $k = 0$

(c) $h = r$ (d) $k = r$

58. A *secant* to a circle is a line that intersects the circle in two distinct points. Find the coordinates of the points of intersection of the secant $x + y = 4$ and the circle $x^2 + y^2 = 10$.

59. Two hyperbolas are *conjugate* to each other if they have the same asymptotes and if the transverse axis of one is equal to the conjugate axis of the other. Find the equation of the hyperbola that is conjugate to $3x^2 - 4y^2 = 12$.

60. Find the length of the side of a square inscribed in an ellipse with semimajor axis of length a and semiminor axis of length b.

For Exercises 61–64, refer to the hyperbola $(x^2/a^2) - (y^2/b^2) = 1$ with point P on the asymptote $y = -\dfrac{b}{a}x$ and foci at $(-c, 0)$ and $(c, 0)$, as shown in the figure.

61. Find the slope of $\overline{FP}$.

62. Find the equation of the line that passes through F and P.

63. Find the coordinates of point P.

64. Find FP, the distance from F to P. What conclusion can you make about the distance between a focus and an asymptote of a hyperbola?

65. Two straight roadways are connected by a circular curve that has central angle $\theta = 90°$ and radius $r = 100$ ft, as

shown in the sketch. A surveyor wants to lay out this circular curve by placing stakes at PC, PT, and at several other points along the curve, such as point A. Take the axes as shown in the figure.

(a) Find the coordinates of PC and PT and the equation of the circular curve.

(b) Determine the coordinates of point A. Round all coordinates to the nearest hundredth of a foot.

66. The arch of a bridge has the shape of the upper half of an ellipse. At the water level, the distance between the supports is 100 feet, and the greatest height above the water level is 20 feet, as shown in the sketch.

(a) Taking the axes as shown in the sketch, find the equation of the ellipse that forms this arch.

(b) Use a graphing calculator to generate the graph of the equation in part (a). Then estimate (to the nearest hundredth of a foot) the height of the arch above the surface of the water at points which are 10 feet apart along the surface of the water, beginning at $x = 10$ ft and ending at $x = 40$ ft.

67. A parabolic satellite dish for receiving television signals has the dimensions shown in the sketch at the top of the next page. At what distance d (in inches) should the receiver be located in order to take advantage of the reflection property of a parabola?

8 in

d

48 inches

Receiver

68. A simply supported beam with a concentrated load at midspan deflects in the shape of a parabola, as shown in the sketch. Determine the equation of the parabola, called the *elastic curve*. Write the equation in the form $y = ax^2 + bx$ with *a* and *b* each rounded to three significant digits.

y

216 in.

Load

0.75 in.

x

Beam

69. The equation of the path of a football kicked from the ground is $x^2 - 174x + 100y = 0$, where *x* and *y* are measured in feet and the *x*-axis is taken along the ground.

(a) Find the horizontal distance traveled by the ball.

(b) Find the greatest height reached by the ball.

70. The roof of a whispering gallery has the shape of an ellipse rotated about its major axis. The distance between the foci is 400 ft and the distance from a vertex to its corresponding focus is 75 ft. If a sound emitted at one focus strikes the roof, find the time before it is heard at the other focus. Assume that sound travels at 1100 ft/s.

71. Ultrahigh-frequency (UHF) sound waves emanating from one focus of a lithotripter are reflected to the other focus of the device to break up a kidney stone. Suppose a lithotripter is formed by rotating the portion of an ellipse below its minor axis about its major axis, as shown in the sketch. How far from the kidney stone (in cm) should the opening of the lithotripter be placed in order to break up the stone?

Major axis

Kidney stone

10 cm

Minor axis

26 cm

Sound source

72. In an engineer's design for a cooling tower in a nuclear power plant, a hyperbola is rotated about the *y*-axis, as shown in the figure. If the distance from *A* to *B* is 12.3 meters, determine the radius of each end of the tower.

y

$y = 2.5x$

10.2 m

A   *B*   *x*

28.2 m

Cumulative Review Exercises for Chapters 4, 5, and 6

1. If the base is 4, find the logarithms of each number.
   (a) 64
   (b) $\frac{1}{16}$
   (c) 32
   (d) $\frac{1}{8}$

2. Find the zeros of the function f defined by
   $f(x) = a + \dfrac{b}{\ln x}$ if a and b are constants.

3. Identify the graph of each equation as a parabola, circle, ellipse, or hyperbola.
   (a) $x^2 + y^2 = 9$
   (b) $x^2 - y^2 = 9$
   (c) $x + y^2 = 9$
   (d) $2x^2 + y^2 = 9$

4. Discuss the position of the vertex of the parabola $y = a(x - h)^2 + k$ for each of the indicated conditions.
   (a) $h = 0$
   (b) $k = 0$
   (c) both $h = 0$ and $k = 0$

5. Factor completely over the integers:
   $x^4 + 2x^3 + x^2 - 8x - 20$

6. Find the remainder when $2x^4 - 3x^2 + 4x - 3$ is divided by $x - a$.

7. Solve for x:
   (a) $\log_x 3 - \log_4 16 - \log_3 \frac{1}{9} - \ln e = 2 \log 1$
   (b) $x^{\ln x} - e^2 x = 0$

8. For $a > 0$, show that $a^x = e^{x \ln a}$. Use this fact to write each of the following expressions as an exponential with base e.
   (a) 10^x
   (b) 3^{-x}
   (c) 5^{3x}
   (d) 2^{x+1}

9. Solve each equation for y.
   (a) $2 \ln x - \frac{1}{2} \ln y = 1$
   (b) $x + 2 = e^{x - \ln y}$
   (c) $2 \log_3(y + 1) = 2 - \log_3(y - 1)$

10. A function f is defined by $f(x) = 2 - e^{-x}$.
    (a) Show that f is one-to-one.
    (b) Find the inverse function f^{-1}, and state its domain.

11. A line is tangent to a circle at the point $(3, 4)$. Find the equation of the line if the equation of the circle is $x^2 + y^2 = 25$.

12. Find the equations of the asymptotes for each of the following hyperbolas.
    (a) $4x^2 - y^2 = 16$
    (b) $y^2 - x^2 + 2x = 5$

13. Explain the behavior of the function f defined by $f(x) = \dfrac{x}{x - 2}$ as x approaches the given value.
    (a) $x \to 2^-$
    (b) $x \to 2^+$

14. Determine the oblique asymptote for the graph of the equation $x^2 - 2xy - y = 0$. Then graph the equation.

15. A sum of $10,000 is deposited in an account at a rate of 9% per year compounded continuously. Find the value of the account after eight years.

16. Given the functions $f(x) = \ln(x + 1)$ and $g(x) = e^{-x} - 1$, find each composite function.
    (a) $f \circ g$
    (b) $g \circ f$

17. Find the coordinates of the intersection points of the ellipses $x^2 + 2y^2 = 18$ and $9x^2 + 4y^2 - 24y = 0$. Round each coordinate to three significant digits.

18. Find the coordinates of the vertex and any x- and y-intercepts for each parabola.
    (a) $x^2 - y - 4x + 3 = 0$
    (b) $y^2 + 4y + x - 1 = 0$

19. Three hundred children are seated in an auditorium in a certain number of rows, with the same number of children seated in each row. Placing two more children in each row would result in five fewer rows being used. In how many rows are the children seated initially?

20. Solve the inequality $x^2 - 4 \le \dfrac{1}{x^2}$ for x.

21. The circumference of a rear wheel on a dragster is 4 feet more than the circumference of a front wheel. A front wheel makes 88 more revolutions than a rear wheel in traveling the quarter-mile track. What is the circumference of each wheel?

22. Find the vertical and horizontal asymptotes for the graph of the function
    $$f(x) = \frac{x^2 - 2}{x^2 - 7x + 6}.$$
    Determine the value of x at the point where the graph of f crosses its horizontal asymptote.

23. The number N of golf balls (in dozens) sold by a certain company is a function of the amount x (in thousand of dollars) that is spent on advertising. This function is given by
    $$N(x) = 2500[1 + \ln(x + 1)].$$
    (a) If no money is spent on advertising, how many dozens of balls are sold?
    (b) If $500,000 is spent on advertising, approximately how many dozens of balls are sold?
    (c) How much money must be spent on advertising to sell 40,000 dozen balls?

393

24. A hot cup of coffee has an initial temperature of $200°F$ and is placed in a room with a temperature of $70°F$. After t minutes, the temperature T of the coffee is $T = 70 + 130e^{-0.38t}$. How many *seconds* does it take for the temperature of the coffee to cool to $120°F$?

25. Find the length of the line segment joining the intersection points of the curves $y = x + 1$ and $x^2 + y^2 = 25$.

26. Discuss the nature of the roots of the equation $x^n - 1 = 0$ if n is a positive integer and (a) n is even or (b) n is odd.

27. Determine the equation of a hyperbola in standard form if its asymptotes are $y = \pm 3x$ and its horizontal transverse axis has length 2.

28. Find the length of the side of a square inscribed in the circle $x^2 + y^2 = 2$.

29. The width of a cardboard box is twice its height, and the length of the box is 2 inches more than its width. Determine the dimensions of the box if its volume is 144 cubic inches.

30. The height of a tin can is 4 inches more than its radius. Determine to three significant digits the height of the can if its volume is 82 cubic inches.

31. Two circular stove pipes that are each 8 inches in diameter are cut at an angle of 45° to form an elbow, as shown in the sketch. Find the length of the major and minor axis of their elliptical intersection.

32. The planet Mercury travels in an elliptical orbit with the Sun at one focus, as shown in the sketch. The eccentricity of the orbit is 0.205, and the distance from the Sun to Mercury at perihelion (shortest distance from the Sun) is 2.85×10^7 miles.

(a) Determine the distance from Mercury to the Sun at aphelion (greatest distance from the Sun).

(b) Find the Cartesian equation of the orbit if the center of the ellipse is at the origin.

The Centrum civic center in Worcester, Massachusetts, sold 12,000 tickets to a recent rock concert. The ticket prices were $12, $18, and $24, and the total income from ticket sales was $201,000. How many tickets of each type were sold if the number of $12 tickets sold was twice the number of $24 tickets sold?

For the solution, see Example 5 in Section 7.1.

Linear Systems and Matrices

7.1   Systems of Linear Equations

◆ **Introductory Comments**

In Section 3.5, we graphed the linear equations

$$A_1 x + B_1 y = C_1 \qquad \text{and} \qquad A_2 x + B_2 y = C_2$$

on the same coordinate plane and determined their points of intersection by using either the substitution method or the elimination method. Figure 7.1 illustrates the three possibilities for the graphs of these two linear equations.

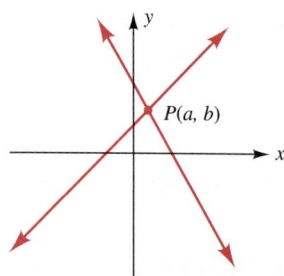

(a) The lines intersect at one point $P(a, b)$.

(b) The lines are parallel and do not intersect.

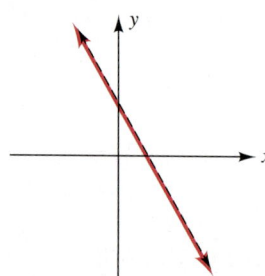

(c) The lines coincide and intersect at an infinite number of points.

FIGURE 7.1

The three possibilities for the intersection of two lines graphed on the same coordinate plane

We say that the pair of linear equations

$$A_1 x + B_1 y = C_1$$
$$A_2 x + B_2 y = C_2$$

represent a **system** of two linear equations in two variables x and y. This system has three possible types of solutions, as illustrated in Figure 7.1:

1. *Exactly one solution,* namely, the ordered pair (a, b). The lines intersect at one point $P(a, b)$, as shown in Figure 7.1(a).

2. *No solution,* an **inconsistent system**. The lines are parallel and do not intersect, as shown in Figure 7.1(b).

3. *Infinitely many solutions,* a **dependent system**. The lines coincide and intersect at an infinite number of points, as shown in Figure 7.1(c).

In this section, we discuss some methods of solving n linear equations in n variables with $n \geq 3$. A system of n linear equations in n variables x_1, x_2, $x_3, \ldots, x_n$ is called an $n \times n$ (read "n by n") linear system:

$$a_{11}x_1 + a_{12}x_2 + a_{13}x_3 + \cdots + a_{1n}x_n = k_1 \quad \leftarrow \boxed{\text{first equation}}$$

$$a_{21}x_1 + a_{22}x_2 + a_{23}x_3 + \cdots + a_{2n}x_n = k_2 \quad \leftarrow \boxed{\text{second equation}}$$

$$a_{31}x_1 + a_{32}x_2 + a_{33}x_3 + \cdots + a_{3n}x_n = k_3 \quad \leftarrow \boxed{\text{third equation}}$$

$$\vdots \qquad \vdots \qquad \vdots \qquad \qquad \vdots \qquad \vdots$$

$$a_{n1}x_1 + a_{n2}x_2 + a_{n3}x_3 + \cdots + a_{nn}x_n = k_n \quad \leftarrow \boxed{n\text{th equation}}$$

To describe the coefficients of each variable we use a *double subscript* notation in which each of the numbers in the subscript has a definite meaning. The first number corresponds to the equation and the second number corresponds to the variable. Thus, a_{32} represents the coefficient in the third equation attached to the second variable (x_2), whereas a_{23} represents the coefficient in the second equation attached to the third variable (x_3).

Although an $n \times n$ linear system with $n \geq 3$ has no graphical representation in the coordinate plane, this system still has three possible types of solutions:

1. Exactly one solution, namely, $x_1 = c_1$, $x_2 = c_2$, . . . , $x_n = c_n$, which is written as an ordered n-tuple $(c_1, c_2, . . . , c_n)$

2. No solution, an inconsistent system

3. Infinitely many solutions, a dependent system

In more advanced math courses, it is shown that the graph of three linear equations in three variables consists of three planes in space. Figure 7.2 shows some ways in which three planes may intersect to obtain exactly one solution, no solution, or infinitely many solutions.

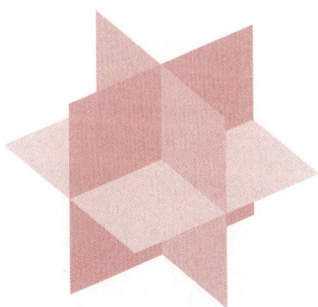

(a) The planes intersect in exactly one point, and the system has exactly one solution.

(b) The planes have no point in common, and the system has no solution.

(c) The planes intersect in a line, and the system has infinitely many solutions.

FIGURE 7.2

Some graphical illustrations of the three types of solutions for a 3×3 linear system

◆ The Elimination Method

We can use the *elimination method* (from Section 3.5) to help solve a 3×3 linear system. The key step in this method is to obtain, for one of the variables, coefficients that are equal in absolute value so that by adding or subtracting the two equations, this variable is eliminated. (See Example 2 in Section 3.5.) For a 3 × 3 linear system, this method of solution involves eliminating one of the variables between a pair of equations, and then eliminating the same variable between a different pair of equations. This procedure yields two linear equations in two variables, which we can solve by using the substitution method or the elimination method. Although the order in which we eliminate the variables from a system of equations is immaterial, we can often reduce the amount of work needed by using some insight and making intelligent choices.

The elimination method relies on generating **equivalent systems**—systems that have the same solution. We can generate equivalent systems by performing any of the following manipulations.

◆ **Generating Equivalent Systems**

> Performing any of the following manipulations on a system of equations generates an equivalent system:
>
> **1.** Interchange the position of two equations.
>
> **2.** Multiply both sides of an equation by a nonzero constant.
>
> **3.** Add a multiple of one equation to another equation.

EXAMPLE 1   Use the elimination method to solve the given system of linear equations.

$$2x - y + 4z = -3$$
$$x + 2y - 3z = 1$$
$$3x + y - 3z = 6$$

◆ **Solution**   If we add the first and third equations, we eliminate y and obtain an equation in x and z:

$$
\begin{array}{r}
2x - y + 4z = -3 \\
3x + y - 3z = 6 \\
\hline
5x \quad\quad + z = 3
\end{array}
$$

Next, we choose another pair of equations and eliminate the same variable. Choosing the first and second equations, we eliminate y and obtain another equation in x and z:

$$
\begin{array}{r}
2x - y + 4z = -3 \\
x + 2y - 3z = 1
\end{array}
\quad
\begin{array}{c}
\xrightarrow{\text{Multiply both sides by 2}} \\
\xrightarrow{\text{No change}}
\end{array}
\quad
\begin{array}{r}
4x - 2y + 8z = -6 \\
x + 2y - 3z = 1 \\
\hline
5x \quad\quad + 5z = -5
\end{array}
$$

We now have two linear equations in the same two variables, x and z. If we subtract one of these equations from the other, we eliminate x:

$$5x + z = 3$$
$$\underline{5x + 5z = -5}$$
$$-4z = 8$$

We find z, x, and y by solving for z, and then substituting back through the equations:

$$-4z = 8: \qquad\qquad\qquad\qquad\qquad \text{implies} \quad z = -2$$
$$5x + z = 3: \qquad\qquad 5x + (-2) = 3 \quad \text{implies} \quad x = 1$$
$$3x + y - 3z = 6: \quad 3(1) + y - 3(-2) = 6 \quad \text{implies} \quad y = -3$$

Thus, the solution of the original system of equations is $x = 1$, $y = -3$, and $z = -2$, which we write as the *ordered triple* $(1, -3, -2)$. We can check this solution by verifying that each of the three original equations is satisfied when we substitute $x = 1$, $y = -3$, and $z = -2$. ◆

In Example 1 we generated the equivalent system

$$3x + y - 3z = 6$$
$$5x + z = 3$$
$$-4z = 8$$

to obtain the solution of the original system of equation. This system is not a unique choice. For the first equation in this equivalent system we can use either of the other original equations, and for the second equation we can use the other equation in x and z, namely, $5x + 5z = -5$.

PROBLEM 1 For the system of equations given in Example 1, show that the solution $(1, -3, -2)$ is also given by the equivalent system

$$2x - y + 4z = -3$$
$$5x + 5z = -5$$
$$-4z = 8$$
◆

In the next example, we solve a dependent system and show how to express its infinitely many solutions as an ordered triple.

EXAMPLE 2 Use the elimination method to solve the given system of linear equations.

$$x - 2y - 3z = 2$$
$$x - 4y - 13z = 14$$
$$-3x + 5y + 4z = 0$$

◆ **Solution**   If we subtract the second equation from the first equation, we eliminate x and obtain an equation in y and z:

$$
\begin{array}{r}
x - 2y - 3z = 2 \\
x - 4y - 13z = 14 \\
\hline
2y + 10z = -12
\end{array}
$$

Next, we choose a different pair of equations and eliminate the same variable. Choosing the first and third equations, we eliminate x and obtain another equation in y and z:

$$
\begin{array}{lcl}
x - 2y - 3z = 2 & \xrightarrow[\text{sides by 3}]{\text{Multiply both}} & 3x - 6y - 9z = 6 \\
-3x + 5y + 4z = 0 & \xrightarrow{\text{No change}} & -3x + 5y + 4z = 0 \\
& & \hline \\[-1.0em]
& & -y - 5z = 6
\end{array}
$$

We now have two linear equations in two unknowns, y and z, which we can try to solve by the elimination method:

$$
\begin{array}{lcl}
2y + 10z = -12 & \xrightarrow{\text{No change}} & 2y + 10z = -12 \\
-y - 5z = 6 & \xrightarrow[\text{sides by 2}]{\text{Multiply both}} & -2y - 10z = 12 \\
& & \hline \\[-1.0em]
& & 0 = 0
\end{array}
$$

Recall from Section 3.5 that if the variables drop out and the resulting statement is true ($0 = 0$), then the lines coincide and the system has infinitely many solutions. Hence, this is a dependent system.

We can express the infinitely many solutions as an ordered triple. First, we solve the equation $-y - 5z = 6$ (or $2y + 10z = -12$) for either y or z. Solving for y is easier, and we obtain

$$ y = -5z - 6. $$

Now, we substitute $y = -5z - 6$ into one of the original equations, say $x - 2y - 3z = 2$, to express x in terms of z:

$$
\begin{aligned}
x - 2y - 3z &= 2 \\
x - 2(-5z - 6) - 3z &= 2 \\
x &= -7z - 10
\end{aligned}
$$

If we let $z = a$, where a is an arbitrary real number, then y and x can also be expressed in terms of a:

$$ y = -5a - 6 \quad \text{and} \quad x = -7a - 10 $$

Hence, the original system of equations has an infinite number of solutions of the form

$$ (-7a - 10, \; -5a - 6, \; a) \quad \text{where } a \text{ is any real number.} $$

We can obtain a *particular solution* of this system by letting a take on any real value. For example, if we let $a = 0$, then we obtain the particular solution $(-10, -6, 0)$. ◆

PROBLEM 2 Use the elimination method to show that the given system of linear equations is an inconsistent system.

$$x - 2y - 3z = -3$$
$$x - 4y - 13z = 14$$
$$-3x + 5y + 4z = 1$$

◆

The elimination method may be extended to an $n \times n$ system of linear equations with $n \geq 4$. For example, with four variables, we eliminate one of the variables between three different pairs of equations. The result is three equations in the remaining three variables, which we can solve by using the procedure illustrated in the previous two examples.

EXAMPLE 3 Use the elimination method to solve the given system of linear equations.

$$x + 2y \quad\quad + t = 3$$
$$y + 2z \quad\quad = 3$$
$$x \quad\quad - 2z \quad\quad = 0$$
$$3y - 4z + t = 2$$

◆ **Solution** If we subtract the first and third equations, we eliminate x and obtain an equation in y, z, and t:

$$x + 2y \quad\quad + t = 3$$
$$\underline{x \quad\quad - 2z \quad\quad = 0}$$
$$2y + 2z + t = 3$$

Joining this equation with the second and fourth equations gives us a 3×3 linear system:

$$2y + 2z + t = 3$$
$$y + 2z \quad\quad = 3$$
$$3y - 4z + t = 2$$

If we subtract the first and third equations of this 3×3 system, we eliminate t and obtain an equation in y and z:

$$2y + 2z + t = 3$$
$$\underline{3y - 4z + t = 2}$$
$$-y + 6z \quad\quad = 1$$

We now have two linear equations in the same two variables, y and z. If we add these equations, we eliminate y:

$$
\begin{aligned}
y + 2z &= 3 \\
-y + 6z &= 1 \\
\hline
8z &= 4
\end{aligned}
$$

We find $z, y, t,$ and x by solving for z and then substituting back through the equations:

$$8z = 4: \qquad\qquad\qquad\qquad\qquad\qquad\qquad \text{implies} \quad z = \tfrac{1}{2}$$

$$y + 2z = 3: \qquad\qquad y + 2\left(\tfrac{1}{2}\right) = 3 \quad \text{implies} \quad y = 2$$

$$2y + 2z + t = 3: \quad 2(2) + 2\left(\tfrac{1}{2}\right) + t = 3 \quad \text{implies} \quad t = -2$$

$$x - 2z = 0: \qquad\qquad x - 2\left(\tfrac{1}{2}\right) = 0 \quad \text{implies} \quad x = 1$$

Thus, the solution of the original system of equations is $x = 1$, $y = 2$, $z = \tfrac{1}{2}$, and $t = -2$, which we write as the *ordered quadruple* $\left(1, 2, \tfrac{1}{2}, -2\right)$. You should check this solution.   ◆

PROBLEM 3   Rework Example 3 by first eliminating the variable t.   ◆

◆ Nonsquare Systems

Thus far we have worked with linear systems in which the number of equations equals the number of variables. Such a system is called a **square system**. If the number of equations does not equal the number of variables, the system is called a **nonsquare system**. A nonsquare system may not have a unique solution.

EXAMPLE 4   Use the elimination method to solve the given system of linear equations.

$$3x + 3y - 2z = 3$$
$$2x + y - z = 2$$

◆ **Solution**   Choosing to eliminate z first, we proceed as follows:

$$
\begin{array}{llll}
3x + 3y - 2z = 3 & \xrightarrow{\text{No change}} & 3x + 3y - 2z = 3 \\
2x + y - z = 2 & \xrightarrow[\text{sides by }-2]{\text{Multiply both}} & \dfrac{-4x - 2y + 2z = -4}{-x + y \phantom{{}-2z{}} = -1}
\end{array}
$$

Since this system does not contain a third equation, we cannot form two equations in x and y. Therefore, we solve the equation $-x + y = -1$ for y in term of x (or for x in terms of y). Solving for y, we obtain

$$y = x - 1.$$

Now, substituting $y = x - 1$ into one of the original equations, we find z in terms of x:

$$2x + y - z = 2: \quad 2x + (x - 1) - z = 2 \quad \text{implies} \quad z = 3x - 3$$

If we let $x = a$, where a is any arbitrary real number, then y and z can also be expressed in terms of a:

$$y = a - 1 \quad \text{and} \quad z = 3a - 3$$

Hence, the original system of equations has an infinite number of solutions of the form

$$(a, a - 1, 3a - 3) \quad \text{where } a \text{ is any real number.} \qquad ◆$$

PROBLEM 4   Solve the system in Example 4 by solving the equation $-x + y = -1$ for x in terms of y (instead of solving for y in terms of x). Then find z in terms of y. Show that the solution of this system of equations can also be written as the ordered triple $(b + 1, b, 3b)$ where b is any real number. ◆

◆ Application: Solving Word Problems

In Section 2.4, we solved word problems by assigning one of the unknowns the variable x and expressing the other unknowns in terms of x. Using this method, it was possible to develop one equation in one unknown and to solve this equation for x. Many word problems, however, are set up more naturally by assigning a different variable to each of the unknowns and then solving a system of equations. The next example illustrates this procedure.

EXAMPLE 5   *Chapter Opening Problem*
The Centrum civic center in Worcester, Massachusetts, sold 12,000 tickets to a recent rock concert. The ticket prices were $12, $18, and $24, and the total income from ticket sales was $201,000. How many tickets of each type were sold if the number of $12 tickets sold was twice the number of $24 tickets sold?

◆ **Solution**   Let

$$x = \text{the number of } \$12 \text{ tickets sold}$$
$$y = \text{the number of } \$18 \text{ tickets sold}$$

and $\qquad\qquad z = \text{the number of } \$24 \text{ tickets sold.}$

Since 12,000 tickets were sold, we have

$$x + y + z = 12{,}000.$$

Since the ticket prices were $12, $18, and $24, and the total income from ticket sales was $201,000, we have

$$12x + 18y + 24z = 201{,}000.$$

Now, since the Centrum sold twice as many $12 tickets as $24 tickets, we have

$$x = 2z, \quad \text{or} \quad x - 2z = 0.$$

Hence, we can write a system of linear equations to describe this problem:

$$
\begin{aligned}
x + y + z &= 12{,}000 \\
12x + 18y + 24z &= 201{,}000 \\
x - 2z &= 0
\end{aligned}
$$

In Problem 5, you are asked to use the elimination method to show that the solution of this system of linear equations is

$$(5000, \ 4500, \ 2500).$$

In summary, the Centrum sold 5000 tickets at $12, 4500 tickets at $18, and 2500 tickets at $24 each.   ◆

PROBLEM 5   Use the elimination method to verify that the solution of the system of linear equations in Example 5 is (5000, 4500, 2500).   ◆

Exercises 7.1

◆ Basic Skills

In Exercises 1–24, use the elimination method to solve each system of linear equations.

1.
$$
\begin{aligned}
x + z &= 11 \\
y + z &= 5 \\
4x + 2y &= 48
\end{aligned}
$$

2.
$$
\begin{aligned}
x + y + z &= 19 \\
2x - y &= 3 \\
y - z &= 10
\end{aligned}
$$

3.
$$
\begin{aligned}
2x + 3y &= 2 \\
x - 2z &= 7 \\
2y - z &= 3
\end{aligned}
$$

4.
$$
\begin{aligned}
3x + y - 2z &= -2 \\
y - 3z &= -1 \\
2x + 3y + 5z &= 3
\end{aligned}
$$

5.
$$
\begin{aligned}
x + 3y - z &= 5 \\
2x + 3y &= 4 \\
-x - y + 3z &= -1
\end{aligned}
$$

6.
$$
\begin{aligned}
x - 2y - 5z &= -1 \\
-3x + y - 2z &= 0 \\
2x + 3z &= 1
\end{aligned}
$$

7.
$$
\begin{aligned}
4x - 3y - z &= 0 \\
x - 3y + 2z &= 7 \\
3x + 9y - z &= -2
\end{aligned}
$$

8.
$$
\begin{aligned}
5x - 2y + z &= -2 \\
3x - y - 3z &= -2 \\
x + 2y - z &= 4
\end{aligned}
$$

9.
$$
\begin{aligned}
3x - 2y + z &= 3 \\
4x + y &= 1 \\
11y - 4z &= -9
\end{aligned}
$$

10.
$$
\begin{aligned}
2x + 3y &= 2 \\
8x - 4z &= 3 \\
3y + z &= -1
\end{aligned}
$$

11.
$$
\begin{aligned}
2x + 3y + z &= 7 \\
5x - y - z &= 2 \\
4x - y + 3z &= 9
\end{aligned}
$$

12.
$$
\begin{aligned}
4x - y + 3z &= 3 \\
3x - 4y + 2z &= 4 \\
2x + 3y - z &= 1
\end{aligned}
$$

13.
$$
\begin{aligned}
x - 2y + z &= 1 \\
3x - y + z &= 6 \\
-2x + y - 4z &= 1
\end{aligned}
$$

14. $5x - 3y + 4z = -1$
$3x + 8y - z = -4$
$2x - 11y + 4z = 2$

15. $6x - 4y - 5z = 4$
$3x - 2y + z = 9$
$\frac{3}{2}x - \frac{2}{3}y + 2z = \frac{41}{6}$

16. $0.2x + 0.3y + 2z = 3$
$3x - 2y + 4z = 6$
$0.5x - y + 3z = -0.5$

17. $x + y + z = 1$
$x + z + t = 2$
$x + y + t = 6$
$y + z + t = 3$

18. $3x + 2y + z = 1$
$x + 3t = -4$
$y - 2z - t = -2$
$2x + y - z = -3$

19. $x - 2y + t = 1$
$2y - z + 3t = 12$
$x - 2z = 2$
$2y + t = 5$

20. $x + y - 2t = 7$
$2x - z + t = -4$
$3x + y + z = 5$
$2y - z + 3t = -2$

21. $x_1 + x_2 = 2$
$x_2 - x_3 + x_4 = 5$
$x_1 + x_3 + x_5 = -1$
$x_2 + x_4 = 0$
$x_4 + x_5 = 2$

22. $x_1 + x_2 + x_3 + x_4 = 12$
$x_1 + x_2 + x_3 + x_5 = 14$
$x_1 + x_2 + x_4 + x_5 = 16$
$x_1 + x_3 + x_4 + x_5 = 18$
$x_2 + x_3 + x_4 + x_5 = 20$

23. $x_1 + 2x_2 + x_5 = 5$
$x_2 - x_3 + x_4 = 1$
$x_1 + 3x_3 - x_5 = -2$
$3x_2 - x_4 + 2x_5 = 8$
$x_3 + 2x_4 - x_5 = -5$

24. $2x_1 + x_2 - x_3 + x_5 = -1$
$x_1 - 2x_2 + x_4 = 5$
$3x_2 - x_3 + 2x_5 = -5$
$3x_1 + 2x_3 - x_4 - x_5 = 7$
$4x_2 + 3x_3 + x_4 - x_5 = -1$

In Exercises 25–32, use the elimination method to solve each non-square system of linear equations.

25. $2x + 2y + z = 4$
$4y - z = 0$

26. $3x - y + 4z = 8$
$y + 2z = 1$

27. $x + 2y - z = 1$
$3x + 2y - z = 5$

28. $x - 3y + 2z = -2$
$2x - 4y + 5z = 1$

29. $3x - 2y + 4t = 4$
$y - 2z + t = 1$
$x - 2y + 8t = 0$

30. $2x - 3y + z = 1$
$3x - y - t = 0$
$y + 3z - 2t = 3$

31. $x + y = 7$
$2x - y = 8$
$x - 3y = -1$

32. $2x - 5y = 8$
$3x + y = 2$
$4x + 3y = 0$

In Exercises 33–44, use a system of linear equations to solve each applied problem.

33. A car rental agency charges a dollars per day to rent a car plus b dollars per mile. Find a and b if the one-day rental charge is \$45 for 100 miles or \$55 for 150 miles.

34. An investor buys two types of stocks—one costs \$40 per share and the other costs \$24 per share. Altogether, the investor purchased 55 shares of stock for a total cost of \$1720. How many shares of each stock were purchased?

35. A total of \$10,000 is invested, part of the money at 10% and the rest at 12%. If the total annual interest is \$1090, how much money is invested at each rate?

36. It takes a ferryboat 4 hours to travel 20 miles upstream and 3 hours to make the return trip downstream. What is the speed of the boat and the speed of the river current?

37. The sum of the three interior angles of any triangle is 180°. Suppose the largest angle is six times as large as the smallest angle and is equal to twice the sum of the two smaller angles. Find the three angles.

38. A contractor buys three parcels of land for \$200,000. Parcel A costs three times as much as parcel B and \$20,000 less than parcels B and C together. Find the cost of each parcel of land.

39. A woman invests \$50,000 at the simple interest rates of 8%, 9%, and $9\frac{1}{2}$% per year. Altogether these investments earn \$4550 per year. How much does she have invested at each rate if she has the same amount invested at 9% and $9\frac{1}{2}$%?

40. A merchant mixes three types of tea, costing \$2.40, \$3.00, and \$3.20 per pound, in order to obtain 50 pounds of a blend that costs \$2.92 per pound. How many pounds of each type does the merchant use if the blend contains 10 pounds more of the \$3.20 type than of the \$2.40 type?

41. The greens fee for a round of golf is \$15 for a child, \$20 for an adult, and \$18 for a senior citizen. On a certain day, 120 rounds of golf were played and the receipts totaled $2220. If twice as many seniors played as children, find the number of adults who played golf that day.

42. Three circular pulleys are tangent to each other, as shown in the sketch. If the center-to-center distances *AB, AC,* and *BC* are 18 cm, 22 cm, and 16 cm, respectively, find the radius of each pulley.

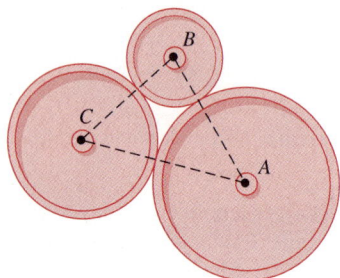

43. A photography studio develops four sizes of prints (in inches); 2 × 3, 3 × 5, 5 × 8, and 8 × 10. After taking senior-class pictures, the studio offers the following four packages to the students.

Package	Number of prints			
	2 × 3	3 × 5	5 × 8	8 × 10
A	4	2	1	—
B	—	2	2	1
C	8	—	4	2
D	12	4	2	2

If package A sells for $12, package B for $24, package C for $46, and package D for $47, determine the cost per print for each size print.

44. The fueling facility at an airport uses three pumps to load each jet with 1500 gallons of fuel before takeoff. When all three pumps are used, each plane can be loaded with fuel in 1 hour 15 minutes. However, on a certain day, the first pump malfunctions and works at only half its normal rate. Fueling now requires 1 hour 40 minutes for each plane. When the malfunctioning pump is closed completely for repair, the fueling takes 2 hours 30 minutes for each plane. What is the normal pumping rate (in gallons per minute) for the malfunctioning pump?

Critical Thinking

45. A system of linear equations has the following equations:

$$x + 3y - 2z = B$$
$$-2x + y + Az = 8$$
$$3x - 5y - z = 12$$

Find values of *A* and *B* for which the system is (a) inconsistent or (b) dependent.

46. Find a system of linear equations whose solution is the ordered triple $(-1, 2, 5)$. (The answer is not unique.)

47. Use a system of linear equations to find the values of *A, B,* and *C* such that

$$A(x - 1)(x + 2) + B(x - 4)(x + 2) + C(x - 4)(x - 1)$$

is equivalent to $x^2 + 8x$.

48. The graph of the circle $x^2 + y^2 + Dx + Ey + F = 0$ passes through the points $(0, 3)$, $(-2, 0)$, and $(4, -4)$. Use a system of linear equations to determine the values for *D, E,* and *F*.

49. The graph of the quadratic function $f(x) = ax^2 + bx + c$ is sketched in the figure. Use a system of linear equations to determine the values for *a, b,* and *c*.

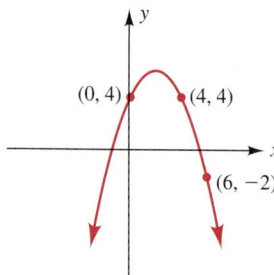

50. The graph of the cubic function

$$f(x) = ax^3 + bx^2 + cx + d$$

is shown in the sketch. Use a system of linear equations to determine the values for *a, b, c,* and *d*.

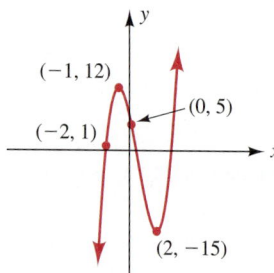

A system of linear equations is said to be **homogeneous** *if all the constant terms are zero. A homogeneous system of equations always has the trivial solution (0, 0, 0, . . . , 0). However, some homogeneous systems also have nontrivial solutions. In Exercises 51 and 52, find the solution to each homogeneous system.*

51.
$$x + 3y - 2z = 0$$
$$3x - 2y + 5z = 0$$
$$-2x - y - z = 0$$

52.
$$3x - 5y + 5z = 0$$
$$x - 4y - 3z = 0$$
$$-4x + 15y + 10z = 0$$

In Exercises 53 and 54, develop a system of linear equation for each applied problem, then solve. [Hint: Each system is dependent and has more than one solution.]

53. A group of 20 skiers spent $260 to ski at Deer Mountain and $180 to ski at Mt. Cabott. The cost of lift tickets at these ski areas are shown in the following table:

	Deer Mountain	Mt. Cabott
Adult	$18	$24
Student	$15	$15
Child	$10	Free

How many adults, students, and children are in this group of 20 skiers?

54. The following table shows the weight composition of four alloys. How many pounds of each alloy must be melted together to form 100 pounds of an alloy that is 20% copper, 46% gold, 21% tin, and 13% zinc?

Alloy	Copper	Gold	Tin	Zinc
A	20%	50%	20%	10%
B	—	60%	40%	—
C	40%	40%	—	20%
D	—	20%	50%	30%

7.2 Matrices and Their Properties

◆ Introductory Comments

A **matrix** is a rectangular array of numbers enclosed by a pair of brackets. Each particular number in the matrix is referred to as an **element** of the matrix. We use a capital letter, such as *A*, *B*, or *C*, to represent a matrix. Consider matrix *A*:

$$A = \begin{bmatrix} a_{11} & a_{12} & a_{13} & \cdots & a_{1j} & \cdots & a_{1n} \\ a_{21} & a_{22} & a_{23} & \cdots & a_{2j} & \cdots & a_{2n} \\ a_{31} & a_{32} & a_{33} & \cdots & a_{3j} & \cdots & a_{3n} \\ \vdots & \vdots & \vdots & & \vdots & & \vdots \\ a_{i1} & a_{i2} & a_{i3} & \cdots & a_{ij} & \cdots & a_{in} \\ \vdots & \vdots & \vdots & & \vdots & & \vdots \\ a_{m1} & a_{m2} & a_{m3} & \cdots & a_{mj} & \cdots & a_{mn} \end{bmatrix}$$

*j*th column

*i*th row

We use double subscript notation for each element. Each of the two numbers in the subscript has a definite meaning: the first number corresponds to the **row** and the second number to the **column** in which the element stands. Thus, a_{ij} represents the element in the *i*th row and *j*th column. We say that matrix *A* has *m* rows and *n* columns and is of **order** $m \times n$ (read "*m* by *n*").

Here are some examples of matrices (plural of matrix):

$$B = \begin{bmatrix} 2 & 0 & 6 \\ 5 & -1 & 3 \end{bmatrix} \qquad C = \begin{bmatrix} 0 & 1 \\ 8 & 4 \\ -2 & 3 \end{bmatrix}$$

Matrix B has two rows and three columns and is of order 2×3. Matrix C has three rows and two columns and is of order 3×2.

For expressing the rules of matrix algebra, we usually write matrix A in the compact form

$$A = [a_{ij}]$$

the matrix with elements a_{ij}

Two matrices are **equal** if and only if their corresponding elements are equal. More formally, we state the following definition of equality of matrices.

◆ **Equality of Matrices**

Two matrices $A = [a_{ij}]$ and $B = [b_{ij}]$ are **equal** if they both are of order $m \times n$ and

$$a_{ij} = b_{ij}$$

for all $i = 1, 2, 3, \ldots, m$ and $j = 1, 2, 3, \ldots, n$.

For example,

$$\begin{bmatrix} a_{11} & a_{12} \\ a_{21} & a_{22} \end{bmatrix} = \begin{bmatrix} -2 & -6 \\ -5 & 7 \end{bmatrix}$$

if and only if $a_{11} = -2$, $a_{12} = -6$, $a_{21} = -5$, and $a_{22} = 7$.

◆ **Using a Matrix to Solve a System of Linear Equations**

We can use a matrix to help solve a system of linear equations. Consider the following nonsquare system of m linear equations in n variables $x_1, x_2, x_3, \ldots, x_n$:

$$
\begin{aligned}
a_{11}x_1 + a_{12}x_2 + a_{13}x_3 + \cdots + a_{1n}x_n &= k_1 & \text{first equation} \\
a_{21}x_1 + a_{22}x_2 + a_{23}x_3 + \cdots + a_{2n}x_n &= k_2 & \text{second equation} \\
a_{31}x_1 + a_{32}x_2 + a_{33}x_3 + \cdots + a_{3n}x_n &= k_3 & \text{third equation} \\
\vdots \qquad\qquad\qquad\qquad & & \\
a_{m1}x_1 + a_{m2}x_2 + a_{m3}x_3 + \cdots + a_{mn}x_n &= k_m & \text{mth equation}
\end{aligned}
$$

For every such $m \times n$ linear system, there corresponds an enlarged or **augmented matrix** of order $m \times (n + 1)$ that is written as follows:

$$\begin{bmatrix} a_{11} & a_{12} & a_{13} & \cdots & a_{1n} & \vdots & k_1 \\ a_{21} & a_{22} & a_{23} & \cdots & a_{2n} & \vdots & k_2 \\ a_{31} & a_{32} & a_{33} & \cdots & a_{3n} & \vdots & k_3 \\ \vdots & \vdots & \vdots & & \vdots & \vdots & \vdots \\ a_{m1} & a_{m2} & a_{m3} & \cdots & a_{mn} & \vdots & k_m \end{bmatrix}$$

coefficients of the variable terms in the linear system | constant terms

For example, the system of linear equations

$$\begin{aligned} x + 3y - 2z &= 1 \\ -2x - y &= 0 \\ 3x \quad - z &= -2 \end{aligned}$$

corresponds to the augmented matrix

$$\begin{bmatrix} 1 & 3 & -2 & \vdots & 1 \\ -2 & -1 & 0 & \vdots & 0 \\ 3 & 0 & -1 & \vdots & -2 \end{bmatrix}$$

We use a vertical dashed line in the augmented matrix to separate the coefficients of the variable terms in the linear system from the constant terms in the system.

An augmented matrix is in **echelon form** if the elements $a_{ij} = 0$ when $i > j$; that is, the element in the first column in every row after the first is 0; the element in the second column in every row after the second is 0; the element in the third column in every row after the third is 0; and so on. For example, the augmented matrix

$$\begin{bmatrix} 1 & 2 & 5 & \vdots & -5 \\ 0 & 2 & -1 & \vdots & 3 \\ 0 & 0 & 2 & \vdots & 4 \end{bmatrix}$$

is in echelon form and corresponds to the system of linear equations

$$\begin{aligned} x + 2y + 5z &= -5 \\ 2y - z &= 3 \\ 2z &= 4 \end{aligned}$$

In the previous section, we discussed three manipulations for generating equivalent systems of equations:

1. Interchange the position of two equations.

2. Multiply both sides of an equation by a nonzero constant.

3. Add a multiple of one equation to another.

Each of these manipulations corresponds to an **elementary row operation** that can be performed on a matrix. If one matrix is obtained from another by a sequence of elementary row operations, we say that the two matrices are **row equivalent**.

◆ **Generating Row-Equivalent Matrices**

Performing any of the following elementary row operations on a matrix generates a row-equivalent matrix:

1. Interchange the position of two rows.

2. Multiply all elements in a row by a nonzero constant.

3. Add a multiple of one row to another row.

If two matrices are row equivalent, we write the symbol ~ between the two matrices.

CAUTION

When working with matrices, do not confuse *equality* (written as =) with *equivalency* (written as ~). For example, when interchanging rows, we write

$$\begin{bmatrix} 2 & 5 \\ -3 & 4 \end{bmatrix} \sim \begin{bmatrix} -3 & 4 \\ 2 & 5 \end{bmatrix}$$

However, since corresponding elements of these matrices are *not* the same, we write

$$\begin{bmatrix} 2 & 5 \\ -3 & 4 \end{bmatrix} \neq \begin{bmatrix} -3 & 4 \\ 2 & 5 \end{bmatrix}$$

In this text, we use the following notation to describe an operation that generates a row-equivalent matrix:

$R_i \leftrightarrow R_j$ Interchange the ith row and jth row of a matrix.

$cR_i \to R_i$ Replace the ith row with a multiple of c times the ith row.

$cR_i + R_j \to R_j$ Replace the jth row with the sum of c times the ith row and the jth row.

In the next example, we solve a system of linear equations by using elementary row operations on an augmented matrix. The scheme is to write the augmented matrix in echelon form, write the corresponding system of equations, and solve for the variables. We refer to this procedure of solving a system of linear equations as the **Gaussian method** (named after German mathematician Carl Friedrich Gauss).

The Gaussian method is the procedure used by most computer programs (and some graphing calculators) to solve systems of linear equations. This method often requires lots of arithmetic with awkward fractions—something that computers do quickly and accurately, but humans do not. To avoid unnecessary fractional computations, we usually begin by generating a row-equivalent matrix in which the element in the first row, first column is 1.

EXAMPLE 1   Use the Gaussian method to solve each system of linear equations.

(a) $13x + 3y = 55$
      $12x + 7y = 55$

(b) $2x - y + 4z = -3$
      $x + 2y - 3z = 1$
      $3x + y - 3z = 6$

◆ **Solution**

(a) Beginning with the augmented matrix that corresponds to this system of linear equations, we use elementary row operations to write the augmented matrix in echelon form:

$$\begin{bmatrix} 13 & 3 & | & 55 \\ 12 & 7 & | & 55 \end{bmatrix} \sim \begin{bmatrix} 1 & -4 & | & 0 \\ 12 & 7 & | & 55 \end{bmatrix} \qquad -R_2 + R_1 \rightarrow R_1$$

$$\sim \begin{bmatrix} 1 & -4 & | & 0 \\ 0 & 55 & | & 55 \end{bmatrix} \qquad -12R_1 + R_2 \rightarrow R_2$$

This last matrix is in echelon form and corresponds to the following system of linear equations:

$$x - 4y = 0$$
$$55y = 55$$

The equation $55y = 55$ implies that $y = 1$. Substituting this value of y back into the other equation, $x - 4y = 0$, we find that $x = 4$. Thus, the solution of the original system is $(4, 1)$. (You should check this solution.)

(b) Beginning with the augmented matrix that corresponds to this system of linear equations, we use elementary row operations to write the augmented matrix in echelon form:

$$\begin{bmatrix} 2 & -1 & 4 & | & -3 \\ 1 & 2 & -3 & | & 1 \\ 3 & 1 & -3 & | & 6 \end{bmatrix} \sim \begin{bmatrix} 1 & 2 & -3 & | & 1 \\ 2 & -1 & 4 & | & -3 \\ 3 & 1 & -3 & | & 6 \end{bmatrix} \qquad R_1 \leftrightarrow R_2$$

$$\sim \begin{bmatrix} 1 & 2 & -3 & | & 1 \\ 0 & -5 & 10 & | & -5 \\ 0 & -5 & 6 & | & 3 \end{bmatrix} \qquad \begin{array}{l} -2R_1 + R_2 \rightarrow R_2 \\ -3R_1 + R_3 \rightarrow R_3 \end{array}$$

$$\sim \begin{bmatrix} 1 & 2 & -3 & | & 1 \\ 0 & -5 & 10 & | & -5 \\ 0 & 0 & -4 & | & 8 \end{bmatrix} \qquad -R_2 + R_3 \rightarrow R_3$$

This last matrix is in echelon form and corresponds to the following system of linear equations:

$$x + 2y - 3z = 1$$
$$-5y + 10z = -5$$
$$-4z = 8$$

The equation $-4z = 8$ implies that $z = -2$. Substituting this value of z back into the other equation, we find that $y = -3$ and $x = 1$. Thus, the solution of the original system is $(1, -3, -2)$.   ◆

Remember, a system of linear equations may have infinitely many solutions or no solution. If we perform row operations on an augmented matrix and obtain a row of zeros except for the last entry in the row, then we conclude that the system is inconsistent (it has no solution).

PROBLEM 1   Use the Gaussian method to show that the given system of linear equations is inconsistent.

$$3x - 9y + 6z = 1$$
$$2x - y - z = 3$$
$$x - 2y + z = 1$$ ◆

◆ Matrix Addition and Scalar Multiples

As we observed in Example 1, matrices can be used to help solve a system of linear equations. Their use, however, far exceeds this single purpose. When English mathematician Arthur Cayley (1821–1895) introduced matrices in 1858, he defined the operations of addition and multiplication with matrices. In doing so, he created a new type of mathematics that has served as a model for various applications, from economics to quantum mechanics.

To add two $m \times n$ matrices, we simply add the elements in corresponding positions. More formally, we state the following rule for matrix addition.

◆ **Matrix Addition**

> If $A = [a_{ij}]$ and $B = [b_{ij}]$ are matrices of order $m \times n$, then their **sum** $A + B$ is a matrix of order $m \times n$ defined by
>
> $$A + B = [a_{ij} + b_{ij}].$$

EXAMPLE 2   Matrices A, B, and C are defined as follows.

$$A = \begin{bmatrix} 6 & 5 & -2 \\ 4 & 0 & -1 \end{bmatrix} \qquad B = \begin{bmatrix} 3 & -5 & 2 \\ 0 & 1 & -3 \end{bmatrix}$$

$$C = \begin{bmatrix} 2 & 3 \\ -1 & 1 \\ 0 & -2 \end{bmatrix}$$

Determine each sum.

(a) $A + B$ **(b)** $C + C$

◆ Solution

(a) $A + B = \begin{bmatrix} 6 + 3 & 5 + (-5) & -2 + 2 \\ 4 + 0 & 0 + 1 & -1 + (-3) \end{bmatrix} = \begin{bmatrix} 9 & 0 & 0 \\ 4 & 1 & -4 \end{bmatrix}$

(b) $C + C = \begin{bmatrix} 2 + 2 & 3 + 3 \\ -1 + (-1) & 1 + 1 \\ 0 + 0 & -2 + (-2) \end{bmatrix} = \begin{bmatrix} 4 & 6 \\ -2 & 2 \\ 0 & -4 \end{bmatrix}$ ◆

CAUTION   We cannot add two matrices of different orders. For the matrices A and C in Example 2, we say that the sum $A + C$ is undefined.

When we add matrix C to itself, as we did in Example 2(b), we obtain a matrix of the same order as C in which each element is twice the corresponding element of C. We refer to the matrix $2C$ as the one obtained by multiplying each element of C by 2. Hence, if C is a matrix, then

$$2C = C + C.$$

When working with matrices, we refer to the number 2 as a **scalar**. In this text, *a scalar is a real number.* To multiply a matrix by a scalar k, we simply multiply each element of the matrix by k.

◆ **Scalar Multiple of a Matrix**

If $A = [a_{ij}]$ is a matrix of order $m \times n$ and k is a scalar, then the **scalar multiple** of A is of order $m \times n$ and is defined by

$$kA = [ka_{ij}].$$

PROBLEM 2   Given the matrices A and B in Example 2, find $-2A + 3B$. ◆

The matrix $-A$ (read "the opposite of A") represents the scalar multiple $-1A$. Thus, $-A$ is a matrix of the same order as A in which each element is the *opposite* of the corresponding element of A. Therefore, for every matrix A we have

$$A + (-A) = \mathbf{0},$$

where $\mathbf{0}$ represents the **zero matrix**, a matrix of the same order as A in which each element is zero. We define matrix subtraction in terms of matrix addition:

$$A - B = A + (-B)$$

EXAMPLE 3   Given the matrices

$$A = \begin{bmatrix} 2 & 5 \\ -3 & 0 \\ 1 & 4 \end{bmatrix} \quad \text{and} \quad B = \begin{bmatrix} 4 & 6 \\ -2 & -5 \\ 0 & 3 \end{bmatrix}$$

determine $2A - B$.

◆ **Solution**   By the definition of matrix subtraction, we have

$$2A - B = 2A + (-B) = \begin{bmatrix} 4 & 10 \\ -6 & 0 \\ 2 & 8 \end{bmatrix} + \begin{bmatrix} -4 & -6 \\ 2 & 5 \\ 0 & -3 \end{bmatrix}$$

$$= \begin{bmatrix} 0 & 4 \\ -4 & 5 \\ 2 & 5 \end{bmatrix}$$

   ◆

PROBLEM 3   Given the matrices A and B in Example 3, find $A - 2B$.

   ◆

◆ **Matrix Multiplication**

The product of two matrices A and B, denoted AB, is found by multiplying each row of matrix A by each column of matrix B. If the product AB is to exist, then the number of columns in matrix A must equal the number of rows in matrix B. The product AB has exactly as many rows as A and exactly as many columns as B:

Matrix A: $m \times n$ Matrix B: $n \times p$

These must be equal for the product AB to be defined.

The order of matrix AB is $m \times p$.

If A and B are matrices such that

$$A = \begin{bmatrix} a_{11} & a_{12} & a_{13} & \cdots & a_{1n} \\ a_{21} & a_{22} & a_{23} & \cdots & a_{2n} \\ \vdots & \vdots & \vdots & & \vdots \\ a_{i1} & a_{i2} & a_{i3} & \cdots & a_{in} \\ \vdots & \vdots & \vdots & & \vdots \\ a_{m1} & a_{m2} & a_{m3} & \cdots & a_{mn} \end{bmatrix} \quad \text{and} \quad B = \begin{bmatrix} b_{11} & b_{12} & \cdots & b_{1j} & \cdots & b_{1p} \\ b_{21} & b_{22} & \cdots & b_{2j} & \cdots & b_{2p} \\ b_{31} & b_{32} & \cdots & b_{3j} & \cdots & b_{3p} \\ \vdots & \vdots & & \vdots & & \vdots \\ b_{n1} & b_{n2} & \cdots & b_{nj} & \cdots & b_{np} \end{bmatrix}$$

then the ith row, jth column element of the matrix AB is

$$a_{i1}b_{1j} + a_{i2}b_{2j} + a_{i3}b_{3j} + \cdots + a_{in}b_{nj}.$$

We now formalize the definition of matrix multiplication.

◆ **Matrix Multiplication**

If $A = [a_{ij}]$ is a matrix of order $m \times n$ and $B = [b_{ij}]$ is a matrix of order $n \times p$, then the **product** AB is a matrix of order $m \times p$ defined by

$$AB = [c_{ij}] \quad \text{where} \quad c_{ij} = a_{i1}b_{1j} + a_{i2}b_{2j} + a_{i3}b_{3j} + \cdots + a_{in}b_{nj}.$$

EXAMPLE 4 Matrices A, B, C, and D are defined as follows.

$$A = [2 \quad 3 \quad -1] \qquad B = \begin{bmatrix} 4 \\ 1 \\ -3 \end{bmatrix} \qquad C = \begin{bmatrix} 2 & -1 \\ 3 & 4 \\ 0 & -2 \end{bmatrix}$$

$$D = \begin{bmatrix} -3 & 2 \\ 1 & 4 \end{bmatrix}$$

Determine each product, if it is defined.

(a) AB **(b)** BA **(c)** CD **(d)** DC

◆ **Solution**

(a) The product AB has exactly as many rows as A and exactly as many columns as B. Thus, the product AB is a 1×1 matrix. Using matrix multiplication, we find

$$AB = [2 \quad 3 \quad -1] \begin{bmatrix} 4 \\ 1 \\ -3 \end{bmatrix} = [(2)(4) + (3)(1) + (-1)(-3)] = [14]$$

(b) The product BA has exactly as many rows as B and exactly as many columns as A. Thus, the product BA is a 3×3 matrix. Using matrix multiplication, we find

$$BA = \begin{bmatrix} 4 \\ 1 \\ -3 \end{bmatrix} [2 \quad 3 \quad -1] = \begin{bmatrix} (4)(2) & (4)(3) & (4)(-1) \\ (1)(2) & (1)(3) & (1)(-1) \\ (-3)(2) & (-3)(3) & (-3)(-1) \end{bmatrix}$$

$$= \begin{bmatrix} 8 & 12 & -4 \\ 2 & 3 & -1 \\ -6 & -9 & 3 \end{bmatrix}$$

(c) Since the number of columns of matrix C is the same as the number of rows of matrix D, we know that the product CD is defined. The product

CD has the same number of rows as C and the same number of columns as D, that is, CD is of order 3×2. Using matrix multiplication, we find

$$CD = \begin{bmatrix} 2 & -1 \\ 3 & 4 \\ 0 & -2 \end{bmatrix} \begin{bmatrix} -3 & 2 \\ 1 & 4 \end{bmatrix} = \begin{bmatrix} (2)(-3)+(-1)(1) & (2)(2)+(-1)(4) \\ (3)(-3)+(4)(1) & (3)(2)+(4)(4) \\ (0)(-3)+(-2)(1) & (0)(2)+(-2)(4) \end{bmatrix}$$

$$= \begin{bmatrix} -7 & 0 \\ -5 & 22 \\ -2 & -8 \end{bmatrix}$$

(d) Note that the number of columns of matrix D is *not* the same as the number of rows of matrix C. Hence, the product DC is not defined. ◆

PROBLEM 4   For the matrices A and C defined in Example 4, find the product AC. ◆

CAUTION   Unlike multiplication of real numbers, matrix multiplication is *not* commutative. Observe from Example 4 that $AB \neq BA$ and $CD \neq DC$.

Although matrix multiplication is not commutative, many of the properties of matrix addition and matrix multiplication are similar to the properties of real numbers:

◆ Properties of Matrix Addition and Matrix Multiplication

For matrices A, B, and C of the appropriate orders so that the operations are defined,

Commutative Property of Addition	$A + B = B + A$
Associative Property of Addition	$(A + B) + C = A + (B + C)$
Associative Property of Multiplication	$A(BC) = (AB)C$
Distributive Property (left)	$A(B + C) = AB + AC$
Distributive Property (right)	$(B + C)A = BA + CA$

◆ Matrices and the Graphing Calculator

We can use the matrix features of a graphing calculator to perform operations with matrices. First, we use the MATRIX key to enter the order and elements of the matrices. We now return to the home (primary) screen on the calculator and perform the indicated operations. Figure 7.3 shows typical displays for finding the product in Example 4(c). For a more detailed explanation about the matrix operations that can be performed on a graphing calculator, consult the manual that came with your graphing calculator or the *Graphing Technology Laboratory Manual* that accompanies this text. After reading this information, use a graphing calculator to verify the products in Example 4(a) and 4(b).

```
MATRIX[C] 3 ×2
[ 2      -1    ]
[ 3       4    ]
[ 0      -2    ]
```

(a) Enter matrix *C*.

```
MATRIX[D] 2 ×2
[ -3      2    ]
[  1      4    ]
```

(b) Enter matrix *D*.

```
[C][D]
        [[-7  0 ]
         [-5 22]
         [-2 -8]]
```

(c) Return to the home screen and perform the indicated operation.

FIGURE 7.3

Typical displays for finding the product in Example 4(c)

◆ **Application: Inventory and Cost**

We conclude this section with an application of matrix multiplication.

EXAMPLE 5 The cost (in dollars) of each size of a certain model of brass bed is given by matrix *C*, and the current inventory of this model of brass bed at two furniture stores in the greater Boston area is given by matrix *B*. Find *CB* and state what it represents.

$$C = \begin{bmatrix} \overset{\text{Twin}}{300} & \overset{\text{Double}}{350} & \overset{\text{Queen}}{400} & \overset{\text{King}}{600} \end{bmatrix} \text{Cost} \qquad B = \begin{bmatrix} \overset{\text{Brookline}}{2} & \overset{\text{Cambridge}}{5} \\ 6 & 0 \\ 0 & 2 \\ 2 & 3 \end{bmatrix} \begin{matrix} \text{Twin} \\ \text{Double} \\ \text{Queen} \\ \text{King} \end{matrix}$$

◀◆ **Solution**

$$CB = \begin{bmatrix} 300 & 350 & 400 & 600 \end{bmatrix} \begin{bmatrix} 2 & 5 \\ 6 & 0 \\ 0 & 2 \\ 2 & 3 \end{bmatrix} = \begin{bmatrix} 3900 & 4100 \end{bmatrix}$$

$300(2) + 350(6) + 400(0) + 600(2)$

$300(5) + 350(0) + 400(2) + 600(3)$

The 1 × 2 matrix

$$CB = \begin{bmatrix} \overset{\text{Brookline}}{3900} & \overset{\text{Cambridge}}{4100} \end{bmatrix} \text{Total cost}$$

represents the total cost (in dollars) of the inventory of this type of bed at each store. ◆

PROBLEM 5 For the matrix *B* given in Example 5, find the matrix [1 1 1 1]*B*, and state what it represents. ◆

Exercises 7.2

Basic Skills

In Exercises 1–16, use the Gaussian method to solve each system of linear equations.

1. $2x - 3y = 1$
$x + 4y = 6$

2. $4x - 3y = 0$
$2x - y = -10$

3. $5x - 20y = -10$
$-2x + 8y = 4$

4. $3x - 12y = 6$
$4x - 16y = -8$

5. $x - 3y = 7$
$2x + z = 5$
$4y + 2z = -2$

6. $4x - 2z = -1$
$2x + 2y = 1$
$x - 3y + z = -7$

7. $2x - y - 2z = 0$
$3x + 3y + z = 6$
$y + 3z = -5$

8. $5x - y + z = 3$
$2x + 3z = -5$
$2x - 3y - 4z = 4$

9. $2x + 3y - 2z = 3$
$3x + 5y + z = 2$
$x + 2y - 5z = 1$

10. $3x - 4y + 19z = 6$
$x - y + 4z = 2$
$5x - 2y - z = 10$

11. $4x - 3y - z = 5$
$3x + 2y - 3z = 2$
$-2x + y - z = 0$

12. $2x - y + z = 7$
$4x - 3y - 3z = 11$
$5x + y - z = 14$

13. $x + 2y + t = 3$
$y + 2z = 3$
$x - 2z = 0$
$3y - 4z + t = 2$

14. $2x - y + t = 9$
$3y + 2z = -6$
$y - 2z + 4t = -14$
$3x - z + t = -4$

15. $x + 3y + 2t = 3$
$3x + 2z - t = 3$
$3y - z + 2t = 1$
$x + 2y - 3z = -4$

16. $5x - y + 2z = -3$
$2x - 5y - t = -3$
$4y - 3z + 4t = -3$
$-x + 4z - 2t = -3$

In Exercises 17–56, perform the indicated matrix operations given that A, B, C, D, E, F, G, and H are defined as follows. If an operation is not defined, state the reason. Use the matrix features of a graphing calculator to verify each answer.

$$A = \begin{bmatrix} 1 & 2 & 0 \\ -2 & 3 & -4 \end{bmatrix} \qquad B = \begin{bmatrix} 2 & -3 & -5 \\ -1 & 4 & 4 \end{bmatrix}$$

$$C = \begin{bmatrix} 2 & -1 \\ 3 & 2 \\ -1 & 4 \end{bmatrix} \qquad D = \begin{bmatrix} 6 & 0 \\ -3 & 9 \\ 0 & 3 \end{bmatrix}$$

$$E = \begin{bmatrix} 2 & -4 & -5 \end{bmatrix} \qquad F = \begin{bmatrix} 1 \\ -3 \\ 2 \end{bmatrix}$$

$$G = \begin{bmatrix} -1 & 1 \\ 1 & -1 \end{bmatrix} \qquad H = \begin{bmatrix} 5 & 5 \\ -2 & -2 \end{bmatrix}$$

17. $2A$

18. $-5G$

19. $-\frac{1}{3}D$

20. $\frac{3}{2}E$

21. $A + B$

22. $C + D$

23. $E + F$

24. $F + F$

25. $3G + H$

26. $2C + 5G$

27. $-4A + 3B$

28. $-\frac{1}{3}D + 2C$

29. $H - B$

30. $D - C$

31. $3G - 2H$

32. $\frac{1}{2}A - \frac{3}{2}B$

33. $-2(3E)$

34. $3A - 2A$

35. $\frac{1}{2}F - \frac{3}{2}F$

36. $2A + 2(B - A)$

37. AC

38. CB

39. GB

40. ED

41. DH

42. HD

43. EF

44. FE

45. HG

46. GH

47. $C(A + B)$

48. $CA + CB$

49. $(A - B)D$

50. $AD - BD$

51. $(EC)A$

52. $E(CA)$

53. $(BD)G$

54. $B(DG)$

55. $F(2E)$

56. $2(FE)$

57. $37.5CB$

58. $[1 \quad 1 \quad 1](37.5CB)$

59. $37.5(B - A)$

60. $37.5C(B - A)$

An engineering firm employs three types of union workers: technicians, engineers, and senior engineers. The company and the union have agreed to a new contract, which states that the hourly wages (in dollars) increase from the figures given in matrix A to those given in matrix B.

Old hourly wage

$$A = \begin{bmatrix} 8.35 \\ 13.45 \\ 22.15 \end{bmatrix} \begin{matrix} \text{Technician} \\ \text{Engineer} \\ \text{Senior engineer} \end{matrix}$$

New hourly wage

$$B = \begin{bmatrix} 9.65 \\ 16.75 \\ 28.25 \end{bmatrix} \begin{matrix} \text{Technician} \\ \text{Engineer} \\ \text{Senior engineer} \end{matrix}$$

and that each person works 37.5 hours per week. The company maintains offices in Boston, Atlanta, and Seattle, and the number of technicians, engineers, and senior engineers at each facility is given in matrix C:

Technician	Engineer	Senior engineer	

$$C = \begin{bmatrix} 81 & 67 & 19 \\ 57 & 42 & 16 \\ 122 & 88 & 31 \end{bmatrix} \begin{matrix} \text{Boston} \\ \text{Atlanta} \\ \text{Seattle} \end{matrix}$$

In Exercises 57–60, find the indicated matrix, and describe what it represents.

A car rental agency rents three types of cars: subcompact, midsize, and large cars. The daily rental rates (in dollars) charged by the agency are given in matrix C.

	Subcompact	Mid-size	Large	
$C =$	[20	32	40]	Cost

Within a certain city, the agency has three terminals. The number of subcompact, midsize, and large cars rented at each terminal on Monday and Tuesday are given in matrices M and T, respectively.

	Terminal 1	Terminal 2	Terminal 3	
$M =$	12	14	18	Subcompact
	8	14	10	Midsize
	2	5	6	Large

	Terminal 1	Terminal 2	Terminal 3	
$T =$	10	8	9	Subcompact
	2	4	10	Midsize
	0	1	6	Large

In Exercises 61–64, find the indicated matrix, and state what it represents.

61. $(M + T)\begin{bmatrix} 1 \\ 1 \\ 1 \end{bmatrix}$

62. $[1 \quad 1 \quad 1](M + T)$

63. $C(M + T)$

64. $C(M + T)\begin{bmatrix} 1 \\ 1 \\ 1 \end{bmatrix}$

◆ **Critical Thinking**

65. The zero product property (Section 2.2) states that if a and b are real numbers and $ab = 0$, then either $a = 0$ or $b = 0$. This property does not hold for matrices: Find a pair of 2×2 matrices A and B for which $AB = 0$, but $A \neq 0$ and $B \neq 0$.

66. Determine the values of x and y such that the following product is true.

$$\begin{bmatrix} 2 & -3 \\ x & y \end{bmatrix}\begin{bmatrix} 4 & 3 \\ -6 & 5 \end{bmatrix} = \begin{bmatrix} 26 & -9 \\ 24 & -1 \end{bmatrix}$$

67. If A is an $m \times m$ matrix, then we define the *square* of matrix A by writing $A^2 = AA$.

(a) Given

$$A = \begin{bmatrix} 1 & -2 & 3 \\ 0 & 3 & -2 \\ 3 & 1 & 4 \end{bmatrix}$$

$$B = \begin{bmatrix} 2 & 0 & -1 \\ 1 & 3 & 4 \\ 0 & -1 & -1 \end{bmatrix}$$

perform each matrix operation:

(i) A^2

(ii) B^2

(iii) $A^2 - B^2$

(iv) $(A + B)(A - B)$

Exercise 67 continues

(b) Compare the answers to (iii) and (iv) of part (a). Does $A^2 - B^2 = (A + B)(A - B)$?

68. Suppose matrices A, X, and B are defined as follows.

$$A = \begin{bmatrix} 1 & -1 & 0 \\ 2 & -1 & 1 \\ 0 & 4 & 5 \end{bmatrix}$$

$$X = \begin{bmatrix} x \\ y \\ z \end{bmatrix} \qquad B = \begin{bmatrix} 2 \\ 9 \\ 23 \end{bmatrix}$$

(a) Express the product AX as a 3×1 matrix.

(b) Use the result of part (a) and equality of matrices to determine x, y, and z if $AX = B$.

In Exercises 69 and 70, use matrices A, B, A^T, and B^T defined as follows:

$$A = \begin{bmatrix} a_{11} & a_{12} \\ a_{21} & a_{22} \end{bmatrix} \qquad B = \begin{bmatrix} b_{11} & b_{12} \\ b_{21} & b_{22} \end{bmatrix}$$

$$A^T = \begin{bmatrix} a_{11} & a_{21} \\ a_{12} & a_{22} \end{bmatrix} \quad and \quad B^T = \begin{bmatrix} b_{11} & b_{21} \\ b_{12} & b_{22} \end{bmatrix}$$

Show that the given statement is true. (The matrices A^T and B^T are called the transposes of matrices A and B, respectively. The transpose of a matrix is formed by switching its rows and columns.)

69. $(AB)^T = B^T A^T$

70. $(A + B)^T = A^T + B^T$

71. Cayley defined matrix multiplication to help simplify double systems of equation such as

$$u = a_{11}r + a_{12}s \qquad x = b_{11}u + b_{12}v$$
$$v = a_{21}r + a_{22}s \qquad y = b_{21}u + b_{22}v$$

(a) Use substitution to find x and y in terms of r and s.

(b) Express the product

$$\begin{bmatrix} b_{11} & b_{12} \\ b_{21} & b_{22} \end{bmatrix} \begin{bmatrix} a_{11} & a_{12} \\ a_{21} & a_{22} \end{bmatrix} \begin{bmatrix} r \\ s \end{bmatrix}$$

as a 2×1 matrix. What do the elements in this 2×1 matrix represent?

72. Use an appropriate substitution to change the given system of equations into a linear system. Apply the Gaussian method to the linear system of equations, and then state the solution of the original system.

(a) $\dfrac{2}{x} + \dfrac{3}{y} = 15$

$\dfrac{4}{x} - \dfrac{5}{y} = -14$

(b) $\begin{aligned} e^x - e^y + 2e^z &= 0 \\ 2e^x - e^y + 3e^z &= 3 \\ 5e^x - 2e^y - 3e^z &= -1 \end{aligned}$

◆ 7.3 Determinants and Inverses of Matrices

◆ Introductory Comments

A matrix of order $n \times n$ is called a **square matrix**. In a square matrix, the elements $a_{11}, a_{22}, a_{33}, \ldots, a_{nn}$ are called the **main diagonal** elements. Associated with every square matrix A is a real number called the *determinant of A*, which we denote by writing $|A|$. The 2×2 matrix $A = \begin{bmatrix} a_{11} & a_{12} \\ a_{21} & a_{22} \end{bmatrix}$

has a **second-order determinant**

$$|A| = \begin{vmatrix} a_{11} & a_{12} \\ a_{21} & a_{22} \end{vmatrix}$$

whose value is found by taking the product of the main diagonal elements and subtracting the product of the elements on the other diagonal.

◆ **Determinant of a 2 × 2 Matrix**

The **determinant** of the 2 × 2 matrix $A = \begin{bmatrix} a_{11} & a_{12} \\ a_{21} & a_{22} \end{bmatrix}$ is

$$|A| = \begin{vmatrix} a_{11} & a_{12} \\ a_{21} & a_{22} \end{vmatrix} = a_{11}a_{22} - a_{21}a_{12}$$

Here are some illustrations of evaluating a second-order determinant:

$$\begin{vmatrix} 8 & -4 \\ 3 & -2 \end{vmatrix} = (8)(-2) - (3)(-4) = -16 + 12 = -4$$

$$\begin{vmatrix} 6 & -1 \\ 0 & 7 \end{vmatrix} = (6)(7) - (0)(-1) = 42 - 0 = 42$$

The determinant of any $n \times n$ matrix in echelon form is the product of the elements along its main diagonal.

◆ **Determinant of a Matrix in Echelon Form**

If A is an $n \times n$ matrix in echelon form, then $|A|$ is the product of the elements along its main diagonal.

Here are some illustrations of evaluating a determinant in echelon form:

$$\begin{vmatrix} 3 & -1 & 4 \\ 0 & 4 & -2 \\ 0 & 0 & -5 \end{vmatrix} = (3)(4)(-5) = -60$$

$$\begin{vmatrix} 2 & -3 & 1 & 0 \\ 0 & 2 & 4 & -7 \\ 0 & 0 & -3 & 1 \\ 0 & 0 & 0 & -1 \end{vmatrix} = (2)(2)(-3)(-1) = 12$$

In this section, we discuss some methods that may be used to evaluate the determinant of an $n \times n$ matrix. We then introduce the inverse of a matrix and show how it may be used to help solve a system of linear equations. We begin by looking at the effects of elementary row operations on a determinant.

◆ **Evaluating the Determinant of a Matrix**

Consider the 2 × 2 matrix $A = \begin{bmatrix} a_{11} & a_{12} \\ a_{21} & a_{22} \end{bmatrix}$ and the effects of elementary row operations on its determinant

$$|A| = \begin{vmatrix} a_{11} & a_{12} \\ a_{21} & a_{22} \end{vmatrix} = a_{11}a_{22} - a_{21}a_{12}$$

1. If the rows of matrix A are interchanged to form the row-equivalent matrix

$$B = \begin{bmatrix} a_{21} & a_{22} \\ a_{11} & a_{12} \end{bmatrix},$$

then

$$\begin{aligned} |B| &= a_{21}a_{12} - a_{11}a_{22} \\ &= -(a_{11}a_{22} - a_{21}a_{12}) \\ &= -|A| \end{aligned}$$

2. If a row of matrix A is replaced by c times that row to form the row-equivalent matrix

$$B = \begin{bmatrix} ca_{11} & ca_{12} \\ a_{21} & a_{22} \end{bmatrix},$$

then

$$\begin{aligned} |B| &= ca_{11}a_{22} - ca_{21}a_{12} \\ &= c(a_{11}a_{22} - a_{21}a_{12}) \\ &= c|A| \end{aligned}$$

3. If a row of matrix A is replaced by the sum of that row and c times another row to form the row equivalent matrix

$$B = \begin{bmatrix} a_{11} & a_{12} \\ ca_{11} + a_{21} & ca_{12} + a_{22} \end{bmatrix},$$

then

$$\begin{aligned} |B| &= a_{11}(ca_{12} + a_{22}) - a_{12}(ca_{11} + a_{21}) \\ &= a_{11}a_{22} - a_{21}a_{12} \\ &= |A| \end{aligned}$$

We now generalize these observations for any $n \times n$ matrix A.

◆ **Effects of Row Operations on a Determinant**

Suppose A is an $n \times n$ matrix.

1. If two rows of A are interchanged to form the row-equivalent matrix B, then $|B| = -|A|$.

2. If a row of A is replaced by c times that row to form the row-equivalent matrix B, then $|B| = c|A|$.

3. If a row of A is replaced by the sum of that row and c times another row to form the row-equivalent matrix B, then $|B| = |A|$.

Operation 2 has the effect of factoring out a common factor from a row. For example,

$$\begin{vmatrix} 8 & -4 \\ 3 & -2 \end{vmatrix} = 4\begin{vmatrix} 2 & -1 \\ 3 & -2 \end{vmatrix} = 4[-4-(-3)] = -4.$$

Factor 4 from row 1.

We can evaluate the determinant of any $n \times n$ matrix by using elementary row operations. The procedure is illustrated in the next example.

EXAMPLE 1   Use elementary row operations to evaluate the determinant of matrix A.

$$A = \begin{bmatrix} 2 & -2 & -3 \\ 0 & 6 & 3 \\ 1 & -2 & -5 \end{bmatrix}$$

◆ **Solution**   We write the determinant in echelon form, and record the effects of the row operations on the determinant:

$$|A| = \begin{vmatrix} 2 & -2 & -3 \\ 0 & 6 & 3 \\ 1 & -2 & -5 \end{vmatrix} = -\begin{vmatrix} 1 & -2 & -5 \\ 0 & 6 & 3 \\ 2 & -2 & -3 \end{vmatrix} \quad \mathbf{R_1 \leftrightarrow R_3}$$

determinant changes sign

$$= -\begin{vmatrix} 1 & -2 & -5 \\ 0 & 6 & 3 \\ 0 & 2 & 7 \end{vmatrix} \quad \mathbf{-2R_1 + R_3 \rightarrow R_3}$$

$$= -3\begin{vmatrix} 1 & -2 & -5 \\ 0 & 2 & 1 \\ 0 & 2 & 7 \end{vmatrix} \quad \mathbf{\tfrac{1}{3}R_2 \rightarrow R_2}$$

Factor out 3 from R_2.

$$= -3\begin{vmatrix} 1 & -2 & -5 \\ 0 & 2 & 1 \\ 0 & 0 & 6 \end{vmatrix} \quad \mathbf{-R_2 + R_3 \rightarrow R_3}$$

Now, multiplying the elements along the main diagonal, we have

$$|A| = -3[(1)(2)(6)] = -36 \qquad ◆$$

We can use elementary row operations to show that any third-order determinant can be expressed in terms of second-order determinants (see Exercise 49):

$$\begin{vmatrix} a_{11} & a_{12} & a_{13} \\ a_{21} & a_{22} & a_{23} \\ a_{31} & a_{32} & a_{33} \end{vmatrix} = a_{11}\begin{vmatrix} a_{22} & a_{23} \\ a_{32} & a_{33} \end{vmatrix} - a_{12}\begin{vmatrix} a_{21} & a_{23} \\ a_{31} & a_{33} \end{vmatrix} + a_{13}\begin{vmatrix} a_{21} & a_{22} \\ a_{31} & a_{32} \end{vmatrix}$$

This formula is called the **expansion of a third-order determinant by its first row**. Using this formula on matrix A in Example 1, we have

$$\begin{vmatrix} 2 & -2 & -3 \\ 0 & 6 & 3 \\ 1 & -2 & -5 \end{vmatrix} = 2\begin{vmatrix} 6 & 3 \\ -2 & -5 \end{vmatrix} - (-2)\begin{vmatrix} 0 & 3 \\ 1 & -5 \end{vmatrix} + (-3)\begin{vmatrix} 0 & 6 \\ 1 & -2 \end{vmatrix}$$

$$= 2[-30 - (-6)] + 2[0 - 3] - 3[0 - 6]$$

$$= -36,$$

which agrees with our result in Example 1.

We can also evaluate the determinant of an $n \times n$ matrix by selecting the appropriate line from the matrix menu on a graphing calculator. Figure 7.4 shows typical displays for evaluating the determinant of matrix A in Example 1.

When we perform elementary row operations on a determinant, it is important to watch for a pair of rows in which corresponding elements are proportional. If we add a certain multiple of one of these rows to the other, we generate a row in which every element is zero. *If every element in a row of matrix A is zero, then $|A| = 0$.*

(a) Enter matrix A

(b) Select "det" from the matrix menu

FIGURE 7.4

Typical displays for evaluating the determinant in Example 1

PROBLEM 1   Explain why the determinant of the given matrix A is zero. Then enter this matrix into a graphing calculator and select the "det" command from the matrix menu to confirm this answer.

$$A = \begin{bmatrix} 1 & -3 & -2 & 6 \\ 2 & 4 & 2 & -1 \\ 0 & 2 & 5 & -6 \\ 6 & 12 & 6 & -3 \end{bmatrix}$$

◆ **Inverse of a Matrix**

A square matrix with the digit 1 along its main diagonal and zeros elsewhere is called the $n \times n$ **identity matrix** and is denoted by

$$I_n = \begin{bmatrix} 1 & 0 & 0 & \dots & 0 \\ 0 & 1 & 0 & \dots & 0 \\ 0 & 0 & 1 & \dots & 0 \\ \vdots & \vdots & \vdots & \vdots & \vdots \\ 0 & 0 & 0 & \dots & 1 \end{bmatrix}$$

Anytime we multiply an $n \times n$ matrix by the identity matrix I_n, we obtain the original $n \times n$ matrix from which we started. We refer to this fact as the **identity property of matrix multiplication**.

◆ **Identity Property of Matrix Multiplication**

If A is a square matrix of order $n \times n$ and I_n is the $n \times n$ identity matrix, then

$$AI_n = I_nA = A.$$

We say that two square $n \times n$ matrices A and B are **inverses** of each other if their product is the identity matrix I_n, that is, if

$$AB = BA = I_n.$$

For example, the square matrices

$$A = \begin{bmatrix} 2 & -5 \\ -1 & 3 \end{bmatrix} \quad \text{and} \quad B = \begin{bmatrix} 3 & 5 \\ 1 & 2 \end{bmatrix}$$

are inverses of each other, since

$$AB = \begin{bmatrix} 2 & -5 \\ -1 & 3 \end{bmatrix}\begin{bmatrix} 3 & 5 \\ 1 & 2 \end{bmatrix} = \begin{bmatrix} 6 + (-5) & 10 + (-10) \\ -3 + 3 & -5 + 6 \end{bmatrix} = \begin{bmatrix} 1 & 0 \\ 0 & 1 \end{bmatrix}$$

and

$$BA = \begin{bmatrix} 3 & 5 \\ 1 & 2 \end{bmatrix}\begin{bmatrix} 2 & -5 \\ -1 & 3 \end{bmatrix} = \begin{bmatrix} 6 + (-5) & -15 + 15 \\ 2 + (-2) & -5 + 6 \end{bmatrix} = \begin{bmatrix} 1 & 0 \\ 0 & 1 \end{bmatrix}$$

identity matrix of order 2×2

CAUTION Two nonsquare matrices A and B cannot be inverses of each other, since their products AB and BA (if they exist) are always of different orders and, therefore, are not equal. In general, a nonsquare matrix cannot have an inverse.

We usually denote the inverse of matrix A by using the notation A^{-1} (read "A inverse"). We now state the **inverse property of matrix multiplication**.

◆ **Inverse Property of Matrix Multiplication**

If A is a square $n \times n$ matrix and A^{-1} is its inverse, then

$$AA^{-1} = A^{-1}A = I_n$$

where I_n is the identity matrix of order $n \times n$.

CAUTION If a is a real number, then $a^{-1} = 1/a$; however, if A is a matrix, the notation A^{-1} does *not* mean $1/A$. In fact, matrix division is *not* defined.

An $n \times n$ matrix A has an inverse if and only if the determinant of A is not zero. If $|A| \neq 0$, then we say that matrix A is **invertible**. If $|A| = 0$, then we say that matrix A is **singular**.

◆ **Procedure for Finding A^{-1}**

If A is an $n \times n$ matrix and $|A| \neq 0$, then its inverse A^{-1} may be found as follows.

1. Write the $n \times 2n$ matrix $\begin{bmatrix} A & \vdots & I_n \end{bmatrix}$ consisting of matrix A on the left-hand side and the identity matrix I_n on the right-hand side of a dashed line.

2. Use elementary row operations on the matrix $\begin{bmatrix} A & \vdots & I_n \end{bmatrix}$ to obtain the matrix $\begin{bmatrix} I_n & \vdots & B \end{bmatrix}$. The matrix B is the inverse matrix of A, that is, $B = A^{-1}$.

EXAMPLE 2   Find the inverse of the given matrix, if it exists.

(a) $A = \begin{bmatrix} -3 & -1 \\ 2 & 2 \end{bmatrix}$ **(b)** $A = \begin{bmatrix} 1 & -1 & 0 \\ 2 & -1 & 1 \\ 0 & 4 & 5 \end{bmatrix}$

◆ **Solution**

(a) The determinant of matrix A is $(-6) - (-2) = -4 \neq 0$. Thus, we know that A is invertible.

Beginning with the 2×4 matrix $\begin{bmatrix} A & \vdots & I_2 \end{bmatrix}$, we perform elementary row operations until we obtain the matrix $\begin{bmatrix} I_2 & \vdots & B \end{bmatrix}$.

$$\begin{bmatrix} -3 & -1 & \vdots & 1 & 0 \\ 2 & 2 & \vdots & 0 & 1 \end{bmatrix} \sim \begin{bmatrix} 1 & 3 & \vdots & 1 & 2 \\ 2 & 2 & \vdots & 0 & 1 \end{bmatrix} \qquad \mathbf{2R_2 + R_1 \rightarrow R_1}$$

$$\sim \begin{bmatrix} 1 & 3 & \vdots & 1 & 2 \\ 0 & -4 & \vdots & -2 & -3 \end{bmatrix} \qquad \mathbf{-2R_1 + R_2 \rightarrow R_2}$$

$$\sim \begin{bmatrix} 1 & 3 & \vdots & 1 & 2 \\ 0 & 1 & \vdots & \frac{1}{2} & \frac{3}{4} \end{bmatrix} \qquad \mathbf{-\frac{1}{4}R_2 \rightarrow R_2}$$

$$\sim \begin{bmatrix} 1 & 0 & \vdots & -\frac{1}{2} & -\frac{1}{4} \\ 0 & 1 & \vdots & \frac{1}{2} & \frac{3}{4} \end{bmatrix} \qquad \mathbf{-3R_2 + R_1 \rightarrow R_1}$$

Therefore, the inverse of matrix A is

$$A^{-1} = \begin{bmatrix} -\frac{1}{2} & -\frac{1}{4} \\ \frac{1}{2} & \frac{3}{4} \end{bmatrix} = \frac{1}{4} \begin{bmatrix} -2 & -1 \\ 2 & 3 \end{bmatrix}$$

To avoid fractional elements, factor the scalar $\frac{1}{4}$ from each element of the matrix.

(b) Using an algebraic procedure for evaluating a third-order determinant, or selecting "det" from the matrix menu on a graphing calculator, we find that $|A| = 1 \neq 0$. Thus, we know that A is invertible.

Beginning with the 3×6 matrix $\left[A \mid I_3\right]$, we perform elementary row operations until we obtain the matrix $\left[I_3 \mid B\right]$.

$$\begin{bmatrix} 1 & -1 & 0 & 1 & 0 & 0 \\ 2 & -1 & 1 & 0 & 1 & 0 \\ 0 & 4 & 5 & 0 & 0 & 1 \end{bmatrix} \sim \begin{bmatrix} 1 & -1 & 0 & 1 & 0 & 0 \\ 0 & 1 & 1 & -2 & 1 & 0 \\ 0 & 4 & 5 & 0 & 0 & 1 \end{bmatrix}$$

$-2R_1 + R_2 \to R_2$

$$\sim \begin{bmatrix} 1 & -1 & 0 & 1 & 0 & 0 \\ 0 & 1 & 1 & -2 & 1 & 0 \\ 0 & 0 & 1 & 8 & -4 & 1 \end{bmatrix}$$

$-4R_2 + R_3 \to R_3$

$$\sim \begin{bmatrix} 1 & -1 & 0 & 1 & 0 & 0 \\ 0 & 1 & 0 & -10 & 5 & -1 \\ 0 & 0 & 1 & 8 & -4 & 1 \end{bmatrix}$$

$-R_3 + R_2 \to R_2$

$$\sim \begin{bmatrix} 1 & 0 & 0 & -9 & 5 & -1 \\ 0 & 1 & 0 & -10 & 5 & -1 \\ 0 & 0 & 1 & 8 & -4 & 1 \end{bmatrix}$$

$R_2 + R_1 \to R_1$

Therefore, the inverse matrix of A is

$$A^{-1} = \begin{bmatrix} -9 & 5 & -1 \\ -10 & 5 & -1 \\ 8 & -4 & 1 \end{bmatrix}$$ ◆

We can use the matrix features of a graphing calculator to find the inverse of a square matrix. First, we use the MATRIX key and enter the elements of the matrix whose inverse we wish to find. We now return to the home (primary) screen on the calculator and find the inverse matrix by using the inverse function key. Figure 7.5 shows typical displays for finding the inverse matrix in Example 2(a). For a more detailed explanation about finding the inverse of a matrix by using a graphing calculator, consult the manual that came with your calculator or the *Graphing Technology Laboratory Manual* that accompanies this text. After reading this information, use a graphing calculator to verify the inverse in Example 2(b).

(a) Enter the elements of matrix *A*.

(b) Return to the home screen to find the inverse matrix.

(c) Express the inverse matrix with fractional elements (optional).

FIGURE 7.5
Typical displays for finding the inverse in Example 2(a)

PROBLEM 2   Show that the given matrix A is singular.

$$A = \begin{bmatrix} 1 & 2 & -3 \\ 1 & 0 & 2 \\ 2 & 2 & -1 \end{bmatrix}$$

Try to find A^{-1} by using the matrix features on a graphing calculator. What error message appears? ◆

◆ Using Inverses to Solve a System of Linear Equations

We can use the inverse of a matrix to help solve a system of linear equations. Consider a system of n linear equations in n unknowns:

$$\begin{aligned} a_{11}x_1 + a_{12}x_2 + a_{13}x_3 + \cdots + a_{1n}x_n &= k_1 \\ a_{21}x_1 + a_{22}x_2 + a_{23}x_3 + \cdots + a_{2n}x_n &= k_2 \\ a_{31}x_1 + a_{32}x_2 + a_{33}x_3 + \cdots + a_{3n}x_n &= k_3 \\ &\vdots \\ a_{n1}x_1 + a_{n2}x_2 + a_{n3}x_3 + \cdots + a_{nn}x_n &= k_n \end{aligned}$$

We can represent this $n \times n$ linear system of equations by the matrix equation

$$AX = B$$

where

$$A = \begin{bmatrix} a_{11} & a_{12} & a_{13} & \cdots & a_{1n} \\ a_{21} & a_{22} & a_{23} & \cdots & a_{2n} \\ a_{31} & a_{32} & a_{33} & \cdots & a_{3n} \\ \vdots & \vdots & \vdots & & \vdots \\ a_{n1} & a_{n2} & a_{n3} & \cdots & a_{nn} \end{bmatrix} \quad X = \begin{bmatrix} x_1 \\ x_2 \\ x_3 \\ \vdots \\ x_n \end{bmatrix} \quad B = \begin{bmatrix} k_1 \\ k_2 \\ k_3 \\ \vdots \\ k_n \end{bmatrix}$$

We refer to matrix A as the **coefficient matrix**, matrix X as the **variable matrix**, and matrix B as the **constant matrix**. If the coefficient matrix A is invertible, then we can solve the equation $AX = B$ for X as follows:

$$AX = B$$
$$A^{-1}(AX) = A^{-1}B \quad \text{Multiply both sides by } A^{-1}$$
$$(A^{-1}A)X = A^{-1}B \quad \text{Apply associative property of matrix multiplication}$$
$$I_nX = A^{-1}B \quad \text{Apply inverse property of matrix multiplication}$$
$$X = A^{-1}B \quad \text{Apply identity property of matrix multiplication}$$

The product $A^{-1}B$ represents the solution to the $n \times n$ linear system of equations. We refer to this procedure of solving a system of linear equations as the **inverse method**.

CAUTION Since matrix division is undefined, we cannot solve the matrix equation $AX = B$ for X by dividing both sides by A. Also, since matrix multiplication is not commutative, $X \neq BA^{-1}$.

EXAMPLE 3 Use the inverse method to solve the given system of linear equations.

$$
\begin{aligned}
x - y &= 2 \\
2x - y + z &= 9 \\
4y + 5z &= 23
\end{aligned}
$$

◆ **Solution** If we let

$$
A = \begin{bmatrix} 1 & -1 & 0 \\ 2 & -1 & 1 \\ 0 & 4 & 5 \end{bmatrix} \qquad X = \begin{bmatrix} x \\ y \\ z \end{bmatrix} \qquad B = \begin{bmatrix} 2 \\ 9 \\ 23 \end{bmatrix}
$$

coefficient matrix   variable matrix   constant matrix

then the matrix equation

$$AX = B$$

$$
\begin{bmatrix} 1 & -1 & 0 \\ 2 & -1 & 1 \\ 0 & 4 & 5 \end{bmatrix}\begin{bmatrix} x \\ y \\ z \end{bmatrix} = \begin{bmatrix} 2 \\ 9 \\ 23 \end{bmatrix}
$$

is equivalent to the given system of three equations in three unknowns. From Example 2(b), we know that the coefficient matrix A is invertible and that

$$
A^{-1} = \begin{bmatrix} -9 & 5 & -1 \\ -10 & 5 & -1 \\ 8 & -4 & 1 \end{bmatrix}
$$

Thus, this system of equations has a unique solution given by

$$
X = A^{-1}B = \begin{bmatrix} -9 & 5 & -1 \\ -10 & 5 & -1 \\ 8 & -4 & 1 \end{bmatrix}\begin{bmatrix} 2 \\ 9 \\ 23 \end{bmatrix}
$$

$$
= \begin{bmatrix} -18 + 45 + (-23) \\ -20 + 45 + (-23) \\ 16 + (-36) + 23 \end{bmatrix}
$$

$$
= \begin{bmatrix} 4 \\ 2 \\ 3 \end{bmatrix}
$$

The solution of this system of linear equations is the ordered triple (4, 2, 3). You can check the solution by substituting $x = 4$, $y = 2$, and $z = 3$ into each of the original equations in the system. ◆

This inverse method is one of the most efficient procedures we have for solving a system of linear equations with a graphing calculator. Figure 7.6 shows typical displays for solving the linear system in Example 3.

(a) Enter the dimension and elements of the coefficient matrix A.

(b) Enter the dimension and elements of the constant matrix B.

(c) Return to the home screen, and compute the product $A^{-1}B$.

FIGURE 7.6

Typical displays for solving the linear system in Example 3

If the coefficient matrix is singular, then the system of linear equations does not have a unique solution.

PROBLEM 3   Show that the given system of linear equations does not have a unique solution. Is this system inconsistent or dependent?

$$x + 2y - 3z = -3$$
$$x + 3z = 4$$
$$2x + 2y - z = 1$$
◆

Exercises 7.3

◆ Basic Skills

In Exercises 1–12, evaluate the determinant of each matrix by inspection.

1. $\begin{bmatrix} 1 & 4 \\ 6 & 2 \end{bmatrix}$

2. $\begin{bmatrix} 9 & 0 \\ 1 & 2 \end{bmatrix}$

3. $\begin{bmatrix} 8 & -4 \\ -1 & 0 \end{bmatrix}$

4. $\begin{bmatrix} 5 & -2 \\ 3 & -4 \end{bmatrix}$

5. $\begin{bmatrix} -10 & -12 \\ 4 & -17 \end{bmatrix}$

6. $\begin{bmatrix} 12 & 6 \\ -15 & -18 \end{bmatrix}$

7. $\begin{bmatrix} 2 & 3 & -1 \\ 0 & 3 & 9 \\ 0 & 0 & -2 \end{bmatrix}$

8. $\begin{bmatrix} -3 & 9 & 0 \\ 0 & -4 & 5 \\ 0 & 0 & 4 \end{bmatrix}$

9. $\begin{bmatrix} 1 & -2 & 3 & 0 \\ 0 & 2 & 8 & 0 \\ 0 & 0 & -8 & 12 \\ 0 & 0 & 0 & -13 \end{bmatrix}$

10. $\begin{bmatrix} -4 & 0 & -8 & 4 \\ 0 & -4 & 5 & 10 \\ 0 & 0 & 13 & 0 \\ 0 & 0 & 0 & -5 \end{bmatrix}$

11. $\begin{bmatrix} 4 & -2 & 10 & 8 \\ 0 & 0 & 0 & 0 \\ -2 & 4 & 12 & 9 \\ 2 & 9 & 4 & -7 \end{bmatrix}$

12. $\begin{bmatrix} 4 & -2 & 6 & 14 \\ 1 & 2 & 3 & 1 \\ 2 & -1 & 3 & 7 \\ 1 & 3 & -4 & 9 \end{bmatrix}$

In Exercises 13–20, evaluate the determinant of each matrix by using the following two methods.

(i) Elementary row operations.

(ii) The formula for the expansion of a third-order determinant by its first row.

13. $\begin{bmatrix} 1 & 0 & 2 \\ 2 & -3 & 4 \\ -3 & 1 & 9 \end{bmatrix}$

14. $\begin{bmatrix} 1 & 2 & -3 \\ 3 & 0 & 3 \\ -4 & 1 & -2 \end{bmatrix}$

15. $\begin{bmatrix} 4 & 0 & -3 \\ 3 & -7 & 2 \\ 1 & -5 & 0 \end{bmatrix}$

16. $\begin{bmatrix} 2 & -3 & 5 \\ -1 & 7 & 3 \\ 6 & -9 & 15 \end{bmatrix}$

17. $\begin{bmatrix} 3 & 2 & 4 \\ 0 & 1 & 3 \\ 0 & -2 & 2 \end{bmatrix}$

18. $\begin{bmatrix} 3 & -3 & 2 \\ 2 & 0 & 1 \\ 2 & 0 & -4 \end{bmatrix}$

19. $\begin{bmatrix} 2 & 6 & 3 \\ 0 & 4 & 4 \\ 2 & -1 & 1 \end{bmatrix}$

20. $\begin{bmatrix} 8 & 2 & -6 \\ 1 & 4 & 8 \\ -6 & 3 & 6 \end{bmatrix}$

In Exercises 21–24, evaluate the determinant of each matrix by using elementary row operations. Check each answer by entering the matrix into a graphing calculator and selecting the "det" command from the matrix menu.

21. $\begin{bmatrix} 1 & -2 & 0 & 4 \\ 3 & 0 & 8 & -2 \\ 1 & 0 & -4 & 5 \\ 2 & -3 & -4 & 1 \end{bmatrix}$.

22. $\begin{bmatrix} 2 & -4 & 5 & 9 \\ 1 & -3 & 4 & 7 \\ 2 & -3 & 0 & 1 \\ 0 & -1 & 2 & 6 \end{bmatrix}$

23. $\begin{bmatrix} 2 & 3 & 0 & 4 \\ 0 & 1 & 5 & 2 \\ 2 & -1 & 3 & 4 \\ -4 & 2 & -2 & 0 \end{bmatrix}$

24. $\begin{bmatrix} 3 & 2 & -1 & 4 \\ 2 & -2 & 4 & 0 \\ -2 & 3 & 6 & 1 \\ 1 & 4 & -3 & 0 \end{bmatrix}$

In Exercises 25–38, find the inverse of matrix A. If A does not have an inverse, state that it is singular. Check each answer by using the matrix features of a graphing calculator.

25. $A = \begin{bmatrix} 3 & 4 \\ -2 & -3 \end{bmatrix}$

26. $A = \begin{bmatrix} 6 & 7 \\ 5 & 6 \end{bmatrix}$

27. $A = \begin{bmatrix} 1 & 4 \\ 2 & 3 \end{bmatrix}$

28. $A = \begin{bmatrix} 4 & 2 \\ -1 & 2 \end{bmatrix}$

29. $A = \begin{bmatrix} 2 & 3 \\ -6 & -9 \end{bmatrix}$

30. $A = \begin{bmatrix} 5 & -2 \\ -10 & 4 \end{bmatrix}$

31. $A = \begin{bmatrix} 1 & 2 & -3 \\ 0 & 2 & -1 \end{bmatrix}$

32. $A = \begin{bmatrix} 3 & 5 \\ 0 & -2 \\ -2 & 3 \end{bmatrix}$

33. $A = \begin{bmatrix} 1 & 2 & -3 \\ 2 & 3 & -4 \\ -1 & 0 & 3 \end{bmatrix}$

34. $A = \begin{bmatrix} 1 & 0 & 2 \\ 0 & 1 & 4 \\ -1 & 0 & 2 \end{bmatrix}$

35. $A = \begin{bmatrix} 2 & 0 & 5 \\ 1 & -1 & 4 \\ -3 & 2 & -8 \end{bmatrix}$

36. $A = \begin{bmatrix} -3 & 9 & 0 \\ 4 & -20 & 1 \\ 1 & -5 & 0 \end{bmatrix}$

37. $A = \begin{bmatrix} -2 & -1 & -8 \\ 3 & 2 & 10 \\ 4 & 4 & 6 \end{bmatrix}$

38. $A = \begin{bmatrix} 3 & -8 & 12 \\ -2 & 6 & -8 \\ 5 & -8 & 12 \end{bmatrix}$

Exercises 39–44, use the inverse method to solve each system of linear equations. Use the matrix features of a graphing calculator to confirm your work.

39. $\begin{aligned} x - y + z &= -3 \\ 2x + y &= 1 \\ y - 3z &= 7 \end{aligned}$ **40.** $\begin{aligned} x + 3y &= 4 \\ y - 2z &= 3 \\ x - y + 3z &= -3 \end{aligned}$

41. $\begin{aligned} x - 3y + z &= -3 \\ 2x - 3z &= 7 \\ -x + y + 2z &= -4 \end{aligned}$

42. $\begin{aligned} 3x - y - 2z &= 0 \\ 2x + 8y &= -1 \\ x - y + 10z &= 27 \end{aligned}$

43. $\begin{aligned} 2x - y + z &= 4 \\ 3x + 2y - 3z &= 2 \\ 4x + 5y - 6z &= -3 \end{aligned}$

44. $\begin{aligned} -2x - y + 4z &= -7 \\ 3x + 2y - z &= 4 \\ 4x - 3y + 2z &= -1 \end{aligned}$

45. Two cables support an 800-lb weight, as shown in the figure. The tensions T_1 and T_2 (in pounds) in each cable may be found by solving the following system of linear equations:

$$0.4226T_1 + 0.7431T_2 = 800$$

$$0.9063T_1 - 0.6691T_2 = 0$$

Use the inverse method in conjunction with the matrix features of a graphing calculator to find T_1 and T_2. Round each answer to three significant digits.

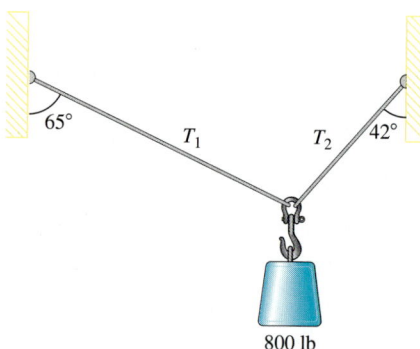
65° T_1 T_2 42°
800 lb

46. For the electrical circuit shown in the figure, the branch currents i_1, i_2, and i_3 (in amperes) may be found by solving the following system of linear equations:

$$i_1 - i_2 - i_3 = 0$$

$$68i_1 + 48i_3 = 12.5$$

$$32i_2 - 48i_3 = -3.5$$

Use the inverse method in conjunction with the matrix features of a graphing calculator to find i_1, i_2, and i_3. Round each answer to three significant digits.

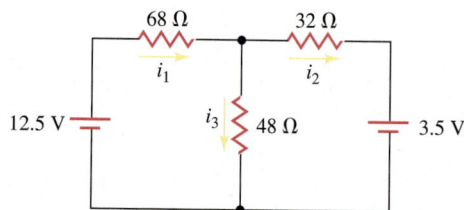
68 Ω   32 Ω
i_1 i_2
12.5 V i_3 48 Ω   3.5 V

47. When an object is propelled vertically *upward*, its distance s (in feet) above the ground after t second(s) is given by $s = -16t^2 + v_0 t + s_0$, where v_0 is the initial velocity of the object and s_0 is the initial distance above the ground. Use the inverse method in conjunction with the matrix features of a graphing calculator to find v_0 and s_0 if $s = 98.3$ ft when $t = 1.2$ s and $s = 123.2$ ft when $t = 2.1$ s. Round each answer to three significant digits.

48. The forces F_1, F_2, and F_3 (in pounds), which act on the beam shown in the sketch, may be found by solving the following system of linear equations:

$$0.7547F_1 \qquad\qquad - 0.4848F_3 = \qquad 0$$
$$0.6561F_1 - \qquad F_2 + 0.8746F_3 = \quad 200$$
$$\qquad\qquad - 18.0F_2 + \quad 21.0F_3 = 2800$$

Use the inverse method in conjunction with the matrix features of a graphing calculator to find F_1, F_2, and F_3. Round each answer to three significant digits.

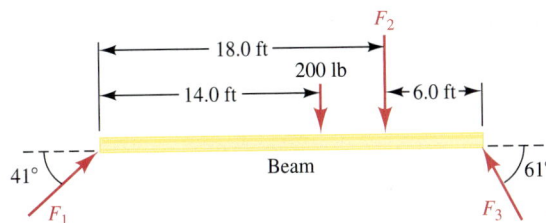

◆ **Critical Thinking**

49. Use elementary row operations to show that

$$\begin{vmatrix} a_{11} & a_{12} & a_{13} \\ a_{21} & a_{22} & a_{23} \\ a_{31} & a_{32} & a_{33} \end{vmatrix} =$$

$$a_{11}\begin{vmatrix} a_{22} & a_{23} \\ a_{32} & a_{33} \end{vmatrix} - a_{12}\begin{vmatrix} a_{21} & a_{23} \\ a_{31} & a_{33} \end{vmatrix} + a_{13}\begin{vmatrix} a_{21} & a_{22} \\ a_{31} & a_{32} \end{vmatrix}$$

50. Find the value(s) of x for which the given determinant is equal to zero.

(a) $\begin{vmatrix} x & -2 \\ -3 & 18 \end{vmatrix}$ (b) $\begin{vmatrix} x & 0 & 3 \\ 4 & -2 & x \\ 6 & 1 & -1 \end{vmatrix}$

51. The triangle shaded in the figure has vertices $(0, 0)$, (x_1, y_1), and (x_2, y_2).

(a) Find the area of this triangle by subtracting the area of the three right triangles from the area of the rectangle.

(b) Evaluate

$$\tfrac{1}{2}\begin{vmatrix} x_1 & y_1 \\ x_2 & y_2 \end{vmatrix},$$

and compare your answer with part (a). What does this determinant represent?

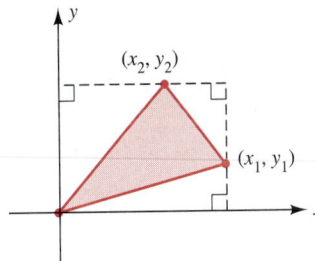

52. Use the determinant in Exercise 51(b) to find the area of the triangle with the given vertices.

(a) $A(0, 0)$, $B(6, 2)$, $C(4, 4)$

(b) $A(0, 0)$, $B(3\sqrt{2}, 3)$, $C(-\sqrt{2}, 3)$

53. Suppose A, B, and C are $n \times n$ matrices and $AB = AC$. Is it possible to "cancel" matrix A from both sides of this equation and conclude that $B = C$? Explain.

54. Describe the inverse of an $n \times n$ matrix having non-zero elements $a_{11}, a_{22}, a_{33}, \ldots, a_{nn}$ along its main diagonal and zeros elsewhere.

55. Given the 2×2 matrix $A = \begin{bmatrix} a & b \\ c & d \end{bmatrix}$, where a, b, c, and d are real numbers, find a formula for A^{-1} and list any restriction(s) on the formula.

56. Use the formula from Exercise 55 to find the inverse of each of the following matrices.

(a) $\begin{bmatrix} 2 & 1 \\ 3 & 7 \end{bmatrix}$ (b) $\begin{bmatrix} 4 & -5 \\ -9 & 4 \end{bmatrix}$

(c) $\begin{bmatrix} 3 & 2 \\ -4 & -3 \end{bmatrix}$ (d) $\begin{bmatrix} 15 & 12 \\ -9 & -8 \end{bmatrix}$

7.4 Systems of Linear Inequalities and Linear Programming

◆ Introductory Comments

Each of the following statements is a **linear inequality** in two variables x and y:

$$Ax + By > C \qquad Ax + By < C$$
$$Ax + By \geq C \qquad Ax + By \leq C$$

If a linear inequality in two variables is true when x is replaced with x_1 and y with y_1, then the ordered pair (x_1, y_1) is a **solution** of the linear inequality. For example, the ordered pair $(2, -1)$ is a solution of the linear inequality $3x - 2y > 7$ because

$$3(2) - 2(-1) > 7$$
$$8 > 7 \quad \text{is true.}$$

However, the ordered pair $(0, 1)$ is *not* a solution of this linear inequality because

$$3(0) - 2(1) > 7$$
$$-2 > 7 \quad \text{is false.}$$

Linear inequalities occur when solving certain types of decision-making problems, called *linear programming problems*. Since these types of problems involve working with the graphs of linear inequalities, we begin with a discussion of graphing linear inequalities.

◆ Graphing Linear Inequalities

One way to illustrate *all* the ordered pairs that are solutions of a linear inequality is to draw its graph. We begin by graphing the equation $Ax + By = C$ using a dashed line if the inequality symbol is $<$ or $>$, and a solid line if the inequality symbol is $\leq$ or $\geq$. This dashed or solid line divides the xy-coordinate plane into two half-planes. We then select any ordered pair form either half-plane as a *test point* in order to determine whether it is a solution of the inequality. If the ordered pair we select is a solution, then every ordered pair in that half-plane is also a solution. Shading this half-plane gives us the *graph of the linear inequality*. The procedure is illustrated in the next example.

EXAMPLE 1   Graph the solution set of each linear inequality.

(a) $3x + y > 0$ **(b)** $3x - 2y \geq 6$

◀◆ **Solution**

(a) Since the inequality symbol is >, the graph of the corresponding equation $3x + y = 0$ is sketched as a dashed line, as shown in Figure 7.7. Selecting $(1, 0)$ as an arbitrary test point for this inequality, we find that

$$3(1) + (0) > 0$$

$$3 > 0 \quad \text{is true}$$

Shading the half-plane that contains the point $(1, 0)$ gives us the graph of $3x + y > 0$ (see Figure 7.7). Every ordered pair in the shaded region, but not on the dashed line, is a solution of the inequality $3x + y > 0$.

(b) Since the inequality symbol is ≥, the graph of the corresponding equation $3x - 2y = 6$ is shown as a solid line in Figure 7.8. Selecting $(0, 0)$ as an arbitrary test point for this inequality, we find that

$$3(0) - 2(0) \geq 6$$

$$0 \geq 6 \quad \text{is false}$$

FIGURE 7.7

Graph of the solution set for the inequality $3x + y > 0$

Shading the half-plane that does *not* contain the point $(0, 0)$ gives us the graph of $3x - 2y \geq 6$ (see Figure 7.8). Every ordered pair in the shaded region or on the solid line is a solution of the inequality $3x - 2y \geq 6$. ◆

Each of the following statements is a linear inequality in two variables x and y written in slope-intercept form:

$$y < ax + b \qquad y > ax + b$$

$$y \leq ax + b \qquad y \geq ax + b$$

FIGURE 7.8

Graph of the solution set for the inequality $3x - 2y \geq 6$

The graph of the solution set of a linear inequality in slope-intercept form is the half-plane *above* the line $y = ax + b$ if the inequality symbol is > or ≥, or the half-plane *below* the line $y = ax + b$ if the inequality symbol is < or ≤. Thus, an alternate method of graphing the solution set of the inequalities in Example 1 is to write the inequality in slope-intercept form as follows:

$$3x + y > 0 \qquad\qquad 3x - 2y \geq 6$$
$$y > -3x \qquad\qquad -2y \geq -3x + 6$$
$$\qquad\qquad\qquad y \leq \tfrac{3}{2}x - 3$$

> Reverse the inequality when multipying or dividing both sides by a negative number.

The graph of the solution set for $y > -3x$ is the half-plane *above* the line $y = -3x$, as shown in Figure 7.7. The graph of the solution set for $y \leq \tfrac{3}{2}x - 3$ is the half-plane *below* the line $y = \tfrac{3}{2}x - 3$, as shown in Figure 7.8.

PROBLEM 1   Graph the solution set for each linear inequality.

(a) $y < -2$ **(b)** $y \leq -2x + 5$

◆ **Graphing a System of Linear Inequalities**

When we have several linear inequalities in two unknowns x and y to be solved simultaneously, we say that we have a **system of linear inequalities**. If an ordered pair (x_1, y_1) satisfies *every* inequality in the system, then (x_1, y_1) is a solution of the system. *All* ordered pairs that satisfy the system can be illustrated graphically by shading the region in the *xy*-coordinate plane that is *common* to the graphs of all the inequalities in the system.

EXAMPLE 2   Graph the solution set of each system of linear inequalities.

(a) $3x + y > 0$ **(b)** $y \geq x - 3$
$\quad\;\; 3x - 2y \geq 6$ $\qquad x + 2y \leq 12$
$\qquad\qquad\qquad\qquad\qquad\quad\; x \geq 0$
$\qquad\qquad\qquad\qquad\qquad\quad\; y \geq 0$

◆ **Solution**

(a) The shaded regions in Figures 7.7 and 7.8 represent the solution sets of the individual inequalities $3x + y > 0$ and $3x - 2y \geq 6$, respectively. Now, the solution set of this system of linear inequalities consists of all points common to both these shaded regions, as illustrated in Figure 7.9. We find the intersection point of the two lines, $\left(\frac{2}{3}, -2\right)$, by solving the equations $3x + y = 0$ and $3x - 2y = 6$ *simultaneously*. Every ordered pair within the shaded region of Figure 7.9 or along the portion of the solid line $3x - 2y = 6$ that touches this region, *except* the point $\left(\frac{2}{3}, -2\right)$, is a solution of this system of inequalities.

(b) We begin by graphing each of the four individual inequalities on the same coordinate plane. The solution set of this system of linear inequalities is the region that is common to all these inequalities. The shaded region in Figure 7.10 shows the solution set. Each of the points where the sides of the shaded region intersect is called a *vertex*, and we find these points (the *vertices* of the graph) as follows:

1. Solving $y = x - 3$ and $y = 0$ simultaneously yields the vertex $(3, 0)$.

2. Solving $y = x - 3$ and $x + 2y = 12$ simultaneously yields the vertex $(6, 3)$.

3. Solving $x + 2y = 12$ and $x = 0$ simultaneously yields the vertex $(0, 6)$.

4. Solving $x = 0$ and $y = 0$ simultaneously yields the vertex $(0, 0)$.

Every ordered pair in the shaded region of Figure 7.10 or along the portions of the solid lines that touch this region is a solution of this system.

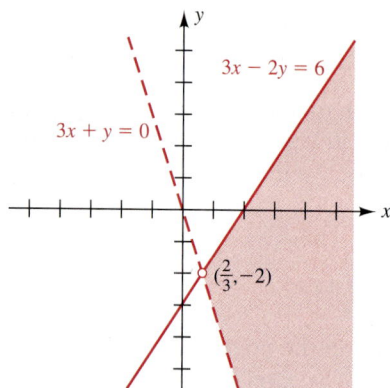

FIGURE 7.9

Graph of the solution set for the system

$3x + y > 0$
$3x - 2y \geq 6$

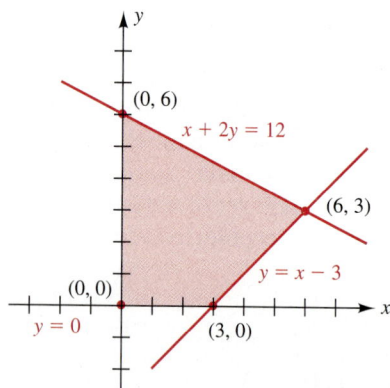

FIGURE 7.10

Graph of the solution set for the system

$y \geq x - 3$
$x + 2y \leq 12$
$x \geq 0$
$y \geq 0$

Most graphing calculators have the capability of generating the graph of a system of linear inequalities. To graph the system in Example 2(a), we first generate the graphs of the corresponding equations $3x + y = 0$ (or $y = -3x$) and $3x - 2y = 6$ (or $y = \frac{3}{2}x - 3$) in the viewing rectangle [see Figure 7.11(a)]. Next, we select "shade" from the DRAW menu to shade the area above $y = -3x$ and below $y = \frac{3}{2}x - 3$ [see Figure 7.11(b)]. Finally, using the built-in program for finding the intersection points of two graphs (or the TRACE key and ZOOM key), we find the vertex of the region [see Figure 7.11(c)].

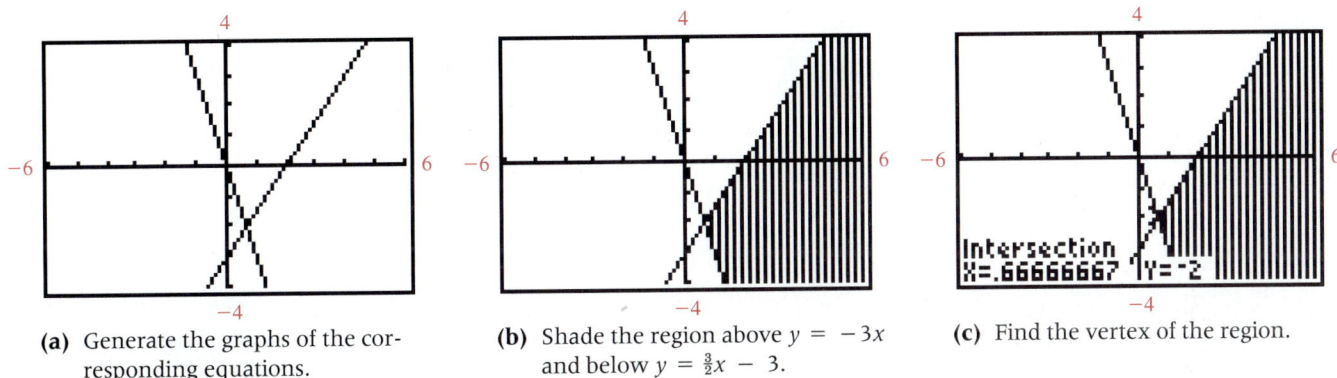

(a) Generate the graphs of the corresponding equations.

(b) Shade the region above $y = -3x$ and below $y = \frac{3}{2}x - 3$.

(c) Find the vertex of the region.

FIGURE 7.11

Typical displays for graphing the system of inequalities in Example 2(a)

PROBLEM 2 Use a graphing calculator to generate the graph of the solution set of the system of inequalities given in Example 2(b). Verify the vertices of the shaded region. ◆

◆ An Introduction to Linear Programming

In Figure 7.10, every line segment joining any two points in the shaded region is contained in that region. A region with this characteristic is said to be **convex**. In **linear programming** (a branch of mathematics), we maximize or minimize a linear function f in two variables x and y over a convex region. The function f, defined by

$$f(x, y) = Ax + By,$$

is called the **objective function** with slope $-A/B$. The convex region over which we maximize or minimize the objective function is formed from a system of linear inequalities called the **constraints**.

Note: The linear programming problems that we discuss in this section represent an introduction to this branch of mathematics. More advanced linear programming problems use more than two unknowns and require sophisticated matrices and computer programming to find maximum and minimum values.

Suppose we wish to find the maximum and minimum values of an objective function f, defined by

$$f(x, y) = 2x + 3y,$$

over the convex region in Figure 7.10. Any line of this form has slope $-\frac{2}{3}$. Figure 7.12 displays the convex region for the constraints given in Example 2(b). It also shows several lines with slope $-\frac{2}{3}$ that pass through some points of the convex region. From the information shown in the figure, we can state the following facts about values of this objective function:

Each of these lines has slope $-\frac{2}{3}$

(0, 6)

(2, 3)

(6, 3)

Maximum value

$2x + 3y = 21$

$2x + 3y = 18$

$2x + 3y = 13$

$2x + 3y = 6$

$2x + 3y = 0$

Minimum value

(0, 0)   (3, 0)

FIGURE 7.12

The maximum value of $f(x, y) = 2x + 3y$ over the given convex region is 21, and this occurs at the vertex (6, 3). The minimum value of $f(x, y) = 2x + 3y$ over the given convex region is 0, and this occurs at the vertex (0, 0).

1. The value of $f(x, y)$ is constant along each line and this functional value may be determined by using any point on the line. Note that $f(0, 0) = 0$, $f(3, 0) = 6$, $f(2, 3) = 13$, $f(0, 6) = 18$, and $f(6, 3) = 21$.

2. The value of $f(x, y)$ increases as lines with slope $-\frac{2}{3}$ move from left to right through the convex region.

3. The maximum value of $f(x, y)$ over the convex region is 21, and this occurs at a vertex of the region, namely, (6, 3).

4. The minimum value of $f(x, y)$ over the convex region is 0, and this also occurs at a vertex of the region, namely, (0, 0).

In summary, we state the following **fundamental principle of linear programming**.

◆ **Fundamental Principle of Linear Programming**

The maximum or minimum value (if it exists) of an objective function f, defined by

$$f(x, y) = Ax + By,$$

always occurs at a vertex of the convex region that is formed from a system of linear inequalities, called constraints.

EXAMPLE 3 Find the maximum and minimum values (if they exist) of the objective function f defined by $f(x, y) = 2x + 5y$, subject to the following set of constraints:

$$y \geq 7 - 2x$$

$$x + 3y \geq 7$$

$$x + y \geq 5$$

$$x \geq 0$$

$$y \geq 0$$

◆ **Solution** We begin by graphing the system of linear inequalities given as constraints (see Figure 7.13). Note that the constraints form an *unbounded* convex region—at least one side of this region is not bounded by a line of constraint.

We find the vertices of this convex region as follows:

1. Solving $y = 7 - 2x$ and $x = 0$ simultaneously yields the vertex $(0, 7)$.
2. Solving $y = 7 - 2x$ and $x + y = 5$ simultaneously yields the vertex $(2, 3)$.
3. Solving $x + y = 5$ and $x + 3y = 7$ simultaneously yields the vertex $(4, 1)$.
4. Solving $x + 3y = 7$ and $y = 0$ simultaneously yields the vertex $(7, 0)$.

According to the fundamental principle of linear programming, the maximum and minimum values of f (if they exist) occur at vertices of this convex region. Thus, we evaluate $f(x, y)$ at each vertex, as shown in the following table.

Vertex	Value of $f(x, y) = 2x + 5y$
$(0, 7)$	$f(0, 7) = 2(0) + 5(7) = 35$
$(2, 3)$	$f(2, 3) = 2(2) + 5(3) = 19$
$(4, 1)$	$f(4, 1) = 2(4) + 5(1) = 13$ ← minimum value
$(7, 0)$	$f(7, 0) = 2(7) + 5(0) = 14$

Using this table and Figure 7.13, we observe that the minimum value of f is $f(4, 1) = 13$.

We cannot conclude from the table that the maximum value of f is $f(0, 7) = 35$. Since values of $f(x, y)$ increase without bound as lines with slope $-\frac{2}{5}$ move from left to right through this unbounded convex region, we must conclude that a maximum value of f does not exist. ◆

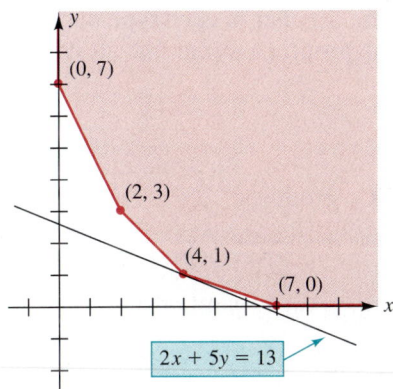

FIGURE 7.13

The minimum value of $f(x, y) = 2x + 5y$ over this unbounded convex region is 13, and it occurs when $x = 4$ and $y = 1$. The function f has no maximum value.

PROBLEM 3 Find the maximum and minimum values (if they exist) of the objective function f, defined by $f(x, y) = 2x - 5y$, subject to the set of constraints given in Example 3. ◆

◆ **Application: Maximizing a Profit**

Linear programming has many applications in business and economics. We conclude this section with an example showing how we can maximize a profit.

EXAMPLE 4   A ski company manufactures a slalom ski and a racing ski. The profit on each pair of slalom skis is $30 and on each pair of racing skis is $50. The company can produce at most 60 pairs of slalom skis per day and at most 40 pairs of racing skis per day. Production of a pair of slalom skis requires 2 hours of labor, and production of a pair of racing skis requires 3 hours of labor. If the maximum number of hours available for the production of skis is 150 hours per day, determine the number of each type of ski that should be manufactured to maximize the profit. (Assume that all skis can be sold.)

◆ **Solution**   We begin by letting

$$x = \text{number of slalom skis produced}$$

and $$y = \text{number of racing skis produced.}$$

If P represents the profit, then the objective function P, defined by

$$P(x, y) = 30x + 50y,$$

is the function we wish to maximize, subject to the following constraints:

$$x \geq 0$$
$$y \geq 0$$
$$x \leq 60$$
$$y \leq 40$$
$$2x + 3y \leq 150$$

The graph of the constraints is shown in Figure 7.14. According to the fundamental principle of linear programming, the maximum value of P must occur at a vertex of this convex region. Thus, we evaluate $P(x, y)$ at each vertex, as shown in the following table.

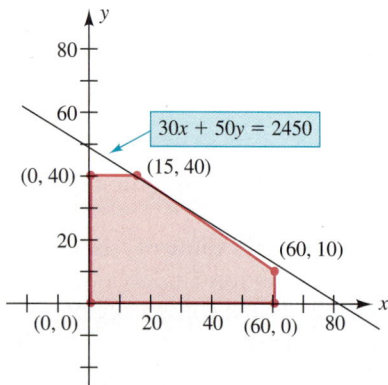

FIGURE 7.14

The maximum value of
$P(x, y) = 30x + 50y$ over
this convex region is $P(15, 40) = 2450$.

Vertex	Value of $P(x, y) = 30x + 50y$	
(0, 0)	$P(0, 0) = 30(0) + 50(0) = 0$	
(0, 40)	$P(0, 40) = 30(0) + 50(40) = 2000$	
(15, 40)	$P(15, 40) = 30(15) + 50(40) = 2450$	← maximum value
(60, 10)	$P(60, 10) = 30(60) + 50(10) = 2300$	
(60, 0)	$P(60, 0) = 30(60) + 50(0) = 1800$	

From the table, we observe that the maximum value of P is $P(15, 40) = \$2450$ (see Figure 7.14). Thus, to maximize profit the company should manufacture 15 pairs of slalom skis and 40 pairs of racing skis per day. ◆

PROBLEM 4 Suppose the ski manufacturer in Example 4 makes some changes in production methods. The company's profit is still \$30 on each pair of slalom skis but is now \$45 on each pair of racing skis. Show that we can find more than one way to maximize the profit subject to these same constraints. ◆

Exercises 7.4

◆ Basic Skills

In Exercises 1–12, graph the solution set of each linear inequality.

1. $3x + 2y \geq 12$

2. $2x - 5y < 10$

3. $2x - 7y < 10$

4. $6x + 4y > 18$

5. $x > 3$

6. $y \leq -2$

7. $2y - 5 > 0$

8. $3x + 4 > 0$

9. $y \leq 2x - 3$

10. $y \geq 4 - 3x$

11. $y < \dfrac{2 - 3x}{4}$

12. $y \geq \dfrac{2x + 8}{3}$

In Exercises 13–30, graph the solution set of each system of linear inequalities. Label the vertices of the shaded region.

13. $x \geq 2$
$y \leq 1$

14. $x + y < 4$
$x > 3$

15. $x + 2y < 7$
$x - y < 1$

16. $x - y \leq 2$
$y < x + 3$

17. $x + 2y < 6$
$x \geq 0$
$y \geq 0$

18. $y < x$
$x < 4$
$y > -2$

19. $5x - 2y > -6$
$x - 2y \leq 2$
$x + 2y \leq 6$

20. $3x + 2y \geq -6$
$11x - 2y < 34$
$x - 4y \geq -16$

21. $y > x$
$x + 2y \geq 6$
$2x + y \leq 15$

22. $3y \geq x$
$x - 3y \geq -12$
$x \leq 6$

23. $x + 3y \leq 15$
$2y \geq x$
$y > 1$
$x \geq 0$

24. $2y \leq x + 9$
$y \leq -3x + 15$
$x > -3$
$y \geq 0$

25. $3x + 2y \geq 6$
$5x + 4y \leq 20$
$x \geq 0$
$y \geq 0$

26. $x + y \geq 10$
$x + y \leq 15$
$y \leq 20$
$x \leq 25$

27. $y \geq 2x - 8$
$x + y \leq 7$
$x \leq 4$
$x \geq 0$
$y \geq 0$

28. $x + 2y \geq 3$
$x - 2y \geq -3$
$3x + 2y \leq 23$
$y \geq x - 6$
$x \geq 1$

29. $2x + y \geq 4$
$x + 2y \geq 5$
$x + 4y \geq 7$
$x \geq 0$
$y \geq 0$

30. $x + 4y \geq 14$
$x + 2y \geq 10$
$x + y \geq 7$
$x \geq 2$
$y \geq 0$

In Exercises 31–44, find the maximum and minimum values (if they exist) of the objective function subject to the given constraints.

Objective function	Constraints
31. $f(x, y) = 3x + y$	$x + y \leq 6$ $x \leq 3$ $x \geq 0$ $y \geq 0$
32. $f(x, y) = 2x - 3y$	$x + y \leq 6$ $x \leq 3$ $x \geq 0$ $y \geq 0$
33. $f(x, y) = 4x - 5y$	$y \leq x + 1$ $y \geq x - 2$ $y \leq 7$ $y \geq 2$

Objective function   *Constraints*

34. $f(x, y) = 3x + 2y$
$$y \le x + 1$$
$$y \ge x - 2$$
$$y \le 7$$
$$y \ge 2$$

35. $f(x, y) = 2x - y$
$$2x + 3y \le 16$$
$$x - 4y \le -3$$
$$y \le 3x - 2$$

36. $f(x, y) = x + 5y$
$$2x + 3y \le 16$$
$$x - 4y \le -3$$
$$y \le 3x - 2$$

37. $f(x, y) = x + y$
$$x + 2y \le 10$$
$$y \le 6 - x$$
$$x \le 4$$
$$x \ge 0$$
$$y \ge 0$$

38. $f(x, y) = 5x + 2y$
$$x + 2y \le 10$$
$$y \le 6 - x$$
$$x \le 4$$
$$x \ge 0$$
$$y \ge 0$$

39. $f(x, y) = 5x + 2y$
$$y \le 27 - 3x$$
$$3x - 2y \le 18$$
$$3x + 4y \ge 18$$
$$3x - 5y \ge -9$$

40. $f(x, y) = 3x + y$
$$y \le 27 - 3x$$
$$3x - 2y \le 18$$
$$3x + 4y \ge 18$$
$$3x - 5y \ge -9$$

41. $f(x, y) = 3x + 4y$ See Exercise 29
42. $f(x, y) = 2x - y$ See Exercise 29
43. $f(x, y) = 7x - 2y$ See Exercise 30
44. $f(x, y) = 2x + 7y$ See Exercise 30

In Exercises 45–52, use linear programming to solve each problem.

45. A doctor is going to purchase bookcases for his office. One type of bookcase costs $100, holds 30 cubic feet of books, and takes up 6 square feet of floor space. Another type costs $200, holds 48 cubic feet of books, and takes up 9 square feet of floor space. The doctor wants to spend no more than $1200 for the bookcases and use no more than 66 square feet of office floor space. How many of each type bookcase should he purchase to *maximize* storage capacity?

46. A sand and gravel company wishes to purchase six-wheel and ten-wheel dump trucks for delivery vehicles. Each six-wheel dump truck sells for $20,000, has an average monthly fuel cost of $100, and has a carrying capacity of 12 cubic yards. Each ten-wheel dump truck sells

for $30,000, has an average monthly fuel cost of $300, and has a carrying capacity of 17 cubic yards. Suppose the company wishes to spend no more than $240,000 for the trucks and no more than $1500 per month for fuel. How many trucks of each type should the company purchase to *maximize* their delivery capacity?

47. A company manufactures oak tables and chairs. Each table requires 2 hours 15 minutes to assemble, 1 hour to finish, and 30 minutes to pack for shipping. Each chair requires 2 hours to assemble, 2 hours to finish, and 10 minutes to pack for shipping. The maximum number of labor hours available each day for assembling is 90 hours, for finishing is 80 hours, and for packing is 15 hours. If the manufacturer makes a profit of $80 on each table and $30 on each chair, determine the number of each that should be produced each day to *maximize* profit. (Assume that all tables and chairs can be sold.)

48. A kennel raises Doberman pinschers and German shepherds. The kennel can raise no more than 40 dogs and wishes to have no more than 24 Doberman pinschers. The cost of raising a Doberman pinscher is $50, the cost of raising a German shephard is $30, and the kennel can invest no more than $1500 for this purpose. If the kennel makes a profit of $150 per pinscher and $100 per shephard, how many dogs of each type should be raised in order to *maximize* profit? (Assume that all dogs can be sold.)

49. A grain company stocks two types of wild-bird seed: sunflower and cracked corn. The company buys the sunflower seed for $10 per bag and the cracked corn for $5 per bag. They sell the sunflower seed for $16 per bag and the cracked corn for $10 per bag. From past experience, the company knows that they will sell at least twice as many bags of sunflower seed as bags of cracked corn. If the company does not wish to order more than $600 worth of seed, how many bags of each type should be ordered to *maximize* profit? (Assume that all bags of seed can be sold.)

50. A woman inherits $48,000 from her grandfather's estate, subject to the following conditions: Part or all of the money must be invested either in treasury bills that pay 8¢ on each dollar or in bonds that pay 10¢ on each dollar. At least $10,000 must be invested in treasury bills and at most $30,000 in bonds. The amount invested in treasury bills must be no more than twice the amount invested in bonds. How much should she invest in each to *maximize* her investment?

51. A company makes two brands of cereal P and Q, both of which are enriched with vitamins A and B. Cereal P cost 10¢ per ounce and cereal Q cost 15¢ per ounce. The number of milligrams of each vitamin that is contained in each ounce of cereal is given in the following table.

	Cereal *P*	Cereal *Q*
Vitamin A	1.2	2.3
Vitamin B	3.3	1.4

Suppose a woman needs at least 8.0 mg of vitamin A and at least 9.0 mg of vitamin B to satisfy her daily requirement. How many ounces of each type of cereal should she consume if she wishes to *minimize* her food cost? Round each answer to the nearest tenth of an ounce.

52. A farmer has 200 acres of land available and wishes to plant at least 40 acres of corn and beans. The cost of the corn seed is $115 per acre and the cost of the bean seed is $185 per acre. Past experience has shown that the cost of maintaining and harvesting the corn crop is $260 per acre, and the cost of maintaining and harvesting the bean crop is $145 per acre. The farmer wants to spend no more than $12,500 for seed and no more than $16,500 on maintenance and harvest. If the expected profits from the corn crop and bean crop are $550 per acre and $520 per acre, respectively, how many acres of each crop should be planted to *maximize* profit? Round each answer to the nearest tenth of an acre.

◆ **Critical Thinking**

53. Given the convex region shown in the figure, find the slope *m* of an objective function $f(x, y) = Ax + By$ that has a maximum value at the given point(s).

   (a) at both points *P* and *Q*
   (b) at both points *Q* and *R*
   (c) at point *P* only   (d) at point *Q* only
   (e) at point *R* only

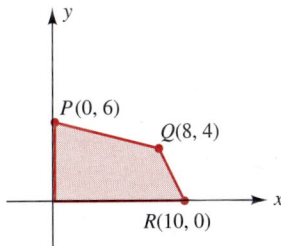

$P(0, 6)$
$Q(8, 4)$
$R(10, 0)$

54. Given the convex region shown in the figure, find the slope *m* of an objective function $f(x, y) = Ax + By$ that has a minimum value at the given point(s).

   (a) at both points *P* and *Q*
   (b) at both points *Q* and *R*
   (c) at point *P* only   (d) at point *Q* only
   (e) at point *R* only

$P(0, 8)$
$Q(2, 3)$
$R(11, 0)$

55. A landscape gardening company has two brands of lawn fertilizer. The nutrient contents in kilograms per bag is shown in the following table:

	Brand A	Brand B
Nitrogen (N)	30	20
Phosphoric acid (P_2O_5)	2	4
Potash (K_2O)	1	4

It has been determined that a certain lawn needs at least 120 kg of nitrogen, at least 16 kg of phosphoric acid, and at least 12 kg of potash. If brand A costs $22 per bag and brand B costs $18 per bag, how many bags of each brand should be used to *minimize* the cost? What is the minimum cost?

56. A computer company has two distribution centers, one in Boston and one in Providence. The company receives an order to ship 40 computers to college A, and 60 computers to college B. The cost of shipping each computer is shown in the following table:

	College A	College B
From **Boston**	$9	$8
From **Providence**	$8	$10

Suppose that the distribution center in Boston has 80 computers available for shipping and the distribution center in Providence has 30 computers available. How should the order be filled in order to *minimize* the total shipping cost? What is the minimum cost?

Chapter 7 Review

◆ Questions for Writing or Group Discussion

1. State the three possible types of solutions for a system of n linear equations in n variables.
2. Explain how the *elimination method* may be used to solve a 3×3 linear system.
3. Can a *nonsquare system* of linear equations have a unique solution? Explain.
4. What is meant by the *order* of a matrix? Give an example of a matrix with order 4×3.
5. What is an *augmented matrix*? How is it formed?
6. List three *elementary row operations* that can be performed on a matrix to generate a row-equivalent matrix. Illustrate each operation with an example.
7. Give an example of two matrices that are equivalent but not equal.
8. Is it possible to find the *sum of two matrices* having different orders? Explain.
9. Explain the procedure for finding the *product of two matrices*. Under what conditions does the product exist?
10. If matrix A has order 3×2 and matrix B has order 2×5, what is the order of AB?
11. Is matrix multiplication commutative? associative? Illustrate with examples.
12. Suppose A and B are $n \times n$ matrices and $AB = BA = I_n$, where I_n is the *identity matrix* of order $n \times n$. What is the relationship between A and B?
13. Explain the procedure for finding the *inverse* of a square matrix. Does every square matrix have an inverse?
14. What is the value of $|A|$ if matrix A is *singular*? Explain.
15. When solving the matrix equation $AX = B$ for X, what restrictions must be placed on A and B for the solution to exist?
16. What is the procedure for finding the *determinant* of a 2×2 matrix? of a 3×3 matrix?
17. What are the effects of elementary row operations on the determinant of an $n \times n$ matrix?
18. List some conditions for which the determinant of an $n \times n$ matrix is zero.
19. Explain the procedure for graphing a *system of linear inequalities*. Illustrate with an example.
20. What is *linear programming*? Where does the maximum or minimum value of the objective function always occur?

◆ Review Exercises

In Exercises 1–18 solve each system of linear equations by using any convenient method.

1. $\begin{aligned} x - y &= 5 \\ 2x + 3y - z &= -4 \\ y + 3z &= -6 \end{aligned}$

2. $\begin{aligned} 2x + y &= 1 \\ x + z &= -4 \\ 3y - 2z &= 11 \end{aligned}$

3. $3x - 2y + z = -3$
$\quad 2x + y - 3z = 2$
$\quad\quad\quad 4y - 2z = 4$

4. $5x - 2y + 3z = 11$
$\quad 2x - y + 4z = 5$
$\quad 4x - 3y + 2z = 9$

5. $2x + 6y - 5z = 1$
$\quad 3x + 4y - 5z = -1$
$\quad\; x + 8y - 5z = 3$

6. $-4x - 3y + z = 2$
$\quad 3x - y + 4z = -3$
$\quad 4x + 3y - z = 0$

7. $\frac{1}{2}x - 3y + 5z = 18$
$\quad 2x - \frac{3}{8}y - z = 15$
$\quad 3x + 5y + 6z = 4$

8. $0.2x - 1.6y - 3.2z = -12$
$\quad 1.4x + 2.3y - 0.7z = 104$
$\quad 9.8x - 2.4y + 1.1z = 113$

9. $13x - 12y - 5z = 6$
$\quad 4x - 11y + 15z = 28$
$\quad 6x - 10y = 9$

10. $3x + 10y + 4z = -8$
$\quad 8x + 5y - 9z = 16$
$\quad 5x - 7y - 11z = 16$

11. $\quad x - 2y + 3z = -9$
$\quad 2x + y - z = 13$
$\quad 3x + 4y - z = 23$

12. $2x + 3y + z = -3$
$\quad 5x - 2y + 3z = 13$
$\quad 3x - y - 4z = -10$

13. $\quad x - 3y + z = 6$
$\quad 4x - 2y - z = -1$

14. $2x + y - 3z = 4$
$\quad 3x + 2y - z = 7$

15. $x + y + t = -1$
$\quad\quad 2y - t = 2$
$\quad x - z + 3t = 11$
$\quad\quad 4y + 3z = 4$

16. $2x - y + 3z - t = 1$
$\quad x + 3y + t = -5$
$\quad\quad 2y + 4z + 5t = 1$
$\quad 4x - z + 2t = 2$

17. $3x - y + z = 0$
$\quad\quad 4y - 4z + 3t = 12$
$\quad 2x - 2y - t = 0$
$\quad 5x - 3y - 2z + 5t = 12$

18. $x_1 + 2x_2 - x_5 = -7$
$\quad\quad x_2 - x_3 + x_4 = -6$
$\quad x_1 + 2x_3 + x_5 = 11$
$\quad\quad 3x_2 - 2x_4 = -4$
$\quad\quad 4x_3 - x_4 - 2x_5 = 5$

In Exercises 19–34, perform the indicated matrix operations given that A, B, C, D, E, F, G, and H are defined as follows.

$$A = \begin{bmatrix} 4 & 5 \\ 3 & 4 \end{bmatrix} \qquad B = \begin{bmatrix} -2 & 3 \\ 1 & 0 \end{bmatrix}$$

$$C = \begin{bmatrix} 1 & 0 & -3 \\ -2 & 4 & 5 \end{bmatrix}$$

$$D = \begin{bmatrix} 4 & 2 & -1 \\ 0 & -3 & -6 \end{bmatrix}$$

$$E = \begin{bmatrix} 2 & -3 \\ -1 & 5 \\ 3 & 2 \end{bmatrix} \qquad F = \begin{bmatrix} 3 & 5 \\ -2 & 7 \\ 1 & 0 \end{bmatrix}$$

$$G = \begin{bmatrix} 1 & 4 & -2 \end{bmatrix} \qquad H = \begin{bmatrix} -6 \\ 3 \\ -9 \end{bmatrix}$$

19. $-3G$

20. $\frac{1}{3}H$

21. $A + B$

22. $F - E$

23. $\frac{1}{2}A - \frac{3}{2}A$

24. $\frac{1}{3}(-3D)$

25. $2C - 3D$

26. $3(A + 2B)$

27. AB

28. CE

29. GH

30. HG

31. $(CF)A$

32. $-2(DH)$

33. $G(2E + F)$

34. $\frac{1}{2}(A^2 + B)$

In Exercises 35–40, find the inverse of each matrix, if it exists.

35. $\begin{bmatrix} 4 & 3 \\ 2 & 2 \end{bmatrix}$

36. $\begin{bmatrix} -6 & 3 \\ 5 & -2 \end{bmatrix}$

37. $\begin{bmatrix} 1 & 4 & 0 \\ 2 & 6 & -1 \\ 3 & 2 & 5 \end{bmatrix}$

38. $\begin{bmatrix} 2 & -1 & 3 \\ 3 & -5 & 8 \\ 4 & 6 & -2 \end{bmatrix}$

39. $\begin{bmatrix} 2 & 8 & -3 & 7 \\ 4 & 1 & -6 & 0 \\ 0 & 3 & 0 & -3 \\ -6 & 9 & 9 & -7 \end{bmatrix}$

40. $\begin{bmatrix} 4 & -9 & -12 & -3 \\ 0 & 4 & 9 & -4 \\ -3 & 7 & 10 & 2 \\ 2 & -2 & 0 & 7 \end{bmatrix}$

In Exercises 41–48, evaluate the determinant of each matrix.

41. $\begin{bmatrix} 2 & -3 \\ 5 & -2 \end{bmatrix}$ **42.** $\begin{bmatrix} 12 & 15 \\ -7 & 9 \end{bmatrix}$

43. $\begin{bmatrix} 1 & -2 & 0 \\ 2 & -3 & 1 \\ 6 & 0 & -2 \end{bmatrix}$

44. $\begin{bmatrix} 3 & -2 & 4 \\ 8 & 2 & -3 \\ -1 & 7 & 3 \end{bmatrix}$

45. $\begin{bmatrix} 3 & -4 & 7 \\ 0 & -2 & 3 \\ 0 & 0 & 5 \end{bmatrix}$

46. $\begin{bmatrix} 1 & 9 & 6 & -3 \\ 0 & -4 & 5 & 1 \\ 0 & 0 & -4 & 2 \\ 0 & 0 & 0 & -2 \end{bmatrix}$

47. $\begin{bmatrix} 2 & -3 & 0 & 1 \\ 1 & -4 & 2 & -5 \\ 0 & -1 & 0 & 2 \\ 2 & 6 & -4 & 0 \end{bmatrix}$

48. $\begin{bmatrix} 1 & 3 & -2 & 4 \\ 0 & 6 & 2 & -4 \\ 2 & 3 & -5 & 10 \\ 3 & 1 & 0 & 6 \end{bmatrix}$

In Exercises 49–52, graph the solution set of each system of linear inequalities. Label the vertices of the shaded region.

49.
$$y < x + 2$$
$$3x + 2y < 6$$
$$x \le 6$$

50.
$$3x + 4y \ge 12$$
$$6x + 5y \le 30$$
$$x \ge 0$$
$$y \ge 0$$

51.
$$x \le 18 - 3y$$
$$3x + 2y \le 19$$
$$2x + y \le 12$$
$$x \ge 0$$
$$y \ge 0$$

52.
$$4x + y \ge 8$$
$$y \ge 5 - x$$
$$x + 5y \ge 13$$
$$x \ge 0$$
$$y \ge 0$$

In Exercises 53 and 54, find the maximum and minimum values (if they exist) of the objective function subject to the given constraints.

53. Objective function: $f(x, y) = 2x + 3y$
Constraints: See Exercise 51

54. Objective function: $f(x, y) = 2x + y$
Constraints: See Exercise 52

55. Find k such that the determinant is equal to the given value.

(a) $\begin{vmatrix} k & 3 \\ 5 & 7 \end{vmatrix} = 6$ (b) $\begin{vmatrix} k & 2 \\ 5 & k \end{vmatrix} = 15$

(c) $\begin{vmatrix} 0 & k & 0 \\ -1 & 2 & 4 \\ 2 & 5 & k \end{vmatrix} = -12$

(d) $\begin{vmatrix} 2 & k & 3 \\ 0 & 3 & k \\ 5 & 1 & 1 \end{vmatrix} = 0$

56. The rows and columns of $A = \begin{bmatrix} a_{11} & a_{12} \\ a_{21} & a_{22} \end{bmatrix}$ are interchanged to form the matrix A^T (called the *transpose* of A). What is the relationship between $|A|$ and $|A^T|$?

57. Recall from Section 3.2 that the equation of a nonvertical line passing through two distinct points (x_1, y_1) and (x_2, y_2) is given by

$$y - y_1 = m(x - x_1)$$

where m is the slope of the line and

$$m = \frac{y_2 - y_1}{x_2 - x_1}.$$

(a) Replace m in the equation $y - y_1 = m(x - x_1)$ with $\dfrac{y_2 - y_1}{x_2 - x_1}$ and solve for y. Express the answer as a single fraction with parentheses removed from the numerator.

(b) Solve the determinant equation

$$\begin{vmatrix} 1 & x & y \\ 1 & x_1 & y_1 \\ 1 & x_2 & y_2 \end{vmatrix} = 0$$

for y and express the answer as a single fraction. Compare your answer to part (a). What does this determinant equation represent?

58. Use the determinant equation in Exercise 57(b) to find the equation of the line passing through the given points A and B. Write your answer in slope-intercept form, $y = mx + b$.

(a) $A(-2, 5)$, $B(2, -3)$
(b) $A(\sqrt{2}, -\sqrt{3})$, $B(-\sqrt{3}, \sqrt{2})$

In Exercises 59–64, use a system of linear equations to solve each problem.

59. A refinery has a "regular" gasoline that sells for 83¢ per gallon and a "super" gasoline that sells for 99¢ per gallon. How many gallons of each type gasoline should be mixed together to form 12,000 gallons of a gasoline that sells for 86¢ per gallon?

60. A man has invested $10,000 in two bank accounts. One account has an annual yield of 8% and the other, an annual yield of 9%. The interest he earns from these accounts is taxable at the rate of 30%. After taxes, he earns $602 from these accounts for the year. How much is invested in each account?

61. Three types of tickets are available for baseball games: box seats for $16, grandstand seats for $10, and bleacher seats for $5. The seating capacity at the ball park is 32,500, and the park has 2500 more grandstand seats than bleacher seats. When a game is sold out, the revenue from ticket sales is $343,500. How many seats of each type are in the ball park?

62. At the baseball game, Kristen buys 2 hot dogs, 1 bag of peanuts, and 2 sodas for $8.50. Caryn buys 1 hot dog, 2 bags of peanuts, 1 ice cream bar, and 1 soda for $7.00. David buys 4 hot dogs and 4 sodas for $16.00. Regina buys 1 hot dog, 2 ice cream bars, and 1 bag of peanuts for $7.00. What is the price of each item: a hot dog? a bag of peanuts? a soda? an ice cream bar?

63. A hiking trail, connecting campsite A to campsite B, has a pond $2\frac{1}{2}$ miles from campsite A. Starting from campsite A, the trail goes downhill for 1 mile, is level for 6 miles, and goes uphill for 2 miles. Suppose a woman can hike from campsite A to campsite B in 3 hours 15 minutes; from campsite B to campsite A in 3 hours; and from campsite A to the pond and back again to campsite A in 1 hour 45 minutes. Find her rates of walking (in miles per hour) downhill, on level ground, and uphill.

64. Find the equation of the parabola $x = ay^2 + by + c$ that passes through the points $(k_1, 1)$, $(k_2, -2)$, and $(k_3, 0)$, given the following information.
   (a) $k_1 = 4$, $k_2 = 1$, and $k_3 = -1$
   (b) $k_1 = -2$, $k_2 = 1$, and $k_3 = -3$
   (c) $k_1 = 3$, $k_2 = 8$, and $k_3 = 4$
   (d) $k_1 = 2$, $k_2 = 5$, and $k_3 = 0$

An appliance store has a two-day sale for three styles of microwave ovens. The sale prices (in dollars) are given in matrix C:

	Model	
400-watt	500-watt	600-watt

$C = [120 \quad 160 \quad 220]$ Price

Within the Boston area, the store has three outlets. The number of 400-watt, 500-watt, and 600-watt models of microwave ovens sold at each outlet during the first and second days of the sale are given in matrices A and B, respectively.

	Outlet 1	Outlet 2	Outlet 3	
$A =$	8	10	16	400-watt
	12	8	9	500-watt
	6	5	6	600-watt

	Outlet 1	Outlet 2	Outlet 3	
$B =$	5	12	8	400-watt
	7	3	6	500-watt
	8	3	6	600-watt

In Exercises 65–68, find the indicated matrix, and state what it represents.

65. $A \begin{bmatrix} 1 \\ 1 \\ 1 \end{bmatrix}$

66. $[1 \quad 1 \quad 1](A + B)$

67. CB

68. $C(A + B)\begin{bmatrix} 1 \\ 1 \\ 1 \end{bmatrix}$

69. Two cables support a 750-lb weight, as shown in the figure. The tensions T_1 and T_2 (in pounds) on each cable may be found by solving the following system of linear equations:

$$0.5150T_1 + 0.7880T_2 = 750$$
$$0.8572T_1 - 0.6157T_2 = 0$$

Use the inverse method in conjunction with the matrix features of a graphing calculator to find T_1 and T_2. Round each answer to three significant digits.

750 lb

70. The *Wheatstone bridge circuit*, shown in the sketch, is used in electrical measurement applications. The currents i_1, i_2, and i_m (in amperes) may be found by solving the following system of linear equations:

$$(10.2 + 31.5)i_1 \qquad\qquad - 31.5i_m = 125$$
$$(20.6 + 41.2)i_2 + \qquad 41.2i_m = 125$$
$$10.2i_1 + \qquad 41.2i_2 + (50.0 + 41.2)i_m = 125$$

Exercise 70 continues

Use the inverse method in conjunction with the matrix features of a graphing calculator to find i_1, i_2, and i_m. Round each answer to three significant digits.

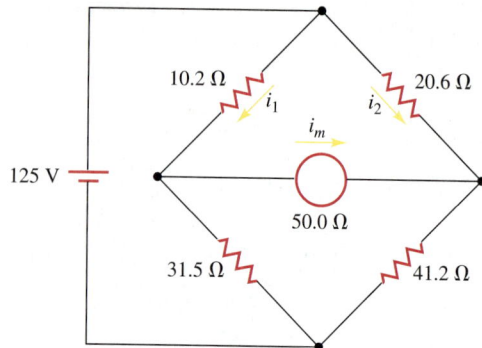

In Exercises 71 and 72, use linear programming to solve each problem.

71. A college football stadium has 27,500 seats and, of these, 7500 are midfield seats. Tickets for a midfield seat cost $20 each, and tickets for all other seats cost $12 each. To make a profit, the college must make at least $300,000 from ticket sales. What is the minimum number of tickets that the college must sell to make a profit?

72. A manufacturer produces two models of log-homes kits: a one-bedroom model and a two-bedroom model. The profit on each one-bedroom kit is $4500 and on each two-bedroom kit, $6000. The company can produce at most 4 one-bedroom kits per week and at most 3 two-bedroom kits per week. To produce a one-bedroom kit requires 200 hours of labor, and to produce a two-bedroom kit requires 300 hours of labor. If the maximum number of hours available for the production of log-home kits is 1100 hours per week, determine the number of each model that should be manufactured to maximize the profit. (Assume all log-home kits can be sold.)

A father decides to deposit $150 each month in a savings account for his daughter's college education. The savings account earns interest at the rate of 9% per year compounded monthly. How much will be in the account at the end of eight years if no money is in the account today?

For the solution, see Example 6 in Section 8.4.

Sequences and Series

8.1 An Introduction to Sequences and Series

◆ Introductory Comments

The phrase "a sequence of events" means the following of one event after another in a given order of succession. In mathematics, we refer to a **sequence** as an ordered list of numbers, and each number in the list is called an **element** of the sequence. When writing a sequence, we use commas to separate the elements and an ellipsis (three dots, . . .) to show an omission of elements that are obviously understood. For example, we write the sequence that represents the first eleven odd, positive integers as

$$1, \ 3, \ 5, \ 7, \ 9, \ 11, \ 13, \ 15, \ 17, \ 19, \ 21$$

or, more compactly, as

$$1, \ 3, \ 5, \ 7, \ . \ . \ . \ , \ 21$$

We refer to this sequence as a *finite sequence* because the sequence stops after a finite number of elements. In a sequence such as

$$1, \ 3, \ 5, \ 7, \ . \ . \ .$$

the ellipsis at the end indicates that the sequence continues forever. We refer to this type of sequence as an *infinite sequence.*

The indicated sum of the elements in a sequence is called a **series**. Associated with the finite sequence $1, 3, 5, 7, \ldots, 21$ is the *finite series*

$$1 + 3 + 5 + 7 + \cdots + 21$$

and associated with the infinite sequence $1, 3, 5, 7, \ldots$ is the *infinite series*

$$1 + 3 + 5 + 7 + \cdots$$

Each number in a series is called a **term** of the series. The terms in a series are identical to the elements in its corresponding sequence.

In this chapter, we study sequences, series, and their applications. We begin by using functional notation to define a sequence. If we think of a sequence in terms of functional notation, then we can use a graphing calculator to support our work with sequences.

◆ Using Functional Notation to Define a Sequence

We define a *finite sequence function* as a function whose domain is the set of positive integers from 1 to k and an *infinite sequence function* as a function whose domain is the set of all positive integers. In either case, the elements in the range of the function form the sequence.

◆ **Definitions of Finite and Infinite Sequences**

A function u, whose domain is the set of positive integers from 1 to k, is a *finite sequence function*. The elements in the range of u, taken in the order

$$u(1),\ u(2),\ u(3),\ .\ .\ .\ ,\ u(k)$$

form a **finite sequence**.

A function v, whose domain is the set of *all* positive integers, is an *infinite sequence function*. The elements in the range of v, taken in the order

$$v(1),\ v(2),\ v(3),\ .\ .\ .$$

form an **infinite sequence**.

EXAMPLE 1   Find the sequence represented by each sequence function.

(a) $u(n) = 2n - 1$ for domain $\{1, 2, 3, 4\}$

(b) $v(n) = (-1)^{n+1}\, 2^n$ for domain $\{1, 2, 3, 4, .\ .\ .\}$

◆ **Solution**

(a) The function u is a finite sequence function. The elements in the range of u are as follows:

$$u(1) = 2(1) - 1 = 1 \qquad u(2) = 2(2) - 1 = 3$$
$$u(3) = 2(3) - 1 = 5 \qquad u(4) = 2(4) - 1 = 7$$

The elements in the range of u, written in the order

$$1,\ 3,\ 5,\ 7,$$

form the finite sequence. This sequence is the first four odd, positive integers.

(b) The function v is an infinite sequence function. The first four elements in the range of v are as follows:

$$v(1) = (-1)^{1+1}\, 2^1 = 2 \qquad v(2) = (-1)^{1+2}\, 2^2 = -4$$
$$v(3) = (-1)^{1+3}\, 2^3 = 8 \qquad v(4) = (-1)^{1+4}\, 2^4 = -16$$

The elements in the range of v, written in the order

$$2,\ -4,\ 8,\ -16,\ .\ .\ .$$

ending with an ellipsis, form the infinite sequence. Note that the signs of the elements in this sequence alternate between positive and negative values. We refer to a sequence with this characteristic as an *alternating sequence.*   ◆

The graph of a sequence function u is formed by plotting the points $(n, u(n))$ in the coordinate plane. Since a sequence function has only integers in its domain, we do *not* connect these points. The graph of the sequence function u, defined in Example 1(a), is shown in Figure 8.1(a).

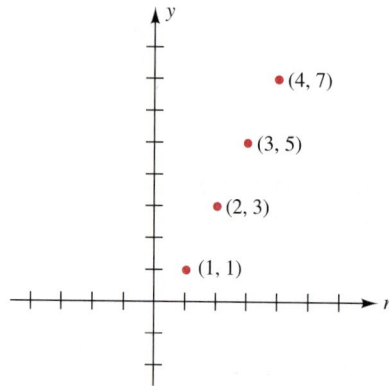

FIGURE 8.1

Graph of the sequence function
$u(n) = 2n - 1$ for domain {1, 2, 3, 4}

(a) Sketched by plotting points

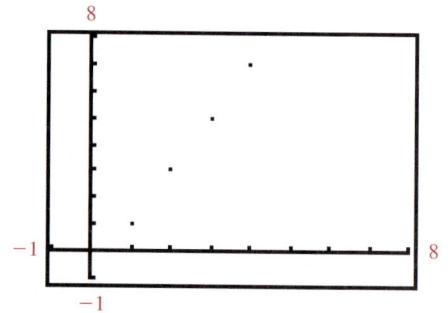

(b) Generated in the viewing rectangle of a graphing calculator

We can generate the graph of a sequence function on most graphing calculators. First, we select "sequence" mode and "dot" mode. We then press the $\boxed{y =}$ key and enter the sequence function we wish to graph. The $\boxed{\text{WINDOW}}$ key (or $\boxed{\text{RANGE}}$ key) is used to select the size of the viewing rectangle as well as other "window" values, such as the value of n at which calculation begins (n-start), the value of $u(n)$ when $n = n$-start [$u(n)$-start], and the values of n at which plotting begins and ends (n-min and n-max, respectively). Figure 8.1(b) shows the graph of $u(n) = 2n - 1$ for domain {1, 2, 3, 4} in the viewing rectangle of a graphing calculator. For this sequence we select these "window" values:

$$n\text{-start} = 1 \qquad u(n)\text{-start} = 1 \qquad n\text{-min} = 1 \qquad n\text{-max} = 4$$

PROBLEM 1   Find the sequence represented by the sequence function

$$u(n) = (-1)^n \, 2^{n-1} \quad \text{for domain } \{1, 2, 3, 4\}$$

Then sketch the graph of this function.   ◆

◆ General Element of a Sequence

It is customary to denote the first element of a sequence as a_1, the second element as a_2, the third element as a_3, and so on. The nth element of a sequence, denoted as a_n, is called the **general element** of the sequence. Thus, for a sequence function u, we have

$$u(1) = a_1 \qquad \text{First element}$$

$$u(2) = a_2 \qquad \text{Second element}$$

$$u(3) = a_3 \qquad \text{Third element}$$

$$\vdots \qquad \vdots \qquad \qquad \vdots$$

$$u(n) = a_n \qquad \text{General element}$$

$$\vdots \qquad \vdots \qquad \qquad \vdots$$

The general element of a sequence may be given **explicitly** in terms of *n*, or it may be expressed in terms of preceding elements. When it is expressed in terms of preceding elements, we say that the sequence is defined **recursively**.

We can generate a sequence in the viewing rectangle of most graphing calculators by selecting the variable u_n (which corresponds to the general element a_n) from the "*y*-vars" menu. First, we enter the general element and store it in u_n. Next, we select the appropriate "window" values for *n*-start and u_n-start. Now, we can generate the sequence defined by this general element by entering u_n (*n*-start, *n*-stop). Figure 8.2 shows a typical display of the sequence defined in Example 1(a). For a more detailed explanation about the sequence features of a graphing calculator, consult the manual that came with your calculator or the *Graphing Technology Laboratory Manual* that accompanies this text. After reading this information, use a graphing calculator to generate the first four elements of the infinite sequence in Example 1(b).

FIGURE 8.2

Typical display of the sequence defined by $u(n) = 2n - 1$, with domain {1, 2, 3, 4}

EXAMPLE 2 Find the first four elements of a sequence whose general element a_n is given.

(a) $a_n = \dfrac{1 + n}{n}$ **(b)** $a_n = \begin{cases} 1 & \text{if } n = 1 \\ na_{n-1} & \text{if } n \geq 2 \end{cases}$

◆ **Solution**

(a) This general element is defined explicitly in terms of *n*. For $n = 1, 2, 3, 4$, we have

$$a_1 = \frac{1+1}{2} = 2 \qquad a_2 = \frac{1+2}{2} = \frac{3}{2}$$

$$a_3 = \frac{1+3}{3} = \frac{4}{3} \qquad a_4 = \frac{1+4}{4} = \frac{5}{4}$$

FIGURE 8.3

Typical display of the first four elements of the sequence with general element

$$a_n = \frac{1 + n}{n}$$

Thus, the first four elements of this sequence are

$$2, \tfrac{3}{2}, \tfrac{4}{3}, \tfrac{5}{4}$$

The display in Figure 8.3 confirms our work. We have chosen *n*-start = 1 and u_n-start = 2 as "window" values.

(b) Note that when $n \geq 2$, the general element a_n is defined in terms of the preceding element a_{n-1}. Hence, this is a recursively defined sequence. For $n = 1, 2, 3, 4$, we have

$$a_1 = 1 \qquad\qquad a_2 = 2a_1 = 2 \cdot 1 = 2$$
$$a_3 = 3a_2 = 3 \cdot 2 = 6 \qquad a_4 = 4a_3 = 4 \cdot 6 = 24$$

Thus, the first four elements of this sequence are

$$1, 2, 6, 24$$

We can confirm our answer by using the sequence features on a graphing calculator. For recursively defined functions, we define u_n by using the variable u_{n-1} on the calculator. Figure 8.4 shows a typical display of this sequence. We have chosen n-start = 1 and u_n-start = 1 as "window" values.   ◆

Referring to Example 2(b), the numbers in the sequence 1, 2, 6, 24 are called the *factorials* of 1, 2, 3, and 4, respectively. A **factorial** of a given positive integer is the product of that integer and all lesser positive integers. In Section 8.6, we discuss factorials and their applications in more detail.

PROBLEM 2   Use the general element in Example 2(b) to find the factorials of 5 and 6.   ◆

FIGURE 8.4

Typical display of the first four elements of the sequence with the general element

$$a_n = \begin{cases} 1 & \text{if } n = 1 \\ na_{n-1} & \text{if } n \geq 2 \end{cases}$$

Given the first few elements of a sequence, we cannot *uniquely* determine the general element. For example, consider the finite sequence

$$2, 4, 6$$

Note that 2, 4, and 6 are even numbers or multiples of 2. Thus, one possibility for the general element is simply

$$a_n = 2n.$$

However, another possibility for the general element is

$$a_n = \begin{cases} 2n & \text{if } n \text{ is odd} \\ 2a_{n-1} & \text{if } n \text{ is even} \end{cases}$$

In the next example, we use **pattern recognition** to determine the general element of a given sequence. Keep in mind that the answers we give are not unique. They do, however, represent the most obvious choices.

EXAMPLE 3   Use pattern recognition to find the general element of each sequence.

(a) $1, 3, 5, 7, \ldots, 21$ **(b)** $1, 3, 9, 27, 81, 243$

(c) $\frac{2}{3}, \frac{3}{4}, \frac{4}{5}, \frac{5}{6}, \ldots$ **(d)** $-x, x^5, -x^9, x^{13}, \ldots$

◆◆ Solution

(a) We recognize the elements of the finite sequence 1, 3, 5, 7, . . . , 21 as odd, positive integers. Since these odd numbers are one less than the even numbers, which have the general element $2n$, we may write the general element for these odd, positive integers as

$$a_n = 2n - 1.$$

(b) The elements in the finite sequence 1, 3, 9, 27, 81, 243 are the powers of three. Since we must obtain 1 when $n = 1$, we can write the general element as

$$a_n = 3^{n-1}.$$

(c) For the infinite sequence $\frac{2}{3}, \frac{3}{4}, \frac{4}{5}, \frac{5}{6}, \ldots$, each denominator is one more than its corresponding numerator. Since we must obtain $\frac{2}{3}$ when $n = 1$, we can write the general element as

$$a_n = \frac{n + 1}{n + 2}.$$

(d) In the infinite sequence $-x, x^5, -x^9, x^{13}, \ldots$, each exponent after the first is four more than the preceding exponent. This suggests an exponent containing $4n$. However, since we must obtain an exponent of 1 when $n = 1$, we write these exponents as $4n - 3$ for $n = 1, 2, 3, \ldots$. Also, since the *odd-numbered* elements are negative, we must include a factor of $(-1)^n$. Thus, we may write the general element for this sequence as

$$a_n = (-1)^n \, x^{4n-3}. \qquad \qquad ◆$$

PROBLEM 3   Verify that each general element found for each part of Example 3 does give the first four elements of its sequence.   ◆

Note: It is not possible to find a formula that defines the general element of every sequence. For the sequence of prime numbers

$$2, 3, 5, 7, 11, 13, 17, 19, 23, 29, \ldots, a_n, \ldots$$

where a_n is the nth prime number, we cannot write a formula that defines a_n.

◆ Sigma Form of a Series

In a sequence, we refer to a_n as the general element, but in a series we refer to a_n as the **general term** of the series.

The finite series

$$a_1 + a_2 + a_3 + \cdots + a_n$$

and the infinite series

$$a_1 + a_2 + a_3 + \cdots + a_n + \cdots$$

are written here in **expanded form**. These series may also be written more compactly by using **sigma form**. The Greek letter Σ (read "sigma") is used as the "summation symbol" in sigma form.

◆ **Sigma Form**

The finite series

$$a_1 + a_2 + a_3 + \cdots + a_n$$

may be written in sigma form as

$$\sum_{i=1}^{n} a_i$$

The infinite series

$$a_1 + a_2 + a_3 + \cdots + a_n + \cdots$$

may be written in sigma form as

$$\sum_{i=1}^{\infty} a_i$$

The letter i in the sigma form is called the **index of summation** or the **summation variable**. We often refer to the summation variable as a *dummy variable* because any other letter can be used without changing the series. The letters $i, j,$ and k are commonly used as summation variables.

To change a series from sigma form to expanded form, we replace the summation variable (index) with successive integers, starting with the integer written below Σ and ending with the integer written above Σ. If the series is infinite, ∞ is written above Σ to indicate that the summation has no ending value.

EXAMPLE 4   Write each series in expanded form.

(a) $\displaystyle\sum_{i=1}^{4} (5i - 2)$ **(b)** $\displaystyle\sum_{j=1}^{\infty} (-1)^j \, j^2$

◆ **Solution**

(a) To express $\displaystyle\sum_{i=1}^{4} (5i - 2)$ [read "the sum of $(5i - 2)$ from $i = 1$ to $i = 4$"] in expanded form, we let the index i take on successive integers from $i = 1$ to $i = 4$. Thus,

$$\sum_{i=1}^{4} (5i - 2) = [5(1) - 2] + [5(2) - 2] + [5(3) - 2] + [5(4) - 2]$$
$$= \underbrace{3 + 8 + 13 + 18}_{\text{expanded form}}$$

FIGURE 8.5

Typical display of the terms in the series

$$\sum_{i=1}^{4} (5i - 2)$$

FIGURE 8.6

Typical display of the first three terms in

the series $\sum_{j=1}^{\infty} (-1)^j j^2$

Since the terms in a series are identical to the corresponding elements in its associated sequence, we can confirm our answer by using the sequence features of a graphing calculator, as shown in Figure 8.5.

(b) To express $\sum_{j=1}^{\infty} (-1)^j j^2$ in expanded form, we let the index j take on *all* successive integers starting with $j = 1$. To do this, we write the first few terms of the series, followed by an ellipsis, then the nth term, followed by an ellipsis:

$$\sum_{j=1}^{\infty} (-1)^j j^2 = (-1)^1\, 1^2 + (-1)^2\, 2^2 + (-1)^3\, 3^2 + \cdots + (-1)^n n^2 + \cdots$$
$$= \underbrace{-1 + 4 - 9 + \cdots + (-1)^n n^2 + \cdots}_{\text{expanded form}}$$

The display in Figure 8.6 shows the first three terms of this infinite series and confirms our work.   ◆

If a finite series contains many terms and we wish to write the series in expanded form, then we usually write the first few terms followed by an ellipsis, the nth term followed by an ellipsis, and the last term. For example,

$$\sum_{i=1}^{40} (5i - 2) = 3 + 8 + 13 + \cdots + (5n - 2) + \cdots + \underbrace{198}_{}$$

We find the last term by evaluating $[5(40) - 2]$.

PROBLEM 4   Write the series $\sum_{k=1}^{27} \dfrac{k}{k + 1}$ in expanded form.   ◆

If the general term of a series in expanded form is known, then we can write the series in sigma form. The letter we use for the summation variable is immaterial—only the general term and the beginning and ending values of the summation variable are significant.

EXAMPLE 5   Use pattern recognition to find the general term of each series. Then write each series in sigma form.

(a) $1 + 3 + 5 + 7 + 9$ **(b)** $2 + 5 + 8 + 11 + \cdots + 83$

(c) $\frac{1}{2} - \frac{1}{4} + \frac{1}{8} - \frac{1}{16} + \cdots$ **(d)** $2x^2 + 3x^3 + 4x^4 + 5x^5 + \cdots$

◆ **Solution**

(a) We recognize the terms in the series $1 + 3 + 5 + 7 + 9$ as the first five odd, positive integers. Thus, from Example 3(a), we know that the general term of this finite series is

$$a_n = 2n - 1.$$

In our definition of sigma form for a finite series, the summation variable i begins at a value of 1 and ends at a value of n, where n is the number of terms in the finite series. Since this series contains 5 terms, we write

$$1 + 3 + 5 + 7 + 9 = \sum_{i=1}^{5} (2i - 1)$$

sigma form

(b) In the finite series $2 + 5 + 8 + 11 + \cdots + 83$, each term after the first is three more than the preceding term. This suggests a general term containing $3n$. However, since we must obtain 2 when $n = 1$, we write the general term of this series as

$$a_n = 3n - 1.$$

Now we must find the upper value of the summation variable. To do this, we set the general term equal to the last term in the series and solve for the number of terms n, as follows:

$$3n - 1 = 83$$
$$3n = 84$$
$$n = 28$$

Thus, using i as the summation variable, we write

$$2 + 5 + 8 + 11 + \cdots + 83 = \sum_{i=1}^{28} (3i - 1)$$

sigma form

(c) The terms of the infinite series $\frac{1}{2} - \frac{1}{4} + \frac{1}{8} - \frac{1}{16} + \cdots$ are powers of one-half with alternating signs. Powers of one-half can be written as $\left(\frac{1}{2}\right)^n$ for $n = 1, 2, 3, \ldots$. Since the even-numbered terms are negative, we must also include the factor $(-1)^{n+1}$. Thus, the general term for this series is

$$a_n = (-1)^{n+1} \left(\frac{1}{2}\right)^n \quad \text{or} \quad \frac{(-1)^{n+1}}{2^n}.$$

The upper value of the summation variable for an infinite series is always ∞. Thus, using i as the summation variable, we have

$$\frac{1}{2} - \frac{1}{4} + \frac{1}{8} - \frac{1}{16} + \cdots = \sum_{i=1}^{\infty} \frac{(-1)^{i+1}}{2^i}$$

sigma form

(d) The coefficients of x in the infinite series $2x^2 + 3x^3 + 4x^4 + 5x^5 + \cdots$ are consecutive integers. Since we must obtain 2 when $n = 1$, we can write these coefficients as $n + 1$ for $n = 1, 2, 3, \ldots$. In each term, the exponent of x is the same as its coefficient. Thus, the general term for this series is

$$a_n = (n + 1)x^{n+1}$$

The upper value of the summation variable for an infinite series is always ∞. Thus, using i as the summation variable, we have

$$2x^2 + 3x^3 + 4x^4 + 5x^5 + \cdots = \sum_{i=1}^{\infty} (i + 1)\, x^{i + 1}$$

sigma form

Our definition of sigma form uses $i = 1$ as the lower value of the summation variable. We can use other lower values to describe a series. Consider the series in Example 5(a):

$$1 + 3 + 5 + 7 + 9 = \sum_{i=1}^{5} (2i - 1)$$

If we write the upper value as $i = 5$, instead of 5, and then perform a *shift of index* by letting $i = k + 1$, we obtain

$$\sum_{i=1}^{i=5} (2i - 1) = \sum_{k+1=1}^{k+1=5} [2(k + 1) - 1) = \sum_{k=0}^{k=4} (2k + 1).$$

Observe that

$$\sum_{k=0}^{4} (2k + 1) = 1 + 3 + 5 + 7 + 9 = \sum_{i=1}^{5} (2i - 1).$$

Note: Unless otherwise instructed, we shall use $i = 1$ as the lower value of the summation variable when changing from expanded form to sigma form.

PROBLEM 5 Perform a shift of index on the sigma form $\sum_{i=1}^{\infty} (i + 1)x^{i+1}$ by letting $i = k - 1$.

Exercises 8.1

Basic Skills

In Exercises 1–6, find the sequence represented by each sequence function. Then sketch the graph of each function.

1. $u(n) = 6 - 3n$ for domain {1, 2, 3, 4, 5}

2. $u(n) = n^2 - 1$ for domain {1, 2, 3, 4, 5, 6}

3. $u(n) = (-1)^{n+1}n$ for domain {1, 2, 3, 4}

4. $u(n) = \dfrac{n}{n + 1}$ for domain {1, 2, 3, 4, 5, 6, 7}

5. $v(n) = \dfrac{2^{2n-1}}{n^2}$ for domain $\{1, 2, 3, 4, \ldots\}$

6. $v(n) = \dfrac{(-1)^n}{2^n - 1}$ for domain $\{1, 2, 3, 4, \ldots\}$

In Exercises 7–20, find the first six elements of a sequence with the given general element. When possible, use the sequence features of a graphing calculator to confirm your answer.

7. $a_n = 5n - 4$

8. $a_n = 4 - 3n$

9. $a_n = 3(-2)^n$

10. $a_n = 9\left(\dfrac{2}{3}\right)^{n-1}$

11. $a_n = 2 + \dfrac{1}{n}$

12. $a_n = \dfrac{n}{2^n}$

13. $a_n = 1 + (-1)^{n+1}$

14. $a_n = \dfrac{(-1)^n}{3n - 4}$

15. $a_n = \begin{cases} n & \text{if } n \text{ is odd} \\ 2n - 1 & \text{if } n \text{ is even} \end{cases}$

16. $a_n = \begin{cases} 0 & \text{if } n \text{ is odd} \\ n^2 - 1 & \text{if } n \text{ is even} \end{cases}$

17. $a_n = \begin{cases} 3 & \text{if } n = 1 \\ 2a_{n-1} & \text{if } n \geq 2 \end{cases}$

18. $a_n = \begin{cases} -1 & \text{if } n = 1 \\ (a_{n-1})^{n-1} & \text{if } n \geq 2 \end{cases}$

19. $a_n = \begin{cases} 1 & \text{if } n = 1 \\ 2 & \text{if } n = 2 \\ a_{n-1} - a_{n-2} & \text{if } n \geq 3 \end{cases}$

20. $a_n = \begin{cases} -5 & \text{if } n = 1 \\ 2 & \text{if } n = 2 \\ a_{n-1}a_{n-2} & \text{if } n \geq 3 \end{cases}$

In Exercises 21–32, use pattern recognition to find the general element of each sequence. (Answers are not unique.)

21. $1, 4, 9, 16, 25, \ldots$

22. $1, 8, 27, 81, 243, \ldots$

23. $2, -4, 6, -8, 10, -12$

24. $-1, 3, -5, 7, -9, \ldots, 27$

25. $3, 4, 5, 6, 7, \ldots, 29$

26. $1, 4, 7, 10, 13, \ldots$

27. $\frac{1}{2}, \frac{3}{4}, \frac{5}{6}, \frac{7}{8}, \frac{9}{10}, \ldots$

28. $\frac{1}{2}, \frac{3}{5}, \frac{5}{8}, \frac{7}{11}, \frac{9}{14}, \frac{11}{17}, \frac{13}{20}$

29. $2, 0.2, 0.02, 0.002, 0.0002, \ldots$

30. $0.31, 0.0031, 0.000031, 0.00000031, \ldots$

31. $-x, \dfrac{x^2}{3}, -\dfrac{x^3}{5}, \dfrac{x^4}{7}, -\dfrac{x^5}{9}, \ldots$

32. $x, -\dfrac{x^3}{5}, \dfrac{x^5}{9}, -\dfrac{x^7}{13}, \dfrac{x^9}{17}, \ldots$

In Exercises 33–42, write each series in expanded form. When possible, use the sequence features of a graphing calculator to confirm your answer.

33. $\displaystyle\sum_{i=1}^{5} 2i$

34. $\displaystyle\sum_{i=1}^{4} (7i - 3)$

35. $\displaystyle\sum_{n=1}^{10} (-1)^n\, n^3$

36. $\displaystyle\sum_{j=1}^{7} (-1)^{j+1} (j^2 - 1)$

37. $\displaystyle\sum_{k=1}^{\infty} k^k$

38. $\displaystyle\sum_{k=1}^{\infty} \dfrac{1}{k}$

39. $\displaystyle\sum_{i=0}^{50} \dfrac{2i - 3}{2}$

40. $\displaystyle\sum_{i=2}^{25} \dfrac{i}{i^2 + 1}$

41. $\displaystyle\sum_{i=1}^{\infty} \dfrac{(-1)^i\, (2x^{i-1})}{i}$

42. $\displaystyle\sum_{i=1}^{\infty} \dfrac{(-1)^{i+1}\, x^i}{3i + 1}$

In Exercises 43–54, write the series in sigma form using $i = 1$ as the lower value of the summation variable.

43. $2 + 4 + 6 + 8 + 10 + 12 + 14$

44. $1 - 3 + 5 - 7 + 9 - 11 + 13 - 15$

45. $2 + 7 + 12 + 17 + \cdots + 147$

46. $3 + 8 + 13 + 18 + \cdots + 98$

47. $1 - 1 + 1 - 1 + 1 - 1 + \cdots$

48. $-1 + 4 - 9 + 16 - 25 + \cdots$

49. $-1 + \frac{1}{2} - \frac{1}{4} + \frac{1}{8} - \frac{1}{16} + \cdots + \frac{1}{512}$

50. $2 + \frac{3}{2} + \frac{4}{3} + \frac{5}{4} + \cdots + \frac{54}{53}$

51. $2 + \frac{4}{3} + \frac{8}{9} + \frac{16}{27} + \frac{32}{81} + \cdots$

52. $1 + \frac{4}{3} + \frac{16}{5} + \frac{64}{7} + \frac{256}{9} + \cdots$

53. $3x + 4x^3 + 5x^5 + 6x^7 + 7x^9$

54. $\dfrac{1}{x^4} - \dfrac{2}{x^{10}} + \dfrac{4}{x^{16}} - \dfrac{8}{x^{22}} + \cdots$

In Exercises 55–60, perform the indicated shift of index.

55. Rewrite $\displaystyle\sum_{i=1}^{4} (2i - 5)$ by letting $i = k + 1$.

56. Rewrite $\displaystyle\sum_{i=0}^{50} (-1)^{i+1} (i + 1)$ by letting $i = k - 1$.

57. Rewrite $\displaystyle\sum_{i=1}^{\infty} \dfrac{(-1)^i\, (2x^{i-1})}{i}$ using $k = 0$ as the lower value of the summation variable.

58. Rewrite $\displaystyle\sum_{i=1}^{\infty} \dfrac{(-1)^{i+1}\, x^i}{3i + 1}$ using $k = 2$ as the lower value of the summation variable.

59. Rewrite $\displaystyle\sum_{i=2}^{25} \dfrac{i}{i^2 + 1}$ using $k = 0$ as the lower value of the summation variable.

60. Rewrite $\displaystyle\sum_{i=0}^{10} \dfrac{(-1)^{i+1}\, 3^{i+2}}{2i + 4}$ using $k = 2$ as the lower value of the summation variable.

━━━ ◆ **Critical Thinking** ━━━━━━━━━━━━

61. One possibility for the general element of a sequence that begins with 1, 3, 5, is $a_n = 2n - 1$. Find another possibility for the general element of a sequence that begins with 1, 3, 5.

62. How does the graph of the sequence function $u(n) = 2n - 3$ differ from the graph of the linear function $f(x) = 2x - 3$?

63. Find a finite sequence function u indicated by each graph, and state its domain.

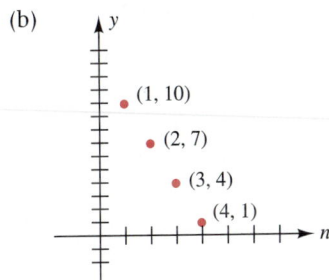

(a)

(b)

64. Give a recursive definition for the general element of each infinite sequence.

(a) 6, 10, 14, 18, 22, . . .   (b) 8, 4, 2, 1, $\frac{1}{2}$, . . .

(c) 1, 4, 10, 22, 46, 94, . . .   (d) 1, 2, 5, 26, 677, . . .

65. The general element of the *Fibonacci sequence* [named after Italian mathematician Leonardo Fibonacci (1180?–1250?)] is given by

$$a_n = \begin{cases} 1 & \text{if } n = 1 \\ 1 & \text{if } n = 2 \\ a_{n-1} + a_{n-2} & \text{if } n \geq 3 \end{cases}$$

(a) Find the first eight elements of Fibonacci sequence.

(b) The elements in a Fibonacci sequence occur frequently in the geometry of nature. Explain how the branching of the tree shown in the figure forms a Fibonacci sequence.

(c) Investigate the branchings of trees in your area. Do they follow a Fibonacci sequence?

66. Each element in the sequence with general element

$$a_n = \begin{cases} \dfrac{N}{2} & \text{if } n = 1 \\ \dfrac{1}{2}\left(a_{n-1} + \dfrac{N}{a_{n-1}} \right) & \text{if } n \geq 2 \end{cases}$$

gives a better approximation of $\sqrt{N}$ than does the element that precedes it. For each of the given square roots, find the fifth element, a_5, rounding your answer to five significant digits. Compare a_5 to the value you obtain by using the $\boxed{\sqrt{}}$ key on your calculator.

(a) $\sqrt{2}$ (b) $\sqrt{3}$
(c) $\sqrt{22}$ (d) $\sqrt{85}$

━━━━━━━━━━━━━━━━━━━━━━━━━━━━

8.2 ◆ **The Sum of a Series and Mathematical Induction**

◆ **Introductory Comments**

To every finite series, a number can be assigned as its *sum*. For example, we find the sum of the finite series

$$\sum_{i=1}^{4} (2i - 1)$$

by adding the terms in the series, as follows:

$$\sum_{i=1}^{4} (2i - 1) = 1 + 3 + 5 + 7 = 16 \quad \boxed{\text{sum of the first four odd integers}}$$

Some, but not all, infinite series can be assigned a sum. For example, in Section 8.5, we show that

$$\sum_{i=1}^{\infty} (-1)^{i+1} \left(\frac{1}{2}\right)^i = \frac{1}{2} - \frac{1}{4} + \frac{1}{8} - \frac{1}{16} + \cdots = \frac{1}{3} \quad \boxed{\text{sum of the infinite series}}$$

However, other infinite series such as

$$\sum_{i=1}^{\infty} i = 1 + 2 + 3 + 4 + \cdots$$

do not appear to sum to any finite number. The sums of infinite series are usually studied in calculus. In this section, we limit our discussion to methods of finding the sums of finite series.

◆ **Finding the Sum of a Series by Direct Computation**

If the number of terms in the series is small, then we can find the sum of the series by direct computation and adding the terms in the series.

EXAMPLE 1   Find the sum of each series.

(a) $\displaystyle\sum_{i=1}^{8} 7$ (b) $\displaystyle\sum_{i=1}^{6} \frac{i}{2^i}$

◆ **Solution**

(a) $\displaystyle\sum_{i=1}^{8} 7 = 7 + 7 + 7 + 7 + 7 + 7 + 7 + 7 = 8(7) = 56$

$\boxed{\text{8 terms}}$

(b) $\displaystyle\sum_{i=1}^{6} \frac{i}{2^i} = \frac{1}{2} + \frac{2}{4} + \frac{3}{8} + \frac{4}{16} + \frac{5}{32} + \frac{6}{64}$

$\qquad = \dfrac{1(32) + 2(16) + 3(8) + 4(4) + 5(2) + 6}{64}$

$\qquad = \dfrac{120}{64} = \dfrac{15}{8}$

To avoid this awkward fractional computation, we can use the sequence features of a graphing calculator in conjunction with its sum feature. We usually access the sum feature of a graphing calculator from the "List" menu. Most graphing calculators have the capability of finding the sum

```
"n/2^n"→Un
                    Done
sum Un(1,6)►Frac
                    15/8
```

FIGURE 8.7

Typical display of the sum of the series in Example 1(b)

of up to 99 numbers in a list. Figure 8.7 shows a typical display of the sum of this series and confirms our work. ◆

The series given in Example 1(a) has the form $\sum_{i=1}^{n} c$ where c is constant. For any series of this form, we can find the sum by the following formula:

$$\sum_{i=1}^{n} c = \underbrace{c + c + c + \cdots + c}_{n \text{ terms}} = nc$$

Formulas like this one enable us to quickly find the sum of a series containing a large number of terms.

PROBLEM 1 Find the sum of the series $\sum_{i=1}^{1000} 4$. ◆

◆ Proof by Mathematical Induction

The direct computation procedure for finding the sum of a series may be tedious if the series contains a large number of terms. To find the sum of a series with a large number of terms, it is advantageous for us to develop and apply a formula. In mathematics, we often use pattern recognition to help develop formulas. For example, suppose we wish to find a formula for the sum of the first n odd, positive integers. We can represent this sum by the series

$$\sum_{i=1}^{n} (2i - 1) = 1 + 3 + 5 + \cdots + (2n - 1).$$

We begin by calculating the sum of the odd, positive integers for the first few values of n and observe any pattern that develops. Table 8.1 shows the sum of the first n odd, positive integers from $n = 1$ to $n = 6$.

The sums in Table 8.1 form the sequence

$$1, 4, 9, 16, 25, 36.$$

Do you detect the pattern? From this sequence, we are led to believe that the sum of the first n odd, positive integers is n^2 or, in series notation,

$$\sum_{i=1}^{n} (2i - 1) = 1 + 3 + 5 + \cdots + (2n - 1) = n^2.$$

TABLE 8.1

Sum of the first n odd, positive integers for $n = 1$ to $n = 6$

n	Sum of the first n odd positive integers
1	$1 = 1$
2	$1 + 3 = 4$
3	$1 + 3 + 5 = 9$
4	$1 + 3 + 5 + 7 = 16$
5	$1 + 3 + 5 + 7 + 9 = 25$
6	$1 + 3 + 5 + 7 + 9 + 11 = 36$

However, *choosing a formula based on a few observations does not guarantee the validity of the formula for all integers n.* What we need is a "chain reaction" procedure that will guarantee that once the formula is valid for a particular integer n, then the formula is valid for the next integer, and the next, and the next, and so on, indefinitely.

Consider the row of dominoes placed on end, as shown in Figure 8.8.

FIGURE 8.8

A row of dominoes placed on end

If we know that

1. the first domino is knocked over, and
2. if any domino in the row is knocked over, then the next one in line is also knocked over,

then we can conclude that *all* dominoes in the row will be knocked over. By analogy with the domino problem, we can state that a *mathematical statement,* such as

$$\sum_{i=1}^{n} (2i - 1) = 1 + 3 + 5 + \cdots + (2n - 1) = n^2$$

is true for all positive integers *n* provided that

1. the statement is true when $n = 1$, and
2. if the statement is true for $n = k$, where k is an arbitrary positive integer, then the statement is true for the next integer, namely, $n = k + 1$.

We refer to this chain reaction procedure as *proof by mathematical induction.*

◆ **Proof by Mathematical Induction**

The following two-step procedure, called proof by mathematical induction, may be used to show that a mathematical statement is true for all positive integers *n*:

Step 1   Show that the statement is true when $n = 1$.

Step 2   Assume that the statement is true for an arbitrary positive integer k and show, by using this assumption, that the statement is also true when $n = k + 1$.

EXAMPLE 2   Use mathematical induction to prove that

$$\sum_{i=1}^{n} (2i - 1) = 1 + 3 + 5 + \cdots + (2n - 1) = n^2$$

for all positive integers *n*.

◆ **Solution**

Step 1   We must first show that the statement is true for $n = 1$:

$$1 = 1^2 \quad ?$$
$$1 = 1 \quad \text{is } \textit{true.}$$

Step 2   Assuming that the statement is true for $n = k$, we must show that, on the basis of that assumption, the statement is true for $n = k + 1$.

Assume: $\qquad\qquad 1 + 3 + 5 + \cdots + (2k - 1) = k^2$

Prove: $\quad 1 + 3 + 5 + \cdots + (2k - 1) + [2(k + 1) - 1] = (k + 1)^2$

Proof:

$$1 + 3 + 5 + \cdots + (2k - 1) + [2(k + 1) - 1] = \{1 + 3 + 5 + \cdots + (2k - 1)\} + [2(k + 1) - 1]$$

Replace these terms with the *assumed* statement.

$$= k^2 \qquad\qquad + [(2k + 2) - 1]$$

$$= k^2 + 2k + 1 \qquad \textbf{Combine like terms}$$

$$= (k + 1)^2 \qquad \textbf{Factor}$$

Thus, the truth of the statement when $n = k$ implies the truth of the statement when $n = k + 1$. Hence, by mathematical induction, we can state that

$$\sum_{i=1}^{n} (2i - 1) = 1 + 3 + 5 + 7 + \cdots + (2n - 1) = n^2$$

for all positive integers n. ◆

PROBLEM 2   Find the sum of the series $\displaystyle\sum_{i=1}^{50} (2i - 1)$ by using the formula we proved in Example 2. Verify the answer by using the sequence and sum features of a graphing calculator. ◆

EXAMPLE 3   Use mathematical induction to prove that the sum of the first n positive integers is $\dfrac{n(n + 1)}{2}$ by showing that

$$\sum_{i=1}^{n} i = 1 + 2 + 3 + \cdots + n = \frac{n(n + 1)}{2}$$

for all positive integers n.

◆ **Solution**

Step 1   We must first show that the statement is true for $n = 1$:

$$1 = \frac{1(1 + 1)}{2} \quad ?$$

$$1 = 1 \qquad \text{is } true.$$

Step 2   Assuming that the statement is true for $n = k$, we must show that, on the basis of that assumption, the statement is true for $n = k + 1$.

Assume: $1 + 2 + 3 + 4 + \cdots + k = \dfrac{k(k + 1)}{2}$

Prove: $1 + 2 + 3 + 4 + \cdots + k + (k + 1) = \dfrac{(k + 1)[(k + 1) + 1]}{2}$

Proof:

$$1 + 2 + 3 + 4 + \cdots + k + (k + 1) = [1 + 2 + 3 + 4 + \cdots + k] + (k + 1)$$

Replace these terms with the *assumed* statement.

$$= \frac{k(k + 1)}{2} + (k + 1)$$

$$= \frac{k(k + 1) + 2(k + 1)}{2} \qquad \textbf{Write as a single fraction}$$

$$= \frac{(k + 1)(k + 2)}{2} \qquad \textbf{Factor}$$

$$= \frac{(k + 1)[(k + 1) + 1]}{2} \qquad \textbf{Rewrite}$$

Thus, the truth of the statement when $n = k$ implies the truth of the statement when $n = k + 1$. Hence, by mathematical induction, we can state that

$$\sum_{i=1}^{n} i = 1 + 2 + 3 + \cdots + n = \frac{n(n + 1)}{2}$$

for all positive integers n.   ◆

   The formula for the sum of positive integers (from Example 3), along with formulas for the *sums of powers of positive integers,* are used in calculus to help find the area under a curve. Here we list the formulas for sums of powers up to and including 3. Mathematical induction may be used to prove formulas 2 and 3 as well (see Exercises 13 and 14).

◆ **Formulas for the Sums of Powers of Positive Integers**

1. $\displaystyle\sum_{i=1}^{n} i = 1 + 2 + 3 + \cdots + n = \frac{n(n+1)}{2}$

2. $\displaystyle\sum_{i=1}^{n} i^2 = 1^2 + 2^2 + 3^2 + \cdots + n^2 = \frac{n(n+1)(2n+1)}{6}$

3. $\displaystyle\sum_{i=1}^{n} i^3 = 1^3 + 2^3 + 3^3 + \cdots + n^3 = \left[\frac{n(n+1)}{2}\right]^2$

PROBLEM 3 Find the sum of each series by using the formulas for the sums of powers of positive integers. Verify each answer by using the sequence and sum features of a graphing calculator.

(a) $\displaystyle\sum_{i=1}^{19} i$ (b) $\displaystyle\sum_{i=1}^{19} i^2$ (c) $\displaystyle\sum_{i=1}^{19} i^3$ ◆

◆ **Summation Properties**

We now state three fundamental **summation properties**. Each property can be proved by mathematical induction or by writing the property in expanded form and then applying the properties of real numbers (see Exercises 62–64).

◆ **Summation Properties**

1. $\displaystyle\sum_{i=1}^{n} ca_i = c\sum_{i=1}^{n} a_i$

2. $\displaystyle\sum_{i=1}^{n} (a_i + b_i) = \sum_{i=1}^{n} a_i + \sum_{i=1}^{n} b_i$

3. $\displaystyle\sum_{i=1}^{n} (a_i - b_i) = \sum_{i=1}^{n} a_i - \sum_{i=1}^{n} b_i$

We use these summation properties in conjunction with the formulas for the sums of powers of integers to find the sums of several other series. For example, suppose we wish to find a formula for the sum of the first n even, positive integers. We can represent this sum by the series

$$\sum_{i=1}^{n} 2i = 2 + 4 + 6 + \cdots + 2n.$$

Now, applying summation property 1, we have

$$\sum_{i=1}^{n} 2i = 2\sum_{i=1}^{n} i.$$

However, from Example 3, we know that

$$\sum_{i=1}^{n} i = \frac{n(n+1)}{2}.$$

Hence,

$$\sum_{i=1}^{n} 2i = 2 \sum_{i=1}^{n} i = 2 \frac{n(n+1)}{2} = n(n+1).$$

Thus, a formula for the sum of the first n even, positive integers is

$$\sum_{i=1}^{n} 2i = 2 + 4 + 6 + \cdots + 2n = n(n+1).$$

EXAMPLE 4   Use the summation properties and the formulas for the sums of powers of integers to find the sum of the following series:

$$\sum_{i=1}^{9} (i^2 + 3i - 4)$$

◆ **Solution**   Using the summation properties, we write

$$\sum_{i=1}^{9} (i^2 + 3i - 4) = \sum_{i=1}^{9} i^2 + 3 \sum_{i=1}^{9} i - \sum_{i=1}^{9} 4$$

Now, using the fact that

$$\sum_{i=1}^{n} i^2 = \frac{n(n+1)(2n+1)}{6}, \quad \sum_{i=1}^{n} i = \frac{n(n+1)}{2}, \quad \text{and} \quad \sum_{i=1}^{n} c = nc,$$

we have

$$\sum_{i=1}^{9} (i^2 + 3i - 4) = \sum_{i=1}^{9} i^2 + 3 \sum_{i=1}^{9} i - \sum_{i=1}^{9} 4$$

$$= \frac{9(9+1)[2(9)+1]}{6} + 3 \left[\frac{9(9+1)}{2} \right] - 9(4)$$

$$= 285 + 135 - 36$$

$$= 384$$

The display in Figure 8.9 confirms our work.   ◆

FIGURE 8.9

Typical display of the sum of the series in Example 4

```
"n²+3n-4"→Un
               Done
sum Un(1,9)
               384
```

PROBLEM 4   Use the methods of Example 4 to find the sum of the series $\sum_{i=1}^{20} (5i - 2)$. Verify the answer by using the sequence and sum features of a graphing calculator.   ◆

◆ Other Applications of Mathematical Induction

Some mathematical statements are *not* true for the first $(j - 1)$ positive integers, but *are* true for integers greater than or equal to j. To prove such statements for $n \geq j$, we can *extend* mathematical induction by showing that

1. the statement is true when $n = j$, and
2. the truth of the statement $n = k$ for $k \geq j$ implies the truth of the statement $n = k + 1$.

EXAMPLE 5 Use mathematical induction to prove that the sum of the interior angles of an n-sided convex polygon is $(n - 2)180°$ for all positive integers $n \geq 3$.

◆ **Solution**

Step 1 We must first show that the statement is true for $n = 3$:

$$(3 - 2)180° = 180° \quad \text{— sum of the interior angles of a triangle (a 3-sided polygon).}$$

Hence, the statement is true for $n = 3$.

Step 2 Assuming that the statement is true for $n = k$ for $k \geq 3$, we must show that, on the basis of that assumption, the statement is true for $n = k + 1$.

Assume: $(k - 2)180°$ is the sum of the interior angles of a k-sided convex polygon.

Prove: $[(k + 1) - 2]180°$ is the sum of the interior angles of a $(k + 1)$-sided convex polygon.

Proof: By attaching a triangle to a k-sided convex polygon, as shown in Figure 8.10, we form a $(k + 1)$-sided convex polygon. Hence, the sum of the interior angles of the $(k + 1)$-sided polygon is

sum of the interior angles of the k-sided convex polygon + sum of the interior angles of the triangle

$$(k - 2)180° \quad + \quad 180°$$

$$[(k - 2) + 1]180° \quad \text{Factor out } 180°$$

$$[(k + 1) - 2]180° \quad \text{Rewrite}$$

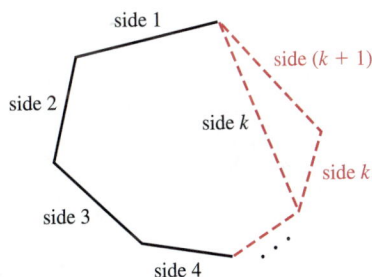

FIGURE 8.10
Attaching a triangle to a k-sided convex polygon forms a $(k + 1)$-sided convex polygon.

Thus, the truth of the statement when $n = k$ for $k \geq 3$ implies the truth of the statement when $n = k + 1$. Therefore, by mathematical induction, we can state that the sum of the interior angles of an n-sided convex polygon is $(n - 2)180°$ for all positive integers $n \geq 3$. ◆

PROBLEM 5 Use the result of Example 5 to determine the sum of the interior angles of a 12-sided convex polygon. ◆

Exercises 8.2

Basic Skills

In Exercises 1–10, use direct computation to find the sum of each series. When possible, use the sequence and sum features of a graphing calculator to confirm each answer.

1. $\displaystyle\sum_{i=1}^{10} 15$

2. $\displaystyle\sum_{i=1}^{24} \frac{5}{6}$

3. $\displaystyle\sum_{i=1}^{100} \pi$

4. $\displaystyle\sum_{i=1}^{250} \sqrt{2}$

5. $\displaystyle\sum_{i=1}^{8} (2 - i)$

6. $\displaystyle\sum_{i=1}^{5} 3(-2)^i$

7. $\displaystyle\sum_{i=1}^{7} (2^i - 1)$

8. $\displaystyle\sum_{i=1}^{6} (i^2 + 1)$

9. $\displaystyle\sum_{i=1}^{5} \frac{(-1)^{i+1}\, 2^i}{i}$

10. $\displaystyle\sum_{i=1}^{4} \frac{(-1)^i\, i^2}{i + 1}$

In Exercises 11–28, prove each statement by using mathematical induction.

11. $2 + 6 + 10 + 14 + \cdots + (4n - 2) = 2n^2$

12. $1 + 4 + 7 + 10 + \cdots + (3n - 2) = \dfrac{n(3n - 1)}{2}$

13. $1^2 + 2^2 + 3^2 + \cdots + n^2 = \dfrac{n(n + 1)(2n + 1)}{6}$

14. $1^3 + 2^3 + 3^3 + \cdots + n^3 = \left[\dfrac{n(n + 1)}{2}\right]^2$

15. $1^2 + 3^2 + 5^2 + 7^2 + \cdots + (2n - 1)^2$
    $= \dfrac{n(2n - 1)(2n + 1)}{3}$

16. $2^2 + 4^2 + 6^2 + 8^2 + \cdots + (2n)^2$
    $= \dfrac{2n(n + 1)(2n + 1)}{3}$

17. $2 \cdot 4 + 4 \cdot 6 + 6 \cdot 8 + 8 \cdot 10 + \cdots + 2n(2n + 2)$
    $= \dfrac{4n(n + 1)(n + 2)}{3}$

18. $1 \cdot 3 + 2 \cdot 5 + 3 \cdot 7 + 4 \cdot 9 + \cdots + n(2n + 1)$
    $= \dfrac{n(n + 1)(4n + 5)}{6}$

19. $\dfrac{1}{1 \cdot 2} + \dfrac{1}{2 \cdot 3} + \dfrac{1}{3 \cdot 4} + \dfrac{1}{4 \cdot 5} + \cdots + \dfrac{1}{n(n + 1)}$
    $= \dfrac{n}{n + 1}$

20. $\dfrac{1}{1 \cdot 3} + \dfrac{1}{3 \cdot 5} + \dfrac{1}{5 \cdot 7} + \cdots + \dfrac{1}{(2n - 1)(2n + 1)}$
    $= \dfrac{n}{2n + 1}$

21. $2 + 2^2 + 2^3 + 2^4 + \cdots + 2^n = 2(2^n - 1)$

22. $3 + 3^2 + 3^3 + 3^4 + \cdots + 3^n = \frac{3}{2}(3^n - 1)$

23. $4 + 4^2 + 4^3 + 4^4 + \cdots + 4^n = \frac{4}{3}(4^n - 1)$

24. $5 + 5^2 + 5^3 + 5^4 + \cdots + 5^n = \frac{5}{4}(5^n - 1)$

25. $1 + 2 \cdot 2 + 3 \cdot 2^2 + 4 \cdot 2^3 + \cdots + n \cdot 2^{n-1}$
    $= 1 + (n - 1)2^n$

26. $1 + 2 \cdot 3 + 3 \cdot 3^2 + 4 \cdot 3^3 + \cdots + n \cdot 3^{n-1}$
    $= \dfrac{1 + (2n - 1)3^n}{4}$

27. $\dfrac{1}{2} + \dfrac{1}{4} + \dfrac{1}{8} + \dfrac{1}{16} + \cdots + \dfrac{1}{2^n} = 1 - \dfrac{1}{2^n}$

28. $\dfrac{3}{2} + \dfrac{5}{2^2} + \dfrac{7}{2^3} + \dfrac{9}{2^4} + \cdots + \dfrac{2n + 1}{2^n}$
    $= 5 - \dfrac{2n + 5}{2^n}$

In Exercises 29–46, use the formulas of Exercises 11–28 to find the sum of the given series. When possible, use the sequence and sum features of a graphing calculator to confirm your answer.

29. $\displaystyle\sum_{i=1}^{20} (4i - 2)$

30. $\displaystyle\sum_{i=1}^{26} (3i - 2)$

31. $\displaystyle\sum_{i=1}^{36} i^2$

32. $\displaystyle\sum_{i=1}^{24} i^3$

33. $\displaystyle\sum_{i=1}^{25} (2i - 1)^2$

34. $\displaystyle\sum_{i=1}^{18} (2i)^2$

35. $\displaystyle\sum_{i=1}^{15} 2i(2i + 2)$

36. $\displaystyle\sum_{i=1}^{18} i(2i + 1)$

37. $\displaystyle\sum_{i=1}^{499} \frac{1}{i(i + 1)}$

38. $\displaystyle\sum_{i=1}^{100} \frac{1}{(2i - 1)(2i + 1)}$

39. $\displaystyle\sum_{i=1}^{10} 2^i$

40. $\displaystyle\sum_{i=1}^{8} 3^i$

41. $\displaystyle\sum_{i=1}^{6} 4^i$

42. $\displaystyle\sum_{i=1}^{5} 5^i$

43. $\displaystyle\sum_{i=1}^{12} i \cdot 2^{i-1}$

44. $\displaystyle\sum_{i=1}^{9} i \cdot 3^{i-1}$

45. $\displaystyle\sum_{i=1}^{8} \frac{1}{2^i}$

46. $\displaystyle\sum_{i=1}^{10} \frac{2i+1}{2^i}$

In Exercises 47–54, use the summation properties and the formulas for the sums of powers of integers to find the sum of the given series. When possible, use the sequence and sum features of a graphing calculator to confirm your answer.

47. $\displaystyle\sum_{i=1}^{50} 3i$

48. $\displaystyle\sum_{i=1}^{36} 5i^2$

49. $\displaystyle\sum_{i=1}^{24} (2i^2 - 5)$

50. $\displaystyle\sum_{i=1}^{10} (4 - i^3)$

51. $\displaystyle\sum_{i=1}^{20} i^2(i - 3)$

52. $\displaystyle\sum_{i=1}^{30} i(i + 4)$

53. $\displaystyle\sum_{i=1}^{18} (2i^2 - i + 1)$

54. $\displaystyle\sum_{i=1}^{12} (3i^3 - 2i - 6)$

In Exercises 55–58, use the sequence and sum features of a graphing calculator to find the sum of each series.

55. $\displaystyle\sum_{i=1}^{84} (9.75 - 1.02i)$

56. $\displaystyle\sum_{i=1}^{75} 0.04(5 + 0.08i)$

57. $\displaystyle\sum_{i=1}^{42} (3.6i^2 - 5.6i + 8.9)$

58. $\displaystyle\sum_{i=1}^{25} (2.78i^3 - 4.32i)$

The series in Exercises 59 and 60 are used in calculus to approximate the area of the shaded region that is shown in the figure. Use the sequence and sum features of a graphing calculator to find the shaded area to the nearest tenth of a square unit.

59. $\displaystyle\sum_{i=1}^{50} 0.04[4 - (0.04i)^2]$

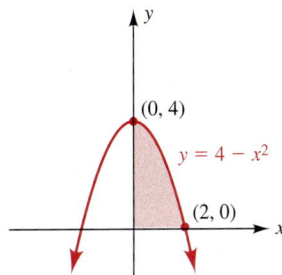

60. $\displaystyle\sum_{i=1}^{50} 0.04[8 - (0.04i)^3]$

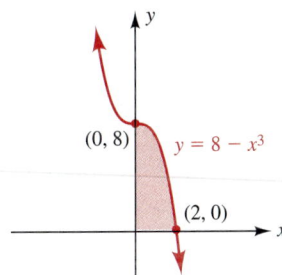

Critical Thinking

61. Find the sum of each series, and express the answer as a single logarithm.

    (a) $\displaystyle\sum_{i=1}^{5} \log i$ (b) $\displaystyle\sum_{i=1}^{5} \log \frac{i+1}{i}$

In Exercises 62–64, prove each summation property by writing it in expanded form and then applying the properties of real numbers.

62. $\displaystyle\sum_{i=1}^{n} ca_i = c \sum_{i=1}^{n} a_i$

63. $\displaystyle\sum_{i=1}^{n} (a_i + b_i) = \sum_{i=1}^{n} a_i + \sum_{i=1}^{n} b_i$

64. $\displaystyle\sum_{i=1}^{n} (a_i - b_i) = \sum_{i=1}^{n} a_i - \sum_{i=1}^{n} b_i$

In Exercises 65–67,

    (a) *use pattern recognition to state the general term of the series.*

    (b) *find the sum of the first 20 terms of the series by using the summation properties and the formulas for the sums of powers of integers.*

65. $1 \cdot 2 + 3 \cdot 4 + 5 \cdot 6 + 7 \cdot 8 + \cdots$

66. $1 \cdot 3 + 3 \cdot 5 + 5 \cdot 7 + 7 \cdot 9 + \cdots$

67. $1 \cdot 2 \cdot 3 + 4 \cdot 5 \cdot 6 + 7 \cdot 8 \cdot 9 + 10 \cdot 11 \cdot 12 + \cdots$

68. On the basis of the results from Exercises 21–24, make a conjecture about the sum of the series

$$\sum_{i=1}^{n} r^i \quad \text{for } r \neq 1.$$

Then prove it by using mathematical induction.

In Exercises 69–76, use mathematical induction to prove the given statement.

69. A regular convex polygon with n sides ($n \geq 4$) has $\frac{1}{2}n(n-3)$ diagonals.

70. If n people are in a room ($n \geq 2$) and every person shakes hands with every other person, then $\frac{1}{2}n(n-1)$ handshakes take place.

71. $x - 1$ is a factor of $x^n - 1$ for all positive integers n. [*Hint:* For step 2, add and subtract x.]

72. $a^n - b^n$ is divisible by $a - b$ for all positive integers n. [*Hint:* For step 2, add and subtract ab^k.]

73. The product of any two consecutive, positive integers n and $n + 1$ is an even number.

74. The sum of cubes of any three consecutive, positive integers n, $n + 1$, and $n + 2$ is a multiple of 9.

75. $2^n > 2n$ for all positive integers $n \geq 3$.

76. $2^n > n^2$ for all positive integers $n \geq 5$.

8.3 Arithmetic Sequences and Series

◆ **Introductory Comments**

Consider the sequence

$$3, 7, 11, 15, 19, 23, \ldots$$

Each element after the first is obtained by adding 4 to the preceding element. A sequence in which every element after the first is obtained by *adding* a fixed number d to the preceding element is called an *arithmetic sequence*.

◀ **Arithmetic Sequence**

> The sequence $a_1, a_2, a_3, \ldots, a_n, \ldots$ is called an **arithmetic sequence** if there exists a fixed number d such that, for all positive integers k,
>
> $$a_{k+1} = a_k + d.$$

It follows from the definition of an arithmetic sequence that the difference between any two successive elements must be the constant d, that is,

$$a_{k+1} - a_k = d.$$

We refer to the constant d as the **common difference** of the sequence.

If the first element and the common difference of an arithmetic sequence are known, then we can generate the sequence. For example, if the first element is $a_1 = -5$ and the common difference is $d = 3$, then the first five elements of this arithmetic sequence are

$$-5, -2, 1, 4, 7$$

In this section we study arithmetic sequences and series and their applications.

◆ **The General Element of an Arithmetic Sequence**

Our definition of an arithmetic sequence allows us to define the general element a_n in terms of the element that precedes it, namely, a_{n-1}. Thus, we can define the general element a_n by the recursive formula

$$a_n = a_{n-1} + d$$

where d is a fixed number (the common difference). We now develop a formula that defines the general element a_n *explicitly*. If the first element in an arithmetic sequence is a_1 and the common difference is d, then the first six elements of the sequence are as follows:

$$
\begin{aligned}
\text{First element:} \quad & a_1 \\
\text{Second element:} \quad & a_2 = a_1 + d \\
\text{Third element:} \quad & a_3 = a_2 + d = (a_1 + d) + d = a_1 + 2d \\
\text{Fourth element:} \quad & a_4 = a_3 + d = (a_1 + 2d) + d = a_1 + 3d \\
\text{Fifth element:} \quad & a_5 = a_4 + d = (a_1 + 3d) + d = a_1 + 4d \\
\text{Sixth element:} \quad & a_6 = a_5 + d = (a_1 + 4d) + d = a_1 + 5d
\end{aligned}
$$

Do you see the pattern? Each element is the sum of a_1 and a multiple of d, where the coefficient of d is one less than the number of the element. Thus, it appears that the general element of an arithmetic sequence may be written as

$$a_n = a_1 + (n - 1)d.$$

We now verify this formula by using mathematical induction (see Section 8.2).

Step 1   We must first show that the statement is true for $n = 1$.

$$a_1 = a_1 + (1 - 1)d \quad ?$$

$$a_1 = a_1 \quad \text{is } true.$$

Step 2   Assuming that the statement is true for $n = k$, we must show that, on the basis of that assumption, the statement is true for $n = k + 1$.

Assume: $\qquad\qquad a_k = a_1 + (k - 1)d$

Prove: $\qquad\qquad a_{k+1} = a_1 + [(k + 1) - 1]d$

Proof: Beginning with the definition of an arithmetic sequence, we have

$$a_{k+1} = \quad a_k \quad + d$$

Rewrite using the *assumed* statement.

$$= [a_1 + (k - 1)d] + d$$

$$= a_1 + [(k - 1)d + d] \qquad \textbf{\textcolor{red}{Apply associative property}}$$

$$= a_1 + [(k - 1) + 1]d \qquad \textbf{\textcolor{red}{Factor}}$$

$$= a_1 + [(k + 1) - 1]d \qquad \textbf{\textcolor{red}{Apply commutative and associative properties}}$$

Thus, the truth of the statement when $n = k$ implies the truth of the statement when $n = k + 1$. Hence, by mathematical induction, we have the following formula.

◆◆ **General Element of an Arithmetic Sequence**

The **general element** of an arithmetic sequence with first term a_1 and common difference d is given by

$$a_n = a_1 + (n - 1)d.$$

EXAMPLE 1   Find the general element of the arithmetic sequence 23, 17, 11, 5, ...

◆ **Solution**   The first element of this sequence is $a_1 = 23$. We can find the common difference d by selecting any two successive elements and computing the difference $a_{k+1} - a_k$. Choosing the third and fourth elements, we find

$$d = a_4 - a_3 = 5 - 11 = -6.$$

Thus, the general element of this arithmetic sequence is

$$a_n = a_1 + (n - 1)d$$
$$= 23 + (n - 1)(-6)$$
$$= 29 - 6n$$

We can use the sequence features of a graphing calculator to confirm our answer. Figure 8.11 shows a typical display of the first four elements of a sequence with general element $a_n = 29 - 6n$. ◆

PROBLEM 1   Find the general element of the arithmetic sequence 3, 7, 11, 15, ... ◆

```
"29-6n"→Un
                    Done
Un(1,4)
        {23 17 11 5}
```

FIGURE 8.11

The first four elements of a sequence with general element $a_n = 29 - 6n$ are 23, 17, 11, 5.

Once the general element of an arithmetic sequence is known, we can find any element in the sequence. For example, to find the 25th element (a_{25}) of the arithmetic sequence given in Example 1, we simply replace n with 25 in the general element $a_n = 29 - 6n$ as follows:

$$a_{25} = 29 - 6(25) = -121.$$

We can also use the sequence features of a graphing calculator to find the 25th element, as shown in Figure 8.12.

FIGURE 8.12
The 25th element of the arithmetic sequence with general element $a_n = 29 - 6n$ is -121.

EXAMPLE 2 Find the 30th element a_{30} of an arithmetic sequence whose first element a_1 is 3 and whose common difference d is 4.

◆ **Solution** For this arithmetic sequence, $a_1 = 3$ and $d = 4$. Thus, the general element of the arithmetic sequence is

$$a_n = 3 + (n - 1)4 = -1 + 4n$$

Therefore, the 30th element is

$$a_{30} = -1 + 4(30) = 119.$$

The display in Figure 8.13 confirms our work. ◆

FIGURE 8.13
The 30th element of the arithmetic sequence with general element $a_n = -1 + 4n$ is 119.

PROBLEM 2 Find the first element a_1 of an arithmetic sequence whose 60th element a_{60} is -98 and whose common difference d is -4. ◆

EXAMPLE 3 Find the common difference d and the first element a_1 of an arithmetic sequence whose 5th element a_5 is 20 and whose 11th element a_{11} is 62.

◆ **Solution** Using the formula $a_n = a_1 + (n - 1)d$, we have

$$a_5 = a_1 + (5 - 1)d \qquad \text{and} \qquad a_{11} = a_1 + (11 - 1)d$$

$$20 = a_1 + 4d \qquad\qquad\qquad 62 = a_1 + 10d$$

To find d and a_1, we solve these equations simultaneously as follows:

$$
\begin{array}{lll}
a_1 + 4d = 20 & \xrightarrow{\text{Multiply by } -1} & -a_1 - 4d = -20 \\
a_1 + 10d = 62 & \xrightarrow{\text{No change}} & \underline{a_1 + 10d = 62} \\
 & & 6d = 42 \quad \textbf{Add and solve for } d \\
 & & d = 7
\end{array}
$$

Substituting $d = 7$ into the equation $a_1 + 4d = 20$, we find

$$a_1 + 4(7) = 20$$

$$a_1 = -8$$

Thus, the common difference is 7 and the first element is -8. ◆

PROBLEM 3 Find the general element of the arithmetic sequence described in Example 3. Then use the general element to verify that $a_5 = 20$ and $a_{11} = 62$. ◆

◆ **Arithmetic Series**

The indicated sum of the elements of an arithmetic sequence is called an **arithmetic series**. The arithmetic series associated with the finite arithmetic sequence

$$a_1, a_1 + d, a_1 + 2d, a_1 + 3d, \ldots, a_1 + (n - 1)d$$

can be written in expanded form as

$$a_1 + (a_1 + d) + (a_1 + 2d) + (a_1 + 3d) + \cdots + [a_1 + (n - 1)d]$$

or in sigma form as

$$\sum_{i=1}^{n} [a_1 + (i - 1)d].$$

We now develop a formula that we can use to find the sum of a finite arithmetic series. If we represent the sum of these n terms by S_n, then

$$S_n = a_1 + (a_1 + d) + (a_1 + 2d) + (a_1 + 3d) + \cdots + [a_1 + (n - 1)d].$$

Alternatively, we can find the sum by starting with the nth term a_n and subtracting the common difference d each time to get to the next term as follows:

$$S_n = a_n + (a_n - d) + (a_n - 2d) + (a_n - 3d) + \cdots + [a_n - (n - 1)d].$$

If we add these two versions of S_n, term by term, the multiples of d drop out and we obtain a formula for the sum of an arithmetic series:

$$
\begin{aligned}
S_n &= a_1 + (a_1 + d) + (a_1 + 2d) + (a_1 + 3d) + \cdots + [a_1 + (n - 1)d] \\
+\ S_n &= a_n + (a_n - d) + (a_n - 2d) + (a_n - 3d) + \cdots + [a_n - (n - 1)d] \\
\hline
2S_n &= (a_1 + a_n) + (a_1 + a_n) + (a_1 + a_n) + (a_1 + a_n) + \cdots + (a_1 + a_n)
\end{aligned}
$$

n terms

$$2S_n = n(a_1 + a_n)$$

$$S_n = n\left(\frac{a_1 + a_n}{2}\right)$$

number of terms

average value of the first and last terms

◄◆ **Sum of a Finite Arithmetic Series**

The **sum S_n of a finite arithmetic series** with n terms is given by

$$S_n = n\left(\frac{a_1 + a_n}{2}\right)$$

where a_1 is the first term and a_n is the last term in the series.

EXAMPLE 4 Find the sum of each arithmetic series.

(a) $\displaystyle\sum_{i=1}^{40} (4i - 3)$ **(b)** $3 + 9 + 15 + \cdots + 273$

◄◆ **Solution**

(a) Since the summation variable i starts at 1 and ends at 40, the number of terms in this series is $n = 40$. Writing this series in expanded form, we have

$$\sum_{i=1}^{40} (4i - 3) = 1 + 5 + 9 + \cdots + 157$$

This is an arithmetic series with common difference $d = 4$, first term $a_1 = 1$, and last term $a_{40} = 157$. The average value of the first and last terms is

$$\frac{a_1 + a_{40}}{2} = \frac{1 + 157}{2} = \frac{158}{2} = 79$$

Thus, the sum of the arithmetic series $\displaystyle\sum_{i=1}^{40} (4i - 3)$ is

$$S_{40} = 40(79) = 3160$$

number of terms ⎫   ⎫ average value of the first and last terms

The display in Figure 8.14 confirms our work.

(b) The first term of the arithmetic series $3 + 9 + 15 + \cdots + 273$ is $a_1 = 3$ and the common difference is $d = 6$. Hence, the general term of this series is

$$a_n = 3 + (n - 1)6 = 6n - 3$$

To find the number of terms n in this series, we let $a_n = 273$ and solve for n as follows:

$$a_n = 6n - 3$$
$$273 = 6n - 3$$
$$276 = 6n$$
$$n = 46$$

```
"4n-3"→Un
                Done
sum Un(1,40)
                3160
```

FIGURE 8.14

The sum of the first 40 terms of a series with general term $a_n = 4n - 3$ is 3160.

The average value of the first and last terms is

$$\frac{a_1 + a_{46}}{2} = \frac{3 + 273}{2} = \frac{276}{2} = 138$$

Thus, the sum of the arithmetic series $3 + 9 + 15 + \cdots + 273$ is

$$S_{46} = \underbrace{46}_{\substack{\text{number} \\ \text{of terms}}}(\underbrace{138}_{\substack{\text{average value of the} \\ \text{first and last terms}}}) = 6348$$

The display in Figure 8.15 supports our work.   ◆

If the summation variable i starts at a number other than 1, we must be certain to determine the correct number of terms in the series. For instance, the arithmetic series $\sum_{i=0}^{40}(4i - 3)$ has 41 terms, not 40.

PROBLEM 4   Find the sum of the arithmetic series $\sum_{i=0}^{40}(4i - 3)$.   ◆

FIGURE 8.15

The sum of the first 46 terms of a series with general term $a_n = 6n - 3$ is 6348.

If we replace a_n with $a_1 + (n - 1)d$ in the formula $S_n = n\left(\dfrac{a_1 + a_n}{2}\right)$, we obtain an alternative formula for the sum of a finite arithmetic series:

$$S_n = n\left(\frac{a_1 + [a_1 + (n - 1)d]}{2}\right) = \frac{n}{2}[2a_1 + (n - 1)d]$$

◆ **Sum of a Finite Arithmetic Series (Alternative Formula)**

The **sum S_n of a finite arithmetic series** with n terms is given by

$$S_n = \frac{n}{2}[2a_1 + (n - 1)d]$$

where d is the common difference and a_1 is the first term in the series.

EXAMPLE 5   Find the sum of the first 50 terms of an arithmetic series whose first term is 24 and whose second term is 20.

◆ **Solution**   The first term is $a_1 = 24$ and the second term is $a_2 = 20$. Thus, the common difference is

$$d = a_2 - a_1 = 20 - 24 = -4$$

and the general term of this arithmetic series is

$$a_n = 24 + (n - 1)(-4) = 28 - 4n$$

FIGURE 8.16
The sum of the first 50 terms of a series with general term $a_n = 28 - 4n$ is -3700.

Now, using the formula $S_n = \dfrac{n}{2}[2a_1 + (n-1)d]$, we find the sum of the first 50 terms:

$$S_{50} = \frac{50}{2}[2(24) + (50-1)(-4)]$$
$$= 25[48 + (49)(-4)]$$
$$= -3700$$

Hence, the sum of the first 50 terms of an arithmetic series with first term 24, and second term 20 is -3700.
The display in Figure 8.16 supports our work. ◆

PROBLEM 5 The first term a_1 of an arithmetic series is 8 and the sum of the first 10 terms is 215. Find the common difference d. ◆

◆ **Application: Installment Plan**

Arithmetic sequences and series are used in many types of applied problems. We conclude this section with an application concerning the payment of a debt by using an *installment plan* (a system for paying a debt by making payments at successive fixed times).

EXAMPLE 6 A stereo system is purchased for $1200 on an installment plan. The plan consists of 24 monthly payments in the amount of $50 plus $1\frac{1}{2}\%$ interest on the unpaid balance for that month.

(a) Find the amount of the first payment, the second payment, and the third payment.

(b) Find the general element of the sequence of payments.

(c) Find the total amount paid for the stereo system.

◆ **Solution**

(a) The first monthly payment a_1 is $50 plus $1\frac{1}{2}\%$ of $1200, that is,

$$a_1 = \$50 + 0.015(\$1200) = \$68.00$$

The unpaid balance after the first payment is $1200 - $50 = $1150. Thus, the second payment a_2 is $50 + $1\frac{1}{2}\%$ of $1150, or

$$a_2 = \$50 + 0.015(\$1150) = \$67.25$$

The unpaid balance after the second payment is $1150 - $50 = $1100. Thus, the third payment a_3 is $50 + $1\frac{1}{2}\%$ of $1100, or

$$a_3 = \$50 + 0.015(\$1100) = \$66.50$$

(b) The first three payments form the arithmetic sequence

$$\$68.00, \quad \$67.25, \quad \$66.50$$

The common difference d of this sequence is

$$d = 1\tfrac{1}{2}\% \text{ of } (-\$50) = -\$0.75.$$

Since $a_1 = \$68.00$ and $d = -\$0.75$, the general element (in dollars) is

$$a_n = 68.00 + (n-1)(-0.75) = 68.75 - 0.75n$$

(c) Since the plan consists of 24 monthly payments, the total amount paid (in dollars) for the stereo system is represented by the arithmetic series

$$\sum_{i=1}^{24} (68.75 - 0.75i)$$

Using the formula $S_n = \dfrac{n}{2}[2a_1 + (n-1)d]$ or the sequence and sum features of a graphing calculator (see Figure 8.17), we find that the total amount paid for the stereo system with this installment plan is \$1,425. ◆

```
"68.75-.75n"→Un
            Done
sum Un(1,24)
            1425
```

FIGURE 8.17

The sum of the first 24 terms of a series with general term $a_n = 68.75 - 0.75n$ is 1425.

PROBLEM 6 Find the amount of the 24th payment for the installment plan described in Example 6. Then use the formula $S_n = n\left(\dfrac{a_1 + a_n}{2}\right)$ to show that the total amount paid for this stereo system is \$1425. ◆

Exercises 8.3

◆ Basic Skills

In Exercises 1–6, write the first five elements of an arithmetic sequence whose first element a_1 and common difference d are given.

1. $a_1 = 3, \ d = 4$ **2.** $a_1 = -28, \ d = 12$

3. $a_1 = 2, \ d = -\tfrac{2}{3}$ **4.** $a_1 = -\tfrac{1}{2}, \ d = -\tfrac{3}{4}$

5. $a_1 = -1, \ d = \pi$ **6.** $a_1 = \sqrt{2}, \ d = -\sqrt{2}$

In Exercises 7–16, find the general element of the given arithmetic sequence. When possible, use the sequence features of a graphing calculator to verify your answer.

7. $3, 9, 15, 21, \ldots$ **8.** $1, 8, 15, 22, \ldots$

9. $11, 4, -3, -10, \ldots$

10. $-2, -7, -12, -17, \ldots$

11. $-\tfrac{2}{3}, 1, \tfrac{8}{3}, \tfrac{13}{3}, \ldots$ **12.** $\tfrac{3}{4}, -2, -\tfrac{19}{4}, -\tfrac{15}{2}, \ldots$

13. $\dfrac{\pi}{6}, -\dfrac{\pi}{6}, -\dfrac{\pi}{2}, -\dfrac{5\pi}{6}, \ldots$

14. $\dfrac{\sqrt{3}}{2}, 2\sqrt{3}, \dfrac{7\sqrt{3}}{2}, 5\sqrt{3}, \ldots$

15. $x - 4, x - 2, x, x + 2, \ldots$

16. $x, y, 2m - x, 3y - 2x, \ldots$

In Exercises 17–26, information is given about an arithmetic sequence $a_1, a_2, a_3, \ldots, a_n$ with common difference d. Find the indicated unknown.

17. $a_1 = 4, \ d = 7; \ a_{28} = ?$

18. $a_1 = 12$, $d = -9$; $a_{32} = ?$

19. $a_{82} = 17$, $d = 4$; $a_1 = ?$

20. $a_{31} = -5$, $d = -\frac{2}{5}$; $a_1 = ?$

21. $a_2 = 9$, $a_{14} = 3$; $d = ?$

22. $a_{13} = 46$, $a_4 = 100$; $a_1 = ?$

23. $a_6 = 2.25$, $a_{16} = 6.25$; $a_{80} = ?$

24. $a_5 = 72$, $a_{21} = -8$; $a_{62} = ?$

25. $a_{19} - a_5 = -36$; $d = ?$

26. $a_5 - a_{24} = 9$; $d = ?$

In Exercises 27–36, find the sum of the arithmetic series. When possible, use the sequence and sum features of a graphing calculator to verify your answer.

27. $\displaystyle\sum_{i=1}^{35} 4i$

28. $\displaystyle\sum_{i=1}^{18} (2i - 5)$

29. $\displaystyle\sum_{i=1}^{58} (7 - 5i)$

30. $\displaystyle\sum_{i=1}^{100} \frac{3i + 2}{2}$

31. $\displaystyle\sum_{i=0}^{25} \left(\frac{i}{2} - 1 \right)$

32. $\displaystyle\sum_{i=3}^{30} (0.6i + 0.8)$

33. $8 + 11 + 14 + 17 + \cdots + 170$

34. $-5 + 1 + 7 + 13 + \cdots + 283$

35. $2.4 + 1.8 + 1.2 + 0.6 + \cdots + (-18)$

36. $\frac{3}{2} + \frac{9}{4} + 3 + \frac{15}{4} + \cdots + 78$

In Exercises 37–46, information is given about an arithmetic series $a_1 + a_2 + a_3 + \cdots + a_n$ with sum S_n and common difference d. Find the indicated unknown.

37. $a_1 = 12$, $d = 5$; $S_{31} = ?$

38. $a_1 = \frac{5}{4}$, $d = -\frac{1}{2}$; $S_{10} = ?$

39. $S_{32} = 448$, $a_1 = 2$; $a_{32} = ?$

40. $S_{28} = -98$, $a_{28} = -181$; $a_1 = ?$

41. $S_{54} = 11{,}583$, $a_1 = 2.5$; $d = ?$

42. $S_{18} = -45$, $d = -9$; $a_1 = ?$

43. $a_1 = -9$, $a_{40} = 147$; $S_{62} = ?$

44. $a_5 = 15$, $a_{19} = 43$; $S_{101} = ?$

45. $S_{24} = 432$, $a_1 = -28$; $a_{33} = ?$

46. $S_{65} = -1560$, $d = -3$; $a_{23} = ?$

In Exercises 47 and 48, use the sequence features of a graphing calculator to find the 79th element of the given arithmetic sequence.

47. -8.32, -6.36, -4.40, -2.44, $\ldots$

48. 837.6, 736.4, 635.2, 534.0, $\ldots$

In Exercises 49–52, use the sequence and sum features of a graphing calculator to find the sum of each arithmetic series.

49. $\displaystyle\sum_{i=1}^{35} (2.78i - 19.4)$

50. $\displaystyle\sum_{i=2}^{54} (437.3 - 17.6i)$

51. $0.0824 + 0.0044 + (-0.0736) + \cdots + (-1.5556)$

52. $1176.2 + 1989.1 + 2802.0 + \cdots + 22{,}311.6$

53. Find the sum of the first 50 even, positive integers.

54. Find the sum of the first 80 odd, positive integers.

55. Find the number of multiples of three between 100 and 599.

56. Find the sum of the integers between 1 and 1000 whose last digit is 9.

57. While making out his will, a man decides to leave $300,000 to his six children in such a way that each child receives $10,000 less than the next oldest child. How much is willed to the oldest child?

58. To drill an artesian well, a company charges $5 per foot until it reaches bedrock and, for each foot thereafter, 10¢ more than the preceding foot. How much does it cost to drill a well 350 feet deep if bedrock is reached at 150 feet?

59. An auditorium has 32 seats in the first row, and each row thereafter has 4 more seats than the preceding row. If the auditorium has 25 rows, find the number of seats

(a) in the last row.   (b) in the auditorium.

60. An object falling freely from a position of rest travels 16 ft during the first second, 48 ft during the second second, 80 ft during the third second, and so on.

(a) Find the distance fallen during the tenth second.

(b) Find the total distance fallen after 10 seconds.

61. A sailboat costing $36,000 is purchased under an agreement to pay $600 per month for 60 months plus 1.25% interest on the unpaid balance for that month.

(a) Find the amount of the first payment, the second payment, and the third payment.

(b) Find the general element of the sequence of payments.

(c) Find the total amount paid for the sailboat.

62. A woman buys a new automobile and agrees to pay $10,929.60 in 24 monthly installments that form an arithmetic sequence. After making the 12th payment, one-third of the debt remains unpaid.

(a) Find the amount of the first payment.

(b) Find the amount of the 24th payment.

◆ Critical Thinking

63. Is the sequence $\ln a$, $\ln a^2$, $\ln a^3$, $\ln a^4$, . . . an arithmetic sequence? If it is an arithmetic sequence, state the common difference.

64. Write the first five elements of the sequence defined recursively by

$$a_n = \begin{cases} 4 & \text{if } n = 1 \\ \dfrac{2a_{n-1} - 1}{2} & \text{if } n \geq 2 \end{cases}$$

Is this an arithmetic sequence? Express a_n explicitly in terms of n.

65. Find x if 10^{-2}, x, 10^{-4} are the first three elements of an arithmetic sequence.

66. Find x if $5x + 3$, $x + 4$, $2x - 5$ are the first three elements of an arithmetic sequence.

67. If the numbers a_1, a_2, a_3, . . . , a_{n-1}, a_n form an arithmetic sequence, then a_2, a_3, . . . , a_{n-1} are called the $(n - 2)$ *arithmetic means* between a_1 and a_n. Insert four arithmetic means between 19 and 32.

68. The sum of the first four terms of an arithmetic series is 80. The fourth term is ten more than twice the second term. Find the first term and the common difference for this series.

69. A person borrows \$20,000 from a loan shark and agrees to repay \$34,800 by making payments of \$300 the first month and, for each month thereafter, \$100 more than the preceding month.

    (a) How many payments are required to repay this loan?

    (b) How much is the last payment?

70. A contractor has agreed to build an in-ground swimming pool in 20 days. If he does not complete the project on time, he must forfeit \$150 for the first day beyond the given time allotment and, for each additional day thereafter, he must forfeit \$20 more than the preceding day. How many days are required to put in the swimming pool if the contractor forfeits \$1760?

8.4 Geometric Sequences and Series

◆ Introductory Comments

Consider the sequence

$$2, \ 6, \ 18, \ 54, \ 162, \ 486, \ . \ . \ .$$

Each element after the first is obtained by multiplying the preceding element by 3. A sequence in which every element after the first is obtained by *multiplying* the preceding element by a fixed, nonzero number r is called a *geometric sequence*.

◆ **Geometric Sequence**

> The sequence a_1, a_2, a_3, . . . , a_n, . . . is called a **geometric sequence** if there exists a fixed, nonzero number r such that for all positive integers k
>
> $$a_{k+1} = a_k r.$$

It follows from the definition of a geometric sequence that the ratio of any two successive elements must be the constant r, that is,

$$\frac{a_{k+1}}{a_k} = r$$

We refer to the constant r as the **common ratio** of the sequence.

If the first element and the common ratio of a geometric sequence are known, then we can generate the sequence. For example, if the first element is $a_1 = 8$, and the common ratio is $-\frac{1}{2}$, then the first five elements of this geometric sequence are

$$8, \; -4, \; 2, \; -1, \; \tfrac{1}{2}$$

In this section we study geometric sequences and series and their applications.

◆ The General Element of a Geometric Sequence

Our definition of a geometric sequence allows us to define the general element a_n in terms of the element that precedes it, namely, a_{n-1}. That is, we can define the general element a_n by the recursive formula

$$a_n = a_{n-1}r$$

where r is a fixed number (the common ratio). We now proceed to find a formula that defines the general element a_n *explicitly*. If the first element in a geometric sequence is a_1 and the common ratio is r, then the first six elements of the sequence are as follows:

First element: a_1

Second element: $a_2 = a_1 r$

Third element: $a_3 = a_2 r = (a_1 r)r = a_1 r^2$

Fourth element: $a_4 = a_3 r = (a_1 r^2)r = a_1 r^3$

Fifth element: $a_5 = a_4 r = (a_1 r^3)r = a_1 r^4$

Sixth element: $a_6 = a_5 r = (a_1 r^4)r = a_1 r^5$

Do you see the pattern? Each element is the product of a_1 and a power of r, where the exponent of r is one less than the number of the element. Thus, it appears that the general element of a geometric sequence can be written as

$$a_n = a_1 r^{n-1}$$

We now verify this formula by using mathematical induction (see Section 8.2).

Step 1   We must first show that the statement is true for $n = 1$.

$$a_1 = a_1 r^{1-1} \quad ?$$

$$a_1 = a_1 r^0 \quad ?$$

$$a_1 = a_1 \quad \text{is } true$$

Step 2   Assuming that the statement is true for $n = k$, we must show that, on the basis of that assumption, the statement is true for $n = k + 1$.

Assume: $a_k = a_1 r^{k-1}$

Prove: $a_{k+1} = a_1 r^{(k+1)-1}$

Proof: Beginning with the definition of a geometric sequence, we have

$$a_{k+1} = \qquad\qquad a_k \, r$$

Rewrite using the *assumed* statement.

$$= [a_1 \, r^{k-1}] \, r$$
$$= a_1 \, r^{(k-1)+1} \qquad \textbf{Add exponents}$$
$$= a_1 \, r^{(k+1)-1} \qquad \textbf{Apply commutative and associative properties}$$

Thus, the truth of the statement when $n = k$ implies the truth of the statement when $n = k + 1$. Hence, by mathematical induction, we have the following formula.

◆ **General Element of a Geometric Sequence**

The **general element** of a geometric sequence with first element a_1 and common ratio r is given by

$$a_n = a_1 r^{n-1}.$$

EXAMPLE 1   Find the general element of the geometric sequence $2, \, -3, \, \frac{9}{2}, \, -\frac{27}{4}, \, \ldots$.

◆ **Solution**   The first element of this sequence is $a_1 = 2$. We can find the common ratio r by selecting any two successive elements and computing the ratio $\frac{a_{k+1}}{a_k}$. Choosing the first and second elements, we find

$$r = \frac{a_2}{a_1} = \frac{-3}{2} = -\frac{3}{2}$$

Thus, the general element of this geometric sequence is

$$a_n = a_1 r^{n-1}$$
$$= 2\left(-\tfrac{3}{2}\right)^{n-1}$$

We can use the sequence features of a graphing calculator to confirm our answer. Figure 8.18 shows a typical display of the first four elements of a sequence with general element $a_n = 2\left(-\frac{3}{2}\right)^{n-1}$. ◆

FIGURE 8.18

The first four elements of a sequence with general element $a_n = 2\left(-\frac{3}{2}\right)^{n-1}$ are $2, -3, \frac{9}{2}, -\frac{27}{4}$.

PROBLEM 1 Find the general element of the geometric sequence 5, 15, 45, 135, ...

Once the general element of a geometric sequence is known, we can find any element in the sequence. For example, to find the tenth element a_{10} of the geometric sequence in Example 1, we simply replace n with 10 in the general element $a_n = 2\left(-\frac{3}{2}\right)^{n-1}$, as follows:

$$a_{10} = 2\left(-\frac{3}{2}\right)^{10-1} = \frac{(-3)^9}{2^8} = -\frac{19{,}683}{256}$$

We can also use the sequence features of a graphing calculator to find the tenth element, as shown in Figure 8.19.

EXAMPLE 2 Find the 13th element a_{13} of a geometric sequence whose first element a_1 is 3 and whose common ratio r is 4.

◆ **Solution** For this geometric sequence, $a_1 = 3$ and $r = 4$. Thus, the general element of the geometric sequence is

$$a_n = 3(4)^{n-1}.$$

Therefore, the 13th element is

$$a_{13} = 3(4)^{13-1} = 50{,}331{,}648.$$

The display in Figure 8.20 confirms our work.

FIGURE 8.19
The tenth element of the geometric sequence with general element $a_n = 2\left(-\frac{3}{2}\right)^{n-1}$ is $-\frac{19683}{256}$.

PROBLEM 2 Find the first element a_1 of a geometric sequence whose ninth element a_9 is $\frac{1}{81}$ and whose common ratio r is $\frac{1}{3}$.

EXAMPLE 3 Find the common ratio r and the first element a_1 of a geometric sequence if the third element a_3 is 24 and the sixth element a_6 is 3.

◆ **Solution** Using the formula $a_n = a_1 r^{n-1}$, we have

$$a_3 = a_1 r^{3-1} \quad \text{and} \quad a_6 = a_1 r^{6-1}$$
$$24 = a_1 r^2 \qquad\qquad 3 = a_1 r^5$$

FIGURE 8.20
The 13th element of the geometric sequence with general element $a_n = 3(4)^{n-1}$ is 50,331,648.

To find a_1 and r, we solve the equations simultaneously by using substitution. Solving for a_1 in the equation $24 = a_1 r^2$, we obtain $a_1 = 24/r^2$. Now, substituting $24/r^2$ for a_1 in the other equation, $3 = a_1 r^5$, we obtain

$$3 = \left(\frac{24}{r^2}\right) r^5$$
$$3 = 24r^3$$
$$\tfrac{1}{8} = r^3$$
$$r = \tfrac{1}{2}$$

Substituting $r = \frac{1}{2}$ into the equation $3 = a_1 r^5$, we find

$$3 = a_1 \left(\tfrac{1}{2}\right)^5$$

$$3 = \frac{a_1}{32}$$

$$a_1 = 96$$

Thus, the common ratio is $\frac{1}{2}$ and the first element is 96.   ◆

PROBLEM 3   Find the general element of the geometric sequence described in Example 3. Then use the general element to verify that $a_3 = 24$ and $a_6 = 3$.   ◆

◆ Geometric Series

The indicated sum of the elements of a geometric sequence is called a **geometric series**. The geometric series associated with the finite geometric sequence

$$a_1, \ a_1 r, \ a_1 r^2, \ a_1 r^3, \ \ldots, \ a_1 r^{n-1}$$

can be written in expanded form as

$$a_1 + a_1 r + a_1 r^2 + a_1 r^3 + \cdots + a_1 r^{n-1}$$

or in sigma form as

$$\sum_{i=1}^{n} a_1 r^{i-1}.$$

We now develop a formula that can be used to find the sum of a finite geometric series. If we represent the sum of these n terms by S_n, then

$$S_n = a_1 + a_1 r + a_1 r^2 + a_1 r^3 + \cdots + a_1 r^{n-2} + a_1 r^{n-1}.$$

Multiplying both sides of this equation by $-r$, we obtain

$$-rS_n = -a_1 r - a_1 r^2 - a_1 r^3 - a_1 r^4 - \cdots - a_1 r^{n-1} - a_1 r^n.$$

If we add these two equations, term by term, all terms on the right-hand side drop out except for a_1 and $-a_1 r^n$, and we obtain a formula for the sum of a finite geometric series:

$$
\begin{aligned}
S_n &= a_1 + a_1 r + a_1 r^2 + a_1 r^3 + \cdots + a_1 r^{n-2} + a_1 r^{n-1} \\
+ \quad -rS_n &= \quad\;\; - a_1 r - a_1 r^2 - a_1 r^3 - a_1 r^4 - \cdots - a_1 r^{n-1} - a_1 r^n \\
\hline
S_n - rS_n &= a_1 \qquad\qquad\qquad\qquad\qquad\qquad\qquad\qquad - a_1 r^n
\end{aligned}
$$

$$S_n(1 - r) = a_1(1 - r^n) \qquad \textcolor{red}{\text{Factor both sides}}$$

$$S_n = \frac{a_1(1 - r^n)}{1 - r} \quad \text{for } r \neq 1 \qquad \textcolor{red}{\text{Divide both sides by } (1 - r)}$$

◆ **Sum of a Finite Geometric Series**

The **sum S_n of a finite geometric series** with a common ratio r, provided that $r \neq 1$, is given by

$$S_n = \frac{a_1(1 - r^n)}{1 - r}$$

where a_1 is the first term and n is the number of terms in the series.

Note: If the common ratio of a finite geometric series is $r = 1$, then the sum of the series $\sum_{i=1}^{n} a_1 r^{i-1}$ is

$$a_1 + a_1 + a_1 + \cdots + a_1 = na_1.$$

n terms

EXAMPLE 4 Find the sum of each finite geometric series.

(a) $\displaystyle\sum_{i=1}^{10} 5(-2)^{i-1}$ **(b)** $\displaystyle\sum_{i=0}^{8} \left(\frac{1}{2}\right)^i$

◆ **Solution**

(a) Since the summation variable i starts at 1 and ends at 10, the number of terms in this series is $n = 10$. Writing this series in expanded form, we have

$$\sum_{i=1}^{10} 5(-2)^{i-1} = 5 - 10 + 20 - 40 + \cdots - 2560.$$

This is a geometric series with common ratio $r = -2$ and first term $a_1 = 5$. Thus, the sum of this geometric series is

$$S_{10} = \frac{a_1(1 - r^{10})}{1 - r}$$

$$= \frac{5[1 - (-2)^{10}]}{1 - (-2)} = \frac{5(-1023)}{3} = -1705$$

The display in Figure 8.21 confirms our work.

(b) For this series the summation variable i ranges from 0 to 8, so the number of terms in this series is $n = 9$. Writing this series in expanded form, we have

$$\sum_{i=0}^{8} \left(\frac{1}{2}\right)^i = 1 + \frac{1}{2} + \frac{1}{4} + \cdots + \frac{1}{256}$$

```
"5(-2)^(n-1)"→Un
                Done
sum Un(1,10)
            -1705
```

FIGURE 8.21

The sum of the first ten terms of a series with general term $a_n = 5(-2)^{n-1}$ is -1705.

This is a geometric series with common ratio $r = \frac{1}{2}$ and first term $a_1 = 1$. Thus, the sum of this geometric series is

$$S_9 = \frac{a_1(1 - r^9)}{1 - r}$$

$$= \frac{1[1 - (\frac{1}{2})^9]}{1 - \frac{1}{2}} = \frac{\frac{511}{512}}{\frac{1}{2}} = \frac{511}{256}$$

The display in Figure 8.22 confirms our work.   ◆

FIGURE 8.22

The sum of the first nine terms of a series with general term $a_n = (\frac{1}{2})^n$ is $\frac{511}{256}$.

PROBLEM 4 Find the first term a_1 of a geometric series if the sum of the first 12 terms is 8190 and the common ratio r is 2.   ◆

The formula for the sum of a finite geometric series can be rewritten to obtain an alternative formula:

$$S_n = \frac{a_1(1 - r^n)}{1 - r} = \frac{a_1 - a_1 r^n}{1 - r} = \frac{a_1 - r(a_1 r^{n-1})}{1 - r} \quad \text{Rewrite the numerator}$$

$$= \frac{a_1 - r a_n}{1 - r} \quad \text{Replace } a_1 r^{n-1} \text{ with } a_n$$

◆ **Sum of a Finite Geometric Series (Alternative Formula)**

The **sum S_n of a finite geometric series** with a common ratio r, provided that $r \neq 1$, is given by

$$S_n = \frac{a_1 - r a_n}{1 - r}$$

where a_1 is the first term and a_n is the last term in the series.

EXAMPLE 5 Find the sum of the geometric series:
$\frac{2}{3} + 2 + 6 + \cdots + 1458$

◆ **Solution**   The common ratio for this geometric series is $r = 3$. The first term is $a_1 = \frac{2}{3}$ and the last term is $a_n = 1458$. Using the alternative formula for the sum of a finite geometric series, we have

$$S_n = \frac{a_1 - r a_n}{1 - r}$$

$$= \frac{\frac{2}{3} - 3(1458)}{1 - 3}$$

$$= \frac{\frac{2}{3}}{-2} - \frac{3(1458)}{-2}$$

$$= -\frac{1}{3} + 2187$$

$$= \frac{6560}{3}$$

We can find the number of terms n in the series

$$\tfrac{2}{3} + 2 + 6 + \cdots + 1458$$

by using the formula $a_n = a_1 r^{n-1}$ with $a_1 = \tfrac{2}{3}$, $a_n = 1458$, and $r = 3$. The equation becomes

$$1458 = \tfrac{2}{3}(3)^{n-1} \qquad \text{or} \qquad 3^{n-1} = 2187.$$

To solve this exponential equation (Section 5.4) for n, we begin by taking the natural logarithm of both sides as follows:

$$\ln 3^{n-1} = \ln 2187$$
$$(n-1)\ln 3 = \ln 2187 \qquad \text{**Apply log property 3**}$$
$$n = 1 + \frac{\ln 2187}{\ln 3} = 8 \qquad \text{**Solve for } n\text{**}$$

Hence, the geometric series $\tfrac{2}{3} + 2 + 6 + \cdots + 1458$ has eight terms and may be written in sigma form as follows:

$$\sum_{i=1}^{8} \frac{2}{3}(3)^{n-1}$$

The display in Figure 8.23 confirms that the sum of the series in Example 5 is $\tfrac{6560}{3}$.

```
"(2/3)(3)^(n-1)"
→Un
              Done
sum Un(1,8)▶Frac
          6560/3
```

FIGURE 8.23

The sum of the first eight terms of the series with general term $a_n = \tfrac{2}{3}(3)^{n-1}$ is $\tfrac{6560}{3}$.

PROBLEM 5 Find the number of terms in the geometric series

$3 + 12 + 48 + \cdots + 201{,}326{,}592$ ◆

◆ Application: Simple Annuities

Recall from Section 5.1 that if a certain principal P is deposited in a savings account at an interest rate r per year and interest is compounded n times per year, then the amount A in the savings account after t years is given by the formula

$$A = P\left(1 + \frac{r}{n}\right)^{nt}.$$

In this formula, r/n is the *interest rate per time interval*, which we denote by i, and nt is the *total number of compounding intervals*, which we denote by N. Thus, the compound interest formula is often written as

$$A = P(1 + i)^N.$$

A **simple annuity** is a sequence of equal periodic payments of R dollars that are made over N equal time intervals at an interest rate per time interval of i. By applying the formula $A = P(1 + i)^N$, we can find the amount accumulated immediately after each periodic payment. Beginning with the Nth payment, we have the following information:

Payment number	Amount accumulated on each payment	Comment
N	R	Earns no interest
$N - 1$	$R(1 + i)$	Earns interest on one time interval
$N - 2$	$R(1 + i)^2$	Earns interest on two time intervals
$N - 3$	$R(1 + i)^3$	Earns interest on three time intervals
$\vdots$	$\vdots$	$\vdots$
1 or $N - (N - 1)$	$R(1 + i)^{N-1}$	Earns interest on $N - 1$ time intervals

The sequence

$$R,\ R(1 + i),\ R(1 + i)^2,\ R(1 + i)^3,\ \ldots,\ R(1 + i)^{N-1}$$

is a geometric sequence with first term $a_1 = R$, common ratio $r = 1 + i$, and general element

$$a_n = R(1 + i)^{n-1}$$

To find the total amount S of an annuity immediately after the Nth payment, we simply sum the amounts accumulated on each payment:

$$S = R + R(1 + i) + R(1 + i)^2 + R(1 + i)^3 + \cdots + R(1 + i)^{N-1}$$

Using the formula for the sum of a finite geometric series, we have

$$S = \frac{a_1(1 - r^n)}{1 - r} = \frac{R[1 - (1 + i)^N]}{1 - (1 + i)} = \frac{R[(1 + i)^N - 1]}{i}.$$

◆ **Amount of an Annuity**

If equal period payments of R dollars are made over N equal time intervals at an interest rate per time interval of i, then the total **amount S of an annuity** immediately after the Nth payment is

$$S = \frac{R[(1 + i)^N - 1]}{i}.$$

Saving plans, rent payments, and life-insurance payments are examples of an annuity. We conclude this section with an example concerning a savings plan.

EXAMPLE 6   *Chapter Opening Problem*

A father decides to deposit $150 each month in a savings account for his daughter's college education. The savings account earns interest at the rate of 9% per year compounded monthly. How much money will be in the account at the end of eight years if no money is in the account today?

◆◆ **Solution**   For this annuity we have equal monthly payments of $150 made over 12(8), or 96, months at an interest rate *per month* of $0.09 \div 12$, or 0.0075. Using the formula for the amount of an annuity with $R = 150$, $N = 96$, and $i = 0.0075$, we obtain

$$S = \frac{R[(1 + i)^N - 1]}{i}$$

$$= \frac{150[(1 + 0.0075)^{96} - 1]}{0.0075}$$

$$\approx 20{,}978.42$$

Thus, after eight years, the account will contain \$20,978.42.
    The general element of the sequence of payments is

$$a_n = 150(1 + 0.0075)^{n-1}.$$

Using this general element with the sequence and sum features of a graphing calculator, we can verify the amount in the account after eight years, as shown in Figure 8.24.   ◆

```
"150(1+.0075)^(n
-1)"→Un
            Done
sum Un(1,96)
       20978.42456
```

FIGURE 8.24

The sum of the first 96 terms with general element $a_n = 150(1 + 0.0075)^{n-1}$ is approximately 20978.42.

PROBLEM 6   Suppose the father's savings account described in Example 6 already contains \$5000. How much will be in the account after eight years?   ◆

Exercises 8.4

◆ **Basic Skills**

In Exercises 1–6, write the first five elements of a geometric sequence whose first element a_1 and common ratio r are given.

1. $a_1 = 3$, $r = 4$

2. $a_1 = -2$, $r = 5$

3. $a_1 = \frac{1}{4}$, $r = -\frac{2}{3}$

4. $a_1 = -16$, $r = -\frac{3}{4}$

5. $a_1 = -1$, $r = \pi$

6. $a_1 = 1 + \sqrt{2}$, $r = 1 - \sqrt{2}$

In Exercises 7–16, find the general element of the given geometric sequence. When possible, use the sequence features of a graphing calculator to verify your answer.

7. 1, 3, 9, 27, . . .

8. 1, −8, 64, −512, . . .

9. −64, 16, −4, 1, . . .

10. 24, 36, 54, 81, . . .

11. $\frac{1}{4}, \frac{1}{6}, \frac{1}{9}, \frac{2}{27}, \ldots$

12. $-\frac{3}{16}, \frac{1}{4}, -\frac{1}{3}, \frac{4}{9}, \ldots$

13. $-\sqrt{2}, \sqrt{6}, -3\sqrt{2}, 3\sqrt{6}, \ldots$

14. $-3\sqrt{3}, -3, -\sqrt{3}, -1, \ldots$

15. $3, 6x^2, 12x^4, 24x^6, \ldots$

16. $2x^3, 10x^2y^2, 50xy^4, 250y^6, \ldots$

In Exercises 17–26, information is given about a geometric sequence $a_1, a_2, a_3, \ldots, a_n$ with common ratio r. Find the indicated unknown.

17. $a_1 = 4$, $r = 3$; $a_9 = ?$

18. $a_1 = 12$, $r = -2$; $a_{12} = ?$

19. $a_6 = 36$, $r = -6$; $a_1 = ?$

20. $a_7 = -5$, $r = -\frac{5}{2}$; $a_1 = ?$

21. $a_2 = -4$, $a_6 = -\frac{81}{4}$; $r = ?$

22. $a_9 = \frac{3}{32}$, $a_4 = 3$; $a_1 = ?$

23. $a_5 = \frac{8}{27}$, $a_8 = -1$; $a_{13} = ?$

24. $a_5 = 0.5$, $a_9 = 312.5$; $a_2 = ?$

25. $\dfrac{a_4}{a_8} = 16$; $r = ?$ **26.** $\dfrac{a_5}{a_8} = -\dfrac{1}{27}$; $r = ?$

In Exercises 27–36, find the sum of the geometric series. When possible, use the sequence and sum features of a graphing calculator to verify your answer.

27. $\displaystyle\sum_{i=1}^{6} 2(-3)^{i-1}$ **28.** $\displaystyle\sum_{i=1}^{9} -3(2)^{i-1}$

29. $\displaystyle\sum_{i=1}^{5} \dfrac{5}{3^{i+1}}$ **30.** $\displaystyle\sum_{i=0}^{10} (-2)^i$

31. $\displaystyle\sum_{i=0}^{8} 42(0.1)^i$ **32.** $\displaystyle\sum_{i=2}^{7} \dfrac{4^i}{3^{i-1}}$

33. $1 + 4 + 16 + 64 + \cdots + 1{,}048{,}576$

34. $-243 + 162 - 108 + 72 - \cdots + \frac{128}{9}$

35. $192 - 96 + 48 - 24 + \cdots - \frac{3}{8}$

36. $2 + \sqrt{2} + 1 + \dfrac{\sqrt{2}}{2} + \cdots + \frac{1}{8}$

In Exercises 37–46, information about a geometric series $a_1 + a_2 + a_3 + \cdots + a_n$ *with sum* S_n *and common ratio* r *is given. Find the indicated unknown.*

37. $a_1 = 16$, $r = -\frac{1}{2}$; $S_{11} = ?$

38. $a_1 = -\frac{5}{4}$, $r = 2$; $S_{10} = ?$

39. $S_6 = 728$, $r = 3$; $a_1 = ?$

40. $S_3 = 93$, $a_1 = 3$; $r = ?$

41. $a_3 = \frac{3}{4}$, $r = \frac{2}{3}$; $S_6 = ?$

42. $a_6 = 2048$, $r = -4$; $S_{10} = ?$

43. $a_1 = -64$, $a_4 = 27$; $S_7 = ?$

44. $a_3 = 24$, $a_6 = 3$; $S_{11} = ?$

45. $S_3 = 57$, $a_1 = 12$; $a_6 = ?$

46. $S_7 = 57\frac{7}{8}$, $r = -\frac{3}{2}$; $a_4 = ?$

In Exercises 47 and 48, use the sequence features of a graphing calculator to find the 12th element of the given geometric sequence. Round your answer to three significant digits.

47. -8.36, -2.926, -1.0241, -0.358435, . . .

48. 6.4, -11.52, 20.736, -37.3248, . . .

In Exercises 49 and 50, use the sequence and sum features of a graphing calculator to find the sum of each geometric series. Round your answer to three significant digits.

49. $\displaystyle\sum_{i=1}^{18} 2.28(0.32)^{i-1}$ **50.** $\displaystyle\sum_{i=1}^{21} 13.3(-1.25)^{i-1}$

51. Find the number of elements in the geometric sequence $0.375, 1.5, 6, \ldots, 6{,}291{,}456$.

52. The first three terms of a geometric series are 6, 12, 24, and the sum of the series is 3066. How many terms are in the series?

53. The population of a city that now has 125,400 inhabitants is increasing by $6\frac{1}{2}\%$ each year. What will the population be six years from now?

54. A car radiator currently contains 3 gallons of the original antifreeze solution. One gallon of the original antifreeze is removed and replaced with one gallon of new antifreeze. One year later, one gallon of this mixture is removed and again replaced with one gallon of new antifreeze. If this process of removing a gallon and replacing a gallon is repeated each year for three more years, how many gallons of the original antifreeze will remain in the radiator?

55. An automobile that costs $16,000 when new depreciates 50% in value each year. What is it worth at the end of five years?

56. A union contract states that the workers shall receive a 6% pay increase per year for the next n years. Suppose that at the beginning of the first year of the contract, each worker is paid $28,540 per year. Express a worker's annual pay P for each succeeding year as a function of n.

57. Suppose you have a choice between two scholarships. One pays $10,000 per year for four years. The other pays $10 the first month, $20 the second month, $40 the third month, and so on, doubling the amount each month for one year only. Which offer should you accept?

58. How many ancestors (parents, grandparents, great-grandparents, and so on) do you have if you trace ten generations?

In Exercises 59 and 60, find the amount accumulated under each annuity. Round your answer to the nearest cent.

59. $100 per month for 120 months, at the rate of 8% per year compounded monthly.

60. $2500 per year for 20 years, at the rate of 9% per year compounded yearly.

61. Suppose your parents had deposited $10,000 in an account on your first birthday and then deposited $1000 on each birthday thereafter. Furthermore, suppose that the account had earned interest at the rate of 8.75% compounded yearly. How much money would be in the account on your 18th birthday?

62. A woman decides to deposit $500 each month in an account that earns interest at the rate of 8.25% per year compounded monthly. How much will she have in the account at the end of 5 years if the account presently contains $22,420?

━━━ ◆ **Critical Thinking** ━━━━━━━━━━━━━━━━━━━━━━━━━

63. Suppose $2x, 5x, 8x, 11x, \ldots$ is an arithmetic sequence.

(a) What type of sequence is $e^{2x}, e^{5x}, e^{8x}, e^{11x}, \ldots$?

(b) State the general element of the sequence in part (a).

64. Suppose $2x, 4x, 8x, 16x, \ldots$ is a geometric sequence.

(a) What type of sequence is $\ln 2x, \ln 4x, \ln 8x, \ln 16x, \ldots$?

(b) State the general element of the sequence in part (a).

65. Find x if $10^{-2}, x, 10^{-4}$ are the first three elements of a geometric sequence.

66. Find x if $x - 4, 5 - 2x, 4x - 1$ are the first three elements of a geometric sequence.

67. If the numbers $a_1, a_2, a_3, \ldots, a_{n-1}, a_n$ form a geometric sequence, then $a_2, a_3, \ldots, a_{n-1}$ are called the $(n - 2)$ *geometric means* between a_1 and a_n. Insert three geometric means between 16 and 81.

68. The sum of the first three terms of a geometric series is 63, and the third term is 45 more than the first. What is the first term of the series?

69. The product of the first three elements of a geometric sequence is -125 and their sum is 21. What is the common ratio of this sequence?

70. A ball is dropped from a height of 12 meters and rebounds after hitting the ground each time one-third the height from which it last fell. Find the *total distance* the ball has traveled when it hits the ground for the sixth time.

━━━

8.5 Infinite Geometric Series

◆ **Introductory Comments**

The series associated with the infinite geometric sequence

$$a_1, \ a_1 r, \ a_1 r^2, \ a_1 r^3, \ \ldots, \ a_1 r^{n-1}, \ \ldots$$

is called an **infinite geometric series**. We can write this series in expanded form as

$$a_1 + a_1 r + a_1 r^2 + a_1 r^3 + \cdots + a_1 r^{n-1} + \cdots$$

or in sigma form as

$$\sum_{i=1}^{\infty} a_1 r^{i-1}$$

To understand what is meant by the sum of an infinite geometric series, we first consider some finite sums:

$$S_1 = a_1$$

$$S_2 = a_1 + a_1 r$$

$$S_3 = a_1 + a_1 r + a_1 r^2$$

$$S_4 = a_1 + a_1 r + a_1 r^2 + a_1 r^3$$

$$\vdots$$

$$S_n = a_1 + a_1 r + a_1 r^2 + a_1 r^3 + \cdots + a_1 r^{n-1} = \frac{a_1(1 - r^n)}{1 - r} \quad \text{provided } r \neq 1$$

We refer to the real number S_n as the **nth partial sum** of the infinite series. For the infinite geometric series

$$\sum_{i=1}^{\infty} \left(\frac{1}{2}\right)^{n-1} = 1 + \tfrac{1}{2} + \tfrac{1}{4} + \tfrac{1}{8} + \cdots$$

the first four partial sums are

$$S_1 = 1 \qquad\qquad S_2 = 1 + \tfrac{1}{2} = \tfrac{3}{2}$$
$$S_3 = 1 + \tfrac{1}{2} + \tfrac{1}{4} = \tfrac{7}{4} \qquad S_4 = 1 + \tfrac{1}{2} + \tfrac{1}{4} + \tfrac{1}{8} = \tfrac{15}{8}$$

The display in Figure 8.25 shows the sums S_{10} and S_{20}. Note how these sums appear to be approaching a value of 2.

The nth partial sum of the series $\sum_{i=1}^{\infty} \left(\frac{1}{2}\right)^{n-1}$ is

$$S_n = \frac{a_1(1 - r^n)}{1 - r} = \frac{1\left[1 - \left(\frac{1}{2}\right)^n\right]}{1 - \frac{1}{2}} = 2\left[1 - \left(\tfrac{1}{2}\right)^n\right].$$

FIGURE 8.25
The 10th and 20th partial sums of the infinite geometric series $\sum_{i=1}^{\infty} \left(\frac{1}{2}\right)^{n-1}$.

The values given in Table 8.2 indicate that as the number of terms n gets larger and larger, the value of $\left(\frac{1}{2}\right)^n$ gets smaller and smaller. Hence, we conclude that as the number of terms n increases without bound ($n \to \infty$), the nth partial sum approaches 2:

$$S_n = 2\left[1 - \left(\tfrac{1}{2}\right)^n\right] \approx 2[1 - 0] = 2$$

TABLE 8.2
Values of $\left(\frac{1}{2}\right)^n$ as n gets larger and larger

n	1	2	3	4	5
$\left(\frac{1}{2}\right)^n$	$\frac{1}{2}$	$\frac{1}{4}$	$\frac{1}{8}$	$\frac{1}{16}$	$\frac{1}{32}$

Since the nth partial sum approaches 2 as $n \to \infty$, we write

$$\sum_{i=1}^{\infty} \left(\frac{1}{2}\right)^{n-1} = 2$$

We can find the sum of some, but not of every infinite geometric series. As we will show, the value of the common ratio r determines whether an infinite geometric series can be assigned a sum.

◆ **Sum of an Infinite Geometric Series**

If the nth partial sum of a geometric series approaches a finite number S as $n \to \infty$, then the infinite series is said to **converge** to the sum S, and we write

$$\sum_{i=1}^{\infty} a_1 r^{i-1} = \lim_{n\to\infty} S_n = S$$

"the limit of S_n as n approaches ∞"

If the nth partial sum approaches positive infinity, negative infinity, or no specific value as $n \to \infty$, then the infinite geometric series cannot be assigned a sum and is said to **diverge**.

Let's examine what happens to the nth partial sum of an infinite geometric series as $n \to \infty$. We must consider four cases for the value of the common ratio r.

Case 1 If an infinite geometric series has a common ratio r such that $|r| < 1$, then $|r^n|$ approaches 0 ($|r^n| \to 0$) as $n \to \infty$. Hence,

$$\sum_{i=1}^{\infty} a_1 r^{i-1} = \lim_{n\to\infty} S_n = \lim_{n\to\infty} \frac{a_1(1 - r^n)}{1 - r}$$

$$= \frac{a_1(1 - 0)}{1 - r}$$

$$= \frac{a_1}{1 - r} \quad \text{if } |r| < 1$$

Case 2 If an infinite geometric series has a common ratio r such that $|r| > 1$, then $|r^n| \to \infty$ as $n \to \infty$. Hence, $\lim_{n\to\infty} S_n$ does not exist and the series diverges.

Case 3 If an infinite geometric series has common ratio $r = 1$, then the sequence of partial sums is

$$a_1, \ 2a_1, \ 3a_1, \ \ldots, \ na_1, \ \ldots$$

Hence, $\lim_{n\to\infty} S_n$ does not exist and the series diverges.

Case 4 If an infinite geometric series has common ratio $r = -1$, then the sequence of partial sums is

$$a_1, \ 0, \ a_1, \ 0, \ a_1, \ 0, \ \ldots$$

Hence, $\lim_{n\to\infty} S_n$ does not exist and the series diverges.

In summary, it is the value of the common ratio r that determines whether an infinite geometric series converges or diverges.

◆ **Sum of an Infinite Geometric Series**

The infinite geometric series

$$\sum_{i=1}^{\infty} a_1 r^{i-1} = a_1 + a_1 r + a_1 r^2 + a_1 r^3 + \cdots + a_1 r^{n-1} + \cdots$$

converges if $|r| < 1$ and has the finite **sum**

$$S = \frac{a_1}{1 - r}$$

or *diverges* if $|r| \geq 1$ and has no finite sum.

EXAMPLE 1 Determine whether the infinite geometric series converges or diverges. If the series converges, find its sum.

(a) $\displaystyle\sum_{n=0}^{\infty} \left(\frac{3}{2} \right)^n$ **(b)** $\displaystyle\sum_{i=1}^{\infty} (-1)^{i+1} \left(\frac{1}{2} \right)^i$

◆ **Solution**

(a) Writing this series in expanded form, we have

$$\sum_{n=0}^{\infty}\left(\frac{3}{2}\right)^{n} = 1 + \frac{3}{2} + \frac{9}{4} + \frac{27}{8} + \cdots$$

This is an infinite geometric series with first term $a_1 = 1$ and common ratio $r = \frac{3}{2}$. Since

$$|r| = \left|\tfrac{3}{2}\right| = \tfrac{3}{2} \geq 1,$$

the series diverges and has no finite sum.

(b) Writing this series in expanded form, we have

$$\sum_{i=1}^{\infty}(-1)^{i+1}\left(\frac{1}{2}\right)^{i} = \frac{1}{2} - \frac{1}{4} + \frac{1}{8} - \frac{1}{16} + \cdots$$

This is an infinite geometric series with first term $a_1 = \frac{1}{2}$ and common ratio $r = -\frac{1}{2}$. Since

$$|r| = \left|-\tfrac{1}{2}\right| = \tfrac{1}{2} < 1,$$

the infinite geometric series converges and the sum S of the series is

$$S = \frac{a_1}{1 - r} = \frac{\frac{1}{2}}{1 - \left(-\frac{1}{2}\right)} = \frac{\frac{1}{2}}{\frac{3}{2}} = \frac{1}{3}.$$

The display in Figure 8.26 shows the sums S_{10} and S_{20}. These sums appear to be approaching a value of $\frac{1}{3}$.   ◆

PROBLEM 1 Determine whether the following infinite geometric series converges or diverges.

$$1 + 0.1 + 0.01 + 0.001 + \cdots$$

If the series converges, find its sum.   ◆

```
"(-1)^(n+1)*(1/2
)^n"→Un
              Done
sum Un(1,10)
       .3330078125
sum Un(1,20)
       .3333330154
```

FIGURE 8.26

The 10th and 20th partial sums of the infinite geometric series $\sum_{i=1}^{\infty}(-1)^{i+1}\left(\frac{1}{2}\right)^{i}$

◆ **Repeating Decimal Numbers**

In Section 1.1 we stated that all decimal numbers that either terminate or repeat the same block of digits are rational numbers. We can express the terminating decimal 0.25 in *fractional form* by writing $\frac{25}{100}$ or in *reduced fractional form* by writing $\frac{1}{4}$. To find the reduced fractional form of a repeating decimal such as

$$0.\overline{12} = 0.121212 \ldots$$

The block of digits under the bar repeats indefinitely.

we can use the formula for the sum of an infinite geometric series. The procedure is outlined in the next example.

EXAMPLE 2   Find the reduced fractional form of each repeating decimal number.

(a) $0.\overline{12} = 0.121212\ldots$ (b) $6.3\overline{450} = 6.3450450450\ldots$

◆ Solution

(a) The decimal number $0.\overline{12}$ can be written as the infinite series

$$0.12 + 0.0012 + 0.000012 + \cdots$$

This is an infinite geometric series with first term $a_1 = 0.12$ and common ratio $r = 0.01$. Since $r < 1$, the series converges to the sum

$$S = \frac{a_1}{1 - r} = \frac{0.12}{1 - 0.01} = \frac{0.12}{0.99} = \frac{4}{33}.$$

Hence, the reduced fractional form of the repeating decimal number $0.\overline{12}$ is $\frac{4}{33}$.

(b) The decimal number $6.3\overline{450}$ can be written as

$$6.3 + [0.0450 + 0.0000450 + 0.0000000450 + \cdots]$$

The expression within brackets is an infinite geometric series with first term $a_1 = 0.0450$ and common ratio $r = 0.001$. Since $r < 1$, the series converges to the sum

$$S = \frac{a_1}{1 - r} = \frac{0.0450}{1 - 0.001} = \frac{45}{999} = \frac{5}{111}.$$

Hence, the reduced fractional form of the repeating decimal number $6.3\overline{450}$ is

$$6.3 + \frac{5}{111} = \frac{63}{10} + \frac{5}{111} = \frac{7043}{1110}$$

Most graphing calculators have the capability of converting a repeating decimal number to reduced fractional form, as shown in Figure 8.27.   ◆

PROBLEM 2   Find the reduced fractional form of the repeating decimal number $0.88888\ldots$. Use a graphing calculator to verify the answer.   ◆

FIGURE 8.27

Converting the repeating decimal numbers in Example 2 to reduced fractional form.

◆ Interval of Convergence

In some infinite series, each term contains a variable. Such series occur in calculus and its applications. For example, consider the infinite series

$$1 + (cx) + (cx)^2 + (cx)^3 + \cdots$$

where x is a variable and c a constant such that $c \neq 0$. Obviously, if $x = 0$, this series converges to 1. If $x \neq 0$, we can think of this series as an infinite geometric series with first term $a_1 = 1$ and common ratio $r = cx$. Depending on the real number that we substitute for x, the infinite geometric series either converges if $|cx| < 1$, or diverges if $|cx| \geq 1$. For an infinite geometric series with variable terms, the subset of real numbers for which the series converges is called its **interval of convergence**.

EXAMPLE 3   Find the interval of convergence and the sum of the infinite series

$$1 + 2x + 4x^2 + 8x^3 + \cdots$$

◆ **Solution**   If $x = 0$, the series converges to 1. If $x \neq 0$, we can think of this series as an infinite geometric series with first term $a_1 = 1$ and common ratio $r = 2x$. The infinite geometric series converges whenever

$$|2x| < 1$$
$$-1 < 2x < 1$$
$$-\tfrac{1}{2} < x < \tfrac{1}{2}$$

Thus, the interval of convergence for this infinite series is $\left(-\tfrac{1}{2}, \tfrac{1}{2}\right)$.

We can find the sum S of this infinite series by using the formula for the sum of an infinite geometric series. Thus,

$$S = \frac{a_1}{1 - r} = \frac{1}{1 - 2x} \quad \text{provided } -\tfrac{1}{2} < x < \tfrac{1}{2}. \qquad ◆$$

PROBLEM 3   Use polynomial long division (Section 4.2) to show that

$$\frac{1}{1 - 2x} = 1 + 2x + 4x^2 + 8x^3 + \cdots \qquad ◆$$

◆ **Application: Hammering a Nail**

Infinite geometric series occur in many types of applied problems. We conclude this section with an application.

EXAMPLE 4   A nail $2\tfrac{1}{2}$ inches long is being hammered into a board. The first impact drives the nail $\tfrac{3}{4}$ inch into the board, and each additional impact drives the nail two-thirds the distance of the preceding impact. If the nail is hammered indefinitely, will its head ever be flush with the board?

◆ **Solution**   The first impact drives the nail $\tfrac{3}{4}$ inch into the board. The second impact drives the nail two-thirds of $\tfrac{3}{4}$ inch, or $\tfrac{1}{2}$ inch into the board. The third impact drives the nail two-thirds of $\tfrac{1}{2}$ inch, or $\tfrac{1}{3}$ inch into the board, and so on.

The total distance that the nail is driven into the board can be represented by the infinite geometric series

$$\tfrac{3}{4} + \tfrac{1}{2} + \tfrac{1}{3} + \cdots$$

where $a_1 = \tfrac{3}{4}$ and $r = \tfrac{2}{3}$. Since $|r| < 1$, the sum S of this infinite geometric series is

$$S = \frac{a_1}{1 - r} = \frac{\tfrac{3}{4}}{1 - \tfrac{2}{3}} = \frac{\tfrac{3}{4}}{\tfrac{1}{3}} = \frac{9}{4} \text{ or } 2\tfrac{1}{4}.$$

Thus, after infinitely many hits, the nail is driven into the board $2\tfrac{1}{4}$ inches. Since the nail is $2\tfrac{1}{2}$ inches long, $\tfrac{1}{4}$ inch of the nail is not hammered into the board. Hence, the head of the nail will never be flush with the board under this condition. ◆

PROBLEM 4 If an initial hammer impact drives a $2\tfrac{1}{2}$-inch nail into the board $\tfrac{7}{8}$ inch, and each successive impact drives the nail two-thirds the distance of the preceding impact, will the head of the nail ever be flush with the board? ◆

Exercises 8.5

◆ Basic Skills

In Exercises 1–8, find the first four partial sums and the nth partial sum for each infinite geometric series.

1. $\displaystyle\sum_{i=1}^{\infty} \left(\frac{2}{3} \right)^{i-1}$

2. $\displaystyle\sum_{i=1}^{\infty} \left(-\frac{3}{4} \right)^{i}$

3. $\displaystyle\sum_{i=0}^{\infty} 2(-3)^{i}$

4. $\displaystyle\sum_{i=2}^{\infty} 3(0.1)^{i-2}$

5. $-243 + 162 - 108 + 72 - \cdots$

6. $1 + 4 + 16 + 64 + \cdots$

7. $\tfrac{9}{8} + \tfrac{3}{4} + \tfrac{1}{2} + \tfrac{1}{3} + \cdots$

8. $15 - 5 + \tfrac{5}{3} - \tfrac{5}{9} + \cdots$

In Exercises 9–20, determine whether the infinite geometric series converges or diverges. If the series converges, find its sum.

9. $\displaystyle\sum_{i=1}^{\infty} 4(3)^{i-1}$

10. $\displaystyle\sum_{i=0}^{\infty} 2\left(-\frac{3}{5} \right)^{i}$

11. $\displaystyle\sum_{i=2}^{\infty} (-0.2)^{i}$

12. $\displaystyle\sum_{i=1}^{\infty} 8\left(\frac{3}{2} \right)^{i-1}$

13. $\displaystyle\sum_{i=1}^{\infty} (-1)^{i+1} \left(\frac{5}{6} \right)^{i}$

14. $\displaystyle\sum_{i=1}^{\infty} \frac{2}{3^{i+1}}$

15. $1 - 2 + 4 - 8 + \cdots$

16. $3.375 + 2.25 + 1.5 + 1 + \cdots$

17. $-6 + 3 - \tfrac{3}{2} + \tfrac{3}{4} - \cdots$

18. $1 - \dfrac{1}{\pi} + \dfrac{1}{\pi^2} - \dfrac{1}{\pi^3} + \cdots$

19. $5\sqrt{10} + 5\sqrt{2} + \sqrt{10} + \sqrt{2} + \cdots$

20. $-\sqrt{3} - \sqrt{6} - 2\sqrt{3} - 2\sqrt{6} - \cdots$

In Exercises 21–28, information is given about an infinite geometric series $a_1 + a_2 + a_3 + \cdots$ with sum S and common ratio r. Find the indicated unknown.

21. $S = -36, \ a_1 = -12; \quad r = ?$

22. $S = 24, \ r = -\tfrac{1}{2}; \quad a_1 = ?$

23. $a_3 = \tfrac{3}{4}, \ r = \tfrac{2}{3}; \quad S = ?$

24. $a_6 = 1.28, \ r = 0.4; \quad S = ?$

25. $a_1 = -64, \ a_4 = 27; \quad S = ?$

26. $a_3 = 24, \ a_6 = 3; \quad S = ?$

27. $S = 18, \ a_2 = 4; \quad a_1 = ?$

28. $S = 32, \ a_2 = 6; \quad r = ?$

In Exercises 29–38, find the reduced fractional form of each repeating decimal number.

29. $0.33333\ldots$

30. $0.77777\ldots$

31. $0.090909\ldots$

32. $0.272727\ldots$

33. $2.\overline{108}$

34. $34.\overline{303}$

35. $15.020312031\ldots$

36. $2.00010101\ldots$

37. $0.8\overline{015}$

38. $5.25\overline{144}$

$\frac{\Delta y}{\Delta x}$ *In Exercises 39–46, find the interval of convergence and the sum S of each infinite series.*

39. $x - \dfrac{x^2}{5} + \dfrac{x^3}{25} - \dfrac{x^4}{125} + \cdots$

40. $3 + 9x + 27x^2 + 81x^3 + \cdots$

41. $1 + 4x^2 + 16x^4 + 64x^6 + \cdots$

42. $x - 8x^4 + 64x^7 - 512x^{10} + \cdots$

43. $1 - (x - 2) + (x - 2)^2 - \cdots$

44. $2(x + 1) + 4(x + 1)^2 + 8(x + 1)^3 - \cdots$

45. $2x + 1 + \dfrac{1}{2x} + \dfrac{1}{4x^2} + \cdots$

46. $1 - \dfrac{2}{x} + \dfrac{4}{x^2} - \dfrac{8}{x^3} + \cdots$

47. A golf ball is dropped from a height of 18 meters and rebounds each time to two-thirds of the height from which it last fell, as shown by the arrows in the figure.

(a) Find the distance the ball travels *downward* before coming to rest.

(b) Find the distance the ball travels *upward* before coming to rest.

(c) Find the *total distance* the ball travels before coming to rest.

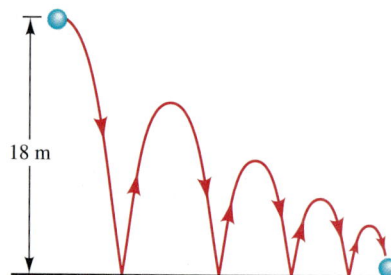

18 m

48. A lawn-mower blade makes 450 revolutions during the first second after the power is shut off and one-sixteenth as many revolutions during each succeeding second. How many revolutions does the blade make before coming to rest?

49. A man runs the 26-mile Boston Marathon at the rate of 10 miles the first hour, 6 miles the second hour, 3.6 miles the third hour, and so on. Does he ever finish the race?

50. A 24-foot steel pile is driven vertically into the soil to form part of a foundation. The first impact drives the pile 5 feet into the soil, and each additional impact drives the pile four-fifths the distance of the preceding impact. Will the top of the pile ever be flush with the ground?

51. A pendulum swings through an arc 22.54 cm long. On each succeeding swing, the pendulum travels an arc that is 0.84 the length of the preceding swing. How far does the pendulum travel before coming to rest?

52. A person with cancer is given 485.0 mg of an experimental drug the first week of treatment, and each succeeding week the patient is given 85% of the previous week's dosage. How many milligrams of this drug is the patient given? Round your answer to the nearest tenth of a milligram.

◆ **Critical Thinking** ◆

53. Determine the value of x if $\displaystyle\sum_{i=1}^{\infty} (3x)^{i-1} = \dfrac{2}{3}$.

54. Find the sum of each infinite series.

(a) $\displaystyle\sum_{i=1}^{\infty} 2\left[\left(\dfrac{2}{3}\right)^{i-1} + \left(\dfrac{3}{4}\right)^i\right]$

(b) $\displaystyle\sum_{i=0}^{\infty} \left(\dfrac{1}{2^i} - \dfrac{1}{3^i} + \dfrac{1}{4^i}\right)$

55. Find the sum of each infinite series, if it exists.

(a) $1 - 1 + 1 - 1 + 1 - 1 + \cdots$

(b) $(1 - 1) + (1 - 1) + (1 - 1) + \cdots$

(c) $1 + (-1 + 1) + (-1 + 1) + \cdots$

Does the associative property of real numbers seem to apply to infinite series?

56. Is it possible to have an infinite geometric series with first term 7 and sum 3? Explain.

57. Find the common ratio of an infinite geometric series in which each term is twice the sum of all the terms that follow it.

58. Find an infinite geometric series in which the first term is 4 and each term is 5 times the sum of all the terms that follow it.

59. A square has sides of length 24 cm. Inside this square, a second square is constructed by joining the midpoints of the sides of the first square. Inside the second square, a third square is constructed by joining the midpoints of the sides of the second square, and so on indefinitely, as shown in the sketch.

(a) What is the sum of the areas of all the squares?

(b) What is the sum of the perimeters of all the squares?

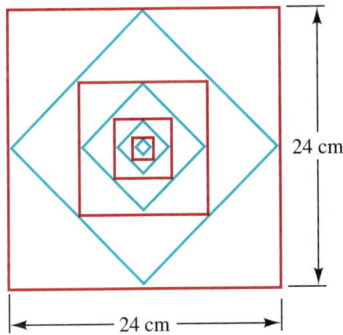

60. In a right triangle ABC, angle A is $30°$, angle B is $60°$, angle C is $90°$, and the length of $\overline{AB}$ is 32 inches. Suppose $\overline{CD}$ is drawn perpendicular to $\overline{AB}$, $\overline{DE}$ is drawn perpendicular to $\overline{AC}$, $\overline{EF}$ is drawn perpendicular to $\overline{AB}$, $\overline{FG}$ is drawn perpendicular to $\overline{AC}$, and so on indefinitely, as shown in the sketch.

(a) What is the sum of all the perpendiculars to side $\overline{AC}$ if $\overline{BC}$ is considered the first of these perpendiculars?

(b) What is the sum of all the perpendiculars to side $\overline{AB}$?

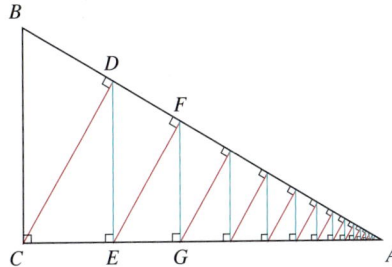

8.6 Factorials and Their Applications

◆ **Introductory Comments**

The sequence defined recursively by

$$a_n = \begin{cases} 1 & \text{if } n = 1 \\ na_{n-1} & \text{if } n \geq 2 \end{cases}$$

has the following elements:

$a_1 = 1$

$a_2 = 2a_1 = 2 \cdot 1$

$a_3 = 3a_2 = 3 \cdot 2 \cdot 1$

$a_4 = 4a_3 = 4 \cdot 3 \cdot 2 \cdot 1$

$a_5 = 5a_4 = 5 \cdot 4 \cdot 3 \cdot 2 \cdot 1$

$a_n = na_{n-1} = n \cdot (n-1) \cdot (n-2) \cdot (n-3) \cdot \ldots \cdot 2 \cdot 1$

The nth element a_n represents the product of all consecutive, positive integers from 1 up to and including n. This special product is called **n factorial** and is denoted by **$n!$**.

◆ ***n* factorial**

> If *n* is a positive integer, then *n* factorial (*n*!) is given by
>
> $$n! = n \cdot (n - 1) \cdot (n - 2) \cdot (n - 3) \cdot \ldots \cdot 2 \cdot 1.$$

We usually extend this definition to include **zero factorial**, 0!. As we will observe in this section, it is convenient to have

$$0! = 1$$

To simplify many factorial expressions, we apply the fact that if *n* and *n* − 1 are consecutive positive integers, then

$$n \cdot (n - 1)! = n!$$

Consider these illustrations of simplified factorial expressions:

$$8 \cdot 7! = 8!$$

$$15 \cdot 14 \cdot 13 \cdot 12 \cdot 11! = 15!$$

$$\frac{10!}{7!} = \frac{10 \cdot 9 \cdot 8 \cdot \cancel{7!}}{\cancel{7!}} = 10 \cdot 9 \cdot 8 = 720$$

In this section, we discuss factorial and their applications to permutations, combinations, and the binomial theorem.

◆ **Permutations**

An arrangement of a set of *n* distinct objects in a given order is called a **permutation**. Any two objects *a* and *b* have two permutations:

$$ab \qquad \text{and} \qquad ba$$

We have two choices for the first position—either *a* or *b*. Once we have made a choice for this position, we have one choice for the last position. Note that the product

$$2 \cdot 1 \qquad \text{or} \qquad 2!$$

gives us the number of permutations of two distinct objects.

The three objects *a*, *b*, and *c* have six permutations:

$$abc, \quad acb, \quad bac, \quad bca, \quad cab, \quad cba$$

We have three choices for the first position—either *a*, *b*, or *c*. Once we have made a choice for this position, we have two choices for the middle position, and once we have made a choice for the middle position, we have one choice for the last position. The product

$$3 \cdot 2 \cdot 1 \quad\text{or}\quad 3!$$

gives us the number of permutations of three distinct objects. In general, the number of permutations of n distinct objects is

$$n \cdot (n - 1) \cdot (n - 2) \cdot \ldots \cdot 2 \cdot 1 \quad\text{or}\quad n!$$

◆ **Permutations of n Objects**

> The number of permutations of n distinct objects is $n!$

Suppose we wish to determine the number of permutations of n distinct objects taken r at a time, where $r \leq n$. Consider the following r boxes:

1st 2d 3d rth
□ □ □ ⋯ □

The 1st box can be filled in n ways.

the 2d box in $(n - 1)$ ways.

the 3d box in $(n - 2)$ ways.

⋮

the rth box in $[n - (r - 1)]$ ways.

Thus, the number of permutations of n distinct objects taken r at a time, where $r \leq n$, is

$$n \cdot (n - 1) \cdot (n - 2) \cdot \ldots \cdot [n - (r - 1)].$$

We can rewrite $[n - (r - 1)]$ as $[(n - r) + 1]$. Now, the next smaller integer after $[(n - r) + 1]$ is $(n - r)$. Thus, multiplying $n \cdot (n - 1) \cdot (n - 2) \cdot \ldots \cdot [n - (r - 1)]$ by $(n - r)!$ gives us $n!$. Hence,

$$n \cdot (n - 1) \cdot (n - 2) \cdot \ldots \cdot [n - (r - 1)] \cdot \frac{(n - r)!}{(n - r)!} = \frac{n!}{(n - r)!}.$$

◆ **Permutations of n Objects Taken r at a Time**

> The number of permutations of n distinct objects taken r at a time, where $r \leq n$, is denoted by ${}_nP_r$ and
>
> $$ {}_nP_r = \frac{n!}{(n - r)!}.$$

Note: If $r = n$, we obtain

$$_nP_n = \frac{n!}{(n - n)!} = \frac{n!}{0!} = \frac{n!}{1} = n!$$

Recall that $n!$ is the number of permutations of n distinct objects.

EXAMPLE 1   At a track meet, eight contestants are entered in the 100-yard dash. Find the number of possible orders of finish for

(a) all eight contestants   **(b)** the first three positions.

◆ **Solution**

(a) The number of permutations of eight distinct objects taken 8 at a time is $_8P_8$. Thus, the number of possible orders of finish for the race is

$$_8P_8 = 8!   \text{or}   40,320 \text{ orders.}$$

(b) The number of permutations of eight distinct objects taken 3 at a time is $_8P_3$. Thus, the number of possible orders of finish for the first three positions is

$$_8P_3 = \frac{8!}{(8 - 3)!} = \frac{8!}{5!}   \text{or}   336 \text{ orders.}$$

Most graphing calculators have a probability menu for finding the number of permutations of n items taken r at a time. Figure 8.28 shows a typical display of evaluating the permutations in this Example.   ◆

```
8 nPr 8
             40320
8 nPr 3
               336
```

FIGURE 8.28

Evaluating the permutations in Example 1

PROBLEM 1   For the race described in Example 1, find the number of possible orders of finish for the first four positions. Use the permutation feature of a graphing calculator to confirm the answer.   ◆

Suppose we wish to find the number of permutations that can be made from the letters of the word PAPA. If we distinguish between the P's by putting one in boldface type (**P**), and between the A's by putting one in boldface type (**A**), then we have 4! or 24 permutations:

PAPA **PA**PA **PA**PA **PA**PA
PPAA **P**PAA **P**PAA **P**PAA
PAAP PAAP **PA**AP PAAP
A**P**PA A**P**PA A**P**PA A**P**PA
AA**P**P AA**P**P AA**P**P AA**P**P
A**P**AP A**P**AP A**P**AP A**P**A**P**.

For each of these six groups, we have 2! ways to arrange the P's and 2! ways to arrange the A's.

However, if we write all the letters in the same style of type, only six of these permutations are *distinguishable*:

PAPA   PPAA   PAAP   APPA   AAPP   APAP.

Thus, for four letters, two of which are P's and two of which are A's, we have permutations of 4 letters

$$\frac{4!}{2!\ 2!} = 6 \text{ distinguishable permutations}$$

number of ways to arrange the P's

number of ways to arrange the A's

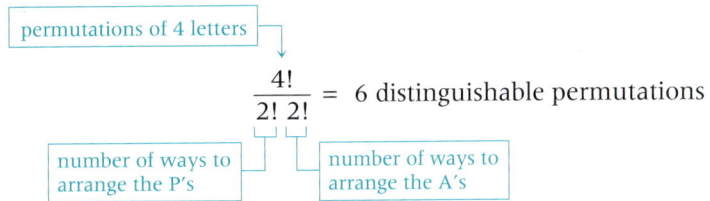

In general, we have the following rule for distinguishable permutations.

◆ **Distinguishable Permutations**

If n objects are made up of n_1 of one kind, n_2 of a second kind, . . . , n_k of a kth kind, such that $n_1 + n_2 + \cdots + n_k = n$, then the number of distinguishable permutations of these n objects is given by

$$\frac{n!}{n_1!n_2! \cdots n_k!}$$

EXAMPLE 2 How many orders of wins and losses are possible for a college football team to end the season with six wins and four losses?

◆ **Solution** The team plays ten games; six of which are wins (W), and four of which are losses (L). The number of distinguishable permutations for WWWWWWLLLL is

$$\frac{10!}{6!\ 4!} = 210$$

Thus, a football team can end the season with six wins and four losses in 210 ways. ◆

PROBLEM 2 How many orders of wins, losses, and ties are possible for a team to end the season with three wins, five losses, and two ties? ◆

◆ **Combinations**

The selection of r objects from a set of n distinct objects *without regard to order* is called a **combination** of n objects taken r at a time. We denote the number of such combinations by $_nC_r$.

We can choose two objects from a set of four distinct objects, a, b, c, and d, in $_4P_2$, or 12 permutations:

ab   ba
ac   ca
ad   da   Once 2 objects are chosen, we
bc   cb   have 2! ways of arranging them.
bd   db
cd   dc

Now, if we disregard order, then each of these groups is counted only once. Hence, the combinations of four objects taken two at a time are

$$ab \quad ac \quad ad \quad bc \quad bd \quad cd$$

The number of such combinations is

permutations of 4 objects taken 2 at a time

$$\frac{{}_4P_2}{2!} = 6$$

permutations of 2 objects

In general, we have the following rule for combinations.

◆ **Combinations of** *n* **Objects Taken** *r* **at a Time**

The number of combinations of *n* distinct objects taken *r* at a time, with $r \leq n$, is denoted by ${}_nC_r$, and

$${}_nC_r = \frac{{}_nP_r}{r!} = \frac{n!}{r!(n-r)!}.$$

CAUTION Do not confuse permutations and combinations. Remember, in a permutation, order is considered. In a combination, order does not matter. For example, a signal made up of three flags that may be obtained from eight differently colored flags is a permutation, because a rearrangement of the three flags makes a different signal (order matters). However, a committee of three members that may be obtained from eight people is a combination, because a rearrangement of the three members is the same committee (the order of the member selection does not change the committee).

EXAMPLE 3 In a certain poker game, five cards are dealt to a player from an ordinary deck of 52 cards.

(a) How many different possible poker hands can be dealt?

(b) How many poker hands consist of all hearts?

(c) How many poker hands consist of three aces?

◆ **Solution** The order in which the five cards are dealt does not matter. Hence, we are finding combinations. A standard deck of 52 cards has four suits (hearts, diamonds, clubs, spades), and each suit contains thirteen cards (2, 3, 4, . . ., 10, jack, queen, king, ace).

(a) The number of possible poker hands that can be dealt is the number of combinations of 52 objects taken 5 at a time:

$${}_{52}C_5 = \frac{52!}{5!(52-5)!} = \frac{52!}{5!47!} = 2,598,960$$

Thus, 2,598,960 poker hands can be dealt.

(b) The number of poker hands consisting of all hearts is the number of combinations of 13 objects taken 5 at a time:

$$_{13}C_5 = \frac{13!}{5!\,(13 - 5)!} = \frac{13!}{5!\,8!} = 1287 \text{ combinations}$$

Thus, a hand of all hearts can be dealt in 1287 ways.

(c) The number of ways of being dealt three of the four aces in the deck is

$$_4C_3 = \frac{4!}{3!\,1!} = 4 \text{ combinations.}$$

The number of ways of being dealt the remaining two cards from the 48 non-ace cards is

$$_{48}C_2 = \frac{48!}{2!\,46!} = 1128$$

Thus, a five-card poker hand consisting of three aces can be dealt in

$$_4C_3 \cdot {}_{48}C_2 = 4 \cdot 1128 = 4512 \text{ combinations.}$$

Most graphing calculators have the capability of finding the number of combinations of n items taken r at a time. Figure 8.29 shows a typical display of evaluating the combinations in this Example. ◆

```
52 nCr 5
            2598960
13 nCr 5
               1287
(4 nCr 3)(48 nCr
2)
               4512
```

FIGURE 8.29

Evaluating the combinations in Example 3

PROBLEM 3 Find the number of five-card poker hands that consist of two queens and three kings. Use the combination feature of a graphing calculator to confirm your answer. ◆

◆ **Binomial Coefficients and the Binomial Theorem**

We usually denote the factorial expression for $_nC_r$ by writing the symbol

$$\binom{n}{r}$$

and refer to this as a *binomial coefficient*.

◆◆ **Binomial Coefficient**

The symbol $\binom{n}{r}$ is called a **binomial coefficient** and is defined by

$$\binom{n}{r} = \frac{n!}{r!\,(n - r)!}$$

where r is a nonnegative integer such that $r \leq n$.

From the following illustrations of evaluating binomial coefficients,

$$\binom{5}{0} = \frac{5!}{0!\,(5-0)!} = \frac{5!}{0! \cdot 5!} = \frac{1}{0!} = \frac{1}{1} = 1$$

$$\binom{5}{1} = \frac{5!}{1!\,(5-1)!} = \frac{5!}{1! \cdot 4!} = \frac{5 \cdot 4!}{1! \cdot 4!} = \frac{5}{1} = 5$$

$$\binom{5}{2} = \frac{5!}{2!\,(5-2)!} = \frac{5!}{2! \cdot 3!} = \frac{5 \cdot 4 \cdot 3!}{2! \cdot 3!} = \frac{20}{2} = 10$$

$$\binom{5}{3} = \frac{5!}{3!\,(5-3)!} = \frac{5!}{3! \cdot 2!} = \frac{5 \cdot 4 \cdot 3!}{3! \cdot 2!} = \frac{20}{2} = 10$$

$$\binom{5}{4} = \frac{5!}{4!\,(5-4)!} = \frac{5!}{4! \cdot 1!} = \frac{5 \cdot 4!}{4! \cdot 1!} = \frac{5}{1} = 5$$

$$\binom{5}{5} = \frac{5!}{5!\,(5-5)!} = \frac{5!}{5! \cdot 0!} = \frac{1}{0!} = \frac{1}{1} = 1$$

we can make two general observations about binomial coefficients:

1. We observe the following equalities:

$$\binom{5}{0} = \binom{5}{5} \qquad \binom{5}{1} = \binom{5}{4} \qquad \binom{5}{2} = \binom{5}{3}$$

In general, if r and n are nonnegative integers such that $r \le n$, then

$$\binom{n}{r} = \binom{n}{n-r}.$$

2. The binomial coefficients $\binom{5}{0}$ and $\binom{5}{5}$ both equal 1. In general, for any nonnegative integer n,

$$\binom{n}{0} = \binom{n}{n} = 1.$$

These are two *fundamental properties of binomial coefficients.*

◆ **Fundamental Properties of Binomial Coefficients**

If r and n are nonnegative integers such that $r \le n$, then

1. $\binom{n}{r} = \binom{n}{n-r}$ 2. $\binom{n}{0} = \binom{n}{n} = 1$

In Section 1.5 we developed formulas for the square and cube of a binomial. Recall that

$$(A + B)^2 = A^2 + 2AB + B^2$$

and
$$(A + B)^3 = A^3 + 3A^2B + 3AB^2 + B^3.$$

Note that the coefficients of the terms in the expansion of $(A + B)^2$ are equivalent to the binomial coefficients

$$\binom{2}{0} = 1, \quad \binom{2}{1} = 2, \quad \text{and} \quad \binom{2}{2} = 1$$

and the coefficients of the terms in the expansion of $(A + B)^3$ are equivalent to the binomial coefficients

$$\binom{3}{0} = 1, \quad \binom{3}{1} = 3, \quad \binom{3}{2} = 3, \quad \text{and} \quad \binom{3}{3} = 1.$$

Let's see if we can use pattern recognition to derive a formula for the expansion of $(A + B)^n$ where n is any positive integer.

Observation	*Conjecture*
1. The expansion of $(A + B)^2$ contains 3 terms. The first term is A^2 and the last term is B^2. The expansion of $(A + B)^3$ contains 4 terms. The first term is A^3 and the last term is B^3.	**1.** The expansion of $(A + B)^n$ should contain $(n + 1)$ terms. The first term should be A^n and the last term should be B^n.
2. In the expansion of $(A + B)^2$, the exponents of A decrease by 1 in each succeeding term after the first, and the sum of the exponents on A and B together in any term is 2. In the expansion of $(A + B)^3$, the exponents on A decrease by 1 in each succeeding term after the first, and the sum of the exponents on A and B together in any term is 3.	**2.** In the expansion of $(A + B)^n$, the exponents on A should decrease by 1 in each succeeding term after the first, and the sum of the exponents on A and B together in any term should be n.
3. The coefficients of the terms in the expansion of $(A + B)^2$ are equivalent to $$\binom{2}{0}, \quad \binom{2}{1}, \quad \binom{2}{2}$$ The coefficients of the terms in the expansion of $(A + B)^3$ are equivalent to $$\binom{3}{0}, \quad \binom{3}{1}, \quad \binom{3}{2}, \quad \binom{3}{3}$$	**3.** The coefficients of the terms in the expansion of $(A + B)^n$ should be equivalent to $$\binom{n}{0}, \quad \binom{n}{1}, \quad \binom{n}{2}, \quad \binom{n}{3}, \ldots, \binom{n}{n}$$

From our preceding conjectures, we write

$$(A + B)^n = \binom{n}{0} A^n + \binom{n}{1} A^{n-1}B + \binom{n}{2} A^{n-2}B^2 + \binom{n}{3} A^{n-3}B^3 + \binom{n}{4} A^{n-4}B^4 + \cdots + \binom{n}{n} B^n$$

We can use mathematical induction (Section 8.2) to prove this formula. The proof, however, is quite cumbersome and we shall simply accept its validity. We refer to this formula as the **binomial theorem** and state it more compactly, using sigma form with lower value $i = 0$:

◆◆ **Binomial Theorem**

If n is a positive integer, then

$$(A + B)^n = \sum_{i=0}^{n} \binom{n}{i} A^{n-i} B^i$$

EXAMPLE 4   Expand $(x - 2y)^5$ by using the binomial theorem.

◆ **Solution**   For $(x - 2y)^5$, we have $A = x$, $B = -2y$, and $n = 5$. Thus, by the binomial theorem, we obtain

$$(x - 2y)^5 = \sum_{i=0}^{5} \binom{5}{i} (x)^{5-i} (-2y)^i = \binom{5}{0} x^5 + \binom{5}{1} x^4(-2y)$$

$$+ \binom{5}{2} x^3(-2y)^2 + \binom{5}{3} x^2(-2y)^3 + \binom{5}{4} x(-2y)^4 + \binom{5}{5} (-2y)^5$$

Now, from our previous discussion, we know that

$$\binom{5}{0} = \binom{5}{5} = 1, \quad \binom{5}{1} = \binom{5}{4} = 5, \quad \text{and} \quad \binom{5}{2} = \binom{5}{3} = 10.$$

Hence,

$$(x - 2y)^5 = 1x^5 + 5x^4(-2y) + 10x^3(4y^2) + 10x^2(-8y^3) + 10x(16y^4) + 1(-32y^5)$$
$$= x^5 - 10x^4y + 40x^3y^2 - 80x^2y^3 + 160xy^4 - 32y^5 \qquad ◆$$

The **kth term** $(1 < k < n + 1)$ in the expansion of $(A + B)^n$ can be written as follows.

◆◆ **kth Term of $(A + B)^n$**

$$\binom{n}{k - 1} A^{n-(k-1)} B^{k-1}$$

PROBLEM 4   Find the fifth term in the expansion of $(x + y)^9$.   ◆

Exercises 8.6

◆ Basic Skills

In Exercises 1–6, simplify each factorial expression.

1. (a) $10 \cdot 9 \cdot 8!$ (b) $k(k-1)(k-2)!$

2. (a) $\dfrac{15!}{14!}$ (b) $\dfrac{(k+1)!}{k!}$

3. (a) $\dfrac{21!}{21}$ (b) $\dfrac{k!}{k}$

4. (a) $\dfrac{12!}{11 \cdot 9!}$ (b) $\dfrac{(k+1)!}{k \cdot (k-2)!}$

5. (a) $\dfrac{1}{10!} + \dfrac{1}{9!}$ (b) $\dfrac{1}{k!} + \dfrac{1}{(k-1)!}$

6. (a) $\dfrac{1}{7! \cdot 11!} + \dfrac{1}{6! \cdot 12!}$

(b) $\dfrac{1}{k! \cdot (r-1)!} + \dfrac{1}{(k-1)! \cdot r!}$

In Exercises 7–20, use the formulas for permutations and combinations to help answer each question. Use the permutation and combination features of a graphing calculator to verify each answer.

7. A family of five has bought seats to a baseball game.
   (a) In how many different ways can they be seated in the five seats?
   (b) In how many ways can they be seated if the father must sit in the middle seat?

8. Signals are sent from a battleship by placing four flags on a vertical pole. How many different signals can be formed by selecting the four flags from
   (a) four differently colored flags?
   (b) two identical red and two identical white flags?

9. How many distinct permutations can be made from the word ALGEBRA?

10. How many distinct permutations starting with M can be made from the word MISSISSIPPI?

11. How many ways can five apple trees, three pear trees, and two peach trees be planted along a driveway if it is not possible to distinguish between trees of the same type?

12. In how many ways is it possible for seven coin tosses to land as four heads and three tails?

13. From nine equally qualified students, how many ways may three of them be selected for scholarships?

14. A hockey lineup consists of two forwards, two defensemen, one center, and one goalie. How many different lineups can a coach form from a roster that has six forwards, six defensemen, three centers, and three goalies?

15. From a collection of records consisting of three hard-rock records, three soft-rock records, and two jazz records, a disk jockey selects five records at random to play.
    (a) How many different selections are possible?
    (b) How many selections are possible with two soft-rock records?
    (c) How many selections are possible with two hard-rock records and two soft-rock records if a certain jazz record must be played.

16. The Greek alphabet contains 24 letters. How many fraternity names can be formed using
    (a) three different Greek letters?
    (b) three Greek letters?

17. A college baseball team has 15 players. How many different ways can the nine starting positions be filled if each player can play any position?

18. In how many ways can a kindergarten teacher assign seats to her 12 students if the classroom has 15 chairs?

19. How many ways can 14 professors be assigned to teach nine sections of college algebra if no professor is assigned more than one section?

20. Suppose 54 Republicans and 46 Democrats sit in the United States Senate. How many ways can a committee of 4 Republicans and 3 Democrats be chosen?

In Exercises 21–34, evaluate each binomial coefficient expression.

21. $\dbinom{42}{1}$ 22. $\dbinom{75}{74}$

23. $\dbinom{19}{19}$ 24. $\dbinom{25}{25}$

25. $\dbinom{31}{0}$ 26. $\dbinom{15}{0}$

27. $\dbinom{6}{2}$ 28. $\dbinom{8}{3}$

29. $\begin{pmatrix} 9 \\ 6 \end{pmatrix}$

30. $\begin{pmatrix} 10 \\ 7 \end{pmatrix}$

31. $\begin{pmatrix} 9 \\ 7 \end{pmatrix} \begin{pmatrix} 7 \\ 3 \end{pmatrix}$

32. $\begin{pmatrix} 11 \\ 8 \end{pmatrix} \begin{pmatrix} 10 \\ 1 \end{pmatrix} \begin{pmatrix} 8 \\ 8 \end{pmatrix}$

33. $\begin{pmatrix} 12 \\ 5 \end{pmatrix} + \begin{pmatrix} 12 \\ 4 \end{pmatrix} - \begin{pmatrix} 13 \\ 5 \end{pmatrix}$

34. $\begin{pmatrix} 9 \\ 4 \end{pmatrix} - \begin{pmatrix} 8 \\ 4 \end{pmatrix} - \begin{pmatrix} 8 \\ 3 \end{pmatrix}$

In Exercises 35–44, use the binomial theorem to expand each expression.

35. $(x + y)^5$ **36.** $(m - n)^7$

37. $(2x - 1)^6$ **38.** $(1 + 3y)^4$

39. $(3a + 2b)^4$ **40.** $(5x - 2y)^5$

41. $(2x^2 - 1)^7$ **42.** $(1 + m^3)^8$

43. $(\sqrt{x} + 2)^8$

44. $(2x^{1/3} - 3y)^6$

In Exercises 45–52, find the indicated term of the expansion.

45. Fourth term of $(x + y)^{12}$

46. Third term of $(a - b)^{16}$

47. Third term of $(3n - 2)^7$

48. Fourth term of $(2p + 1)^9$

49. Fifth term of $(x + 3y^2)^8$

50. Fifth term of $(x - 2y^{-1/2})^8$

51. Seventh term of $(\sqrt{t} - 2)^{10}$

52. Next to last term of $(9x^2 - y^5)^{12}$

◆ Critical Thinking

In Exercises 53–58, solve for n.

53. $_nP_2 = 182$ **54.** $_nP_3 = 1320$

55. $_nC_2 = 36$ **56.** $_nC_3 = 120$

57. $_nC_3 = {}_nP_2$ **58.** $_nC_3 - {}_{n-1}P_2 = 0$

59. Show that $_nC_r = {}_nC_{n-r}$.

60. Show that $_nC_r + {}_nC_{r+1} = {}_{n+1}C_{r+1}$.

61. In how many different ways can k distinguishable trees be planted in a circle if $k = 2$? 3? 4? 5? . . . n? [*Hint:* Two circular permutations are different only if corresponding objects are preceded or followed by different objects. Hence, simply rotating the objects clockwise or counterclockwise one or more positions does not make a different circular permutation.]

In Exercises 62 and 63, show that the given statement is true for all positive integers n.

62. $\begin{pmatrix} n \\ 0 \end{pmatrix} + \begin{pmatrix} n \\ 1 \end{pmatrix} + \begin{pmatrix} n \\ 2 \end{pmatrix} + \cdots + \begin{pmatrix} n \\ n \end{pmatrix} = 2^n$

63. $\begin{pmatrix} n \\ 0 \end{pmatrix} - \begin{pmatrix} n \\ 1 \end{pmatrix} + \begin{pmatrix} n \\ 2 \end{pmatrix} - \cdots + (-1)^n \begin{pmatrix} n \\ n \end{pmatrix} = 0$

64. Use the binomial theorem to find the indicated power of the complex number.

     (a) $(2 + 3i)^4$ (b) $(2 - i)^6$

65. Use the binomial theorem to expand each trinomial.

     (a) $(x^2 + x + 1)^5$ (b) $(x^2 - x - 1)^5$

66. For each function f, defined as indicated, find the difference quotient $\dfrac{f(x + h) - f(x)}{h}$.

     (a) $f(x) = x^4$

     (b) $f(x) = x^n$ for n a positive integer

67. In the 17th century, French mathematician Blaise Pascal discovered a triangular pattern for the coefficients of the terms in the expansion of $(A + B)^n$. The first seven rows of *Pascal's triangle* are given:

Row													Coefficients in the expansion of:
1						1							$(A + B)^0$
2					1		1						$(A + B)^1$
3				1		2		1					$(A + B)^2$
4			1		3		3		1				$(A + B)^3$
5		1		4		6		4		1			$(A + B)^4$
6	1		5		10		10		5		1		$(A + B)^5$
7	1	6		15		20		15		6		1	$(A + B)^6$

Each row in the triangle starts and ends with a coefficient of 1. The other coefficients in each row can be obtained from the coefficients in the preceding row. Do you see the pattern? Use your observation to determine the coefficients for each row:

     (a) row 8   **(b)** row 9   **(c)** row 10

68. Find the sum of the numbers in each row of Pascal's triangle (see Exercise 67). Do you observe a pattern to these sums? Guess a formula for the sum of the numbers in the nth row of Pascal's triangle, and verify your answer.

69. In calculus, it is shown that the expansion of $(1 + x)^n$, where n is a negative integer or a rational number and $|x| < 1$, is given by the *infinite binomial series*

$$1 + nx + \frac{n(n-1)}{2!}x^2 + \frac{n(n-1)(n-2)}{3!}x^3 + \cdots$$

Write the first four terms in the expansion of the given expression.

(a) $(1 + x)^{-1}$ provided $|x| < 1$

(b) $(1 - y)^{-2}$ provided $|y| < 1$

(c) $(1 - z)^{-1/2}$ provided $|z| < 1$

(d) $(1 + x)^{1/3}$ provided $|x| < 1$

70. Use mathematical induction to prove that $n! > 2^n$ for all integers $n \geq 4$.

71. In calculus, it is shown that the *power series* for e^x is given by

$$e^x = 1 + x + \frac{x^2}{2!} + \frac{x^3}{3!} + \frac{x^4}{4!} + \cdots$$

Use the first ten terms of this series to estimate the value of e. Compare this value of e to the value you obtain by using the $\boxed{e^x}$ key on your calculator with $x = 1$.

72. *Stirling's formula* states that a good approximation to $n!$, especially for large values of n, is given by

$$\sqrt{2\pi n}\left(\frac{n}{e}\right)^n.$$

(a) Use Stirling's formula and a calculator to estimate 5!, 10!, and 15!.

(b) Use the $\boxed{n!}$ key on your calculator to find the values of 5!, 10!, and 15!.

(c) Compute the error in the approximations of 5!, 10!, and 15! when using Stirling's formula.

Chapter 8 Review

◆ **Questions for Writing or Group Discussion**

1. What is the domain of an *infinite sequence function*?

2. Explain the similarities and differences between a *sequence* and a *series*.

3. Explain what is meant by defining a sequence *recursively* and *explicitly*.

4. Explain the procedure for finding the fifth element of a sequence whose *general element* is known.

5. How can the process of *mathematical induction* be used to show that a baby can climb a flight of stairs?

6. What is the upper value of the *summation variable* for an infinite series?

7. State the procedure for changing a finite series from *expanded form* to *sigma form*. Illustrate with an arithmetic series.

8. List three *summation properties*. Give an example of each that illustrates its use.

9. Explain the procedure for finding the *common ratio* of a geometric sequence and the *common difference* of an arithmetic sequence.

10. Does there exist a sequence that is both an *arithmetic sequence* and a *geometric sequence*? If so, give an example.

11. Do the opposites of the elements in an arithmetic sequence also form an arithmetic sequence? Explain.

12. Do the reciprocals of the elements in a geometric sequence also form a geometric sequence? Explain.

13. What is a *simple annuity*?

14. What can we conclude about an infinite geometric series whose *nth partial sum* approaches a finite value as n increases without bound ($n \to \infty$)?

15. Under what conditions does an infinite geometric series *converge* or *diverge*?

16. Is the sequence log 2, log 4, log 8, log 16, . . . arithmetic, geometric, or neither? Explain.

17. Give an example of an *infinite geometric series* that
(a) converges and has a sum of 3.   **(b)** diverges.

18. Explain the procedure for changing a *repeating decimal* to reduced fractional form.

19. What is meant by the *interval of convergence* of an infinite geometric series with variable terms?

20. State another *binomial coefficient* that is equal to $\begin{pmatrix} n \\ k \end{pmatrix}$.

21. What is the difference between a *permutation* and a *combination*? Illustrate with examples.

22. State the *binomial theorem*. Illustrate with an example how it is used to expand a binomial.

◆ **Review Exercises**

In Exercises 1 and 2, list the first six elements of each sequence function, then sketch its graph.

1. $u(n) = \dfrac{n+1}{n}$ **2.** $v(n) = (-1)^{n+1} (2n)$

In Exercises 3–8, find the first six elements of a sequence with the given general element.

3. $a_n = 3n - 7$ **4.** $a_n = 6\left(-\frac{1}{2}\right)^{n-1}$

5. $a_n = (-1)^{n+1} n!$ **6.** $a_n = \dfrac{n}{n+2}$

7. $a_n = \begin{cases} n - 1 & \text{if } n \text{ is odd} \\ (a_{n-1})^2 & \text{if } n \text{ is even} \end{cases}$

8. $a_n = \begin{cases} 3 & \text{if } n = 1 \\ 5 & \text{if } n = 2 \\ a_{n-1} + a_{n-2} & \text{if } n \geq 3 \end{cases}$

In Exercises 9–22,

(a) determine whether each sequence appears to be arithmetic, geometric, or neither.

(b) use pattern recognition to find the general element of each sequence.

9. $-1, 3, -5, 7, -9, 11, -13, 15, \ldots$

10. $2, -4, 6, -8, 10, -12, \ldots$

11. $3, 9, 15, 21, 27, 33, 39, \ldots$

12. $49, 38, 27, 16, 5, -6, -17, -28, \ldots$

13. $768, 192, 48, 12, 3, \ldots$

14. $5, 15, 45, 135, 405, 1215, \ldots$

15. $\frac{1}{2}, -\frac{3}{4}, \frac{5}{6}, -\frac{7}{8}, \frac{9}{10}, \ldots$

16. $3, \frac{4}{3}, 1, \frac{6}{7}, \frac{7}{9}, \ldots$

17. $-4, 2, -1, \frac{1}{2}, -\frac{1}{4}, \frac{1}{8}, \ldots$

18. $2500, 25, 0.25, 0.0025, 0.000025, \ldots$

19. $\frac{13}{3}, 3, \frac{5}{3}, \frac{1}{3}, -1, \ldots$

20. $-\frac{3}{2}, -\frac{1}{2}, \frac{1}{2}, \frac{3}{2}, \frac{5}{2}, \frac{7}{2}, \ldots$

21. $0, 3, 8, 15, 24, 35, 48, 63, 80, \ldots$

22. $1, 1, 2, 6, 24, 120, 720, \ldots$

In Exercises 23–34,

(a) determine whether each series is arithmetic, geometric, or neither.

(b) find the sum of each series.

23. $\displaystyle\sum_{i=1}^{35} 2i$ **24.** $\displaystyle\sum_{i=1}^{20} (7 - 3i)$

25. $\displaystyle\sum_{i=2}^{10} 8\left(-\frac{3}{2}\right)^{i-1}$ **26.** $\displaystyle\sum_{i=0}^{12} (-3)^i$

27. $\displaystyle\sum_{i=1}^{18} 3i^2$ **28.** $\displaystyle\sum_{i=1}^{24} i(i - 3)$

29. $\displaystyle\sum_{i=1}^{\infty} 5\left(\frac{1}{2}\right)^{i-1}$ **30.** $\displaystyle\sum_{i=2}^{\infty} \dfrac{1}{(-2)^i}$

31. $\displaystyle\sum_{i=0}^{30} \left(\dfrac{i}{3} - 2\right)$ **32.** $\displaystyle\sum_{i=0}^{32} \dfrac{5i + 4}{6}$

33. $\displaystyle\sum_{i=1}^{16} (2i^3 - 3i + 5)$ **34.** $\displaystyle\sum_{i=1}^{6} \dfrac{(-1)^{i+1} \, 2^i}{i}$

In Exercises 35–40, prove each statement using mathematical induction.

35. $2 + 8 + 14 + 20 + \cdots + (6n - 4) = 3n^2 - n$

36. $6 + 11 + 16 + 21 + \cdots + (5n + 1) = \dfrac{n(5n + 7)}{2}$

37. $1 \cdot 2 + 2 \cdot 3 + 3 \cdot 4 + 4 \cdot 5 + \cdots + n(n + 1)$
$= \dfrac{n(n + 1)(n + 2)}{3}$

38. $1 \cdot 2 \cdot 3 + 2 \cdot 3 \cdot 4 + 3 \cdot 4 \cdot 5 + 4 \cdot 5 \cdot 6 + \cdots$
$+ n(n + 1)(n + 2) = \dfrac{n(n + 1)(n + 2)(n + 3)}{4}$

39. $1^3 + 3^3 + 5^3 + 7^3 + \cdots + (2n - 1)^3 = n^2(2n^2 - 1)$

40. $\dfrac{1}{2} + \dfrac{2}{2^2} + \dfrac{3}{2^3} + \dfrac{4}{2^4} + \cdots + \dfrac{n}{2^n} = 2 - \dfrac{n + 2}{2^n}$

In Exercises 41–44, use the results of Exercises 35–40 to find the sum of the given series.

41. $\displaystyle\sum_{i=1}^{28} i(i + 1)$

42. $\displaystyle\sum_{i=1}^{36} i(i + 1)(i + 2)$

43. $\displaystyle\sum_{i=1}^{15} (2i - 1)^3$

44. $\displaystyle\sum_{i=1}^{12} \dfrac{i}{2^i}$

In Exercises 45 and 46, write each arithmetic series in sigma form using (a) $i = 1$ and (b) $i = 0$ as the lower value of the summation variable.

45. $5 + 9 + 13 + \cdots + 1001$

46. $88 + 76 + 64 + 52 + \cdots$

In Exercises 47 and 48, write each geometric series in sigma form using $i = 1$ as the lower value of the summation variable.

47. $243 - 162 + 108 - 72 + \cdots$

48. $2 + 6 + 18 + 54 + \cdots + 9{,}565{,}938$

In Exercises 49 and 50,

(a) *find the first four partial sums and the nth partial sum for each infinite geometric series.*

(b) *find the sum of the series, if it exists.*

49. $384 - 96 + 24 - 6 + \cdots$

50. $\displaystyle\sum_{i=1}^{\infty} \left(\dfrac{5}{4}\right)^{i-1}$

51. What is the 26th element of an arithmetic sequence whose first element is 92 and whose common difference is -7?

52. What is the general element of an arithmetic sequence whose 6th element is 34 and whose 12th element is 52?

53. What is the sum of the first 60 terms of an arithmetic series whose first term is 9 and whose second term is 24?

54. The first term of an arithmetic series is -19, and the sum of the first 30 terms is 300. What is the 15th term of the series?

55. What is the first element of a geometric sequence whose 11th element is $\frac{3}{32}$ and whose common ratio is $\frac{1}{2}$?

56. What is the general element of a geometric sequence whose second element is 12 and whose eighth element is 8748?

57. Find x so that $x - 3, 2x - 3, 6x + 1$ are the first three terms of a geometric series.

58. The sum of the first three terms of a geometric series is 26, and the third term is 16 more than the first term. What is the general term of the series?

59. What is the sum of an infinite geometric series whose first term is $\frac{27}{76}$ and whose fifth term is $\frac{1}{3}$?

60. What is the common ratio of an infinite geometric series in which each term is three times the sum of all the terms that follow it?

61. Find the reduced fractional form represented by each repeating decimal:
(a) $3.030303\ldots$ (b) $2.7\overline{306}$

62. Find the interval of convergence and the sum of each infinite series:
(a) $9 + 6x + 4x^2 + \cdots$
(b) $3 - 9(x - 1) + 27(x - 1)^2 - \cdots$

63. A Christmas tree farm has 20 trees planted in the first row, and each row thereafter has 6 more trees than the preceding row. If the farm has 116 rows of trees, how many trees are planted altogether?

64. A car costing $12,000 is purchased under an agreement to pay $250 per month plus $\frac{3}{4}$% interest on the unpaid balance for that month. Find the total amount paid for the car.

65. To save for a down payment on a home, a married couple deposits $300 each month in an account that earns interest at the rate of 8% per year compounded monthly. Determine how much money will be in the account at the end of 3 years given the indicated account balance.
(a) The account is empty today.
(b) The account has $6000 today.

66. The enrollment at a certain college is presently 5411 students, but the enrollment is decreasing by 4% each year. Determine the enrollment five years from now.

67. A boy pedals his bicycle up a 105-ft hill at the rate of 36 ft the first second, 24 ft the second second, 16 ft the third second, and so on. Does he ever reach the top of the hill?

68. A tennis ball is dropped from a height of 16 feet, and rebounds each time one-quarter the height from which it last fell. Find the total distance the ball travels before coming to rest.

In Exercise 69–74, evaluate each binomial coefficient expression.

69. $\dbinom{16}{1}$

70. $\dbinom{22}{0}$

71. $\dbinom{13}{11}$

72. $\dbinom{20}{4}$

73. $\dbinom{15}{5} + \dbinom{15}{4} - \dbinom{16}{5}$

74. $\dfrac{\dbinom{7}{2}\dbinom{3}{1}}{\dbinom{10}{3}}$

In Exercises 75 and 76, use the binomial theorem to expand each expression.

75. $(3x - 2)^6$

76. $(2 + a^{1/2})^8$

77. Find the fourth term of $(a^{-1} + b)^{10}$.

78. Find the tenth term of $(4x^2 - 1)^{12}$.

79. How many ways can six sedans, three coupes, and four station wagons be lined up in a row at a dealership, if one does not distinguish between vehicles of the same type?

80. A college athletic director must hire 3 assistant football coaches from 12 equally qualified candidates. How many ways can he select the 3 coaches?

81. In the game of bridge, 13 cards are dealt to each of 4 players from a deck of 52 cards.

    (a) How many different bridge hands are possible?

    (b) How many bridge hands consist of 4 aces?

    (c) How many bridge hands consist of 8 spades?

82. How many ways can 10 people be assigned to play the four infield positions of a baseball team if each person can play any infield position?

1. Find $f(1) + f(2) + f(3) + \cdots + f(20)$ for the given function f.

   (a) $f(x) = 2x$ (b) $f(x) = 2^x$

2. The sum of a finite arithmetic series with n terms is n^2. Find the sum of the first and last terms of the series.

3. (a) Use mathematical induction to show that

$$\sum_{i=1}^{n} i^4 = 1^4 + 2^4 + 3^4 + \cdots + n^4$$
$$= \frac{n(n + 1)(2n + 1)(3n^2 + 3n - 1)}{30}.$$

   (b) Find the sum of the series $1^4 + 2^4 + 3^4 + \cdots + 20^4$.

4. (a) Use mathematical induction to show that

$$\sum_{i=1}^{n} i^5 = 1^5 + 2^5 + 3^5 + \cdots + n^5$$
$$= \frac{n^2(n + 1)^2(2n^2 + 2n - 1)}{12}.$$

   (b) Find the sum of the series $1^5 + 2^5 + 3^5 + \cdots + 20^5$.

5. The nth partial sum of an infinite series is $S_n = 4[1 - (2/3)^n]$. What is the sum of the series?

6. How many consecutive multiples of 6, beginning with 6, must be taken so that their sum is 7650?

7. Solve each system of linear equations.

   (a) $\begin{aligned} 2x + y + z &= 6 \\ x - 2y + 2z &= 10 \\ 3x - y - z &= 4 \end{aligned}$

   (b) $\begin{aligned} x - y \quad\;\; - 2w &= 2 \\ 2y + 3z + 2w &= -8 \\ x \quad\;\; + 2z - 4w &= 1 \\ x + 4y + z \quad\;\; &= 6 \end{aligned}$

8. Given that the sequence a^2, b^2, c^2 is an arithmetic sequence, show that

$$\frac{1}{b + c}, \quad \frac{1}{a + c}, \quad \frac{1}{a + b}$$

   is also an arithmetic sequence.

9. Given the matrices

$$A = \begin{bmatrix} 7 & -2 & 6 \\ 2 & 3 & 5 \\ 5 & -1 & 4 \end{bmatrix}$$

$$X = \begin{bmatrix} x \\ y \\ z \end{bmatrix} \qquad B = \begin{bmatrix} 5 \\ 2 \\ 5 \end{bmatrix}$$

   find A^{-1}, and solve the matrix equation $AX = B$ for X.

10. Find x and y if $x + 3y, 2x - y, x - 8y, 8x + 6$ are the first four elements in an arithmetic sequence.

11. Find x and y if $4, x, y$ are the first three elements in an arithmetic sequence and $x, y, 18$ are the first three elements in a geometric sequence.

12. Solve for x.

   (a) $\begin{vmatrix} x & 2 \\ 1 & 3 \end{vmatrix} = 7$

   (b) $\begin{vmatrix} 2 & x & 3 \\ -2 & 4 & x \\ 5 & -3 & -2 \end{vmatrix} = -51$

13. Determine the nature of the sequence

$$\log a_1, \log a_2, \log a_3, \log a_4, \ldots$$

   if $a_1, a_2, a_3, a_4, \ldots$ is a geometric sequence with positive elements and common ratio r.

14. If $a_1, a_2, a_3, a_4, \ldots$ is an arithmetic sequence with common difference d, what type of sequence is $10^{a_1}, 10^{a_2}, 10^{a_3}, 10^{a_4}, \ldots$?

15. Perform the indicated operations, given that matrices A and B are defined as follows.

$$A = \begin{bmatrix} 2 & 6 & -1 & 0 \\ 5 & 5 & -4 & 2 \\ 0 & 8 & 2 & -3 \\ 6 & 8 & 2 & -3 \end{bmatrix}$$

$$B = \begin{bmatrix} 5 & -1 & 4 & -7 \\ 3 & -1 & 0 & 3 \\ 9 & 2 & 4 & 0 \\ -6 & -3 & 7 & 2 \end{bmatrix}$$

   (a) $A + B$ (b) $3B - 2A$
   (c) AB (d) BA

16. Find the interval of convergence and the sum of the infinite geometric series.

$$\frac{1}{x + 1} + \frac{x}{(x + 1)^2} + \frac{x^2}{(x + 1)^3} + \cdots$$

17. The product of the first seven terms in a geometric series is 0.0002187. What is the fourth term of the series?

18. Find the sum of each infinite geometric series.

   (a) $2, 2 - \sqrt{2}, \dfrac{\sqrt{2} - 1}{\sqrt{2} + 1}, \ldots$

   (b) $\displaystyle\sum_{i=1}^{\infty} \left[\left(\tfrac{1}{3}\right)^{i-1} + \left(\tfrac{1}{2}\right)^{i} \right]$

19. Evaluate each series.

(a) $\displaystyle\sum_{k=0}^{6} \binom{6}{k}$

(b) $\displaystyle\sum_{k=0}^{6} \left[\binom{6}{k} + \binom{6}{6-k} \right]$

(c) $\displaystyle\sum_{k=0}^{6} \left[\binom{6}{k} - \binom{6}{6-k} \right]$

(d) $\displaystyle\sum_{k=0}^{6} \left[\binom{6}{k} \binom{6}{6-k} \right]$

20. Find the middle term in the expansion of $(x - y)^{20}$.

21. The sides of a right triangle form an arithmetic sequence with common difference 6. Find the lengths of the sides of the triangle.

22. Fifty thousand dollars is invested—part at 5%, part at 6%, and part at 7%—yielding an annual interest of $3080. The income from the 6% investment yields $60 more than the income from the 7% investment. How much money is invested at each percentage rate?

23. Suppose 12 points lie in a plane, but no two points lie on the same straight line.

(a) How many different lines are determined by the twelve points?

(b) How many different triangles are determined by the twelve points?

24. It is estimated that a new car costing $12,000 depreciates in value by 25% each year. What is its approximate value after seven years?

25. The sum of the first four terms of a series is $\frac{25}{12}$. The first term is twice the second term, and the second term is twice the fourth term. Six times the third term is one less than twice the sum of the first two terms. Determine the first four terms of the series, and then use pattern recognition to describe the general term.

26. An equilateral triangle has sides of length 12 cm. Inside this triangle, a second equilateral triangle is constructed by joining the midpoints of the sides of the first triangle. Inside the second triangle, a third equilateral triangle is constructed by joining the midpoints of the sides of the second triangle, and so on, indefinitely.

(a) What is the sum of the perimeters of all the equilateral triangles?

(b) What is the sum of the areas of all the equilateral triangles?

27. Given that

$$A = \begin{bmatrix} 3 & 4 \\ 2 & 3 \end{bmatrix} \qquad B = \begin{bmatrix} -6 & 3 \\ 1 & 2 \end{bmatrix}$$

$$C = \begin{bmatrix} 2 & 3 \\ -1 & 4 \end{bmatrix}$$

solve the matrix equation $2AX + B = C$ for the matrix X.

28. Use appropriate substitutions to change the given system of equations to a linear system. Solve the linear system and then state the solution of the original system.

$$\frac{2}{x} + \frac{3}{y} + \frac{1}{z} = 4$$

$$\frac{3}{x} - \frac{5}{y} + \frac{2}{z} = -5$$

$$\frac{4}{x} - \frac{6}{y} + \frac{3}{z} = -7$$

29. A cross-country ski trail connects condominium A to condominium B, with a warming hut located 4 miles from A. Starting from condominium A, the trail goes uphill for 3 miles, level for 4 miles, and then downhill for 6 miles. Suppose a man can ski from A to B in 2 hours; from B to A in 2 hours 45 minutes; and from A to the warming hut and back again to A in 1 hour 30 minutes. What are this skier's rates of skiing uphill, on level ground, and downhill (in miles per hour, mi/h).

30. When two resistors R_a and R_b are connected in parallel, as shown in the figure, the total resistance R_t between points A and B may be found by using the equation

$$\frac{1}{R_t} = \frac{1}{R_a} + \frac{1}{R_b}.$$

Find the values of three resistors R_1, R_2, and R_3 if the total resistance is 60 ohms (Ω) when R_1 and R_2 are connected in parallel, 100 Ω when R_2 and R_3 are connected in parallel, and 75 Ω when R_1 and R_3 are connected in parallel.

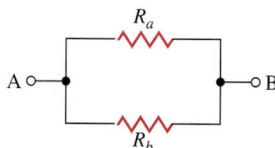

31. How long does it take for an annuity to amount to a million dollars if $600 is invested each month into an account that earns interest at the rate of 9% per year compounded monthly?

32. A company manufactures different types of prefinished oak flooring and packs the flooring in 30-square-foot cartons for shipping. The company receives an order for 65 cartons of a parquet flooring from customer A and an order for 90 cartons of the same parquet flooring from customer B. The company has 100 cartons of this product at warehouse W_1 and 70 cartons at warehouse W_2. The shipping costs from the warehouses to the customers are given in the following table.

(a) How should the order be filled in order to minimize the shipping cost?

(b) What is the minimum shipping cost?

Warehouse	Customer	Shipping cost per carton
W_1	A	$1.25
W_2	A	$1.40
W_1	B	$1.30
W_2	B	$1.50

APPENDIX ◆ Significant Digits

Consider measuring the width of the wooden block in Figure A.1 with a ruler marked in intervals of 0.1-inch. The width of the block appears to be 1.8 inches. However, does the end of the block fall exactly in the middle of the marking for 1.8 inches, or slightly to the left or right of this marking? If we use a powerful magnifying glass, we might attempt to answer this question, but we could never determine the *exact* width of the block. For this reason, we say that every number found by a measuring process is an *approximate number*.

The number 8 in the tenths position of the number "1.8 inches" does have some *significance*, since the width seems to be closer to 1.8 inches than to either 1.7 inches or 1.9 inches. If a digit contributes to our knowledge of how good an approximation is, it is called a **significant digit**. Throughout this text, you are asked to round answers to a certain number of significant digits. The following may be used as a guide for this purpose.

FIGURE A.1

Measuring the width of a block with a ruler marked in 0.1-inch intervals.

◆ **Significant Digits**

1. All nonzero digits are significant.

   Examples: 475 has 3 significant digits.
   12.827 has 5 significant digits.

2. Zeros between nonzero digits are significant.

   Examples: 6506 has 4 significant digits.
   42.0072 has 6 significant digits.

3. Zeros appearing at the end of a decimal number are significant.

   Examples: 7.00 has 3 significant digits.
   76.40 has 4 significant digits.

4. Zeros at the beginning of a decimal number are *not* significant and serve only to locate the decimal point correctly.

   Examples: 0.002 has 1 significant digit.
   0.023 has 2 significant digits.

5. Zeros at the end of an integer are *not* significant unless a tilde (˜) is placed above one of the zeros. The tilde is placed over the last zero that is significant.

   Examples: 22,000 has 2 significant digits.
   22,0̃00 has 4 significant digits.

If a number is written in scientific notation as

$$k \times 10^n, \quad \text{where } 1 \le |k| < 10 \text{ and } n \text{ is an integer,}$$

then the number of significant digits in that number is the same as the number of significant digits in *k*.

Examples: 6.2×10^3 = 6200 has 2 significant digits.

6.20×10^3 = 62̃00 has 3 significant digits.

6.2×10^{-3} = 0.0062 has 2 significant digits.

6.20×10^{-3} = 0.00620 has 3 significant digits.

To avoid making a *rounding error* when working with approximate numbers, we carry along a few extra significant digits through the calculating process and then round the final answer to the desired accuracy. For example, suppose we wish to find the area of the gable end of the house shown in Figure A.2 and round this answer to three significant digits. First, we find the area of the rectangular part of the gable, without rounding the answer, as follows:

$$A = lw = (24.26 \text{ ft})(9.78 \text{ ft}) \approx 237.2628 \text{ sq ft.}$$

Next, we find the area of the triangular part of the gable, without rounding the answer, as follows:

$$A = \tfrac{1}{2}bh = \tfrac{1}{2}(24.26 \text{ ft})(8.52 \text{ ft}) \approx 103.3476 \text{ sq ft.}$$

Finally, we add these areas and then round to the desired accuracy of three significant digits. Hence, the area of the gable end of the house is

$$237.2628 \text{ sq ft} + 103.3476 \text{ sq ft} = 340.6104 \text{ sq ft} \approx 341 \text{ sq ft.}$$

round to 3 significant digits

FIGURE A.2

Dimensions of the gable end of a house

If we round the areas of the rectangular and triangular parts to three significant digits and then add these areas, we accumulate a *rounding error*. Note that

$$237 \text{ sq ft} + 103 \text{ sq ft} = 34\tilde{0} \text{ sq ft.}$$

This answer does not have the desired accuracy. When working with approximate numbers, we usually use a calculator to perform the basic operations. It is best to let the calculator store all the digits and then round the final display to the desired accuracy. By using this procedure, you will be sure to avoid rounding errors.

Some exercises in this text that deal with approximate numbers may not specify a required number of significant digits in the final answer. For exercises of this nature, look at the data given in the problem and determine which approximate number has the *fewest* number of significant digits. It is common practice to round the final answer to the same number of significant digits as the approximate number with the fewest number of significant digits. For example, note in Figure A.2 that both 8.52 ft and 9.78 ft have the fewest number of significant digits, namely, three. Hence, it is acceptable to round the area of the gable to three significant digits as well.

◆ Solutions to Problems and Answers to Odd-Numbered Exercises

◆ Chapter 1

SECTION 1.1

◆ Problems

1. (a) $a \leq 8$ (b) $-4 \leq b < 0$

2. $|c - (-1)| \geq 5$ or $|c + 1| \geq 5$

3. (a)

   (b)

4. (a)

   (b)

◆ Exercises

1. $x < 0$ 3. $a \leq 7$ 5. $2 < p \leq 10$ 7. $0 < c < 8$

9. $-2 \leq t \leq 0$ 11. 15   13. 1.9   15. $\frac{29}{24}$ 17. $9\frac{1}{6}$

19. $\pi - \sqrt{2}$ 21. $|a - 7| \geq 3$ 23. $|x + 4| < 2$

25. $|a + 1| > 4$ 27. $|1 - y| < |y|$

29. $|b + 2| > |b - 5|$ 31. $[-4, -2]$ 33. $(2, 6)$

35. $[-1, 4)$ 37. $(-\infty, -1]$ 39. $(10, \infty)$

41. $(-\infty, 0) \cup (0, \infty)$ 43. $(-\infty, -7] \cup [-4, \infty)$

45. $[0, 2) \cup (2, \infty)$ 47. $(-\infty, 1) \cup (1, 6) \cup (6, \infty)$

49. $(-\infty, 0) \cup [2, 3) \cup (3, \infty)$

51.

53.

55.

57.

59.

61. $x - \pi$ 63. 1   65. $-5 < x < 5$

67. The interval notation $[a, a]$ represents the single point a on the real number line. The interval notation (a, a) is meaningless.

69. $(-\infty, -1) \cup (0, 1)$ 71. $\frac{157}{50}$, π, $\frac{22}{7}$, 3.145, $\sqrt{10}$, 3.2

SECTION 1.2

◆ Problems

1. $\dfrac{4^{-2}x^{-3}y^4}{4x^{-4}y^{-3}} = \dfrac{x^4y^3y^4}{4 \cdot 4^2x^3} = \dfrac{xy^7}{4^3}$ or $\dfrac{xy^7}{64}$

2. $\dfrac{9.3 \times 10^7 \text{ mi}}{1.86 \times 10^5 \text{ mi/s}} = 5 \times 10^2 \text{ s}$

   $= 500 \text{ s} \cdot \dfrac{1 \text{ min}}{60 \text{ s}} \approx 8.3 \text{ min}$

3. $P = \dfrac{(6 \times 10^5)^2}{3 \times 10^1} = \dfrac{36 \times 10^{10}}{3 \times 10^1}$

   $= 12 \times 10^9 \text{ W} = 12{,}000 \times 10^6 \text{ W} = 12{,}000 \text{ MW}$

◆ Exercises

1. 36   3. -16 5. $\frac{1}{64}$ 7. $-\frac{1}{64}$ 9. $\frac{81}{16}$ 11. $\frac{1}{81}$ 13. $\frac{2}{9}$

15. $-\frac{1}{4}$ 17. $\frac{7}{12}$ 19. $\frac{6}{5}$ 21. $-\dfrac{8}{xy^3}$ 23. $\dfrac{10}{(1 - 2x)^6}$

25. $-50x^4y^{13}$ 27. $\dfrac{16}{y^4}$ 29. $2x^3y^4$ 31. $\dfrac{1}{x + 3}$

33. $(x - 2)^8$ 35. $(2y + 3)^2$ 37. $\dfrac{1}{18mn^2}$

39. $8p^3(q - r)^7$ 41. 0.0000000069

43. $-175{,}000{,}000{,}000$ **45.** 0.000392

47. 5.43×10^4 **49.** 1.3×10^{-7}

51. 2.40×10^5 **53.** 2×10^3 **55.** 3.1×10^{-18}

57. 3.125×10^{48} **59.** 2.4×10^{-6} **61.** 1.119

63. $42{,}3\tilde{0}0$ **65.** 1.07×10^{-23} **67.** 1.99×10^{-30}

69. 12 yr   **71.** $6.7 \, \mu A$ **73.** $33{,}100$ sq m   **75.** 1

77. Since $a/0$ is undefined, the expression $1 = (a/0)^0$ is meaningless.

79. (a) $\$643.70$ (b) Approximately $\$231{,}732$

SECTION 1.3

◆ Problems

1. (a) $\sqrt[5]{(x^2 + 2)^2} = (x^2 + 2)^{2/5}$
(b) $(2x^2)^{3/7} = \sqrt[7]{(2x^2)^3} = \sqrt[7]{8x^6}$

2. (a) $[(-2)^4 x^4]^{1/4} = [(-2)^4]^{1/4}(x^4)^{1/4} = |-2|x = 2x$
(b) $[(-2)^4]^{1/4}(x^4)^{1/4} = |-2||x| = 2|x|$

3. (a) $\sqrt[3]{32(x+y)^3} = \sqrt[3]{8(x+y)^3 \cdot 4} = 2(x+y)\sqrt[3]{4}$
(b) $\sqrt[4]{\sqrt[3]{x^2y^6}} = \sqrt[12]{x^2y^6} = \sqrt[12]{(xy^3)^2} = \sqrt[6]{xy^3}$

4. $\dfrac{\sqrt[3]{5x^2}}{\sqrt[3]{2y}} = \dfrac{\sqrt[3]{5x^2}}{\sqrt[3]{2y}} \cdot \dfrac{\sqrt[3]{4y^2}}{\sqrt[3]{4y^2}} = \dfrac{\sqrt[3]{20x^2y^2}}{2y}$

◆ Exercises

1. 2 **3.** -2 **5.** 343 **7.** 25 **9.** $\frac{1}{1000}$

11. $-\frac{1}{27}$ **13.** $\frac{1}{4}$ **15.** $\frac{2}{3}$ **17.** -1 **19.** 8

21. 2.290 **23.** 0.001123 **25.** 16.58 **27.** 0.6553

29. $(2a)^{1/2}$ **31.** $(y^2 - 3)^{4/7}$ **33.** $x(x^2 + y^2)^{-1/2}$

35. $x^{3/2}$ **37.** $\sqrt{x}$ **39.** $\sqrt[4]{27m^6}$ or $m\sqrt[4]{27m^2}$

41. $\sqrt[5]{(x+y)^4}$ **43.** $\dfrac{2}{\sqrt[3]{x^2}}$ **45.** $4x$ **47.** $\dfrac{4}{x^2}$

49. $(x+y)^{17/6}$ **51.** $\dfrac{y^2}{16x^2}$ **53.** $\dfrac{x^{4/3}}{(x^2+4)^{1/4}}$ **55.** $\dfrac{y^2}{2}$

57. $\dfrac{x}{(x^2 - 1)^{1/2}}$ **59.** $2|x| + 2x$ **61.** $2\sqrt{6}$

63. $x + 1$ **65.** $x^4\sqrt{x}$ **67.** $2xy\sqrt[3]{2x}$

69. $7(a + b)\sqrt{a^2 + b^2}$ **71.** $\dfrac{3xy\sqrt[3]{2y}}{5z}$

73. $\sqrt[3]{5xy^2}$ **75.** $\dfrac{x\sqrt{2x}}{y}$ **77.** $|x + 1|$

79. $6\sqrt{2}$ **81.** $\dfrac{\sqrt[3]{12x^2y}}{3xy}$ **83.** $\dfrac{42a}{5\sqrt{6a}}$

85. $\dfrac{x + 2}{\sqrt[3]{3x(x+2)}}$ **87.** $\sqrt[6]{675}$ **89.** $\sqrt[6]{x^3y^4}$

91. (a) $2 - 2\sqrt[3]{2} + 2\sqrt[3]{5}$
(b) $a^2b\sqrt{c} + b^2c\sqrt{a} + ac^2\sqrt{b}$

93. $\dfrac{\pi\sqrt{2L}}{4}$ **95.** Approximately 8.00 ft

SECTION 1.4

◆ Problems

1. $\left(3 + \sqrt{-25}\right) - \left(3 - \sqrt{-36}\right)$
$= (3 + 5i) - (3 - 6i) = 11i$

2. $(1 + 3i)(2 + 5i) = 2 + 11i + 15i^2 = -13 + 11i$

3. $\dfrac{1 + \sqrt{-9}}{1 - \sqrt{-9}} = \dfrac{(1 + 3i)(1 + 3i)}{(1 - 3i)(1 + 3i)}$

$= \dfrac{1 + 6i + 9i^2}{1 - 9i^2}$

$= \dfrac{-8 + 6i}{10} = -\dfrac{4}{5} + \dfrac{3}{5}i$

4. $\left(2 - \sqrt{-9}\right)^2 = (2 - 3i)^2 = (2 - 3i)(2 - 3i)$
$= 4 - 12i + 9i^2 = -5 - 12i$

◆ Exercises

1. $3 + 8i$ **3.** $15 + 3i$ **5.** $24 + 16i$ **7.** $39 + 33i$

9. 6 **11.** $2 - 4i$ **13.** $\frac{21}{29} + \frac{9}{29}i$ **15.** $-\frac{3}{5} - \frac{4}{5}i$

17. $9i$ **19.** -18 **21.** $3i\sqrt{5}$ **23.** $-2 + 33i$

25. 3 **27.** $30 + 19i$ **29.** $-\dfrac{\sqrt{6}}{5} + \dfrac{2}{5}i$ **31.** -5

33. $-2i$ **35.** $-27i$ **37.** $108i$ **39.** $5 - 12i$

41. $-\frac{9}{1681} - \frac{40}{1681}i$ **43.** 0

45. Since $(1 + 0i)(a + bi) = a + bi$, the complex number $1 + 0i$ is the multiplicative identity of $a + bi$.

47. (a) 0 (b) 0 **49.** (a) 0 (b) 0 **51.** $\pm 3i$

53. (a) $4.24i$ (b) $22.5i$ (c) $12.4i$ (d) $0.936i$

55. 1.07 A

SECTION 1.5

◆ Problems

1. $(5x^2 - 3x) + (5x - x^2) - (2x + 4x^2)$
   $= 5x^2 - 3x + 5x - x^2 - 2x - 4x^2$
   $= 0$

2.
$$
\begin{array}{l}
m^3 - m^2 + 3m \;\; + 4 \\
\qquad\qquad m^2 + 2 \\
\hline
m^5 - m^4 + 3m^3 + 4m^2 \\
\qquad\quad\; 2m^3 - 2m^2 + 6m + 8 \\
\hline
m^5 - m^4 + 5m^3 + 2m^2 + 6m + 8
\end{array}
$$

3. (a) $(2x - 3)(2x + 3) = (2x)^2 - (3)^2 = 4x^2 - 9$
   (b) $(3x - 5y)^2 = (3x)^2 + 2(3x)(-5y) + (-5y)^2$
   $= 9x^2 - 30xy + 25y^2$
   (c) $(x^4 + 4)^3 = (x^4)^3 + 3(x^4)^2(4) + 3(x^4)(4)^2 + (4)^3$
   $= x^{12} + 12x^8 + 48x^4 + 64$

4. $x^2 - [y(x + 2y) - (x + y)(2y - x)]$
   $= x^2 - [xy + 2y^2 - (2y^2 + xy - x^2)]$
   $= x^2 - [xy + 2y^2 - 2y^2 - xy + x^2]$
   $= x^2 - x^2 = 0$

5. $(x^{-4} + 4)^3 = (x^{-4})^3 + 3(x^{-4})^2(4) + 3(x^{-4})(4)^2 + (4)^3$
   $= x^{-12} + 12x^{-8} + 48x^{-4} + 64$

◆ Exercises

1. $-x^2$ 3. $-5x^2y$ 5. $-2x^{-1}$ 7. $10\sqrt{3y}$

9. $\sqrt{10x}$ 11. polynomial of degree 6

13. not a polynomial, because y has a negative exponent

15. polynomial of degree 2   17. polynomial of degree 8

19. not a polynomial, because variables appear in the radicand

21. $7m^2 - 5m - 9$ 23. $2xy - x^2$

25. $-32m^5n + 8m^4 - 4m^3 + 4m^2$

27. $3x^3 + 7x^2 - 27xy - 63y$ 29. $12x^2 - xy - y^2$

31. $9x^2 - y^2$ 33. $2n^3 + n^2 - 12n + 4$

35. $x^3 + 27$ 37. $8x^3 - 1$ 39. $x^3\sqrt{3} - 9$

41. $m^2 + 12m + 36$ 43. $4a^2 - 12ab^2 + 9b^4$

45. $9x^6 + 30x^3 + 25$ 47. $x^3 + 9x^2 + 27x + 27$

49. $64x^3 - 144x^2y^2 + 108xy^4 - 27y^6$

51. $-4n^2 - 9n + 4$ 53. m^3

55. $32x^3 - 32x^2 + 8x$

57. $a^4 - 32a^2 + 256$ 59. $162x^4 - 162x^3 + 51x^2$

61. $4x^{-2}$ 63. $x - 3y$ 65. $4x - 28y\sqrt{x} + 49y^2$

67. $x^{1/3} - x^{2/3} - 8$

69. degree of $P + Q$ is m; degree of $P - Q$ is m; degree of PQ is $m + n$; and degree of P^2 is $2m$

71. $5x^2 + 12x - 10$ 73. $10\pi r - 25\pi$

75. (a) 19   (b) -5 77. $1.06x$

79. (a) $(6.32 \times 10^{-3})x^2 - (3.46 \times 10^{-4})x^3$
   $+ (4.72 \times 10^{-6})x^4$
   (b) 0.53 inch

SECTION 1.6

◆ Problems

1. $(9 - 3y) + (2xy - 6x) = 3(3 - y) - 2x(3 - y)$
   $= (3 - y)(3 - 2x)$

2. $x^2 + 10x + 25 = (x + 5)(x + 5) = (x + 5)^2$

3. (a) $(x + y)^2 - 25y^4 = [(x + y) + 5y^2][(x + y) - 5y^2]$
   $= (x + y + 5y^2)(x + y - 5y^2)$
   (b) $64 - x^3 = (4 - x)(16 + 4x + x^2)$

◆ Exercises

1. $ab(3a - 1)$ 3. $3xy(2x^2 - 3x + y)$

5. $(x - 1)(5 - x)$ 7. $7m(m + 3)^3$

9. $(a + b)(c + d)$ 11. $(x - 3)(x^2 + 2)$

13. $(2 - x)(3 - y)$ 15. $(5 - x^2)(3 - 2y)$

17. $(a + 5)(a - 2)$ 19. $(x - 9y)(x - 2y)$

21. $(12 + x)(2 + x)$ 23. $(x^2 + 6)(x^2 - 3)$

25. $(2x + 3)(x + 1)$ 27. $(3a - 4b)(a + 5b)$

29. $(4x + 1)(x - 3)$ 31. $(5 - 2t)(3 - 2t)$

33. $(4x - 5y)(3x - 2y)$ 35. $(2n^2 + 5)(n^2 + 1)$

37. $(x^3 - 2y^2)(6x^3 - 5y^2)$ 39. $(y^6 + 2)(8y^6 - 3)$

41. $(x + 9)(x - 9)$ 43. $(5x + y)(5x - y)$

45. $(2x^2 - 5x + 10)(2x^2 + 5x + 10)$

47. $(t + 2)(t^2 - 2t + 4)$

49. $(3x + 2y^2)(9x^2 - 6xy^2 + 4y^4)$

51. $(2 - 5n)(4 + 10n + 25n^2)$

53. $(t + 4)^2$ **55.** $(2a^2 - 3b^2)^2$

57. $4xy(x - 3)(x - 4)$ **59.** $2x(3x + 2)(x - 2)$

61. $5a(a + 3)(a - 3)$ **63.** $2xy(2x + 1)(4x^2 - 2x + 1)$

65. $2(n + 2)(n - 2)(n^2 + 4)$

67. $(2x + 3)(2x - 3)(x + 1)(x - 1)$

69. $3a(a^3 + 3)(a - 1)(a^2 + a + 1)$

71. $2y(x + 1)(2x + 3)(2x - 3)$

73. $(x + 1)(x - 1)(x^2 + x + 1)(x^2 - x + 1)$

75. $(x + y - 2)(x - y - 2)$

77. $(2x + 2y + 3)(2x - 2y + 3)$

79. $\left(x + \sqrt{10}\right)\left(x - \sqrt{10}\right)$

81. $\left(a + b\sqrt[3]{5}\right)\left(a^2 - ab\sqrt[3]{5} + b^2\sqrt[3]{25}\right)$

83. $(x + 5i)(x - 5i)$

85. $\left(t + \sqrt{6}\right)\left(t - \sqrt{6}\right)\left(t + i\sqrt{6}\right)\left(t - i\sqrt{6}\right)$

87. $2\pi rt(h + r)$ **89.** $(3.2 + A)(3.2 - A)(1 + A^2)$

SECTION 1.7

◆ Problems

1. $\dfrac{16 - x^2}{2x^3 - 5x^2 - 12x} = \dfrac{(4 + x)(4 - x)}{x(x - 4)(2x + 3)}$

$\qquad = \dfrac{-(4 + x)}{x(2x + 3)}$

2. $\dfrac{x^2}{2x^2 - 5x + 2} \div \dfrac{4x}{4 - x^2}$

$\qquad = \dfrac{x^2}{(2x - 1)(x - 2)} \cdot \dfrac{(2 + x)(2 - x)}{4x}$

$\qquad = \dfrac{x(2 + x)}{4(1 - 2x)}$

3. $\dfrac{2x^2 - 9}{2x^2 + 11x + 5} - \dfrac{x^2 + 16}{2x^2 + 11x + 5}$

$\qquad = \dfrac{(2x^2 - 9) - (x^2 + 16)}{2x^2 + 11x + 5}$

$\qquad = \dfrac{x^2 - 25}{(2x + 1)(x + 5)} = \dfrac{x - 5}{2x + 1}$

4. $\dfrac{x + 1}{x^2 - 2x + 1} - \dfrac{1}{x - 1} = \dfrac{x + 1}{(x - 1)^2} - \dfrac{x - 1}{(x - 1)^2}$

$\qquad = \dfrac{(x + 1) - (x - 1)}{(x - 1)^2} = \dfrac{2}{(x - 1)^2}$

5. $\dfrac{\sqrt{(x + h) + 1} - \sqrt{x + 1}}{h} \cdot \dfrac{\sqrt{(x + h) + 1} + \sqrt{x + 1}}{\sqrt{(x + h) + 1} + \sqrt{x + 1}}$

$\qquad = \dfrac{[(x + h) + 1] - (x + 1)}{h\left(\sqrt{(x + h) + 1} + \sqrt{x + 1}\right)}$

$\qquad = \dfrac{1}{\sqrt{(x + h) + 1} + \sqrt{x + 1}}$

6. $\dfrac{\dfrac{3}{x + h} - \dfrac{3}{x}}{h} = \dfrac{\dfrac{3x - 3(x + h)}{x(x + h)}}{h}$

$\qquad = \dfrac{-3h}{x(x + h)} \cdot \dfrac{1}{h} = \dfrac{-3}{x(x + h)}$

7. $x[-2(1 - 3x)^{-3}(-3)] + (1 - 3x)^{-2}$

$\qquad = \dfrac{6x}{(1 - 3x)^3} + \dfrac{1}{(1 - 3x)^2}$

$\qquad = \dfrac{6x + (1 - 3x)}{(1 - 3x)^3}$

$\qquad = \dfrac{3x + 1}{(1 - 3x)^3}$

◆ Exercises

1. $\dfrac{1}{1 - 2x}$ **3.** $-\dfrac{1}{3}$ **5.** $\dfrac{2x + 1}{x - 1}$ **7.** $-\dfrac{n + 1}{n + 6}$

9. $\dfrac{1}{x + 3}$ **11.** $\dfrac{3 + x + y}{3 - x + y}$ **13.** $\dfrac{3(x + 1)}{x - 1}$

15. $\dfrac{2 + n}{1 - n}$ **17.** $\dfrac{2x}{x + y}$ **19.** $-\dfrac{3}{3b + a}$ **21.** $n + 2$

23. $\dfrac{x}{3}$ **25.** $\dfrac{m}{m^2 - 9}$ **27.** $\dfrac{x}{(x - 1)^3}$ **29.** $\dfrac{4 - 2x^2}{x(x + 2)^2}$

31. $\dfrac{-2}{(1 + a)(1 - a)}$ **33.** $\dfrac{1}{a - 1}$ **35.** $\dfrac{x^2 + y^2}{xy}$

37. $-\dfrac{a + b + c}{bc}$ **39.** $\dfrac{-1}{3 + \sqrt{t + 3}}$

41. $\dfrac{2}{\sqrt{2(x + h)} + \sqrt{2x}}$

43. $\dfrac{2x + h}{\sqrt{(x + h)^2 + 1} + \sqrt{x^2 + 1}}$ **45.** $\dfrac{a}{1 - a}$

47. $\dfrac{-1}{x(x + h)}$ **49.** $\dfrac{-(2x + h)}{x^2(x + h)^2}$ **51.** $(8x + 6)(x + 3)^2$

53. $\dfrac{2 + 6t}{(2 - 3t)^4}$ **55.** $\dfrac{5 - 16n}{5(1 - 4n)^{6/5}}$ **57.** $\dfrac{3 - 5t}{(3 - 4t)^{3/4}}$

59. $\dfrac{5t^2 - 1}{t^{4/3}}$

61. The expressions are *not* equal when $x = 1$ because $x^2 - 1 = 0$, and division by 0 is undefined.

63. (a) $y + 7 - \dfrac{5}{y}$ (b) $\dfrac{14}{t + 7} - t - 7$

65. (a) $x + \sqrt{x^2 - 1}$ (b) $\sqrt{a^2 + b^2} - b$

67. $\dfrac{2E}{2R + r}$ **69.** $\dfrac{120 - 60\sqrt{n} - 4n + n^2}{4 - n}$

SECTION 1.8

◆ Problems

1. Adding -3 to both sides, we obtain $|4t - 3| = -1$. However, since the absolute value of a real number is always *nonnegative*, we conclude that this equation has no solution.

2. Multiplying both sides by 8, we obtain

$$4(y + 1) - 5(y - 1) > 8$$
$$4y + 4 - 5y + 5 > 8$$
$$-y > -1$$
$$y < 1$$

Hence, the solution set is $(-\infty, 1)$.

3. $ax + b = cx + d$
$b - d = cx - ax$
$b - d = x(c - a)$
$x = \dfrac{b - d}{c - a} \cdot \dfrac{-1}{-1} = \dfrac{d - b}{a - c}$

4. $A = \dfrac{1}{2}h(b_1 + b_2)$

$\dfrac{2A}{h} = b_1 + b_2$

$b_1 = \dfrac{2A}{h} - b_2$

◆ Exercises

1. -7 **3.** -4 **5.** -4 **7.** $-\frac{3}{2}$ **9.** 5

11. 2 **13.** $\frac{2}{5}$ **15.** $\frac{7}{6}$ **17.** ± 4 **19.** $1, \frac{1}{5}$

21. $-\frac{1}{3}, \frac{19}{3}$ **23.** $\pm\sqrt{3}$ **25.** ± 4 **27.** $\pm\sqrt{2}$

29. $-6, 8$ **31.** $(-\infty, 3)$ **33.** $(-\infty, 2]$ **35.** $(-7, \infty)$

37. $\left(-\frac{3}{2}, \infty\right)$ **39.** $(-\infty, 10]$ **41.** $(-3, 3)$

43. $\left(-\infty, -\frac{3}{2}\right) \cup \left(\frac{3}{2}, \infty\right)$ **45.** $(-\infty, 1) \cup (4, \infty)$

47. $\left[\frac{5}{2}, \frac{11}{2}\right]$ **49.** $\left(-\infty, -\frac{14}{3}\right] \cup \left[\frac{10}{3}, \infty\right)$

51. $x = \dfrac{bc}{a}$, $a \neq 0$, $b \neq 0$

53. $x = \dfrac{c}{a - b}$, $a \neq b$

55. $x = \dfrac{ak - a}{b - 3k}$, $k \neq 0$, $b \neq 3k$

57. $x = \dfrac{-b \pm 1}{a}$, $a \neq 0$

59. $x \leq \dfrac{-b}{a - c}$ **61.** $x \geq \dfrac{abc - ay}{b}$ **63.** $15°C$

65. 5 cm   **67.** 9π ft   **69.** (a) $E = IR$ (b) $R = \dfrac{E}{I}$

71. (a) $L_0 = \dfrac{L}{1 + \mu\Delta t}$ (b) $\Delta t = \dfrac{L - L_0}{L_0\mu}$

73. (a) $r = \dfrac{A - P}{Pt}$ (b) $P = \dfrac{A}{1 + rt}$

75. (a) $t = \dfrac{mv - mv_0}{F}$ (b) $m = \dfrac{Ft}{v - v_0}$

77. (a) $R_1 = \dfrac{P - I^2R_2}{I^2}$ (b) $I = \pm\sqrt{\dfrac{P}{R_1 + R_2}}$

79. $A - B = 0$ is an equivalent equation; $A/B = 1$ and $A^2 = B^2$ are *not* equivalent equations.

81. $\left(-\infty, -\frac{8}{3}\right)$

83. If $x > 5$, then $5 - x < 0$. Hence, when dividing both sides by $5 - x$, we must reverse the direction of the inequality.

85. $48.6°F$

CHAPTER 1 REVIEW EXERCISES

1. $a \geq 7$ **3.** $-10 < c < 0$ **5.** $|a - 3| \leq 2$

7. 15 **9.** $[3, 7]$ **11.** $(-1, \infty)$

13. $(-\infty, -6] \cup [-1, \infty)$ **15.** $[0, 4) \cup (4, \infty)$

17. $-50x^6y^{13}$ **19.** $\dfrac{1}{(2x + 1)^4}$ **21.** $x - 1$

23. $x + 1$ **25.** $\dfrac{y^6}{8x^{3/2}}$ **27.** 3×10^{-24} **29.** $(x + 2)^{2/3}$

31. $\dfrac{1}{\sqrt[6]{x+4}}$ **33.** $4(x+y)\sqrt{2(x+y)}$

35. $2y\sqrt[4]{4x^3}$ **37.** $3|xy|$ **39.** $22i\sqrt{5}$

41. -64 **43.** $12-i$ **45.** $47+i$

47. $2a^2+a$ **49.** $2\sqrt{10x}$

51. $2x^6+4x^5-3x^4-13x^3-2x^2+9x+3$

53. $30x^4-13x^2y-10y^2$ **55.** $4t^2-25$

57. x^3-64 **59.** $9x^2-42xy+49y^2$

61. $8x^3-60x^2+150x-125$

63. $x^6-27x^4+243x^2-729$ **65.** $-3y^2$

67. x^2+3 **69.** $3y(y-2)(y+2)$

71. $(m-3)(m^2+1)$ **73.** $(3x+4y)(2x-3y)$

75. $(2x^2-3)(x^2-5)$ **77.** $(x+5)(x-3)$

79. $(5x+3y^2)(5x-3y^2)$ **81.** $(2x-1)(4x^2+2x+1)$

83. $(t+2)^3$ **85.** $2xy(4x-3)(x+1)$

87. $2xy(1+2x)(1+3x)(1-3x)$ **89.** $\dfrac{n+9}{3n-2}$

91. $\dfrac{1}{x}$ **93.** $-6x$ **95.** $\dfrac{21y^2-10x^3}{24xy}$

97. $\dfrac{-x(2x+3)}{3y(x+3)}$ **99.** $\dfrac{-3}{x+3}$ **101.** $\dfrac{12}{x(x+4)^2}$

103. $\dfrac{2x}{(x+2y)(x-2y)^2}$ **105.** a

107. $\dfrac{x-5}{\sqrt[3]{3x(x-5)}}$ **109.** $2x+\sqrt{4x^2-1}$ **111.** 2

113. $-\frac{1}{2}$ **115.** $-4,\ \frac{20}{3}$ **117.** ± 4 **119.** $L=\dfrac{Rd^2}{\mu+\mu t}$

121. $(-\infty,-3)$ **123.** $[2,\infty)$

125. $(-\infty,-1)\cup\left(\frac{13}{3},\infty\right)$ **127.** 40 m

129. 103,000 sq ft

131. (a) $r=\sqrt{\dfrac{3V}{\pi h}}$ (b) 6 inches

◆ **Chapter 2**

SECTION 2.1

◆ **Problems**

1.

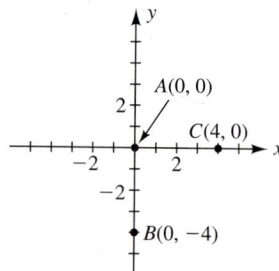

2. $PQ=\sqrt{[1-(-3)]^2+(-2-1)^2}=\sqrt{25}=5$

3. $\left(\dfrac{-4+5}{2},\dfrac{-3+(-1)}{2}\right)=\left(\dfrac{1}{2},-2\right)$

4. $AB=\sqrt{[2-(-2)]^2+(5-3)^2}=\sqrt{20}=2\sqrt{5}$
$AC=\sqrt{(4-2)^2+(1-5)^2}=\sqrt{20}=2\sqrt{5}$
$BC=\sqrt{[4-(-2)]^2+(1-3)^2}=\sqrt{40}=2\sqrt{10}$

Since $AB=AC$, we conclude that triangle ABC is isosceles. Also, since $(AB)^2+(AC)^2=(BC)^2$, we conclude that triangle ABC is a right triangle with hypotense $\overline{BC}$.

Area $=\frac{1}{2}(2\sqrt{5})(2\sqrt{5})=10$ square units.

◆ **Exercises**

1. III   **3.** positive; negative   **5.** x-axis

7. $A(2,1),\ B(-4,4),\ C(-3,-1),\ D(4,-2),\ E(3,0),$
$F(0,-3)$

9. (a) 5   (b) $\left(\frac{5}{2},4\right)$

11. (a) $\sqrt{34}$ (b) $\left(-\frac{1}{2},\frac{11}{2}\right)$

13. (a) $\sqrt{113}$ (b) $\left(-1,-\frac{3}{2}\right)$

15. (a) $\frac{5}{12}$ (b) $\left(-\frac{5}{8},\frac{5}{6}\right)$

17. (a) $3\sqrt{2}$ (b) $\left(-2,\dfrac{\sqrt{2}}{2}\right)$

19. $d=\sqrt{x^2+y^2}$ **21.** $3\sqrt{13}+3\sqrt{5}$

23. $(AB)^2=(AC)^2+(BC)^2$, where $AB=5\sqrt{2}$,
$AC=2\sqrt{10},\ BC=\sqrt{10}$; area $=10$ square units

25. $AB=AC=BC=2\sqrt{2}$; area $=2\sqrt{3}$ square units

27. $\left(-1,\frac{3}{2}\right),\ (2,-3),\ \left(5,-\frac{15}{2}\right)$

29. $2\sqrt{17}, 5\sqrt{2}, \sqrt{74}$ **31.** $\left(0, \frac{7}{9}\right)$ **33.** $\left(\frac{a}{2}, \frac{b}{2}\right)$

35. $\dfrac{\sqrt{a^2 + b^2}}{2}$ **37.** $(a + b, c)$ **39.** $\left(\dfrac{a + b}{2}, \dfrac{c}{2}\right)$

41. $\left(\frac{14}{3}, 2\right)$ and $\left(\frac{10}{3}, 0\right)$

43. radius ≈ 4.33 units, area ≈ 59.0 square units

SECTION 2.2

◆ Problems

1.

2.

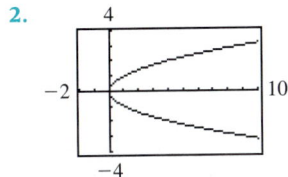

3. The graph of $y^2 + x = 4$ is symmetric with respect to the x-axis, since replacing y with $-y$ yields an equivalent equation:

$$(-y)^2 + x = 4$$
$$y^2 + x = 4$$

To find the x-intercept, let $y = 0$:

$$(0)^2 + x = 4$$
$$x = 4$$

To find the y-intercepts, let $x = 0$:

$$y^2 + (0) = 4$$
$$y = \pm 2$$

4. The graph of $y = x^3 - 3x$ is symmetric with respect to the origin, since replacing x with $-x$ and y with $-y$ yields an equivalent equation:

$$-y = (-x)^3 - 3(-x)$$
$$-y = -x^3 + 3x$$
$$y = x^3 - 3x$$

The graph of $x^2 + 3y^2 - 4x - 12 = 0$ is symmetric with respect to the x-axis, since replacing y with $-y$ yields an equivalent equation:

$$x^2 + 3(-y)^2 - 4x - 12 = 0$$
$$x^2 + 3y^2 - 4x - 12 = 0$$

5.

Setting up the direct proportion,

$$\frac{1}{0.7} = \frac{32}{x},$$

we find $x = 22.4$. Therefore, for $t = 3.7$ min, $T \approx 262 - 22.4 = 239.6°$F

◆ Exercises

1.

3.

5.

7.

9.

11.

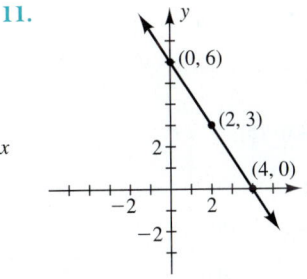

13. Symmetric with respect to the y-axis; x-intercepts ± 3; y-intercept $\frac{9}{4}$

15. Symmetric with respect to the origin; no x-intercept; no y-intercept

17. Symmetric with respect to the y-axis; x-intercepts ± 3; y-intercept 3

19. Symmetric with respect to the origin; no x-intercept; no y-intercept

21. Symmetric with respect to the x-axis; x-intercept 2; no y-intercept

23. None of these types of symmetry; x-intercept $\frac{5}{3}$; y-intercept 2

25. Symmetric with respect to the y-axis; x-intercepts ± 6; y-intercept 12

27. None of these types of symmetry; x-intercepts 0, 2; y-intercept 0

29. Symmetric with respect to the x-axis; x-intercepts 8, -2; y-intercepts ± 4

31. Symmetric with respect to the origin; x-intercepts $\pm\sqrt{2}$; no y-intercept

33. None of these types of symmetry; x-intercepts $-\frac{3}{2}$, $-\frac{15}{2}$; y-intercept 1

35. Symmetric with respect to the origin; x-intercepts 0, ± 1; y-intercept 0

37. Symmetric with respect to the x-axis; x-intercept -2; y-intercepts ± 2

39. Symmetric with respect to the y-axis; no x-intercept; y-intercept 1

41. (a)

(b) 59.8°F

43. (a)

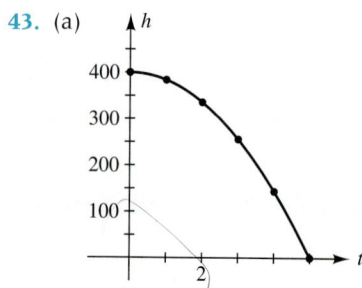

(b) 211.2 ft

45. The graph of $y = |x| + c$, with $c > 0$, is the same as the graph of $y = |x|$, but shifted vertically upward c units.

47. The graph of $y = \sqrt{x - c}$, with $c > 0$, is the same as the graph of $y = \sqrt{x}$, but shifted horizontally to the right c units.

49. (a)

C (in °C)	3.9	23.3	-19.4	12	31	-6
F (in °F)	39	74	-3	53.6	87.8	21.2

(b)

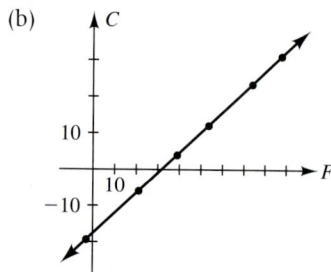

(c) F-intercept 32, the Fahrenheit temperature when the Celsius temperature is 0. C-intercept $-\frac{160}{9}$, the Celsius temperature when the Fahrenheit temperature is 0.
(d) $-40°F = -40°C$. Replace C with F in the equation $5F - 9C = 160$ and solve for F.

SECTION 2.3

◆ **Problems**

1. Solving the equation for y, we obtain $y = \pm\sqrt{4 - 2x}$. Since there are two outputs for each input value $x < 2$, we conclude that the equation $2x + y^2 = 4$ does *not* define y as a function of x. Also, note that the graph of this equation fails the vertical line test.

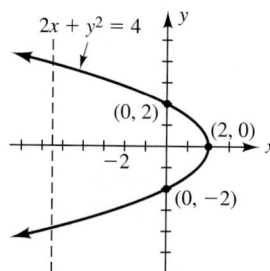

2. $f(\pi) = \pi^2 - 4\pi + 3 \approx 0.303$

3. Step 1: $g(x + h) = 4 - 3(x + h)$
$$= 4 - 3x - 3h$$
Step 2: $g(x + h) - g(x)$
$$= (4 - 3x - 3h) - (4 - 3x)$$
$$= -3h$$
Step 3: $\dfrac{g(x + h) - g(x)}{h} = \dfrac{-3h}{h} = -3$

4. $h(1) = (1) - 1 = 0$

5. The radicand $2 - x$ must be positive. Thus,

$$2 - x > 0$$
$$-x > -2$$
$$x < 2$$

Hence, the domain is $(-\infty, 2)$.

◆ **Exercises**

1. Defines y as a function of x

3. Defines y as a function of x

5. Defines y as a function of x

7. Does not define y as a function of x

9. Defines y as a function of x

11. Defines y as a function of x

13. Defines y as a function of x **15.** 13   **17.** 5

19. $9 - 2\sqrt{2}$ **21.** $|2ab + 3|$ **23.** $t - 4$

25. $n^2 - n + 1$ **27.** $\sqrt{x}$ **29.** $\sqrt{1 + x^2} - 4$

31. $4x - 2\sqrt{x - 2} - 7$ **33.** $4x - 34\sqrt{x} + 73$

35. (a) 9   (b) -11 (c) 4   (d) 0

37. (a) 7   (b) 3   (c) 5   (d) $\begin{cases} x + 1 & \text{if } x > -1 \\ 0 & \text{if } x = -1 \\ -(x + 1) & \text{if } x < -1 \end{cases}$

39. 2   **41.** $2x + h$ **43.** $2x + 2 + h$ **45.** $\dfrac{-1}{x(x + h)}$

47. $(-\infty, \infty)$ **49.** $(-\infty, \infty)$ **51.** $(-\infty, 4]$

53. $[-4, 4]$ **55.** $(-\infty, -2) \cup (-2, \infty)$

57. $(-\infty, -2) \cup (-2, 2) \cup (2, \infty)$

59. $(-\infty, -5) \cup (-5, 2) \cup (2, \infty)$

61. $\left[\frac{1}{2}, 4\right) \cup (4, \infty)$

63. $r(25) = 7$ meters, which represents the radius of the oil spill after 25 minutes

65. $C(12.4) = \$19.48$, which represents the cost of a taxi-cab fare when the cab is driven 12.4 miles

67. No; several values of y correspond to each input value of x.

69. (a) Yes; for each element in set X there corresponds exactly one element in set Y.

(b) No; the element 2 in set X is not assigned a value.

(c) Yes; for each element in set X there corresponds exactly one element in set Y.

(d) No; the element 2 in set X corresponds to two elements in set Y.

71. (a) $A = 4$ (b) $A = 134$

73. The population is one-fourth as large.

SECTION 2.4

◆ **Problems**

1. (a) $F(0) = -1 + \sqrt{15} \approx 2.87$

(b) Since 6 is not an element in the domain of F, the functional value $F(6)$ is undefined.

2. The function g has domain $(-\infty, -2] \cup [2, \infty)$ and range $[0, \infty)$. The function h has domain $(-\infty, \infty)$ and range $(-\infty, \infty)$.

3.
$$\sqrt{2x - 1} + 2 = x$$
$$\sqrt{2x - 1} = x - 2$$
$$2x - 1 = x^2 - 4x + 4$$
$$x^2 - 6x + 5 = 0$$
$$(x - 1)(x - 5) = 0$$
$$x = 1, 5$$

However, 1 is an extraneous root. Thus, the only solution is 5.

4.
$$\frac{x + 1}{2} - \frac{5(x - 1)}{8} \geq 1$$
$$4(x + 1) - 5(x - 1) \geq 8$$
$$-x + 9 \geq 8$$
$$-x \geq -1$$
$$x \leq 1$$

Hence, the solution set is $(-\infty, 1]$, which agrees with the graphical result in Example 4(b).

5.
$$(x + 6)^2 = x^2 + 18^2$$
$$x^2 + 12x + 36 = x^2 + 324$$
$$12x = 288$$
$$x = 24$$

◆ **Exercises**

1. (a) $(-\infty, \infty)$ (b) $[-2, \infty)$ (c) ± 2 (d) Even

3. (a) $(-\infty, \infty)$ (b) $[-1, \infty)$ (c) $0, -2$
(d) Neither

5. (a) $[-\sqrt{5}, \sqrt{5}]$ (b) $[0, \sqrt{5}]$ (c) $\pm\sqrt{5}$
(d) Even

7. (a) $(-\infty, \infty)$ (b) $[-2, 0)$ (c) None   (d) Even

9. (a) $[-2, 2]$ (b) $[-22, 22]$ (c) $0, \pm\sqrt[4]{5}$
   (d) Odd

11. Odd   13. Even   15. Odd   17. Even

19. Neither   21. $\frac{5}{3}$ 23. $8, -5$ 25. $3, \frac{1}{2}$ 27. $\frac{3}{2}$

29. ± 2 31. 5 33. $-\frac{1}{2}, \frac{3}{2}$ 35. $17, -25$ 37. $\frac{7}{9}$

39. -3 41. $0, -2$ 43. $(-\infty, 2]$ 45. $(-\infty, 10]$

47. $(-\infty, 1) \cup (4, \infty)$

49. $(-\infty, -3] \cup [-\sqrt{3}, \sqrt{3}] \cup [3, \infty)$ 51. 13 inches

53. 9,728 box seats

55. 441 sq ft for the square room; 459 sq ft for the rectangular room

57. 170 ft   59. 1, 3, 5 or 3, 5, 7   61. $(2, 3)$

63. It has no real solution

65. One y-intercept; infinitely many x-intercepts

67. No, it would fail the vertical line test.

69. $6\frac{1}{2}$ years

SECTION 2.5

◆ Problems

1. The graph of $h(x) = x^3 - 1$ is the same as the graph of $f(x) = x^3$, but shifted vertically *downward* 1 unit:

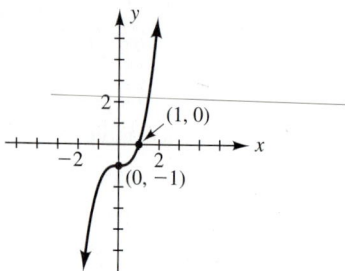

2. The graph of $h(x) = (x - 1)^3$ is the same as the graph of $f(x) = x^3$, but shifted horizontally *to the right* 1 unit:

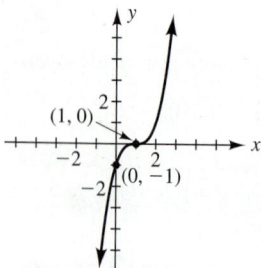

3. The graph of $F(x) = (x + 1)^2 - 4$ is obtained by shifting the graph of $f(x) = x^2$ to the left 1 unit and downward 4 units:

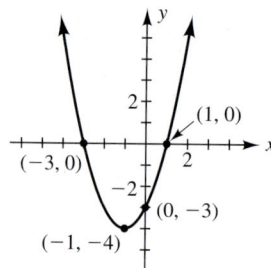

4. The graph of $F(x) = -|x + 2|$ is obtained by shifting the graph of $f(x) = |x|$ to the left 2 units and then reflecting this graph about the x-axis:

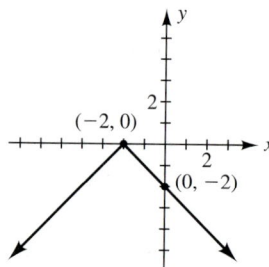

5. The graph of $F(x) = 4x - 1$ is formed by stretching the graph of $f(x) = x$ by a factor of 4 and then shifting this graph downward 1 unit:

6.

◆ Exercises

1.

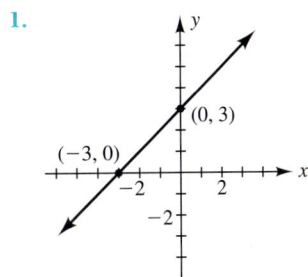

Line through $(-3, 0)$ and $(0, 3)$.

3.

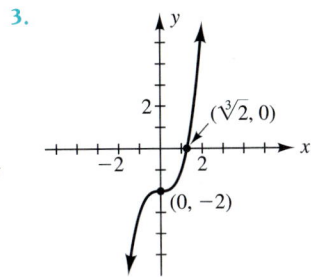

$(\sqrt[3]{2}, 0)$, $(0, -2)$.

5.

$(0, 3)$.

7.

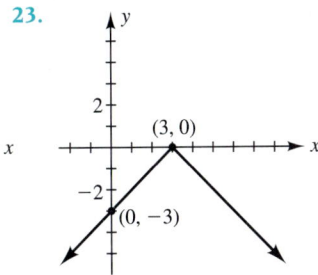

$(0, 1)$, $(1, 0)$.

9.

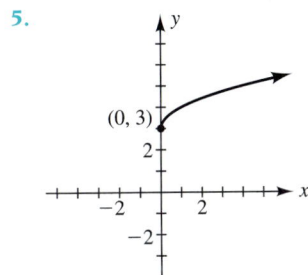

$(0, 4)$, $(-2, 0)$.

11.

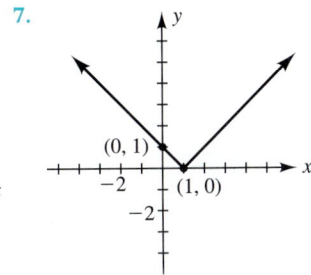

$x = -2$, $(0, \frac{1}{2})$.

13.

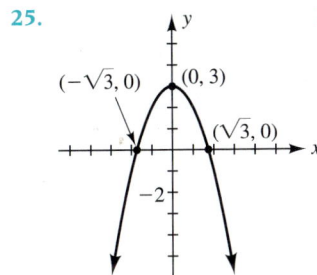

$(-1, 1)$, $(1, -1)$.

15.

$(0, 0)$.

17.

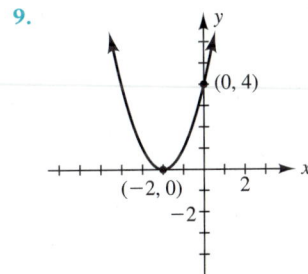

$(0, 1)$, $(1, 0)$.

19.

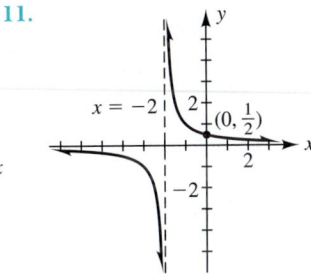

$(0, 4)$, $(1, 3)$.

21.

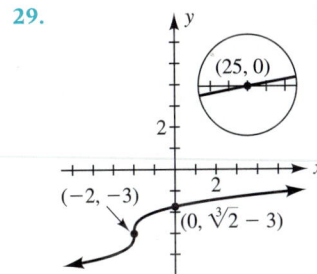

$(0, \sqrt{2} - 1)$, $(-1, 0)$, $(-2, -1)$.

23.

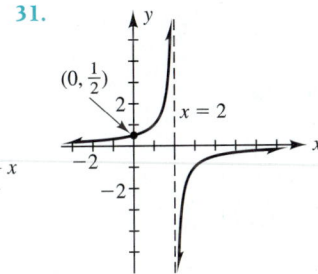

$(3, 0)$, $(0, -3)$.

25.

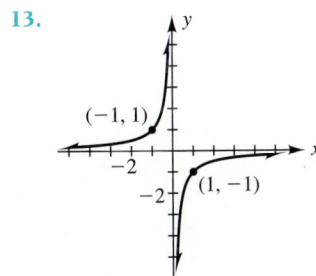

$(-\sqrt{3}, 0)$, $(0, 3)$, $(\sqrt{3}, 0)$.

27.

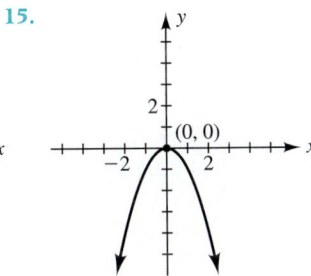

$(-4, 2)$, $(-4 + \sqrt[3]{2}, 0)$, $(0, -62)$.

29.

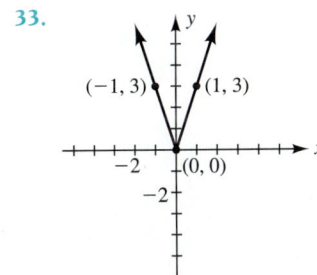

$(25, 0)$, $(-2, -3)$, $(0, \sqrt[3]{2} - 3)$.

31.

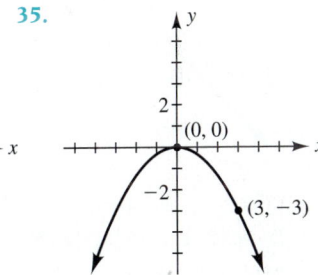

$(0, \frac{1}{2})$, $x = 2$.

33.

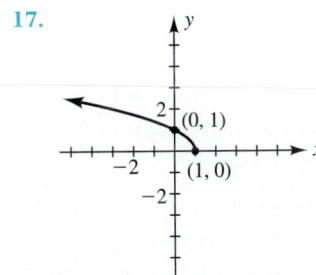

$(-1, 3)$, $(1, 3)$, $(0, 0)$.

35.

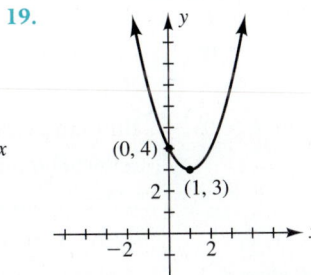

$(0, 0)$, $(3, -3)$.

37.

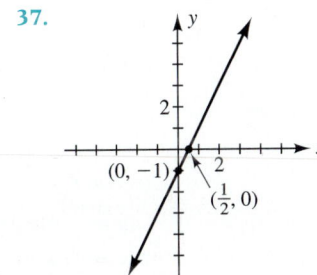

$(0, -1)$, $(\frac{1}{2}, 0)$.

39.

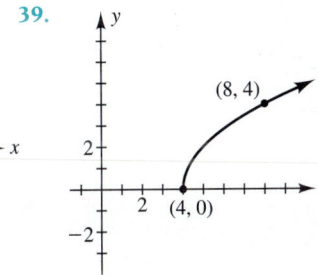

$(8, 4)$, $(4, 0)$.

41.

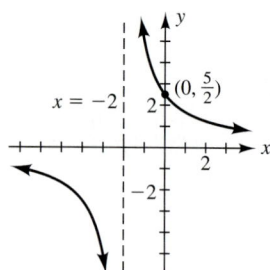

43. *Increasing functions:* Identity, cubing, square root, cube root; *decreasing function:* Reciprocal

45. Increasing   **47.** Decreasing   **49.** Neither

51. Increasing

53.

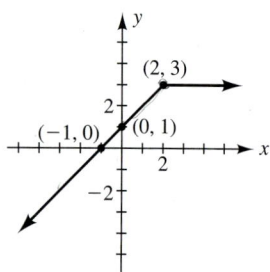

Increasing: $(-\infty, 3)$
Constant: $(3, \infty)$

55.

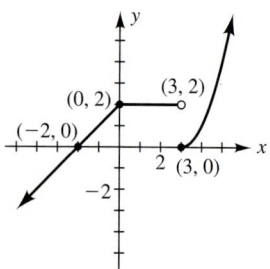

Increasing: $(-\infty, 0) \cup (3, \infty)$
Constant: $(0, 3)$

57.

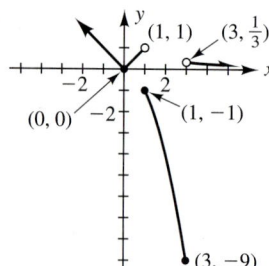

Increasing: $(0, 1)$
Decreasing: $(-\infty, 0) \cup (1, 3) \cup (3, \infty)$

59.

61.

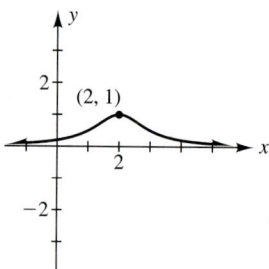

63. $y = 1 + \sqrt{4 - x^2}$

65. $y = -2 + \sqrt{4 - (x + 1)^2}$

67. Vertical shift rule

69. x- and y-axis reflection rules

71.

To graph:	Draw the graph of $y = f(x)$ and then
$y = f(x) + c$, for $c > 0$	shift vertically upward c units
$y = f(x) - c$, for $c > 0$	shift vertically downward c units
$y = f(x + c)$, for $c > 0$	shift horizontally to the left c units
$y = f(x - c)$, for $c > 0$	shift horizontally to the right c units
$y = -f(x)$	reflect about the x-axis
$y = f(-x)$	reflect about the y-axis
$y = cf(x)$, for $c > 1$	stretch vertically by a factor of c
$y = cf(x)$, for $0 < c < 1$	compress vertically by a factor of $1/c$

73. (a)

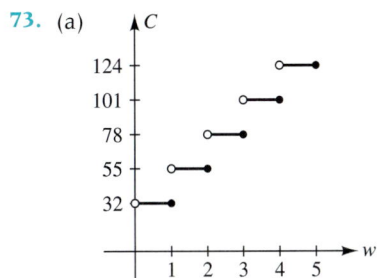

(b) $C(2.4) = 78¢$

SECTION 2.6

◆ Problems

1. $(f \circ g)(0) = f(g(0)) = f(0) = -2$

2. (a) $f(9) = 3$. Hence, $(g \circ f)(9) = g(f(9))$
$$= g(3)$$
$$= 3^2 - 4 = 5$$

(b) $(g \circ f)(9) = 9 - 4 = 5$

3. $H(a) = H(b)$ implies
$$a^2 - 1 = b^2 - 1$$
$$a^2 = b^2$$
$$a = b \quad \text{for } a \geq 0, \, b \geq 0$$

Also, note that the graph of $H(x) = x^2 - 1$ with $x \geq 0$ passes the horizontal line test.

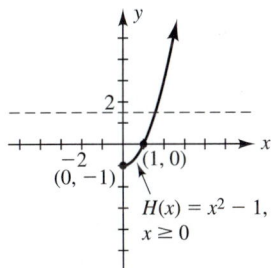

4. As shown in the figure, the functions f and g are one-to-one functions and their domains and ranges are the set of all real numbers. The graphs of f and g appear to be reflections of one another in the line $y = x$, which indicates that these functions might be inverses of each other.

Note that
$$f(g(x)) = f\left(\sqrt[5]{x} + 2\right) = \left[\left(\sqrt[5]{x} + 2\right) - 2\right]^5$$
$$= \left(\sqrt[5]{x}\right)^5 = x$$
and $\quad g(f(x)) = g[(x-2)^5] = \sqrt[5]{(x-2)^5} + 2$
$$= (x-2) + 2 = x.$$

Hence, we conclude that f and g are inverses of each other.

5. Replacing $f(x)$ with y, we solve for x as follows:

$$y = \frac{1}{x+2}$$
$$x + 2 = \frac{1}{y}$$
$$x = \frac{1}{y} - 2$$

Interchanging x and y, we obtain $y = \dfrac{1}{x} - 2$.

Hence, the inverse of f is $f^{-1}(x) = \dfrac{1}{x} - 2$.

◆ Exercises

1. 0   **3.** Undefined   **5.** 2   **7.** 1

9. $(f \circ g)(x) = 2x^2 + 3; \; (-\infty, \infty)$

11. $(G \circ f)(x) = \dfrac{1}{2x+1}; \; \left(-\infty, -\frac{1}{2}\right) \cup \left(-\frac{1}{2}, \infty\right)$

13. $(F \circ h)(x) = \sqrt{x-4}; \; [4, \infty)$

15. $(G \circ H)(x) = \dfrac{x-1}{2}; \; (-\infty, 1) \cup (1, \infty)$

17. $(H \circ f)(x) = \dfrac{1}{x}; \; (-\infty, 0) \cup (0, \infty)$

19. $(f \circ f)(x) = 4x + 3; \; (-\infty, \infty)$ **21.** $\frac{1}{2}$

23. (a) One-to-one

25. (a) Not one-to-one   (b) $[0, \infty)$ is one restriction.

27. (a) One-to-one

29. (a) Not one-to-one   (b) $[-2, \infty)$ is one restriction.

31. (a) Not one-to-one   (b) $[2, \infty)$ is one restriction.

33. f and g are inverses.   **35.** f and g are not inverses.

37. f and g are inverses.   **39.** f and g are inverses.

41. f and g are inverses.   **43.** f and g are inverses.

45. (a) $f^{-1}(x) = \dfrac{x}{2}$ (b)

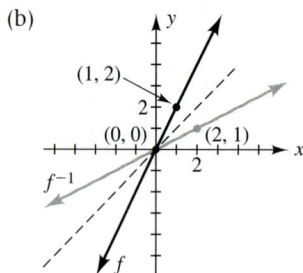

47. (a) $f^{-1}(x) = 3 - x$ (b)

49. (a) $f^{-1}(x) = \sqrt[3]{2 - x}$ (b)

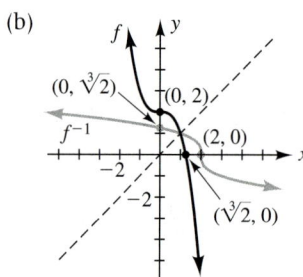

51. (a) $f^{-1}(x) = x^3 + 4$ (b)

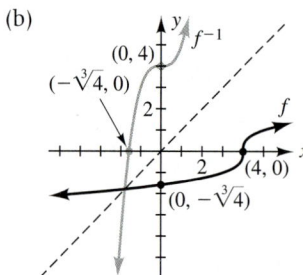

53. (a) $f^{-1}(x) = \dfrac{1}{x} - 1$ (b)

55. (a) $f^{-1}(x) = (x - 1)^2 + 4,\ x \geq 1$
(b)

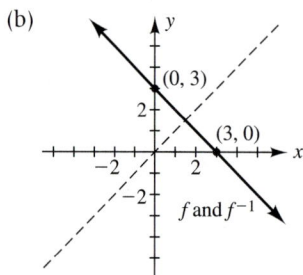

57. (a) $f^{-1}(x) = \sqrt{x} + 1$ (b)

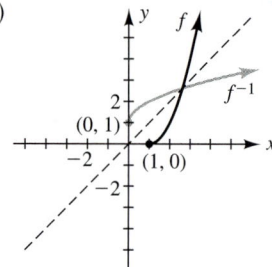

59. (a) $f^{-1}(x) = \sqrt{9 - x^2},\ 0 \leq x \leq 3$
(b)

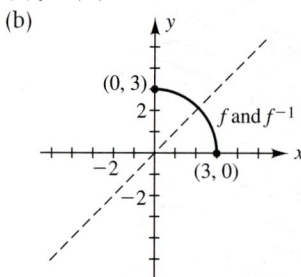

61. (a) $f^{-1}(x) = -\sqrt{x - 3}$
(b)

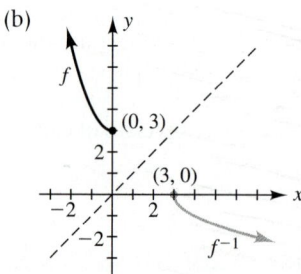

63. 4   **65.** 3   **67.** 2

69.

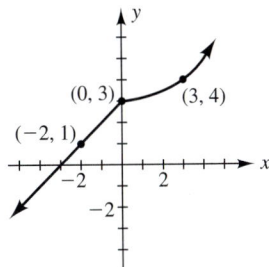

x	-2	0	3
$f^{-1}(x)$	1	3	4

71.

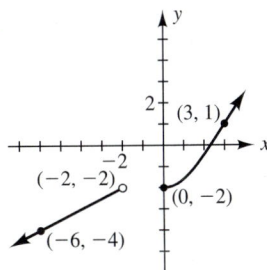

x	-2		0	3
$f^{-1}(x)$	undefined		-2	1

73. $(f \circ g)^{-1}(x) = (g^{-1} \circ f^{-1})(x) = \dfrac{3x + 1}{3}$

75. $(-\infty, 0) \cup (0, \infty)$ **77.** $\left(-\infty, \frac{2}{3}\right) \cup \left(\frac{2}{3}, \infty\right)$

79. $f(x) = \frac{1}{3}x - 2$

81. (a) $(V \circ r)t = 36\pi(t + 1)^3$
(b) Volume of the balloon as a function of time after the inflation process begins
(c) 2304π cubic inches   (d) 9 seconds

SECTION 2.7

◆ **Problems**

1. If $x = 50$ ft and $y = 100 - x$, then $y = 50$ ft. Maximum area occurs when the rectangle is a square.

2. From the similar triangles, we have $h = 3r$. Hence,

$$V = \frac{1}{3}\pi r^2 h = \frac{1}{3}\pi r^2(3r) = \pi r^3$$

3. $T = \dfrac{168{,}000}{40} = \$4200$

4. $t = \dfrac{240}{50} = 4.8$ h

5. Since the electrical resistance R of a wire varies directly as its length L and inversely as the square of its radius r, we write $R = \dfrac{kL}{r^2}$. Doubling both L and r, we obtain

$$R = \frac{k(2L)}{(2r)^2} = \frac{1}{2} \cdot \frac{kL}{r^2}.$$

Thus, the resistance becomes one-half as large.

◆ **Exercises**

1. $d = \dfrac{C}{\pi}$ **3.** $C = 40 + \dfrac{n}{5}$

5. (a) $s = \dfrac{d}{\sqrt{2}}$ (b) $A = \dfrac{d^2}{2}$ (c) $P = 2\sqrt{2}\,d$

7. $W = 125{,}000 + 4000t$

9. (a) $V = \frac{4}{3}\pi h^3$ (b) $V = \frac{1}{6}\pi r^3$

11. (a) $d = \sqrt{2 - 2x}$ (b) $[-1, 1]$

13. (a) $A = \dfrac{x\sqrt{x}}{2}$ (b) $P = x + \sqrt{x} + \sqrt{x^2 + x}$

15. $S = 2x^2 + \dfrac{256}{x}$

17. (a) $V = 9h^2$
(b) Domain $[0, 4]$
(c)

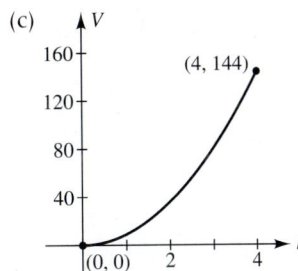

19. (a) $d = 30\sqrt{9 + t^2}$
(b) Domain $[0, 3]$; range $\left[90, 90\sqrt{2}\right]$

21. (a) $A = \dfrac{x\sqrt{262.44 - x^2}}{2}$
(b) Approximately 64.6 sq in.

23. (a) $d = \dfrac{F}{50}$

(b)

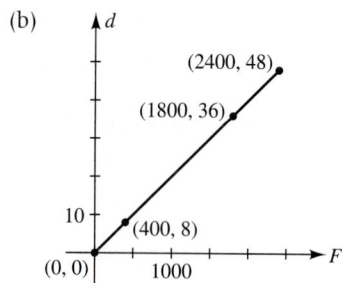

(c) 36 cm

25. (a) $W = 125l$

(b)

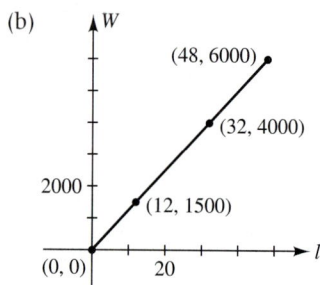

(c) 4000 lb

27. (a) $N = \dfrac{10,000}{d}$

(b)

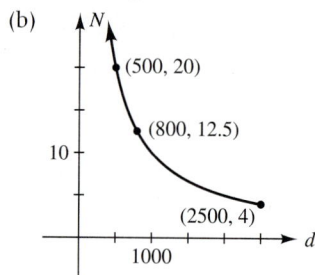

(c) 12.5 calls

29. (a) $W = \dfrac{300}{f}$

(b)

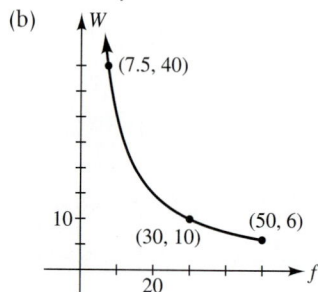

(c) 10 m

31. (a) $F = \dfrac{160,000}{d^2}$

(b)

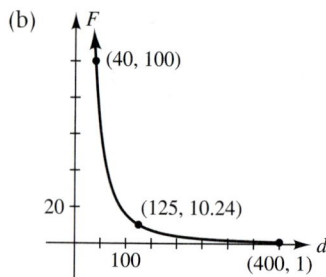

(c) 10.24 N

33. (a) $T \approx 2.007\sqrt{l}$

(b)

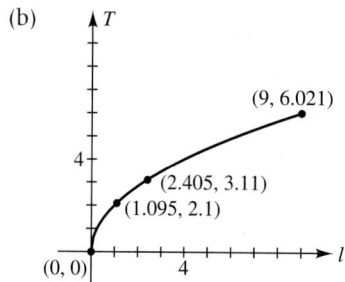

(c) Approximately 3.11 s

35. The force is tripled.

37. (a) $A = 1000x - 2x^2$ (b) $(0, 500)$
(c) 250 ft by 500 ft
(d) The ratio of x to y is 1 to 2.

39. (a) $A = 30r - 2r^2 - \dfrac{\pi r^2}{2}$ (b) $r \approx 4.20$ ft
(c) $y = 4.20$ ft; the ratio of r to y is 1 to 1.

41. $d = \begin{cases} 30t & \text{if } t \le 3 \\ 30\sqrt{9 + (t-3)^2} & \text{if } 3 < t \le 6 \end{cases}$

43. 25 seconds

CHAPTER 2 REVIEW EXERCISES

1. (a) 5   (b) $\left(2, \frac{9}{2}\right)$

3. (a) $2\sqrt{5}$ (b) $(0, 2)$

5. (a) $\sqrt{97}$ (b) $\left(\frac{3}{2}, -3\right)$

7. $AB = BC = CD = DA = 2\sqrt{2}$;
perimeter is $8\sqrt{2}$; area is 8

9. $\dfrac{a}{2}$; half as long as OA

11. $\dfrac{\sqrt{(a-b)^2 + c^2}}{2}$; half as long as AB

13. (a) *x*-intercept 0; *y*-intercept 0
(b) Symmetric with respect to origin
(c)

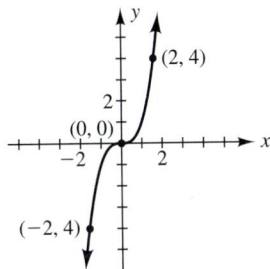

(d) Defines *y* as a function of *x*

15. (a) *x*-intercept 0; *y*-intercept 0
(b) Symmetric with respect to *x*-axis
(c)

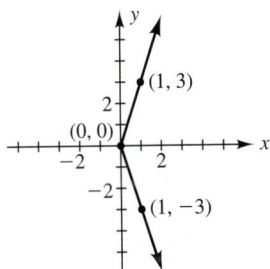

(d) Does not define *y* as a function of *x*

17. (a) No axis intercepts
(b) Symmetric with respect to *y*-axis
(c)

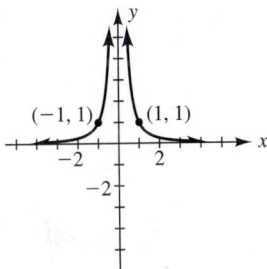

(d) Defines *y* as a function of *x*

19. (a) *x*-intercept 3; *y*-intercept 2
(b) No symmetry
(c)

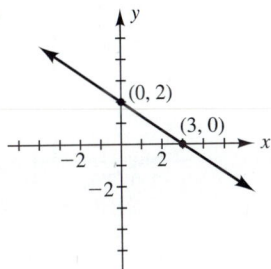

(d) Defines *y* as a function of *x*

21. (a) *x*-intercepts ± 3; *y*-intercept -3
(b) Symmetric with respect to *y*-axis
(c)

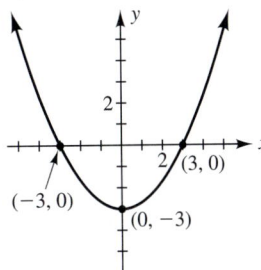

(d) Defines *y* as a function of *x*

23. (a) *x*-intercepts ± 2; *y*-intercepts ± 1
(b) Symmetric with respect to *x*-axis, *y*-axis and origin
(c)

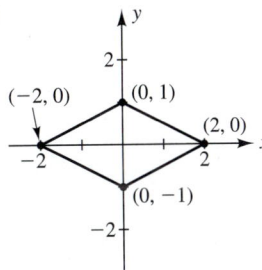

(d) Does not define *y* as a function of *x*

25. 1   **27.** 5   **29.** $2x^2 - 13x + 21$ **31.** -5

33. 5   **35.** $18x^4 + 45x^2 + 28$ **37.** $\dfrac{3x + 1}{x - 1}$

39. $\dfrac{x - 2}{3}$ **41.** 2   **43.** $4x - 1 + 2h$

45. (a) $(-\infty, \infty)$
(b) -3
(c) Neither
(d)

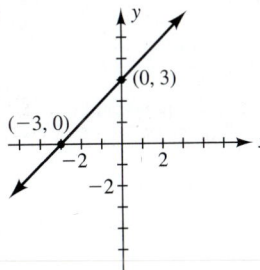

(e) Increasing $(-\infty, \infty)$
(f) One-to-one
(g) $(-\infty, \infty)$

47. (a) $(-\infty, \infty)$
(b) 0

(c) Odd

(d)

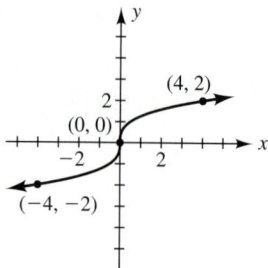

(e) Increasing $(-\infty, \infty)$ (f) One-to-one   (g) $(-\infty, \infty)$

49. (a) $(-\infty, \infty)$ (b) ± 3 (c) Even

(d)

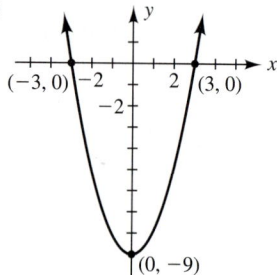

(e) Decreasing on $(-\infty, 0)$; increasing on $(0, \infty)$

(f) Not one-to-one   (g) $[-9, \infty)$

51. (a) $(-\infty, \infty)$ (b) -3 (c) Neither

(d)

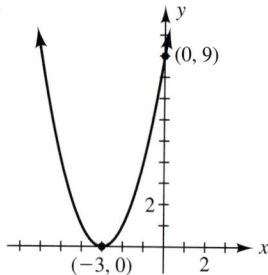

(e) Decreasing on $(-\infty, -3)$; increasing on $(-3, \infty)$

(f) Not one-to-one   (g) $[0, \infty)$

53. (a) $(-\infty, \infty)$ (b) -2 (c) Neither

(d)

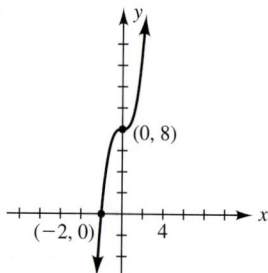

(e) Increasing $(-\infty, \infty)$ (f) One-to-one   (g) $(-\infty, \infty)$

55. (a) $(-\infty, \infty)$ (b) 2 and 4   (c) Neither

(d)

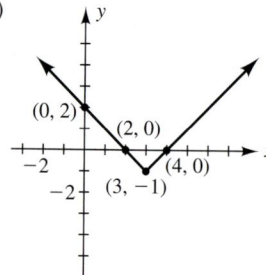

(e) Decreasing on $(-\infty, 3)$; increasing on $(3, \infty)$

(f) Not one-to-one   (g) $[-1, \infty)$

57. (a) $(-\infty, \infty)$ (b) 0 and 6   (c) Neither

(e) Increasing on $(-\infty, 3)$; decreasing on $(3, \infty)$

(f) Not one-to-one   (g) $(-\infty, 9]$

59. (a) $[-10, 10]$ (b) ± 10 (c) Even

(e) Increasing on $(-10, 0)$; decreasing on $(0, 10)$

(f) Not one-to-one   (g) $[0, 10]$

61. (a) $(-\infty, \infty)$ (b) 0 and $\pm\sqrt{2}$ (c) Even

(e) Increasing on $(-\infty, -1) \cup (0, 1)$; decreasing on $(-1, 0) \cup (1, \infty)$ (f) Not one-to-one

(g) $(-\infty, 1]$

63. (a) $(-\infty, 3]$ (b) None   (c) Neither

(d)

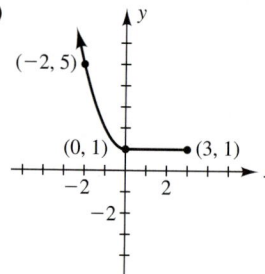

(e) Decreasing on $(-\infty, 0)$; constant on $(0, 3)$

(f) Not one-to-one   (g) $[1, \infty)$

65.

67.

69.

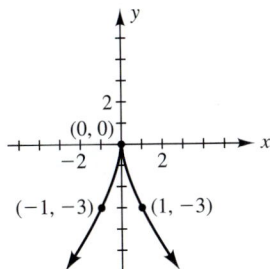

71. (a) $f^{-1}(x) = -\sqrt{x}$ (b)

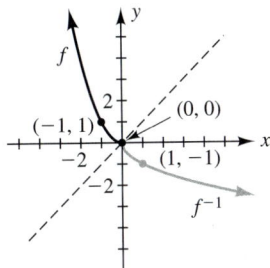

73. (a) $f^{-1}(x) = \sqrt[3]{3 - x}$ (b)

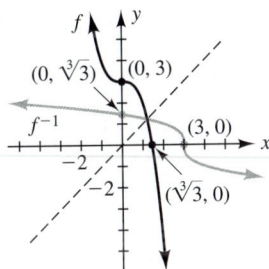

75. (a) $f^{-1}(x) = x^2 + 2$, $x \geq 0$
(b)

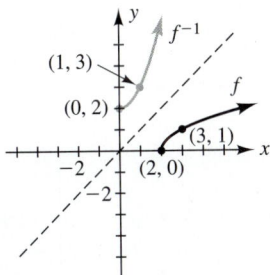

77. (a) $f^{-1}(x) = \begin{cases} \sqrt[3]{x} & \text{if } x < 1 \\ 2x - 1 & \text{if } x \geq 1 \end{cases}$
(b)

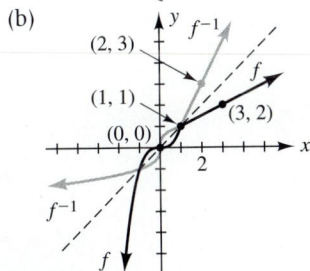

79. $(-\infty, 3]$ **81.** 3   **83.** 4

85.

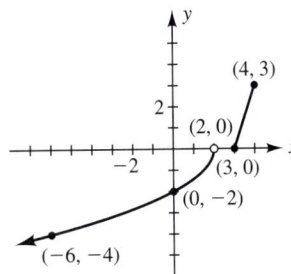

87. $-\frac{1}{2}$ **89.** $\frac{1}{9}$ **91.** $\left(-\infty, \frac{7}{4}\right]$

93. $8400 at 8%; $4100 at 9%

95. (a) $63.43   (b) 425.6 mi

97. $C = \begin{cases} 0.015n & \text{if } n \leq 1000 \\ 15 + 0.02(n - 1000) & \text{if } n > 1000 \end{cases}$

99. (a) $A = 3x - x^3$ (b) $\left(0, \sqrt{3}\right)$ (c) $P(1, 2)$

101. (a) $T = \dfrac{V}{25}$

(b)

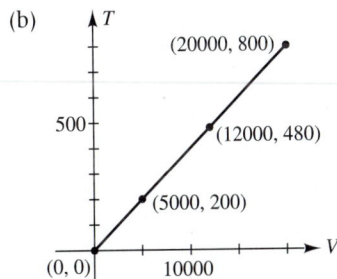

(c) $480

103. (a) $G = \dfrac{n(2000 - n)}{19,200}$

(b) Approximately 39 field mice per acre

◆ Chapter 3

SECTION 3.1

◆ Problems

1. (a) $m = \dfrac{2 - 4}{-1 - 3} = \dfrac{-2}{-4} = \dfrac{1}{2}$

(b) $m = \dfrac{-3 - 4}{2 - (-2)} = \dfrac{-7}{4} = -\dfrac{7}{4}$

(c) $m = \dfrac{3 - 3}{-2 - 3} = \dfrac{0}{-5} = 0$

Relabeling the points does not change the value of the slope.

2.

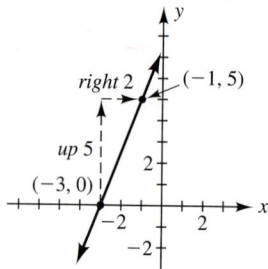

3. slope is 3; y-intercept is $-\frac{1}{2}$

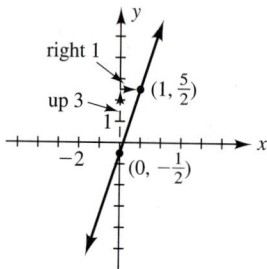

4. *x-intercept:* $3x + 8(0) = 12$
$$3x = 12$$
$$x = 4$$

y-intercept: $3(0) + 8y = 12$
$$8y = 12$$
$$y = \frac{3}{2}$$

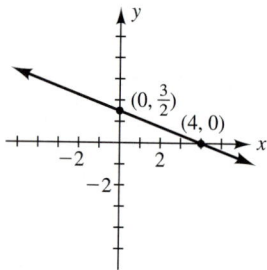

5. $0.33x - 4656.50 = 15000$
$$0.33x = 19656.50$$
$$x = \$59,565.15$$

◆ **Exercises**

1. 2   **3.** $-\frac{2}{3}$ **5.** Undefined   **7.** -4 **9.** $\dfrac{3 - b}{4}$

11. -1

13.

15.

17.

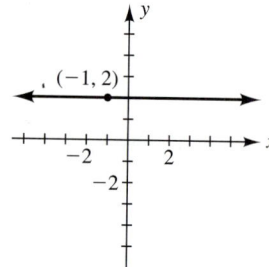

19. $m = 3,\ b = 0$

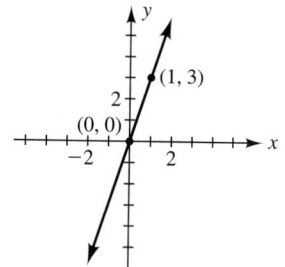

21. $m = -\frac{1}{3},\ b = 0$

23. $m = 2,\ b = 3$

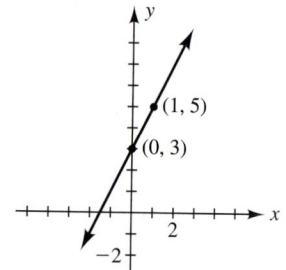

25. $m = -4,\ b = 1$

27. $m = \frac{1}{2},\ b = -1$

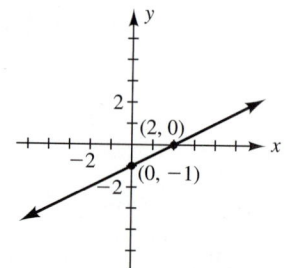

29. $m = -\frac{3}{4}$, $b = \frac{1}{4}$

31.

33.

35.

37.

39.

41.

43.

45.

47.

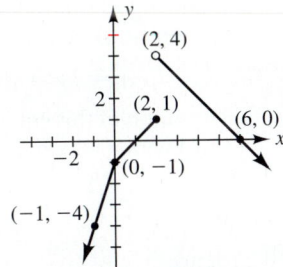

49. *x*-intercept 3.43; *y*-intercept 6.00

51. *x*-intercept -4.07; *y*-intercept 3.17

53. *x*-intercept 5.35; *y*-intercept -3.94

55. *x*-intercept -12.0; *y*-intercept 1.88

57. (a) $m = \frac{9}{5}$, $b = 32$

(b)

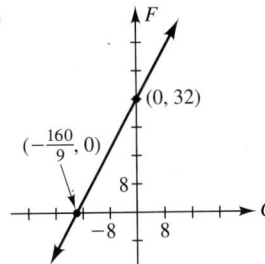

59. (a) $m = -3$, $b = 44$

(b)

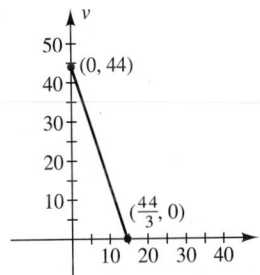

(c) $t = \frac{44}{3}$ s is the time when car comes to rest.

61. (a) $C = \begin{cases} 3 & \text{if } x \le 30 \\ 0.06x + 1.2 & \text{if } 30 < x \le 100 \\ 0.04x + 3.2 & \text{if } 100 < x \le 300 \\ 0.01x + 12.2 & \text{if } x > 300 \end{cases}$

(b)

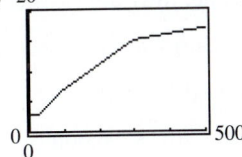

(c) \$12.80   (d) 272 kilowatt-hours

63. $(-5, 5)$ and $(3, -1)$ **65.** (a) 1   (b) 9   (c) -5

67. $f^{-1}(x) = \dfrac{x - b}{m}$

69. $\dfrac{x}{C/A} - \dfrac{y}{C/B} = 1$; C/A is the *x*-intercept of the line and C/B is the *y*-intercept of the line.

SECTION 3.2

◆ Problems

1. $y - 0 = \frac{1}{2}[x - (-3)]$, or $y = \frac{1}{2}x + \frac{3}{2}$

2. $y - 1 = 1(x - 2)$
   $y - 1 = x - 2$
   $y = x - 1$

3. $y - 0 = \frac{2}{3}[x - (-3)]$
   $y = \frac{2}{3}x + 2$

4. Slope of line through $(3, 0)$ and $(2, -3)$:

   $$m = \frac{0 - (-3)}{3 - 2} = 3$$

   Equation of the perpendicular line through $(3, -4)$:

   $$y - (-4) = -\frac{1}{3}(x - 3)$$
   $$y = -\frac{1}{3}x - 3$$

5. The V-intercept ($\$11,000$) represents the value of the car when new.

◆ Exercises

1. $y = \frac{3}{4}x + 3$ 3. $y = 3x + 5$ 5. $y = -\frac{7}{3}x + \frac{28}{3}$

7. $y = -3x$ 9. $y = -3x + 11$ 11. $y = -\frac{7}{15}x + \frac{2}{5}$

13. $y = -\frac{23}{18}x + \frac{1}{6}$ 15. $y = \frac{2}{3}x - 2$

17. $y = -24x + 6$ 19. $y = \frac{3}{2}$ 21. $x - 3 = 0$

23. $y = \frac{1}{5}x - 2$ 25. $y = -x + 5$

27. $y = -\frac{1}{2}x - 1$ 29. $y = 3x + 8$

31. $y = \frac{2}{3}x - \frac{13}{3}$ 33. $y = \frac{1}{4}x + 3$ 35. $y = \frac{3}{4}x + \frac{3}{4}$

37. $y = 2x$

39. Both lines have slope 1.72 and, thus, are parallel.

41. (a) $r = \frac{5}{2}T - 100$ (b) 0 chirps/min

43. (a) $V = 60,000 - 9000x$ (b) $\$24,000$

45. (a) $V = 15,000x + 50,000$ (b) $\$230,000$

47. (a) $w = -0.040625T + 4.57185$
    (b) Approximately 3.15 mm
    (c) Approximately 112.5°F

49. (a) $y = -\frac{1}{3}x - \frac{16}{3}$ or $x + 3y + 16 = 0$
    (b) $y = \frac{4}{3}x - \frac{11}{6}$ or $8x - 6y - 11 = 0$

51. $f(x) = \frac{4}{3}x + \frac{13}{3}$ 53. $g(x) = 3x - 7$

55. $\frac{12}{5}$ 57. (a) -2 (b) 0   (c) -6 (d) 4

59. $y = -6x + 4$ 61. $y = -\frac{A}{B}x + \left(\frac{A}{B}x_1 + y_1\right)$

SECTION 3.3

◆ Problems

1. Vertex: $(3, -1)$
   x-intercepts: $0 = (x - 3)^2 - 1$
   $$1 = (x - 3)^2$$
   $$\pm 1 = x - 3$$
   $$x = 4 \quad \text{or} \quad x = 2$$
   y-intercept: $f(0) = (0 - 3)^2 - 1 = 8$

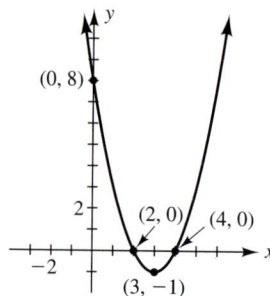

2. The basic form is $y = a(x - 4)^2 + 1$. To find a, we substitute $x = 2$ and $y = 3$:

   $$3 = a(2 - 4)^2 + 1$$
   $$3 = 4a + 1$$
   $$a = \frac{1}{2}$$

   Hence, the equation is $y = \frac{1}{2}(x - 4)^2 + 1$.

3. To find the x-intercepts, we solve the equation $f(x) = 0$:

   $$x^2 - 3x - 4 = 0$$
   $$(x - 4)(x + 1) = 0$$
   $$x - 4 = 0 \quad \text{or} \quad x + 1 = 0$$
   $$x = 4 \qquad\qquad x = -1$$

   Hence, the x-intercepts are 4 and -1.

4. Since $a = -3 < 0$, the parabola opens downward with a maximum value at

   $$x = -\frac{0}{2(-3)} = 0.$$

   Hence, the maximum value is $y = 3 - 3(0)^2 = 3$, and this occurs when $x = 0$.

5. If fencing all four sides, then $100 = 2l + 2w$, or $l = 50 - w$. Hence,

   $$A = lw = (50 - w)(w) = 50w - w^2.$$

   The graph of this equation is a parabola that opens downward with a maximum value at

   $$w = -\frac{50}{2(-1)} = 25 \text{ ft.}$$

   Hence,

   $$l = 50 - 25 = 25 \text{ ft.}$$

Thus, the largest area that can be enclosed is

$$(25 \text{ ft})(25 \text{ ft}) = 625 \text{ sq ft.}$$

6. We let $s = 4$ and solve for t:

$$4 = -16t^2 + 96t + 4$$
$$16t^2 - 96t = 0$$
$$16t(t - 6) = 0$$
$$t = 0 \quad \text{or} \quad t = 6$$

Hence, the ball is in flight 6 seconds.

◆ **Exercises**

1.

3.

5.

7.

9.

11.

13.

15.

17.

19.

21.

23.

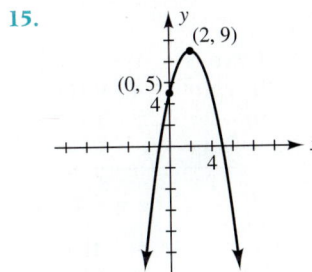

25. $y = \frac{1}{4}x^2$ 27. $y = -\frac{1}{2}x^2 + 2$

29. $y = -\frac{2}{9}x^2 + \frac{4}{3}x$ 31. $y = \frac{2}{3}x^2 + \frac{4}{3}x + 2$

33. Maximum value of 16   35. Minimum value of $\frac{3}{4}$

37. Maximum value of 9   39. (0.744, 3.78)

41. (2.63, 1.51)

43. (a) $-16t^2 + 64t + 80$ (b) 2 s
(c) 144 ft   (d) 5 s

45. (a) 50 pairs   (b) $2300   47. 62 and 62

49. 11 in. by 11 in.   51. $33\frac{1}{3}$ ft by 100 ft

53. Approximately 343 watts

55. (a) $12   (b) $3600   57. $\dfrac{a + b}{2}$

59. $y = \frac{1}{2}(x - 1)^2 - 2$ 61. $c = 0$

63. No effect; the axis of symmetry of
$y = x^2 + 4x + c$ is $x = -2$ for *all* values of c.

SECTION 3.4

◆ **Problems**

1. $x^2 - x - 12 = 3(x + 2)(x - 2)$
$x^2 - x - 12 = 3x^2 - 12$
$2x^2 + x = 0$ (standard form)
$x(2x + 1) = 0$
$x = 0 \quad \text{or} \quad x = -\frac{1}{2}$

2. *Check* $1 + 2i$:

$$(1 + 2i)^2 + 5 \overset{?}{=} 2(1 + 2i)$$
$$(1 + 4i + 4i^2) + 5 \overset{?}{=} 2 + 4i$$
$$(-3 + 4i) + 5 \overset{?}{=} 2 + 4i$$
$$2 + 4i = 2 + 4i$$

Check $1 - 2i$:

$$(1 - 2i)^2 + 5 \overset{?}{=} 2(1 - 2i)$$
$$(1 - 4i + 4i^2) + 5 \overset{?}{=} 2 - 4i$$
$$(-3 - 4i) + 5 \overset{?}{=} 2 - 4i$$
$$2 - 4i = 2 - 4i$$

3.
$$x^2 < 16x$$
$$x^2 - 16x < 0$$
$$x(x - 16) < 0$$

Boundary numbers: 0, 16

Algebraic signs: $++++$ $----$ $++++$

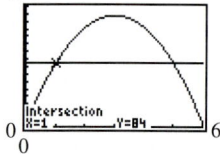

Solution set: $(0, 16)$

4. Let x = uniform width (in feet) of the additional parking area. Then

$$(220 + x)(100 + x) = 2(220)(100)$$
$$x^2 + 320x - 22{,}000 = 0.$$

So, $x = \dfrac{-320 \pm \sqrt{(320)^2 - 4(1)(-22{,}000)}}{2(1)}$

$$x = \dfrac{-320 \pm \sqrt{190{,}400}}{2}$$

$$x \approx 58.2 \text{ ft}$$

5.

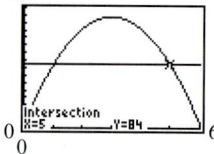

◆ Exercises

1. $-4, 2$ **3.** $0, 2$ **5.** ± 6 **7.** $-1, \frac{5}{2}$

9. $-6, -3$ **11.** $-4, 2$

13. $\dfrac{11 \pm 5\sqrt{5}}{2}$ **15.** $\frac{3}{4}, -\frac{4}{5}$ **17.** $\dfrac{2 \pm i}{5}$

19. $\dfrac{-5 \pm 3\sqrt{5}}{2}$ **21.** $\dfrac{\sqrt{7} \pm 5i}{4}$ **23.** $\pm \dfrac{\sqrt{5}}{2}$

25. $\pm \dfrac{\sqrt{15}}{2}$ **27.** $\pm \frac{7}{2}i$ **29.** $\dfrac{-3 \pm 4\sqrt{2}}{4}$

31. $5 \pm 3i\sqrt{2}$ **33.** $-6 \pm 4\sqrt{3}$ **35.** $6, -2$

37. $\dfrac{3 \pm 3\sqrt{5}}{2}$ **39.** $\frac{7}{5}, -\frac{3}{5}$ **41.** $(-3, 3)$

43. $(-\infty, 0] \cup [12, \infty)$ **45.** $[-1, 9]$

47. $\left(-\infty, -\frac{7}{2}\right) \cup \left(\frac{1}{2}, \infty\right)$

49. $\left(\dfrac{11 - 5\sqrt{5}}{2}, \dfrac{11 + 5\sqrt{5}}{2}\right)$

51. $\left[-4, -\frac{2}{3}\right]$ **53.** $x \approx -0.991, 3.34$

55. $x \approx -0.927, 1.12$ **57.** $(-\infty, -1.82) \cup (1.82, \infty)$

59. $[-1.27, 1.70]$ **61.** 18, 14

63. Base is 9 in., height is 12 in.

65. Width is 5 ft, length is 7 ft

67. Approximately 3.57 ft   **69.** $700

71. Between 2 m and 11 m

73. 0, 2, 4 or 2, 4, 6 or 4, 6, 8   **75.** When $p > \$3$

77. $5 \text{ s} < t < 15 \text{ s}$ **79.** $0, -\dfrac{b}{a}$ **81.** $-a \pm b$

83. $a, -(a + 2b)$ **85.** $k = 64$ **87.** $k = \pm 2\sqrt{6}$

89. $k = 8$ **91.** $(-4, -2) \cup (3, 5)$ **93.** $[3, 8]$

95. $(-2, 4)$ **97.** $k \leq -2\sqrt{6}$ or $k \geq 2\sqrt{6}$

99. $k \leq 0$ or $k \geq 4$

SECTION 3.5

◆ Problems

1. $y = 3 - 5x$: $-\frac{31}{13} \overset{?}{=} 3 - 5\left(\frac{14}{13}\right)$

$$-\tfrac{31}{13} \overset{?}{=} \tfrac{39}{13} - \tfrac{70}{13}$$
$$-\tfrac{31}{13} = -\tfrac{31}{13}$$

$y = \frac{3}{2}x - 4$: $-\frac{31}{13} \overset{?}{=} \frac{3}{2}\left(\frac{14}{13}\right) - 4$

$$-\tfrac{31}{13} \overset{?}{=} \tfrac{21}{13} - \tfrac{52}{13}$$
$$-\tfrac{31}{13} = -\tfrac{31}{13}$$

2. (a)
$$2x - y = 3$$
$$2x - (4x + 5) = 3$$
$$-2x = 8$$
$$x = -4$$

Hence, $y = 4(-4) + 5 = -11$. The lines intersect at the point $(-4, -11)$.

(b)
$$2x - y = 3$$
$$\underline{-4x + y = 5}$$
$$\overline{-2x \quad\quad = 8}$$
$$x = -4$$

Hence, $y = 4(-4) + 5 = -11$. The lines intersect at the point $(-4, -11)$.

3. Substituting $x^2 - 2x + 3$ for y, we obtain

$$4x - (x^2 - 2x + 3) = 6$$
$$x^2 - 6x + 9 = 0$$
$$(x - 3)^2 = 0$$
$$x = 3$$

Hence, $y = (3)^2 - 2(3) + 3 = 6$. The line is tangent to the parabola at the intersection point $(3, 6)$.

4. Substituting x^2 for y, we obtain

$$x^2 = 4x - x^2$$
$$2x^2 - 4x = 0$$
$$2x(x - 2) = 0$$
$$x = 0, 2$$

Hence, $y = 0^2 = 0$ and $y = 2^2 = 4$. The parabolas intersect at $(0, 0)$ and $(2, 4)$.

5.

$$x^6 + 4x^2 - 16 = 0$$
$$(x^6 - 8) + 4x^2 - 8 = 0$$
$$(x^2 - 2)(x^4 + 2x^2 + 4) + 4(x^2 - 2) = 0$$
$$(x^2 - 2)(x^4 + 2x^2 + 8) = 0$$

$$x^2 - 2 = 0 \quad \text{or} \quad x^4 + 2x^2 + 8 = 0$$

$$x = \pm\sqrt{2} \qquad \text{No real solution}$$

Hence, the exact coordinates of the intersection points are $\left(-\sqrt{2}, -2\sqrt{2}\right)$ and $\left(\sqrt{2}, 2\sqrt{2}\right)$.

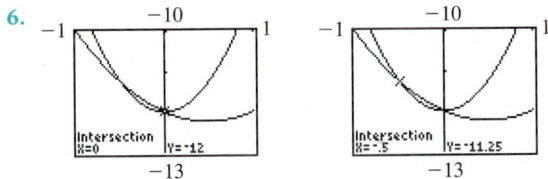

6.

◆◇ Exercises

1. $\left(2, \frac{2}{5}\right)$ **3.** $\left(-\frac{1}{2}, -1\right)$ **5.** $\left(-\frac{1}{2}, \frac{11}{2}\right)$

7. $(-3, -23)$ **9.** No intersection point   **11.** $\left(\frac{7}{3}, \frac{5}{6}\right)$

13. $(-3, -6), (1, 2)$ **15.** $(2, 5), \left(-\frac{3}{2}, \frac{3}{2}\right)$

17. $\left(\frac{1}{2}, -\frac{7}{4}\right), (5, 5)$ **19.** $(-2, -3), \left(\frac{1}{4}, \frac{21}{8}\right)$

21. $\left(\dfrac{-11 + \sqrt{41}}{4}, \dfrac{-11 + \sqrt{41}}{8}\right),$

$\left(\dfrac{-11 - \sqrt{41}}{4}, \dfrac{-11 - \sqrt{41}}{8}\right)$

23. $\left(5 + \sqrt{13}, \dfrac{19 + 5\sqrt{13}}{2}\right),$

$\left(5 - \sqrt{13}, \dfrac{19 - 5\sqrt{13}}{2}\right)$

25. $(0, 3), (3, 0)$ **27.** $\left(4, 2\sqrt{5}\right), \left(4, -2\sqrt{5}\right)$

29. $(3, 4), \left(-\frac{24}{5}, \frac{7}{5}\right)$ **31.** $\left(-\frac{1}{2}, -2\right), (1, 1)$

33. $(4, 0), (8, 2)$ **35.** No intersection point

37. $\left(1, 3\sqrt{3}\right), \left(1, -3\sqrt{3}\right), \left(-1, 3\sqrt{3}\right),$
$\left(-1, -3\sqrt{3}\right)$

39. $(0, 1), \left(-\frac{8}{5}, -\frac{3}{5}\right)$ **41.** $\left(2, \sqrt{2}\right)$

43. $(0, 2), (0, -2)$ **45.** $\left(\frac{5}{4}, -\frac{3}{4}\right)$

47. $(9, 3), (9, -3)$ **49.** $(1, 2), \left(\frac{2}{5}, -\frac{8}{5}\right)$

51. $(-1.879, -6.638), (0.347, 0.042), (1.532, 3.596)$

53. $(-1.195, 3.817), (2.930, 2.723)$

55. $(2.000, 1.000), (0.400, -0.600)$

57. $(-1.328, 6.220), (0.816, 0.886)$

59. -4 **61.** 2 **63.** $-1, \frac{5}{2}$ **65.** $-3, -6$

67. $-4, 2$ **69.** No real solution

71. Approximately $0.854, -5.854$

73. No real solution   **75.** $(5, -1), (2, 2), (-1, -7)$

77. $(1, 5), (9, 1), (-1, 1)$ **79.** $4\sqrt{5}$

81. $P\left(-1, \frac{9}{2}\right), Q(4, 2), R\left(4, -\frac{7}{4}\right), S\left(-1, -\frac{1}{2}\right)$

83. $P(2, 1), Q(4, 5), R(4, -5)$

85. $P(-2, 5), Q(1, 2), R(-2, -3), S\left(\frac{7}{2}, -3\right)$

CHAPTER 3 REVIEW EXERCISES

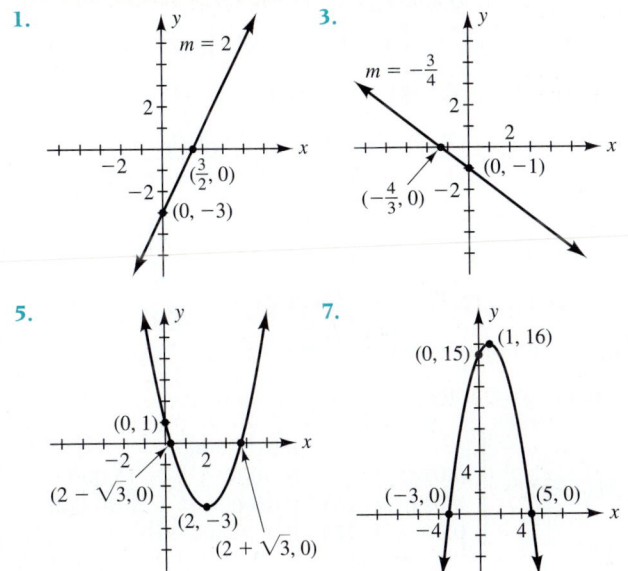

1.

3.

5.

7.

9.

11.

13.

73. 200 ft along tar road by 125 ft along gravel road

75. 30 m   **77.** 1.5 in.   **79.** $\pm 10\sqrt{26}$

15. $y = \frac{3}{4}x - 2$ **17.** $y = 3x + 10$

19. $y = -\frac{3}{2}x - 2$ **21.** $y = x^2 - 4$

23. $y = \frac{3}{4}(x + 2)^2 - 3$ **25.** $8, -2$ **27.** $\pm 4i$

29. $2 \pm \sqrt{14}$ **31.** $1 \pm 2i$ **33.** $\sqrt{6}, -\dfrac{\sqrt{6}}{2}$

35. $\left[-3, \frac{2}{3}\right]$ **37.** $\left(-\infty, -1 - \sqrt{5}\right) \cup \left(-1 + \sqrt{5}, \infty\right)$

39. $\left(-\infty, -\frac{2}{3}\right) \cup (2, \infty)$ **41.** -2 **43.** $-1, \frac{1}{9}$

45. Approximately $0.764, 5.236$ **47.** $\left(\frac{1}{2}, -3\right)$

49. $(4, 6), (1, 3)$ **51.** $\left(\sqrt{11}, 5\right), \left(-\sqrt{11}, 5\right), (0, -6)$

53. $\left(1, \sqrt{2}\right), \left(1, -\sqrt{2}\right), \left(-1, \sqrt{2}\right), \left(-1, -\sqrt{2}\right)$

55. $\left(6, \frac{3}{2}\right), \left(-\frac{10}{3}, -\frac{5}{6}\right)$ **57.** $(1.104, 1.344)$

59. (a) $x = \dfrac{C_1 B_2 - C_2 B_1}{A_1 B_2 - A_2 B_1}$ (b) $y = \dfrac{A_1 C_2 - A_2 C_1}{A_1 B_2 - A_2 B_1}$

61. (a) $y = \dfrac{B}{A}x$ (b) $\left(\dfrac{-AC}{A^2 + B^2}, \dfrac{-BC}{A^2 + B^2}\right)$

63. $(-2, 2), \left(\frac{4}{3}, \frac{16}{3}\right), \left(-\frac{1}{3}, \frac{7}{6}\right), \left(3, \frac{9}{2}\right)$

65. Minimum value of $\frac{11}{3}$ **67.** $a = -2$

69. (a) $V = -400x + 2400$ (b) 6 years

71. $C(x) = \begin{cases} 120 & \text{if } x \le 5000 \\ 0.018x + 30 & \text{if } 5000 < x \le 15{,}000 \\ 0.011x + 135 & \text{if } 15{,}000 < x \le 30{,}000 \\ 0.003x + 375 & \text{if } x > 30{,}000 \end{cases}$

◆ **Cumulative Review: Chapters 1, 2, 3**

1. (a) $(x + 1)(x - 4)(2x + 3)$
    (b) $(x - 2a)(x + 2a - 4y)$

3. 8

5. (a) $k = 0$ (b) $k = 9$ (c) $k < 9$ (d) $k > 9$

7. $2i$ **9.** $A = \dfrac{x^2}{18}$

11. (a) $\dfrac{1}{b - c}$ (b) $\dfrac{x^2 - x + 1}{x^2}$

13. (a) $v = 2t + 10$ (b)

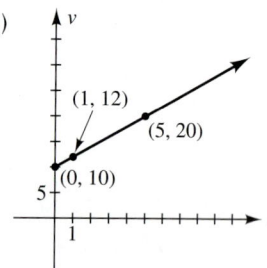

The v-intercept represents the initial velocity.

15. $f(0)$ **17.** At least $7\frac{1}{2}$ quarts   **19.** 70 square units

21. (a) Symmetric with respect to the y-axis
    (b) Symmetric with respect to the origin
    (c) Symmetric with respect to the x-axis
    (d) Symmetric with respect to the x-axis, y-axis, and origin

23. (a) $y = 1$ (b) $y = x - 2$
    (c) $3x + 2y = 11$ (d) $y = -3x + 10$

25. (a) 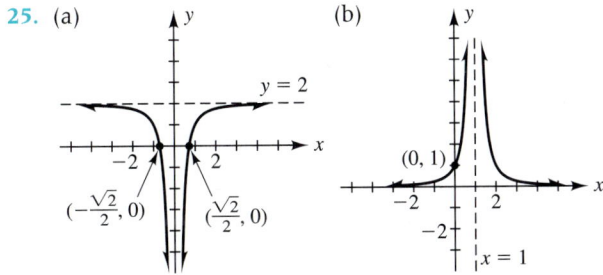 (b)

27. $m = \frac{5}{3}$

29. All squares from 9 in. by 9 in. to 11 in. by 11 in.

31. (a) $A = 6000 + 100x - 20x^2$ (b) in $2\frac{1}{2}$ weeks

◆ Chapter 4

SECTION 4.1

◆ Problems

1. *x*-intercept:
$$1 - (x + 2)^6 = 0$$
$$(x + 2)^6 = 1$$
$$x + 2 = \pm 1$$
$$x = -2 \pm 1$$
$$x = -1, -3$$
y-intercept: $P(0) = 1 - (0 + 2)^6 = -63$

2.

3. (a)

(b)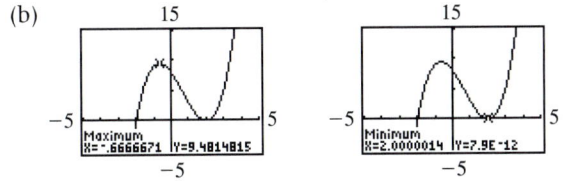

4. *M* is an odd function:
$$M(-x) = 100(-x) - (-x)^3$$
$$= -100x + x^3$$
$$= -M(x)$$

◆ Exercises

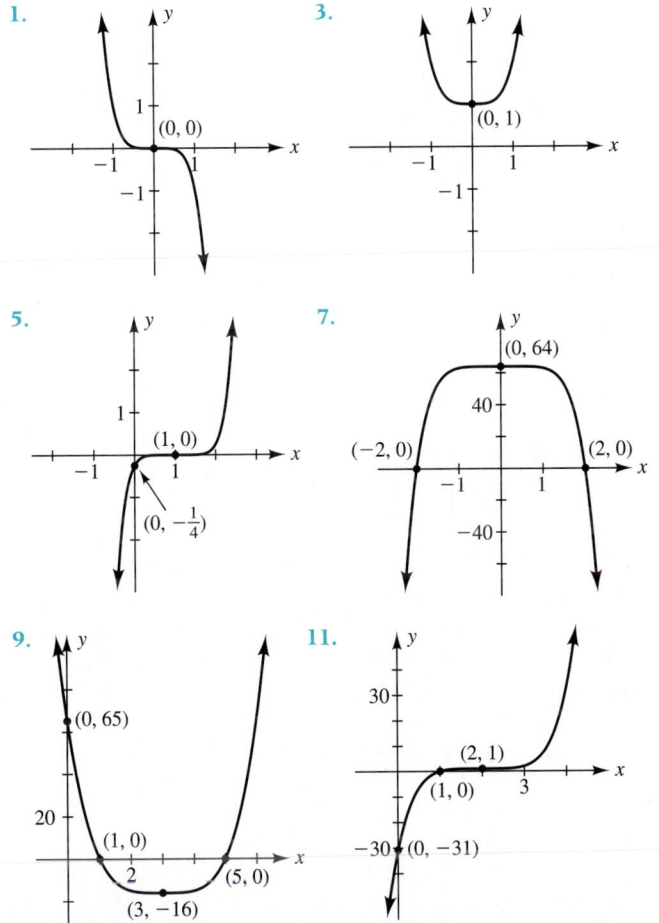

1.

3.

5.

7.

9.

11.

13. (a) Down to the left as $x \to -\infty$; up to the right as $x \to \infty$

(b) Two or none

15. (a) Down to the left as $x \to -\infty$; down to the right as $x \to \infty$

(b) Three or one

17. (a) Up to the left as $x \to -\infty$; down to the right as $x \to \infty$
    (b) Six, four, two, or none

19. (a) Up to the left as $x \to -\infty$; up to the right as $x \to \infty$
    (b) Five, three, or one

21. Maximum $(0.239, -1.66)$; minimum $(2.09, -8.05)$

23. Maxima $(-0.951, 11.0)$, $(0.824, 7.48)$; minimum $(0.128, 5.87)$

25. Maximum $(0.734, -0.605)$; minimum $(-0.734, -5.40)$

27. Minimum $(0.000, 1.000)$

29.

31.

33.

35.

37.

39.

41.

43.

45.
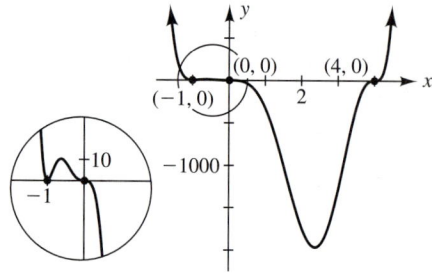

47. 13 years

49. (a) Domain $(0, 6)$
    (b)
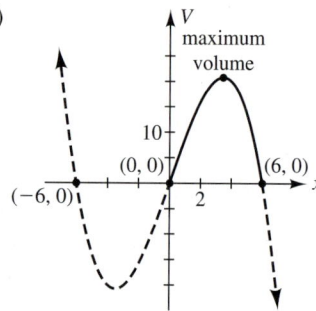
    (c) $x \approx 3.464$ ft

51. (a) $V = 180x^2 - 4x^3$; domain $(0, 45)$
    (b)
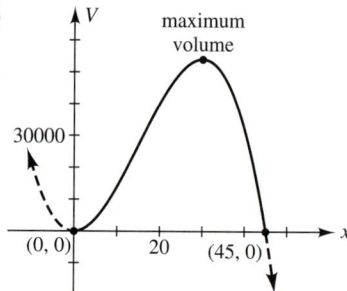
    (c) $x \approx 30$ inches

53. (a) $V = 64x - 28x^2 + 3x^3$
    (b)
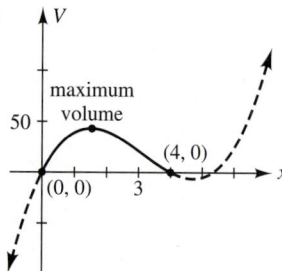
    (c) $x \approx 1.51$ cm

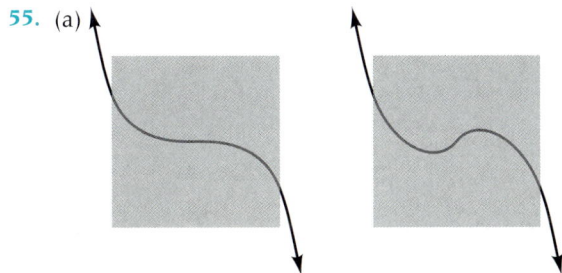

55. (a)

(b) $y = -x^3 - 2x$; $y = -x^3 + 2x$

57. (a)

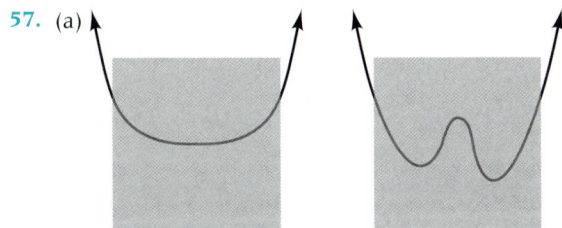

(b) $y = x^4 + 2x^2$; $y = x^4 - 2x^2$

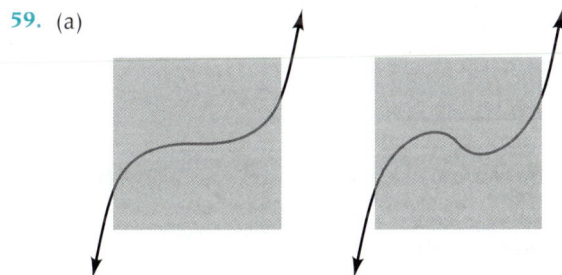

59. (a)

(b) $y = x^5 + 3x^3$; $y = x^5 - 3x^3$; $y = x^5 - 3x^3 + 2x$

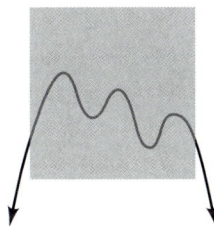

61. (a)

(b) $y = -x^6 - 2x^4$; $y = -x^6 + 2x^4$; $y = -x^6 + 2x^4 - x^2$

63. (a) $(1, 0)$, $\left(\sqrt{3}, 3\sqrt{3} - 3\right)$, $\left(-\sqrt{3}, -3\sqrt{3} - 3\right)$
(b) $(3, -12)$, $\left(\sqrt{5}, \sqrt{5} - 15\right)$, $\left(-\sqrt{5}, -\sqrt{5} - 15\right)$

65.

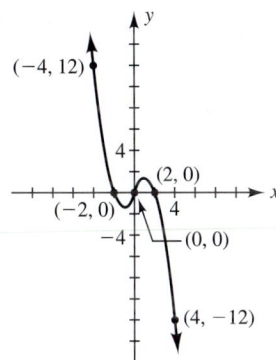

67. No; its graph is never symmetric with respect to the y-axis.

◆ **Problems**

1. $(x - 2)(2x^2 - x + 1) + (-2)$
$= (2x^3 - x^2 + x - 4x^2 + 2x - 2) + (-2)$
$= 2x^3 - 5x^2 + 3x - 4$

2.

$$3x^2 + 0x - 2 \overline{)\begin{array}{l} 2x \quad - 1 \\ 6x^3 - 3x^2 + 0x + 0 \end{array}}$$

$$\underline{6x^3 + 0x^2 - 4x}$$

$$-3x^2 + 4x$$

$$\underline{-3x^2 + 0x + 2}$$

$$4x - 2$$

Hence, $\dfrac{6x^3 - 3x^2}{3x^2 - 2} = 2x - 1 + \dfrac{4x - 2}{3x^2 - 2}$.

3.

$$1\overline{)\begin{array}{rrrr} 1 & -5 & 7 & -3 \\ & 1 & -4 & 3 \\ \hline 1 & -4 & 3 & 0 \end{array}}$$

Hence, $\dfrac{x^3 - 5x^2 + 7x - 3}{x - 1} = x^2 - 4x + 3$.

4. $(x + 3)(2x^3 - 6x^2 + 3x) + 1$
$= (2x^4 - 6x^3 + 3x^2 + 6x^3 - 18x^2 + 9x) + 1$
$= 2x^4 - 15x^2 + 9x + 1$

5.
$$
\begin{array}{r|rrrrr}
3) & 1 & -3 & -4 & 13 & -8 \\
& & 3 & 0 & -12 & 3 \\
\hline
& 1 & 0 & -4 & 1 & \boxed{-5} \text{ remainder}
\end{array}
$$

6.
$$
\begin{array}{r|rrrrr}
\tfrac{1}{2}) & 4 & -6 & 0 & 4 & -3 \\
& & 2 & -2 & -1 & \tfrac{3}{2} \\
\hline
& 4 & -4 & -2 & 3 & \boxed{-\tfrac{3}{2}} \text{ remainder}
\end{array}
$$

Hence, $P\left(\tfrac{1}{2}\right) = -\tfrac{3}{2}$.

7.
$$
\begin{array}{r|rrrr}
1 - i) & 3 & -8 & 10 & -4 \\
& & 3 - 3i & -8 + 2i & 4 \\
\hline
& 3 & -5 - 3i & 2 + 2i & 0
\end{array}
$$

Since the remainder is 0, we conclude that $1 - i$ is a zero of P.

◆ **Exercises**

1. $3x^2 + x + 2 + \dfrac{-1}{x - 1}$

3. $3x^2 - 4x - 1 + \dfrac{-4x + 6}{x^2 - 2x + 3}$

5. $2x^2 + x + 2 + \dfrac{-1}{2x - 1}$

7. $x^2 + 1 + \dfrac{-3x^2 + 4}{2x^3 + 1}$

9. $x^2 - x + 1 + \dfrac{-1}{x + 1}$

11. $-3x^3 - x - 1$ 13. $-3 + \dfrac{6}{2x + 1}$

15. $2x + 2 + \dfrac{6x - 2}{(2x - 1)^2}$

17. $2x - 1 + \dfrac{3}{x - 1}$

19. $x^2 + 3x - 8 + \dfrac{20}{x + 2}$

21. $2x^3 + x^2 + x + 4$

23. $4x^3 + x^2 - 6x + 4 + \dfrac{-24}{x + 6}$

25. $-x^4 - 2x^3 - 4x^2 - 8x - 16 + \dfrac{-29}{x - 2}$

27. $3x^2 - 6x + 15$

29. $2x^3 - 4x^2 + 10x - 44 + 118/\left(x + \tfrac{5}{2}\right)$

31. $14.5x^2 + 2.5x - 45.5$

33. $275x^3 + 495x^2 + 765x + 750$

35. $P(1) = 8$ 37. $F(-2) = 7$ 39. $P(4) = 25$

41. $Q\left(\tfrac{1}{2}\right) = 0$ 43. $f\left(-\tfrac{1}{3}\right) = 0$

45. $h(1 - i) = 7 - 4i$ 47. $P\left(\sqrt{2}\right) = 12$

49. $P(1.7) = -5.0582$ 51. No   53. Yes   55. Yes

57. Yes   59. No   61. Yes   63. Yes   65. Yes

67. 455 m   69. $P(x) = x^4 + 7x^3 + 7x^2 + 11x + 18$

71. $2x^2 - 3x + 2$ 73. $8x^3 + x^2 + 2x - 2$

75. $-x^3 - 2x + 3$ 77. $k = 16$ 79. $k = 2$

SECTION 4.3

◆ **Problems**

1.
$$
\begin{array}{r|rrrr}
2i) & 2 & -3 & 8 & -12 \\
& & 0 + 4i & -8 - 6i & 12 \\
\hline
& 2 & -3 + 4i & -6i & 0
\end{array}
$$

Since $P(2i) = 0$, we conclude that $x - 2i$ is a factor of $P(x) = 2x^3 - 3x^2 + 8x - 12$.

2. $P(x) = a(x - 0)^3(x + 3)$
$8 = a(1 - 0)^3(1 + 3)$
$8 = 4a$
$a = 2$

Hence, the polynomial function P with these characteristics is

$$P(x) = 2(x - 0)^3(x + 3)$$
$$= 2x^4 + 6x^3$$

3. The possible rational zeros of $P(x) = 3x^3 + 22x^2 + 25x + 6$ are

$$\frac{\pm 1}{\pm 1}, \frac{\pm 2}{\pm 1}, \frac{\pm 3}{\pm 1}, \frac{\pm 6}{\pm 1}, \quad \text{and} \quad \frac{\pm 1}{\pm 3}, \frac{\pm 2}{\pm 3}, \frac{\pm 3}{\pm 3}, \frac{\pm 6}{\pm 3}$$

These simplify to $\pm 1, \pm 2, \pm 3, \pm 6, \pm \tfrac{1}{3}, \pm \tfrac{2}{3}$. Testing -1 using synthetic division, we find

$$
\begin{array}{r|rrr}
-1) 3 & 22 & 25 & 6 \\
& -3 & -19 & -6 \\
\hline
3 & 19 & 6 & 0
\end{array}
$$

Hence, $P(x) = (x + 1)(3x^2 + 19x + 6)$
$= (x + 1)(x + 6)(3x + 1)$.
By the factor theorem, the zeros of P are -1, -6, and $-\tfrac{1}{3}$.

4. $(x - 3)^2[x - (1 + 2i)][x - (1 - 2i)]$
   $= (x^2 - 6x + 9)[x^2 - (1 + 2i)x - (1 - 2i)x$
      $+ (1 + 2i)(1 - 2i)]$
   $= (x^2 - 6x + 9)(x^2 - 2x + 5)$
   $= x^4 - 8x^3 + 26x^2 - 48x + 45$

5.

Hence, the irrational zero in the interval $(-2, -1)$ is approximately -1.634.

6.

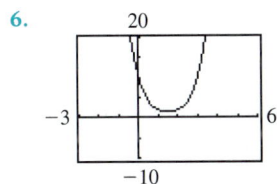

Since the graph does not intersect the x-axis, we conclude that this function has no real zero.

◆ Exercises

1. No   **3.** No   **5.** Yes   **7.** Yes   **9.** No   **11.** Yes

13. $P(x) = 2x^3 - 4x^2 - 2x + 4$

15. $P(x) = -4x^4 + 16x^3 - 12x^2 - 16x + 16$

17. $P(x) = x^4 - 1$ **19.** $P(x) = -5x^5 + 10x^3$

21. (a) $P(x) = (x - 1)(x - 1)(x + 3)$
   (b) 1 of multiplicity two, -3

23. (a) $P(x) = (2x + 1)(3x - 4)(x + 1)$
   (b) $-\frac{1}{2}, \frac{4}{3}, -1$

25. (a) $P(x) = (x - 2)^2(x + i)(x - i)$
   (b) 2 of multiplicity two, $\pm i$

27. (a) $P(x) = (2x - 3)(x - 3)(x - 2)^2$
   (b) $\frac{3}{2}, 3, 2$ of multiplicity two

29. (a) $P(x) = (x + 2i)(x - 2i)(x - 4)(x + 1)$
   (b) $\pm 2i, 4, -1$

31. (a) $P(x) = [x - (3 - i)][x - (3 + i)]$
      $\times [x - (-1 + \sqrt{6})][x - (-1 - \sqrt{6})]$
   (b) $3 \pm i, -1 \pm \sqrt{6}$

33. (a) 1, 2, 3   (b) $P(x) = (x - 1)(x - 2)(x - 3)$

35. (a) $4, -2, \frac{1}{2}$ (b) $f(x) = (x - 4)(x + 2)(2x - 1)$

37. (a) $-\frac{1}{4}$ (b) $H(x) = (4x + 1)(x^2 + 2x + 6)$

39. (a) $\frac{1}{2}, \frac{2}{3}, -\frac{3}{2}$ (b) $P(x) = (2x - 1)(3x - 2)(2x + 3)$

41. (a) $-2, \frac{3}{2}$ (b) $f(x) = (x + 2)(2x - 3)(x^2 + 3)$

43. (a) $\frac{8}{3}$ (b) $g(x) = (3x - 8)(x^3 - 2x^2 - 1)$

45. (a) -1 of multiplicity two, $-2, -\frac{1}{2}$
   (b) $P(x) = (x + 1)^2(x + 2)(2x + 1)$

47. (a) ± 2
   (b) $f(x) = (x - 2)(x + 2)(x^3 + x^2 - 2x + 4)$

49. (a) 1 of multiplicity two, -2 of multiplicity two, $-\frac{1}{2}$
   (b) $F(x) = (x - 1)^2(x + 2)^2(2x + 1)$

51. For $f(x) = x^n - a^n$, $f(-a) = 0$ when n is even, but $f(-a) = -2a^n$ when n is odd. Hence, $x + a$ is a factor of $x^n - a^n$ only when n is even.

53. $k = -7$ **55.** $k = 4, -2$ **57.** $\frac{1}{3}$ **59.** $\frac{2}{3}, -2$

61. 2   **63.** $f(x) = -2x^3 + 4x^2 + 10x - 12$

65. $f(x) = 6x^4 + 6x^3 - 18x^2 - 30x - 12$

67. Approximately 5 h 55 min after high tide.

SECTION 4.4

◆ Problems

1. $(x^2 + 4)(x^2 - 3x - 4) = 0$
   $x^2 + 4 = 0$ or $x^2 - 3x - 4 = 0$
   $x^2 = -4$ $(x + 1)(x - 4) = 0$
   $x = \pm 2i$ $x = -1, 4$

Hence, the four solutions are $\pm 2i, -1$, and 4.

2. Writing the equation in standard form, we obtain $x^3 - x^2 - 10x - 8 = 0$. The possible rational roots are $\pm 1, \pm 2, \pm 4$, and ± 8. The graph in the figure indicates that $-2, -1$, and 4 may be rational roots.

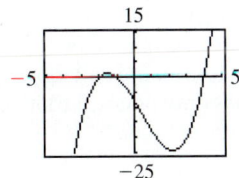

Test -1: $-1)\overline{\begin{array}{cccc} 1 & -1 & -10 & -8 \end{array}}$
   $\begin{array}{ccc} & -1 & 2 & 8 \end{array}$
   $\overline{\begin{array}{cccc} 1 & -2 & -8 & ⓪ \end{array}}$ remainder

Since the remainder is zero, -1 is a root. Thus, $x + 1$ is a factor:

$$x^3 - x^2 - 10x - 8 = 0$$
$$(x + 1)(x^2 - 2x - 8) = 0$$
$$(x + 1)(x + 2)(x - 4) = 0$$
$$x = -1, -2, 4$$

Hence, the three solutions are -1, -2, and 4.

3. Writing the equation in standard form, we obtain
$x^4 + 2x^3 + x^2 - 2x - 8 = 0$. The possible rational
roots are ± 1, ± 2, ± 4, and ± 8. The graph in the figure
indicates that -2 may be a rational root:

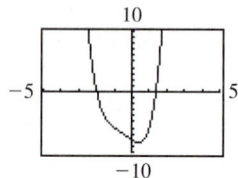

Test -2:
$$-2 \overline{)\,\begin{array}{ccccc} 1 & 2 & 1 & -2 & -8 \\ & -2 & 0 & -2 & 8 \\ \hline 1 & 0 & 1 & -4 & \boxed{0} \end{array}} \text{ remainder}$$

The root in the interval $(1, 2)$ must be irrational. Tracing
to this x-intercept, we find that this root is approximately 1.379. Hence, the two real roots of this equation
are -2 and 1.379 (nearest thousandth).

4.
$$x^3 + 6 \geq 7x$$
$$x^3 - 7x + 6 \geq 0$$
$$(x - 1)(x^2 + x - 6) \geq 0$$
$$(x - 1)(x - 2)(x + 3) \geq 0$$

Boundary numbers: $1, 2, -3$
Algebraic signs: $---+++++++++++--+++$

Solution set: $[-3, 1] \cup [2, \infty)$

5. To find the radius x when the volume is 4000 cubic feet,
we solve the equation

$$4000 = \pi x^2(60 - x) + \frac{2}{3}\pi x^3$$

$$\pi x^3 - 180\pi x^2 + 12000 = 0$$

The graph in the figure indicates that
$y = \pi x^3 - 180\pi x^2 + 12000$ intersects the x-axis
only once in the interval $(0, 60)$. Hence, the equation
$\pi x^3 - 180\pi x^2 + 12000 = 0$ has only one real root.
As shown in this figure, the radius $x \approx 4.6675$ ft.

6. Using the result of the synthetic division in Example 6,
we factor completely over the set of integers:

$$2x^3 - 23x^2 + 60x - 45$$
$$(x - \tfrac{3}{2})(2x^2 - 20x + 30)$$
$$(2x - 3)(x^2 - 10x + 15)$$

The boundary number in the interval $(1, 2)$ is the positive
root of the equation $x^2 - 10x + 15 = 0$. To find the exact value of this boundary number, we apply the quadratic formula:

$$x = \frac{-(-10) \pm \sqrt{(-10)^2 - 4(1)(15)}}{2(1)}$$

$$= \frac{10 \pm \sqrt{40}}{2} = \frac{10 \pm 2\sqrt{10}}{2}$$

$$= 5 \pm \sqrt{10}$$

Hence, the exact value of the boundary number in the
interval $(1, 2)$ is $5 - \sqrt{10}$.

◆ **Exercises**

1. 0. ± 1 **3.** 0 of multiplicity two, $\dfrac{3 \pm \sqrt{17}}{2}$

5. 0 of multiplicity three, $\dfrac{3 \pm \sqrt{17}}{4}$ **7.** $\tfrac{3}{2}$, $\pm\sqrt{2}\,i$

9. 0, -2, -5 of multiplicity two   **11.** 1, $-\tfrac{8}{3}, \tfrac{3}{2}$

13. $\tfrac{1}{3}$ of multiplicity two, $\pm i$ **15.** 0, -3, $\dfrac{3 \pm 3i\sqrt{3}}{2}$

17. ± 2, $-1 \pm 2i$ **19.** $-\tfrac{1}{2}, \tfrac{2}{3}, 3, 4$ **21.** 1.38

23. -1.55 **25.** $-1.88, 0.78, 4.11$

27. $2.09, -1.15, -0.28$ **29.** $0.69, 1.78$ **31.** $-1, 1.13$

33. $-\tfrac{3}{2}, -0.20$ **35.** $-1, 2, 1.22$

37. ± 2, $-3.26, -0.28, 0.54$ **39.** $1, -1.25$

41. $(-\infty, 2 - 2\sqrt{3}\,] \cup [0, 2 + 2\sqrt{3}\,]$

43. $(-\infty, -3) \cup (-\sqrt{2}, 0) \cup (\sqrt{2}, \infty)$

45. $(-\sqrt{5}, \sqrt{5})$ **47.** $[-2, \tfrac{1}{2}] \cup [\tfrac{2}{3}, \infty)$

49. $[-1, \tfrac{5}{2}]$ **51.** $(-\infty, 1) \cup (1, 3)$ **53.** $[\tfrac{1}{3}, \tfrac{1}{2}]$

55. $(-\infty, -\sqrt{3}) \cup [1, \sqrt{3}\,] \cup [5, \infty)$

57. Approximately 5.1 years   **59.** 8 units or more

61. Approximately 4.44 in. by 4.44 in. by 2.44 in.

63. 4 ft by 3 ft by $1\tfrac{1}{2}$ ft   **65.** 5 inches

67. Positive and less than 0.177 cm, or greater than $1\frac{1}{2}$ cm but less than 2 cm

69. $1.3788,\ -0.6894 \pm 1.5575i$

71. (a) $x^3 - 12x + 16 = 0$ (b) 3 of multiplicity two, -3

73. $(2, 8), \left(\frac{1}{2}, \frac{1}{8}\right), \left(-\frac{5}{2}, -\frac{125}{8}\right)$

75. One negative real root and two imaginary roots

77. One positive real root, one negative real root, and two imaginary roots

SECTION 4.5

◆ Problems

1. (a) The zero of the denominator $D(x) = x - 1$ is 1. Hence, the domain of F is $(-\infty, 1) \cup (1, \infty)$.
(b) The zeros of the denominator $D(x) = x^2 - 2x - 8 = (x + 2)(x - 4)$ are -2 and 4. Thus, the domain of f is $(-\infty, -2) \cup (-2, 4) \cup (4, \infty)$.

2. For $F(x) = 3$, we have

$$\frac{3x - 2}{x - 1} = 3$$
$$3x - 2 = 3(x - 1)$$
$$3x - 2 = 3x - 3$$
$$-2 = -3,$$

which is a contradiction.

3. We solve the equation $g(x) = 2x$:

$$\frac{2x^3}{x^2 + 3} = 2x$$
$$2x^3 = 2x(x^2 + 3)$$
$$2x^3 = 2x^3 + 6x$$
$$6x = 0$$
$$x = 0$$

Hence, the rational function g intersects its oblique asymptote at $(0, 0)$.

4. From the graph in the figure, we conclude that the coordinates of the relative maximum point are approximately $(-0.354, -0.369)$.

5. The graph is shown in the figure. We cannot evaluate the function at $x = 1$. This suggests that a hole occurs in the graph at $(1, 3)$.

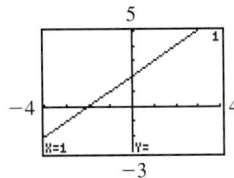

6. To determine the number of kits that are produced if the average cost per kit is \$16,640, we solve the equation

$$16,640 = \frac{141,120 + 12,000x + 20x^2}{x}$$
$$16,640x = 141,120 + 12,000x + 20x^2$$
$$20x^2 - 4640x + 141,120 = 0$$
$$x^2 - 232x + 7056 = 0$$

So, $x = \dfrac{-(-232) \pm \sqrt{(-232)^2 - 4(1)(7056)}}{2(1)}$

$= 36$ or   196 kits.

◆ Exercises

1. (a) $x = -2$
(b) As $x \to -2^-$, $f(x) \to \infty$; as $x \to -2^+$, $f(x) \to -\infty$.

3. (a) None

5. (a) $x = \frac{1}{2}$, $x = -3$
(b) As $x \to \frac{1}{2}^-$, $f(x) \to \infty$; as $x \to \frac{1}{2}^+$, $f(x) \to -\infty$. As $x \to -3^-$, $f(x) \to -\infty$; as $x \to -3^+$, $f(x) \to \infty$.

7. (a) $x = 2$
(b) As $x \to 2^-$, $f(x) \to -\infty$; as $x \to 2^+$, $f(x) \to \infty$.

9. (a) $x = 2$, $x = 4$, $x = -1$
(b) As $x \to 2^-$, $f(x) \to -\infty$; as $x \to 2^+$, $f(x) \to \infty$. As $x \to 4^-$, $f(x) \to \infty$; as $x \to 4^+$, $f(x) \to -\infty$. As $x \to -1^-$, $f(x) \to -\infty$; as $x \to -1^+$, $f(x) \to \infty$.

11. (a) $y = 1$
(b) As $x \to \infty$, $f(x) \to 1^+$; as $x \to -\infty$, $f(x) \to 1^-$.
(c) Does not cross.

13. (a) $y = \frac{1}{2}$
(b) As $x \to \infty$, $f(x) \to \frac{1}{2}^-$; as $x \to -\infty$, $f(x) \to \frac{1}{2}^+$.
(c) Crosses at $\left(\frac{13}{7}, \frac{1}{2}\right)$

15. None exist

17. (a) $y = -3$
(b) As $x \to \infty$, $f(x) \to -3^-$; as $x \to -\infty$, $f(x) \to -3^+$.
(c) Crosses at $\left(-\frac{1}{10}, -3\right)$

19. (a) $y = 0$

(b) As $x \to \infty$, $f(x) \to 0^+$; as $x \to -\infty$, $f(x) \to 0^+$.

(c) Crosses at $\left(\frac{1}{3}, 0\right)$

37. (a) $y = x$ (b) Does not cross

(c)

21.

23.

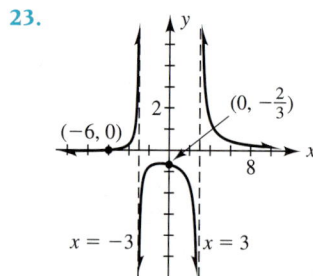

39. (a) $y = x + 3$ (b) Does not cross

(c)

25.

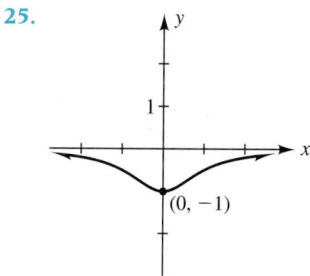

41. (a) $y = x - 6$ (b) Does not cross

(c)

27.

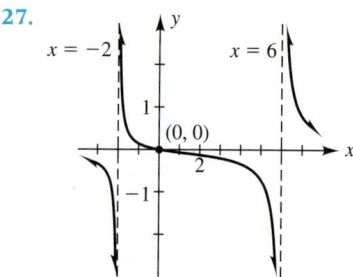

43. (a) $y = x + 3$ (b) Crosses at $\left(-\frac{20}{13}, \frac{19}{13}\right)$

(c)

29.

31.

33.

35.

45.

47.

49.

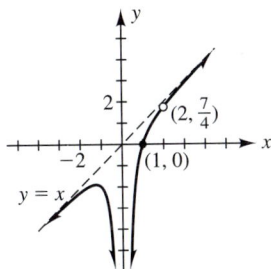

51. (b) $x = 18$ in. by $y = 9$ in.

53. (a) $C(x) = \dfrac{81{,}000 + (400x + 0.1x^2)}{x}$

(b) Approximately 900 stoves

55. No; the procedure of dividing numerator and denominator by the highest power of x that appears in the rational function yields at most one horizontal asymptote.

57. One example is $f(x) = \dfrac{x^2 - 6x + 8}{x^3}$; its graph crosses the horizontal asymptote $y = 0$ at $(2, 0)$ and at $(4, 0)$.

59. No. If $P(1) = 0$, then the graph of f has no vertical asymptote but, instead, has a hole at $x = 1$.

61. (a)

(b)

63. (a) $C(x) \to \infty$, a prohibitive cost

◆ **Problems**

1. The LCD is $2(m + 2)$:

$$f(m) = \frac{3}{2} - \frac{1}{m + 2} - \frac{m}{2m + 4}$$

$$= \frac{3(m + 2) - 1(2) - m}{2(m + 2)} = \frac{2m + 4}{2m + 4}$$

Reducing to lowest terms, gives us a new function F:

$$F(m) = \frac{2m + 4}{2m + 4} = 1 \quad \text{for } m \neq -2.$$

2.

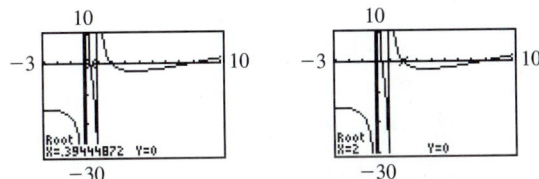

3.

$$\frac{x^2 + 2x - 4}{x + 2} \geq 1$$

$$\frac{x^2 + 2x - 4}{x + 2} - 1 \geq 0$$

$$\frac{(x^2 + 2x - 4) - (x + 2)}{x + 2} \geq 0$$

$$\frac{(x + 3)(x - 2)}{x + 2} \geq 0$$

Boundary numbers: $-3, -2, 2$
Algebraic signs: $- - + + - - - - + + + +$

Solution set: $[-3, -2) \cup [2, \infty)$

4. Let $x =$ crew's speed (in mi/h) in still water.
Then $\quad x + 3 =$ crew's speed downstream
and $\quad x - 3 =$ crew's speed upstream

Since it takes the same time to row 4 mi downstream as it does to row 3 mi upstream, we write the following equation:

$$\frac{4}{x + 3} = \frac{3}{x - 3}$$

$$4(x - 3) = 3(x + 3)$$

$$x = 21 \text{ mi/h}$$

Hence, the team will qualify for the NCAA trials.

5. Let $x =$ amount of time (in hours) to flood the bog using both spillways.

$$\frac{1}{48} + \frac{1}{36} = \frac{1}{x}$$
$$3x + 4x = 144$$
$$x = 20\tfrac{4}{7} \text{ hours}$$

◆ **Exercises**

1. 8   **3.** $-\frac{1}{5}$ **5.** 1   **7.** No solution   **9.** 5, -4

11. 2, $-\frac{3}{7}$ **13.** 4, 2   **15.** 2   **17.** 4   **19.** $\pm\dfrac{\sqrt{5}}{2}$

21. -3 **23.** $1 \pm \sqrt{3}$ **25.** 2, $\pm\dfrac{\sqrt{3}}{3}$

27. 0, -1, $\dfrac{3 \pm \sqrt{321}}{6}$ **29.** $\left(0, \tfrac{1}{2}\right)$ **31.** $(-3, 1)$

33. $\left[-\tfrac{7}{5}, -\tfrac{1}{2}\right)$ **35.** $(-\infty, -4] \cup [1, 2)$

37. $\left(-3 - \sqrt{7}, -5\right) \cup \left(-3 - \sqrt{7}, 5\right)$

39. $(-8, -2) \cup (0, 2)$ **41.** $(-3, 2)$ **43.** $(-3, 4]$

45. $(-2, 2)$ **47.** $\left[-\tfrac{5}{2}, -2\right) \cup \left(-1, -\tfrac{1}{2}\right] \cup (2, 3]$

49. 4 and 12   **51.** 6 mi/h

53. Approximately 10.4 knots, 12.6 knots   **55.** $37\tfrac{1}{2}$ hours

57. Large pump, 40 min; small pump, 60 min

59. 12 students   **61.** $x = \dfrac{2mn}{n - m}$ **63.** $x = -\dfrac{k}{5}$

65. 0, $\dfrac{a + b}{b}$ **67.** a, $-\dfrac{1}{a}$ **69.** $(-\infty, -3] \cup [-1, \infty)$

71. $\left(-\infty, -\tfrac{3}{2}\right] \cup (1, \infty)$

73. We cannot multiply both sides by $x - 1$ because we don't know whether $x - 1$ is positive or negative. The correct solution set is $(-1, 0)$.

75. $0\ \Omega < R_2 \leq 36.2\ \Omega$

CHAPTER 4 REVIEW EXERCISES

1. $3x + 2 + \dfrac{-6x + 2}{x^2 + 1}$

3. $2x^3 - 2x^2 + 4x - 5 + \dfrac{9}{x + 2}$

5. $6x^3 + 9x + 3 + \dfrac{-3}{x - \tfrac{1}{3}}$

7. $x^3 - x - 1 + \dfrac{x^2 + 2x + 2}{x^3 + x + 1}$

9. 6   **11.** 0   **13.** Yes   **15.** No   **17.** Yes

19. (a) $P(x) = (x - 2)(x - 4)(x + 1)$ (b) 2, 4, -1

21. (a) $P(x) = (3x - 4)\left[x - \left(1 + \sqrt{6}\right)\right]$
$\times \left[x - \left(1 - \sqrt{6}\right)\right]$
(b) $\tfrac{4}{3}$, $1 + \sqrt{6}$, $1 - \sqrt{6}$

23. (a) $P(x) = (2x - 1)(x + 3)(x + 2i)(x - 2i)$
(b) $\tfrac{1}{2}$, -3, $\pm 2i$

25. (a) $P(x) = (x - 1)(x + 1)(x - 2)(2x - 1)(2x + 3)$
(b) ± 1, 2, $\tfrac{1}{2}$, $-\tfrac{3}{2}$

27. (a) $P(x) = (x + i)(x - i)[x - (2 + 2i)]$
$\times [x - (2 - 2i)]$
(b) $\pm i$, $2 \pm 2i$

29. (a) $P(x) = [x - (2 + i)][x - (2 - i)]$
$\times (2x - 1)(x + 2i)(x - 2i)$
(b) $2 \pm i$, $\tfrac{1}{2}$, $\pm 2i$

31. 1.67   **33.** $\tfrac{5}{2}$, -1.15 **35.** -5.98, -0.92

37. 2 of multiplicity two, 4.17   **39.** $\tfrac{1}{6}$ **41.** 7

43. $3 \pm \sqrt{5}$ **45.** $\left[-1 - \sqrt{5}, 0\right] \cup \left[-1 + \sqrt{5}, \infty\right)$

47. $(2, \infty)$ **49.** $(-\infty, -2) \cup [4, \infty)$

51. $(-\infty, -4) \cup \left(-\tfrac{2}{3}, 0\right) \cup (2, \infty)$

53.

55.

57.

59.

61.

63.

65.

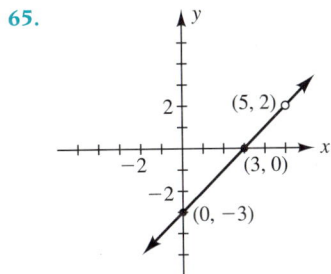

67. (a) $P(x) = \frac{1}{2}(x^3 - 2x^2 - 5x + 6)$
(b) $P(x) = 2(x^3 - 2x^2 - 5x + 6)$

69.

71. $y = \frac{2}{3}$; crosses at $\left(\frac{5}{3}, \frac{2}{3}\right)$ **73.** $2\frac{1}{2}$ ft **75.** 4 cm

77. (a) $V = 4x^3 - 70x^2 + 300x$
(b) 5 cm or approximately 1.10 cm
(c) About 2.83 cm

79. (b) 5 m by 5 m by 4 m **81.** 24 mi/h

◆ Chapter 5

SECTION 5.1

◆ Problems

1. The domain and range of H may be determined from its graph (see Figure 5.4): domain $(-\infty, \infty)$; range $(-\infty, 3)$.

2.

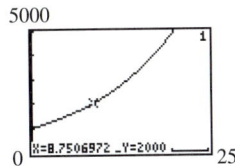

3. $\dfrac{125^{x-2}}{25^{2x-3}} = \dfrac{(5^3)^{x-2}}{(5^2)^{2x-3}} = \dfrac{5^{3x-6}}{5^{4x-6}} = 5^{-x} = \left(\dfrac{1}{5}\right)^x$

4. As illustrated in the figure, in about 8.75 years the amount in the account will be $2000.

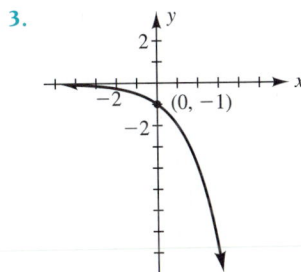

5. Continuous compounding yields \$5.87 more than quarterly compounding over the five-year period.

◆ Exercises

1. (a) $(0, 1)$ (b) Largest $y = 2^x$; smallest $y = 3^x$
(c) Largest $y = 3^x$; smallest $y = 2^x$

3.

5.

7.

9.

11.

13.

15.

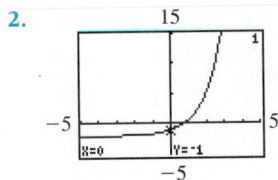

17. 1   **19.** 36^{-x} **21.** 1   **23.** e^x **25.** 9^x **27.** $\frac{1}{3}$ **29.** 5

31. (a) $A = 5000(1.09)^t$ (b) \$7057.91

33. (a) $A = 10,000(1.02125)^{4t}$ (b) \$27,437.17

35. (a) $A = 2000e^{0.06t}$ (b) \$3432.01

37. (a) $A = 120,000e^{0.08t}$ (b) \$124,897.29

39. (a) $k = 15$ (b) 6.74 psi

41. (a) 18 bacteria   (b) 133 bacteria

43. (a) \$12,000   (b) \$4,466
    (c) The machine's scrap value is \$2000.

45. (a) $4^x - 4^{-x}$ (b) 4   **47.** 2

49. 2   **51.** ± 1 **53.** $a = \frac{1}{4}$

55.

x	0	1	2	3	4	-1	-2	-3	-4
$f(x)$	1	-2	4	-8	16	$-\frac{1}{2}$	$\frac{1}{4}$	$-\frac{1}{8}$	$\frac{1}{16}$

The function f is not defined when $x = 1/n$, where n is an even integer. Thus, we cannot connect these points to form the graph.

57. (ii) 8.0% compounded monthly

59. (b)

h	1	0.1	0.01	0.001	0.0001
$\dfrac{e^h - 1}{h}$	1.7183	1.0517	1.0050	1.0005	1.0001

(c) $\dfrac{e^h - 1}{h} \to 1$ as $h \to 0^+$.

SECTION 5.2

◆ Problems

1. (a) $\log 1 = \log_{10} 1 = 0$ (b) $\ln e = \log_e e = 1$

2.

```
log 1          0
ln e^1
               1
```

3. The graph of $g(x) = \log_{1/4} x = \dfrac{\ln x}{\ln \frac{1}{4}}$:

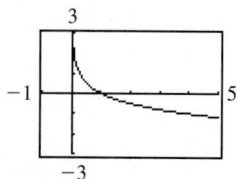

4. The graph of $F(x) = \log_4(x + 2) = \dfrac{\ln(x + 2)}{\ln 4}$:

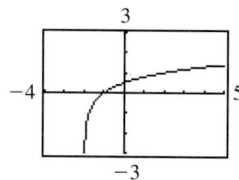

The graph of $G(x) = 1 + \log_4(-x) = 1 + \dfrac{\ln(-x)}{\ln 4}$:

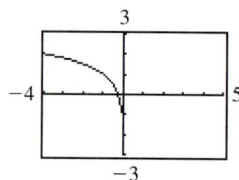

5. $\log_3 9^{2x-3} = \log_3 (3^2)^{2x-3} = \log_3 3^{4x-6} = 4x - 6$
As shown in the figure, the graph of $y = \log_3 9^{2x-3}$ is identical to the graph of $y = 4x - 6$:

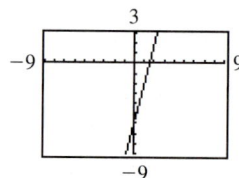

6. $R(5) = 100 - 29 \log[27(5) + 1]$
$ = 100 - 29 \log 136 \approx 38.1\%$

◆ Exercises

1. 2   **3.** 3   **5.** 1   **7.** -4 **9.** -2

11. 2   **13.** 0   **15.** 2

17. Undefined   **19.** $\frac{3}{2}$ **21.** $-\frac{4}{3}$ **23.** 3

25. 5   **27.** 10   **29.** 0   **31.** 1.447   **33.** 2.079

35. 3.585   **37.** 2.757   **39.** -0.3691 **41.** -8.697

43. (a) $(1, 0)$ (b) Largest $y = \log x$; smallest $y = \log_2 x$
    (c) Largest $y = \log_2 x$; smallest $y = \log x$

45.

47.

49.

51.

(b)

53.

55.

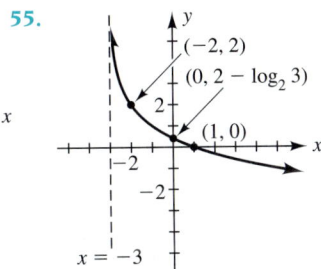

87. (a) $f^{-1}(x) = 8^x$

(b)

57.

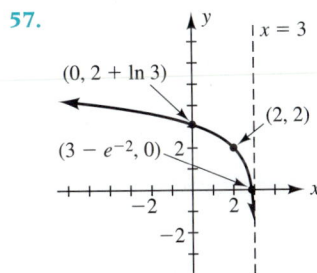

89. (a) It appears that $\ln(xy) = \ln x + \ln y$.

(b) It appears that $\ln \dfrac{x}{y} = \ln x - \ln y$.

(c) It appears that $\ln x^y = y \ln x$.

91. 2.01

93. (a) $(0, \infty)$ (b) $x = 1$ (c) $f(x) \to 0$

(d) $f(x) \to -\infty$

(e) $x = e$

59. $x - 3$ **61.** $x + 2$ **63.** $x^2 + 4$ **65.** $x^2 + 2x$

67. 1   **69.** $3x + 6$ **71.** xe^x **73.** x^2

75. pH ≈ 6.4

77. (a) 100%   (b) 69%

(c) About 4 days

79. (a) 1000   (b) 5615

(c) About \$19,000

81. (a) 3   (b) 4

83. (a) $(0, e) \cup (e, \infty)$ (b) $(0, 10^{10}]$

85. (a) $f^{-1}(x) = \log_6 x$

SECTION 5.3

◆ Problems

1. For example, $(\log_2 8)^2 \neq 2 \log_2 8$. Note that
$(\log_2 8)^2 = 3^2 = 9$, whereas $2 \log_2 8 = 2 \cdot 3 = 6$.

2. The graph of $H(x) = \log_4 \dfrac{1}{x - 1} = \dfrac{\ln[1/(x - 1)]}{\ln 4}$:

3.
```
log 50+2*log 4-3
*log 2
                  2
```

4. (a)
```
ln 2/ln 12+ln 6/
ln 12
                  1
```
(b)
```
ln 16/ln 3*ln 27
/ln 2
                 12
```

5. Let I_a = the intensity of the 1933 earthquake

and I_b = the intensity of the 1989 earthquake.

Then

$$8.9 = \log I_a - \log I_0$$
$$\underline{7.1 = \log I_b - \log I_0}$$
$$1.8 = \log I_a - \log I_b$$

Hence, $\dfrac{I_a}{I_b} = 10^{1.8} \approx 63$ times as intense.

◆ Exercises

1. $\log_2(8 + 8) = \log_2 16 = 4$ and

   $\log_2 8 + \log_2 8 = 3 + 3 = 6$

3. $\dfrac{\log_2 8}{\log_2 2} = \dfrac{3}{1} = 3$ and $\log_2 \dfrac{8}{2} = \log_2 4 = 2$

5. 10   **7.** $3 + \log_2 x + \log_2(x + 2)$

9. $2 + \ln x - \ln 10$ **11.** $\frac{2}{3}\log_3 x - 1$

13. $2\log_b x - \frac{2}{3}\log_b(x + 1)$

15. $-\left(2 + \frac{3}{2}\log_3 x + \frac{1}{3}\log_3 y\right)$

17. $2x^2 - 2\ln(e^x + 1)$

19. $\ln 24$ **21.** 2   **23.** -2

25. $\log_5 \dfrac{x^2 + 2x - 3}{x^2}$

27. $\ln \dfrac{x - 2}{x + 2}$ **29.** $\ln \dfrac{9\sqrt{x^2 - 9}}{x^2}$

31.

33.

35.

37.

39. 1   **41.** 4   **43.** 2   **45.** 1

47. About 32 times more intense

49. About 1.32×10^{10} times brighter

51. 10^9 times more intense

53. No; the domain of f is $(-\infty, 0) \cup (0, \infty)$, whereas the domain of g is $(0, \infty)$.

55. (a) $\frac{1}{2}$ (b) $\frac{1}{3}$ (c) 1   (d) $\frac{1}{6}$
     (e) $\frac{1}{5}$ (f) $-\frac{1}{5}$ (g) -1 (h) 1

57. $\log(6.2 \times 10^k) = \log 6.2 + \log 10^k = \log 6.2 + k$

59. (a) 1   (b) $\ln e = 1$

SECTION 5.4

◆ Problems

1.
$$5^{x-1} = 325$$
$$(x - 1)\ln 5 = \ln 325$$
$$x - 1 = \frac{\ln 325}{\ln 5}$$
$$x = 1 + \frac{\ln 325}{\ln 5} \approx 4.594$$

2.
$$e^{x-1} = 3^x$$
$$(x - 1)\ln e = x\ln 3$$
$$x - 1 = x\ln 3$$
$$x - x\ln 3 = 1$$
$$x(1 - \ln 3) = 1$$
$$x = \frac{1}{1 - \ln 3} \approx -10.14$$

3. $2\ln x - \ln 9 = 4$
$$\ln \frac{x^2}{9} = 4$$
$$e^4 = \frac{x^2}{9}$$
$$x^2 = 9e^4$$
$$x = 3e^2$$

4.
$$x + \ln y = \ln c$$
$$\ln y - \ln c = -x$$
$$\ln \frac{y}{c} = -x$$
$$e^{-x} = \frac{y}{c}$$
$$y = ce^{-x}$$

5.
$$\log \sqrt{x} = \sqrt{\log x}$$
$$\tfrac{1}{2} \log x = \sqrt{\log x}$$
$$\tfrac{1}{4} (\log x)^2 = \log x$$
$$(\log x)^2 - 4 \log x = 0$$
$$\log x (\log x - 4) = 0$$
$$\log x = 0 \quad \text{or} \quad \log x - 4 = 0$$
$$x = 1 \qquad\qquad x = 10^4$$

6.

7.
$$2 = e^{r(5.5)}$$
$$r(5.5) = \ln 2$$
$$r = \frac{\ln 2}{5.5} \approx 0.126 \quad \text{or} \quad 12.6\%$$

8.
$$540 = 18e^{0.1354t}$$
$$30 = e^{0.1354t}$$
$$0.1354t = \ln 30$$
$$t = \frac{\ln 30}{0.1354} \approx 25 \text{ minutes}$$

9.
$$27 = 30e^{-0.02476t}$$
$$0.9 = e^{-0.02476t}$$
$$-0.02476t = \ln 0.9$$
$$t = \frac{\ln 0.9}{-0.02476} \approx 4.26 \text{ years}$$

◆ **Exercises**

1. $\tfrac{3}{2}$ **3.** $-\tfrac{1}{3}$ **5.** $\dfrac{\ln 35}{\ln 7} \approx 1.827$

7. $\dfrac{\ln \tfrac{2}{3}}{3} \approx -0.135$

9. $\dfrac{3 - \log 28}{2} \approx 0.7764$ **11.** $\dfrac{\ln 120}{\ln 9} \approx 2.179$

13. $-\dfrac{\ln \tfrac{15}{4}}{7} \approx -0.1888$ **15.** $-\log 60 \approx -1.778$

17. $\dfrac{1}{2 + \log 4} \approx 0.384$ **19.** $\dfrac{\ln 4}{\ln 4 - 1} \approx 3.589$

21. $\dfrac{\ln \tfrac{3}{4}}{\ln 3} \approx -0.2619$ **23.** $\dfrac{\ln 3072}{\ln 16} \approx 2.896$ **25.** $\tfrac{1}{100}$

27. 8   **29.** $2e$ **31.** $\tfrac{1}{2}, -3$ **33.** 2   **35.** $3e - 1$

37. 4   **39.** 2   **41.** 4   **43.** 1   **45.** $y = ce^{-2x}$

47. $y = \dfrac{10}{(x + a)^2}$ **49.** $y = \dfrac{kx - x}{1 + k}$ for $k = e^c$

51. $y = \dfrac{ka}{1 + kb}$ for $k = e^{ac}$ **53.** $1, e^2$ **55.** $\dfrac{\ln 4}{2}$

57. 0   **59.** 2.123   **61.** 1.202   **63.** $-1.637, 1.000$

65. (a) 5.8 yr   (b) 25.6 yr   (c) 65.8 yr

67. 11%

69. (a) $A(t) = 24e^{0.1792t}$ (b) 1996

71. Approximately 2140 B.C.

73. (a) $P = P_0 2^{-(1/3)t}$ (b) 25%

75. $t = \dfrac{\ln(A/P)}{n \ln[1 + (r/n)]}$ **77.** (b) A line

79. (a) $h^{-1}(x) = 2 \ln x$
(b)

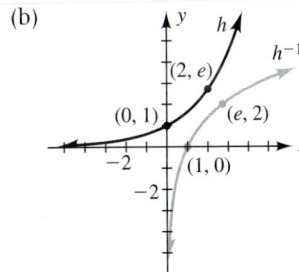

81. (a) $G^{-1}(x) = \tfrac{1}{2} \log(1 - x)$
(b)

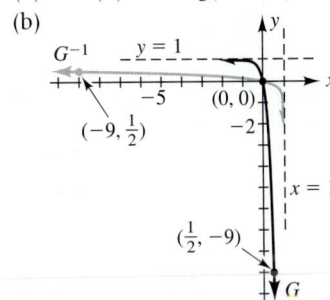

83. (a) $h^{-1}(x) = 2e^{-x}$
(b)

85. (a) $G^{-1}(x) = \frac{1}{3}(10^{(x-1)/2})$
(b)

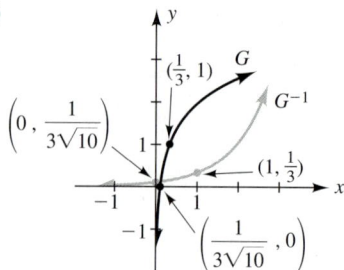

87. $\ln\left(10 \pm 3\sqrt{11}\,\right) \approx \pm 2.993$ **89.** 3, 9

91. (a) 50 ft   (b) $50\ln\left(\dfrac{3 \pm \sqrt{5}}{2}\right) \approx \pm 48.1$ ft

93. (a) $i(t) \to 20$ amp   (b) Approximately 79 ms

95. (a) $T = 70 + 330e^{-0.1805t}$ (b) 262°F
(c) About 7.85 minutes

CHAPTER 5 REVIEW EXERCISES

1. 1   **3.** e **5.** 2^{-x} **7.** 2   **9.** $\frac{2}{3}$ **11.** -3

13. $\frac{1}{3}$ **15.** $-\frac{2}{3}$ **17.** Approximately 3.465

19. Approximately -6.644 **21.** 4   **23.** 3   **25.** $2x$

27. $x^2 + 1$ **29.** $3 - 3x$ **31.** xe^{2x} **33.** 1

35. $2 + 2\log_3 x + \log_3(x - 1)$

37. $-\log x - \frac{1}{2}\log(2x - 3)$

39. $-x^2 + \ln x - \ln(e^x - 1)$

41. $\ln 8$ **43.** $\log_3(x + 1)^2$ **45.** 0

47.

49.

51.

53.

55.

57.

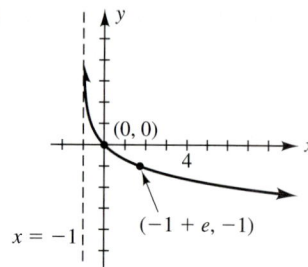

59. $\dfrac{\log 80}{2} \approx 0.9515$ **61.** $-\frac{3}{4}$ **63.** $\dfrac{\ln 4}{\ln 36} \approx 0.3869$

65. $\dfrac{\ln 6}{1 + \ln 9} \approx 0.5604$ **67.** 5   **69.** $-2 \pm e^3$

71. No solution   **73.** 3   **75.** $y = kx^2$ for $k = e^c$

77. ± 1 **79.** 3, $\sqrt{3}$ **81.** $f^{-1}(x) = \log_8 x$

83. $h^{-1}(x) = \dfrac{2 + \ln(x/2)}{3}$ **85.** $G^{-1}(x) = 2^{-(x+1)}$

87. (a) $12,058.57   (b) $12,256.79   (c) $12,298.02

89. Approximately 7 years

91. (a) $A(t) = 10^3 e^{1.535t}$ for t in hours
(b) Approximately 27 minutes

93. About 794 times more intense

95. (a) $k \approx 0.0405$ (b) Approximately 99°F
(c) About 35 minutes

◆ Chapter 6

SECTION 6.1

◆ Problems

1. The graph of the equation $(x - 2)^2 + (y + 1)^2 = 5$ (or $y = -1 \pm \sqrt{5 - (x - 2)^2}\,$) is shown in the figure using a square set viewing rectangle. The circle passes through the point (4, 0) and the coordinates of its center appear to be (2, −1).

2. x-intercepts: $(x + 2)^2 + (0 - 1)^2 = 5$
$(x + 2)^2 = 4$
$x + 2 = \pm 2$
$x = -2 \pm 2$
$x = 0, -4$

y-intercepts: $(0 + 2)^2 + (y - 1)^2 = 5$
$$(y - 1)^2 = 1$$
$$y - 1 = \pm 1$$
$$y = 1 \pm 1$$
$$y = 0, 2$$

3. Completing the square, we have
$(x + 2)^2 + (y + 3)^2 = 0$. Since $r^2 = 0$, the graph of this equation is the single point $(-2, -3)$, so it is not the graph of a circle.

4. The distance from the center $C(0, -56.57)$ to point $A(20, 20.89)$ is 80 ft:
$$CA = \sqrt{(20 - 0)^2 + [20.89 - (-56.57)]^2}$$
$$= \sqrt{6400.0516} \approx 80 \text{ ft}$$

◆ Exercises

1. Radius 4; center $(0, 2)$ 3. Radius 3; center $(0, 0)$

5. Radius 2; center $(0, 1)$ 7. Radius 4; center $(3, -4)$

9. Radius $\frac{4}{3}$; center $\left(-\frac{4}{3}, 0\right)$

11. Radius 2; center $\left(1, -\frac{3}{2}\right)$

13. The equation defines a degenerate circle, and its graph is the single point $(-5, 1)$.

15. The equation does not define a circle and has no graph.

17. x-intercepts $\pm 2\sqrt{3}$; y-intercepts $-2, 6$

19. x-intercept 3; y-intercepts $-4 \pm \sqrt{7}$

21. $x^2 + y^2 = 4$ 23. $x^2 + y^2 - 4x + 6y + 11 = 0$

25. $x^2 + y^2 = 25$ 27. $x^2 + y^2 - 8y + 11 = 0$

29. $x^2 + y^2 + 4x - 10y + 25 = 0$

31. $x^2 + y^2 - 8x - 4y - 5 = 0$

33. Center $(-5.36, 0.00)$; radius 7.49

35. Center $(3.61, 2.42)$; radius 2.56

37. (a) PC$(-88.39, 0)$; PT$(88.39, 0)$;
$x^2 + (y + 88.39)^2 = 125^2$
(b) $A(50, 26.17)$

39. (a) PC$(-27.00, 0)$; PT$(27.00, 0)$;
$x^2 + (y + 46.77)^2 = 54^2$
(b) $A(20, 3.39)$

41. (a) $(x - 8.2)^2 + y^2 = 14.3^2$ (b) 12.4 ft

43. (a) $x^2 + y^2 - 10x + 6y + 25 = 0$ (b) 27π

45. $(4, 2)$ and $(-3, -5)$

47. (a) $4x + 3y = 25$ (b) $x - 4y = -6$

49. $h = 0$ or 8   51. $m = \pm \frac{4}{3}$

SECTION 6.2

◆ Problems

1. The graph of the equation $x = \frac{1}{2}(y - 3)^2 + 2$ (or $y = 3 \pm \sqrt{2x - 4}$) is shown in the figure. The parabola passes through the point $(4, 1)$ and the coordinates of its vertex appear to be $(2, 3)$.

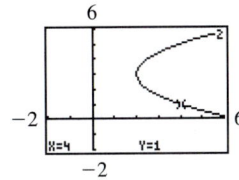

2. x-intercepts of $2x^2 + 3y = 6$:
$$2x^2 + 3(0) = 6$$
$$2x^2 = 6$$
$$x^2 = 3$$
$$x = \pm\sqrt{3}$$

x-intercept of $y^2 - 2x - 4y + 10 = 0$:
$$(0)^2 - 2x - 4(0) + 10 = 0$$
$$-2x + 10 = 0$$
$$-2x = -10$$
$$x = 5$$

3. Solving the equation $3y^2 - 8x = 0$ for x, we obtain $x = \frac{3}{8}y^2$. Hence, $1/(4p) = \frac{3}{8}$, which implies that $p = \frac{2}{3}$. Thus, the coordinates of the focus are $\left(\frac{2}{3}, 0\right)$ and the directrix is the vertical line $x = -\frac{2}{3}$.

4. We can change to general form as follows:
$$x = -\tfrac{1}{14}(y - 1)^2 - \tfrac{3}{2}$$
$$-14x = (y - 1)^2 + 21$$
$$-14x = (y^2 - 2y + 1) + 21$$
$$y^2 + 14x - 2y + 22 = 0$$

5. Replacing p with $\frac{9}{16}$, we obtain
$$x = \frac{1}{4p}y^2 = \frac{1}{4\left(\frac{9}{16}\right)}y^2 = \frac{4}{9}y^2$$

Hence, the equation of the parabola is $x = \frac{4}{9}y^2$.

◆ Exercises

1. Vertex $(0, 0)$; opens to the right

3. Vertex $(0, -8)$; opens upward

5. Vertex $(4, 1)$; opens to the left

7. Vertex $(2, 4)$; opens downward

9. Vertex $\left(-\frac{25}{4}, \frac{3}{2}\right)$; opens to the right

11. Vertex $(-1, 2)$; opens to the right

13. Vertex $(-4, 0)$; opens upward

566   SOLUTIONS AND ANSWERS ◆

15. Vertex $(2, -4)$; opens to the left

17. Focus $\left(0, -\frac{63}{8}\right)$; directrix $y = -\frac{65}{8}$

19. Focus $\left(-6, \frac{3}{2}\right)$; directrix $x = -\frac{13}{2}$

21. Focus $\left(\frac{13}{8}, -4\right)$; directrix $x = \frac{19}{8}$

23. $x = \frac{1}{8}y^2$ 25. $y = 2(x + 1)^2 - 1$

27. $x = -\frac{1}{4}(y - 1)^2 + 1$

29. $y = -\frac{1}{20}(x + 3)^2 + 2$

31. $x = \frac{1}{16}(y + 1)^2 + 3$

33. $(0.744, 7.34)$ 35. $(-1.78, 0.979)$

37. 30 meters   39. $d = 9$ inches   41. $x = 0.152y^2$

43. $\dfrac{a + b}{2}$ 45. $c = 0$

47. Narrower for $0 < p < \frac{1}{4}$; wider for $p > \frac{1}{4}$

49. $\left(0, -\dfrac{E}{4A}\right)$

SECTION 6.3

◆ Problems

1. The graph of the equation $\dfrac{(x + 1)^2}{16} + \dfrac{(y - 3)^2}{4} = 1$
(or $y = 3 \pm \frac{1}{2}\sqrt{16 - (x + 1)^2}$) is shown in the figure. The ellipse appears to have minor axis of length 4 and vertices at $(3, 3)$ and $(-5, 3)$.

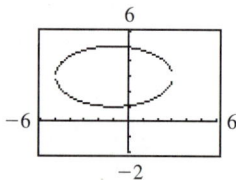

2. Completing the square, we have $3(x - 1)^2 + 2(y + 2)^2 = -4$. Since $-4 < 0$, this equation does not define an ellipse. No point (x, y) with real coordinates satisfies this equation.

3. Writing the equation in standard form, we obtain

$$\frac{x^2}{1^2} + \frac{y^2}{3^2} = 1,$$

which we recognize as an ellipse with $a = 1$, $b = 3$, and center at the origin. To locate the foci, we apply $c^2 = 3^2 - 1^2$, which implies that $c = \pm 2\sqrt{2}$. Hence, the coordinates of the foci are $F_1(0, 2\sqrt{2})$ and $F_2(0, -2\sqrt{2})$. The eccentricity is $e = 2\sqrt{2}/3 \approx 0.94$.

4. To change to general form, we proceed as follows:

$$\frac{(x - 2)^2}{12} + \frac{(y - 1)^2}{16} = 1$$
$$4(x - 2)^2 + 3(y - 1)^2 = 48$$
$$4(x^2 - 4x + 4) + 3(y^2 - 2y + 1) = 48$$
$$4x^2 + 3y^2 - 16x - 6y - 29 = 0$$

◆ Exercises

1. Center $(0, 0)$; vertices $(-4, 0)$, $(4, 0)$

3. Center $(0, 0)$; vertices $(0, 5)$, $(0, -5)$

5. Center $(0, 0)$; vertices $(-3, 0)$, $(3, 0)$

7. Center $(0, 0)$; vertices $\left(-\sqrt{3}, 0\right)$, $\left(\sqrt{3}, 0\right)$

9. Center $(2, -1)$; vertices $(-2, -1)$, $(6, -1)$

11. Center $(0, -2)$; vertices $(0, 1)$, $(0, -5)$

13. Center $(1, -3)$; vertices $(1, -1)$, $(1, -5)$

15. Center $(-5, 2)$; vertices $(-5, 3)$, $(-5, 1)$

17. Foci $\left(0, \sqrt{21}\right)$, $\left(0, -\sqrt{21}\right)$; eccentricity $\sqrt{21}/5 \approx 0.92$

19. Foci $\left(2 + 2\sqrt{3}, -1\right)$, $\left(2 - 2\sqrt{3}, -1\right)$; eccentricity $\sqrt{3}/2 \approx 0.87$

21. Foci $\left(1, \dfrac{-15 + 2\sqrt{21}}{5}\right)$, $\left(1, \dfrac{-15 - 2\sqrt{21}}{5}\right)$; eccentricity $\dfrac{\sqrt{21}}{5} \approx 0.92$

23. $\dfrac{4x^2}{25} + y^2 = 1$ 25. $\dfrac{4(x - 2)^2}{9} + \dfrac{(y + 2)^2}{16} = 1$

27. $\dfrac{(x + 2)^2}{8} + \dfrac{y^2}{9} = 1$ 29. $\dfrac{x^2}{9} + \dfrac{y^2}{5} = 1$

31. $\dfrac{\left(x - \frac{3}{2}\right)^2}{9} + \dfrac{4y^2}{27} = 1$ 33. $(0, 2.11)$, $(0, -2.11)$

35. $(0, 13.2)$, $(0, -4.38)$ 37. $8\sqrt{2} \approx 11.3$ ft

39. $w \approx 8.1$ m

41. (a) 40 ft by 20 ft
(b) Along the major axis, $10\sqrt{3} \approx 17.3$ ft on either side of the center

43. Tack down the ends of the string with the thumbtacks and insert the pencil point into the loop of the string. Now, keeping the string taut, move the pencil to trace out an ellipse.

45. $x^2 + y^2 = a^2$

49. Domain $[-3, 3]$; range $[0, 2]$

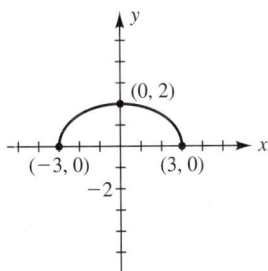

SECTION 6.4

◆ Problems

1. The graph of the equation $\dfrac{x^2}{9} - \dfrac{4(y-1)^2}{9} = 1$

$\left(\text{or } y = 1 \pm \tfrac{1}{2}\sqrt{x^2 - 9}\right)$ and its asymptotes $y = \pm\tfrac{1}{2}x + 1$ are shown in the figure. The hyperbola appears to have transverse axis of length 6 and vertices $(-3, 1)$ and $(3, 1)$.

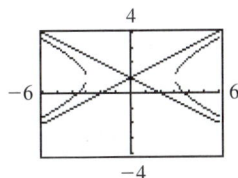

2. Completing the square, we obtain $(x - 3)^2 - 4(y - 1)^2 = 0$. Since the right-hand side of this equation is zero, we conclude this equation does not define a hyperbola. The graph of this equation is the pair of intersecting lines $y = \tfrac{1}{2}x - \tfrac{1}{2}$ and $y = -\tfrac{1}{2}x + \tfrac{5}{2}$.

3. Writing the equation in standard form, we obtain

$$\frac{x^2}{1^2} - \frac{y^2}{3^2} = 1,$$

which we recognize as a hyperbola with $a = 1$, $b = 3$, and center at the origin. To locate the foci, we apply $c^2 = 3^2 + 1^2$, which implies that $c = \pm\sqrt{10}$. Hence, the coordinates of the foci are $F_1\left(\sqrt{10},\, 0\right)$ and $F_2\left(-\sqrt{10},\, 0\right)$. The eccentricity is $e = \sqrt{10}/1 \approx 3.2$.

4. To change to general form, we proceed as follows:

$$\frac{(y-1)^2}{4} - \frac{(x-2)^2}{12} = 1$$
$$3(y-1)^2 - (x-2)^2 = 12$$
$$3(y^2 - 2y + 1) - (x^2 - 4x + 4) = 12$$
$$3y^2 - x^2 + 4x - 6y - 13 = 0$$

◆ Exercises

1. Center $(0, 0)$; vertices $(-2, 0)$, $(2, 0)$

3. Center $(0, 0)$; vertices $(0, 2)$, $(0, -2)$

5. Center $(0, 0)$; vertices $\left(-\tfrac{3}{4}, 0\right)$, $\left(\tfrac{3}{4}, 0\right)$

7. Center $(0, 0)$; vertices $\left(0, \tfrac{2}{5}\right)$, $\left(0, -\tfrac{2}{5}\right)$

9. Center $(0, -2)$; vertices $(-3, -2)$, $(3, -2)$

11. Center $(1, 3)$; vertices $(1, 9)$, $(1, -3)$

13. Center $(2, -3)$; vertices $\left(\tfrac{3}{2}, -3\right)$, $\left(\tfrac{5}{2}, -3\right)$

15. Center $(-4, -1)$; vertices $(-4, 2)$, $(-4, -4)$

17. Foci $\left(0, \sqrt{13}\right)$, $\left(0, -\sqrt{13}\right)$; eccentricity $\sqrt{13}/2 \approx 1.8$

19. Foci $\left(\sqrt{13}, -2\right)$, $\left(-\sqrt{13}, -2\right)$; eccentricity $\sqrt{13}/3 \approx 1.2$

21. Foci $\left(\dfrac{4 + \sqrt{37}}{2}, -3\right)$, $\left(\dfrac{4 - \sqrt{37}}{2}, -3\right)$; eccentricity $\sqrt{37} \approx 6.1$

23. $\dfrac{y^2}{16} - \dfrac{x^2}{256} = 1$ **25.** $\dfrac{x^2}{4} - \dfrac{y^2}{4} = 1$

27. $\dfrac{y^2}{9} - \dfrac{(x+2)^2}{16} = 1$ **29.** $\dfrac{x^2}{1} - \dfrac{y^2}{3} = 1$

31. $4\left(x - \tfrac{3}{2}\right)^2 - \dfrac{y^2}{2} = 1$ **33.** $(1.61, 0)$, $(-1.61, 0)$

35. $(0, -15.2)$, $(0, -4.34)$ **37.** 50 cm

39. (a) $PA - PB = 1430$ ft; P must lie on a branch of a hyperbola because the difference in distances from two fixed points A and B is constant (1430 ft).

(b) $\dfrac{x^2}{511,225} - \dfrac{y^2}{1,738,755} = 1$

(c) $PA = 2340$ ft, $PB = 910$ ft

41. (a) The opening of each branch of the hyperbola is very narrow.

(b) The opening of each branch of the hyperbola is very wide.

43. No effect

45. (a) Degenerate circle; a single point $(3, 4)$

(b) Degenerate ellipse; a single point $(-3, 2)$

(c) Degenerate hyperbola; a pair of intersecting lines, $y = 2x - 6$ and $y = -2x + 2$

(d) Degenerate hyperbola; a pair of intersecting lines, $y = 3x - 1$ and $y = -3x + 11$

47. Domain $(-\infty, -2] \cup [2, \infty)$; range $[0, \infty)$

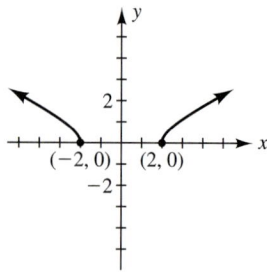

49. Domain $(-\infty, \infty)$; range $(-\infty, -3]$

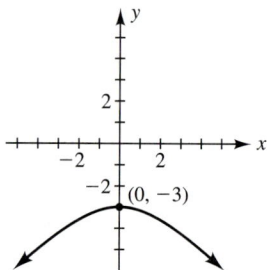

CHAPTER 6 REVIEW EXERCISES

1. Circle; radius 4; center $(0, 0)$

3. Parabola; vertex $(0, 0)$; opens to the right

5. Ellipse; center $(0, 0)$; vertices $(-4, 0)$, $(4, 0)$

7. Hyperbola; center $(0, 0)$; vertices $(-5, 0)$, $(5, 0)$

9. Circle; radius $\frac{1}{2}$; center $(-2, -1)$

11. Parabola; vertex $(2, -3)$; opens upward

13. Ellipse; center $(0, 3)$; vertices $(0, 6)$, $(0, 0)$

15. Hyperbola; center $(0, 1)$; vertices $(-4, 1)$, $(4, 1)$

17. Circle; radius 5, center $(5, 4)$

19. Parabola; vertex $(16, 1)$; opens to the left

21. Ellipse; center $(1, -2)$; vertices $(1, 2)$, $(1, -6)$

23. Hyperbola; center $(0, 1)$; vertices $\left(-\frac{1}{3}, 1\right)$, $\left(\frac{1}{3}, 1\right)$

25. Degenerate hyperbola; a pair of intersecting lines $y = 2x - 2$ and $y = -2x + 2$

27. Degenerate ellipse; a single point $(1, -2)$

29. Focus $(1, 0)$; directrix $x = -1$

31. Focus $\left(\frac{63}{4}, 1\right)$; directrix $x = \frac{65}{4}$

33. Foci $\left(\pm 5\sqrt{2}, 0\right)$; eccentricity $\sqrt{2} \approx 1.4$

35. Foci $\left(1, -2 \pm 2\sqrt{3}\right)$; eccentricity $\sqrt{3}/2 \approx 0.87$

37. $x^2 + y^2 = 49$ **39.** $(x - 2)^2 + \left(y - \frac{9}{2}\right)^2 = \frac{25}{4}$

41. $(x - 2)^2 + (y - 3)^2 = 25$ **43.** $y = x^2 - 4$

45. $x = \frac{2}{9}(y + 3)^2 - 2$ **47.** $y = \frac{1}{2}(x - 1)^2 - \frac{1}{2}$

49. $\dfrac{(x - 3)^2}{9} + \dfrac{y^2}{4} = 1$ **51.** $\dfrac{(x - 2)^2}{16} + \dfrac{y^2}{12} = 1$

53. $\dfrac{y^2}{9} - \dfrac{4x^2}{81} = 1$ **55.** $\dfrac{x^2}{9} - \dfrac{(y + 1)^2}{7} = 1$

57. (a) Center on the y-axis   (b) Center on the x-axis
(c) Tangent to the y-axis   (d) Tangent to the x-axis

59. $3y^2 - 4x^2 = 12$ **61.** a/b

63. $\left(\dfrac{-a^2}{c}, \dfrac{ab}{c}\right)$

65. (a) PC$(-70.71, 0)$; PT$(70.71\ 0)$;
$x^2 + (y + 70.71)^2 = 10,000$
(b) $A(30, 24.68)$

67. $d = 18$ in.   **69.** (a) 174 ft   (b) 75.69 ft

71. 24 cm

◆ **Cumulative Review: Chapters 4, 5, 6**

1. (a) 3   (b) -2 (c) $\frac{5}{2}$ (d) $-\frac{3}{2}$

3. (a) Circle   (b) Hyperbola   (c) Parabola   (d) Ellipse

5. $(x + 2)(x - 2)(x^2 + 2x + 5)$

7. (a) 3   (b) e^2, e^{-1}

9. (a) $y = e^{-2}x^4$ (b) $y = \dfrac{e^x}{x + 2}$ (c) $y = 2$

11. $3x + 4y = 25$ **13.** (a) $f(x) \to -\infty$ (b) $f(x) \to \infty$

15. $\$20,544.33$ **17.** $(-1.99, 2.65)$, $(1.99, 2.65)$

19. 30 rows   **21.** Rear wheel 10 ft; front wheel 6 ft

23. (a) 2500 dozen   (b) 35,306 dozen
(c) $\$3,269,016.00$

25. $7\sqrt{2}$ **27.** $9x^2 - y^2 = 9$

29. Height 3 in.; width 6 in.; length 8 in.

31. Length of major axis $8\sqrt{2} \approx 11.3$ in.; length of minor axis 8 inches

Chapter 7

◆ Problems

1.
$$-4z = 8 \quad \text{implies} \quad z = -2$$
$$z = -2 \text{ and } 5x + 5z = -5 \quad \text{implies} \quad x = 1$$
$$z = -2, \ x = 1, \ 2x - y + 4z = -3 \quad \text{implies} \quad y = -3$$

2. Subtracting the second equation from the first, we eliminate x:

$$\begin{array}{r} x - 2y - 3z = -3 \\ x - 4y - 13z = 14 \\ \hline 2y + 10z = -17 \end{array}$$

Adding 3 times the first equation to the third equation, we again eliminate x:

$$\begin{array}{r} 3x - 6y - 9z = -9 \\ -3x + 5y + 4z = 1 \\ \hline -y - 5z = -8 \end{array}$$

We now have two linear equations in two unknowns y and z, which we can try to solve in the usual manner.

$$2y + 10z = -17 \xrightarrow{\text{No change}} 2y + 10z = -17$$
$$-y - 5z = -8 \xrightarrow{\text{Multiply by 2}} \begin{array}{r} -2y - 10z = -16 \\ \hline 0 = -33 \end{array}$$

Recall from Section 3.5 that if the variables drop out and the resulting statement is false ($0 = -33$), then the system has no solution. Hence, this is an inconsistent system.

3. Subtracting the fourth equation from the first equation, we eliminate t:

$$\begin{array}{r} x + 2y + t = 3 \\ 3y - 4z + t = 2 \\ \hline x - y + 4z = 1 \end{array}$$

Adding this equation to the second equation in the system ($y + 2z = 3$), we eliminate y:

$$\begin{array}{r} x - y + 4z = 1 \\ y + 2z = 3 \\ \hline x + 6z = 4 \end{array}$$

Subtracting the third equation in the system ($x - 2z = 0$) from this equation, we eliminate x:

$$\begin{array}{r} x + 6z = 4 \\ x - 2z = 0 \\ \hline 8z = 4 \quad \text{implies} \quad z = \tfrac{1}{2} \end{array}$$

Finally, we find x, y, and t by substituting back through the system.

$$z = \tfrac{1}{2} \text{ and } x - 2z = 0 \quad \text{implies} \quad x = 1$$
$$z = \tfrac{1}{2} \text{ and } y + 2z = 3 \quad \text{implies} \quad y = 2$$
$$x = 1, \ y = 2, \text{ and } x + 2y + t = 3 \quad \text{implies} \quad t = -2$$

4. Solving $-x + y = -1$ for x, we obtain $x = y + 1$. Using this equation, we express z in terms of y:

$$2x + y - z = 2 \quad \text{implies}$$
$$2(y + 1) + y - z = 2 \quad \text{or} \quad z = 3y$$

Now, if we let $y = b$, where b is any real number, we obtain

$$x = b + 1 \qquad \text{and} \qquad z = 3b$$

Hence, $(b + 1, \ b, \ 3b)$ describes the solution of this system.

5. Adding -18 times the first equation to the second equation, we eliminate y:

$$\begin{array}{r} -18x - 18y - 18z = -216{,}000 \\ 12x + 18y + 24z = 201{,}000 \\ \hline -6x + 6z = -15{,}000 \end{array}$$

Adding $-6x + 6z = -15{,}000$ to 3 times the third equation, we eliminate z:

$$\begin{array}{r} -6x + 6z = -15{,}000 \\ 3x - 6z = 0 \\ \hline -3x = -15{,}000 \quad \text{implies } x = 5000 \end{array}$$

Finally, we find z and y by substituting back through the system: $z = 2500$, $y = 4500$.

◆ Exercises

1. $(10, 4, 1)$ **3.** $(1, 0, -3)$ **5.** $(-1, 2, 0)$

7. $\left(\tfrac{2}{3}, -\tfrac{1}{9}, 3\right)$

9. Dependent system; infinite number of solutions of the form $\left(\dfrac{5 - a}{11}, \dfrac{4a - 9}{11}, a\right)$, where a is any real number

11. $(1, 1, 2)$ **13.** $\left(\tfrac{5}{2}, 0, -\tfrac{3}{2}\right)$ **15.** $(1, -2, 2)$

17. $(1, 2, -2, 3)$ **19.** $(0, 1, -1, 3)$

21. Dependent system; infinite number of solutions of the form $(4 - a, a - 2, -5, 2 - a, a)$, where a is any real number

23. $(1, 2, -1, -2, 0)$

25. $(2 - 3a, a, 4a)$, where a is any real number

27. $(2, a, 2a + 1)$, where a is any real number

29. $(2a + 2, 5a + 1, 3a, a)$, where a is any real number

31. $(5, 2)$ **33.** $a = \$25$, $b = \$0.20$

35. \$5500 at 10%, \$4500 at 12%   **37.** 120°, 40°, 20°

39. \$6000 at 8%, \$22,000 at 9%, \$22,000 at $9\frac{1}{2}$%

41. 20 children, 60 adults, 40 seniors

43. \$0.75 for 2×3 print, \$2.00 for 3×5 print, \$5.00 for 5×8 print, \$10.00 for 8×10 print

45. (a) $A = \frac{3}{2}$, B is any real number except -28
(b) $A = \frac{3}{2}$, $B = -28$

47. $A = \frac{8}{3}$, $B = -1$, $C = -\frac{2}{3}$

49. $a = -\frac{1}{2}$, $b = 2$, $c = 4$

51. $(-a, a, a)$, where a is any real number

53. No adults, 12 students, 8 children or 5 adults, 4 students, 11 children

SECTION 7.2

◆ Problems

1. $\begin{bmatrix} 3 & -9 & 6 & | & 1 \\ 2 & -1 & -1 & | & 3 \\ 1 & -2 & 1 & | & 1 \end{bmatrix}$

$\sim \begin{bmatrix} 1 & -2 & 1 & | & 1 \\ 2 & -1 & -1 & | & 3 \\ 3 & -9 & 6 & | & 1 \end{bmatrix}$ $R_1 \leftrightarrow R_3$

$\sim \begin{bmatrix} 1 & -2 & 1 & | & 1 \\ 0 & 3 & -3 & | & 1 \\ 0 & -3 & 3 & | & -2 \end{bmatrix}$ $-2R_1 + R_2 \rightarrow R_2$
                                  $-3R_1 + R_3 \rightarrow R_3$

$\sim \begin{bmatrix} 1 & -2 & 1 & | & 1 \\ 0 & 3 & -3 & | & 1 \\ 0 & 0 & 0 & | & -1 \end{bmatrix}$ $R_2 + R_3 \rightarrow R_3$

Since all elements in the last row except the last entry are zero, we conclude the system is inconsistent and has no solution.

2. $-2A + 3B$

$= -2\begin{bmatrix} 6 & 5 & -2 \\ 4 & 0 & -1 \end{bmatrix} + 3\begin{bmatrix} 3 & -5 & 2 \\ 0 & 1 & -3 \end{bmatrix}$

$= \begin{bmatrix} -12 & -10 & 4 \\ -8 & 0 & 2 \end{bmatrix} + \begin{bmatrix} 9 & -15 & 6 \\ 0 & 3 & -9 \end{bmatrix}$

$= \begin{bmatrix} -3 & -25 & 10 \\ -8 & 3 & -7 \end{bmatrix}$

3. $A - 2B = A + (-2B)$

$= \begin{bmatrix} 2 & 5 \\ -3 & 0 \\ 1 & 4 \end{bmatrix} + (-2)\begin{bmatrix} 4 & 6 \\ -2 & -5 \\ 0 & 3 \end{bmatrix}$

$= \begin{bmatrix} 2 & 5 \\ -3 & 0 \\ 1 & 4 \end{bmatrix} + \begin{bmatrix} -8 & -12 \\ 4 & 10 \\ 0 & -6 \end{bmatrix}$

$= \begin{bmatrix} -6 & -7 \\ 1 & 10 \\ 1 & -2 \end{bmatrix}$

4. $AC = \begin{bmatrix} 2 & 3 & -1 \end{bmatrix}\begin{bmatrix} 2 & -1 \\ 3 & 4 \\ 0 & -2 \end{bmatrix}$

$= \begin{bmatrix} 4 + 9 + 0 & -2 + 12 + 2 \end{bmatrix}$

$= \begin{bmatrix} 13 & 12 \end{bmatrix}$

5. $\begin{bmatrix} 1 & 1 & 1 & 1 \end{bmatrix}\begin{bmatrix} 2 & 5 \\ 6 & 0 \\ 0 & 2 \\ 2 & 3 \end{bmatrix}$

$= \begin{bmatrix} 2 + 6 + 0 + 2 & 5 + 0 + 2 + 3 \end{bmatrix}$

$= \begin{bmatrix} 10 & 10 \end{bmatrix}$

The 1×2 matrix $\begin{bmatrix} 10 & 10 \end{bmatrix}$ represents the total inventory of this type bed at each outlet.

◆ Exercises

1. $(2, 1)$

3. Dependent system; infinite number of solutions of the form $(4a - 2, a)$, where a is any real number

5. $(1, -2, 3)$ **7.** $(-1, 4, -3)$ **9.** $\left(\frac{23}{4}, -3, -\frac{1}{4}\right)$

11. $\left(\frac{1}{10}, -\frac{11}{10}, -\frac{13}{10}\right)$ **13.** $\left(1, 2, \frac{1}{2}, -2\right)$

15. $\left(\frac{2}{3}, -\frac{1}{3}, \frac{4}{3}, \frac{5}{3}\right)$ **17.** $\begin{bmatrix} -2 & 4 & 0 \\ -4 & 6 & -8 \end{bmatrix}$

19. $\begin{bmatrix} -2 & 0 \\ 1 & -3 \\ 0 & -1 \end{bmatrix}$ **21.** $\begin{bmatrix} 3 & -1 & -5 \\ -3 & 7 & 0 \end{bmatrix}$

23. Undefined; matrices of different orders cannot be added.

25. $\begin{bmatrix} 2 & 8 \\ 1 & -5 \end{bmatrix}$ **27.** $\begin{bmatrix} 2 & -17 & -15 \\ 5 & 0 & 28 \end{bmatrix}$

29. Undefined; matrices of different orders cannot be subtracted.

31. $\begin{bmatrix} -13 & -7 \\ 7 & 1 \end{bmatrix}$ **33.** $\begin{bmatrix} -12 & 24 & 30 \end{bmatrix}$

35. $\begin{bmatrix} -1 \\ 3 \\ -2 \end{bmatrix}$ **37.** $\begin{bmatrix} 8 & 3 \\ 9 & -8 \end{bmatrix}$

39. $\begin{bmatrix} -3 & 7 & 9 \\ 3 & -7 & -9 \end{bmatrix}$ **41.** $\begin{bmatrix} 30 & 30 \\ -33 & -33 \\ -6 & -6 \end{bmatrix}$

43. $\begin{bmatrix} 4 \end{bmatrix}$ **45.** $\begin{bmatrix} 0 & 0 \\ 0 & 0 \end{bmatrix}$

47. $\begin{bmatrix} 9 & -9 & -10 \\ 3 & 11 & -15 \\ -15 & 29 & 5 \end{bmatrix}$ **49.** $\begin{bmatrix} -21 & 60 \\ -3 & -33 \end{bmatrix}$

51. $\begin{bmatrix} 57 & -96 & 120 \end{bmatrix}$ **53.** $\begin{bmatrix} -63 & 63 \\ 66 & -66 \end{bmatrix}$

55. $\begin{bmatrix} 4 & -8 & -10 \\ -12 & 24 & 30 \\ 8 & -16 & -20 \end{bmatrix}$

57. $\begin{bmatrix} 91,524.375 \\ 63,958.125 \\ 132,264.375 \end{bmatrix}$, new weekly payroll (in dollars) at each facility

59. $\begin{bmatrix} 48.75 \\ 123.75 \\ 228.75 \end{bmatrix}$, weekly pay increase (in dollars) for each technician, engineer, and senior engineer

61. $\begin{bmatrix} 71 \\ 48 \\ 20 \end{bmatrix}$, total number of subcompact, midsize, and large cars that were rented on Monday and Tuesday

63. $\begin{bmatrix} 840 & 1256 & 1660 \end{bmatrix}$, total revenue (in dollars) from cars that were rented at each terminal on Monday and Tuesday

65. One example is $A = \begin{bmatrix} a & a \\ b & b \end{bmatrix}$,

$B = \begin{bmatrix} -1 & 1 \\ 1 & -1 \end{bmatrix}$,

where a and b are any real numbers.

67. (a) (i) $\begin{bmatrix} 10 & -5 & 19 \\ -6 & 7 & -14 \\ 15 & 1 & 23 \end{bmatrix}$

(ii) $\begin{bmatrix} 4 & 1 & -1 \\ 5 & 5 & 7 \\ -1 & -2 & -3 \end{bmatrix}$

(iii) $\begin{bmatrix} 6 & -6 & 20 \\ -11 & 2 & -21 \\ 16 & 3 & 26 \end{bmatrix}$

(iv) $\begin{bmatrix} 5 & -2 & 34 \\ -1 & 2 & -22 \\ 6 & 0 & 27 \end{bmatrix}$

(b) From parts (iii) and (iv), we see that $A^2 - B^2 \neq (A + B)(A - B)$.

71. (a) $x = (a_{11}b_{11} + a_{21}b_{12})r + (a_{12}b_{11} + a_{22}b_{12})s$
$y = (a_{11}b_{21} + a_{21}b_{22})r + (a_{12}b_{21} + a_{22}b_{22})s$

(b) $\begin{bmatrix} (a_{11}b_{11} + a_{21}b_{12})r + (a_{12}b_{11} + a_{22}b_{12})s \\ (a_{11}b_{21} + a_{21}b_{22})r + (a_{12}b_{21} + a_{22}b_{22})s \end{bmatrix}$

The elements represent the values of x and y.

SECTION 7.3

◆ Problems

1. Corresponding elements in rows 2 and 4 are proportional. If we add $-3R_2$ to R_4, we obtain a row in which every element is zero. If every element in a row of a matrix is zero, then the determinant of the matrix is 0.

2. $\begin{bmatrix} 1 & 2 & -3 & | & 1 & 0 & 0 \\ 1 & 0 & 2 & | & 0 & 1 & 0 \\ 2 & 2 & -1 & | & 0 & 0 & 1 \end{bmatrix}$

$\sim \begin{bmatrix} 1 & 2 & -3 & | & 1 & 0 & 0 \\ 0 & -2 & 5 & | & -1 & 1 & 0 \\ 0 & -2 & 5 & | & -2 & 0 & 1 \end{bmatrix}$

$\sim \begin{bmatrix} 1 & 2 & -3 & | & 1 & 0 & 0 \\ 0 & -2 & 5 & | & -1 & 1 & 0 \\ 0 & 0 & 0 & | & -1 & -1 & 1 \end{bmatrix}$

Since we obtain a row of zeros on the A portion of this matrix, we conclude that A is singular.

3. The coefficient matrix for this system of equations is the same as matrix A defined in Problem 2. Since matrix A is singular (see Problem 2), A^{-1} does not exist and we cannot solve the matrix equation

$$\begin{bmatrix} 1 & 2 & -3 \\ 1 & 0 & 2 \\ 2 & 2 & -1 \end{bmatrix} \begin{bmatrix} x \\ y \\ z \end{bmatrix} = \begin{bmatrix} -3 \\ 4 \\ 1 \end{bmatrix}$$

uniquely for x, y, and z. The system of equations in Problem 3 is a dependent system with infinitely many solutions of the form $\left(4 - 2a, \dfrac{5a - 7}{2}, a \right)$.

◆ **Exercises**

1. -22 **3.** -4 **5.** 218 **7.** -12 **9.** 208

11. 0 **13.** -45 **15.** 64 **17.** 24 **19.** 40

21. 368 **23.** 304 **25.** $\begin{bmatrix} 3 & 4 \\ -2 & -3 \end{bmatrix}$

27. $\dfrac{1}{5} \begin{bmatrix} -3 & 4 \\ 2 & -1 \end{bmatrix}$ **29.** Singular   **31.** Singular

33. $\dfrac{1}{4} \begin{bmatrix} -9 & 6 & -1 \\ 2 & 0 & 2 \\ -3 & 2 & 1 \end{bmatrix}$

35. $\dfrac{1}{5} \begin{bmatrix} 0 & -10 & -5 \\ 4 & 1 & 3 \\ 1 & 4 & 2 \end{bmatrix}$

37. $\begin{bmatrix} -14 & -13 & 3 \\ 11 & 10 & -2 \\ 2 & 2 & -\frac{1}{2} \end{bmatrix}$

39. $(0, 1, -2)$ **41.** $\left(\frac{1}{2}, \frac{1}{2}, -2 \right)$ **43.** $(1, -5, -3)$

45. $T_1 = 56\tilde{0}$ lb, $T_2 = 758$ lb

47. $v_0 = 80.5$ ft/s, $s_0 = 24.8$ ft

51. (a) $\dfrac{x_1 y_2 - x_2 y_1}{2}$

(b) The area of a triangle with vertices $(0, 0)$, (x_1, y_1), and (x_2, y_2)

53. Matrix A can be cancelled only if it is invertible.

55. $A^{-1} = \dfrac{1}{ad - bc} \begin{bmatrix} d & -b \\ -c & a \end{bmatrix}$ for $ad - bc \neq 0$

SECTION 7.4

◆ **Problems**

1. (a)

(b)

2. 8

3. The objective function has no maximum or minimum value when lines with slope $\frac{2}{5}$ pass through the unbounded region shown in Figure 7.13.

4. The profit P is now given by $P = 30x + 45y$.

Vertex	Values of $P = 30x + 45y$
$(0, 0)$	0
$(0, 40)$	1800
$(15, 40)$	2250 ←
$(60, 10)$	2250 ←
$(60, 0)$	1800

maximum value

Hence, to maximize profits, the company should manufacture either 15 pairs of slalom skis and 40 pairs of racing skis, or 60 pairs of slalom skis and 10 pairs of racing skis.

◆ **Exercises**

1.

3.

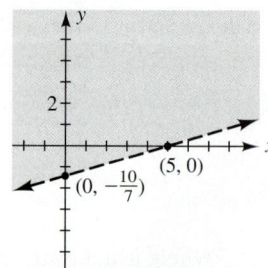

5. (3, 0)

7. $(0, \frac{5}{2})$

25. (0, 5) (0, 3) (2, 0) (4, 0)

27. (0, 7) (4, 3) (0, 0) (4, 0)

9. $(\frac{3}{2}, 0)$ (0, −3)

11. $(0, \frac{1}{2})$ $(\frac{2}{3}, 0)$

29. (0, 4) (1, 2) (3, 1) (7, 0)

13. (2, 1)

15. (3, 2)

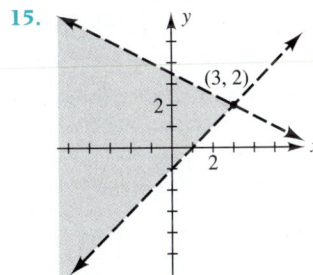

31. Maximum value 12; minimum value 0

33. Maximum value 6; minimum value −11

35. Maximum value 8; minimum value 0

37. Maximum value 6; minimum value 0

39. Maximum value 47; minimum value 16

41. No maximum value; minimum value 11

43. No maximum value; no minimum value

45. Eight $100-cases, two $200-cases

47. 24 tables, 18 chairs

17. (0, 3) (0, 0) (6, 0)

19. (0, 3) (4, 1) (−2, −2)

49. 24 bags of cracked corn, 48 bags of sunflower seeds

51. 1.6 oz. of cereal P, 2.6 oz. of cereal Q

53. (a) $m = -\frac{1}{4}$ (b) $m = -2$ (c) $-\frac{1}{4} < m \le 0$
(d) $-2 < m < -\frac{1}{4}$ (e) $m < -2$ or $m > 0$

55. 2 bags of brand A and 3 bags of brand B; minimum cost $98

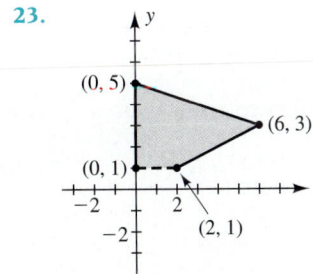

21. (5, 5) (2, 2)

23. (0, 5) (6, 3) (0, 1) (2, 1)

CHAPTER 7 REVIEW EXERCISES

1. (2, −3, −1) **3.** $\left(-\frac{1}{3}, \frac{2}{3}, -\frac{2}{3}\right)$

5. Dependent system; infinite number of solutions of the form $(2a − 2, a, 2a − 1)$, where a is any real number

7. $\left(7, -4, \frac{1}{2}\right)$ **9.** $\left(\frac{1}{4}, -\frac{3}{4}, \frac{5}{4}\right)$ **11.** (4, 2, −3)

13. $(a + 1, a, 2a + 5)$, where a is any real number

15. $(-11, 4, -4, 6)$ **17.** $\left(\frac{3}{4}, \frac{1}{4}, -2, 1\right)$

19. $\begin{bmatrix} -3 & -12 & 6 \end{bmatrix}$ **21.** $\begin{bmatrix} 2 & 8 \\ 4 & 4 \end{bmatrix}$

23. $\begin{bmatrix} -4 & -5 \\ -3 & -4 \end{bmatrix}$ **25.** $\begin{bmatrix} -10 & -6 & -3 \\ -4 & 17 & 28 \end{bmatrix}$

27. $\begin{bmatrix} -3 & 12 \\ -2 & 9 \end{bmatrix}$ **29.** $\begin{bmatrix} 24 \end{bmatrix}$ **31.** $\begin{bmatrix} 15 & 20 \\ 18 & 27 \end{bmatrix}$

33. $\begin{bmatrix} -23 & 59 \end{bmatrix}$ **35.** $\begin{bmatrix} 1 & -\frac{3}{2} \\ -1 & 2 \end{bmatrix}$

37. $\frac{1}{20}\begin{bmatrix} -32 & 20 & 4 \\ 13 & -5 & -1 \\ 14 & -10 & 2 \end{bmatrix}$ **39.** Singular

41. 11 **43.** -14 **45.** -30 **47.** -16

49.

51.

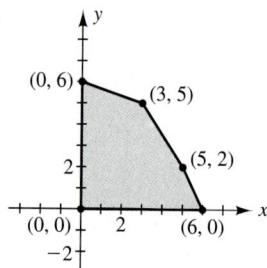

53. Minimum value 0; maximum value 21

55. (a) $k = 3$ (b) $k = \pm 5$ (c) $k = 6, -2$
(d) $k = 3, -\frac{13}{5}$

57. (a) $y = \dfrac{x_2 y_1 - x_1 y_2 + x y_2 - x y_1}{x_2 - x_1}$

(b) The equation of a nonvertical line passing through two distinct points (x_1, y_1), and (x_2, y_2)

59. 9750 gallons of regular; 2250 gallons of super

61. 11,000 box seats; 12,000 grandstand seats; 9500 bleacher seats

63. 4 mi/h downhill; 3 mi/h on level ground; 2 mi/h uphill

65. $\begin{bmatrix} 34 \\ 29 \\ 17 \end{bmatrix}$, total number of 400-watt, 500-watt, and 600-watt microwave ovens sold during first day of sale

67. $\begin{bmatrix} 3480 & 2580 & 3240 \end{bmatrix}$, revenue (in dollars) from sales of microwave ovens at each outlet on second day of sale

69. $T_1 = 465$ lb, $T_2 = 648$ lb   **71.** 20,000 seats

◆ **Chapter 8**

SECTION 8.1

◆ **Problems**

1. $u(1) = (-1)^1 2^0 = -1,$ $u(2) = (-1)^2 2^1 = 2,$
$u(3) = (-1)^3 2^2 = -4,$ $u(4) = (-1)^4 2^3 = 8$

The elements $-1, 2, -4, 8$ form an alternating sequence. The graph is shown in the figure.

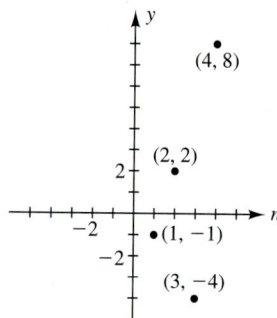

2. $a_5 = 5a_4 = 5(24) = 120;$
$a_6 = 6a_5 = 6(120) = 720$
Hence, 5 factorial is 120 and 6 factorial is 720.

3. (a) For $a_n = 2n - 1$, $a_1 = 2(1) - 1 = 1$,
$a_2 = 2(2) - 1 = 3$, $a_3 = 2(3) - 1 = 5$,
$a_4 = 2(4) - 1 = 7$
(b) For $a_n = 3^{n-1}$, $a_1 = 3^0 = 1$,
$a_2 = 3^1 = 3$, $a_3 = 3^2 = 9$, $a_4 = 3^3 = 27$
(c) For $a_n = \dfrac{n + 1}{n + 2}$, $a_1 = \dfrac{1 + 1}{1 + 2} = \dfrac{2}{3}$,
$a_2 = \dfrac{2 + 1}{2 + 2} = \dfrac{3}{4}$, $a_3 = \dfrac{3 + 1}{3 + 2} = \dfrac{4}{5}$,
$a_4 = \dfrac{4 + 1}{4 + 2} = \dfrac{5}{6}$
(d) For $a_n = (-1)^n x^{4n-3}$, $a_1 = (-1)^1 x^{4-3} = -x$,
$a_2 = (-1)^2 x^{8-3} = x^5$, $a_3 = (-1)^3 x^{12-3} = -x^9$,
$a_4 = (-1)^4 x^{16-3} = x^{13}$

4. $\displaystyle\sum_{k=1}^{27} \dfrac{k}{k + 1} = \dfrac{1}{2} + \dfrac{2}{3} + \dfrac{3}{4} + \cdots + \dfrac{n}{n + 1} + \cdots + \dfrac{27}{28}$

5. Replacing i with $k - 1$, we obtain

$$\sum_{i=1}^{\infty} (i + 1)x^{i+1} = \sum_{k-1=1}^{k=\infty} [(k - 1) + 1]x^{(k-1)+1}$$

$$= \sum_{k=2}^{\infty} kx^k$$

◆ Exercises

1. 3, 0, −3, −6, −9   **3.** 1, −2, 3, −4

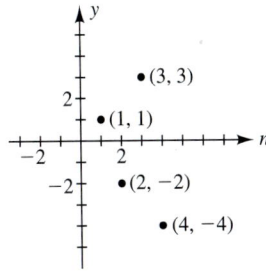

5. 2, 2, $\frac{32}{9}$, 8, $\frac{512}{25}$, · · ·

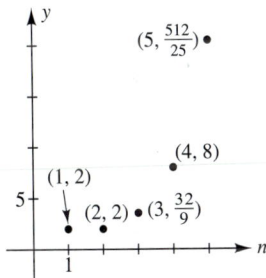

7. 1, 6, 11, 16, 21, 26   **9.** −6, 12, −24, 48, −96, 192

11. 3, $\frac{5}{2}$, $\frac{7}{3}$, $\frac{9}{4}$, $\frac{11}{5}$, $\frac{13}{6}$ **13.** 2, 0, 2, 0, 2, 0

15. 1, 3, 3, 7, 5, 11   **17.** 3, 6, 12, 24, 48, 96

19. 1, 2, 1, −1, −2, −1   **21.** $a_n = n^2$

23. $a_n = (-1)^{n+1}(2n)$ **25.** $a_n = n + 2$

27. $a_n = \dfrac{2n - 1}{2n}$ **29.** $a_n = 2(0.1)^{n-1}$

31. $a_n = (-1)^n \dfrac{x^n}{2n - 1}$ **33.** $2 + 4 + 6 + 8 + 10$

35. $-1 + 8 - 27 + 64 - 125 + 216 - 343 + 512$
$\qquad - 729 + 1000$

37. $1 + 4 + 27 + 256 + 3125 + \cdots$

39. $-\dfrac{1}{2} + \dfrac{1}{2} + \dfrac{3}{2} + \dfrac{5}{2} + \cdots + \dfrac{2n - 3}{2} + \cdots + \dfrac{97}{2}$

41. $-2 + x - \dfrac{2x^2}{3} + \dfrac{x^3}{2} - \dfrac{2x^4}{5} + \cdots$

43. $\displaystyle\sum_{i=1}^{7} 2i$ **45.** $\displaystyle\sum_{i=1}^{30} (5i - 3)$ **47.** $\displaystyle\sum_{i=1}^{\infty} (-1)^{i+1}$

49. $\displaystyle\sum_{i=1}^{10} \dfrac{(-1)^i}{2^{i-1}}$ **51.** $\displaystyle\sum_{i=1}^{\infty} \dfrac{2^i}{3^{i-1}}$

53. $\displaystyle\sum_{i=1}^{5} (i + 2)x^{2i-1}$ **55.** $\displaystyle\sum_{k=0}^{3} (2k - 3)$

57. $\displaystyle\sum_{k=0}^{\infty} \dfrac{(-1)^{k+1}(2x^k)}{k + 1}$ **59.** $\displaystyle\sum_{k=0}^{23} \dfrac{k + 2}{(k + 2)^2 + 1}$

61. One possibility is
$a_n = (2n - 1) + (n - 1)(n - 2)(n - 3)$, which
generates the sequence 1, 3, 5, 13, 33, ...

63. (a) $u(n) = 4n - 6$ for domain {1, 2, 3, 4, 5}
(b) $u(n) = -3n + 13$ for domain {1, 2, 3, 4}

65. (a) 1, 1, 2, 3, 5, 8, 13, 21
(b) Horizontal lines through the tree intersect the
branches in the sequence 1, 1, 2, 3, 5, 8, as shown in the
figure.

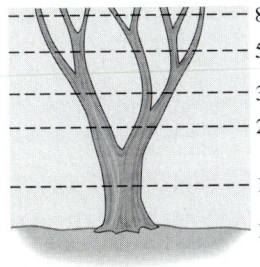

SECTION 8.2

◆ Problems

1. $\displaystyle\sum_{i=1}^{1000} 4 = 1000(4) = 4000$

2. $\displaystyle\sum_{i=1}^{50} (2i - 1) = (50)^2 = 2500$

3. (a) $\displaystyle\sum_{i=1}^{19} i = \dfrac{19(19 + 1)}{2} = 190$

(b) $\displaystyle\sum_{i=1}^{19} i^2 = \dfrac{19(19 + 1)[2(19) + 1]}{6} = 2470$

(c) $\displaystyle\sum_{i=1}^{19} i^3 = \left[\dfrac{19(19 + 1)}{2}\right]^2 = 36{,}100$

4. $\displaystyle\sum_{i=1}^{20} (5i - 2) = 5 \sum_{i=1}^{20} i - \sum_{i=1}^{20} 2$
$\qquad\qquad = 5\left[\dfrac{20(20 + 1)}{2}\right] - 20(2) = 1010$

5. The sum of the interior angles of a 12-sided convex polygon is $(12 - 2)180° = 1800°$

◆ Exercises

1. 150   **3.** 100π **5.** -20 **7.** 247   **9.** $\frac{76}{15}$

29. 800   **31.** 16,206   **33.** 20,825   **35.** 5440

37. $\frac{499}{500}$ **39.** 2046   **41.** 5460   **43.** 45,057   **45.** $\frac{255}{256}$

47. 3825   **49.** 9680   **51.** 35,490   **53.** 4065

55. -2822.4 **57.** 87,423

59. Approximately 5.3 square units

61. (a) log 120   (b) log 6

65. (a) $a_n = (2n - 1)(2n) = 4n^2 - 2n$
(b) 11,060

67. (a) $a_n = (3n - 2)(3n - 1)(3n)$
$\qquad = 27n^3 - 27n^2 + 6n$
(b) 1,114,470

SECTION 8.3

◆ Problems

1. For the sequence 3, 7, 11, 15, . . . , we have $a_1 = 3$ and $d = 4$. Hence, the general element is

$$a_n = a_1 + (n - 1)d$$
$$= 3 + (n - 1)4$$
$$= 4n - 1$$

2. Using $a_n = a_1 + (n - 1)d$ with $n = 60$, we have

$$a_{60} = a_1 + 59d$$

Hence,

$$a_1 = a_{60} - 59d = -98 - (59)(-4) = 138$$

3. If $a_1 = -8$ and $d = 7$, then

$$a_n = a_1 + (n - 1)d$$
$$= -8 + (n - 1)7$$
$$= 7n - 15$$

Hence,

$$a_5 = 7(5) - 15 = 20 \quad \text{and} \quad a_{11} = 7(11) - 15 = 62.$$

4. $\displaystyle\sum_{i=0}^{40} (4i - 3) = 41\left(\frac{-3 + 157}{2}\right) = 3157$

5. Using $S_n = \dfrac{n}{2}[2a_1 + (n - 1)d]$ with $n = 10$, we have

$$S_{10} = \frac{10}{2}[2a_1 + (10 - 1)d]$$
$$215 = 5[2(8) + 9d]$$
$$43 = 16 + 9d$$
$$27 = 9d$$
$$d = 3$$

6. Using $a_n = a_1 + (n - 1)d$, we find

$$a_{24} = \$68 + (24 - 1)(-\$0.75) = \$50.75.$$

Hence, $S_{24} = 24\left(\dfrac{\$68.00 + \$50.75}{2}\right) = \$1425.$

◆ Exercises

1. 3, 7, 11, 15, 19   **3.** 2, $\frac{4}{3}$, $\frac{2}{3}$, 0, $-\frac{2}{3}$

5. $-1, -1 + \pi, -1 + 2\pi, -1 + 3\pi, -1 + 4\pi$

7. $a_n = 6n - 3$ **9.** $a_n = 18 - 7n$

11. $a_n = \dfrac{5n - 7}{3}$ **13.** $a_n = \dfrac{(3 - 2n)\pi}{6}$

15. $a_n = x + (2n - 6)$ **17.** $a_{28} = 193$

19. $a_1 = -307$ **21.** $d = -\frac{1}{2}$ **23.** $a_{80} = 31.85$

25. $d = -\frac{18}{7}$ **27.** 2520   **29.** -8149 **31.** $\frac{273}{2}$

33. 4895   **35.** -273 **37.** $S_{31} = 2697$ **39.** $a_{32} = 26$

41. $d = 8$ **43.** $S_{62} = 7006$ **45.** $a_{33} = 100$

47. 144.56   **49.** 1072.4   **51.** -16.2052 **53.** 2550

55. 166   **57.** \$75,000

59. (a) 128 seats   (b) 2000 seats

61. (a) $a_1 = \$1050$, $a_2 = \$1042.50$, $a_3 = \$1035$
(b) $a_n = 1057.50 - 7.50n$ (c) \$49,725

63. Yes; common difference is ln a **65.** $x = 0.00505$

67. $\frac{108}{5}$, $\frac{121}{5}$, $\frac{134}{5}$, $\frac{147}{5}$

69. (a) 24 payments   (b) \$2600

SECTION 8.4

◆ Problems

1. For the sequence 5, 15, 45, 135, . . . we have $a_1 = 5$ and $r = 3$. Thus, the general element is

$$a_n = a_1 r^{n-1} = 5(3)^{n-1}$$

2. Using $a_n = a_1 r^{n-1}$ with $n = 9$, we have $a_9 = a_1 r^8$. Hence,

$$a_1 = \frac{a_9}{r^8} = \frac{\frac{1}{81}}{\left(\frac{1}{3}\right)^8} = \frac{\left(\frac{1}{3}\right)^4}{\left(\frac{1}{3}\right)^8} = 81$$

3. If $a_1 = 96$ and $r = \frac{1}{2}$, then

$$a_n = a_1 r^{n-1} = 96\left(\tfrac{1}{2}\right)^{n-1}$$

Hence,

$$a_3 = 96\left(\tfrac{1}{2}\right)^2 = 24 \quad \text{and} \quad a_6 = 96\left(\tfrac{1}{2}\right)^5 = 3$$

4. Using $S_n = \dfrac{a_1(1 - r^n)}{1 - r}$ with $n = 12$, we have

$$S_{12} = \dfrac{a_1(1 - r^{12})}{1 - r}$$

Hence,

$$a_1 = \dfrac{S_{12}(1 - r)}{1 - r^{12}} = \dfrac{8190(1 - 2)}{1 - 2^{12}} = 2$$

5. Using $a_n = a_1 r^{n-1}$ with $a_1 = 3$, $r = 4$, and $a_n = 201,326,592$, we find

$$201,326,592 = 3(4)^{n-1}$$
$$67,108,864 = 4^{n-1}$$
$$\ln 67,108,864 = (n - 1) \ln 4$$
$$n - 1 = \dfrac{\ln 67,108,864}{\ln 4}$$
$$n = \dfrac{\ln 67,108,864}{\ln 4} + 1 = 14$$

Hence, the series has 14 terms.

6. The amount of \$5000 earns interest at the rate of 9% compounded monthly, for 10 years. Using the compound interest formula (Section 5.1), we find

$$A = P\left(1 + \dfrac{r}{n}\right)^{nt} = \$5000\left(1 + \dfrac{0.09}{12}\right)^{120}$$
$$= \$5000(1.0075)^{120}$$
$$= \$12,257$$

Adding this amount to the amount of the annuity (\$29,027) yields a total amount of \$41,284.

◆ **Exercises**

1. 3, 12, 48, 192, 768   **3.** $\frac{1}{4}, -\frac{1}{6}, \frac{1}{9}, -\frac{2}{27}, \frac{4}{81}$

5. $-1, -\pi, -\pi^2, -\pi^3, -\pi^4$ **7.** $a_n = 3^{n-1}$

9. $a_n = -64\left(-\frac{1}{4}\right)^{n-1}$ **11.** $a_n = \frac{1}{4}\left(\frac{2}{3}\right)^{n-1}$

13. $a_n = -\sqrt{2}\,(-\sqrt{3})^{n-1}$ **15.** $a_n = 3(2x^2)^{n-1}$

17. $a_9 = 26,244$ **19.** $a_1 = -\frac{1}{216}$ **21.** $r = \pm\frac{3}{2}$

23. $a_{13} = \frac{243}{32}$ **25.** $r = \pm\frac{1}{2}$ **27.** -364 **29.** $\frac{605}{729}$

31. 46.66666662   **33.** 1,398,101   **35.** $\frac{1023}{8}$

37. $S_{11} = \frac{683}{64}$ **39.** $a_1 = 2$ **41.** $S_6 = \frac{665}{144}$

43. $S_7 = -\frac{2653}{64}$ **45.** $a_6 = \frac{729}{8}, -\frac{9375}{8}$ **47.** -0.0000807

49. 3.35   **51.** 13   **53.** Approximately 182,976   **55.** \$500

57. Choose the scholarship that doubles the amount each month for one year. It pays \$40,950, whereas the other scholarship pays only \$40,000.

59. \$18,294.60   **61.** \$81,397.92

---

63. (a) Geometric sequence with $r = e^{3x}$ (b) $a_n = e^{(3n-1)x}$

65. $x = 10^{-3}$ **67.** 24, 36, 54   **69.** $r = -\frac{1}{5}, -5$

SECTION 8.5

◆ **Problems**

1. The infinite series $1 + 0.1 + 0.01 + 0.001 + \cdots$ has $a_1 = 1$ and $r = 0.1$. Since $|r| < 1$, this series converges and its sum S is

$$S = \dfrac{a_1}{1 - r} = \dfrac{1}{1 - 0.1} = \dfrac{1}{0.9} = \dfrac{10}{9}$$

2. The repeating decimal $0.88888\ldots$ can be written as an infinite series with $a_1 = 0.8$, and $r = 0.1$. Thus,

$$S = \dfrac{0.8}{1 - 0.1} = \dfrac{8}{9}$$

3.

$$
\require{enclose}
\begin{array}{r}
1 + 2x + 4x^2 + 8x^3 + \cdots \\[2pt]
1 - 2x\,\enclose{longdiv}{1} \\
\underline{1 - 2x} \\
2x \\
\underline{2x - 4x^2} \\
4x^2 \\
\underline{4x^2 - 8x^3} \\
8x^3 \\
\underline{8x^3 - 16x^4} \\
16x^4
\end{array}
$$

4. If the initial hammer impact drives the nail $\frac{7}{8}$ inch, then the total distance the nail can be driven is given by

$$\dfrac{7}{8} + \dfrac{7}{12} + \dfrac{7}{18} + \dfrac{7}{27} + \cdots$$

The sum of this infinite geometric series is

$$S = \dfrac{a_1}{1 - r} = \dfrac{\frac{7}{8}}{1 - \frac{2}{3}} = \dfrac{21}{8} = 2\dfrac{5}{8} \text{ inches.}$$

Since the nail is $2\frac{1}{2}$ inches long and $2\frac{5}{8} > 2\frac{1}{2}$, it is possible for the head of the nail to be flush with the board. In fact, this will occur during the eighth impact.

◆ **Exercises**

1. $1, \frac{5}{3}, \frac{19}{9}, \frac{65}{27}; 3\left[1 - \left(\frac{2}{3}\right)^n\right]$

3. $2, -4, 14, -40; \frac{1}{2}[1 - (-3)^n]$

5. $-243, -81, -189, -117; -\frac{729}{5}\left[1 - \left(-\frac{2}{3}\right)^n\right]$

7. $\frac{9}{8}, \frac{15}{8}, \frac{19}{8}, \frac{65}{24}; \frac{27}{8}\left[1 - \left(\frac{2}{3}\right)^n\right]$

9. Diverges   **11.** Converges to the sum $\frac{1}{30}$

13. Converges to the sum $\frac{5}{11}$ **15.** Diverges

17. Converges to the sum -4

19. Converges to the sum $\frac{25}{4}\left(\sqrt{10} + \sqrt{2}\right)$

21. $r = \frac{2}{3}$ **23.** $S = \frac{81}{16}$ **25.** $S = -\frac{256}{7}$ **27.** $a_1 = 6, 12$

29. $\frac{1}{3}$ **31.** $\frac{1}{11}$ **33.** $\frac{78}{37}$ **35.** $\frac{500627}{33330}$ **37.** $\frac{2669}{3330}$

39. Interval of convergence $(-5, 5)$; $S = \dfrac{5x}{x + 5}$

41. Interval of convergence $\left(-\frac{1}{2}, \frac{1}{2}\right)$; $S = \dfrac{1}{1 - 4x^2}$

43. Interval of convergence $(1, 3)$; $S = \dfrac{1}{x - 1}$

45. Interval of convergence $\left(-\infty, -\frac{1}{2}\right) \cup \left(\frac{1}{2}, \infty\right)$;
$S = \dfrac{4x^2}{2x - 1}$

47. (a) 54 m   (b) 36 m   (c) 90 m

49. No; he runs only 25 miles.   **51.** 140.875 cm

53. $x = -\frac{1}{6}$

55. (a) Does not exist   (b) 0   (c) 1;
The associative property of real numbers does not apply to infinite series.

57. $r = \frac{1}{3}$

59. (a) 1152 sq cm   (b) $\left(192 + 92\sqrt{2}\right)$ cm

SECTION 8.6

◆ **Problems**

1. The number of possible orders of finish for the first four positions is
$$_8P_4 = \frac{8!}{(8 - 4)!} = \frac{8!}{4!} = 1680 \text{ orders}$$

2. A team that plays ten games can end the season with 3 wins, 5 losses, and 2 ties in $\dfrac{10!}{3!\, 5!\, 2!} = 2520$ permutations.

3. A five-card poker hand consisting of two queens and three kings can be dealt in $_4C_2 \cdot {}_4C_3 = \dfrac{4!}{2!\, 2!} \cdot \dfrac{4!}{3!\, 1!} = 24$ combinations.

4. The fifth term in the expansion of $(x + y)^9$
is $\begin{pmatrix} 9 \\ 4 \end{pmatrix} x^{9-4}y^4 = 126x^5y^4$.

◆ **Exercises**

1. (a) 10!   (b) $k!$ **3.** (a) 20!   (b) $(k - 1)!$

5. (a) $\dfrac{11}{10!}$ (b) $\dfrac{1 + k}{k!}$ **7.** (a) 120   (b) 24

9. 2520   **11.** 2520   **13.** 84

15. (a) 56   (b) 30   (c) 9

17. 1,816,214,400   **19.** 726,485,760   **21.** 42   **23.** 1

25. 1   **27.** 15   **29.** 84   **31.** 1260   **33.** 0

35. $x^5 + 5x^4y + 10x^3y^2 + 10x^2y^3 + 5xy^4 + y^5$

37. $64x^6 - 192x^5 + 240x^4 - 160x^3 + 60x^2 - 12x + 1$

39. $81a^4 + 216a^3b + 216a^2b^2 + 96ab^3 + 16b^4$

41. $128x^{14} - 448x^{12} + 672x^{10} - 560x^8 + 280x^6$
   $- 84x^4 + 14x^2 - 1$

43. $x^4 + 16x^3\sqrt{x} + 112x^3 + 448x^2\sqrt{x} + 1120x^2$
   $+ 1792x\sqrt{x} + 1792x + 1024\sqrt{x} + 256$

45. $220x^9y^3$ **47.** $20,412n^5$ **49.** $5670x^4y^8$

51. $13,440t^2$ **53.** $n = 14$ **55.** $n = 9$ **57.** $n = 8$

61. $(n - 1)!$

65. (a) $x^{10} + 5x^9 + 15x^8 + 30x^7 + 45x^6 + 51x^5$
   $+ 45x^4 + 30x^3 + 15x^2 + 5x + 1$
(b) $x^{10} - 5x^9 + 5x^8 + 10x^7 - 15x^6 - 11x^5$
   $+ 15x^4 + 10x^3 - 5x^2 - 5x - 1$

67. (a) 1, 7, 21, 35, 35, 21, 7, 1
(b) 1, 8, 28, 56, 70, 56, 28, 8, 1
(c) 1, 9, 36, 84, 126, 126, 84, 36, 9, 1

69. (a) $1 - x + x^2 - x^3 + \cdots$
(b) $1 + 2y + 3y^2 + 4y^3 + \cdots$
(c) $1 + \dfrac{z}{2} + \dfrac{3z^2}{8} + \dfrac{5z^3}{16} + \cdots$
(d) $1 + \dfrac{x}{3} - \dfrac{x^2}{9} + \dfrac{5x^3}{81} - \cdots$

71. Power series yields $e \approx 2.718281526$; $\boxed{e^x}$ key yields $e \approx 2.718281828$

CHAPTER 8 REVIEW EXERCISES

1. $2, \frac{3}{2}, \frac{4}{3}, \frac{5}{4}, \frac{6}{5}, \frac{7}{6}$

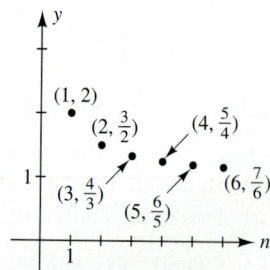

3. $-4, -1, 2, 5, 8, 11$ **5.** $1, -2, 6, -24, 120, -720$

7. $0, 0, 2, 4, 4, 16$

9. (a) Neither   (b) $a_n = (-1)^n(2n - 1)$

11. (a) Arithmetic   (b) $a_n = 6n - 3$

13. (a) Geometric   (b) $a_n = 768\left(\frac{1}{4}\right)^{n-1}$

15. (a) Neither   (b) $a_n = \dfrac{(-1)^{n+1}(2n - 1)}{2n}$

17. (a) Geometric   (b) $a_n = -4\left(-\frac{1}{2}\right)^{n-1}$

19. (a) Arithmetic   (b) $a_n = \dfrac{17 - 4n}{3}$

21. (a) Neither   (b) $a_n = n^2 - 1$

23. (a) Arithmetic   (b) 1260

25. (a) Geometric   (b) $-\dfrac{12{,}117}{64}$

27. (a) Neither   (b) 6327

29. (a) Geometric   (b) 10

31. (a) Arithmetic   (b) 93

33. (a) Neither   (b) 36,664

41. 8120   **43.** 101,025

45. (a) $\displaystyle\sum_{i=1}^{250} (4i + 1)$ (b) $\displaystyle\sum_{i=0}^{249} (4i + 5)$

47. $\displaystyle\sum_{i=1}^{\infty} 243\left(-\frac{2}{3}\right)^{i-1}$

49. (a) $384, 288, 312, 306;\ \frac{1536}{5}\left[1 - \left(-\frac{1}{4}\right)^n\right]$ (b) $\frac{1536}{5}$

51. -83 **53.** 27,090   **55.** 96

57. $x = 4, -\frac{3}{2}$ **59.** $\frac{81}{16}$ or $\frac{81}{80}$

61. (a) $\frac{100}{33}$ (b) $\frac{3031}{1110}$

63. 42,340   **65.** (a) \$12,160.67   (b) \$19,782.09

67. Yes, during the 9th second   **69.** 16   **71.** 78   **73.** 0

75. $729x^6 - 2916x^5 + 4860x^4 - 4320x^3 + 2160x^2 - 576x + 64$

77. $120a^{-7}b^3$ **79.** 60,060

81. (a) 635,013,559,600   (b) 1,677,106,640
(c) 740,999,259

◆ Cumulative Review: Chapters 7, 8

1. (a) 420   (b) 2,097,150

3. (b) 722,666   **5.** 4

7. (a) $(2, -1, 3)$ (b) $\left(1, 2, -3, -\frac{3}{2}\right)$

9. $A^{-1} = \dfrac{1}{17}\begin{bmatrix} -17 & -2 & 28 \\ -17 & 2 & 23 \\ 17 & 3 & -25 \end{bmatrix}$;

$X = \begin{bmatrix} 3 \\ 2 \\ -2 \end{bmatrix}$

11. $x = 8, y = 12$ or $x = \frac{1}{2}, y = -3$

13. Arithmetic sequence with common difference $\log r$

15. (a) $\begin{bmatrix} 7 & 5 & 3 & -7 \\ 8 & 4 & -4 & 5 \\ 9 & 10 & 6 & -3 \\ 0 & 5 & 9 & -1 \end{bmatrix}$

(b) $\begin{bmatrix} 11 & -15 & 14 & -21 \\ -1 & -13 & 8 & 5 \\ 27 & -10 & 8 & 6 \\ -30 & -25 & 17 & 12 \end{bmatrix}$

(c) $\begin{bmatrix} 19 & -10 & 4 & 4 \\ -8 & -24 & 18 & -16 \\ 60 & 5 & -13 & 18 \\ 90 & -1 & 11 & -24 \end{bmatrix}$

(d) $\begin{bmatrix} -37 & 1 & -7 & 7 \\ 19 & 37 & 7 & -11 \\ 28 & 96 & -9 & -8 \\ -15 & 21 & 36 & -33 \end{bmatrix}$

17. 0.3

19. (a) 64   (b) 128   (c) 0   (d) 924

21. 18, 24, 30

23. (a) 66   (b) 220

25. $1 + \dfrac{1}{2} + \dfrac{1}{3} + \dfrac{1}{4} + \cdots + \dfrac{1}{n} + \cdots$

27. $\begin{bmatrix} 16 & -4 \\ -11 & 3 \end{bmatrix}$

29. 3 mi/h uphill, 8 mi/h on level ground, 12 mi/h downhill

31. About 29 years

INDEX ◆

TRIGONOMETRIC DEFINITIONS

$$\sin \theta = \frac{y}{r} \qquad \csc \theta = \frac{r}{y}$$

$$\cos \theta = \frac{x}{r} \qquad \sec \theta = \frac{r}{x}$$

$$\tan \theta = \frac{y}{x} \qquad \cot \theta = \frac{x}{y}$$

$$r = \sqrt{x^2 + y^2}$$

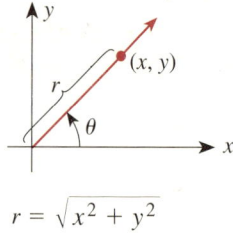

TRIGONOMETRIC RATIOS FOR RIGHT TRIANGLES

$$\sin \theta = \frac{\text{opp}}{\text{hyp}} \qquad \csc \theta = \frac{\text{hyp}}{\text{opp}}$$

$$\cos \theta = \frac{\text{adj}}{\text{hyp}} \qquad \sec \theta = \frac{\text{hyp}}{\text{adj}}$$

$$\tan \theta = \frac{\text{opp}}{\text{adj}} \qquad \cot \theta = \frac{\text{adj}}{\text{opp}}$$

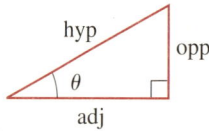

GRAPHS OF THE TRIGONOMETRIC FUNCTIONS

1. Sine Function
$y = \sin x$

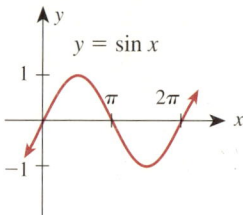

2. Cosine Function
$y = \cos x$

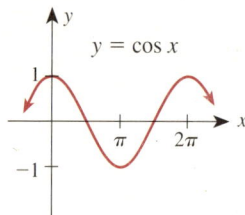

3. Tangent Function
$y = \tan x$

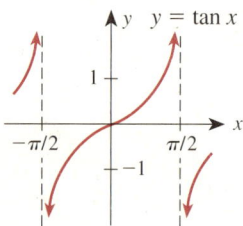

4. Cotangent Function
$y = \cot x$

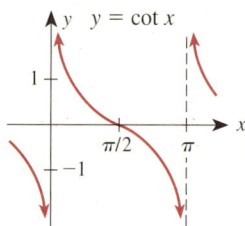

5. Cosecant Function
$y = \csc x$

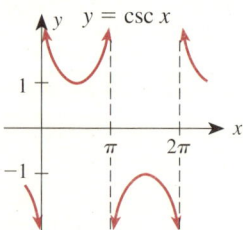

6. Secant Function
$y = \sec x$

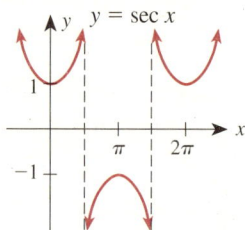

GRAPHS OF THE INVERSE TRIGONOMETRIC FUNCTIONS

1. Inverse Sine Function
$y = \sin^{-1} x$

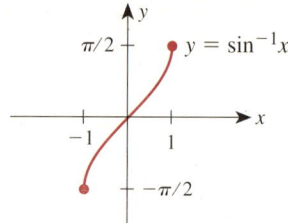

2. Inverse Cosine Function
$y = \cos^{-1} x$

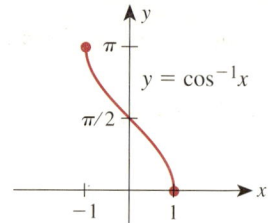

3. Inverse Tangent Function
$y = \tan^{-1} x$

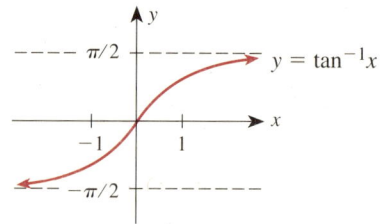

OBLIQUE TRIANGLES

1. The Law of Sines

$$\frac{a}{\sin A} = \frac{b}{\sin B} = \frac{c}{\sin C}$$

2. The Law of Cosines

$$a^2 = b^2 + c^2 - 2bc \cos A$$
$$b^2 = a^2 + c^2 - 2ac \cos B$$
$$c^2 = a^2 + b^2 - 2ab \cos C$$

3. Hero's Formula

$$\text{Area} = \sqrt{s(s-a)(s-b)(s-c)},$$

where s is half the perimeter of the triangle.

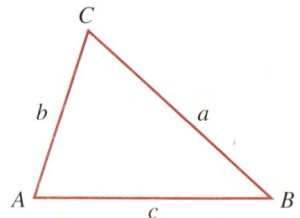

VECTORS

For the nonzero vector $\mathbf{A} = \langle x, y \rangle$ with magnitude $|\mathbf{A}|$ and direction angle θ,

1. $|\mathbf{A}| = \sqrt{x^2 + y^2}$

2. $\tan \theta = \dfrac{y}{x}$

3. $x = |\mathbf{A}| \cos \theta$

4. $y = |\mathbf{A}| \sin \theta$